Tracking Environmental Change Using Lake Sediments

Developments in Paleoenvironmental Research

VOLUME 5

Aims and Scope:

Paleoenvironmental research continues to enjoy tremendous interest and progress in the scientific community. The overall aims and scope of the *Developments in Paleoenvironmental Research* book series is to capture this excitement and document these developments. Volumes related to any aspect of paleoenvironmental research, encompassing any time period, are within the scope of the series. For example, relevant topics include studies focused on terrestrial, peatland, lacustrine, riverine, estuarine, and marine systems, ice cores, cave deposits, palynology, isotopes, geochemistry, sedimentology, paleontology, etc. Methodological and taxonomic volumes relevant to paleoenvironmental research are also encouraged. The series will include edited volumes on a particular subject, geographic region, or time period, conference and workshop proceedings, as well as monographs. Prospective authors and/or editors should consult the **Series Editor John P. Smol** for more details. Any comments or suggestions for future volumes are welcomed.

For further volumes:
http://www.springer.com/series/5869

H. John B. Birks • André F. Lotter • Steve Juggins
John P. Smol
Editors

Tracking Environmental Change Using Lake Sediments

Data Handling and Numerical Techniques

Editors

H. John B. Birks
Department of Biology
University of Bergen
PO Box 7803
N-5020 Bergen
Norway

André F. Lotter
Department of Geobiology
Laboratory of Palaeobotany and Palynology
University of Utrecht
Budapestlaan 4
3584 CD Utrecht
The Netherlands

Steve Juggins
School of Geography, Politics and Sociology
Newcastle University
Daysh Building 14b
Newcastle-upon-Tyne, NE1 7 RU
United Kingdom

John P. Smol
Department of Biology
Paleoecological Environmental
Assessment and Research Laboratory
(PEARL)
Queen's University
Kingston, Ontario, K7L 3N6
Canada

Additional material to this book can be downloaded from http://extras.springer.com

ISSN 1571-5299
ISBN 978-94-007-2744-1 ISBN 978-94-007-2745-8 (eBook)
DOI 10.1007/978-94-007-2745-8
Springer Dordrecht Heidelberg New York London

Library of Congress Control Number: 2012933106

Printed on acid-free paper

Springer is part of Springer Science+Business Media (www.springer.com)

This book is dedicated to Cajo ter Braak, whose work on quantifying species-environmental relationships underpins many of the recent advances in quantitative palaeolimnology described in this book

Preface

Palaeoenvironmental research has been thriving for several decades, with innovative methodologies being developed at a frenetic rate to help answer a myriad of scientific and policy-related questions. This burst in activity was the impetus for the establishment of the *Developments in Paleoenvironmental Research* (*DPER*) book series over a decade ago. The first four *DPER* volumes dealt primarily with methodologies employed by palaeolimnologists. Subsequent volumes addressed a spectrum of palaeoenvironmental applications, ranging from ice cores to dendrochronology to the study of sedimentary deposits from around the globe.

This book does not deal with the collection and synthesis of primary data, but instead it discusses the key role of data handling and numerical and statistical approaches in analysing palaeolimnological data. As summarised in our introductory chapter, palaeoenvironmental research has steadily moved from studies based on one or a few types of proxy data to large, data-rich, multi-disciplinary studies. In addition, there has been a rapid shift from simply using qualitative interpretations based on indicator species to more quantitative assessments. Although there remains an important place in palaeoenvironmental research for qualitative analyses, the reality is that many researchers now employ a wide range of numerical and statistical methodologies. It is time to review critically some of these approaches and thereby make them more accessible to the wider research community. We hope the 21 chapters making up this volume meet these goals.

Many people helped with the planning, development, and final production of this book. We would like to acknowledge the hard work and professionalism of our many reviewers, who provided constructive comments on earlier drafts of the manuscripts. We would also like to acknowledge the assistance we received from our publishers, and especially the efforts and encouragement from our main Springer colleagues—Tamara Welschot and Judith Terpos. We are grateful to Irène Hofmann for her work in the early stages of the book. We are particularly indebted to the enormous amount of work that Cathy Jenks has done in the processing and editing of the chapters, compiling and checking bibliographies and the glossary, and in the overall production of this book. Thanks are also due to our host institutions and our various funding sources, which helped facilitate our academic endeavours. We also gratefully acknowledge a variety of publishers and authors who allowed us to reproduce previously published figures. Foremost, we would like to thank the authors for their hard work and especially for their patience with the delays in completing this book. We hope that the final product was worth the wait.

Structure of the Book

This book consists of 21 chapters arranged in four parts. Part I is introductory and Chap. 1 considers the rapid development and ever-increasing application of numerical and statistical techniques in palaeolimnology. Chapter 2 provides an overview of the basic numerical and statistical approaches used in palaeolimnology in the context of data collection and assessment, data summarisation, data analysis, and data interpretation. Many of these techniques are described in more detail in chapters in Parts II and III but some important approaches such as classification, assignment and identification, and regression analysis and statistical modelling are described in Chap. 2 as they are not specifically covered elsewhere in the book. Chapter 3 describes the modern and stratigraphical data-sets that are used in some of the later chapters.

Part II considers numerical approaches that can be usefully applied to the two major types of palaeolimnological data, namely modern surface-sediment data-sets and core sediment data-sets. These approaches are exploratory data analysis and data display (Chap. 5), assessment of uncertainties associated with laboratory methods and microfossil analysis (Chap. 6), clustering and partitioning (Chap. 7), classical indirect and canonical direct ordination (Chap. 8), and a battery of techniques grouped together as statistical-learning methods in Chap. 9. These include classification and regression trees, multivariate regression trees, other types of tree-based methods, artificial neural networks and self-organising maps, Bayesian networks and genetic algorithms, principal curves, and shrinkage methods. Some other numerical techniques are not covered in these five chapters (e.g., estimating compositional turnover, richness, and species optima and tolerances, and comparing clusterings and ordinations) because the topics are not sufficiently large to warrant individual chapters. They are outlined in Chap. 4 as an introduction to Part II.

Part III contains seven chapters. They describe numerical techniques that are only applicable to the quantitative analysis of stratigraphical data-sets. Chapter 11 discusses numerical techniques for zoning or partitioning stratigraphical sequences and for detecting patterns within stratigraphical data-sets. Chapter 12 considers the essential task of establishing age-depth relationships that provide the basis for estimating rates of change and temporal patterns within and between stratigraphical sequences. Chapter 13 discusses an important but rarely used approach to core correlation by sequence-slotting. Chapter 14 discusses the quantitative reconstruction of environmental variables such as lake-water pH from, for example, fossil diatom assemblages. This general topic of environmental reconstruction has been a central focus of many palaeolimnological investigations in the last 20 years and Chap. 14 highlights the assumptions and limitations of such reconstructions, and the testing, evaluation, and validation of reconstructions. Chapter 15 considers modern analogue methods in palaeolimnology as a procedure for quantitative environmental reconstructions and for comparing fossil and modern assemblages as a tool in lake restoration and management. Chapter 16 concludes Part III by presenting new approaches to assessing temporal patterns in palaeolimnological temporal-series where the major assumptions of conventional time-series analysis are not met. Other numerical techniques such as palaeopopulation analysis, stratigraphical changes in richness, and approaches to temporal-series analysis such as LOESS smoothing and the SiZer (Significant Zero crossings of the derivative) approach and its relatives BSiZer and SiNos that are not discussed in Chaps. 11, 12, 13, 14, 15 and 16 are outlined briefly in Chap. 10, which also provides an overview and introduction to Part III.

Part IV consists of five chapters. Chapter 17 provides an introduction and overview to this Part. Three chapters (Chaps. 18, 19, 20) describe case studies where some of the numerical methods presented in Parts II and III are used to answer particular palaeolimnological research questions and to test palaeolimnological hypotheses. Chapter 18 considers limnological responses to environmental changes at inter-annual to decadal time-scales. Chapter 19 reviews the application of numerical techniques to evaluate surface-water acidification and eutrophication. Chapter 20 discusses tracking Holocene climatic change using stratigraphical palaeolimnological data and numerical techniques. The last chapter, Chap. 21, discusses eight areas of research that represent future challenges in the improved numerical analysis of palaeolimnological data.

Data-sets, figures, software, and R scripts used or mentioned in this book, links to important websites relevant to this book and its contents are available from Springer's Extras website (http://extras.springer.com).

About The Editors

H. John B. Birks is Professor in Quantitative Ecology and Palaeoecology at the Department of Biology, University of Bergen and the Bjerknes Centre for Climate Research (Norway), Emeritus Professor and Fellow at the Environmental Change Research Centre, University College London (UK), and Distinguished Visiting Fellow at the School of Geography and the Environment, University of Oxford (UK). He was Adjunct Professor in the Department of Biology at Queen's University (Kingston, Ontario, Canada) until 2011. His teaching and research include ecology, biogeography, palaeoecology, palaeolimnology, pollen analysis, numerical methods, and alpine botany.

André F. Lotter is Professor in Palaeoecology at the Institute of Environmental Biology of Utrecht University (The Netherlands), where he is head of the Laboratory of Palaeobotany and Palynology.

Steve Juggins is a Senior Lecturer and Head of Department in the School of Geography, Politics and Sociology, Newcastle University (UK), where he teaches and researches issues of aquatic pollution, diatom analysis, palaeolimnology, and quantitative palaeoecology.

John P. Smol is a Professor in the Department of Biology at Queen's University (Kingston, Ontario, Canada), where he also holds the *Canada Research Chair in Environmental Change*. He co-directs the Paleoecological Environmental Assessment and Research Lab (PEARL). John Smol was the founding editor of the international *Journal of Paleolimnology* (1987–2007) and is current editor of the journal *Environmental Reviews*, and editor of the *Developments in Paleoenvironmental Research* book series.

Contents

Contributors

N. John Anderson Department of Geography, Loughborough University, Loughborough, UK

H. John B. Birks Department of Biology and Bjerknes Centre for Climate Research, University of Bergen, Bergen, Norway

Environmental Change Research Centre, University College London, London, UK

School of Geography and the Environment, University of Oxford, Oxford, UK

Maarten Blaauw School of Geography, Archaeology & Palaeoecology, Queen's University Belfast, Belfast, UK

Geoffrey S. Boulton School of GeoSciences, University of Edinburgh, Edinburgh, UK

R. Malcolm Clark School of Mathematical Sciences, Monash University, Clayton, VIC, Australia

Brian F. Cumming Paleoecological Environmental Assessment and Research Laboratory (PEARL), Department of Biology, Queen's University, Kingston, ON, Canada

Pierre Dutilleul Laboratory of Applied Statistics, Department of Plant Science, Faculty of Agricultural and Environmental Sciences, McGill University, Ste-Anne-de-Bellevue, QC, Canada

Sherylyn C. Fritz Department of Earth and Atmospheric Sciences, University of Nebraska, Lincoln, NE, USA

Roland I. Hall Department of Biology, University of Waterloo, Waterloo, ON, Canada

Einar Heegaard Department of Biology and Bjerknes Centre for Climate Research, University of Bergen, Bergen, Norway

Norwegian Forest and Landscape Institute, Fana, Norway

Oliver Heiri Institute of Plant Sciences and Oeschger Centre for Climate Change Research, University of Bern, Bern, Switzerland

Vivienne J. Jones Environmental Change Research Centre, University College London, London, UK

Steve Juggins School of Geography, Politics & Sociology, Newcastle University, Newcastle-upon-Tyne, UK

Kathleen R. Laird Paleoecological Environmental Assessment and Research Laboratory (PEARL), Department of Biology, Queen's University, Kingston, ON, Canada

Pierre Legendre Département de sciences biologiques, Université de Montréal, Montréal, QC, Canada

Melinda Lontoc-Roy Laboratory of Applied Statistics, Department of Plant Science, Faculty of Agricultural and Environmental Sciences, McGill University, Ste-Anne-de-Bellevue, QC, Canada

André F. Lotter Laboratory of Palaeobotany and Palynology, Department of Biology, University of Utrecht, Utrecht, The Netherlands

Louis J. Maher Jr. Geology and Geophysics, University of Wisconsin, Madison, WI, USA

Gavin L. Simpson Environmental Change Research Centre, University College London, London, UK

John P. Smol Paleoecological Environmental Assessment and Research Lab (PEARL), Department of Biology, Queen's University, Kingston, ON, Canada

Richard J. Telford Department of Biology and Bjerknes Centre for Climate Research, University of Bergen, Bergen, Norway

Roy Thompson School of GeoSciences, University of Edinburgh, Edinburgh, UK

Dirk Verschuren Limnology Unit, Department of Biology, Ghent University, KL Ledeganckstraat 35, B-9000, Ghent, Belgium

Part I
Introduction, Numerical Overview, and Data-Sets

Chapter 1
The March Towards the Quantitative Analysis of Palaeolimnological Data

John P. Smol, H. John B. Birks, André F. Lotter, and Steve Juggins

Abstract We outline the aims of palaeolimnology and describe the major types of palaeolimnological data. The distinction between biological data derived from stratigraphical studies of cores and modern surface-sediment samples with associated environmental data is discussed. A brief history of the development of quantitative palaeolimnology is presented, starting with early applications of principal component analysis in 1975. Major developments occurred in the late 1980s, thanks to the work of Cajo ter Braak and others. The structure of the book in terms of four parts is explained. Part I is introductory and presents an overview of numerical methods and of the data-sets used. Part II presents numerical approaches

J.P. Smol (✉)
Paleoecological Environmental Assessment and Research Lab (PEARL), Department of Biology, Queen's University, Kingston, ON K7L 3N6, Canada
e-mail: smolj@queensu.ca

H.J.B. Birks
Department of Biology and Bjerknes Centre for Climate Research, University of Bergen, PO Box 7803, Bergen N-5020, Norway

Environmental Change Research Centre, University College London, Pearson Building, Gower Street, London WC1E 6BT, UK

School of Geography and the Environment, University of Oxford, Oxford OX1 3QY, UK
e-mail: john.birks@bio.uib.no

A.F. Lotter
Laboratory of Palaeobotany and Palynology, Department of Biology, University of Utrecht, Budapestlaan 4, Utrecht NL-3584 CD, The Netherlands
e-mail: a.f.lotter@uu.nl

S. Juggins
School of Geography, Politics & Sociology, University of Newcastle, Newcastle-upon-Tyne NE1 7RU, UK
e-mail: stephen.juggins@ncl.ac.uk

H.J.B. Birks et al. (eds.), *Tracking Environmental Change Using Lake Sediments*, Developments in Paleoenvironmental Research 5, DOI 10.1007/978-94-007-2745-8_1,

appropriate to the analysis of modern and stratigraphical palaeolimnological data. Part III considers numerical techniques that are only applicable to stratigraphical data, and Part IV presents three case-studies and concludes with a discussion of future challenges.

Keywords Calibration • Calibration functions • Data-sets • Numerical methods • Palaeolimnology • Temporal scales • Transfer functions

Palaeolimnology

Palaeolimnology can broadly be defined as the study of the physical, chemical, and biological information stored in lake and river sediments. As such, palaeolimnology is a multi-disciplinary science with many diverse applications. The questions posed by palaeolimnologists can vary widely, ranging from applied issues, such as tracking the effects of lake acidification, eutrophication, chemical contamination, and erosion, to more fundamental scientific subjects, such as examining hypotheses regarding biogeography, evolution, natural modes of climatic change, and theoretical ecology. Common questions posed by palaeolimnologists include: Have lakes changed? If so, when and by how much? What was the cause of the change? How have species distributions and abundances changed over long time frames? Not surprisingly, given the growing interest and concern in environmental problems, and the general lack of reliable long-term monitoring data, a large portion of recent palaeolimnological research has been directed to applied issues (Smol 2008).

The overall palaeolimnological approach is relatively straightforward. The raw materials used are lake sediments[1] which, under ideal circumstances, accumulate at the bottom of a basin in an orderly and undisturbed manner. In a typical study, sediment cores can be retrieved using a variety of sampling devices, after which the sediment profiles can be sectioned into appropriate time slices, and the age-depth profile can be established using geochronological techniques (Last and Smol 2001a). Incorporated in these sediments is a diverse array of indicators and other proxy data (Last and Smol 2001b; Smol et al. 2001a, b); the palaeolimnologist's job is then to interpret these proxy data in a defendable and rigorous manner that is of interest to other scientists and the public at large.

Several recent textbooks (e.g., Cohen 2003; Smol 2008) have been published on palaeolimnology, synthesising this rapidly growing discipline. Furthermore, many palaeolimnological approaches and methods have been standardised, at least

---

[1]In this chapter and in this book as a whole, we will refer primarily to lake sediments; however, palaeolimnologists can also use pond, river, wetland, estuarine, and other types of sediment profiles, assuming reliable and undisturbed stratigraphic sequences can be retrieved.

at a general level (e.g., Last and Smol 2001a, b; Smol et al. 2001a, b). Over the last two decades, the amount, diversity, and quality of data generated by palaeolimnologists have been increasing steadily (Pienitz et al. 2009), with many studies producing large and complex data-sets. Parallel with these advances in data generation have been research and developments on quantifying these data in a numerically or statistically robust fashion. Not surprisingly many numerical and statistical techniques are now standard components of the palaeolimnologist's tool-kit. This book summarises some of these numerical approaches.

Types of Palaeolimnological Data

The quantity, quality, and diversity of palaeolimnological proxy data grow steadily. In the 1970s most palaeolimnological studies were largely restricted to some geochemical data, perhaps coupled with analyses of fossil diatoms and pollen grains. Today, a typical palaeolimnological study may include ten or more different types of proxy data. A scan of papers published in the international *Journal of Paleolimnology* shows a clear trend of larger and more multi-authored papers since the journal's inception in 1988.

Smol (2008) provides summaries of the commonly used palaeolimnological indicators, whilst previous volumes in this *DPER* book series (e.g., Last and Smol 2001a, b; Smol et al. 2001a, b; Battarbee et al. 2004; Francus 2004; Pienitz et al. 2004; Leng 2006) contain more detailed reviews. Sediment components are typically categorised, at least at a broad level, by their source. Allochthonous components originate from outside the lake basin, such as soil particles and pollen grains from terrestrial vegetation. Autochthonous components originate from the lake itself, including algal and aquatic invertebrates or chemical precipitates. The list of physical, chemical, and biological indicators continues to grow steadily (Pienitz et al. 2009). Amongst the biological indicators, it is true that diatom valves, chironomid head capsules, and pollen grains are most frequently used, but virtually every organism living in a lake system leaves some sort of morphological or biogeochemical (e.g., fossil pigments, lipids) indicator. These biotic indicators are either used to reconstruct past environmental conditions (e.g., lake-water pH or phosphorus, temperature, oxygen availability) in a qualitative or quantitative way or their reaction to different stressors (e.g., climate change, nutrient enrichment, heavy metal pollution) is studied using numerical methods. Similarly, a broad spectrum of inorganic and organic chemical and physical markers (e.g., metals, isotopes, persistent organic pollutants) can be used to interpret lake histories (e.g., Coolen and Gibson 2009; Francus et al. 2009; Heiri et al. 2009; Weijers et al. 2009). Although most of our examples will deal with biological indicators, numerical approaches are often equally applicable to chemical- and physical-based studies (Birks 1985, 1987; Rosén et al. 2000, 2010; Grosjean et al. 2009).

Different Temporal Scales: From Surface-Sediment Calibration Sets to Detailed Sediment-Core Studies

Palaeolimnologists use sediment cores of appropriate length (i.e., temporal range) and sectioned into time slices to answer the research questions under study (Last and Smol 2001a) and to track long-term changes within a specific lake or other aquatic system (e.g., a bay of a river). Proxy indicators (e.g., diatom valves, geochemical markers, isotopic data) contained in these sediment time-slices are isolated, identified, and counted or analysed in various ways (Last and Smol 2001b; Smol et al. 2001a, b). In studies conducted before the 1980s, most biological palaeolimnologists would have little choice but to interpret the stratigraphical changes in bio-indicators (or other types of data) qualitatively using whatever ecological data were available in the scientific literature at that time. However, the increased use of surface-sediment, modern calibration data-sets (also known as modern training-sets), beginning primarily in the late 1980s, was a major step forward in quantifying and aiding the interpretation of information preserved in sediment cores.

The concepts and assumptions underpinning surface-sediment calibration data-sets are fairly straightforward (Smol 2008), although the statistical treatment of these data is not so simple (Birks 1998, 2010; Birks et al. 2010). For example, suppose a palaeolimnologist wishes to reconstruct lake-water pH using diatoms preserved in sediment cores for a particular region. The first question one might have is how would the palaeolimnologist provide any palaeoenvironmental interpretations based on these assemblages, which may easily encompass several hundred diatom taxa, in a quantifiable and statistically defendable manner? Surface-sediment calibration sets have made this possible. A suite of calibration lakes (typically 40 or more in number) are first carefully chosen to reflect the limnological conditions that are likely to be encountered (and therefore need to be reconstructed) from the down-core sediment assemblages. For example, if this is an acidification study, and past pH may have fluctuated approximately over a pH range of about 5.5–7.0, it would be prudent to choose a calibration-set that encompasses lakes with current pH levels of about 5.0–7.5, or so. Recent limnological data for the calibration lakes are collated, which should include the limnological variables that may most likely influence assemblages (e.g., lake-water pH, nutrients, and other physical, chemical, and possibly biological factors). This represents the first data matrix: the environmental data. The second challenge is to characterise the recent biological assemblages present in the calibration lakes (in this example, diatom species composition and abundance). Because surface sediments (i.e., the top 1 or 0.5 cm) contain assemblages that lived in the lake's recent past (i.e., last few years), these data are used for the second data matrix: the species data. Numerical and statistical techniques, as described in this book, are then used to explore, define, and quantify the relationships between the two data matrices, and develop calibration or transfer functions whereby the palaeolimnologist can reconstruct past environmental conditions based on the indicators preserved in sediment profiles.

A large portion of this book addresses various numerical or statistical approaches that have been developed to deal with these complex data. Although almost all of this calibration work has focused on biological indicators, similar approaches can be used for quantifying other types of proxy data (Pienitz et al. 2009).

Palaeolimnologists, however, use a variety of temporal frameworks. Surface-sediment calibration-sets represent only one of many types of palaeolimnological applications that require numerical analyses. For example, a broad spectrum of different types of data can be generated for the down-core portion of the study. The complexity and diversity of proxy data may at times be staggering, especially in multi-proxy studies (Lotter 2003; Birks and Birks 2006). Moreover, the data can be collected and presented in different ways, including relative frequencies, concentrations, accumulation rates, and various ratios. In some cases, simple presence and absence data can be useful (e.g., Sweetman and Smol 2006). Each approach may provide some important palaeoenvironmental insights, but also requires a careful assessment and evaluation of various assumptions.

The adoption and widespread use of numerical methods in palaeolimnology has been surprisingly rapid. Early work (e.g., pre-1990s) was typically restricted to qualitative interpretations of species distributions, or the development of simple indices and ratios (for a historical review, see Battarbee et al. 1986). Some of the earliest numerical work was by Pennington and Sackin (1975), who used principal component analysis on down-core pollen and geochemical data. By the late 1970s and early 1980s, some researchers were using, for example, simple agglomerative cluster analyses to group fossil samples (e.g., Davis and Norton 1978; Jatkar et al. 1979; Norton et al. 1981; Carney 1982), or stratigraphically constrained cluster analysis to derive fossil assemblage zones (Binford 1982). Several early studies showed the value of applying numerical methods such as ordinations or clusterings to summarise patterns in modern sediment geochemistry (Dean and Gorham 1976a, b) and in modern assemblages of, for example, plant macrofossils (Birks 1973), cladocerans (Beales 1976; Hofmann 1978; Synerholm 1979), diatoms (Brugam 1980; Bruno and Lowe 1980), and ostracods (Kaesler 1966; Maddocks 1966). Similarly, palaeolimnologists began to apply numerical partitioning, clustering, or ordination techniques to sediment lithostratigraphical (Brown 1985), biological, and geochemical data (Sergeeva 1983; Peglar et al. 1984).

In his pioneering study, Binford (1982) used the results of numerical analyses of ostracod and cladoceran assemblages to reconstruct, in a semi-quantitative way, changes in water-level, salinity, and substrate over the past 12,000 years at Lake Valencia, Venezuela. Other early and creative applications of multivariate data analysis in palaeolimnology include Whiteside (1970) who used multiple discriminant analysis (= canonical variates analysis) to relate fossil assemblages of cladocerans in Danish lake sediments to modern assemblages, and Elner and Happey-Wood (1980) who used principal component and correspondence analyses to compare diatom, pollen, and geochemical stratigraphies in Holocene sedimentary records from two lakes in North Wales.

In the related field of Quaternary pollen analysis, Maher (1972a, b, 1980, 1981) emphasised the importance of considering the inherent errors in counting microfossils and showed how robust confidence intervals could be calculated for both relative percentage and concentration and accumulation-rate data (see Maher et al. 2012: Chap. 6). Maher's pioneering work on analytical and counting errors will hopefully become more used as research questions in palaeolimnology become increasingly more refined and more demanding in terms of data precision, accuracy, and uncertainties.

By the late 1980s, several multivariate techniques were being used to summarise patterns in palaeolimnological data but the techniques used were often not optimal for the research problems being addressed (e.g., unconstrained cluster analysis applied to stratigraphical time-constrained data). Several important publications in the late 1980s changed the way that many palaeolimnologists analysed their data numerically. These publications included (1) ter Braak (1986) where canonical correspondence analysis was introduced as a means of analysing species compositional and environmental data simultaneously, (2) Birks and Gordon (1985) where numerical techniques developed specifically for the analysis of Quaternary pollen-stratigraphical data were synthesised, and (3) ter Braak and Prentice (1988) where a unified theory of gradient analysis was presented with the first explicit distinction between gradient analytical techniques (e.g., regression, calibration, ordination, constrained ordination) appropriate for species data with linear or monotonic responses to the environment and for species data with unimodal responses to the environment. The scene was then set for the seminal paper by ter Braak and van Dam (1989) where two-way weighted-averaging regression and calibration were shown to be an effective and robust way of inferring lake-water pH from diatom assemblages. A year later Birks et al. (1990) demonstrated how some of these approaches could be applied in a rigorous manner to lake acidification studies, developed numerical methods for estimating sample-specific errors of prediction, and presented various numerical approaches for evaluating environmental reconstructions.

At about the same time as these publications and developments, the INQUA Commission for the Study of the Holocene, under its President-elect Brigitta Ammann, established a working group on data-handling methods in 1987. An annual or 6-monthly newsletter containing details of software, relevant literature, methodological developments, and on-going research was produced and widely distributed amongst numerical palaeoecologists and palaeolimnologists. It was originally edited by JC Ritchie (1988–1990) and then by LJ Maher (1990–1997). The newsletter was then edited by KD Bennett and the last issue was in 2003. Besides providing useful newsletters with many articles of direct interest to the rapidly evolving cohort of quantitative palaeoecologists and palaeolimnologists, Lou Maher created an invaluable file boutique of useful software for estimating confidence intervals, sequence-slotting, rarefaction analysis, etc. Although these programs were written to run under MS-DOS, they can, with care, be run under Microsoft Windows® and other operating systems. They represent a rich array of software, much of which is as relevant to the quantitative palaeolimnologist in the twenty-first century as they were to the pioneering palaeolimnologists of the mid 1980s. For details of all the newsletters, software, publications, etc., go to http://www.geology.wisc.edu/~maher/inqua.html or http://www.chrono.qub.ac.uk/inqua/index.htm. The working group on data-handling methods made a significant contribution to the spread of numerical ideas, literature, and software amongst the palaeolimnological research community in the late 1980s-late 1990s.

Opportunities and Challenges

Palaeolimnologists can justifiably be proud of their accomplishments, but there also remain many challenges and opportunities. Of course, no amount of 'statistical finesse' will ever compensate for a poor data-set or a poorly designed sampling programme. Although interpretations based on multi-proxy studies typically provide much stronger environmental and ecological interpretations, they are not without their problems (Lotter 2003; Birks and Birks 2006). Nonetheless, exciting opportunities are available as more, high quality data-sets become available. For example, as many palaeolimnological approaches are now standardised to a certain level, and because a large number of studies have been completed for some regions, it is now possible to undertake meta-analyses of large data-sets to probe various hypotheses (e.g., Smol et al. 2005; Rühland et al. 2008). Much scope remains for these types of syntheses. Moreover, in the quest for more robust calibration functions, much of the hard-earned ecological and biogeographical data contained in modern calibration-sets (e.g., Telford et al. 2006; Vyverman et al. 2007; Vanormelingen et al. 2008; Bennett et al. 2010) often remains under-utilised. It is important to keep in mind that, even in a book on numerical analyses, detailed statistical interpretations are not always needed in palaeoecological studies, and in fact elegant ecological work can often be done at a qualitative level. For example, as certain *Chaoborus* species cannot co-exist with fish, the simple presence of mandibles from one species of this taxon in a lake's profile may indicate fishless conditions (e.g., Uutala 1990).

Outline of the Book

This book focuses on numerical and statistical methods that have been widely applied in palaeolimnology (e.g., ordination methods, weighted averaging) or that have considerable potential in palaeolimnology (e.g., classification and regression trees). All the methods presented are robust and can often take account of the numerical properties of many palaeolimnological data-sets (e.g., closed percentage data, many variables, many zero values). They can all be used to answer specific research questions in palaeolimnology that can contribute to our understanding of lake history, development, and responses to a range of environmental factors. The emphasis throughout is on numerical thinking (ideas, reasons, and potentialities of numerical techniques), rather than numerical arithmetic (the actual numerical manipulations involved).

The book aims to provide an understanding of the most appropriate numerical methods for the quantitative analysis of complex multivariate palaeolimnological data. There has in the last 15–20 years been a maturation of many methods. The book provides information to what these methods can and cannot do and some guidance as to when and when not to use particular methods. It attempts to outline the major assumptions, limitations, strengths, and weaknesses of different methods.

The book is divided into four parts and contains 21 chapters. Part I is introductory and contains this chapter, an overview (Chap. 2) of the basic numerical and statistical methods (e.g., regression analysis and statistical modelling) used in palaeolimnology, and Chap. 3 that describes the data-sets used in some of the chapters. Part II includes chapters on numerical methods that can usefully be applied to the analysis of modern surface-sediment data and of core sediment data. Chapter 4 gives an introduction and overview of the methods in this part. Chapter 5 explores the essential first step of exploratory data analysis and graphical display and the important questions of identifying potential outlying data-points in large complex data-sets. Chapter 6 considers a topic that is surprisingly rarely considered in palaeolimnology, namely the assessment of uncertainties associated with laboratory methods and microfossil analysis. Chapter 7 outlines the basic techniques available for the clustering and partitioning of multivariate palaeolimnological data to detect groups or clusters and to establish the relationships of biologically defined clusters and environmental variables. Chapter 8 discusses the range of ordination techniques currently available to palaeolimnologists to detect and summarise patterns in both modern and core data. These include classical techniques such as principal component analysis and correspondence analysis and canonical or constrained techniques such as canonical correspondence analysis and redundancy analysis. The chapter also presents new and improved techniques for partitioning variation in data-sets and in detecting spatial or temporal structures at a range of scales. Chapter 9 outlines statistical-learning techniques such as the various types of classification and regression trees and artificial neural networks as useful tools in the exploration and mining of very large, heterogeneous data-sets and in developing robust, simple predictive models. It also considers powerful techniques such as principal curves and surfaces as a means of summarising patterns in complex multivariate data and shrinkage techniques such as ridge regression, the lasso, and the elastic net to help develop robust regression models based on large data-sets.

Part III is devoted to numerical techniques that are only applicable to stratigraphical data-sets and an overview of these techniques is given in Chap. 10. Chapter 11 describes basic techniques for summarising patterns in stratigraphical data by partitioning, clustering, or ordination methods, and for estimating rates of change within stratigraphical sequences. Chapter 12 discusses the critical problem of establishing age-depth relationships. It outlines procedures for calibrating radiocarbon dates first because age-depth models based on uncalibrated dates are meaningless. It reviews various age-depth modelling procedures and discusses the difficult problem of deciding which model to accept and to use. Age-depth modelling is an area where considerable advances are being made, in particular by adopting a Bayesian approach to age calibration and age-depth modelling. Chapter 13 outlines robust numerical procedures for correlating two or more cores on the basis of some measured properties (e.g., loss-on-ignition, magnetic susceptibility). These procedures can be very useful in correlating multiple undated cores with a master dated core from a lake. Chapter 14 discusses the quantitative reconstruction of past environmental variables from fossil assemblages using calibration or transfer functions. A range of such calibration-function methods is discussed, the question of which model(s) to select is explored, and the ecological problems and pitfalls of interpreting palaeoenvironmental reconstructions derived from calibration functions are considered. Chapter 15 outlines the modern analogue technique, a useful technique not only in environmental reconstruction but also in identifying potential 'reference lakes' in restoration programmes. Chapter 16 provides an introduction to robust techniques for analysing patterns in temporally ordered palaeolimnological data. The chapter cautions against the use of several conventional time-series analysis techniques (e.g., spectral analysis) that are designed for evenly spaced samples in the time-series, a property that is very rarely realised in palaeolimnology.

Part IV begins with an introduction (Chap. 17) and considers three different case studies where many of the numerical methods described in Parts II and III are used to answer particular research questions in palaeolimnology such as limnological responses to environmental changes (Chap. 18), human impacts of acidification and eutrophication (Chap. 19), and tracking climatic change using palaeolimnological techniques (Chap. 20). These three chapters highlight that the numerical methods discussed in Parts II and III are not ends in themselves but are a means to an end. In the case of palaeolimnology, the end is to improve our understanding of the timing, rates, magnitudes, and causes of limnological changes over a range of time scales. Part IV concludes with some views on where quantitative palaeolimnology has reached, what the future challenges are, and what are the limitations of our current data-sets and our numerical methods (Chap. 21).

This book makes no attempt to review the now vast literature on the application of numerical methods in palaeolimnology. Almost every paper in palaeolimnology involves at least one numerical analysis. This book similarly makes no attempt to be a complete textbook on numerical methods in ecology, environmental sciences, or palaeontology. Instead it concentrates on numerical techniques of primary interest and relevance to palaeolimnologists. For palaeolimnologists interested in textbooks on numerical ecology, environmental sciences, or palaeontology, we recommend the following:

1. Numerical ecology

Jongman RHG, ter Braak CJF, van Tongeren OFR (1987) Data analysis in community and landscape ecology. Pudoc, Wageningen, 299 pp

Legendre P, Legendre L (1998) Numerical ecology, 2nd English edn. Elsevier, Amsterdam, 853 pp

Lepš J, Šmilauer P (2003) Multivariate analysis of ecological data using CANOCO. Cambridge University Press, Cambridge, 269 pp

McCune B, Grace, JB, Urban DL (2002) Analysis of ecological communities. MjM Software Design, Gleneden Beach, Oregon, 300 pp
Zuur AF, Ieno EN, Smith GM (2007) Analysing ecological data. Springer, New York, 672 pp

2. Environmental sciences

Fielding AH (2007) Cluster and classification techniques for the biosciences. Cambridge University Press, Cambridge, 246 pp
Hanrahan G (2009) Environmental chemometrics. CRC, Boca Raton, 292 pp
Manly BFJ (2009) Statistics for environmental science and management, 2nd edn. CRC, Boca Raton, 295 pp
Qian SS (2010) Environmental and ecological statistics with R. CRC Press, Boca Raton, 421 pp
Quinn GP, Keough MJ (2002) Experimental design and data analysis for biologists. Cambridge University Press, Cambridge, 537 pp
Shaw PJA (2003) Multivariate statistics for the environmental sciences. Arnold, London, 233 pp
Sparks T (ed) (2000) Statistics in ecotoxicology. Wiley, Chichester, 320 pp
Varmuza K, Filzmoser P (2009) Introduction to multivariate statistical analysis in chemometrics. CRC Press, Boca Raton, 321 pp

3. Palaeontology

Davis JC (2002) Statistics and data analysis in geology, 3rd edn. Wiley, New York, 646 pp
Hammer Ø, Harper DAT (2006) Paleontological data analysis. Blackwell, Oxford, 351 pp
Haslett SK (ed) (2002) Quaternary environmental micropalaeontology. Arnold, London, 340 pp
Reyment RA, Savazzi E (1999) Aspects of multivariate statistical analysis in geology. Elsevier, Amsterdam, 285 pp

Because the development of user-friendly Windows® type software for implementing specialised numerical and statistical analyses lags far behind the development of the actual numerical techniques, we give limited references in the chapters to available software for particular analyses. We believe that there will inevitably be future developments in numerical palaeolimnology but that these developments will only be available to researchers as scripts and packages in the R programming language and its vast libraries for statistical and numerical procedures. Useful introductions to R as well as to basic statistics and statistical modelling include:

Aitkin M, Francis B, Hinde J, Darnell R (2009) Statistical modelling with R. Oxford University Press, Oxford, 576 pp
Borcard D, Gillet F, Legendre P (2011) Numerical ecology with R. Springer, New York, 206 pp

Cohen Y, Cohen JY (2008) Statistics and data with R. An applied approach through examples. Wiley, Chichester, 599 pp
Crawley MJ (2005) Statistics. An introduction using R. Wiley, Chichester, 327 pp
Crawley MJ (2007) The R book. Wiley, Chichester, 942 pp
Dalgaard P (2008) Introductory statistics with R, 2nd edn. Springer, New York, 364 pp
Everitt BS (2005) An R and S-PLUS companion to multivariate analysis. Springer, London, 222 pp
Everitt BS, Hothorn T (2009) A handbook of statistical analyses using R. Chapman & Hall/CRC, London, 376 pp
Everitt BS, Hothorn T (2011) An introduction to applied multivariate analysis with R. Springer, New York, 283 pp
Fox J (2002) An R and S-PLUS companion to applied regression. Sage, Thousand Oaks, 312 pp
Good PI (2005) Introduction to statistics through resampling methods and R/S-PLUS. Wiley, Hoboken, 229 pp
Logan M (2010) Biostatistical design and analysis using R. A practical guide. Wiley-Blackwell, Chichester, 546 pp
Reiman C, Filzmoser P, Garrett R, Dutter R (2008) Statistical data analysis explained – applied environmental statistics with R. Wiley, Chichester, 343 pp
Torgo L (2011) Data mining with R. CRC Press, Boca Raton, 289 pp
Verzani J (2005) Using R for introductory statistics. Chapman & Hall/CRC, Boca Raton, 414 pp
Wehrens R (2011) Chemometrics with R. Springer, New York, 285 pp
Wright DB, London K (2009) Modern regression techniques using R. A practical guide for students and researchers. Sage, London, 204 pp
Zuur AF, Ieno EN, Meesters EHWG (2009) A beginner's guide to R. Springer, New York, 218 pp

This book, by necessity, assumes a basic knowledge of statistics and modern data-analytical procedures. There are many excellent and clearly written books on basic statistics and multivariate analysis relevant to ecology (and hence palaeoecology and palaeolimnology). These include, in addition to those mentioned above, the following:

Crawley MJ (2002) Statistical computing. An introduction to data analysis using S-PLUS. Wiley, Chichester, 761 pp
Everitt BS, Dunn G (2001) Applied multivariate data analysis, 2nd edn. Arnold, London, 342 pp
Fox J (1997) Applied regression analysis, linear models, and related methods. Sage, Thousand Oaks, 597 pp
Fox J (2008) Applied regression analysis and generalized linear models, 2nd edn. Sage, Thousand Oaks, 665 pp
Gotelli NJ, Ellison AM (2004) A primer of ecological statistics. Sinauer Associates, Sunderland, MA, 510 pp

Gotelli NJ, Graves GR (1996) Null models in ecology. Smithsonian Institution Press, Washington, 368 pp
Manly BFJ (2005) Multivariate statistical methods. A primer, 3rd edn. Chapman & Hall/CRC, Boca Raton, 214 pp
Manly BFJ (2007) Randomization, bootstrap, and Monte Carlo methods in biology, 3rd edn. Chapman & Hall/CRC, Boca Raton, 326 pp
Roff DA (2006) Introduction to computer-intensive methods of data analysis in biology. Cambridge University Press, Cambridge, 368 pp
Scheiner SM, Gurevitch J (eds) (2001) Design and analysis of ecological experiments. Oxford University Press, Oxford, 415 pp
Sokal RR, Rohlf FJ (1995) Biometry, 2nd edn. WH Freeman, New York, 887 pp
van Belle G (2008) Statistical rules of thumb. Wiley, Hoboken, 272 pp
Waite S (2000) Statistical ecology in practice. Prentice Hall, London, 414 pp
Whitlock M, Schluter D (2009) The analysis of biological data. Roberts & Company, Greenwood Village, Colorado, 700 pp

In addition, the reader may find BS Everitt's (2002) *The Cambridge Dictionary of Statistics* (2nd edition, Cambridge University Press, Cambridge, 410 pp) a useful reference work for clear, simple definitions of particular statistical terms and concepts used in this book. Nomenclature of numerical terms and methods follows, wherever possible, Everitt (2002).

Acknowledgements We acknowledge Hilary Birks for useful comments and Cathy Jenks for invaluable help in the preparation of this chapter. This is publication number A207 from the Bjerknes Centre for Climate Research.

References

Battarbee RW, Smol JP, Meriläinen J (1986) Diatoms as indicators of pH: a historical review. In: Smol JP, Battarbee RW, Davis RB, Meriläinen J (eds) Diatoms and lake acidity. Junk Publications, Dordrecht, pp 5–14
Battarbee RW, Gasse F, Stickley CE (eds) (2004) Past climate variability through Europe and Africa. Springer, Dordrecht
Beales PW (1976) Palaeolimnological studies of a Shropshire mere. PhD thesis, University of Cambridge
Bennett JR, Cumming BF, Ginn BK, Smol JP (2010) Broad-scale environmental response and niche conservatism in lacustrine diatom communities. Global Ecol Biogeogr 19:724–732
Binford MW (1982) Ecological history of Lake Valencia, Venezuela: interpretation of animal microfossils and some chemical, physical, and geological factors. Ecol Monogr 52:307–333
Birks HH (1973) Modern macrofossil assemblages in lake sediments in Minnesota. In: Birks HJB, West RG (eds) Quaternary plant ecology. Blackwell, Oxford, pp 173–190
Birks HH, Birks HJB (2006) Multi-proxy studies in palaeolimnology. Veg Hist Archaeobot 15: 235–251
Birks HJB (1985) Recent and possible future mathematical developments in quantitative palaeoecology. Palaeogeogr, Palaeoclim, Palaeoecol 50:107–147
Birks HJB (1987) Multivariate analysis of stratigraphic data in geology: a review. Chemometr Intell Lab Syst 2:109–126

Birks HJB (1998) Numerical tools in palaeolimnology – progress, potentialities, and problems. J Paleolimnol 20:307–332

Birks HJB (2010) Numerical methods for the analysis of diatom assemblage data. In: Smol JP, Stoermer EF (eds) The diatoms: applications for the environmental and earth sciences, 2nd edn. Cambridge University Press, Cambridge, pp 23–54

Birks HJB, Gordon AD (1985) Numerical methods in Quaternary pollen analysis. Academic Press, London

Birks HJB, Line JM, Juggins S, Stevenson AC, ter Braak CJF (1990) Diatoms and pH reconstruction. Philos Trans R Soc Lond B 327:263–278

Birks HJB, Heiri O, Seppä H, Bjune AE (2010) Strengths and weaknesses of quantitative climate reconstructions based on late-Quaternary biological proxies. Open Ecol J 3:68–110

Brown AG (1985) Traditional and multivariate techniques applied to the interpretations of floodplain sediment grain size variations. Earth Surf Process Landf 10:281–291

Brugam RB (1980) Postglacial diatom stratigraphy of Kirchner Marsh, Minnesota. Quaternary Res 13:133–146

Bruno MG, Lowe RL (1980) Differences in the distribution of some bog diatoms: a cluster analysis. Am Midl Nat 104:70–79

Carney HJ (1982) Algal dynamics and trophic interactions in the recent history of Frains Lake, Michigan. Ecology 63:1814–1826

Cohen AS (2003) Palaeolimnology: the history and evolution of lake systems. Oxford University Press, Oxford

Coolen MJL, Gibson JAE (2009) Ancient DNA in lake sediment records. PAGES News 17: 104–106

Davis RB, Norton SA (1978) Paleolimnologic studies of human impact on lakes in the United States, with emphasis on recent research in New England. Pol Arch Hydrobiol 25:99–115

Dean WE, Gorham E (1976a) Classification of Minnesota lakes by Q- and R-mode factor analysis of sediment mineralogy and geochemistry. In: Merriam DF (ed) Quantitative techniques for the analysis of sediments. Pergamon Press, Oxford, pp 61–71

Dean WE, Gorham E (1976b) Major chemical and mineral components of profundal surface sediments in Minnesota lakes. Limnol Oceanogr 21:259–284

Elner JK, Happey-Wood CM (1980) The history of two linked but contrasting lakes in North Wales from a study of pollen, diatoms, and chemistry in sediment cores. J Ecol 68:95–121

Everitt BS (2002) The Cambridge dictionary of statistics, 2nd edn. Cambridge University Press, Cambridge

Francus P (ed) (2004) Image analysis, sediments and paleoenvironments. Springer, Dordrecht

Francus P, Lamb H, Nakagawa T, Marshall M, Brown E, Suigetsu 2006 project members (2009) The potential of high-resolution X-ray fluorescence core scanning: applications in paleolimnology. PAGES News 17:93–95

Grosjean M, von Gunten L, Trachsel M, Kamenik C (2009) Calibration-in-time: transforming biogeochemical lake sediment proxies into quantitative climate variable. PAGES News 17:108–110

Heiri O, Wooller MJ, van Hardenbroek M, Wang YV (2009) Stable isotopes in chitinous fossils of aquatic vertebrates. PAGES News 17:100–102

Hofmann W (1978) Analysis of animal microfossils from the Segerberger See (FRG). Archiv für Hydrobiologie 82:316–346

Jatkar SA, Rushforth SR, Brotherson JD (1979) Diatom floristics and succession in a peat bog near Lily Lake, Summit County, Utah. Great Basin Nat 39:15–43

Kaesler RL (1966) Quantitative re-evaluation of ecology and distribution of recent foraminfera and ostracoda of Todos Santos Bay, Baja California, Mexico. University of Kansas Paleontological Contributions 10, pp 1–50

Last WM, Smol JP (eds) (2001a) Tracking environmental change using lake sediments. Volume 1: Basin analysis, coring, and chronological techniques. Kluwer Academic Publishers, Dordrecht

Last WM, Smol JP (eds) (2001b) Tracking environmental change using lake sediments. Volume 2: Physical and geochemical methods. Kluwer Academic Publishers, Dordrecht

Leng MJ (ed) (2006) Isotopes in palaeoenvironmental research. Springer, Dordrecht

Lotter AF (2003) Multi-proxy climatic reconstructions. In: Mackay AW, Battarbee RW, Birks HJB, Oldfield F (eds) Global change in the Holocene. Arnold, London, pp 373–383

Maddocks RF (1966) Distribution patterns of living and subfossil podocopid ostracods in the Nosy Be' area, Northern Madagascar. University of Kansas Paleontological Contributions 12, pp 1–72

Maher LJ (1972a) Nomograms for computing 0.95 confidence limits of pollen data. Rev Palaeobot Palynol 13:85–93

Maher LJ (1972b) Absolute pollen diagram of Redrock Lake, Boulder County, Colorado. Quaternary Res 2:531–553

Maher LJ (1980) The confidence limit is a necessary statistic for relative and absolute pollen data. Proc IV Int Palynol Conf 3:152–162

Maher LJ (1981) Statistics for microfossil concentration measurements employing samples spiked with marker grains. Rev Palaeobot Palynol 32:153–191

Maher LJ, Heiri O, Lotter AF (2012) Chapter 6: Assessment of uncertainties associated with palaeolimnological laboratory methods and microfossil analysis. In: Birks HJB, Lotter AF, Juggins S, Smol JP (eds) Tracking environmental change using lake sediments. Volume 5: Data handling and numerical techniques. Springer, Dordrecht

Norton SA, Davis RB, Brakke DF (1981) Responses of northern New England lakes to atmospheric inputs of acids and heavy metals. Report to Land and Water Resources Center, University of Maine

Peglar SM, Fritz SC, Alapieti T, Saarnisto M, Birks HJB (1984) Composition and formation of laminated sediments in Diss Mere, Norfolk, England. Boreas 13:13–28

Pennington W, Sackin MJ (1975) An application of principal components analysis to the zonation of two Late-Devensian profiles. New Phytol 75:419–453

Pienitz R, Douglas MSV, Smol JP (eds) (2004) Long-term environmental change in Arctic and Antarctic lakes. Springer, Dordrecht

Pienitz R, Lotter AF, Newman L, Kiefer T (eds) (2009) Advances in paleolimnology. PAGES News 17:89–136

Rosén P, Dåbakk E, Renberg I, Nilsson M, Hall RA (2000) Near-infrared spectroscopy (NIRS): a new tool for inferring past climatic changes from lake sediments. Holocene 10:161–166

Rosén P, Vogel H, Cunningham L, Reuss N, Conley DJ, Persson P (2010) Fourier transform infrared spectroscopy, a new method for rapid determination of total organic and inorganic carbon and biogenic silica concentration in lake sediments. J Paleolimnol 43:247–259

Rühland K, Paterson AM, Smol JP (2008) Hemispheric-scale patterns of climate-induced shifts in planktonic diatoms from North American and European lakes. Global Change Biol 14: 2740–2754

Sergeeva LV (1983) Trace element associations as indicators of sediment accumulation in lakes. Hydrobiologia 103:81–84

Smol JP (2008) Pollution of lakes and rivers: a paleoenvironmental perspective, 2nd edn. Blackwell Publishing, Oxford

Smol JP, Birks HJB, Last WM (eds) (2001a) Tracking environmental change using lake sediments. Volume 3: Terrestrial, algal, and siliceous indicators. Kluwer Academic Publishers, Dordrecht

Smol JP, Birks HJB, Last WM (eds) (2001b) Tracking environmental change using lake sediments. Volume 4: Zoological indicators. Kluwer Academic Publishers, Dordrecht

Smol JP, Wolfe AP, Birks HJB et al. (2005) Climate-driven regime shifts in the biological communities of Arctic lakes. Proc Natl Acad Sci USA 102:4397–4402

Sweetman JN, Smol JP (2006) Reconstructing past shifts in fish populations using subfossil *Chaoborus* (Diptera: Chaoboridae) remains. Quaternary Sci Rev 25:2013–2023

Synerholm CC (1979) The chydorid Cladocera from surface lake sediments in Minnesota and North Dakota. Archiv für Hydrobiologie 86:137–151

Telford RJ, Vandvik V, Birks HJB (2006) Dispersal limitations matter for microbial morphospecies. Science 312:1015 ter Braak CJF (1986) Canonical correspondence analysis: a new eigenvector technique for multivariate direct gradient analysis. Ecology 67:1167–1179

ter Braak CJF, Prentice IC (1988) A theory of gradient analysis. Adv Ecol Res 18:271–317

ter Braak CJF, van Dam H (1989) Inferring pH from diatoms: a comparison of old and new calibration methods. Hydrobiologia 178:209–223

Uutala AJ (1990) *Chaoborus* (Diptera: Chaoboridae) mandibles – paleolimnological indicators of the historical status of fish populations in acid-sensitive lakes. J Paleolimnol 4:139–151

Vanormelingen P, Verleyen E, Vyverman W (2008) The diversity and distribution of diatoms: from cosmopolitanism to narrow endemism. Biodiv Cons 17:393–405

Vyverman W, Verleyen E, Sabbe K, Vanhoutte K, Sterken M, Hodgson DA, Mann DG, Juggins S, van de Vijver B, Jones V, Flower R, Roberts D, Chepurnov VA, Kilroy C, Vanormelingen P, de Wever A (2007) Historical processes constrain patterns in global diatom diversity. Ecology 88:1924–1931

Weijers JWH, Blaga CI, Werne JP, Sinninghe Damsté JS (2009) Microbial membrane lipids in lake sediments as a paleothermometer. PAGES News 17:102–104

Whiteside MC (1970) Danish chydorid Cladocera: modern ecology and core studies. Ecol Monogr 40:79–118

Chapter 2
Overview of Numerical Methods in Palaeolimnology

H. John B. Birks

Abstract This chapter presents a general introduction and overview of the numerical and statistical techniques that are most commonly used in quantitative palaeolimnology. After discussing the different types of palaeolimnological data (modern surface-samples and stratigraphical data) and the role of quantification in palaeolimnology, it presents a brief overview of the numerical techniques used in data collection, data assessment, data summarisation, data analysis, and data interpretation. In addition, the chapter describes important numerical and statistical procedures that are not covered elsewhere in the book such as numerical tools in identification, classification, and assignment, and statistical techniques of regression analysis and statistical modelling.

The major techniques discussed are linear discriminant and multiple discriminant analyses for identification, classification, and assignment (data collection); exploratory data analysis, primarily graphical techniques (data assessment and summarisation); error estimation (data assessment); regression analysis and statistical modelling involving general linear models, generalised linear models, mixed-effects models, non-parametric regression models, generalised additive models, classification and regression trees, artificial neural networks and self-organising maps, multivariate reduced-rank regression, and model selection and shrinkage

H.J.B. Birks (✉)
Department of Biology and Bjerknes Centre for Climate Research, University of Bergen, PO Box 7803, Bergen N-5020, Norway

Environmental Change Research Centre, University College London, Pearson Building, Gower Street, London WC1E 6BT, UK

School of Geography and the Environment, University of Oxford, Oxford OX1 3QY, UK
e-mail: john.birks@bio.uib.no

H.J.B. Birks et al. (eds.), *Tracking Environmental Change Using Lake Sediments*, Developments in Paleoenvironmental Research 5, DOI 10.1007/978-94-007-2745-8_2,

(data assessment, data summarisation, data analysis); quantitative environmental reconstructions involving calibration and inverse regression (data analysis); temporal-series analysis (data analysis); and confirmatory data analysis involving permutation tests (data interpretation).

Keywords Accumulation rates • Artificial neural networks • Auto-correlograms • Bootstrap cross-validation • Calibration • Calibration data-sets • Canonical correlation analysis • Canonical correspondence analysis • Classical regression • Classification and regression tress • Closed compositional data • Confirmatory data analysis • Constrained Gaussian ordination • Correlation coefficient • Exploratory data analysis • Gaussian logit regression • General linear models • Generalised additive models • Generalised linear models • Genetic algorithms • Hypothesis testing • Inverse regression • LOESS • Monte Carlo permutation tests • Multivariate regression • Non-linear canonical analysis of principal coordinates • Palaeolimnological data • Partial constrained ordination • Periodograms • Piece-wise regression • Randomisation tests • Reduced-rank regression • Redundancy analysis • Regression analysis • Self-organising maps • Shrinkage • Smoothers • Splines • Stratigraphical data • Surface samples • Temporal autocorrelation • Temporal-series analysis • Time-series analysis • Training-sets • Type I error • Type II error • Variation partitioning • Weighted averaging

Introduction

This chapter presents a general introduction and overview of the numerical and statistical approaches and techniques that are most commonly used in quantitative palaeolimnology. After discussing the different types of palaeolimnological data and the role of quantification in palaeolimnology, it presents a brief overview of the numerical approaches relevant in the various stages of a palaeolimnological study, namely data collection, data assessment, data summarisation, data analysis, and data interpretation.

The major numerical approaches and techniques used in these stages are then outlined to provide a basic background and overview for the more detailed chapters in this book that consider specific approaches (e.g., exploratory data analysis, environmental reconstructions) or particular methods (e.g., canonical correspondence analysis, modern analogue technique), In addition, two important numerical approaches, namely techniques to aid identification, classification, and assignment of biological remains in lakes sediments, and regression analysis and statistical modelling of palaeolimnological data are presented in more detail because they are not covered specifically elsewhere in this book. Both approaches are essential in many palaeolimnological studies.

Types of Palaeolimnological Data

Palaeolimnological data are counts or measurements of a wide range of biological, chemical, and physical elements preserved in lake sediments. The biotic components include the presence or absence, the estimated abundance (e.g., rare, occasional, frequent, abundant), or counts of remains of organisms (fossils or 'sub-fossils') preserved in sediments deposited in lakes, embayments, deltas, and reservoirs (see Smol 2008). The commonest groups of organisms studied by palaeolimnologists are diatoms, chrysophytes (cysts and scales), chironomids, cladocerans, ostracods, plant macrofossils (seeds, fruits, leaves, etc.), pollen and spores, and mollusca, although other groups are more rarely studied such as beetles, oribatid mites, testate amoebae, and biochemical markers such as photosynthetic pigments, lipids, and DNA (see Smol et al. 2001a, b for detailed accounts of the different types of biological remains studied in lake sediment). The sediments are themselves valuable sources of abiotic palaeolimnological data, such as inorganic and organic geochemistry, magnetic properties, sediment composition and grain-size information, stable isotopes of H, C, N, and O, etc. (see Last and Smol 2001a, b; Pienitz et al. 2009 for further details). The sediments and, in the case of radiocarbon-dating using accelerator mass spectrometry (AMS), their contained fossils provide the basic material for radiometric age determinations on which so much of palaeolimnology depends (see Last and Smol 2001b; Walker 2005).

The sediment samples can be from one sediment core collected from the lake under investigation (temporal, stratigraphical data) or from one age or depth at several sites (spatial, geographical data). The most common type of spatial data consists of fossil counts or physical or chemical measurements from the uppermost, surficial (0–1 cm) sediments, so-called modern surface-samples. Stratigraphical and spatial data often contain counts of many (50–500) different fossil taxa or measurements of many geochemical or geophysical properties (20–75) in a large number of samples ($\sim$50–500). Such data are highly multivariate, containing many variables (biological taxa, chemical elements, magnetic properties) and many objects (sediment samples). Counts of different fossils are usually expressed as percentages of individual taxa relative to some calculation sum (e.g., total diatom valves counted in the sample). The sample sum itself is relatively uninformative in many numerical analyses. It is largely determined by the initial research and sampling design, namely the number of individual fossils to be counted per sample. The actual number counted can also be a function of fossil concentrations, preservation, and amount of analytical time. More rarely, the concentrations or accumulation rates of different fossils may be estimated as number of objects per unit volume of sediment or unit weight of sediment (concentrations) or as net number of objects per unit of sediment area or unit weight of sediment per unit time (accumulation rates or flux density) (Birks and Birks 1980; Birks and Gordon 1985). To convert concentrations to estimates of accumulation rates, it is necessary to estimate sediment deposition-times, the

Table 2.1 Types of palaeolimnological data, their relative commonness, and their units

	Biotic variables		Abiotic variables	
Data type	Percentages	Accumulation rates	Percentages	Accumulation rates
Stratigraphical data	Common	Occasional	Common	Frequent
Spatial data (modern surface-samples)	Common	Very rare	Common	Rare
Units	%	Objects cm^{-2} $year^{-1}$ or objects g^{-1} $year^{-1}$	%	Variable cm^{-2} $year^{-1}$ or variable g^{-1} $year^{-1}$

amount of time per unit thickness in years cm^{-1} or unit weight of sediment in years g^{-1} or its reciprocal, sediment matrix accumulation rate, the net thickness or weight of sediment accumulated per unit time after compaction and diagenesis in cm $year^{-1}$ or g $year^{-1}$ (Birks and Birks 1980; Birks and Gordon 1985). Deposition times are derived most commonly by obtaining a series of radiometric dates at different depths through the sediment sequence and estimating the deposition times from the age-depth relationship (see Blaauw and Heegaard 2012: Chap. 12). Counts of the number of annual laminations between samples from laminated sediments can be used to provide estimates of sediment deposition-times and hence fossil accumulation rates (see Lotter and Anderson 2012: Chap. 18). The various types of palaeolimnological data, their relative commonness, and their units are summarised in Table 2.1.

Many biotic palaeolimnological data contain many zero values (taxa absent or not found in many samples). Because of the percentage calculation, the data are 'closed' compositional data and thus there are in-built interrelationships between the variables (Birks and Gordon 1985; Aitchison 1986; Reyment and Savazzi 1999). Closed data require special numerical methods for correct statistical analysis. Reyment and Savazzi (1999) discuss statistical approaches for analysing closed data involving log ratios (cf. O'Hara and Kotze 2010) (see also Baxter 1989, 1991, 1992; Elston et al. 1996; Jackson 1997; Kucera and Malmgren 1998; Reyment 1999; Aitchison and Greenacre 2002). The large numbers of zero values in many palaeolimnological data-sets cause problems in the use of log ratios. Ter Braak and Šmilauer (2002) discuss the statistical properties of correspondence analysis (CA) and its canonical or constrained relative canonical correspondence analysis (CCA) (see Legendre and Birks 2012b: Chap. 8) in relation to generalised linear models and show how CA is appropriate and robust with percentage data containing many zero values (see also ter Braak and Verdonschot 1995).

Palaeolimnological samples are usually in known stratigraphical or temporal order (stratigraphical data) or in a known geographical context (modern surface-samples, spatial data). Modern surface-sample data often have associated contemporary environmental data (e.g., lake-water chemistry, catchment vegetation, climate) and together the modern biological and environmental data-sets comprise 'training-sets' or 'calibration data-sets' that are now so important in many aspects of quantitative palaeolimnology (Smol 2008; Juggins and Birks 2012: Chap. 14; Simpson 2012: Chap. 15; Smol et al. 2012: Chap. 1).

Some palaeolimnological data-sets have many more objects (samples) than variables (e.g., cladoceran, pollen, ostracod, geochemical, magnetic data) whereas other data-sets have many more variables than objects (e.g., diatom data). With the rapid development of palaeolimnology, national and international data-bases of modern and stratigraphical palaeolimnological data are now being assembled (e.g., European Diatom Database Initiative (EDDI) http://craticula.ncl.ac.uk/Eddi/jsp/). Palaeolimnological data can thus also consist of fossil counts or geochemical determinations from many different cores from a large number of sites over a broad geographical area, thereby combining both temporal and spatial data.

The Role of Quantification in Palaeolimnology

In nearly 40 years of research in quantitative palaeoecology and palaeolimnology, some of the most commonly asked questions, at least in the first 10–15 years of this research, were 'why attempt quantification in palaeolimnology?' and 'do we really need numerical methods in palaeolimnology?' There was a feeling that 'statistics is the scientific equivalent of a trip to the dentist' and should be kept to an absolute minimum or be avoided altogether!

There are several answers to these and related questions. First, palaeolimnological data are very time consuming and expensive to collect and they are highly quantitative. It is a waste of time and money to collect such quantitative data and then ignore the quantitative aspects of the data. Second, palaeolimnological data are complex, multivariate, and often stratigraphically ordered and numerical methods can help to summarise, describe, characterise, and interpret such data. Numerical methods can help to identify 'signal' from 'noise' in complex data, detect features that may otherwise escape attention, generate hypotheses, aid in further data collection by identifying previously unsuspected features in the data, and assist in the communication and display of complex data. Third, numerical methods force the investigator to be explicit. Walker (1972) elegantly summarised this "The more orthodox amongst us should at least reflect that many of the same imperfections are implicit in our own cerebrations and welcome the exposure which numbers brings to the muddle which words may obscure". Fourth, numerical methods allow palaeolimnologists to tackle research problems that are not otherwise solvable because the problems require the testing of competing hypotheses. It is often easier to test hypotheses using numerical methods, particularly statistical methods. Numerical techniques can thus hopefully lead to a more rigorous science involving not only hypothesis generation but also hypothesis testing. Fifth, deriving quantitative environmental reconstructions of, for example, lake-water pH or total phosphorus, is important in many branches of environmental science and applied limnology (see Smol 2008), for example in validating hindcasts or back-predictions from ecosystem- or catchment-scale models (e.g., Jenkins et al. 1990; Battarbee et al. 2005).

Overview of Numerical Methods

Numerical methods are useful tools at the many stages in a palaeolimnological investigation (Fig. 2.1). During *data collection*, they can be invaluable in the identification of certain fossil types. Computer-based techniques can also be useful in data capture, data compilation, and data storage. In *data assessment*, statistical techniques are essential in estimating the inherent errors associated with different palaeolimnological analyses (biological, magnetic, geochemical, etc.). Exploratory data analysis is also an essential step in data assessment. For *data summarisation*, there is a range of robust numerical techniques that can be used to detect and summarise the major underlying patterns in palaeolimnological data, both modern and stratigraphical. For single stratigraphical sequences or modern calibration data-sets, the numerical delimitation of sample groups or zones (for stratigraphical data) can be a useful first step in data summarisation. Other numerical techniques such as ordination procedures, can summarise temporal or spatial trends in stratigraphical or modern data-sets. In *data analysis*, numerical methods are essential for detecting temporal patterns such as trends and periodicities and for quantitatively reconstructing past environmental variables from biotic assemblages. The final stage in a palaeolimnological study, *data interpretation*, can be greatly aided by numerical techniques for the reconstruction of past biotic assemblages and lake types and in the testing of competing hypotheses about underlying causative factors such as climate change, human activity, or internal lake dynamics in determining lake development and limnological changes. As broader and increasingly more complex questions are

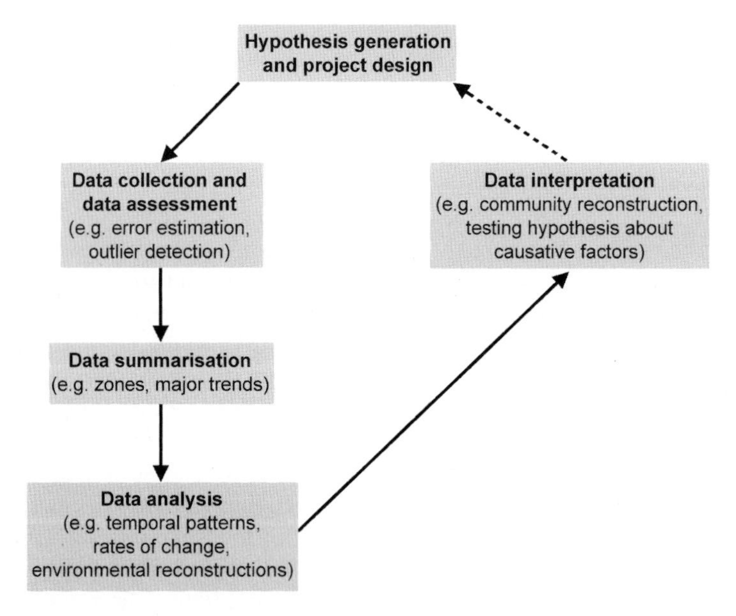

Fig. 2.1 Major stages in a palaeolimnological investigation

being asked of palaeolimnological data, numerical methods are constantly being updated and modified in an attempt to help answer new, challenging, and critical research questions.

The major numerical approaches used in these five main stages of a palaeolimnological study are listed in Table 2.2. They are then briefly described in this chapter so as to provide a basic background and overview for the more detailed chapters about specific approaches and methods in this book. In addition, important numerical approaches that are not covered specifically in this book (e.g., identification, classification, and assignment; regression analysis) are discussed in this chapter. Just as in everyday life, common-sense is an essential attribute, common-sense in statistics is equally important. van Belle (2008) provides an invaluable set of 'statistical rules of thumb' all based on common-sense.

Identification, Classification, and Assignment

Reliable and detailed identification of the biological remains of research interest (e.g., diatoms, chironomids, pollen) preserved in surface or core sediment samples is the first essential step of *data collection* (Fig. 2.1, Table 2.2) in any biologically based palaeolimnological study. Identification of palaeolimnological microfossils most commonly proceeds by comparison of the fossils either with descriptions and illustrations in taxonomic monographs or floras for specific geographical areas or particular ecological systems, as in the case of diatoms and chrysophyte cysts, or with modern reference material prepared and mounted in the same way as the fossil material, as in the case of chironomids, pollen, and plant macrofossils. In critical cases in diatom analysis, comparisons may be made with taxonomic type material. Identification involves assigning fossils to pre-existing categories of modern taxa. With experience, many fossils can be assigned by memory to modern taxonomic categories. However, some numerically dominant and/or ecologically important fossil types may be so similar in overall morphology and appearance that the investigator cannot consistently assign a particular fossil to a modern taxon. In such cases, numerical techniques of classification or assignment can be of value.

It is important to note here the essential distinction between two types of numerical classification: unsupervised classification and supervised classification (Næs et al. 2002; Simpson and Birks 2012: Chap. 9). The first type is termed in this book clustering or partitioning and involves some form of cluster analysis. It is used in situations when there is little or no *a priori* information about group structure within the data. The goal is to identify clusters and to partition the data into groups of samples on the basis of the data themselves without using any *a priori* information (see Fielding 2007; Birks 2012b: Chap. 11; Legendre and Birks 2012a: Chap. 7). Clustering can also be performed visually in an informal way using ordination or scaling techniques (see Legendre and Birks 2012b: Chap. 8).

The other type of classification, supervised classification, is also known as discriminant analysis. This is a powerful class of methods for the analysis of

Table 2.2 Basic numerical approaches and the major stages in a palaeolimnological study

	Major stages in a palaeolimnological study					
Numerical approach	Collection	Assessment	Summarisation	Analysis	Interpretation	Relevant chapters
Identification or assignment procedures	+				+	2
Exploratory data analysis	+	+	+	+		
Univariate and bivariate	+	+	+	+		5, 6
Multivariate	+	+	+	+		7, 8, 9, 10
Regression analysis (statistical modelling)	+	+	+	+	+	2, 5, 6, 8, 9, 12, 13, 14, 18
Quantitative environmental reconstruction (calibration or inverse regression)				+	+	14, 15, 19, 20
Temporal-series analysis				+	+	13, 16, 20
Confirmatory data analysis (hypothesis testing)					+	8, 14, 16, 18, 19, 20

multivariate data similar in some ways to ordination procedures (Prentice 1980), that are primarily used to discriminate between pre-defined groups and build classification rules for a number of *a priori* groups (Hand 1981; Fielding 2007). These rules can later be used for assigning or allocating new and unknown objects to the most probable group (Lachenbruch 1975; James 1985; Webb 1999). Another important application of discriminant analysis is to help in interpreting differences between *a priori* groups of samples (Reyment et al. 1984). Discriminant analysis can be viewed in several different ways – as a form of canonical ordination (Borcard et al. 2011; Legendre and Birks 2012b: Chap. 8) where linear combinations of variables are found to maximise the separation of two or more groups or as a form of qualitative calibration where the quantity to be calibrated is not a continuous quantitative variable like pH (see Juggins and Birks 2012: Chap. 14; Simpson 2012: Chap. 15) but is a categorical group variable (Næs et al. 2002). The overall aim is to assess whether or not a set of variables distinguish or discriminate between two (or more) *a priori* groups of objects.

In the two-group case the most commonly used method is Fisher's linear discriminant function (Davis 2002; Everitt 2005; Hammer and Harper 2006; Fielding 2007; Wehrens 2011) in which a linear combination or transformation (z) of the m variables (x) that gives the maximum separation between the two *a priori* groups is determined

$$z = a_1x_1 + a_2x_2 + a_3x_3 + \cdots + a_mx_m$$

The ratio of the between-group variance of z to the within-group variance is maximised. The solution for the discriminant coefficient $\mathbf{a} = (a_1, \ldots, a_m)$ is

$$\mathbf{a} = \mathbf{S}^{-1}(\mathbf{x}_1 - \mathbf{x}_2)$$

where $\mathbf{S}^{-1}$ is the inverse of the pooled within-groups variance-covariance matrix of the two groups and $\mathbf{x}_1$ and $\mathbf{x}_2$ are the group mean vectors. For palaeoecological examples, see Birks and Peglar (1980), Brubaker et al. (1987), Rose et al. (1996), and Weng and Jackson (2000).

The linear discriminant function provides a classification or allocation rule that can be used to assign or identify unknown objects to one of the two groups (Fielding 2007). The derivation of this linear discriminant function assumes, amongst many assumptions (see Birks and Peglar 1980) that the variance-covariance matrices of the two groups are the same. If they are not, a quadratic discriminant function may be necessary to distinguish between the two groups (Hammer and Harper 2006; Wehrens 2011). Such a function contains powers and cross-products of the variables. In the pattern recognition literature, the sample of objects from which the discriminant function is derived is often called a 'training-set'. This usage of the term 'training-set' is very different from its common usage in palaeolimnology where it is a set of modern biological assemblages and related environmental data used for deriving transfer or calibration functions for palaeoenvironmental reconstructions (see Juggins and Birks 2012: Chap. 14; Simpson 2012: Chap. 15).

There are statistical problems with allocation rules based on a discriminant function that seeks to estimate the proportions of different fossil types by identifying each individual fossil in the sediment sample (Gordon and Prentice 1977; Gordon 1982; Birks and Gordon 1985). The statistical method of maximum likelihood mixtures (Mood et al. 1974; Webb 1999) can be used to model the counts for the fossil types of interest as a mixture of multinomial (Gordon and Prentice 1977) or normal distributions (Gordon 1982). Classification and regression trees and other tree-based methods such as random forests ('decision trees') (Webb 1999; Lindbladh et al. 2002; Fielding 2007; Simpson and Birks 2012: Chap. 9) can also provide robust estimates of the likely proportions of different taxa in fossil assemblages (e.g., Lindbladh et al. 2003, 2007).

Rose et al. (1996) illustrate the use of linear discriminant analysis to characterise the surface chemical differences between modern carbonaceous fly-ash particles formed by the high temperature combustion of oil or coal and preserved in lake sediments. They developed a discriminant function based on modern particles from eight power stations and then used a discrimination rule to identify carbonaceous particles in a lake-sediment core as either being oil- or coal-derived. Their procedure assumes that the carbonaceous particles in the lake sediment have either an oil or coal origin, whereas there are other potential sources for carbonaceous particles such as peat-fired power stations. To avoid potential misclassification of such non-coal or non-oil derived particles, they standardised the discriminant function, calculated simple 95% confidence intervals for the two groups, and defined any sediment particles lying outside the 95% confidence interval for either fuel type as unclassified. They then applied the discriminant function to carbonaceous particles in sediment cores to identify those from oil and from coal.

When there are more than two groups, all with the same variance-covariance matrix, it is possible to determine several linear combinations or functions of the variables for separating the groups. The number of such functions that can be derived is the smaller of m and $g-1$ where m is the number of variables and g is the number of groups. This procedure is called multiple discriminant analysis or canonical variates analysis (CVA) and the linear functions are termed canonical discriminant functions or simply canonical functions (Reyment et al. 1984). Ter Braak (1987a), ter Braak and Verdonschot (1995), and Borcard et al. (2011) discuss the relationship between CVA and canonical correspondence analysis (CCA) and ter Braak and Šmilauer (2002) show how CVA can be implemented using the program CANOCO. CVA has rarely been used in palaeolimnology but has been more commonly used in palaeoecology, in particular pollen analysis. Examples include Whiteside (1970), Birks et al. (1975), Birks (1976, 1977, 1980), Lamb (1984), Liu and Lam (1985), MacDonald (1987), Sugden and Meadows (1989), Liu (1990), Meadows and Sugden (1991), Horrocks and Ogden (1994), Liu et al. (2001), Shen et al. (2008a, b), Catalan et al. (2009), and Mackay et al. (2011).

Canonical variates analysis can be viewed as multiple discriminant analysis, as a dimension-reduction technique where multivariate data consisting of g groups with m variables can be represented in m or $g-1$ dimensions, whichever is the smaller, or as an ordination technique for comparing groups in which the implicit dissimilarity coefficient or distance measure is the square root of Mahalanobis' D^2 between groups of objects (Gower 1966a; Prentice 1980). Canonical variates are linear combinations of variables that maximise the *ratio* of the between-group variance to the within-group variance. In contrast, principal component analysis (PCA, see Legendre and Birks 2012b: Chap. 8) partitions the *total* variance in a data-set into successive components with the maximal concentration of variance in the first few components (Prentice 1980). The first canonical variate is the linear combination of variables that provides the most effective discrimination between the groups, the second contributes the greatest additional discriminating power, and so on (Prentice 1980). The canonical variates are discriminant axes, and their loadings, suitably scaled, are discriminant coefficients. These can be used to derive a series of classification functions or allocation rules for groups with similar variance-covariance matrices (Fielding 2007).

The performance of one or more discriminant functions for two or more groups can be assessed by the misclassification or 'error' rate. This can be estimated by a simple 'plug-in' allocation procedure based on the discriminant index (two groups) or indices (three or more groups). However, this is likely to under-estimate the true error rate as the same data are used to both generate the discriminant functions and to test them (Rose et al. 1996). A more realistic estimate of the error rate can be obtained by some form of cross-validation (Hand 1986; Kohavi 1995; Fielding 2007) such as leave-one-out or the so-called 0.632 bootstrap estimator, which gives superior performance in simulation experiments (McLachlan 1992; Efron and Tibshirani 1993; Kohavi 1995; Molinaro et al. 2005). It is often useful to use a step-wise variable-selection procedure to produce a reduced set of variables with maximum prediction power by eliminating variables that do not contribute significantly to the discrimination. Such variables with little or no discriminating power, either alone or in combination with other variables, add noise to the discrimination model and their inclusion can decrease the predictive ability by increasing the error rate (Hand 1981; Catalan et al. 2009). Step-wise selection in linear discriminant function analysis or CVA can easily be implemented using the program CANOCO (ter Braak and Šmilauer 2002) where the statistical significance of the F-ratio for each variable and of the overall discriminant function model can be assessed using Monte Carlo permutation tests. This test has the advantage over standard tests in linear discriminant analysis and CVA in that it does not require the assumption of normality in the variables (ter Braak and Šmilauer 2002).

Mathematical details of CVA are presented in Klecka (1980), Prentice (1980), Hand (1981), Reyment et al. (1984), ter Braak (1987a), Reyment (1991), Everitt and Dunn (2001), Everitt (2005), Varmuza and Filzmoser (2009), and Wehrens (2011). Campbell and Atchley (1981) provide a simple geometrical presentation of the mathematics behind CVA. Examples of its use in many areas of geology and biology are discussed by Cacoullos (1973) and Reyment et al. (1984).

Linear discriminant analysis and CVA of closed compositional data require a special log-ratio transformation prior to analysis (Aitchison 1986; Reyment and Savazzi 1999; Hammer and Harper 2006).

There are many extensions of the two-group linear discriminant analysis and CVA (Fielding 2007; Wehrens 2011) that are of potential use in palaeolimnology. These include flexible discriminant analysis (Hastie et al. 1994) that uses a non-parametric regression that creates non-linear decision boundaries used in allocation rules. Mixture discriminant analysis (Hastie and Tibshirani 1996) creates a classifier by fitting a Gaussian mixture model to each group. A penalised discriminant analysis (Hastie et al. 1995) attempts to overcome the problems of collinearity created when many variables are correlated by using a penalised regression. Robust quadratic discriminant analysis allows the boundaries between the groups to be curved quadratic surfaces in contrast to the flat boundaries in linear discriminant analysis (Fielding 2007). In addition, in the quadratic method, there is no assumption that the within-group covariance matrices are equal. While this can be an advantage with some data, it comes at a cost by increasing the numbers of parameters fitted in the model, potentially leading to a greater chance of over-fitting the training-set data (Fielding 2007). Because of this risk and the surprisingly robust nature of linear discriminant analysis (e.g., Gilbert 1969; Birks and Peglar 1980), quadratic discriminants are often only useful when the differences between the group covariance matrices are very large (Marks and Dunn 1974). Fielding (2007) and Wehrens (2011) discuss other approaches to supervised classification such as naive Bayes classifiers, logistic (= logit) regression, generalised additive models, and decision trees (e.g., random forests, artificial neural networks – see below and Simpson and Birks 2012: Chap. 9), all of which have potential applications in certain palaeolimnological problems in which *a priori* group structure is an essential property of the data.

Fossil morphological characteristics are often of different data-types (e.g., quantitative size measurements, ordinal or nominal multistate (e.g., smooth, rough, very rough), binary or dichotomous (e.g., presence or absence of setae)). Numerical techniques such as Goodall's (1969) deviant index for mixed morphological data can be useful in assigning fossils to modern taxonomic categories. The index assesses how different a particular fossil is to the mean, median, or mode of reference material of the likely modern taxon, depending on the data-type. Hansen and Cushing (1973) illustrated the use of the deviant index in identifying different species of fossil *Pinus* pollen in New Mexico. Classification and regression trees (CART: see below) (Fielding 2007; Simpson and Birks 2012: Chap. 9) have been shown to provide powerful and robust techniques for identifying different species of *Picea* (Lindbladh et al. 2002) and *Pinus* pollen (Barton et al. 2011) using mixed morphological data. CART provides a simple, non-parametric approach for recursively classifying levels of a dependent variable (e.g., species) using a set of independent variables (e.g., morphology) (Barton et al. 2011). Modern species represented by, for example, pollen are classified based on morphological attributes. In the first step of a CART analysis each morphological variable is tested to find the combination of a variable and a split threshold that separates the entire sample into two sub-sets that are internally as homogenous as possible with respect to species identity. Each of the two sub-sets is then partitioned in turn. The process is repeated recursively through the descending nodes so that a classification tree is 'grown' until a specified level of complexity and certain stopping rules are reached. At this point the resulting tree is 'pruned back' to an optimally fitting version based on n-fold cross-validation (usually $n = 10$) (Molinaro et al. 2005). The end result is a classification or decision tree (Fielding 2007; Simpson and Birks 2012: Chap. 9) much like a standard taxonomic key, allowing the identification of unknown fossils using a small set of morphological variables.

The classification success of the tree constructed in this way can be assessed as the proportion of the sample of known identity that is correctly assigned by the tree. However, this proportion is likely to be biased and over-optimistic and inflated because the tree is constructed via the exhaustive search procedure so as to minimise the classification error. This bias can be controlled via cross-validation pruning but this involves a trade-off coefficient that weights classification accuracy against tree complexity (Breiman et al. 1984; Simpson and Birks 2012: Chap. 9). An independent and reliable assessment of the reliability of the pruned tree can be obtained by making predictions from the model against a reserved independent test-set of objects not used in the construction of the CART model (e.g., a test-set of 100 independent objects to evaluate the prediction accuracy of the tree trained on 1,000 objects) (Kohavi 1995; Lindbladh et al. 2002; Molinaro et al. 2005; Barton et al. 2011).

There is a range of specialised techniques for morphometric analysis, the study of phenotypic variation, and for cladistic and phylogenetic analysis of taxa found in palaeolimnological studies (see Birks 2010 for a brief discussion of these techniques in the context of diatom research, Zuur et al. 2007 for examples of numerical analysis of morphometric data, and van der Meeren et al. 2010 for a palaeolimnological application).

Identification and counting of fossils are very time consuming and much progress has been made in the last 10 years in the automation of pollen identification and counting using texture measures, artificial neural networks, Gabor transforms, digital moments, and other pattern description and recognition techniques (Flenley 2003; Li et al. 2004; Treloar et al. 2004; Zhang et al. 2004; Allen 2006; O'Gorman et al. 2008). It is a rapidly developing research area. With ever-increasing computing power and improved and faster image-capture and image-analytical techniques, major advances can be expected in the coming years that may revolutionise micro-fossil analysis in palaeolimnology (e.g., Allen 2006; Holt et al. 2011).

Exploratory Data Analysis

Exploratory data analysis (EDA) and data display, and the estimation of laboratory and analytical errors are essential parts of *data collection*, *data assessment*, and *data summarisation* and are important preliminary steps in *data analysis* (Fig. 2.1, Table 2.2). EDA is described by Everitt (2002, p. 136) as "an approach to data analysis that emphasizes the use of informal graphical procedures not based on prior assumptions about the structure of the data or on formal models for the data. The essence of this approach is that, broadly speaking, data are assumed to possess the following structure

$$\text{Data} = \text{Smooth} + \text{Rough}$$

where the 'Smooth' is the underlying regularity or pattern in the data. The objective of the exploratory approach is to separate the Smooth from the Rough with minimal use of formal mathematics or statistical methods."

A first step in EDA is what Everitt (2002: p. 189) calls initial data analysis (IDA) which he describes as "The first place in the examination of a data-set which consists of a number of informal steps including

- checking the quality of the data,
- calculating simple summary statistics and constructing appropriate graphs

The general aim is to clarify the structure of the data, obtain a simple descriptive summary, and perhaps get ideas for a more sophisticated analysis (see Chatfield 1988)."

In palaeolimnology the main purposes of EDA and IDA are to provide simple summary statistics of location, dispersion, and skewness and kurtosis for individual variables, to guide the investigator to appropriate transformations, to generate graphical displays of the data for single (univariate), two (bivariate), and many (multivariate) variables, to identify potential outliers in a given data-set, and to separate the smooth (*sensu* Everitt 2002) ('signal') from the rough (sensu Everitt 2002) ('noise') in a data-set. Juggins and Telford (2012: Chap. 5) discuss EDA and data display in detail. The standard texts on EDA are Tukey (1977) and Velleman and Hoaglin (1981). There are several wide-ranging and thought-provoking reviews and texts on graphical data display in EDA, including Chambers et al. (1983), Tufte (1983, 1990), Cleveland (1993, 1994), Jacoby (1997, 1998), and Gelman et al. (2002). Zuur et al. (2010) provide a valuable protocol for data exploration that avoids many common statistical problems, in particular avoiding Type I (the null hypothesis is erroneously rejected, representing a false positive) and Type II (the null hypothesis is erroneously accepted, representing a false negative) errors, thereby reducing the chance of making wrong conclusions.

The commonest tools in EDA and IDA that are of considerable practical value for palaeolimnologists are simple graphical displays such as histograms and frequency or cumulative frequency graphs, and summary statistics for individual variables of location (e.g., mean, weighted mean, median), dispersion (e.g., range, percentiles, variance, standard deviation, standard error of mean), and skewness (a measure of how one tail of a frequency curve is drawn out), and kurtosis (a measure of the peakedness of a frequency curve). Sokal and Rohlf (1995) and Zuur et al. (2010) are excellent references for details of these and related summary statistics. These summary statistics and basic graphical displays are essential tools in deciding if data transformations are required (see O'Hara and Kotze 2010; Juggins and Telford 2012: Chap. 5; Legendre and Birks 2012b: Chap. 8). There are two main reasons for transforming palaeolimnological data – to achieve comparability between variables measured in different units (e.g., lake-water pH, specific conductivity, alkalinity, Ca^{2+}, Na^{+}) by standardising each variable to zero mean and unit variance, and to make particular variables have a better fit to a particular underlying statistical model (e.g., normal, log-normal models) (O'Hara and Kotze 2010) by applying transformations such as square roots, cubic roots, fourth roots, logarithmic power functions (to various bases such as base 10 or base 2), and the Box-Cox transformation (see Sokal and Rohlf 1995 for details of these transformations). Other reasons for transforming palaeolimnological data, especially biological counts, are to stabilise variances and to dampen the effects of very abundant taxa (see Prentice 1980; Birks and Gordon 1985 for discussions of appropriate transformations). Other useful graphical approaches in EDA and IDA include quantile-quantile (Q-Q) plots, box-and-whisker plots, and density estimation procedures applied to histograms (see Hartwig and Dearing 1979; Fox and Long 1990; Crawley 2002, 2007; Fox 2002, 2008; Venables and Ripley 2002; Warton 2008; Juggins and Telford 2012: Chap. 5 for further details of these and other EDA tools).

An important aspect of EDA and IDA is to explore relationships between variables. Scatter-plots of pairs of variables and scatter-plot matrices where scatter-plots of all pair-wise scatter-plots of variables are arranged together and aligned into a matrix with shared scales are invaluable graphical tools (see Fox 2002; Zuur et al. 2010; Juggins and Telford 2012: Chap. 5). Scatter-plots can be enhanced by adding box-plots for each variable, by coding and using colours, by jittering where a small random quantity is added to each value to separate over-plotted points, by applying bivariate kernel-density estimates, and by fitting regression lines or locally weighted scatter-plot smoothers (LOESS or LOWESS) to the scatter-plots (see Fox 2002 and Zuur et al. 2010 for examples). Regression and smoothers are discussed in detail below.

Graphical display of multivariate data, which are the commonest data type in palaeolimnology, involves dimension-reduction techniques such as ordination or scaling (see Legendre and Birks 2012b: Chap. 8) or clustering or partitioning (see Legendre and Birks 2012a: Chap. 7). Clustering or partitioning of data into groups is a useful first step in exploring large, multivariate data-sets and yet retaining the multivariate character of the data (Everitt et al. 2011; Wehrens 2011). The groupings of samples can be constrained in one dimension (e.g., depth or time in stratigraphical data) or two dimensions (e.g., locations in modern spatial data) (see Birks 2012b: Chap. 11; Legendre and Birks 2012a: Chap. 7) and indicator variables can be identified and statistically evaluated for particular groupings (see Legendre and Birks 2012a: Chap. 7 for details and Catalan et al. 2009 for examples in palaeolimnology).

Ordination techniques used in the context of EDA primarily fall in the general category of classical or indirect gradient analysis (ter Braak 1987a; ter Braak and Prentice 1988; Wehrens 2011; Legendre and Birks 2012b: Chap. 8). Techniques such as principal component analysis (PCA) (Jolliffe 2002), correspondence analysis (CA) (Greenacre 1984, 2007; Greenacre and Blasius 2006), detrended correspondence analysis (DCA) (Hill and Gauch 1980), principal coordinate analysis (Gower 1966b), or non-metric multidimensional scaling (NMDS) (Kruskal

1964) provide a convenient low-dimensional (usually two or three) geometric representation of multivariate data in such a way that the configuration of the objects in the resulting low-dimensional ordination plot reflects, as closely as possible, the similarities between the objects in the original multi-dimensional space of the data (ter Braak 1987a; Husson et al. 2011). The full dimensions of the data are either the number of objects or the number of variables, whichever is smaller. Ordination plots, appropriately scaled (see ter Braak 1994; ter Braak and Verdonschot 1995 for details of scalings) can provide very effective summaries of the major patterns or signal within a data-set (a major function of EDA) and can suggest potential outlying objects.

Principal curves (Hastie and Stuetzle 1989; Simpson and Birks 2012: Chap. 9) are generalisations of the first principal component line, being a smooth, one-dimensional curve fitted through the data in m dimensions such that the curve fits the data best in a statistical sense (De'ath 1999; Hastie et al. 2011). Principal curves are a very powerful technique for summarising patterns in complex palaeoecological data, especially stratigraphical data.

The identification of outliers in a data-set is an important aim of EDA. Everitt (2002) defines an outlier as "an observation that appears to deviate markedly from the other members of the sample in which it occurs... More formally the term refers to an observation which appears to be inconsistent with the rest of the data, relative to an assumed model." EDA provides a means of identifying such outliers in terms of their leverage (the potential for influence resulting from unusual values) and influence (an observation is influential if its deletion substantially changes the results). Outliers may result from some abnormality in the measured feature of an object or they may result from an error in measurement, recording, or data entry. It is important to realise that the concept of an outlier is model dependent, so an object may appear to be an outlier in one context but may appear to conform to the rest of the data in another context (Juggins and Birks 2012: Chap. 14; Juggins and Telford 2012: Chap. 5). There are several well-established statistical techniques for assessing leverage and influence, particularly in the context of a regression model (see Rawlings 1988; Hamilton 1992; Crawley 2002, 2007; Fox 2002, 2008). Barnett and Lewis (1978) and Hawkins (1980) provide detailed accounts of detecting outliers in data-sets.

Problems can arise with missing data, for example in environmental data-sets (Allison 2002). The most common approach in dealing with missing data is to delete objects containing missing observations (Nakagawa and Freckleton 2008). This approach reduces statistical power, discards valuable information, and increases model bias. Nakagawa and Freckleton (2008) discuss the powerful but largely underused approach of multiple imputation to allow for missing data.

An important aspect of EDA is graphical representations of the individual variables and their relationships and of the objects in a data-set. Although there is a temptation for the palaeolimnologist, after s/he has completed all the diatom or chironomid analyses, to rush to do constrained ordinations or develop environmental reconstruction models, it cannot be emphasised too much that time spent on EDA is time well spent and that in EDA one should never forget the graphs (Warton 2008; Zuur et al. 2010; Juggins and Telford 2012: Chap. 5).

Another aspect of exploratory data analysis that is often ignored in quantitative palaeolimnology is *data assessment* (Fig. 2.1, Table 2.2), in particular the estimation of analytical errors (Maher et al. 2012: Chap. 6). In the context of biostratigraphical data, a count of, for example, diatom valves in a given sediment sample is hopefully an unbiased sample count of the diatoms preserved in that sediment sample. As in all sampling, there are statistical uncertainties associated with any sample count. The larger the count, the smaller the uncertainties become. As larger counts require more time to complete, there is a trade-off between time and level of uncertainty. It is therefore important to estimate the uncertainty associated with all fossil counts (see Maher et al. 2012: Chap. 6). Methods for estimating counting errors and confidence intervals associated with relative percentage data (see Table 2.1) are well developed (Maher 1972a; Maher et al. 2012: Chap. 6). The errors depend on the size of the count and on the proportion in the count sum of the taxon of interest. Methods for estimating errors and confidence intervals associated with concentrations are also available. Deriving the total error requires careful propagation of errors associated with the various steps involved in estimating concentration values. The total error depends not only on estimating and combining these various errors but also on the number of fossils counted (Maher 1981). Estimating the errors for accumulation rates (Table 2.1) is similar to estimating concentration errors (Maher 1972b; Bennett 1994), but with the additional complexity of incorporating the uncertainties associated with estimating sediment accumulation rates based on radiometric age estimates and, in the case of radiocarbon dates, calibrating radiocarbon ages into calibrated years (Blaauw and Heegaard 2012: Chap. 12). The total error associated with accumulation rates may be 25–40% or more of the estimated values. It is therefore essential to derive and display reliable estimates of the total errors before attempting interpretation of any observed changes in accumulation rates (Maher 1972b; Bennett 1994; Bennett and Willis 2001). Hughes and Hase (2010) provide a thorough account of measurements, uncertainties, and error analysis in the natural sciences.

Regression Analysis and Statistical Modelling

Regression analysis and statistical modelling are important in the *data assessment*, *data summarisation*, *data analysis*, and *data interpretation* stages in a palaeolimnological study (Table 2.2) and they can also be valuable in *data collection* (see above and Table 2.2). Regression analysis is implicit in many widely used approaches and techniques in quantitative palaeolimnology, including palaeoenvironmental reconstructions using calibration functions (Juggins and Birks 2012: Chap. 14), ordination techniques, both classical and canonical (Legendre and Birks 2012b: Chap. 8), classification, regression, and other decision trees, networks, and other statistical-learning techniques (Simpson and Birks 2012: Chap. 9), data assessment and error estimation (Maher et al. 2012: Chap. 6), age-depth modelling (Blaauw and Heegaard 2012: Chap. 12), temporal-series analysis (Dutilleul et al. 2012: Chap. 16), and confirmatory data analysis and hypothesis testing (Cumming et al. 2012: Chap. 20; Lotter and Anderson 2012: Chap. 18; Simpson and Hall 2012: Chap. 19). Because regression analysis is a vast (and complex) topic with very many textbooks devoted to different aspects of modern regression analysis and statistical modelling, the account here is rather general and is designed to give the reader a background of what regression analysis is, what it can be used for, what are the different major types of regression analysis, and in what circumstances should a particular regression approach be used. For detailed, in-depth accounts, the very readable books by Mosteller and Tukey (1977), Kleinbaum et al. (1988), Rawlings (1988), Hamilton (1992), Montgomery and Peck (1992), Crawley (1993, 2002, 2005, 2007), Fox (1997, 2002, 2008), Ramsey and Schafer (1997), Draper and Smith (1998), Zuur et al. (2007), Dalgaard (2008), and the chapter by ter Braak and Looman (1987) are particularly recommended. There are several very useful handbooks in the SAGE Series on Quantitative Applications in the Social Sciences on specific topics in regression analysis, including Achen (1982), Berry and Feldman (1985), Schroeder et al. (1986) and Berry (1993). More detailed reference texts include Harrell (2001), Venables and Ripley (2002), Crawley (2002, 2007), Aitken et al. (2009), Wright and Landon (2009), and Hastie et al. (2011).

Regression modelling is defined by Everitt (2002: pp. 319–320) as "a frequently used applied statistical technique that serves as a basis for studying and characterising a system of interest, by formulating a reasonable mathematical model of the relationship between a response variable, y, and a set of q explanatory variables, x_1, $x_2, \ldots, x_q$. The choice of the explicit form of the model may be based on previous knowledge of the system or on considerations such as 'smoothness' and continuity of y as a function of the x variables. In very general terms all such models can be considered to be of the form

$$y = f\left(x_1, \ldots, x_q\right) + \varepsilon$$

where the function f reflects the true but unknown relationship between y and the explanatory variables. The random additive error ε which is assumed to have mean zero and variance $\sigma_\varepsilon{}^2$ reflects the dependence of y on quantities other than $x_1, \ldots, x_q$. The goal is to formulate a function $f\ (x_1, x_2, \ldots, x_p)$ that is a reasonable approximation of f. If the correct parametric form of f is known, then methods such as least squares estimation or maximum likelihood estimation can be used to estimate the set of unknown coefficients. If f is linear in the parameters, for example, then the model is that of multiple regression. If the experimenter is unwilling to assume a particular parametric form for f, then non-parametric regression modelling can be used, for example, kernel regression, smoothing, recursive partitioning regression, or multivariate adaptive regression splines."

Regression analysis is thus a set of statistical methods that can be used to explore or model statistically the relationships between a response variables (e.g., diatom taxon, geochemical element, isotope value) and one or a set of predictor or explanatory variables on the basis of observations about the response variables and the predictor variable(s) at a series of sites or in a set of samples. The response variable, here assumed to be a biological taxon, may be recorded as abundances (relative or absolute values – see Table 2.1) or as present or absent (dichotomous or binary variable). In contrast to clustering, partitioning, and ordination techniques (Legendre and Birks 2012a, b: Chaps. 7 and 8), it is not possible to analyse all taxa simultaneously in regression techniques except with multivariate and reduced-rank regression (see below). In conventional regression, each taxon must be analysed and modelled separately (ter Braak and Looman 1987). Each regression focuses on a particular taxon and on how this taxon is related to the particular predictor variable(s) (e.g., environmental variables such as lake-water pH, alkalinity). The term 'response variable' comes from the idea that a taxon reacts or responds to the environmental or predictor variable(s) in a causal way; however, as ter Braak and Looman (1987) emphasise, causality cannot be proved from the results of a regression analysis.

The aim of regression analysis is thus more modest (ter Braak and Looman 1987), namely to describe the response variable as a quantitative function of one or more predictor variables. This is also called direct gradient analysis (ter Braak and Prentice 1988) and contrasts with multivariate gradient analysis (e.g., canonical correspondence analysis) where the responses of more than one response variable are described as a function of the predictor variables (ter Braak 1987a; ter Braak and Prentice 1988; Legendre and Birks 2012b: Chap. 8). This function, f, is termed the response function and it usually cannot be estimated so that the function can predict the response without some error. Regression analysis attempts to make the errors as small as possible and to average them to zero. The value predicted by the response function is thus the expected response (Ey), namely the response with the error term averaged out (ter Braak and Looman 1987).

In palaeolimnology, regression analysis is used mainly to

1. estimate parameters of palaeolimnological interest, for example the optimum and ecological amplitude or tolerance of a taxon in relation to an environmental predictor variable (Fig. 2.2), namely estimation and description
2. assess which predictor variables contribute most to a taxon's response and which predictor variables appear to be unimportant statistically. Such assessments require tests of statistical significance and thus involve statistical modelling
3. predict the taxon's responses (presence/absence or abundance) at sites with only observed values of one or more predictor variables, namely statistical prediction
4. predict or infer the values of predictor variables (e.g., environmental values) from observed values of one or more taxa. This is called calibration or environmental reconstruction and is discussed in detail by Juggins and Birks (2012: Chap. 14).

The vast topic of regression analysis and associated statistical modelling is simplified here and discussed as a general introduction to regression modelling; general linear models; generalised linear models; smoothers; generalised additive models; classification and regression trees; artificial neural networks; multivariate regression; and model selection and shrinkage. These accounts draw heavily on

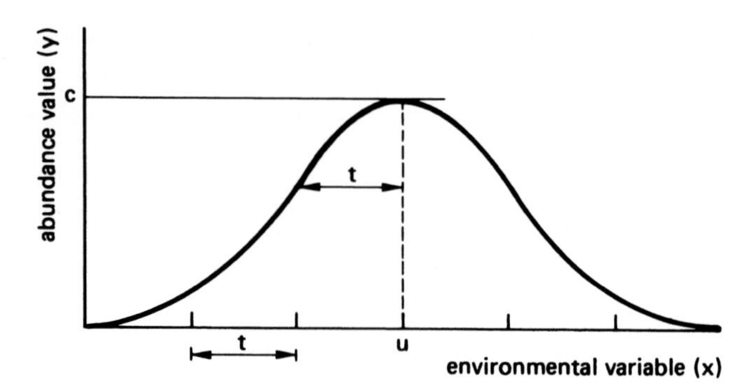

Fig. 2.2 A Gaussian unimodal relationship between the abundance (y) of a taxon and an environmental variable (x). The three important ecological parameters of the model are shown – u optimum of the taxon's response curve, t tolerance of the taxon's response curve, and c maximum of the response curve's height

ter Braak and Looman (1987), Manly (1992, 2007, 2009), Crawley (1993, 2002, 2005, 2007), Fox (1997, 2002, 2008), Ramsey and Schafer (1997), Everitt (2002), Lepš and Šmilauer (2003, especially Chap. 8), Zuur et al. (2007), and Hastie et al. (2011).

Introduction to Regression Analysis and Statistical Modelling

The simplest way to describe any regression model is (Lepš and Šmilauer 2003)

$$y = \mathrm{E}y + \varepsilon$$

where y is the value of the response variable of interest, Ey is the expected values of the response variable for particular values of the predictor variable(s), and ε is the variability of the observed true values of y around the expected values Ey, namely the random additive error (Lepš and Šmilauer 2003).

The expected values of the response variable are described as a function, f, of the predictor variables $x_1, \ldots, x_q$

$$\mathrm{E}y = f\left(x_1, \ldots, x_q\right)$$

The Ey part is called the *systematic part* or systematic component and ε is the stochastic or *error part* or error component of the regression model. The general properties and different roles of these parts of regression models are summarised in Table 2.3 (based on Lepš and Šmilauer 2003).

When a particular regression is fitted to a data-set, assumptions about the error part are fixed, such as the distributional properties, independence of individual observations, and cross-dependence between individual observations, but the

Table 2.3 Summary of the differences between the systematic and error parts of a regression model

Systematic part	Error part
Determined by the research question of interest or the hypothesis to be tested (ter Braak and Looman 1987)	Reflects *a priori* assumptions of the regression model (e.g., Crawley 1993, 2002)
Parameters (regression coefficients) are estimated by fitting the regression model (e.g., Crawley 1993, 2002)	Parameter(s) are estimated during or after model fitting the variance of the response variable (e.g., Crawley 1993, 2002)
Interpreted and tested as a working hypothesis or model. Individual parameters can be tested by further modelling (e.g., Crawley 1993, 2002)	Used to estimate model robustness and goodness-of-fit through regression diagnostics (e.g., ter Braak and Looman 1987; Crawley 1993, 2002)

Modified from Lepš and Šmilauer (2003), Table 8.1

systematic part can be varied depending on the nature of the data and the research questions (Lepš and Šmilauer 2003).

In the simplest case of the simple classical linear regression model with one response variable, y, and one predictor variable, x, the systematic part can be specified as

$$\mathrm{E}y = f(x) = \beta_0 + \beta_1 x$$

where β_0 and β_1 are fixed but unknown regression coefficients, β_0 is the intercept of the fitted line, and β_1 is the slope of the fitted line (ter Braak and Looman 1987).

It is, of course, possible to have a more complex model by modelling the dependence of y on x through a polynomial model. We thus have three possible types of model (Lepš and Šmilauer 2003; see also Blaauw and Heegaard 2012: Chap. 12).

1. $\mathrm{E}y = \beta_0$	The so-called null model
2. $\mathrm{E}y = \beta_0 + \beta_1 x$	Linear model of y linearly dependent on x
3. $\mathrm{E}y = \beta_0 + \beta_1 x + \beta_2 x^2$	Polynomial model of y having a non-linear dependency on x that can be expanded to be a polynomial to the nth degree where n is the number of observations $- 1$.

The most complex polynomial model, the so-called full model, passes exactly through all the observations but it provides no simplification of the data (which is one of the basic aims of any statistical model) and is thus of no practical use (Crawley 1993, 2002, 2007). In contrast, the null model simplifies reality so much that nothing new is gained from such a model. It too is of no practical use in gaining new insights from the data-set of interest via statistical modelling as gaining insights is also a basic aim of such modelling (Lepš and Šmilauer 2003).

Clearly some balance has to be found between the simple null model that provides no new insights and the complex full model that provides no data simplification (Crawley 1993, 2002, 2007; Lepš and Šmilauer 2003). Besides being complex and providing little or no simplifications of the data, complex models are frequently too well fitted to the data-set (so-called over-fitting where the model contains more unknown parameters than can be justified by the data – Everitt 2002), and may produce biased predictions for the part of the underlying statistical population that has not been or cannot be sampled. Thus the aim of modern regression analysis and associated statistical modelling is to try to find a robust compromise between a model that is as simple as possible for it to be useful in modelling and prediction and a model that is too complex (or too simple) to be useful (Lepš and Šmilauer 2003). Such a model is often called a parsimonious model or a minimal adequate model (Crawley 1993, 2002, 2007). Such a model accords with the principle of parsimony in statistical modelling, namely the general principle that among competing models, all of which provide an adequate fit for a particular data-set, the model with the fewest parameters is to be preferred (Everitt 2002). This principle is the statistical implementation of Occam's razor proposed by William of Occam (also spelt Ockham) (1280–1349) that "entia non sumt multiplicanda praetor mecessitatem", namely that "a plurality (of reasons) should not be posited without necessity" (Everitt 2002). Curtis et al. (2009) and Kernan et al. (2009) illustrate the derivation of minimal adequate models with multivariate palaeolimnological data.

Crawley (1993, 2002, 2007), Venables and Ripley (2002), Aitken et al. (2009), and Wright and Landon (2009) provide detailed insights into the scientific philosophy and the underlying techniques for statistical modelling using regression analysis.

General Linear Models

The regression techniques included in the class of general linear models (not to be confused with generalised linear models, discussed below) can all be represented as (Lepš and Šmilauer 2003)

$$Y_i = \beta_0 + \sum_{j=1}^{m} \beta_j \cdot X_{ji} + \varepsilon \qquad (2.1)$$

This general linear model differs from the traditional linear regression model discussed above in that both quantitative variables and qualitative variables or factors can be used as predictors. (A factor is a categorical variable with a small number of levels under investigation used as a possible source of variation – Everitt 2002.) Thus analysis of variance (ANOVA) is part of general linear models as the predictor variables are factors. A factor (e.g., soil moisture: dry, medium, wet) can be replaced by $k-1$ dummy presence/absence variables if the factor has k different levels. In Eq. 2.1 above, a factor is usually represented by more than one predictor X_j and therefore by more than one regression coefficient β_j (ter Braak and Looman

1987; Lepš and Šmilauer 2003). The symbol ε represents the random error part of the regression model. In general linear models, this is usually assumed to be zero and to have a constant variance (Lepš and Šmilauer 2003).

It is important to note that the general linear model is not the same as a generalised linear model (GLM) (Faraway 2005). As Lepš and Šmilauer (2003) emphasise, GLMs are based on the general linear model but represent a major generalisation of the general linear model, as discussed in the next sub-section on GLM.

An important property of a general linear model is that it is additive (Lepš and Šmilauer 2003). The effects of individual predictor variables are mutually independent but this does not mean that the predictors are not correlated (ter Braak and Looman 1987; Lepš and Šmilauer 2003). If the value of a predictor variable is increased by one unit, this has a constant effect, as expressed by the value of the regression coefficient for that predictor, independent of the values that the other predictor variables have and are even independent of the original values of the predictor variable that is increased (Lepš and Šmilauer 2003).

Equation 2.1 considers the theoretical underlying population of all possible observations of Y and X. In practice, we have a sample of this population which provides the observed data-set for analysis. Based on this finite and hopefully unbiased sample, the 'true' values of the regression coefficients β_j are estimated according to some numerical criteria (e.g., least-squares minimisation) and the resulting estimates are usually represented as b_j (Lepš and Šmilauer 2003). Estimating the values of b_j is known as model fitting (Lepš and Šmilauer 2003). Given the observed values of the predictor variables, the fitted or predicted values of the response variable are

$$\mathrm{E}Y = \hat{Y} = b_0 + \sum_{j=1}^{m} b_j \cdot X_j$$

Given the fitted values of the response variable, it is now possible to estimate the random variable representing the stochastic or error part of the model in Eq. 2.1 as (Lepš and Šmilauer 2003)

$$e_i = Y_i - \hat{Y}_i$$

This random variable, the regression residual, is the difference between the observed values of the response variable and the corresponding values predicted by the regression model (Lepš and Šmilauer 2003).

As a result of fitting a general linear model to the data, it is now possible to partition the total variation in the data into the model or regression variation and the residual variation (ter Braak and Looman 1987). The steps are as follows (Lepš and Šmilauer 2003).

1. Estimate the total variation in the values of the response variable as the total sum-of-squares (TSS), as

$$\mathrm{TSS} = \sum_{i=1}^{n} \left(Y_i - \bar{Y}\right)^2$$

where $\bar{Y}$ is the mean of response variable Y (Lepš and Šmilauer 2003).
2. Partition TSS into two components on the basis of the fitted regression model; the variation explained by the fitted model (regression or model sum-of-squares) and the residual variation represented by the residual sum-of-squares (Lepš and Šmilauer 2003).

   The regression or model sum-of-squares (MSS) is estimated as

$$\mathrm{MSS} = \sum_{i=1}^{n} \left(\hat{Y}_i - \bar{Y}\right)^2$$

   and the residual sum-of-squares (RSS) is estimated as

$$\mathrm{RSS} = \sum_{i=1}^{n} \left(Y_i - \hat{Y}\right)^2 = \sum_{i=1}^{n} {e_i}^2$$

   Clearly TSS = MSS + RSS. These statistics can now be used to test the statistical significance of the regression model. Under the null hypothesis that the response variable is independent of the predictor variable(s), MSS will not be different from RSS if both are divided by their respective degrees of freedom (MSS: number of parameters fitted − 1; RSS: number of objects minus number of parameters) to form the model mean square and the residual mean square that can be used in an F-test to evaluate the significance of the overall regression model. Total mean square is TSS divided by its degrees of freedom which are the number of objects − 1. See ter Braak and Looman (1987), Crawley (1993, 2002, 2007), Sokal and Rohlf (1995), Lepš and Šmilauer (2003), and Faraway (2005) for further details of statistical testing of general linear regression models.

An important topic that frequently arises in general linear models (and in other regression modelling) is the question of selecting predictor variables in a multiple linear regression. There are several methods for variable selection (Miller 1990; Ramsey and Schafer 1997; Faraway 2005; Murtaugh 2009; Dahlgren 2010; Simpson and Birks 2012: Chap. 9). These include backward elimination where all predictors are entered and variables are removed, starting with the least important and not statistically significant variables, and continuing until all the remaining variables are statistically significant. Forward selection involves starting with no variables and then adding in the 'best' predictor variable (explaining the largest part of the variation in the response variable). The process is continued as long as the new variables added are all statistically significant in the presence of the other included variables, namely have a statistically significant 'conditional' effect. The best-subset approach tries all possible combinations and the best set of predictors, in a statistical sense of high regression or model sum-of-squares and small number of statistically significant predictors and hence few model parameters, is selected. Olden and Jackson (2000) and Mundry and Nunn (2009) provide detailed analyses of the pros and cons of different procedures for selecting predictor variables in regression modelling and warn against step-wise procedures in hypothesis-testing analyses. Rawlings (1988), Miller (1990), Montgomery and Peck (1992), Sokal and Rohlf (1995), Ramsey and Schafer (1997), Faraway (2005), Murtaugh (2009), and Dahlgren (2010) discuss in detail the intricacies and problems in variable selection. Schielzeth (2010) discusses simple means to help improve the interpretability of regression coefficients in multiple linear regression models. Murray and Conner (2009) present six methods to try to quantify the relative importance of predictor variables in regression models. They show that zero-order correlations performed best at eliminating superfluous variables and that independent effects provided the most reliable ranking of the importance of different predictor variables. The topic of shrinkage of coefficients in regression models aimed at improving the predictive performance of such models is discussed briefly below and is reviewed in detail by Simpson and Birks (2012: Chap. 9).

The Akaike Information Criterion (AIC) is a widely used tool in selecting between competing regression and other statistical models. It takes into account both the statistical goodness-of-fit and the number of parameters that have been estimated to achieve this degree of fit (Crawley 2002, 2007). It imposes a penalty for increasing the number of parameters. The more parameters are used in a regression model, the better the fit and a perfect fit is obtained if there is a parameter for every object, but the model then has no explanatory or predictive power (Crawley 1993). AIC attempts to assess model 'parsimony' by penalising models for their complexity, as measured by the number of model degrees of freedom (Lepš and Šmilauer 2003). There is thus a trade-off between model goodness-of-fit and the number of parameters as required by the principle of parsimony. Low values of AIC indicate the preferred model, that is the one with the fewest parameters that still provides an adequate fit to the data. In multiple regression, AIC is the RSS plus twice the number of regression coefficients including the intercept. AIC can thus be used to compare the fit of alternative models with different numbers of parameters and hence degrees of freedom (model complexity). AIC is an invaluable guide to model selection. A sound general rule is that given alternative competing models, all with approximately the same explanatory power and goodness-of-fit but involving different numbers of regression parameters, the model with the lowest AIC should be selected. Hastie and Tibshirani (1990), Chambers and Hastie (1992), Venables and Ripley (2002), Murtaugh (2009), and Hastie et al. (2011) provide further details of AIC and its relative Bayes (Schwarz) Information Criterion (BIC) (see also Ramsey and Schafer 1997).

General linear models include linear and polynomial regression with one or more quantitative predictor variables, analysis of variance (ANOVA) and analysis of covariance (ANCOVA) with one or more qualitative nominal factors as predictor variables and, in the case of ANCOVA, a quantitative variable as a covariable or concomitant variable. These covariables are predictor variables that are not of primary interest in a study but are measured because it is likely that they may influence the response variable of interest and consequently need to be included in the regression analysis and model building. Multiple regression can also be used to assess the simultaneous effect of both quantitative and nominal predictors (ter Braak and Looman 1987; Hardy 1993). All these techniques involve least-squares estimation and thus they make various assumptions of the data and the model errors. As a result, many things can go wrong in seemingly straightforward regression analysis (Manly 1992; Ramsey and Schafer 1997). Regression diagnostics are a powerful set of techniques to detect such problems (see Belsey et al. 1980; Cook and Weisberg 1982; Hocking and Pendleton 1983; Bollen and Jackman 1990). The critical assumptions in least-squares estimation in regression are that (1) the variance of the random error part is constant for all observations, (2) the errors are normally distributed, and (3) the errors for all observations are independent. Assumption (1) is required to justify choosing estimates of the b parameter so as to minimise the residual sum-of-squares and is needed in significance tests of t and F values. Clearly, in minimising the RSS, it is essential that no residuals should be markedly larger than others. Assumption (2) is needed to justify tests of statistical significance and estimation of confidence intervals.

The first step in regression diagnostics (Bollen and Jackman 1990; Manly 1992) is to plot the residuals of the regression model against the fitted values or against each of the predictor variables and to look for outliers and systematic trends in these plots. Plots of residuals against the fitted values can highlight non-constancy of variances whereas plots against the predictors may detect evidence of curvature within the model. In addition, residuals can be plotted against the sequence of data collection to detect any temporal correlation and normal probability plots constructed where the ordered standardised residuals are plotted against expected values assuming a standard normal distribution to detect non-normality of errors.

Two final topics in general linear models are so-called Model II regression (Legendre and Legendre 1998) and piece-wise regression (Toms and Lesperance 2003). Model II regression is important when both the response and the predictor variables are random and are not controlled by the researcher and thus there is error associated with measures of both sets of variables, as commonly occurs in morphometrics and water chemistry. When there is a need to estimate the parameters of the equation that describes the relationship between pairs of random variables, it is essential to use Model II regression for parameter estimation because the slope parameter estimated by least-squares linear regression (Model I regression) may be biased by the presence of measurement error in the predictor variable. There are various Model II regression methods such as major-axis regression, standard major-axis regression, and ranged major-axis regression (Legendre and Legendre 1998). The strengths and weaknesses of these Model II regression procedures and the problems of statistical testing for Model II models are discussed by McArdle (1988) and Legendre and Legendre (1998).

Piece-wise regression models are 'broken-stick' models where two or more lines are joined at unknown point(s), called 'break-point(s)' representing abrupt changes in the data-set under analysis (Toms and Lesperance 2003). The segments are usually straight lines but the technique can be generalised to segments of other shapes (e.g., polynomial segments) (Seber and Wild 1989). Piece-wise regression models can be difficult to fit because of the number of possible break-points. Palaeolimnological examples of piece-wise regression include Heegaard et al. (2006) and Engels and Cwynar (2011). These examples involve detecting ecotones or thresholds in biological data-sets along altitudinal or depth gradients, respectively.

All the general linear models discussed so far have assumed normally distributed variables and the error part is assumed to be normally distributed. How can data with response variables that are presences or absences or proportions be analysed by regression analysis? We now turn to generalised linear models (GLM) that are more generalised than the restrictive general linear models.

Extending the General Linear Model

Introduction

The general linear model is central to modern regression analysis and statistical modelling. The model takes the form

$$y = \beta_0 + \beta_1 x_1 + \cdots + \beta_m x_m + \varepsilon$$

where y is the response variable, $x_1, \ldots, x_m$ are the predictor variables, and ε is the normally distributed error part.

In this section we present four extensions to this general linear model. the first generalises the y part, the second the ε part, the third the x part of the general linear model, and the fourth part extends the response variable, y, to many response variables, so-called multivariate regression.

The general linear model cannot handle non-normal responses in y such as counts, proportions, or presence-absence data. This has led to the development of generalised linear models (GLM) that can represent binary, categorical, and other response types (McCullagh and Nelder 1989; Faraway 2006).

Mixed-effects models (MEM) allow the modelling of data that have a grouped, nested, or hierarchical structure (Pinheiro and Bates 2000). Repeated measures and longitudinal and multi-level data consist of several observations taken on the same individual or group. This includes a correlation structure in the error part, ε (Faraway 2006).

In the general linear model, the predictors, x, are combined in a linear way to model their effect on the response. Sometimes this linearity is not adequate to capture the structure of the data and more flexibility is required. Methods such as generalised additive models (GAMs), classification and regression trees (CARTs), and artificial neural networks (ANNs) allow a more flexible regression modelling of the responses that combine the predictors in a non-parametric manner (Faraway 2006).

All the approaches discussed (general linear model, GLM, MEM, GAM, CART, ANN) most commonly consider one response variable in relation to one or more predictor variables. Regression techniques are required that allow the modelling of the responses of many variables (e.g., biotic assemblages) in relation to one or many predictors. These approaches are grouped here as multivariate regression techniques. These overlap with canonical or constrained ordination techniques discussed by Legendre and Birks (2012b: Chap. 8) and multivariate regression trees discussed by Simpson and Birks (2012: Chap. 9) and Legendre and Birks (2012a: Chap. 7).

Generalised Linear Models

Generalised linear models (GLM) (McCullagh and Nelder 1989) extend and generalise the general linear model in two critically important ways. Before considering these extensions, it is useful to clarify what GLMs are. They are not models that only show a straight-line relationship between the response variable and the predictor variable(s). A linear model is an equation that contains mathematical variables, parameters, and random variables that is linear in the parameters and the random variables (Crawley 1993), such as

$$y = a + bx$$

$$y = a + bx + cx^2 = a + bx + cz, \text{ where } z = x^2$$

$$y = a + bc^x = a + bz, \text{ where } z = \text{exponential } (x)$$

Some non-linear models can be expressed in a linear form by suitable transformations

$$y = \exp(a + bx) \rightarrow \log_e y = a + bx$$

There are some models that are intrinsically non-linear such as the hyperbolic function or the asymptotic exponential. Nothing can linearise them for all parameters.

Excellent introductions to GLM include ter Braak and Looman (1987), O'Brian (1992), Crawley (1993, 2002, 2005, 2007), Gill (2001), Fox (2002, 2008), Lepš and Šmilauer (2003), and Dunteman and Ho (2006). The underlying theory is presented by McCullagh and Nelder (1989), Dobson (2001), Myers et al. (2002), and Faraway (2006).

The first way that GLMs extend and generalise the general linear model is that the expected values of the response variable (Ey) are not supposed to be always directly equal to a linear combination of the predictor variables. Rather, the scale

Table 2.4 Summary of useful combinations of GLM link functions and types of response-variable reference statistical distributions

Type of response variable	Typical link function	Reference statistical distribution	Variance function (mean-variance relation)
Counts	Log	Poisson	$v \propto \mathrm{E}y$
Proportions	Logit or probit	Binomial	$v \propto \mathrm{E}y \cdot (1 - \mathrm{E}y)$
Ratios	Inverse or log	Gamma	$v \propto (\mathrm{E}y)^2$
Measurements	Identity	Gaussian (normal)	$v = \text{constant}$

Based on O'Brian (1992), Crawley (1993), and Lepš and Šmilauer (2003)
The assumed relation between the variance (v) and the expected values of the response variable (Ey) is also given

of the response depends on the scale of the predictors through a simple parametric function, the so-called link function g (Lepš and Šmilauer 2003),

$$g\,(\mathrm{E}y) = \eta$$

where η is the linear predictor and is defined in exactly the same way as the entire systematic part of a general linear model (Lepš and Šmilauer 2003) as

$$\eta = \beta_0 + \sum_{j=1}^{m} \beta_j x_j$$

The aim of the link function is to allow mapping of values from the whole scale of real values of the linear predictor (ranging from $-\infty$ to $+\infty$) into a specific interval that makes more sense for the response variable such as non-negative counts or values bounded between 0 and 1 for proportions (ter Braak and Looman 1987; Lepš and Šmilauer 2003).

The second way that GLMs generalise and extend the general linear model is that GLMs have less specific assumptions about the error or stochastic part of the regression model compared to general linear models. The variance does not need to be constant and can depend on the expected values of the response variable Ey (Lepš and Šmilauer 2003). The relationship between the mean and variance is usually specified through the statistical distribution assumed for the error part and, therefore, for the response variable (Lepš and Šmilauer 2003). It is important to note that the mean-variance relationship described by the variance function (Table 2.4) is the *essential* part of a GLM specification, not the specific statistical distribution (Lepš and Šmilauer 2003).

A GLM consists of three main parts – the error part, the linear predictor, and the link function. The choices available for the link functions and for the assumed type of distribution of the response variable cannot be combined independently (Table 2.4) (Lepš and Šmilauer 2003). The logit link function (Table 2.4), for example, maps the real scale onto a range from 0 to +1 (ter Braak and Looman 1987) so it is not an appropriate link function for count data with an assumed

Poisson distribution (Lepš and Šmilauer 2003). The logit function is useful mainly for modelling probability as a parameter of the binomial distribution (ter Braak and Looman 1987; Pampel 2000; Menard 2002; Lepš and Šmilauer 2003). Table 2.4 lists the commonest combinations of the assumed link function and the expected distribution of the response variable together with a summary of the types of response variables (e.g., counts, ratios) that match these assumptions. As in classical linear regression and ANOVA within general linear models, it is not assumed that the response variable has a particular distribution (Lepš and Šmilauer 2003).

The commonly used *error parts* in GLMs within palaeolimnology are

- Gaussian or normal;
- Poisson for count data that are non-negative integers whose variance increases with the mean;
- binomial for observed proportions from a total consisting of non-negative values with a maximum value of 1.0 and variance is largest at $p = q = 0.5$ where p is the proportion of individuals that responded in a given way ('successes'), and q ('failures') is $(1 - p)$ (Pampel 2000; Menard 2002);
- gamma for concentration data with non-negative real values, standard deviation increases with the mean, and the data often have many near-zero values and some high values;
- exponential for data on time-to-death in, for example, survival analysis (Crawley 1993).

Clearly the choice of error function depends on the nature of the response variable and on the proportional relationship between the variance and the expected values of the response variable (Table 2.4).

The *linear predictor* η (eta) is defined as

$$\eta = \sum_{j=1}^{m} x_{ij} \beta_j$$

where x_{ij} is the value of the predictor variable j in object i, β_j are the model coefficients, and there are m predictor variables. The right-hand side is called the linear structure. To determine the fit of a given model, a linear predictor is needed for each value of the response variable, and then the predicted value is compared with a transformed value of y, the transformation applied is specified by the link function. The fitted value is computed by the inverse of the link function to get back to the original scale of measurement of the response.

The third part of a GLM is the *link function*. This relates the mean value of y to its linear predictor η by

$$\eta = g(y)$$

where g (.) is the link function. The linear predictor emerges from the linear model as a sum of the terms for each of the m predictor variables. This is not a value of

y (except in the case of the identity link (Table 2.4) where $\eta = \mu$). The value of η is obtained by transforming the values of y by the link function, and the predicted value of y is obtained by applying the inverse link function to η. So

$$\mu = g^{-1}(\eta)$$

and we can combine the link function and linear predictor to form the basic or core GLM equation

$$y = g^{-1}(\eta) + \varepsilon \quad \text{or} \quad g(y) = \eta + \varepsilon$$

where ε is the error part, η is the linear predictor, and g is the link function (Lepš and Šmilauer 2003).

The most frequently used link functions are given in Table 2.4. The most important criterion in selecting a link function is to ensure that the fitted values stay within reasonable bounds (ter Braak and Looman 1987). We need to ensure that all counts are integer values greater than or equal to zero or the responses expressed as proportions have fitted values that lie between 0 and 1. In the first case, a log link is appropriate because the fitted values are antilogs of the linear predictor and all antilogs are greater than or equal to zero. In the second case, the logit link is appropriate because the estimated proportions are calculated from the antilogs of the logs-odd ratio, log (p/q), and must lie between 0 and 1. Crawley (1993, 2002, 2005, 2007) summarises the main types of GLM used in ecology and these include those most appropriate in palaeolimnology. Note that linear (simple or multiple), ANOVA, and ANCOVA within general linear models, are, in fact, members of the GLM family with identity link functions and an assumed normal or Gaussian error reference distribution.

GLMs thus include many types of regression models (Lepš and Šmilauer 2003). These include:

1. 'classical' general linear models including (multiple) linear regression and most types of ANOVA and ANCOVA (Neter et al. 1996);
2. extensions of these classical linear models to include response variables with non-constant variance (counts, relative frequencies, proportions) (Fleiss 1981; Aitchison 1986);
3. analysis of contingency table using log-linear models (Fienberg 1980; Christensen 1990; Everitt 1992; Harrell 2001);
4. models of survival probabilities (probit analysis) used in, for example, ecotoxicology (Piegorsch and Bailer 1997; Barnett 2004).

To summarise GLMs so far, they are mathematical extensions or generalisations of general linear models (Faraway 2006) that do not force real-life data into unnatural, unrealistic, or inappropriate scales. They therefore allow for non-linearities and non-constant variance structures in data. They are based on an assumed relationship (link function) between the mean of the response variable and the linear combination of the predictor variables (linear predictor). The data can be assumed to be several families of reference statistical distributions such as continuous probability distributions (Gaussian (= normal), gamma, exponential, chi-square) or discrete probability distributions (binomial, Poisson, multinomial, hypergeometric) which may approximate the assumed error structure of most real-life data-sets (O'Brian 1992). GLMs are thus more flexible and better suited for analysing many real-life data-sets than 'conventional' regression techniques within general linear models (Lepš and Šmilauer 2003).

GLMs extend the concept of residual sum-of-squares in general linear models. The extent of discrepancy between the observed 'true' values of the response variable and those predicted by the regression model is expressed as the GLM's deviance. Therefore to assess the quality of a model, statistical tests are used that are based on the analysis of deviance which is quite similar in concept to an analysis of variance of a regression model in general linear modelling (ter Braak and Looman 1987).

An important property of the general linear model, namely its linearity, is retained in GLM on the scale of the linear predictor (Lepš and Šmilauer 2003). The effect of a particular predictor is expressed by a single parameter, the linear transformation coefficient or regression coefficient. Similarly the model additivity is kept on the linear predictor scale. In terms of the scale of the predictor variable, there can be major differences (Lepš and Šmilauer 2003). For example, with a log link-function, the additivity on the scale of the linear predictor corresponds to a multiplicative effect on the scale of the response variable (Lepš and Šmilauer 2003).

There are five major stages in the fitting of GLM.

1. Identify the response (y) and the predictor (x) variables
2. Identify and define the model equation
3. Choose an appropriate assumed error function for the response variable under study
4. Use appropriate model parameter estimation procedures
5. Use appropriate model evaluation and criticism procedures.

There are many books that discuss GLMs and the stages in fitting them – Crawley (1993, 2002, 2005, 2007), Ramsey and Schafer (1997), Harrell (2001), Myers et al. (2002), Faraway (2006), Fox (2008), and Wright and Landon (2009) and the interested reader is referred to these books for details.

I will concentrate here on the question of parameter estimation, model development, and model criticism from a practical viewpoint in palaeolimnology.

Given the error function and the link function, it is necessary to formulate the linear predictor term. It is essential to be able to estimate its parameters and to find the linear predictor that minimises the total residual deviance in the model. If a normal error function and an identity link function are used, then least-squares estimation of general linear models is appropriate. Other error functions and link functions require maximum-likelihood estimation.

In maximum-likelihood estimation (Eliason 1993), the aim is to find parameter values that give the 'best fit' to the data. Best in the context of maximum likelihood involves consideration of (1) the data for the response variable, (2) specification of the model, and (3) estimation of the model parameters. The aim is to find the minimal adequate model to describe the data. The 'best' model is the model that produces the minimal total residual deviance subject to the constraint that all the parameters in the model are statistically significant. The model should be minimal because of the principle of parsimony in statistical modelling and adequate because there is no point in retaining an inadequate model that does not describe a significant part of the variation in the response variable data. There may be no one minimal adequate model or there may be several adequate models. Statistical modelling involving GLM is designed to find the minimal adequate model but in real-life cases there may be more than one plausible and acceptable model. In the context of GLM, the principle of parsimony (Occam's Razor) (Crawley 1993) proposes that

1. models should have as few parameters as possible
2. linear models are to be preferred to non-linear models
3. models relying on few assumptions are to be preferred to models with many assumptions
4. models should be simplified until they are minimal adequate models
5. simple explanations are to be preferred to complex explanations.

Maximum-likelihood estimation, given the data, model, link function, and error part, provides values for the model parameters by finding iteratively the parameter values in the specified model that would make the observed data most likely, namely to find model parameters that maximise the likelihood of the data being observed. It involves maximising the likelihood or log-likelihood with respect to the parameters and depends not only on the data but also on the model specification. It is here that the concept of deviance is important as a measure of goodness-of-fit. Fitted or expected values (Ey) are most unlikely to match the observed values perfectly. The magnitude of the discrepancy between fitted and observed values is a measure of the inadequacy of the model and deviance is a measure of this discrepancy.

The aim of statistical modelling within the framework of GLM is to determine the minimal adequate model in which all the parameters are significantly different from zero. This is achieved by a step-wise process of model simplification (Crawley 1993, 2002, 2007) beginning with the saturated model with one parameter for every observation. This model has a perfect fit (deviance is zero), no degrees of freedom, and no explanatory power. The next step is to develop the maximal model with one parameter for all (p) factors, interactions, and variables and covariables that might be of any interest, and to note its residual deviance and the degrees of freedom ($n-p-1$). This model may have some explanatory power but many of the model parameters may not be statistically significant. Model simplification begins by inspecting the parameter estimates and by removing the least significant terms starting with the highest order interactions. If the deletion causes an insignificant increase in deviance, the term is left out of the model, the parameter estimates are inspected, and the least significant terms remaining in the model are removed. If the deletion causes a significant increase in deviance, the term is put back into the model. These are potentially statistically significant and important terms as assessed by their effects on the deviance when deleted from the maximal model. Removal of terms from the model is continued by repeating the previous two steps until the model contains nothing but significant terms. The resulting model is the minimal adequate model. If none of the parameters is significant, then the null model with only one parameter (the overall mean of the response variable y) is the minimal adequate model but with no fit, the residual or error sum-of-squares equals the total sum-of-squares, and thus the model has no explanatory power.

Hopefully there will be at least one minimal adequate model with at least one significant parameter. It will have $0 \leq p' \leq p$ significant parameters, $n-p'-1$ degrees of freedom, a model fit less than the maximal model but not significantly so, and an explanatory power r^2 of residual sum-of-squares/total sum-of-squares (or low residual deviance).

It is not uncommon to find two or more equally plausible or acceptable minimal adequate models. Additional tools for selecting the minimal adequate model are the Akaike Information Criterion (AIC) and the Bayes Information Criterion (BIC) (Crawley 2002, 2005, 2007). The more parameters there are in a model, the better the fit is but the explanatory or predictive power is low. There is thus a trade-off between the goodness-of-fit and the number of model parameters. AIC and BIC penalise any superfluous parameters by adding penalties to the deviance. AIC supplies a relatively light penalty for any lack of parsimony, whereas BIC applies a heavier penalty for lack of parsimony. One should select as the minimal adequate model(s), the model(s) that gives the lowest AIC and/or BIC. Burnham and Anderson (2002) discuss in detail the various approaches and problems in model selection (see also Birks 2012c: Chap. 21; Simpson and Birks 2012: Chap. 9).

The most effective approach to model selection and evaluation involves external cross-validation (Juggins and Birks 2012: Chap. 14) where one data-set is used to develop a statistical model based on GLMs. The regression model is then applied to an independent test data-set to evaluate the predictive power of the GLM statistical model (e.g., Oberdorff et al. 2001). Shao (1993) discusses linear model selection by internal cross-validation, illustrating the limitations of leave-one-out cross-validation, and recommends data-spitting or k-fold cross-validation.

The most widespread use of GLM in palaeolimnology is probably the fitting of response curves or surfaces to the present-day occurrence or abundance of individual taxa (e.g., diatom taxa) to an environmental variable such as lake-water pH (e.g., ter Braak and van Dam 1989; Birks et al. 1990). There are many types of ecological response curves and a compromise is necessary between ecological realism and numerical simplicity (ter Braak and Prentice 1988; ter Braak and van Dam 1989; ter Braak 1996). The Gaussian unimodal response model with a symmetric unimodal curve is a suitable compromise (Fig. 2.2) (ter Braak and Prentice 1988). The Gaussian logit or logistic model within the GLM framework is usually applied to presence-absence data (see e.g., ter Braak and Looman 1986, 1987; Vanderpoorten and Durwael 1999; Oberdorff et al. 2001; Ysebaert et al. 2002; Coudun and Gégout 2006). However, it can be used, as in palaeolimnology, as a quasi-likelihood model for proportions and as an approximation to the more complex multinomial logit model (ter Braak and van Dam 1989; Birks 1995). The multinomial logit (= logistic) model can be difficult to fit and its parameters difficult to interpret because of indeterminacies. Here a Gaussian logit model is used as a quasi-likelihood model because the mean-variance relationship described by the variance function (see Table 2.4) is the essential property of the model specification and not the specific reference distribution (Lepš and Šmilauer 2003).

If the aim is to establish how many taxa in a modern calibration organism-environment data-set (see Juggins and Birks 2012: Chap. 14) have a statistically significant fit to a Gaussian logit model or to the simpler linear or sigmoidal logit model, a Gaussian logit model can be fitted to all those taxa expressed as proportions with at least ten occurrences in the data-set with binomial error structure. From the Gaussian logit regression coefficients, b_0, b_1, and b_2, the optimum ($\hat{u}_k$), tolerance ($\hat{t}_k$), and the height of the curve's peak (c_k) of the fitted Gaussian response curve (Fig. 2.2) can be estimated (ter Braak and Looman 1986, 1987), along with the approximate 95% confidence intervals for the estimated $\hat{u}_k$ and the standard error of the estimated $\hat{t}_k$ (ter Braak and Looman 1986, 1987).

For each taxon, the significance ($\alpha = 0.05$) of the Gaussian logit model (three parameters) can be tested against the simpler linear logit (sigmoidal) model (two parameters) by a residual deviance test. The significance of the Gaussian logit regression coefficient b_2 against the null hypothesis ($b_2 \geq 0$) can be assessed by a one-sided t-test (ter Braak and Looman 1986, 1987). If the null hypothesis is rejected in favour of $b_2 < 0$, the taxon's optimum is considered statistically significant. If either the Gaussian unimodal model or the optimum are not significant, the linear logit model and its regression coefficient b_1 can be tested against the null model that the taxon shows no significant relationship with the environmental variable of interest by using deviance and two-sided t-tests (Birks et al. 1990).

This simple hierarchical set of statistical models has been extended by Huisman et al. (1993) and Oksanen and Minchin (2002) in their HOF modelling procedures and software. Here a set of five response models are fitted (skewed, symmetric, plateau, monotonic, null) with either a Poisson or binomial error function. The most complex model (skewed) is fitted first using maximum-likelihood estimation, and the simpler models are then fitted by backward elimination. Deviance for each model is calculated and if the drop in deviance is greater than 3.84, the extra model parameter is significant at $\alpha = 0.05$ (chi-square distribution). If the data are over-dispersed with the deviance greater than the degrees of freedom, an F-test is used to assess the drops in deviance as model simplification continues. This model simplification procedure provides a means of selecting the most parsimonious or minimal adequate response model for a taxon using statistical criteria. The number of taxa in the data-set showing a statistically significant response to the environmental variable of interest can be established and the most frequent type of response curve can be determined. Palaeolimnological data-sets invariably have symmetric Gaussian responses as the commonest response curve (e.g., Lotter et al. 1997, 1998). Skewed responses and null responses are rare, plateau responses are very rare, and monotonic responses are frequent (Birks et al. 1990; Bjune et al. 2010).

There are simpler heuristic methods based on simple weighted averaging (WA) regression to estimate the optimum and tolerance of a single response variable (e.g., chironomid species) in relation to a predictor environmental variable (e.g., mean July air temperature) (ter Braak and Prentice 1988). When a species shows a unimodal relationship with a predictor variable (Fig. 2.2), the presences of the species will be concentrated around the peak of the occurrence function. A simple estimate of the species' optimum is thus the average of the values of the predictor variable at those sites where the species is present. With quantitative abundance data, WA applies weights proportional to the species' abundances. Absences have zero weight (ter Braak and Prentice 1988). WA estimates the optimum of a Gaussian logit curve as efficiently as the GLM technique of Gaussian logit regression if the site values are equally spaced over the whole range of occurrences of the species along the gradient of the predictor variable and the site values are closely spaced in comparison with the species tolerance (ter Braak and Barendregt 1986; ter Braak and Looman 1986; ter Braak and Prentice 1988). WA is an important part of several techniques used in palaeolimnology for reconstructing environmental variables from fossil assemblages, so-called calibration (see Juggins and Birks 2012: Chap. 14).

Mixed-Effects Models

Mixed-effects models are mentioned only briefly here because they have hardly been used in palaeolimnology (see Heegaard et al. 2005, 2006; Adler et al. 2010; Eggermont et al. 2010 for examples). They are of considerable potential value with certain types of data (Pinheiro and Bates 2000; Zuur et al. 2009).

They are a class of regression and analysis of variance models that allow the usual assumptions that the residual or error terms are independent and identically distributed to be relaxed and generalised. They thus allow generalisations of the error part, ε, in the general linear model (Faraway 2006).

Mixed-effects models can take into account more complicated data structures in a flexible way, for example if the data have a grouped, nested, or hierarchical structure. Repeated measures, longitudinal, spatial, and multilevel data consist of several observations taken on the same individual or group, thereby inducing a correlation structure in the error. Analyses that assume independence of the observations would be inappropriate. Mixed-effects models allow the modelling of such data by either modelling interdependence directly or by introducing random effect terms which induce observations on the same object to be correlated. The major advantages of these types of models include an increase in the precision of estimates and the ability to make wider inferences (Everitt 2002).

Mixed-effects models commonly involve random effects to model grouping structure. A fixed effect is an unknown constant that we attempt to estimate from the data. Fixed effect parameters are commonly used in general and generalised linear models. In contrast, a random effect is a random variable. It does not make sense to estimate a random effect; instead what is done is to estimate the parameters that describe the distribution of this random effect (Faraway 2006).

Consider an experiment to investigate the effect of a pollutant on a sample of aquatic invertebrates. Typically, interest is on specific pollutant treatments and we would treat the pollutant effects as fixed. However, it makes more sense to treat the invertebrate effects as random. It is often reasonable to treat the invertebrate samples as being random samples from a larger collection of invertebrates whose characteristics we wish to emulate. Furthermore, we are not particularly interested in these specific invertebrates but in the whole population of invertebrates. A random effects approach to modelling effects is more ambitious in that it is attempting to infer something about the wider population beyond the available samples. Blocking factors can often be viewed as random effects, because these often arise as a random sample of those blocks potentially available (Faraway 2006).

A mixed-effects model has both fixed and random effects. A simple example of such a model would be a two-way ANOVA

$$y_{ijk} = \mu + \mathrm{T}_i + v_j + \varepsilon_{ijk}$$

where μ and T_i are fixed objects and the error, ε_{ijk}, and the random effects v_j are independent and identically distributed $N(0, \sigma^2)$ and $N(0, \sigma_v^2)$, respectively, where σ^2 and σ_v^2 are variances. We wish to estimate T_i and test the hypothesis $H0: \mathrm{T}_i = 0$ while we estimate σ_v^2 and test $H0: \sigma_v^2 = 0$. Note that we are estimating and testing several fixed-effect parameters but only estimating and testing a single random effect parameter (Faraway 2006). Pinheiro and Bates (2000), Zuur et al. (2007, 2009), Allison (2009), and Bolker et al. (2009) provide excellent introductions to mixed-effects modelling in ecology.

Non-parametric Regression Models

In general and generalised linear models, the predictor variables, x, are combined in a linear way to model their effect on the response variable. In some instances this linearity is inadequate and too restrictive to capture the structure of the data and more flexibility in modelling is required. In this section we consider smoothers, generalised additive models (GAM), classification and regression trees (CART) and related decision trees, and artificial neural networks (ANN) and related techniques. These can result in a more flexible regression modelling of the response by combining the predictors in a non-parametric manner (Faraway 2006). Non-parametric methods are statistical techniques of estimation and inference that are based on a function of the sample observations, the probability distribution of which does not depend on a complete specification of the reference probability distribution of the population from which the sample observations were drawn. Consequently the techniques are valid under relatively general assumptions about the underlying populations.

Smoothers: We first consider smoothers, a term that can be applied to almost all techniques in statistics that involve fitting some model to a set of observations (Goodall 1990; Everitt 2002). Smoothers generally refer to those methods that use computing techniques to highlight structure within a data-set very effectively by taking advantage of the investigator's ability to draw conclusions from well-designed graphs (Everitt 2002). We will consider locally weighted regression (LOESS) (Goodall 1990; Jacoby 1997) and smoothing spline functions (Marsh and Cormier 2002). Many other smoothing techniques exist, including non-parametric regressions such as kernel regression smoothers and multivariate adaptive regression splines (Venables and Ripley 2002; Faraway 2006; Simpson and Birks 2012: Chap. 9).

The basic aim of all smoothers is to derive a non-parametric function from a set of observations. The fitted values are produced by the application of smoothers and are thus less variable than the actual observations, hence the term 'smoother' (Lepš and Šmilauer 2003).

As Lepš and Šmilauer (2003) emphasise, there are very many different types of smoothers, some are not very good but are simple to understand and others are very good but not simple to understand. An example of the former type is the simple moving-average or running-mean smoother. This is used primarily to smooth time-series in which a target observation is replaced by a weighted average of the target observation and its near neighbours with weights being a function of distance or time from the target observation. It is only really appropriate with time-series and less so with temporal-series (Juggins and Telford 2012: Chap. 5). An example of the second type of smoother that is versatile and robust is the LOESS smoother. LOESS is an acronym for **lo**cally weighted regr**ess**ion and is a generalisation of the technique known as LOWESS derived from **lo**cally **we**ighted **s**catterplot **s**moother (Cleveland 1979). Crawley (2002, 2005, 2007) describes LOWESS as a non-parametric curve fitter and LOESS as a modelling tool. Venables and Ripley (2002) describe LOESS as an extension of the ideas of LOWESS which will work in one, two, or more dimensions in a similar way. For consistency with Lepš and Šmilauer (2003) we use the term LOESS for smoothers based on locally weighted linear regression (Cleveland and Devlin 1988; Hastie and Tibshirani 1990).

LOESS does not attempt, in contrast to general or generalised linear modelling, to fit a simple model (e.g., a quadratic model) between Y and X throughout the entire range of X (Efron and Tibshirani 1991; Trexler and Travis 1993). Instead it fits a series of local regressions for values of the predictor, in each case using only data points near the selected values (Jacoby 1997; Fox 2000; Lepš and Šmilauer 2003). Essentially, LOESS performs a series of robust weighted regressions at each of t different locations or target values (v_j with j from 1 to t) along the predictor variable's range. Each regression uses only a subset of observations that fall close to the target value on the horizontal predictor axis. The coefficients from each local regression are used to generate a predicted or fitted value $g\ (v_j)$. The t different points (v_j, $g\ [v_j]$) are plotted, and adjacent points are connected by line segments to produce the final smooth curve (Jacoby 1997). The area (band for a single predictor) around the target value or estimation point which is used to select the data points to be fitted in the local regression model is called the band-width or span and is specified as a proportion of the total available data-set (usually 0.45 or 0.33). Thus a span value α of 0.33 specifies that at each estimation point a third of all the available observations is used in the local regression and this third consists of all the observations closest to the estimation point. The complexity of the local linear regression model is specified by the second parameter of the smoother, the degree (λ). Normally only two degrees are available: 1 for a linear regression model and 2 for a second-order polynomial model (Fox 2000; Lepš and Šmilauer 2003).

The observations used in fitting the local regression model do not all carry the same weight. The weights depend on their distance from the target value in the predictor's space. If a data point has exactly the same values as the predictor, its weight is equal to 1.0. The weight decreases smoothly as a tricubic function to 0.0 at the edge of the smoother's band-width. The procedure (Jacoby 1997) is thus

1. Describe how 'smooth' the fitted relationship should be by selecting the value of band-width or span (α) to use.
2. Each observation is given a weight depending on its proximity to the target value or estimation point. All points within the span are considered and the weights applied are derived from a tricubic distance-decay function.
3. A simple linear regression ($\lambda = 1$) or second-order polynomial regression ($\lambda = 2$) is fitted for all adjacent points using weighted least-squares regression where the weights are defined in step 2.
4. Repeat for all observations in the predictor variable.
5. Calculate the residuals between the observed and fitted values of the response variable.
6. Derive robustness weights based on the residuals so that well-fitted points carry high weight.
7. Repeat steps 3 and 4 but with new weights based on the robustness weights and the distance weights.

The fitted curve is a non-parametric regression estimate because it does not assume a global parametric form (e.g., quadratic) for the regression.

An important feature of a LOESS regression model is that its complexity can be expressed using the same kind of units used in general or generalised linear models, namely the degrees of freedom taken from the data by the fitted model (Lepš and Šmilauer 2003). These are alternatively called the equivalent number of parameters. As a LOESS model produces, as in other regression models, fitted values of the response variable, it is possible to estimate the variability in the values of the response variable accounted for by the fitted model and compare it with the residual sum-of-squares (Lepš and Šmilauer 2003). As the number of degrees of freedom of the model is estimated, the residual degrees of freedom can be calculated along with the sum-of-squares per one degree of freedom, corresponding to the mean square of a general linear regression model (Lepš and Šmilauer 2003). Consequently, LOESS models can be compared using an analysis of variance in the same way as is done for general linear models (Lepš and Šmilauer 2003). It is possible to use generalised cross-validation to find the 'optimal' LOESS model (Jacoby 1997; Venables and Ripley 2002). It is important, however, to remember what Cleveland (1993) wrote about LOESS, namely "in any specific application of LOESS, the choice of the two parameters α and λ must be based on a combination of judgement and trial and error. There is no substitute for the latter".

Smoothing splines (Goodall 1990) have similar aims to locally weighted smoothers and are also used in some other forms of non-parametric regression such as generalised additive models. Given data of x and y variables on the same set of objects, it is possible to connect these points with a smooth continuous line, a so-called spline function. The term spline comes from the flexible drafting spline made from a narrow piece of wood or plastic that can be bent to conform to an irregular shape. Splines are not analytical functions and they do not provide statistical models in contrast to general or generalised linear regression. They are arbitrary functions and have no real theoretical basis except for the theory that defines the characteristics of the lines themselves. Splines are piece-wise polynomials of degree m that are constrained to have continuous derivatives as the joints or knots between the pieces or objects. For example, if $t_1, t_2, \ldots, t_m$ are a set of m values lying in the interval a–b such that $a < t_1 \leq t_2 \leq \ldots < t_m \leq b$, then a cubic spline is a function q such that on each of the intervals (a, t_1), (t_1, t_2), ..., (t_m, b), q is a cubic polynomial and the polynomial pieces fit together at the points t in such a way that q itself and its first and second derivatives are continuous at each t, and hence on the whole of a, b (Everitt 2002). The points t are called knots. The most common type is a cubic spline for the smoother estimation of the function f in modelling the dependence of response variable y on a predictor variable x

$$y = f(x) + \varepsilon$$

where ε represents the error term with an expected value of zero. For further details of smoothing splines and how they are calculated, see Eubank (1988), Marsh and Cormier (2002), Crawley (2002, 2005, 2007), Venables and Ripley (2002), and Faraway (2006).

In palaeolimnology, LOESS smoothers are mainly used to highlight the major trends or 'signal' in scatter-plots (e.g., of predicted versus observed values in organism-environment calibration models (see Juggins and Birks 2012: Chap. 14)) or in stratigraphical plots (e.g., Hammarlund et al. 2002; Helmens et al. 2007; Herzschuh et al. 2010). They can also be used directly in organism-environment modelling (e.g., Correa-Metrio et al. 2011). Splines are less rarely used except in age-depth modelling (see Blaauw and Heegaard 2012: Chap. 12) and as the major smoother in generalised additive models.

Generalised additive models: Generalised additive models (GAMs) are semi-parametric extensions of GLM which use smoothing techniques such as splines to identify and represent possible non-linear relationships between the predictor and the response variables (Efron and Tibshirani 1991; Yee and Mitchell 1991). They are an alternative to considering polynomial terms in GLM or for searching for appropriate transformations of both predictor and response variables (Everitt 2002). GLM require an *a priori* statistical model such as the Gaussian logit model and are thus model-driven. With real-life data, GLM may not be flexible enough to approximate the response adequately as the response may be bimodal, badly skewed, or more complex than the *a priori* model.

GAMs are thus a useful exploratory alternative to GLM (Yee and Mitchell 1991). In GAMs, the link function of the expected values of the response variable is modelled as the sum of a number of smooth functions of the predictor variables rather than in terms of the predictor variables themselves (Everitt 2002).

The similarities and differences between GLM and GAM can be summarised in their general equations (Yee and Mitchell 1991)

$$\text{GLM} \quad g(\mathrm{E}y) = \alpha + \beta x = \beta_0 + \sum_{j=1}^{m} \beta_j x_j$$

where g is the link function, Ey is the modelled abundance of the response variable y, β_0 is the intercept or constant, β_j are the regression coefficients or model parameters, and $x_j, \ldots, x_m$ are predictor variables.

$$\text{GAM} \quad g(\mathrm{E}y) = \alpha + fx = \beta_0 + \sum_{j=1}^{m} f_j(x_j)$$

where f_j are unspecified smoothing functions estimated from the data using smoothers such as cubic splines or LOESS to give maximum explanatory power. In GAMs, the data determine the shape of the response curve rather than being limited by the shapes available in parametric GLM. GAMs can thus detect bimodality and extreme skewness (Efron and Tibshirani 1991; Yee and Mitchell 1991; Acharya et al. 2011).

The additive predictor $f_j(x_j)$ (η_A) in GAM contrasts with the linear predictor (η) in GLM as η_A is represented as a sum of independent contributions of the individual predictors. However, the effect of a particular predictor variable is not summarised using a simple regression coefficient β but as a smooth function f_j for predictor variable j. This describes the transformation from the predictor values to the additive effect of that predictor on the expected values of the response variable

$$\eta_\mathrm{A} = \beta_0 + \sum_{j=1}^{m} f_j(x_j)$$

The additive predictor replaces the linear predictor of a GLM but is related to the scale of the response variable via the link function (Lepš and Šmilauer 2003) as

$$f_j(x_j) = \beta_j \cdot x_j$$

In the more general case, smooth transformation functions (so-called 'smooth terms') are fitted using a LOESS smoother or, more commonly, a cubic spline smoother (Yee and Mitchell 1991; Lepš and Šmilauer 2003).

When fitting a GAM, it is not possible to prescribe the shape of the smooth functions of the individual predictors. Instead it is necessary to specify the complexity of the individual curves in terms of their degrees of freedom. It is also necessary to select the type of smoother to use to find the shape of the smooth transformation functions for the individual predictors (Lepš and Šmilauer 2003).

The best practice is to fit GAMs of different complexity and select the 'best' one. The performance of individual GAMs can be assessed by the AIC, as in GLM. AIC penalises an individual model for its complexity and the number of degrees of freedom used in the model (see Hastie and Tibshirani 1990; Lepš and Šmilauer 2003). The model with the lowest AIC value (highest parsimony) should be selected. It is often important to compare GLM and GAM fits in which the same reference distribution (e.g., Poisson) and associated link function are assumed for the response variables and with the comparable amount of complexity (Heegaard 1997; Acharya et al. 2011). To compare a second-order polynomial GLM with a comparable GAM, the smooth term in the GAM must be set at 2 degrees of freedom (Lepš and Šmilauer 2003). The GLM and GAM results can be compared visually or the fitted values of one can be plotted against the fitted values for the other model. Alternatively, the residuals from the two models can be plotted. These two alternatives provide complementary information because the residuals are the values of the observed response variable minus the fitted values and the GLM and GAM models share the same observed values for the response variable (Lepš and Šmilauer 2003). With GAM it is possible, as in GLM, to do a step-wise model selection including not only a selection of the predictors used in the systematic part of the model but also a selection of complexity in their smooth terms (Lepš and Šmilauer 2003).

GAMs are primarily graphical tools and cannot be easily summarised numerically in contrast to GLMs where their parameters, the regression coefficients, summarise the shape of the response curve and regression model. The fitted GAM is best summarised by plotting the estimated smooth terms representing the relation between the predictor values and their effects on the modelled response variable.

Just as in GLM, interactions between two or more predictor variables can be included in the additive predictor to assess if the effect of one variable depends on the values of the other. An interaction term is thus

$$f_q\left(x_j \cdot x_k\right)$$

just as in a linear predictor in GLM where the interaction term is

$$\beta_q\left(x_j \cdot x_k\right)$$

GAMs can thus be viewed as semi-parametric extensions of GLMs. The main underlying assumptions are that the functions are additive and that the components are smooth. Like GLMs, GAMs use a link function to establish the relationship between the mean of the response variable and a 'smoothed' function of the predictor variable(s). The strength of GAMs is their ability to cope with non-linear and non-monotonic relationships between the response variable and the set of predictor variables. They are very much more data-driven than model-driven as in GLMs. The data determine the nature of the relationship between the response and the predictor variables. GAMs are a very useful exploratory tool as they can handle complex non-linear data structures.

GAMs have an enormous potential in palaeolimnology, for example in modelling species responses in relation to environmental variables – see Šmilauer and Birks (1995), Heegaard (2002), Yuan (2004), Hájkova et al. (2008), and Mackay et al. (2011) for examples. Excellent introductions to GAM are given by Yee and Mitchell (1991), Guisan et al. (2002), and Zuur et al. (2007). Mathematical details are presented in Hastie and Tibshirani (1986, 1990), Faraway (2006), and Wood (2006), and a range of illustrative examples is discussed by Hastie and Tibshirani (1987) and Leathwick (1995).

There is a continuum of regression models from general linear models (including simple linear and multiple linear regression) which are the most restrictive in terms of underlying assumptions (but are the most used and misused!) to generalised linear models that are fairly general but are strictly model-based, to generalised additive models that are the most general as they are strongly data-driven.

Together these three main types of regression analysis and statistical modelling provide the palaeolimnologist with a powerful set of techniques for detecting and modelling relationships between a response variable and one or more predictor variables. We now turn to two very different approaches to regression modelling, namely classification and regression trees and artificial neural networks.

Classification and Regression Trees (CARTs)

These tree-based methods and artificial neural networks (Simpson and Birks 2012: Chap. 9) discussed below are probably the most non-parametric types of regression models that can be used to describe the dependence of the response variable values on the values of the predictor variables (Lepš and Šmilauer 2003).

CARTs are defined by a recursive binary partitioning of a data-set into subgroups that are successively more and more homogeneous in the values of the response variable. The partitioning is very similar to the recursive binary splitting in TWINSPAN (Legendre and Birks 2012a: Chap. 7) and in the binary partitioning of stratigraphical data (Birks 2012b: Chap. 11) but in these methods the multivariate data-set is partitioned and, in the case of TWINSPAN, the criterion for splitting is different (Lepš and Šmilauer 2003). At each partitioning step in a CART, exactly one of the predictors is used to define the binary split. The split that maximises the homogeneity of the two resulting subgroups and the difference between the subgroups is selected (Lepš and Šmilauer 2003). Each split uses only one of the predictor variables and these may be qualitative (as in a classification tree) or quantitative (as in a regression tree) (Lepš and Šmilauer 2003).

The results of the fitting are presented as a 'tree' showing the successive splits. Each branch is described by a specific splitting rule. This rule can take the form of an inequality for a quantitative predictor (e.g., pH < 7) or a set of possible values for a qualitative factor (e.g., moisture has value 1 or 2) (Efron and Tibshirani 1991; Lepš and Šmilauer 2003; Molinaro et al. 2005). The two subgroups defined by such a splitting rule are further subdivided until they are too small or sufficiently homogeneous in the values of the response variable (Lepš and Šmilauer 2003). The terminal groups ('leaves') are identified by a predicted value of the response variable if this is a quantitative variable or by a predicted object membership group if the response variable is a qualitative factor, a categorical variable with a small number of classes (Lepš and Šmilauer 2003).

As a tree-based method is fitted to a data-set, the resulting tree is usually over-complicated (over-fitted) based on the data-set used. It is therefore necessary to follow the principle of parsimony in regression modelling and to try to find an 'optimum' size for the tree for the prediction of the response values (Efron and Tibshirani 1991; Lepš and Šmilauer 2003; Hastie et al. 2011). A cross-validation procedure is used to determine the 'optimal' size of the tree by creating a series of progressively reduced 'pruned' trees using only a subset of the total data-set and then using the remaining part as a test-set to assess the performance of the 'pruned' tree. The test-set is run through the hierarchical set of splitting rules and the predicted value of the response variable is then compared with its known value. For each size and complexity of the tree model, this n-fold cross-validation is repeated several times (Venables and Ripley 2002; Lepš and Šmilauer 2003; Hastie et al. 2011). The data-set is usually split into ten parts of roughly the same size and for each of these parts a tree model of given size is fitted using nine parts and the remaining part or test-set is used to assess the tree's performance (Venables and Ripley 2002; Lepš and Šmilauer 2003). A graph of the tree 'quality' against its complexity or size typically shows a minimum corresponding to the optimal tree size (Venables and Ripley 2002; Lepš and Šmilauer 2003). If a larger tree is used, the model is over-fitted – it provides a close approximation of the data-set but a biased description of the underlying population from which the data-set was sampled (Lepš and Šmilauer 2003; Simpson and Birks 2012: Chap. 9).

Excellent introductions to CARTs are given by Efron and Tibshirani (1991), Michaelsen et al. (1994), De'ath and Fabricius (2000), Fielding (2007), Zuur et al. (2007), and Olden et al. (2008). Breiman et al. (1984) is the standard text on CART (see also Venables and Ripley 2002; Faraway 2006; Hastie et al. 2011).

De'ath (2002) has extended univariate regression trees with a single response variable to consider multivariate responses. In multivariate regression trees (MRTs), the univariate response is replaced by a multivariate assemblage response (see Borcard et al. 2011; Legendre and Birks 2012a: Chap. 7; Simpson and Birks 2012: Chap. 9).

There are other types of related CART techniques such as bagging, boosted trees, random forests, and multivariate adaptive regression splines (Prasad et al. 2006). These are outlined in Simpson and Birks (2012: Chap. 9).

What is the role of CARTs and related techniques in regression modelling? ANOVA and regression are powerful techniques but as the number of predictor variables and the complexity of the data increases with interactions, unbalanced designs, and empty cells, general and generalised linear modelling become less effective. CARTs are simpler and less sensitive to unbalanced designs and zero values (De'ath and Fabricius 2000). The binary splits represent an optimum set of one-degree-of-freedom contrasts. The advantages of CARTs increase as the number of predictor variables and data complexity increase. They are, in many ways, 'data-mining' tools that are most useful with huge heterogeneous and highly variable data-sets (Witten and Frank 2005). De'ath's (2002) multivariate regression tree approach similarly has some advantages over canonical ordination techniques such as canonical correspondence analysis (CCA) and redundancy analysis (RDA) (see Borcard et al. 2011; Legendre and Birks 2012b: Chap. 8 for details of CCA and RDA). MRTs make no assumptions about organism-environment response models, in contrast to CCA and RDA, resulting in considerable robustness of MRTs. They are invariant to monotonic transformation of the predictor variables, they emphasise the local data-structure and interactions whereas canonical ordinations generally consider the global data-structure, and they can be used to predict response variables from the values of predictor variables. MRT is one tree, whereas m univariate CARTs are needed for m response variables. MRTs can match or outperform RDA or CCA in analysing and predicting assemblage composition (De'ath 2002), particularly with large heterogeneous data-sets. MRTs are, as discussed by Legendre and Birks (2012a: Chap. 7), more regression-based than TWINSPAN supplemented by simple discriminant functions.

Although widely used in ecology, biogeography, and environmental science (e.g., Thuiller et al. 2003; Pyšek et al. 2010; Aho et al. 2011), CARTs have rarely been used in palaeoecology and palaeolimnology. Examples of the use of CARTs include Pelánková et al. (2008) and of MRTs include Bjerring et al. (2009), Davidson et al. (2010a, b), and Herzschuh and Birks (2010).

Artificial Neural Networks (ANNs) and Self-Organising Maps (SOMs)

Everitt (2002) describes artificial neural networks (ANNs) as "a mathematical structure modelled on the human neural network and designed to attack many statistical problems, particularly in the areas of pattern recognition, multivariate analysis, learning and memory". The general idea is to use a network to learn some target values or vectors from a set of associated input signals through a set of iterative adjustments of a set of parameters and to minimise the error between the network and the desired output following some learning rule. ANNs have been used in, for example, regression and statistical modelling, supervised classification and assignment, unsupervised classification, and calibration (= inverse regression) (Warner and Misra 1996; Simpson and Birks 2012: Chap. 9).

The essential feature of an ANN structure is a network of simple processing elements (artificial neurons) linked together so that they can co-operate. From a set of 'inputs' and an associated set of parameters, the artificial neurons produce an 'output' that provides a possible solution to the problem under study, for example regression modelling. In many ANNs, the relationship between the input received by a neuron and its output is determined by a simple GLM. Training of ANNs involves two main processes – forward-propagation and back-propagation.

A forward-propagation or feed-forward network is essentially an extension of the idea of the perceptron, a simple classifier, which computes a linear combination of the variables and returns the sign of the result – observations with positive values are classified into one group and those with negative values into another group (Everitt 2002). In a forward-propagation network, the vertices are numbered so that all connections go from a vertex to one with a higher number, and the vertices are arranged in layers with connections only to higher layers (Everitt 2002). The layers are the input layer, one or more hidden layers, and an output layer. The input vector and the output vector are linked directly to the input layer and the output layer, respectively. Each neuron in the hidden layer and output layer receives weighted signals from each neuron in the previous layer. Each neuron sums its inputs to form a total input x_j and applies a function f_j to x_j to give an output y_j. The links have weights w_{ij} which multiply the signals travelling along them by that factor. Forward propagation is performed alone, without the following back-propagation step when running an already trained network (Malmgren and Nordlund 1997; Simpson and Birks 2012: Chap. 9).

In back-propagation the difference or 'error' between the actual output vector resulting from the forward-propagation process and the desired target vector is used to adjust incrementally the weights between the output layer and the last of the hidden layers according to a learning algorithm based on the gradient-descent method (Malmgren and Nordlund 1997). For each layer, going backwards through the network, the values used for adjusting the weights are the error terms, or their derivatives, from the immediately succeeding layer. As the process is going backwards through the network, these terms have already been computed. The size of the incremental adjustments of the weights is determined by the learning rate, which is set between 0 and 1. If the learning rate is too high, the result may be a network that never converges, while too low a learning rate may result in excessively long learning (Malmgren and Nordlund 1997).

Introductions to ANNs and their various types are given by Eberhart and Dobbins (1990), Bishop (1995, 2007), Warner and Misra (1996), Malmgren and Nordlund (1997), Abdi et al. (1999), Venables and Ripley (2002), Faraway (2006), Fielding (2007), Olden et al. (2008), Ripley (2008), and Franklin (2010). ANNs have been widely used in limnology, freshwater biology, earth sciences, ecology, and conservation biology – see Lek and Guégan (2000) for a review. Warner and Misra (1996) discuss ANNs in the context of GLM and conclude that ANNs are only useful when the investigator has no idea of the functional relationship between the response and the predictor variables. If one has some idea of the functional relationship, they advise using GLM. They have rarely been used in palaeolimnology, mainly for the purposes of environmental reconstructions from fossil assemblages (e.g., Racca et al. 2001, 2003, 2004). Their use in palaeoecology and palaeoclimatology are outlined by Birks et al. (2010).

Olden and Jackson (2002) compared the performance of ANNs with GLM, CARTs, and linear discriminant analysis to predict the presence or absence of different fish species in lakes as a function of 13 habitat features. They showed that CARTs and ANNs generally outperformed the other methods and they suggested that this was because these approaches can model both linear and non-linear species responses.

Telford et al. (2004) and Telford and Birks (2005) urge caution in assessing ANN performance by showing the importance of having not only an independent test-set (as in CARTs), but also an independent optimisation set in ANN. ANNs learn by iteratively adjusting a large number of parameters (originally set as random values) to minimise the error between the predicted output and the actual input. If ANNs are trained for too long, they can over-fit the data by learning specific features of the data rather than learning the general rules. It is therefore necessary to have an independent optimisation data-set to optimise the network after training and only when training and optimisation are done should the independent test-set be used to assess the ANN's performance. When this is done, ANNs rarely out-perform more classical regression and statistical modelling procedures, including CARTs and their variants. A further disadvantage of ANN is that it is difficult to interpret network coefficients in any biologically or ecologically meaningful way as ANNs are essentially black boxes.

ANNs can, if used carefully and critically, provide a flexible class of linear and non-linear regression models (Warner and Misra 1996; Ripley 2008). By adding more hidden layers, the complexity of the ANN model can be controlled from relatively simple models with a single hidden layer to models with complex structure and many hidden layers. This specification of the architecture of ANNs in terms of hidden layers is equivalent to specifying a suitable model, particularly the systematic part, in GLM. Training ANNs is equivalent to estimating the parameters of a GLM given a data-set. ANNs seem attractive to use because they require less experience and statistical understanding and insight compared with GLMs or GAMs. However, users of ANNs need to pay careful attention to basic questions of transformations, outliers, and influential observations and the need to develop minimal adequate models (see Racca et al. 2003 for one approach with ANNs). ANNs may be useful for prediction but they are poor for understanding data and model structure as the ANN weights and coefficients are usually un-interpretable. ANNs can introduce complex interactions that may not reflect reality. As Telford et al. (2004) show, ANNs are easy to over-fit and give over-optimistic performance-statistics and unreliable estimates of values of additional independent data. There is no statistical theory for inference, diagnostics, or model selection. ANNs are at best a heuristic tool (Warner and Misra 1996). They do not provide a rigorous approach to regression analysis and statistical modelling as they lack the underlying theory of general linear models, generalised linear models, and generalised additive models.

Self-organising maps (SOMs) (Kohonen 2001) are superficially similar to ANNs, as they were both developed for applications in artificial-intelligence research (Simpson and Birks 2012: Chap. 9). In reality SOMs can be regarded as a constrained version of the κ-means clustering or partitioning method (Legendre and Birks 2012a: Chap. 7). The goal of a basic SOM is to preserve the similarities between samples so that similar samples are mapped onto the same or neighbouring units on a rectangular or hexagonal grid of units with pre-defined dimensions in terms of the numbers of rows and columns. A SOM can be implemented in an unsupervised mode, where it learns features from the data themselves. SOMs can also be constructed in a supervised mode when both dependent and independent predictor variables are available. Such a supervised SOM allows for predictions of the response variable to be made at sites with new values of the predictor variables (see Simpson and Birks 2012: Chap. 9 for further details and a palaeolimnological example).

In the general category of network-based techniques, Simpson and Birks (2012: Chap. 9) discuss two additional techniques – Bayesian networks (also known as belief or Bayesian belief networks) and genetic algorithms, in particular the Genetic Algorithm for Rule-set Prediction (GARP). Bayesian networks can be viewed as a graphical summarisation of the system under study where key features are represented by nodes that are linked together in some way so that the cause-and-effect relationships between nodes are discovered. They are a powerful technique for describing a means by which reasoning in the face of uncertainties can be performed (Jensen and Nielsen 2007; Simpson and Birks 2012: Chap. 9).

Genetic algorithms are one of a number of stochastic optimisation techniques that fall under the general heading of evolutionary computing (Fielding 2007; Olden et al. 2008; Simpson and Birks 2012: Chap. 9). Although they are very flexible optimisation tools, they are not well suited to all problems and CARTs and related methods generally perform as well or better than genetic algorithms for general regression or classification problems. Despite the popularity of GARP in, for example, developing species distribution models (e.g., Elith et al. 2006; Franklin 2010), genetic algorithms do not appear to have any particular advantages over CARTs and related techniques in palaeolimnology (Simpson and Birks 2012: Chap. 9).

Multivariate Regression

Almost all the regression procedures discussed so far consider only one response variable in relation to one or more predictor variables. In palaeolimnology interest is often centred on the relationship between assemblages comprising many species (e.g., diatoms, chironomids) and one or more environmental predictor variables. Here the general techniques of canonical ordination (Legendre and Birks 2012b: Chap. 8) (also known as constrained ordination or multivariate direct gradient analysis – ter Braak and Prentice 1988) are particularly useful.

The basic idea is to search for a weighted sum of predictor variables that fit the data of *all* response variables as well as possible, i.e., that gives the maximum total regression sum-of-squares under the assumption of a linear response model of the response variables to the predictor variables. The resulting technique, redundancy analysis (RDA) (ter Braak and Prentice 1988; ter Braak 1994; Borcard et al. 2011) is an ordination of the response variables in which the axes are constrained to be linear combinations of the predictor variables. Canonical correspondence analysis (CCA) (ter Braak 1986; ter Braak and Prentice 1988; Borcard et al. 2011) is the comparable technique under the assumption of a unimodal species-environment response model.

When many response variables are of interest, each response variable can be analysed either in a separate multiple regression (ter Braak and Looman 1994) or jointly in a multivariate regression. The former approach is justified by the Gauss-Markoff set-up of regression theory, where estimation in multivariate regression reduces to a series of multiple regressions (Finn 1974; ter Braak and Looman 1994; Neter et al. 1996). If a restriction is imposed on the rank (dimensionality) of the matrix of regression coefficients, this restriction can result in a more parsimonious model in which the response variables react to the regressors or predictor variables only through a restricted number of 'latent variables' estimated as canonical variates (ter Braak and Looman 1994). Izenman (1975) introduced the appropriate name of reduced-rank regression for this approach.

RDA is a form of reduced-rank regression (Davies and Tso 1982) and is intermediate between principal component analysis and separate multiple regressions for each of the response variables. It is a constrained ordination and a constrained form of multivariate multiple regression (Davies and Tso 1982; Israëls 1984; ter Braak and Prentice 1988). With two axes, RDA uses $2(q+m)+m$ parameters to describe the response data, whereas multiple regression uses $m(q+1)$ parameters where m is the number of response variables and q is the number of predictor variables (ter Braak and Prentice 1988). RDA results can be displayed in 2- or 3-dimensional ordination plots or biplots (Gabriel 1982) that simultaneously display (1) the main patterns of assemblage variation as far as this variation can be explained statistically by the available predictor variables and (2) the main patterns in the correlation coefficients between the response variables and each of the predictor variables (van der Meer 1991; ter Braak 1994; ter Braak and Looman 1994).

Canonical correlation analysis (CCoA) (Tso 1981; Thompson 1984; Gittins 1985; ter Braak 1990; Borcard et al. 2011) is related to RDA and is the standard linear multivariate technique for relating two sets of variables (responses and predictors). It differs from RDA in its assumptions about the error part – correlated normal errors in canonical correlation analysis and uncorrelated errors with equal variance in RDA (Tso 1981; ter Braak and Prentice 1988; van der Meer 1991). From a practical viewpoint, the most important difference is that RDA can analyse any number of response variables whereas in CCoA the number of responses (m) must be less that $n-q$ where n is the number of objects and q is the number of predictor variables (ter Braak and Prentice 1988). This is particularly restrictive in regressions of diatom assemblages containing 150–200 taxa in 50–100 samples.

Canonical variates analysis (CVA) (= multiple discriminant analysis) (see above) is a special case of canonical correlation analysis in which the predictor variables are a set of dummy variables reflecting a single-factor *a priori* classification of the objects. There is thus a similar restriction on the number of response variables that can be analysed by CVA. RDA with dummy 1/0 variables for class membership provides an attractive alternative to CVA as it avoids this restriction (e.g., Catalan et al. 2009).

Like RDA, CCA results can be displayed as an ordination diagram that displays the main patterns of the response assemblage variation, as far as these reflect variations in the predictor variables. A CCA plot also shows the main patterns in the weighted averages (WA) (not correlations as in RDA) of each of the response variables with respect to the predictor variables (ter Braak 1986, 1987b). CCA is intermediate between correspondence analysis and separate WA calculations for each response variable in relation to each predictor variable. CCA is thus a constrained form of WA regression in which the weighted averages are restricted to lie in a low-dimensional sub-space (ter Braak and Prentice 1988).

RDA and CCA provide the easiest ways of implementing multivariate reduced-rank regression involving two or more response variables and one or more predictor variables. CCA assumes a symmetric Gaussian species response to environmental variables (Fig. 2.2) that can be approximated by simple weighted averaging regression. With the great increase in computing power in the last 20 years, Yee (2004) has followed up ter Braak's (1986) idea of constrained Gaussian ordination (CGO) with estimation by maximum likelihood rather than by the simple WA algorithm used in CCA (ter Braak and Prentice 1988). Quadratic reduced-rank vector-based GLMs (Yee 2004) implement CGO but they assume symmetric unimodal species responses. Yee (2006) has extended this approach further to reduced-rank vector-based GAMs where no response model is assumed (see also Yee and Mackenzie 2002). Instead the data determine the response model as in GAMs. These techniques are still under development, but when mature these methods will give palaeolimnologists powerful ordination and constrained ordination (= multivariate regression) techniques within the GLM/GAM theoretical framework incorporating a mixture of linear, quadratic, and smooth responses.

A related development in constrained ordination and multivariate regression is non-linear canonical analysis of principal coordinates (NCAP) (Millar et al. 2005). It builds on canonical analysis of principal coordinates (CAP) (Anderson and Willis 2003; Hewitt et al. 2005; Anderson 2008b; Oksanen et al. 2011) but it considers intrinsically non-linear relationships between assemblages and non-linear environmental variables. It involves extending the traditional univariate linear model implicit in RDA and CAP to a GLM through the use of a link function and non-linear optimisation procedures as in GLM. The smallest number of axes to retain is determined using the Bayes Information Criterion in the context of NCAP. By means of permutation tests (see below) it is possible to test if the non-linear model fitted is preferable to the linear model. This method is of considerable potential in analysing multivariate non-linear biological systems and non-linear environmental gradients within the framework of multivariate regression (e.g., Anderson et al. 2005a).

Finally, it is important to consider the critical question in all regression analysis, namely when to use constrained ordination (reduced-rank regression) with all response variables considered simultaneously and when to use a series of separate regressions of each response variable? As ter Braak and Prentice (1988) discuss, the answer depends on whether there is an advantage biologically in analysing all the responses simultaneously. In constrained ordination, the responses are assumed to react to the same composite gradients of the predictor variables, whereas in standard multiple regression, a separate composite gradient is constructed for each response variable (ter Braak and Prentice 1988). Regressions can result in more detailed descriptions and more robust predictions and calibrations if the regression modelling is done with regard to the statistical assumptions of the model used (as in GLM) and if sufficient data are available (ter Braak and Prentice 1988). Palaeolimnological data are often collected over a wide range of environmental variation and their analysis necessitates the use of non-linear models. Building robust non-linear models by regression can be difficult and demanding. In CCA the composite gradients are linear combinations of the predictor variables and non-linearity enters through the unimodal response model with respect to a few composite gradients (ter Braak and Prentice 1988). Constrained ordinations are easier to apply and require less data than conventional regression analysis involving GLM, and they provide a graphical summary of the response-predictor variable relationship. Constrained ordinations are thus very useful for the exploratory analysis of large and heterogeneous data-sets and for simple multivariate reduced-rank regression of all the response variables in relation to the predictor variables, whereas regression analysis involving GLM is most useful in analysing specific taxa in relation to particular environmental gradients (e.g., Guisan et al. 1999).

Model Selection and Shrinkage

Selection of alternative statistical regression models has conventionally been based on p-values as the criterion for model selection. However, there is an increasing shift to a greater emphasis on AIC and multi-model inference (Burnham and Anderson 2002; Anderson 2008a). This shift is important in reducing the reliance of the 'truth' of model selected by step-wise procedures such as forward selection and in understanding the error tendencies of conventional variable-selection approaches (Whittingham et al. 2006). However, though this type of multi-model inference is useful for understanding model-based uncertainty, whether it is the best way to predict reliably an outcome remains unclear (Elith and Leathwick 2009). Other model-averaging techniques from computer sciences use a range of approaches to develop concurrently a set of models that together predict well (Hastie et al. 2011). Critical research comparing the conceptual bases and performance of various model averaging or consensus approaches including regression and AIC, Bayesian approaches, and statistical-learning model ensembles (e.g., bagged or boosted trees, random forests; Simpson and Birks 2012: Chap. 9) is needed in quantitative palaeolimnology.

There are several alternative approaches to selecting a single final regression model. The different information criteria (AIC, BIC) provide a range of trade-offs between model complexity and predictive performance and can be used within cross-validation to select a model (Hastie et al. 2011). Some methods focus on simultaneous selection of variables and parameter estimation, for example, by shrinkage of coefficient estimates (e.g., Reineking and Schröder 2006; Simpson and Birks 2012: Chap. 9), using ridge regression, the lasso, or the elastic net (Hastie et al. 2011). These provide alternative methods for selecting a final regression model that are generally more reliable than step-wise methods (Elith and Leathwick 2009). In statistical- and machine-learning methods, these ideas of model selection and tuning are called 'regularisation', as they make the fitted line or surface more regular or smooth by controlling over-fitting (Phillips et al. 2006). Use of these alternative methods in palaeolimnology is still relatively rare but it is likely to increase as research questions become more demanding in terms of model selection. The general topic of model selection is considered more fully in Chap. 9 on statistical learning and in Chap. 21 as a future challenge in quantitative palaeolimnology.

Shrinkage is a common phenomenon in regression analysis (Simpson and Birks 2012: Chap. 9). It is noticed when an equation from, say, a multiple regression model is applied to a new data-set and the model predicts much less well than in the original sample (Everitt 2002). In particular, the value of the multiple correlation coefficient (R) is less, namely it has 'shrunk'. Palaeolimnologists may be most familiar with the concept of shrinkage when two-way weighted averaging regression and calibration (Juggins and Birks 2012: Chap. 14) are used to reconstruct an environmental variable from palaeolimnological data. As averages are taken twice, the resulting estimates are 'shrunk', hence the need for a deshrinking equation by either classical or inverse linear regression (Birks et al. 1990; Juggins and Birks 2012: Chap. 14).

Shrinkage of the coefficients in regression models aims at improving the predictive performance of such models. The best known shrinkage methods (e.g., Brown et al. 2002; Dahlgren 2010; Hastie et al. 2011) are proportional shrinkage (Copas 1983; Breiman and Friedman 1997), and methods based on ridge regression (Hoerl and Kennard 1970), the lasso (Tibshirani 1996), and the elastic net (Zou and Hastie 2005; Hastie et al. 2011) (see Simpson and Birks 2012: Chap. 9). These methods generally share the feature that all coefficients are shrunken regardless of whether the coefficients are small or large in a statistical sense (ter Braak 2006). This form of shrinkage is not necessarily a prerequisite for good predictive power. Alternative shrinkage methods based on Bayesian approaches shrink the small coefficients but shrink the large coefficients only slightly. Such models can have excellent predictive power, and they are based on the prior belief that only a few coefficients contain the major signal in the regression model.

The major approaches to shrinkage of regression models (ridge regression, the lasso, and the elastic net) are discussed more fully by Simpson and Birks (2012: Chap. 9).

Quantitative Environmental Reconstruction, Calibration, and Inverse Regression

The commonest approach to environmental reconstruction from palaeolimnological data is to quantify the observed patterns of occurrence, abundance, and covariance of all p taxa in n modern surface samples ($\mathbf{Y}_m$) in relation to the environmental variable ($\mathbf{X}_m$) of interest and to derive $\hat{\mathbf{U}}_m$, a set of modern empirical calibration functions (also known as transfer functions) (Birks et al. 2010). The underlying model is thus

$$\mathbf{Y}_m = \hat{\mathbf{U}}_m \cdot \mathbf{X}_m$$

Given $\hat{\mathbf{U}}_{\mathrm{m}}$, they can be used to infer the past environment, $\mathbf{X}_0$, from the fossil assemblage ($\mathbf{Y}_0$)

$$\hat{\mathbf{X}}_0 = \hat{\mathbf{U}}_m^{-1}(\mathbf{Y}_0)$$

Although $\hat{\mathbf{U}}_m$ and $\hat{\mathbf{X}}_0$ can be estimated or inferred in several different ways (see Juggins and Birks 2012: Chap. 14), in practice nearly all quantitative environmental reconstructions involve two stages. First, the responses of the modern taxa to the contemporary environmental variable of interest are modelled by some form of regression (ter Braak and Prentice 1988). This involves a modern training-set of taxon assemblages (response variables) from surface-sediment samples with associated environmental data (predictor variables). Second, the modelled modern responses are used to infer values of the environment variable in the past from the composition of the fossil assemblages. This is a calibration problem (ter Braak and Prentice 1988).

There are two conceptually different approaches to quantitative environmental reconstruction, as discussed in detail by ter Braak (1995). These are the classical and the inverse approaches (Osborne 1991). In chemometrics, Martens and Næs (1989) call the classical approach an indirect or reverse approach and the inverse approach a forward or direct approach.

In the classical approach, the empirical calibration functions, $\hat{\mathbf{U}}_m$, are estimated from the modern training-set by regressing $\mathbf{Y}_m$ on $\mathbf{X}_m$. This can be a linear or a non-linear regression and a univariate or a multivariate regression. The estimated $\hat{\mathbf{U}}_m$ is then 'inverted' to estimate the unknown environmental variable $\mathbf{X}_0$ from a fossil sample $\mathbf{Y}_0$. The steps are

$\mathbf{Y}_m = \hat{\mathbf{U}}_m(\mathbf{X}_m) + \text{error}$	classical regression
$\hat{\mathbf{X}}_0 = \hat{\mathbf{U}}_m^{-1}(\mathbf{Y}_0)$	calibration

With the exception of simple linear calibration, the inverse of $\hat{\mathbf{U}}_m$ does not exist and what is attempted instead is to find values of $\hat{\mathbf{X}}_0$ so that the two sides of the calibration equation are as similar as possible in some statistical sense (ter Braak 1995). In practice, values of $\hat{\mathbf{X}}_0$ are sought that have the highest probability of producing the observed fossil assemblages $\mathbf{Y}_0$ if the estimated value of $\hat{\mathbf{X}}_0$ is the true value.

In the inverse approach, the empirical calibration functions $\hat{\mathbf{U}}_m$ are estimated directly from the training-set by the ecologically curious inverse regression of $\mathbf{X}_m$ on $\mathbf{Y}_m$ in contrast to the classical approach where $\mathbf{Y}_m$ is regressed on $\mathbf{X}_m$. In the inverse approach, the past environmental variable $\mathbf{X}_0$, given a fossil assemblage $\mathbf{Y}_0$, is estimated directly from the modern calibration functions. The steps are

$\mathbf{X}_m = \hat{\mathbf{U}}_m(\mathbf{Y}_m) + \text{error}$	inverse regression
$\hat{\mathbf{X}}_0 = \hat{\mathbf{U}}_m(\mathbf{Y}_0)$	calibration

There is a substantial statistical literature on the relative merits of the classical and inverse approaches (ter Braak 1995). In practice, inverse models appear to perform best if the fossil assemblages are similar in composition to samples in the central part of the modern training data-set, whereas classical methods may perform better at the extremes and with some slight extrapolation.

Of the numerical techniques widely used in palaeolimnological environmental reconstructions, Gaussian logit regression (GLR) and maximum-likelihood (ML) calibration fall within the general classical approach, whereas weighted averaging (WA) (regression and calibration) and weighted-averaging partial least squares (WAPLS) lie within the inverse approach (ter Braak 1995).

The conditions under which WA regression approaches GLR were discussed above. In calibration, WA estimates the value of an environmental variable at a site as well as the corresponding ML estimate if the species shows Gaussian curves and has Poisson-distributed abundance values. Further conditions are that the species optima are equally spaced along the environmental gradient over an interval that extends for a sufficient distance in both directions from the true value of the environmental variable of interest; the species have equal tolerances (Fig. 2.2); the species have equal maximum values in their response curves (Fig. 2.2); and the species optima are closely spaced in comparison to their tolerances (ter Braak and Prentice 1988). These conditions mirror a species packing model where the species have equal response breadths and spacing (ter Braak and Prentice 1988).

The numerical methods commonly used in palaeolimnology for environmental reconstructions are discussed by ter Braak (1995), Birks (1995, 1998), Birks et al. (2010), and Juggins and Birks (2012: Chap. 14).

Temporal-Series Analysis

Stratigraphical palaeolimnological data are the values of one or more variables counted or measured in a series of samples in stratigraphical order and hence recorded over the time interval represented by the stratigraphical sequence. They form palaeolimnological temporal-series (Dutilleul et al. 2012: Chap. 16). Reconstructed environmental variables inferred from the fossil assemblages preserved in the stratigraphical sequence by calibration functions (Juggins and Birks 2012: Chap. 14) are also palaeolimnological temporal-series. These values usually vary with depth and hence with time, and this variation may be long-term trends, short-terms fluctuations, cyclical, irregular, or seemingly random.

In many time-series analytical techniques (see, for example, Shumway and Stoffer 2006; Cryer and Chan 2008), the term time-series is reserved for partial realisation of a discrete-time stochastic process, namely that the observations are made repeatedly on the same random variable at equal time intervals (Dutilleul et al. 2012: Chap. 16). There are clear difficulties in meeting this requirement in many palaeolimnological studies unless the observations are from annually laminated sediments (e.g., Tian et al. 2011). Dutilleul et al. (2012: Chap. 16) propose that the vast majority of palaeolimnological stratigraphical data-sets comprise 'temporal-series' rather than time-series in the strict statistical sense.

Numerical analysis of temporal-series can involve looking at the correlation structure within a variable in relation to time, between variables in relation to time, temporal trends within a variable and between variables, and other patterns within and between variables.

Prior to any numerical analysis, it is useful to perform exploratory data analyses on the individual variables or reconstructions within the temporal-series, such as simple graphical displays (Juggins and Telford 2012: Chap. 5) and testing for trend within each variable and assessing the statistical significance of the trends by randomisation tests (Manly 2007). LOESS smoothers (see above) are useful graphical tools for highlighting the 'signal' or major patterns in individual temporal-series (Juggins and Telford 2012: Chap. 5). An alternative to LOESS that combines graphical display, hypothesis testing, and temporal-series analysis is provided by SiZer (**Si**gnificance of **Zer**o crossings of the derivatives) (Chaudhuri and Marron 1999; Holmström and Erästö 2002; Sonderegger et al. 2009). Korhola et al. (2000) introduced the SiZer approach into palaeolimnology (for details, see Birks 2012a: Chap. 10).

There are two main approaches to time-series analysis when the inter-sample time intervals are constant (Dutilleul et al. 2012: Chap. 16). There is the time-domain approach based on the concept of temporal correlation, namely the correlation between objects in the same sequence that are k time intervals apart. The autocorrelation coefficients are measures of the correlation between objects separated by different time intervals and can be plotted as a correlogram to assess the autocorrelation structure, behaviour, and patterns of the palaeolimnological variable of interest over time. Time-series of two different variables can be compared by cross-correlations to detect patterns of temporal variation and temporal relationships between variables (Legendre and Legendre 1998; Davis 2002).

The second general approach involves the frequency domain and focuses on bands of frequency or wavelength over which the variance of a time-series is concentrated (Dutilleul 1995). It estimates the proportion of the variance that can be attributed to each of a continuous series of frequencies. The power spectrum of a time-series can help to detect periodicities within the data and the main tools are spectral density functions, cross-spectral analysis, and the periodogram (see Birks 1998 for palaeolimnological applications).

In the absence of equally spaced samples in time, a common procedure is to interpolate samples to equal time intervals. This is equivalent to a low-pass filter and may result in an under-estimation of the high-frequency components in the spectrum. Although techniques have been developed for spectral, cross-spectral, and wavelet analysis for unevenly sampled temporal series (Schulz and Stattegger 1997; Schulz and Mudelsee 2002; Witt and Schumann 2005), they do not appear to have been applied to palaeolimnological stratigraphical data. Dutilleul et al. (2012: Chap. 16) provide a balanced critique of these techniques. As more robust alternatives, Dutilleul et al. (2012: Chap. 16) present the methodology and application of auto-correlograms using distance classes and a novel frequency-domain technique, so-called multifrequential periodogram analysis, to analyse four different palaeolimnological temporal-series, all of which consist of unequally spaced observations.

Confirmatory Data Analysis

Many factors can influence lakes, their sediments, their hydrology, and their biota, and hence the modern and fossil assemblages of organisms. The living assemblages can be influenced by many factors such as biotic interactions, water-level changes, habitat availability, light, turbulence, resources, water chemistry, growing-season duration, and, in the case of fossil assemblages, transportation, resuspension, and preservation. These and other factors can themselves be influenced by, for example, human activity, climate change, volcanic activity, catchment-soil development and erosion, catchment vegetation, atmospheric contamination, etc. (Smol 2008). Interpretation of palaeolimnological data involves testing hypotheses about possible drivers or 'forcing factors' that may have influenced changes in palaeolimnological assemblages and sediment character and composition.

Birks (1998) discusses the importance of confirmatory data analysis or hypothesis testing in palaeolimnology, where falsifiable working hypotheses are proposed to explain the observed patterns. He proposes that there are two main approaches to testing hypotheses in palaeolimnology – the direct and the indirect approaches.

In the direct approach, palaeolimnological data are collected from lakes in contrasting settings today (e.g., geology, land-use, climate, vegetation) to test a specific hypothesis. This is what Deevey (1969) suggested as "coaxing history to perform experiments". Birks (1998, 2010) outlines examples of this direct approach. This approach is not dependent on statistical data-analysis, although quantitative environmental reconstructions, and quantitative summarisation and comparison of conditions before and after an environmental perturbation are often relevant.

In the indirect approach (Birks 1998) statistical techniques are central, as results from statistical modelling involving regression analysis are used to test alternative hypotheses. Crawley (2002) discusses in general terms the meaning of significance, good and bad hypotheses, p values, and interpretation of the results of statistical models. He emphasises that two kinds of mistakes can be made in interpreting results from statistical models – the null hypothesis is rejected when it is true (Type I error) and the null hypothesis is accepted when it is false (Type II error). Interpretation of model results should carefully consider these two kinds of error.

The basic statistical tests all involve some form of regression analysis (Birks 1997) based on an underlying linear or unimodal response model (ter Braak and Prentice 1988). The appropriate technique depends on the number of response variables and the number of predictor or explanatory variables. With one response variable and one predictor variable, simple linear regression or Gaussian logit or Poisson regression are relevant. With one response variable and two or more predictor variables, multiple linear regression or multiple Gaussian logit or Poisson regression may be appropriate (ter Braak and Looman 1987). With more than one response variable and one or more predictor variables, the appropriate techniques are canonical ordination or reduced-rank regression such as RDA and CCA (see above; Borcard et al. 2011; Legendre and Birks 2012b: Chap. 8). There are partial versions of all these techniques where the effects of covariables or concomitant variables or 'nuisance' predictor variables can be adjusted for and partialled out statistically (ter Braak 1988; ter Braak and Prentice 1988; Borcard et al. 2011). Partial techniques provide a powerful means of testing competing hypotheses as the effects of particular predictor variable can be partialled out and the relative importance of other predictors can be assessed statistically (ter Braak and Prentice 1988; Lotter and Birks 1993; Lotter and Anderson 2012: Chap. 18).

From a numerical viewpoint, the critical question is how to evaluate the statistical significance of the fitted regression model, possibly with covariables, given the complex numerical properties of much palaeolimnological data (closed percentage data, many zero values, many variables, temporally ordered samples). It is not possible to evaluate the statistical significance of regression models based on such data using conventional F-tests. Instead distribution-free Monte Carlo permutation tests (e.g., Legendre and Legendre 1998; ter Braak and Šmilauer 2002; Lepš and Šmilauer 2003; Manly 2007; Legendre et al. 2011) can be used. In these, an empirical distribution of the test statistic of relevance is derived by repeated permutations of the predictor variables or the regression residuals if covariables are present in the regression model (Legendre and Legendre 1998; ter Braak and Šmilauer 2002; Legendre et al. 2011) and a comparison made between the observed test statistics and, say, 999 values of the same statistic based on permuted data to permit a Monte Carlo test of significance and an exact probability for the observed test statistic. Such Monte Carlo tests are distribution-free as they do not assume normality of the error distribution. They do, however, require independence or exchangeability of the samples. The validity of the results from any permutation test rests on how appropriate the type of permutation test used is (ter Braak and Šmilauer 2002; Churchill and Doerge 2008). Temporally ordered stratigraphical data require a special permutation test where the samples are retained in stratigraphical order but the match of the response and predictor variables is permuted (ter Braak and Šmilauer 2002). Such restricted permutation tests provide a powerful non-parametric means of overcoming the problem of performing statistical tests in the presence of temporal autocorrelation in palaeolimnological data. Similar restricted permutation tests are available for spatially autocorrelated data (ter Braak and Šmilauer 2002). Manly (2007) provides a very readable introduction to randomisation and permutation tests as tools for testing hypotheses. More specialised texts include Edgington (1995), Legendre and Legendre (1998), Sprent (1998), Lunneborg (2000), Good (2001, 2005), and Roff (2006). Lotter and Anderson (2012: Chap. 18) review the use of permutation tests in evaluating competing hypotheses about the role of different drivers in influencing palaeolimnological changes.

CCA, RDA, and their partial relatives are increasingly used in palaeolimnology to assess the statistical significance of environmental variables in determining the composition and abundance of biotic assemblages in modern calibration data-sets and in testing specific hypotheses relating to calibration data-sets. Partitioning the variation (see Borcard et al. 1992, 2011; Legendre and Birks 2012b: Chap. 8) in the response variables can help in assessing the relative importance of different predictor variables in explaining the variation in modern assemblages (see Juggins and Birks 2012: Chap. 14). Variation partitioning can also be used with stratigraphical data to try to partition the variation in the response variables into different unique components, their covariation terms, and the unexplained variation. As the statistical significance of some of these components can be assessed by Monte Carlo permutation tests, the relative importance of competing hypotheses can be evaluated and quantified (see Lotter and Anderson 2012: Chap. 18; Simpson and Hall 2012: Chap. 19 for examples).

Other types of permutation-based statistical analyses are being used increasingly by palaeolimnologists and they are outlined here to provide a background for the reader who will encounter them in later chapters (e.g., Cumming et al. 2012: Chap. 20) and in the palaeolimnological literature (e.g., Werner and Smol 2005; Tremblay et al. 2010; Wiklund et al. 2010). These include analysis of similarities (ANOSIM), similarity percentage tests (SIMPER), and permutational multivariate analysis of variance using distance matrices (ADONIS). The vegan package in R (Oksanen et al. 2011) has functions for anosim() and adonis(), and SIMPER is part of the PRIMER package (Clarke and Gorley 2006).

Analysis of similarities (Clarke 1993; Clarke and Warwick 1994, 2001) uses any distance or dissimilarity measure (Legendre and Birks 2012b: Chap. 8) to assess statistically significant differences in species assemblages between two or more groups of samples from different habitats or from *a priori* partitionings or clusters. It is analogous to a standard analysis of variance (ANOVA – see above) but it allows the use of any distance measure. ANOSIM calculates an R statistic (analogous to the F-ratio test in ANOVA) based on the difference of mean ranks between groups and within groups for testing if assemblage composition varies across groups, but it is based on the differences of mean *ranks* between groups and within groups (cf. ANOVA). R scales from -1 to $+1$. A value of 1 indicates that all the most similar samples are within the same groups. When R is near zero, this indicates that there are no differences between groups, whereas an R value of less than zero is unlikely as it implies that similarities between different sites are greater than within sites. The statistical significance of the observed value of R is assessed by permuting the vector representing the sample groupings to obtain an empirical distribution of R under the null model. A global test is calculated first to test the null hypothesis that there are no differences between groups of samples specified *a priori*. If the null hypothesis is rejected, specific pairs of groups of samples can be compared through pair-wise R values that give a measure of how distinct the groups are.

Related to ANOSIM is SIMPER, a similarity percentage test (Clarke and Gorley 2006) that identifies specific taxa that account for the differences between groups assessed by ANOSIM (Sokal et al. 2008). By looking at the overall percentage contribution each species makes to the average dissimilarity between two groups (an average of all possible pairs of dissimilarity coefficients, taking one sample from each group), it is possible to list the species in decreasing order of their importance in distinguishing between the two sets of samples (Clarke and Gorley 2006).

Commenting on anosim() in vegan, Oksanen (in Oksanen et al. 2011) comments "I don't quite trust this method. Somebody should study its performance critically ... Most ANOSIM models could be analysed with adonis() which seems to be a more robust alternative". The `adonis()` function within vegan is for the analysis and partitioning of sums-of-squares using a range of distance matrices. As it partitions sums-of-squares of a multivariate data-set, it is directly analogous to a multivariate analysis of variance (MANOVA). Anderson (2001) and McArdle and Anderson (2001) call it non-parametric MANOVA or permutational MANOVA. It uses permutation tests to derive pseudo-F ratios and it is a robust alternative to parametric MANOVA and to redundancy analysis (Legendre and Anderson 1999). Typical uses of adonis() include the analysis of ecological assemblage data (e.g., Anderson et al. 2005b, c; Langlois et al. 2005) or genetic data where there might be a limited number of samples or individuals but thousands of columns of gene expression data (e.g., Zapala and Schork 2006). ADONIS is a robust method for performing non-parametric analysis of variance but with multivariate response data. It has wide potential in the analysis of palaeolimnological data and it should be used in preference to ANOSIM.

There are many other permutation tests relevant for hypothesis-testing using palaeolimnological data (see Legendre and Legendre 1998; Roff 2006; Manly 2007). For example, Tian et al. (2011) test the statistical significance of the correlations (r) between sedimentary and climatic variables for the last 100 years at Steel Lake, a lake with annually laminated sediments in Minnesota (USA). Tian et al. (2011) used a block bootstrap (cf. h-block cross-validation: Burman et al. 1994; Telford and Birks 2009) to account for temporal autocorrelation in their time-series. This involved resampling with replacement of the time-series in blocks of three consecutive samples (9 years). A two-tailed significance test was performed on the bootstrap distribution of the correlation coefficient based on 10,000 resamplings of the time-series to test for significance of the observed r values. Many palaeolimnologists calculate r between temporal-series (e.g., inferred summer temperature and observed summer temperature) but do not consider the strong autocorrelation in these series. Tian et al. (2011) is a welcome study that considers the problem of temporal autocorrelation present in stratigraphical data.

Conclusions

This chapter provides an overview of the major types of numerical approaches and methods currently used in the analysis of modern and stratigraphical palaeolimnological data-sets. These approaches and methods are presented in terms of their roles in data collection, data assessment, data summarisation, data analysis, and data interpretation (Table 2.2). Some of the approaches and methods such as exploratory data analysis, clustering and partitioning, classical and canonical ordination, weighted-averaging partial least squares, classification and regression trees, age-depth modelling, modern analogue techniques, and temporal-series analysis are only discussed briefly as they are covered more fully in subsequent chapters in this book. Basic numerical techniques that are not covered elsewhere in the book are discussed in more detail here, for example, discriminant analysis and regression analysis and statistical modelling. Regression analysis and statistical modelling are discussed in some detail as they are essential parts of, for example, age-depth modelling, classical and constrained ordinations, and environmental reconstructions.

The numerical approaches and the specific numerical or statistical techniques used within these approaches that are discussed in this chapter and elsewhere in the book represent the great majority of the numerical tools used by palaeolimnologists in the analysis of their data today. New established techniques in applied statistics such as additive modelling (Simpson and Anderson 2009) or newly developed techniques such as significance testing of palaeoenvironmental reconstructions (Telford and Birks 2011) are constantly being added to the ever-expanding palaeolimnologist's numerical tool-kit. Future challenges and the need for new numerical methods for particular research questions are outlined in the last chapter of this book (Birks 2012c: Chap. 21).

Acknowledgements I am greatly indebted to Hilary Birks, Einar Heegaard, Steve Juggins, Gavin Simpson, and Richard Telford for helpful discussions, to John Smol for his valuable comments and suggestions, and to Cathy Jenks for her invaluable help in preparing and editing this chapter. This is publication A346 from the Bjerknes Centre for Climate Research.

References

Abdi H, Valentin D, Edelman B (1999) Neural networks. Sage, Thousand Oaks

Acharya KP, Vetaas OR, Birks HJB (2011) Orchid species richness along Himalayan elevational gradients. J Biogeogr 38:1821–1833

Achen CH (1982) Interpreting and using regression. Sage Publications, Beverly Hills

Adler S, Hübener T, Dressler M, Lotter AF, Anderson NJ (2010) A comparison of relative abundance versus class data in diatom-based quantitative reconstructions. J Environ Manage 91:1380–1388

Aho K, Weaver T, Regele S (2011) Identification and siting of native vegetation types on disturbed land: demonstration of statistical methods. Appl Veg Sci 14:277–290

Aitchison J (1986) The statistical analysis of compositional data. Chapman & Hall, London

Aitchison J, Greenacre M (2002) Biplots of compositional data. Appl Stat 51:375–392

Aitken M, Francis B, Hinde J, Darnell R (2009) Statistical modelling with R. Oxford University Press, Oxford

Allen G (2006) An automated pollen recognition system. Unpublished M. Eng. thesis, Massey University, Plamerston North, New Zealand

Allison PD (2002) Missing data. Sage Publications, Thousand Oaks

Allison PD (2009) Fixed effects regression models. Sage Publications, Thousand Oaks

Anderson DR (2008a) Model based inference in the life sciences. A primer on evidence. Springer, New York

Anderson MJ (2001) A new method for non-parametric multivariate analysis of variance. Aust J Ecol 26:32–46
Anderson MJ (2008b) Animal-sediment relationships re-visited: characterising species' distributions along an environmental gradient using canonical analysis and quantile regression splines. J Exp Mar Biol Ecol 366:16–27
Anderson MJ, Willis TJ (2003) Canonical analysis of principal co-ordinates: a useful method of constrained ordination for ecology. Ecology 84:511–525
Anderson MJ, Millar RB, Blom WM, Diebel CE (2005a) Nonlinear multivariate models of successional change in community structure using the von Bertalanffy curve. Oecologica 146:279–286
Anderson MJ, Connell SD, Gillanders BM, Diebel CE, Blom WM, Saudners JE, Landers TJ (2005b) Relationships between taxonomic resolution and spatial scales of multivariate variation. J Anim Ecol 74:636–646
Anderson MJ, Diebel CE, Blom WM, Landers TJ (2005c) Consistency and variation in kelp holdfast assemblages: spatial patterns of biodiversity for the major phyla at different taxonomic resolutions. J Exp Mar Biol Ecol 320:35–56
Barnett V (2004) Environmental statistics. Methods and applications. Wiley, Chichester
Barnett V, Lewis T (1978) Outliers in statistical data. Wiley, Chichester
Barton AM, Nurse AM, Michaud K, Hardy SW (2011) Use of CART analysis to differentiate pollen of red pine (*Pinus resinosa*) and jack pine (*P. banksiana*) in New England. Quaternary Res 75:18–23
Battarbee RW, Monteith DT, Juggins S, Evans CD, Jenkins A, Simpson GL (2005) Reconstructing pre-acidification pH for an acidified Scottish loch: a comparison of palaeolimnological and modelling approaches. Environ Pollut 137:135–149
Baxter MJ (1989) Multivariate analysis of data on glass compositions: a methodological note. Archaeometry 31:45–53
Baxter MJ (1991) Principal component and correspondence analyses of glass compositions: an empirical study. Archaeometry 33:29–41
Baxter MJ (1992) Statistical analysis of chemical compositional data and the comparison of analyses. Archaeometry 34:267–277
Belsey DA, Kuh E, Welsch RE (1980) Regression diagnostics. Chapman & Hall, London
Bennett KD (1994) Confidence intervals for age estimates and deposition times in late-Quaternary sediment sequences. Holocene 4:337–348
Bennett KD, Willis KJ (2001) Pollen. In: Smol JP, Birks HJB, Last WM (eds) Tracking environmental change using lake sediments. Volume 3: Terrestrial, algal, and siliceous indicators. Kluwer, Dordrecht, pp 5–32
Berry WD (1993) Understanding regression assumptions. Sage Publications, Newbury Park
Berry WD, Feldman S (1985) Multiple regression in practice. Sage Publications, Beverly Hills
Birks HJB (1976) Late-Wisconsinan vegetational history at Wolf Creek, central Minnesota. Ecol Monogr 46:395–429
Birks HJB (1977) Modern pollen rain and vegetation of the St Elias Mountains, Yukon Territory. Can J Bot 55:2367–2382
Birks HJB (1980) Modern pollen assemblages and vegetational history of the moraines of the Klutlan Glacier and its surroundings, Yukon Territory, Canada. Quaternary Res 14:101–129
Birks HJB (1995) Quantitative palaeoenvironmental reconstructions. In: Maddy D, Brew JS (eds) Statistical modelling of Quaternary science data. Volume 5: Technical guide. Quaternary Research Association, Cambridge, pp 161–254
Birks HJB (1997) Reconstructing environmental impacts of fire from the Holocene sedimentary record. In: Clark JS, Cachier H, Goldammer JG, Stocks BJ (eds) Sediment records of biomass burning and global change. Springer, Berlin, pp 295–311
Birks HJB (1998) Numerical tools in palaeolimnology – progress, potentialities, and problems. J Paleolimnol 20:307–332

Birks HJB (2010) Numerical methods for the analysis of diatom assemblage data. In: Smol JP, Stoermer EF (eds) The diatoms: applications for the environmental and earth sciences, 2nd edn. Cambridge University Press, Cambridge, pp 23–54

Birks HJB (2012a) Chapter 10: Introduction and overview of part II. In: Birks HJB, Lotter AF, Juggins S, Smol JP (eds) Tracking environmental change using lake sediments. Volume 5: Data handling and numerical techniques. Springer, Dordrecht

Birks HJB (2012b) Chapter 11: Analysis of stratigraphical data. In: Birks HJB, Lotter AF, Juggins S, Smol JP (eds) Tracking environmental change using lake sediments. Volume 5: Data handling and numerical techniques. Springer, Dordrecht

Birks HJB (2012c) Chapter 21: Conclusions and future challenges. In: Birks HJB, Lotter AF, Juggins S, Smol JP (eds) Tracking environmental change using lake sediments. Volume 5: Data handling and numerical techniques. Springer, Dordrecht

Birks HJB, Birks HH (1980) Quaternary palaeoecology. Arnold, London

Birks HJB, Gordon AD (1985) Numerical methods in Quaternary pollen analysis. Academic, London

Birks HJB, Peglar SM (1980) Identification of *Picea* pollen of late Quaternary age in eastern North-America – a numerical approach. Can J Bot 58:2043–2058

Birks HJB, Webb T, Berti AA (1975) Numerical analysis of pollen samples from central Canada – comparison of methods. Rev Palaeobot Palynol 20:133–169

Birks HJB, Line JM, Juggins S, Stevenson AC, ter Braak CJF (1990) Diatoms and pH reconstruction. Philos Trans R Soc Lond B 327:263–278

Birks HJB, Heiri O, Seppä H, Bjune AE (2010) Strengths and weaknesses of quantitative climate reconstructions based on late-Quaternary biological proxies. Open Ecol J 3:68–110

Bishop CM (1995) Neural networks for pattern recognition. Clarendon, Oxford

Bishop CM (2007) Pattern recognition and machine learning. Springer, New York

Bjerring R, Becares E, Declerck S, Gross EM, Hansson L-A, Kairesalo T, Nykänen M, Halkiewicz A, Kornijów R, Conde-Porcuna JM, Seferlis M, Nõges T, Moss B, Amsinck SL, Odgaard BV, Jeppesen E (2009) Subfossil Cladocera in relation to contemporary environmental variables in 54 pan-European lakes. Freshw Biol 54:2401–2417

Bjune AE, Birks HJB, Peglar S, Odland A (2010) Developing a modern pollen-climate calibration data-set for Norway. Boreas 39:674–688

Blaauw M, Heegaard E (2012) Chapter 12: Estimation of age-depth relationships. In: Birks HJB, Lotter AF, Juggins S, Smol JP (eds) Tracking environmental change using lake sediments. Volume 5: Data handling and numerical techniques. Springer, Dordrecht

Bolker BM, Books ME, Clark CJ, Geange SW, Poulsen JR, Stevens MHH, White J-SS (2009) Generalized linear mixed models: a practical guide for ecology and evolution. Trends Ecol Evol 24:127–135

Bollen KA, Jackman RW (1990) Regression diagnostics. In: Fox J, Long JS (eds) Modern methods of data analysis. Sage, Newbury Park, pp 257–291

Borcard D, Legendre P, Drapeau P (1992) Partialling out the spatial component of ecological variation. Ecology 73:1045–1055

Borcard D, Gillet F, Legendre P (2011) Numerical ecology with R. Springer, New York

Breiman L, Friedman JH (1997) Predicting multivariate responses in multiple linear regression. J R Stat Soc B 59:3–54

Breiman L, Friedman JH, Olshen RA, Stone CJ (1984) Classification and regression trees. Wadsworth, Belmont

Brown PJ, Vannucci M, Fearn T (2002) Bayes model averaging with selection of regressors. J R Stat Soc B 64:519–536

Brubaker LB, Graumlich LJ, Anderson PM (1987) An evaluation of statistical techniques for discriminating *Picea glauca* from *Picea mariana* pollen in northern Alaska. Can J Bot 65: 899–906

Burman P, Chow E, Nolan D (1994) A cross-validatory method for dependent data. Biometrika 81:351–358

Burnham KP, Anderson DR (2002) Model selection and multimodel inference: a practical information-theoretic approach, 2nd edn. Springer, New York
Cacoullos T (ed) (1973) Discriminant analysis and applications. Academic Press, New York
Campbell NA, Atchley WE (1981) The geometry of canonical variates analysis. Syst Zool 30: 268–280
Catalan J, Barbieri MG, Bartumeus F, Bitusik P, Botev I, Brancelj A, Cogalniceau D, Manca M, Marchetto A, Ognjanova-Rumenova N, Pla S, Rieradevall M, Sorvari S, Stefkova E, Stuchlik E, Ventura M (2009) Ecological thresholds in European Alpine lakes. Freshw Biol 54:2494–2517
Chambers JM, Hastie TJ (1992) Statistical models in S. Wadsworth & Brooks, Pacific Grove
Chambers JM, Cleveland WS, Kleiner B, Tukey PA (1983) Graphical methods for data analysis. Wadsworth, Monterey
Chatfield C (1988) Problem solving. A statistician's guide. Chapman & Hall, London
Chaudhuri P, Marron JS (1999) SiZer for exploration of structures in curves. J Am Stat Assoc 94:807–823
Christensen R (1990) Log-linear models. Springer, New York
Churchill GA, Doerge RW (2008) Naive application of permutation testing leads to inflated Type I error rates. Genetics 178:609–610
Clarke KR (1993) Non-parametric multivariate analyses of changes in community structure. Aust J Ecol 18:117–143
Clarke KR, Gorley RN (2006) PRIMER v6: user manual/tutorial. PRIMER-E, Plymouth
Clarke KR, Warwick RM (1994) Similarity-based testing for community patterns: the two-way layout with no replication. Mar Biol 118:167–176
Clarke KR, Warwick RM (2001) Change in marine communities: an approach to statistical analysis and interpretation, 2nd edn. PRIMER-E, Plymouth
Cleveland WA (1979) Robust locally weighted regression and smoothing scatterplots. J Am Stat Assoc 74:829–836
Cleveland WA, Devlin SJ (1988) Locally weighted regression: an approach to regression analysis by local fitting. J Am Stat Assoc 83:596–610
Cleveland WS (1993) Visualizing data. AT&T Bell Laboratories, Murray Hill
Cleveland WS (1994) The elements of graphing data. AT&T Bell Laboratories, Murray Hill
Cook RD, Weisberg S (1982) Residuals and influence in regression. Chapman & Hall, London
Copas JB (1983) Regression, prediction and shrinkage. J R Stat Soc B 45:311–354
Correa-Metrio A, Bush MB, Pérez L, Schwalb A, Cabrera KR (2011) Pollen distribution along climatic and biogeographic gradients in northern Central America. Holocene 21:681–692
Coudun C, Gégout J-C (2006) The derivation of species response curves with Gaussian logistic regression is sensitive to sampling intensity and curve characteristics. Ecol Model 199: 164–175
Crawley MJ (1993) GLM for ecologists. Blackwell Scientific Publications, Oxford
Crawley MJ (2002) Statistical computing. Wiley, Chichester
Crawley MJ (2005) Statistics. An introduction using R. J Wiley & Sons, Chichester
Crawley MJ (2007) The R book. Wiley, Chichester
Cryer JD, Chan K-S (2008) Time series analysis with applications in R, 2nd edn. Springer, New York
Cumming BF, Laird KR, Fritz SC, Verschuren D (2012) Chapter 20: Tracking Holocene climatic change with aquatic biota from lake sediments: case studies of commonly used numerical techniques. In: Birks HJB, Lotter AF, Juggins S, Smol JP (eds) Tracking environmental change using lake sediments. Volume 5: Data handling and numerical techniques. Springer, Dordrecht
Curtis CJ, Juggins S, Clarke G, Battarbee RW, Kernan M, Catalan J, Thompson R, Posch M (2009) Regional influence of acid deposition and climate change in European mountain lakes assessed using diatom transfer functions. Freshw Biol 54:2555–2572
Dahlgren JP (2010) Alternative regression methods are not considered in Murtaugh (2009) or by ecologists in general. Ecol Lett 13:E7–E9
Dalgaard P (2008) Introductory statistics with R. Springer, New York

Davidson TA, Sayer CD, Langdon PG, Burgess A, Jackson M (2010a) Inferring past zooplanktivorous fish and macrophyte density in a shallow lake: application of a new regression tree model. Freshw Biol 55:584–599
Davidson TA, Sayer CD, Perrow M, Bramm M, Jeppesen E (2010b) The simultaneous inference of zooplanktivorous fish and macrophyte density from sub-fossil cladoceran assemblages: a multivariate regression tree approach. Freshw Biol 55:546–564
Davies PT, Tso MK-S (1982) Procedures for reduced-rank regression. Appl Stat 31:244–255
Davis JC (2002) Statistics and data analysis in geology, 3rd edn. Wiley, New York
De'ath G (1999) Principal curves: a new technique for indirect and direct gradient analysis. Ecology 80:2237–2253
De'ath G (2002) Multivariate regression trees: a new technique for modeling species-environment relationships. Ecology 83:1108–1117
De'ath G, Fabricus KE (2000) Classification and regression trees: a powerful yet simple technique for ecological data analysis. Ecology 81:3178–3192
Deevey ES (1969) Coaxing history to conduct experiments. Bioscience 19:40–43
Dobson AJ (2001) An introduction to generalized linear models, 2nd edn. Chapman & Hall, Boca Raton
Draper NR, Smith H (1998) Applied regression analysis, 3rd edn. Wiley, NewYork
Dunteman GH, Ho M-HR (2006) An introduction to generalized linear models. Sage, Thousand Oaks
Dutilleul P (1995) Rhythms and autocorrelation analysis. Biol Rhythm Res 26:173–193
Dutilleul P, Cumming BF, Lontoc-Roy M (2012) Chapter 16: Autocorrelogram and periodogram analyses of palaeolimnological temporal series from lakes in central and western North America to assess shifts in drought conditions. In: Birks HJB, Lotter AF, Juggins S, Smol JP (eds) Tracking environmental change using lake sediments. Volume 5: Data handling and numerical techniques. Springer, Dordrecht
Eberhart RC, Dobbins RW (eds) (1990) Neural networks PC tools. Academic Press, London
Edgington ES (1995) Randomisation tests, 3rd edn. Marcel Dekker, New York
Efron B, Tibshirani RJ (1991) Statistical data analysis in the computer age. Science 253:390–396
Efron B, Tibshirani RJ (1993) An introduction of the bootstrap. Chapman & Hall, London
Eggermont H, Heiri O, Russell J, Vuille M, Audenaert L, Verschuren D (2010) Paleotemperature reconstruction in tropical Africa using fossil Chironomidae (Insecta: Diptera). J Paleolimnol 43:413–435
Eliason SR (1993) Maximum likelihood estimation – logic and practice. Sage, Newbury Park
Elith J, Leathwick JR (2009) Species distribution models: ecological explanation and prediction across space and time. Annu Rev Ecol Evol Syst 40:677–697
Elith J, Graham CH, Anderson RP, Dudík M, Ferrier S, Guisan A et al. (2006) Novel methods improve prediction of species' distributions from occurrence data. Ecography 29:129–151
Elston DA, Illius AW, Gordon IJ (1996) Assessment of preference among a range of options using log ratio analysis. Ecology 77:2538–2548
Engels S, Cwynar LC (2011) Changes in fossil chironomid remains along a depth gradient: evidence for common faunal thresholds within lakes. Hydrobiologia 665:15–38
Eubank RL (1988) Smoothing splines and parametric regression. Marcel Dekker, New York
Everitt BS (1992) The analysis of contingency tables, 2nd edn. Chapman & Hall, London
Everitt BS (ed) (2002) The Cambridge dictionary of statistics, 2nd edn. Cambridge University Press, Cambridge
Everitt BS (2005) An R and S-PLUS® companion to multivariate analysis. Springer, New York
Everitt BS, Dunn G (2001) Applied multivariate data analysis, 2nd edn. Arnold, London
Everitt BS, Landau S, Leese M, Stahl D (2011) Cluster analysis, 5th edn. Academic Press, London
Faraway JJ (2005) Linear models with R. CRC Press, Boca Raton
Faraway JJ (2006) Extending the linear model with R. Generalized linear, mixed effects and nonparametric regression. Chapman & Hall, Boca Raton
Fielding AH (2007) Cluster and classification techniques for the biosciences. Cambridge University Press, Cambridge

Fienberg SE (1980) The analysis of cross-classified categorical data, 2nd edn. The MIT Press, Cambridge
Finn JD (1974) A general model for multivariate analysis. Holt, Reinhart & Winston, New York
Fleiss J (1981) Statistical methods for rates and proportions, 2nd edn. Wiley, Chichester
Flenley J (2003) Some prospects for lake sediment analysis in the 21st century. Quaternary Int 105:77–80
Fox J (1997) Applied regression analysis, linear models, and related methods. Sage, Thousand Oaks
Fox J (2000) Non-parametric simple regression. Smoothing scatterplots. Sage, Thousand Oaks
Fox J (2002) An R and S-PLUS® companion to applied regression. Sage, Thousand Oaks
Fox J (2008) Applied regression analysis and generalized linear models. Sage, Thousand Oaks
Fox J, Long JS (1990) Modern methods of data analysis. Sage, Newbury Park
Franklin J (2010) Mapping species distributions – spatial inference and prediction. Cambridge University Press, Cambridge
Gabriel KR (1982) Biplot. In: Kotz S, Johnson NL (eds) Encyclopedia of statistical sciences 1. Wiley, New York, pp 263–267
Gelman A, Pasarica C, Dodhia R (2002) Let's practice what we preach: turning tables into graphs in statistic research. Am Stat 56:121–130
Gilbert ED (1969) The effect of unequal variance covariance matrices on Fisher's linear discriminant function. Biometrics 25:505–515
Gill J (2001) Generalized linear models. A unified approach. Sage, Thousand Oaks
Gittins R (1985) Canonical analysis: a review with applications in biology. Springer, Berlin
Good PI (2001) Resampling methods. A practical guide to data analysis. Birkhäuser, Boston
Good PI (2005) Introduction to statistics through resampling methods in R/S-PLUS®. Wiley-Interscience, Hoboken
Goodall C (1990) A survey of smoothing techniques. In: Fox J, Long JS (eds) Modern methods of data analysis. Sage, Newbury Park, pp 126–176
Goodall DW (1969) Deviant index: a new tool for numerical taxonomy. Nature 210:216
Gordon AD (1982) Some observations on methods of estimating the proportions of morphologically similar pollen types in fossil samples. Can J Bot 60:1888–1894
Gordon AD, Prentice IC (1977) Numerical methods in Quaternary paleoecology 4. Separating mixtures of morphologically similar pollen taxa. Rev Palaeobot Palynol 23:359–372
Gower JC (1966a) A Q-technique for the calculation of canonical variates. Biometrika 53:588–590
Gower JC (1966b) Some distance properties of latent root and vector methods used in multivariate analysis. Biometrika 53:325–338
Greenacre MJ (1984) Theory and applications of correspondence analysis. Academic Press, London
Greenacre MJ (2007) Correspondence analysis in practice, 2nd edn. Chapman & Hall/CRC, Boca Raton
Greenacre MJ, Blasius J (eds) (2006) Multiple correspondence analysis and related methods. Chapman & Hall/CRC, Boca Raton
Guisan A, Weiss SB, Weiss AD (1999) GLM versus CCA spatial modeling of plant species distribution. Plant Ecol 143:107–122
Guisan A, Edwards TC, Hastie TJ (2002) Generalized linear and generalized additive models in studies of species distributions: setting the scene. Ecol Model 157:89–100
Hájková P, Hájek M, Apostolova I, Zelený D, Díte D (2008) Shifts in the ecological behaviour of plant species between two distant regions: evidence from the base richness gradient in mires. J Biogeogr 35:282–294
Hamilton LC (1992) Regression with graphics. A second course in applied statistics. Brooks, Pacific Grove
Hammarlund D, Barnekow L, Birks HJB, Buchardt B, Edwards TWD (2002) Holocene changes in atmospheric circulation recorded in the oxygen-isotope stratigraphy of lacustrine carbonates from northern Sweden. Holocene 12:339–351
Hammer Ø, Harper DAT (2006) Paleontological data analysis. Blackwell, Oxford

Hand DJ (1981) Discrimination and classification. Wiley, Chichester
Hand DJ (1986) Recent advances in error rate estimation. Pattern Recognit Lett 4:335–346
Hansen BS, Cushing EJ (1973) Identification of pine pollen of late Quaternary age from Chuska mountains, New Mexico. Geol Soc Am Bull 84:1181–1199
Hardy MA (1993) Regression with dummy variables. Sage Publications, Newbury Park
Harrell FE (2001) Regression modeling strategies. Springer, New York
Hartwig F, Dearing BE (1979) Exploratory data analysis. Sage Publications, Beverly Hills
Hastie TJ, Stuetzle W (1989) Principal curves. J Am Stat Assoc 84:502–516
Hastie TJ, Tibshirani RJ (1986) Generalized additive models. Stat Sci 1:297–318
Hastie TJ, Tibshirani RJ (1987) Generalized additive models: some applications. J Am Stat Assoc 82:371–386
Hastie TJ, Tibshirani RJ (1990) Generalized additive models. Chapman & Hall, London
Hastie TJ, Tibshirani RJ (1996) Discriminant analysis by Gaussian mixtures. J R Stat Soc B 58:155–176
Hastie TJ, Tibshirani RJ, Buja A (1994) Flexible discriminant analysis by optimal scoring. J Am Stat Assoc 89:1255–1270
Hastie TJ, Buja A, Tibshirani RJ (1995) Penalized discriminant analysis. Ann Stat 23:73–102
Hastie TJ, Tibshirani RJ, Friedman J (2011) The elements of statistical learning, 2nd edn. Springer, New York
Hawkins DM (1980) Identification of outliers. Chapman & Hall, London
Heegaard E (1997) Ecology of *Andreaea* in western Norway. J Bryol 19:527–636
Heegaard E (2002) The outer border and central border for species environmental relationships estimated by non-parametric generalised additive models. Ecol Model 157:131–139
Heegaard E, Birks HJB, Telford RJ (2005) Relationships between calibrated ages and depth in stratigraphical sequences: an estimation procedure by mixed-effect regression. Holocene 15:612–618
Heegaard E, Lotter AF, Birks HJB (2006) Aquatic biota and the detection of climate change: are there consistent aquatic ecotones? J Paleolimnol 35:507–518
Helmens KF, Bos JAA, Engels S, van Meerbeeck CJ, Bohncke SJP, Renssen H, Heiri O, Brooks SJ, Seppä H, Birks HJB, Wohlfarth B (2007) Present-day temperatures in northern Scandinavia during the last glaciation. Geology 35:987–990
Herzschuh U, Birks HJB (2010) Evaluating the indicator value of Tibetan pollen taxa for modern vegetation and climate. Rev Palaeobot Palynol 160:197–208
Herzschuh U, Birks HJB, Ni J, Zhao Y, Liu H, Liu X, Grosse G (2010) Holocene land-cover changes on the Tibetan Plateau. Holocene 20:91–104
Hewitt JE, Anderson MJ, Thrush SF (2005) Assessing and monitoring ecological community health in marine systems. Ecol Appl 15:942–953
Hill MO, Gauch HG (1980) Detrended correspondence analysis – an improved ordination technique. Vegetatio 42:47–58
Hocking RR, Pendleton OJ (1983) The regression dilemma. Commun Stat Theory Methods 12:497–527
Hoerl AE, Kennard R (1970) Ridge regression: biased estimation for nonorthogonal problems. Technometrics 12:55–67
Holmström L, Erästö P (2002) Making inferences about past environmental change using smoothing in multiple time scales. Comput Stat Data Anal 41:289–309
Holt K, Allen G, Hodgson R, Marsland S, Flenley J (2011) Progress towards an automated trainable pollen location and classifier system for use in the palynology laboratory. Rev Palynol Palaeobot 167:175–183
Horrocks M, Ogden J (1994) Modern pollen spectra and vegetation on Mt Hauhungatahi, central North Island, New Zealand. J Biogeogr 21:637–649
Hughes IG, Hase TPA (2010) Measurements and their uncertainties. Oxford University Press, Oxford
Huisman J, Olff H, Fresco LFM (1993) A hierarchical set of models for species response models. J Veg Sci 4:37–46

Husson F, Lê S, Pagès J (2011) Exploratory multivariate analysis by example using R. CRC Press, Boca Raton

Israëls AZ (1984) Redundancy analysis for qualitative variables. Psychometrika 49:331–346

Izenman AJ (1975) Reduced-rank regression for the multivariate linear model. J Multivar Anal 5:248–264

Jackson DA (1997) Compositional data in community ecology: the paradigm or peril of proportions? Ecology 78:929–940

Jacoby WG (1997) Statistical graphics for univariate and bivariate data. Sage Publications, Thousand Oaks

Jacoby WG (1998) Statistical graphics for visualizing multivariate data. Sage Publications, Thousand Oaks

James M (1985) Classification algorithms. Collins, London

Jenkins A, Whitehead PG, Cosby BJ, Birks HJB (1990) Modelling long-term acidification – a comparison with diatom reconstructions and the implications for reversibility. Philos Trans R Soc Lond B 327:435–440

Jensen FV, Nielsen TD (2007) Bayesian networks and decision graphs, 2nd edn. Springer, New York

Jolliffe IT (2002) Principal component analysis, 2nd edn. Springer, New York

Juggins S, Birks HJB (2012) Chapter 14: Quantitative environmental reconstructions from biological data. In: Birks HJB, Lotter AF, Juggins S, Smol JP (eds) Tracking environmental change using lake sediments. Volume 5: Data handling and numerical techniques. Springer, Dordrecht

Juggins S, Telford RJ (2012) Chapter 5: Exploratory data analysis and data display. In: Birks HJB, Lotter AF, Juggins S, Smol JP (eds) Tracking environmental change using lake sediments. Volume 5: Data handling and numerical techniques. Springer, Dordrecht

Kernan M, Ventura M, Bitušík P, Brancelj A, Clarke G, Velle G, Raddum GG, Stuchlík E, Catalan J (2009) Regionalisation of remote European mountain lake ecosystems according to their biota: environmental versus geographical patterns. Freshw Biol 54:2470–2493

Klecka WR (1980) Discriminant analysis. Sage Publications, Newbury Park

Kleinbaum DG, Kupper LL, Muller KE (1988) Applied regression analysis and other multivariate methods. PWS-Kent, Boston

Kohavi R (1995) A study of cross-validation and bootstrap for accuracy estimation and model selection. Int Joint Conf Artif Intell 14:1137–1143

Kohonen T (2001) Self-organising maps, 3rd edn. Springer, Berlin

Korhola A, Weckström J, Holmström L, Erästö P (2000) A quantitative Holocene climatic record from diatoms in northern Fennoscandia. Quaternary Sci Rev 54:284–294

Kruskal JB (1964) Multidimensional scaling by optimizing goodness of fit to a nonmetric hypothesis. Psychometrika 29:1–27

Kucera M, Malmgren BA (1998) Logratio transformation of compositional data – a resolution of the constant sum constraint. Mar Micropal 34:117–120

Lachenbruch PA (1975) Discriminant analysis. Hafner Press, New York

Lamb HF (1984) Modern pollen spectra from Labrador and their use in reconstructing Holocene vegetational history. J Ecol 72:37–59

Langlois TJ, Anderson MJ, Babcock RC (2005) Reef-associated predators influence adjacent soft-sediment communities. Ecology 86:1508–1519

Last WM, Smol JP (eds) (2001a) Tracking environmental change using lake sediments. Volume 1: Basin analysis, coring, and chronological techniques. Kluwer Academic Publishers, Dordrecht

Last WM, Smol JP (eds) (2001b) Tracking environmental change using lake sediments. Volume 2: Physical and geochemical methods. Kluwer Academic Publishers, Dordrecht

Leathwick JR (1995) Climatic relationships of some New Zealand forest tree species. J Veg Sci 6:237–248

Legendre P, Anderson MJ (1999) Distance-based redundancy analysis: testing multispecies responses in multifactorial ecological experiments. Ecol Monogr 69:1–24

Legendre P, Birks HJB (2012a) Chapter 7: Clustering and partitioning. In: Birks HJB, Lotter AF, Juggins S, Smol JP (eds) Tracking environmental change using lake sediments. Volume 5: Data handling and numerical techniques. Springer, Dordrecht

Legendre P, Birks HJB (2012b) Chapter 8: From classical to canonical ordination. In: Birks HJB, Lotter AF, Juggins S, Smol JP (eds) Tracking environmental change using lake sediments. Volume 5: Data handling and numerical techniques. Springer, Dordrecht

Legendre P, Legendre L (1998) Numerical ecology, 2nd English edn. Elsevier, Amsterdam

Legendre P, Oksanen J, ter Braak CJF (2011) Testing the significance of canonical axes in redundancy analysis. Methods Ecol Evol 2:269–277

Lek S, Guégan J-F (2000) Artificial neuronal networks: application to ecology and evolution. Springer, Berlin

Lepš J, Šmilauer P (2003) Multivariate analysis of ecological data using CANOCO. Cambridge University Press, Cambridge

Li P, Treloar WJ, Flenley JR, Empson L (2004) Towards automation of palynology 2: the use of texture measures and neural network analysis for automated identification of optical images of pollen grains. J Quaternary Sci 19:755–762

Lindbladh M, O'Connor R, Jacobson GL (2002) Morphometric analysis of pollen grains for palaeoecological studies: classification of *Picea* from eastern North America. Am J Bot 89:1459–1467

Lindbladh M, Jacobson GL, Shauffler M (2003) The post-glacial history of tree *Picea* species in New England, USA. Quaternary Res 59:61–69

Lindbladh M, Oswald W, Foster D, Faison E, Hou J, Huang Y (2007) A late-glacial transition from *Picea glauca* to *Picea mariana* in southern New England. Quaternary Res 67:502–508

Liu H, Cui H, Huang Y (2001) Detecting Holocene movements of the woodland-steppe ecotone in northern China using discriminant analysis. J Quaternary Sci 16:237–244

Liu K-B (1990) Holocene palaeoecology of the boreal forest and Great Lakes-St Lawrence forest in Northern Ontario. Ecol Monogr 60:179–212

Liu K-B, Lam NS-N (1985) Paleovegetational reconstruction based on modern and fossil pollen data: an application of discriminant analysis. Ann Assoc Am Geogr 75:115–130

Lotter AF, Anderson NJ (2012) Chapter 18: Limnological responses to environmental changes at inter-annual to decadal time-scales. In: Birks HJB, Lotter AF, Juggins S, Smol JP (eds) Tracking environmental change using lake sediments. Volume 5: Data handling and numerical techniques. Springer, Dordrecht

Lotter AF, Birks HJB (1993) The impact of the Laacher See tephra on terrestrial and aquatic ecosystems in the Black Forest, southern Germany. J Quaternary Sci 8:263–276

Lotter AF, Birks HJB, Hofmann W, Marchetto A (1997) Modern diatom, Cladocera, Chironomid, and Chrysophyte cyst assemblages as quantitative indicators for the reconstruction of past environmental conditions in the Alps. I. Climate. J Paleolimnol 18:395–420

Lotter AF, Birks HJB, Hofmann W, Marchetto A (1998) Modern diatom, Cladocera, Chironomid, and Chrysophyte cyst assemblages as quantitative indicators for the reconstruction of past environmental conditions in the Alps. II. Nutrients. J Paleolimnol 19:443–463

Lunneborg CE (2000) Data analysis by resampling. Duxbury, Pacific Grove

MacDonald GM (1987) Post-glacial development of the subalpine-boreal transition forest of Western Canada. J Ecol 75:303–320

Mackay AW, Davidson TA, Wolski P, Woodward S, Mazebedi R, Masamba WRL, Todd M (2011) Diatom sensitivity to hydrological and nutrient variability in a subtropical, flood-pulse wetland. Ecohydrology. doi:10.1002/eco.242

Maher LJ (1972a) Nomograms for computing 0.95 confidence limits of pollen data. Rev Palaeobot Palynol 13:85–93

Maher LJ (1972b) Absolute pollen diagrams of Redrock Lake, Boulder County, Colorado. Quaternary Res 2:531–553

Maher LJ (1981) Statistics for microfossil concentration measurements employing samples spiked with marker grains. Rev Palaeobot Palynol 32:153–191

Maher LJ, Heiri O, Lotter AF (2012) Chapter 6: Assessment of uncertainties associated with palaeolimnological laboratory methods and microfossil analysis. In: Birks HJB, Lotter AF, Juggins S, Smol JP (eds) Tracking environmental change using lake sediments. Volume 5: Data handling and numerical techniques. Springer, Dordrecht
Malmgren BA, Nordlund W (1997) Application of artificial neural networks to paleoceanographic data. Palaeogeogr, Palaeoclim, Palaeoecol 136:359–373
Manly BFJ (1992) The design and analysis of research studies. Cambridge University Press, Cambridge
Manly BFJ (2007) Randomization, bootstrap, and Monte Carlo methods in biology, 3rd edn. Chapman & Hall/CRC, London
Manly BFJ (2009) Statistics for environmental science and management, 2nd edn. CRC Press, Boca Raton
Marks S, Dunn OJ (1974) Discriminant functions when covariance matrices are unequal. J Am Stat Assoc 69:555–559
Marsh LC, Cormier DR (2002) Spline regression models. Sage, Thousand Oaks
Martens H, Næs T (1989) Multivariate calibration. Wiley, Chichester
McArdle B (1988) The structural relationship: regression in biology. Can J Zool 66:2329–2339
McArdle BH, Anderson MJ (2001) Fitting multivariate models to community data: a comment on distance-based redundancy analysis. Ecology 82:290–297
McCullagh P, Nelder JA (1989) Generalized linear models. Chapman & Hall, London
McLachlan GJ (1992) Discriminant analysis and statistical pattern recognition. Wiley, Chichester
Meadows ME, Sugden JM (1991) The application of multiple discriminant analysis to the reconstruction of the vegetation history of Fynbos, southern Africa. Grana 30:325–336
Menard S (2002) Applied logistic regression analysis, 2nd edn. Sage, Thousand Oaks
Michaelsen J, Schimel DS, Friedl MA, Davis FW, Dubayah RC (1994) Regression tree analysis of satellite and terrain data to guide vegetation sampling and surveys. J Veg Sci 5:673–686
Millar RB, Anderson MJ, Zunun G (2005) Fitting nonlinear environmental gradients to community data: a general distance-based approach. Ecology 86:2245–2251
Miller AJ (1990) Subset selection and regression. Chapman & Hall, London
Molinaro AM, Simon R, Pfeiffer RM (2005) Prediction error estimation: a comparison of resampling methods. Bioinformatics 21:3301–3307
Montgomery DC, Peck EA (1992) Introduction to linear regression analysis, 2nd edn. Duxbury Press, Boston
Mood AM, Graybill FA, Boes DC (1974) Introduction to the theory of statistics. McGraw-Hill, Tokyo
Mosteller F, Tukey JW (1977) Data analysis and regression: a second course in statistics. Addison Wesley, Reading
Mundry R, Nunn CL (2009) Stepwise model fitting and statistical inference: turning noise into signal pollution. Am Nat 173:119–123
Murray K, Conner MM (2009) Methods to quantify variable importance: implications for the analysis of noisy ecological data. Ecology 90:348–355
Murtaugh PA (2009) Performance of several variable-selection methods applied to ecological data. Ecol Lett 12:1061–1068
Myers RH, Montgomery DC, Vining GG (2002) Generalized linear models with applications in engineering and the sciences. Wiley, New York
Nakagawa S, Freckleton RP (2008) Missing in action: the dangers of ignoring missing data. Trends Ecol Evol 23:592–596
Neter J, Kutner MH, Nachtshein CJ, Wasserman W (1996) Applied linear statistical models, 4th edn. WCB (McGraw-Hill), Boston
Næs T, Isaksson T, Fearn T, Davies T (2002) A user-friendly guide to multivariate calibration and classification. NIR Publications, Chichester
Oberdorff T, Pont D, Hugueny B, Chessel D (2001) A probabilistic model characterizing fish assemblages of French rivers: a framework for environmental assessment. Freshw Biol 46: 399–415

O'Brian L (1992) Introducing quantitative geography. Measurement, methods and generalized linear models. Routledge, London
O'Gorman L, Samon MJ, Seul M (2008) Practical algorithms for image analysis, 2nd edn. Cambridge University Press, Cambridge
O'Hara RB, Kotze DJ (2010) Do not log-transform count data. Methods Ecol Evol 1:118–122
Oksanen J, Minchin PR (2002) Continuum theory revisited: what shape are species responses along ecological gradients? Ecol Model 157:119–129
Oksanen J, Blanchet FG, Kindt R, Legendre P, O'Hara RB, Simpson GL, Solymos P, Stevens MHM, Wagner H (2011) vegan: community ecology package. R package version 1.17-8. http://CRAN.R-project.org/package=vegan
Olden JD, Jackson DA (2000) Torturing the data for the sake of generality: how valid are our regression models? Ecoscience 7:501–510
Olden JD, Jackson DA (2002) A comparison of statistical approaches for modelling fish species distributions. Freshw Biol 47:1976–1995
Olden JD, Lawler JJ, Poff NL (2008) Machine learning methods without tears: a primer for ecologists. Quaternary Rev Biol 83:171–193
Osborne C (1991) Statistical calibration: a review. Int Stat Rev 59:309–336
Pampel FC (2000) Logistic regression. A primer. Sage, Thousand Oaks
Pelánková B, Kuneš P, Chytrý M, Jankovská V, Ermakov N, Svobodová-Svitavaská H (2008) The relationships of modern pollen spectra to vegetation and climate along a steppe-forest-tundra transition in southern Siberia, explored by decision trees. Holocene 18:1259–1271
Phillips SJ, Anderson RP, Schapire RE (2006) Maximum entropy modeling of species geographic distributions. Ecol Model 190:231–259
Piegorsch WW, Bailer AJ (1997) Statistics for environmental biology and toxicology. Chapman & Hall, London
Pienitz R, Lotter AF, Newman L, Kiefer T (eds) (2009) Advances in Paleolimnology. PAGES News 17:89–136
Pinheiro J, Bates D (2000) Mixed effects models in S and S-PLUS. Springer, New York
Prasad AM, Iverson LR, Liaw A (2006) Newer classification and regression tree techniques: bagging and random forests for ecological predictions. Ecosystems 9:181–199
Prentice IC (1980) Multidimensional scaling as a research tool in Quaternary palynology – a review of theory and methods. Rev Palaeobot Palynol 31:71–104
Pyšek P, Bacher S, Chytrý M, Jarošik V, Wild J, Celesti-Grapow L, Gassó N, Kenis M, Lambdon PW, Nentwig W, Pergl J, Roques A, Sádlo J, Solarz W, Vilà M, Hiulme PE (2010) Contrasting patterns in the invasions of European terrestrial and freshwater habitats by alien plants, insects and vertebrates. Global Ecol Biogeogr 19:317–331
Racca JMJ, Philibert A, Racca R, Prairie YT (2001) A comparison between diatom-based pH inference models using artificial neural networks (ANN), weighted averaging (WA) and weighted averaging partial least squares (WA-PLS) regressions. J Paleolimnol 26:411–422
Racca JMJ, Wild M, Birks HJB, Prairie YT (2003) Separating wheat from chaff: diatom taxon selection using an artificial neural network pruning algorithm. J Paleolimnol 29:123–133
Racca JMJ, Gregory-Eaves J, Pienitz R, Prairie YT (2004) Tailoring paleolimnological diatom-based transfer functions. Can J Fish Aquat Sci 61:2440–2454
Ramsey FL, Schafer DW (1997) The statistical sleuth – a course in methods of data analysis. Duxbury Press, Belmont
Rawlings JO (1988) Applied regression analysis. A research tool. Wadsworth & Brooks, Pacific Grove
Reineking B, Schröder B (2006) Constrain to perform: regularization of habitat models. Ecol Model 193:675–690
Reyment RA (1991) Multidimensional palaeobiology. Pergamon Press, Oxford
Reyment RA (1999) Multivariate statistical analysis of geochemical data exemplified by Proterozoic dyke swarms in Sweden. GFF (J Geol Soc Sweden) 121:49–55
Reyment RA, Savazzi E (1999) Aspects of multivariate statistical analyses in geology. Elsevier, Amsterdam

Reyment RA, Blackith RE, Campbell NA (1984) Multivariate morphometrics, 2nd edn. Academic Press, London
Ripley BD (2008) Pattern recognition and neural networks. Cambridge University Press, Cambridge
Roff DA (2006) Introduction to computer-intensive methods of data analysis in biology. Cambridge University Press, Cambridge
Rose NL, Juggins S, Watt J (1996) Fuel-type characterisation of carbonaceous fly-ash particles using EDS-derived surface chemistries and its application to particles extracted from lake sediments. Proc R Soc Lond A 452:881–907
Schielzeth H (2010) Simple means to improve the interpretability of regression coefficients. Methods Ecol Evol 1:103–113
Schroeder LD, Sjoquist DL, Stephan PE (1986) Understanding regression analysis. An introductory guide. Sage Publications, Beverly Hills
Schulz M, Mudelsee M (2002) REDFIT: estimating red-noise spectra directly from unevenly spaced paleoclimatic time series. Comput Geosci 28:421–426
Schulz M, Stattegger K (1997) SPECTRUM: spectral analysis of unevenly spaced paleoclimatic time series. Comput Geosci 23:929–945
Seber GAF, Wild CJ (1989) Nonlinear regression. Wiley, New York
Shao J (1993) Linear model selection by cross-validation. J Am Stat Assoc 88:486–494
Shen C, Liu K-B, Morrill C, Overpeck JT, Peng J, Tang L (2008a) Ecotone shift and major droughts during the mid-late Holocene in the central Tibetan Plateau. Ecology 89:1079–1088
Shen C, Liu K-B, Tang L, Overpeck JT (2008b) Numerical analysis of modern and fossil pollen data from the Tibetan Plateau. Ann Assoc Am Geogr 98:755–772
Shumway RH, Stoffer DS (2006) Time series analysis and its applications, 2nd edn. Springer, New York
Simpson GL (2012) Chapter 15: Analogue methods in palaeolimnology. In: Birks HJB, Lotter AF, Juggins S, Smol JP (eds) Tracking environmental change using lake sediments. Volume 5: Data handling and numerical techniques. Springer, Dordrecht
Simpson GL, Anderson NJ (2009) Deciphering the effect of climate change and separating the influence of confounding factors in sediment core records using additive models. Limnol Oceanogr 54:2529–2541
Simpson GL, Birks HJB (2012) Chapter 9: Statistical learning in palaeolimnology. In: Birks HJB, Lotter AF, Juggins S, Smol JP (eds) Tracking environmental change using lake sediments. Volume 5: Data handling and numerical techniques. Springer, Dordrecht
Simpson GL, Hall IR (2012) Chapter 19: Human impacts – applications of numerical methods to evaluate surface-water acidification and eutrophication. In: Birks HJB, Lotter AF, Juggins S, Smol JP (eds) Tracking environmental change using lake sediments. Volume 5: Data handling and numerical techniques. Springer, Dordrecht
Šmilauer P, Birks HJB (1995) The use of generalised additive models in the description of diatom-environment response surfaces. Geological Survey of Denmark (DGU) Service Report 7:42–47
Smol JP (2008) Pollution of lakes and rivers: a paleoenvironmental perspective, 2nd edn. Blackwell, Oxford
Smol JP, Birks HJB, Last WM (eds) (2001a) Tracking environmental change using lake sediments. Volume 3: Terrestrial, algal, and siliceous indicators. Kluwer Academic Publishers, Dordrecht
Smol JP, Birks HJB, Last WM (eds) (2001b) Tracking environmental change using lake sediments. Volume 4: Zoological indicators. Kluwer Academic Publishers, Dordrecht
Smol JP, Birks HJB, Lotter AF, Juggins S (2012) Chapter 1: The march towards the quantitative analysis of palaeolimnological data. In: Birks HJB, Lotter AF, Juggins S, Smol JP (eds) Tracking environmental change using lake sediments. Volume 5: Data handling and numerical techniques. Springer, Dordrecht
Sokal RR, Rohlf FJ (1995) Biometry – the principles and practice of statistics in biological research. WH Freeman, New York

Sokal MA, Hall RI, Wolfe BB (2008) Relationships between hydrological and limnological conditions in lakes of the Slave River Delta (NWT, Canada) and quantification of their roles on sedimentary diatom assemblages. J Paleolimnol 39:533–550
Sonderegger DL, Wang H, Clements WH, Noon BR (2009) Using SiZer to detect thresholds in ecological data. Front Ecol Environ 7:190–195
Sprent P (1998) Data driven statistical methods. Chapman & Hall, London
Sugden JM, Meadows ME (1989) The use of multiple discriminant analysis in reconstructing recent vegetation changes on the Nuweveldberg, South Africa. Rev Palaeobot Palynol 60: 131–147
Telford RJ, Birks HJB (2005) The secret assumption of transfer functions: problems with spatial autocorrelation in evaluating model performance. Quaternary Sci Rev 24:2173–2179
Telford RJ, Birks HJB (2009) Design and evaluation of transfer functions in spatially structured environments. Quaternary Sci Rev 28:1309–1316
Telford RJ, Birks HJB (2011) A novel method for assessing the statistical significance of quantitative reconstructions inferred from biotic assemblages. Quaternary Sci Rev 30:1272–1278
Telford RJ, Andersson C, Birks HJB, Juggins S (2004) Biases in the estimation of transfer function prediction errors. Paleoceanography 19:PA4014
ter Braak CJF (1986) Canonical correspondence analysis: a new eigenvector technique for multivariate direct gradient analysis. Ecology 67:1167–1179
ter Braak CJF (1987a) Ordination. In: Jongman RHG, ter Braak CJF, van Tongeren OFR (eds) Data analysis in community and landscape ecology. Pudoc, Wageningen, pp 91–173
ter Braak CJF (1987b) The analysis of vegetation-environment relationships by canonical correspondence analysis. Vegetatio 69:69–77
ter Braak CJF (1988) Partial canonical correspondence analysis. In: Bock HH (ed) Classification and related methods of data analysis. North-Holland, Amsterdam, pp 551–558
ter Braak CJF (1990) Interpreting canonical correlation analysis through biplots of structure correlations and weights. Psychometrika 55:519–531
ter Braak CJF (1994) Canonical community ordination. Part I: basic theory and linear methods. Ecoscience 1:127–140
ter Braak CJF (1995) Non-linear methods for multivariate statistical calibration and their use in palaeoecology: a comparison of inverse (*k*-nearest neighbours, partial least squares and weighted averaging partial least squares) and classical approaches. Chemometr Intell Lab Syst 28:165–180
ter Braak CJF (1996) Unimodal models to relate species to environment. DLO-Agricultural Mathematics Group, Wageningen
ter Braak CJF (2006) Bayesian sigmoidal shrinkage with improper variance priors and an application to wavelet denoising. Comput Stat Data Anal 51:1232–1242
ter Braak CJF, Barendregt LG (1986) Weighted averaging of species indicator values: its efficiency in environmental calibration. Math Biosci 78:57–72
ter Braak CJF, Looman CWN (1986) Weighted averaging, logit regression and the Gaussian response model. Vegetatio 65:3–11
ter Braak CJF, Looman CWN (1987) Regression. In: Jongman RHG, ter Braak CJF, Tongeren OFR (eds) Data analysis in community and landscape ecology. Pudoc Press, Wageningen, pp 29–90
ter Braak CJF, Looman CWN (1994) Biplots in reduced-rank regression. Biomet J 36:983–1003
ter Braak CJF, Prentice IC (1988) A theory of gradient analysis. Adv Ecol Res 18:271–317
ter Braak CJF, Šmilauer P (2002) CANOCO reference manual and CanoDraw for Windows user's guide: software for canonical community ordination (version 4.5). Microcomputer Power, Ithaca
ter Braak CJF, van Dam H (1989) Inferring pH from diatoms – a comparison of old and new calibration methods. Hydrobiologia 178:209–223
ter Braak CJF, Verdonschot PFM (1995) Canonical correspondence analysis and related multivariate methods in aquatic ecology. Aq Sci 57:255–289
Thompson B (1984) Canonical correlation analysis. Sage, Beverly Hills

Thuiller W, Araújo MB, Lavorel S (2003) Generalized models vs classification tree analysis: predicting spatial distributions of plant species at different scales. J Veg Sci 14:669–680
Tian J, Nelson DM, Hu FS (2011) How well do sediment indicators record past climate? An evaluation using annually laminated sediments. J Paleolimnol 45:73–84
Tibshirani R (1996) Regression shrinkage and selection via the lasso. J R Stat Soc Ser B 58: 267–288
Toms JD, Lesperance ML (2003) Piecewise regression: a tool for identifying ecological thresholds. Ecology 84:2034–2041
Treloar WJ, Taylor GE, Flenley JR (2004) Towards automation of palynology 1: analysis of pollen shape and ornamentation using simple geometric measures, derived from scanning electron microscope images. J Quaternary Sci 18:645–754
Tremblay V, Larocque-Tobler I, Sirois P (2010) Historical variability of subfossil chironomids (Diptera: Chironomidae) in three lakes impacted by natural and anthropogenic disturbances. J Paleolimnol 44:483–495
Trexler JC, Travis J (1993) Non-traditional regression analysis. Ecology 74:1629–1637
Tso MK-S (1981) Reduced-rank regression and canonical analysis. J R Stat Soc B 43:183–189
Tufte ER (1983) The visual display of quantitative information. Graphics Press, Cheshire
Tufte ER (1990) Envisioning information. Graphics Press, Cheshire
Tukey JW (1977) Exploratory data analysis. Addison-Wesley, Reading
van Belle G (2008) Statistical rules of thumb, 2nd edn. Wiley, Hoboken
van der Meer J (1991) Exploring macrobenthos-environment relationships by canonical correlation analysis. J Exp Mar Biol Ecol 148:105–120
van der Meeren T, Verschuren D, Ito E, Martens K (2010) Morphometric techniques allow environmental reconstructions from low-diversity continental ostracode assemblages. J Paleolimnol 44:903–911
Vanderpoorten A, Durwael L (1999) Trophic response curves of aquatic bryophytes in lowland calcareous streams. The Bryologist 102:720–728
Varmuza K, Filzmoser P (2009) Introduction to multivariate statistical analysis in chemometrics. CRC Press, Boca Raton
Velleman PF, Hoaglin DC (1981) Applications, basics, and computing of exploratory data analysis. Duxbury Press, Boston
Venables WN, Ripley BD (2002) Modern applied statistics with S®, 4th edn. Springer, New York
Walker D (1972) Quantification in historical plant ecology. Proc Ecol Soc Austral 6:91–104
Walker M (2005) Quaternary dating methods. Wiley, Chichester
Warner B, Misra M (1996) Understanding neural networks as statistical tools. Am Stat 50:284–293
Warton DI (2008) Raw data graphing: an informative but under-utilized tool for the analysis of multivariate abundances. Austral Ecol 33:290–300
Webb A (1999) Statistical pattern recognition. Arnold, London
Wehrens R (2011) Chemometrics with R. Springer, New York
Weng C, Jackson ST (2000) Species differentiation of North American spruce (*Picea*) based on morphological and anatomical characteristics of needles. Can J Bot 78:1367–1383
Werner P, Smol JP (2005) Diatom-environmental relationships and nutrient transfer functions from contrasting shallow and deep limestone lakes in Ontario, Canada. Hydrobiologia 533:145–173
Whiteside MC (1970) Danish chydorid cladocera: modern ecology and core studies. Ecol Monogr 40:79–118
Whittingham MJ, Stephens PA, Bradbury RB, Freckleton RP (2006) Why do we still use stepwise modelling in ecology and behaviour? J Animal Ecol 75:1182–1189
Wiklund JA, Bozinovski N, Hall RI, Wolfe BB (2010) Epiphytic diatoms as flood indicators. J Paleolimnol 44:25–42
Witt A, Schumann AY (2005) Holocene climate variability on millennial scales recorded in Greenland ice cores. Nonlinear Process Geophys 12:345–352
Witten IH, Frank E (2005) Data mining: practical machine learning tools and techniques. Morgan Kaufmann/Elsevier, Amsterdam
Wood SN (2006) Generalized additive models. Chapman & Hall, Boca Raton

Wright DB, Landon K (2009) Modern regression techniques using R. Sage, London
Yee TW (2004) A new technique for maximum-likelihood canonical Gaussian ordination. Ecol Monogr 74:685–701
Yee TW (2006) Constrained additive ordination. Ecology 97:203–213
Yee TW, Mackenzie M (2002) Vector generalized additive models in plant ecology. Ecol Model 157:141–156
Yee TW, Mitchell ND (1991) Generalized additive models in plant ecology. J Veg Sci 2:587–602
Ysebaert T, Meire P, Herman PMJ, Verbeek H (2002) Macrobenthic species response surfaces along estuarine gradients: prediction by logistic regression. Mar Ecol Prog Ser 225:79–95
Yuan LL (2004) Assigning macroinvertebrate tolerance classifications using generalised additive models. Freshw Biol 49:662–667
Zapala MA, Schork NJ (2006) Multivariate regression analysis of distance matrices for testing associations between gene expression patterns and related variables. Proc Natl Acad Sci USA 103:19430–19435
Zhang Y, Fountain DW, Hodgson RM, Flenley JR, Gunetileke S (2004) Towards automation of palynology 3: pollen pattern recognition using Gabor transforms and digital moments. J Quaternary Sci 19:7673–7768
Zou H, Hastie T (2005) Regularization and variable selection via the elastic net. J R Stat Soc Ser B 67:301–320
Zuur AF, Ieno EN, Smith GM (2007) Analyzing ecological data. Springer, New York
Zuur AF, Ieno EN, Walker NJ, Savelier AA, Smith GM (2009) Mixed effect models and extensions in ecology with R. Springer, New York
Zuur AF, Ieno EN, Elphick CS (2010) A protocol for data exploration to avoid common statistical problems. Methods Ecol Evol 1:3–14

Chapter 3
Data-Sets

H. John B. Birks and Vivienne J. Jones

Abstract The main data-sets used to illustrate particular numerical methods in this book are described. They are a Holocene diatom-stratigraphy from The Round Loch of Glenhead and a modern diatom-pH calibration-set from north-west Europe developed as part of the Surface Waters Acidification Programme.

Keywords Data-sets • Diatoms • Palaeolimnology • pH • SWAP • The Round Loch of Glenhead

Introduction

In various chapters of this book, two standard sets of representative palaeolimnological data are used to assist in illustrating the use of particular numerical methods and to demonstrate the application of these techniques to real data rather than to artificial data.

As many palaeolimnological data are generally of two main types – down-core stratigraphical data and modern surface-sediment data (Smol et al. 2012: Chap. 1; Birks 2012: Chap. 2) – the data-sets used are of these two types. The data-sets

H.J.B. Birks (✉)
Department of Biology and Bjerknes Centre for Climate Research, University of Bergen, PO Box 7803, Bergen N-5020, Norway

School of Geography and the Environment, University of Oxford, Oxford OX1 3QY, UK
e-mail: john.birks@bio.uib.no

V.J. Jones
Environmental Change Research Centre, University College London, Pearson Building, Gower Street, London WC1E 6BT, UK
e-mail: vivienne.jones@ucl.ac.uk

H.J.B. Birks et al. (eds.), *Tracking Environmental Change Using Lake Sediments*, Developments in Paleoenvironmental Research 5, DOI 10.1007/978-94-007-2745-8_3,

are The Round Loch of Glenhead diatom core data-set, called here the RLGH or RLGH3 data-set, and the Surface Waters Acidification Programme (SWAP) modern diatom-pH calibration data-set, called here the SWAP data-set.

The Round Loch of Glenhead Data-Set

The Round Loch of Glenhead (RLGH) data-set consists of fossil diatom relative abundances in 101 samples from 0.3 to 256.5 cm depth in core RLGH3. The core covers the last 10,000 years or more (Jones et al. 1989). All taxa identified to species level or below using the SWAP diatom taxonomic guidelines (Stevenson et al. 1991) and attaining a value of 1% or more in at least two fossil samples are included, giving a total of 139 taxa. Some taxa found in the RLGH core are absent from the modern SWAP diatom data-set, and vice versa. Abundances are expressed as percentages of the total diatom count (c. 500 valves) at RLGH.

The Round Loch of Glenhead is a small (12.5 ha), 13.5 m deep lake situated at 300 m on granite bedrock in Galloway, south-west Scotland (55.095°N; 4.429°W). Its catchment is almost entirely *Molinia caerulea*-dominated blanket mire on deep peat and peaty podsols, open *Calluna vulgaris-Erica cinerea* heath on shallow podsols or skeletal soils, or bare rock (Jones et al. 1989). Average lake-water pH at the time of the palaeolimnological investigations of the RLGH3 core and of SWAP (1984–1990) was 4.7 (based on measurements from 1979 to 1990) with an annual range of 4.6–5.0 (1981–1982) (Battarbee et al. 1989; Jones et al. 1989). The lake's pH has recently been measured to be about 5.2 (April 2003-March 2006) (Battarbee et al. 2008). Details of the ecological setting and the sediment, pollen, and diatom stratigraphies of the site, of the field and laboratory methods used, and of the palaeolimnological interpretation of the core are presented in full by Jones et al. (1986, 1989).

In addition to the diatom data from the RLGH3 core, there is a series of 20 radiocarbon dates based on dating lake-sediment samples from the core (Jones et al. 1989; Stevenson et al. 1990) and a ^{210}Pb chronology based on 11 dates for the uppermost 20 cm of the core (Stevenson et al. 1990). These radiocarbon dates are used to estimate age-depth relationships for the core by Blaauw and Heegaard (2012: Chap. 12).

The RLGH diatom data-set covers the entire Holocene (post-glacial) and it was collected to test the hypothesis of land-use and associated catchment vegetation and soil changes as a cause of recent lake acidification (Jones et al. 1986, 1989). The data were also used to illustrate the application of two-way weighted averaging and Gaussian logit regression and maximum likelihood calibration to reconstruct lake-water pH from fossil diatom assemblages (Birks et al. 1990a), to reconstruct using weighted averaging total aluminium and dissolved organic carbon from fossil diatom assemblages (Birks et al. 1990b), and to illustrate the use of modern analogue techniques in palaeolimnology (Simpson 2007). The palaeolimnology of RLGH has also been intensively studied in connection with understanding the causes of blanket-peat erosion (Stevenson et al. 1990), in the comparison of diatom-inferred pH with hindcast simulations from catchment-based models of lake acidification (Jenkins et al. 1990; Battarbee et al. 2005), in studying early signs of reversibility of recent lake acidification (Allott et al. 1992), in deriving and defining lake restoration goals and reference conditions (Flower et al. 1997; Simpson et al. 2005; Battarbee et al. 2011a), and in assessing the reliability of different diatom-based transfer functions for defining reference pH conditions (Battarbee et al. 2008). In addition, RLGH's water chemistry, epilithic diatoms, aquatic macrophytes, macroinvertebrates, and salmonids have been regularly monitored since 1988 (Monteith and Evans 2005; Monteith et al. 2005; Battarbee 2010; Battarbee et al. 2011b).

The SWAP Data-Set

The SWAP data-set consists of diatoms counts for 167 surface-sediment (0.5 cm) samples from lakes in England (five lakes), Norway (49), Scotland (55), Sweden (28), and Wales (30). It includes all diatom taxa (277) that are present in at least two samples with an abundance of 1% or more in at least one sample and that are identified to species level or below. Abundances are expressed as percentages of the total diatom count (c. 500 valves). In addition the data contain pH determinations for all 167 lakes measured after the water samples had equilibrated to room temperature (20°C). The pH values for each lake are based on the arithmetic mean of $[H^+]$ (Barth 1975), after initial data screening. Many lakes have pH data based on three or more readings (131), though some only have one (24) or two (13) readings. The pH range is 4.33–7.25 (mean = 5.56, median = 5.27, standard deviation = 0.77). Further details of the data-set and the methods used are given by Birks et al. (1990a), Munro et al. (1990), Stevenson et al. (1991), and Simpson (2007). For some of the numerical analyses illustrated in this book (e.g., Legendre and Birks 2012: Chap. 7; Simpson and Hall 2012: Chap. 19), a subset of this data-set is used consisting of the 90 samples from Scotland, Wales, and England. This subset contains 234 taxa and is called the SWAP-UK data-set.

The SWAP diatom-chemistry data-set was compiled as part of the palaeolimnological programme within SWAP to provide a large modern calibration or training data-set that could be used for the quantitative reconstruction of lake-water pH, dissolved organic carbon, and total aluminium from fossil diatom assemblages (Birks et al. 1990a, b). The modern data-set attempted to represent the full range of lake types in the acid-sensitive or recently acidified regions in Sweden, Norway, and the United Kingdom. For details of the palaeolimnology programme within SWAP, see Battarbee et al. (1990), Renberg and Battarbee (1990), and Battarbee (1994).

Data Availability

The data-sets are available from http://extra.springer.com. They are also included in the analogue (Simpson and Oksanen 2009) and rioja (Juggins 2009) packages for R.

Acknowledgements We are indebted to the many diatomists who contributed to the compilation, harmonisation, and refinement of the SWAP modern pH-diatom data-set; and to Cathy Jenks for help in the production of this chapter. This is publication number A208 from the Bjerknes Centre for Climate Change.

References

Allott TEH, Harriman R, Battarbee RW (1992) Reversibility of lake acidification at the Round Loch of Glenhead, Galloway, Scotland. Environ Pollut 77:219–225

Barth EF (1975) Average pH. J Water Pollut Control Fed 47:2191–2192

Battarbee RW (1994) Diatoms, lake acidification, and the Surface Water Acidification Programme (SWAP): a review. Hydrobiologia 274:1–7

Battarbee RW (2010) Are our acidified upland waters recovering? Freshw Biol Assoc News 52:5–6

Battarbee RW, Stevenson AC, Rippey B, Fletcher C, Natkanski J, Wik M, Flower RJ (1989) Causes of lake acidification in Galloway, south-west Scotland: a palaeoecological evaluation of the relative roles of atmospheric contamination and catchment change for two acidified sites with non-afforested catchments. J Ecol 77:651–672

Battarbee RW, Mason J, Renberg I, Talling JF (1990) Palaeolimnology and lake acidification. Royal Society, London

Battarbee RW, Monteith DT, Juggins S, Evans CD, Jenkins A, Simpson GL (2005) Reconstructing pre-acidification pH for an acidified Scottish loch: a comparison of palaeolimnological and modelling approaches. Environ Pollut 137:135–149

Battarbee RW, Monteith DT, Juggins S, Simpson GL, Shilland EM, Flower RJ, Kreiser AM (2008) Assessing the accuracy of diatom-based transfer functions in defining reference pH conditions for acidified lakes in the UK. Holocene 18:57–67

Battarbee RW, Simpson GL, Bennin H, Curtis C (2011a) A reference typology of low alkalinity lakes in the UK based on pre-acidification diatom assemblages from lake sediment cores. J Paleolimnol 45:489–505

Battarbee RW, Curtis CJ, Shilland EW (2011b) The Round Loch of Glenhead: recovering from acidification, climate change monitoring and future threats. Scottish Natural Heritage Commissioned Report No. 469

Birks HJB (2012) Chapter 2: Overview of numerical methods in palaeolimnology. In: Birks HJB, Lotter AF, Juggins S, Smol JP (eds) Tracking environmental change using lake sediments. Volume 5: Data handling and numerical techniques. Springer, Dordrecht

Birks HJB, Line JM, Juggins S, Stevenson AC, ter Braak CJF (1990a) Diatoms and pH reconstruction. Philos Trans R Soc Lond B 327:263–278

Birks HJB, Juggins S, Line JM (1990b) Lake surface-water chemistry reconstructions from palaeolimnological data. In: Mason BJ (ed) The Surface Waters Acidification Programme. Cambridge University Press, Cambridge, pp 301–313

Blaauw M, Heegaard E (2012) Chapter 12: Estimation of age-depth relationships. In: Birks HJB, Lotter AF, Juggins S, Smol JP (eds) Tracking environmental change using lake sediments. Volume 5: Data handling and numerical techniques. Springer, Dordrecht

Flower RJ, Juggins S, Battarbee RW (1997) Matching diatom assemblages in lake sediment cores and modern surface sediment samples: the implications for lake conservation and restoration with special reference to acidified systems. Hydrobiologia 344:27–40

Jenkins A, Whitehead PG, Cosby BJ, Birks HJB (1990) Modelling long-term acidification – a comparison with diatom reconstructions and the implications for reversibility. Philos Trans R Soc Lond B 327:435–440

Jones VJ, Stevenson AC, Battarbee RW (1986) Lake acidification and the land-use hypothesis: a mid-post-glacial analogue. Nature 322:157–158

Jones VJ, Stevenson AC, Battarbee RW (1989) Acidification of lakes in Galloway, south west Scotland: a diatom and pollen study of the post-glacial history of the Round Loch of Glenhead. J Ecol 77:1–23

Juggins S (2009) rioja: analysis of Quaternary science data. R package version 0.5-6. http://www.staff.ncl.ac.uk/staff/stephen.juggins/

Legendre P, Birks HJB (2012) Chapter 7: Clustering and partitioning. In: Birks HJB, Lotter AF, Juggins S, Smol JP (eds) Tracking environmental change using lake sediments. Volume 5: Data handling and numerical techniques. Springer, Dordrecht

Monteith DT, Evans CD (2005) The United Kingdom Acid Waters Monitoring Network: a review of the first 15 years and introduction to the special issue. Environ Pollut 137:3–13

Monteith DT, Hildrew AG, Flower RJ, Raven PJ, Beaumont WRB, Collen P, Kreiser AM, Shilland EM, Winterbottom JM (2005) Biological responses to the chemical recovery of acidified fresh waters in the UK. Environ Pollut 137:83–101

Munro MAR, Kreiser AM, Battarbee RW, Juggins S, Stevenson AC, Anderson DS, Anderson NJ, Berge F, Birks HJB, Davis RB, Flower RJ, Fritz SC, Haworth EY, Jones VJ, Kingston JC, Renberg I (1990) Diatom quality control and data handling. Philos Trans R Soc Lond B 327:257–261

Renberg I, Battarbee RW (1990) The SWAP Palaeolimnology Programme: a synthesis. In: Mason BJ (ed) The Surface Waters Acidification Programme. Cambridge University Press, Cambridge, pp 281–300

Simpson GL (2007) Analogue methods in palaeoecology: using the analogue package. J Stat Software 22:1–29

Simpson GL, Hall RI (2012) Chapter 19: Human impacts – implications of numerical methods to evaluate surface-water acidification and eutrophication. In: Birks HJB, Lotter AF, Juggins S, Smol JP (eds) Tracking environmental change using lake sediments. Volume 5: Data handling and numerical techniques. Springer, Dordrecht

Simpson GL, Oksanen J (2009) analogue: analysis and weighted averaging methods for palaeoecology. R package version 0.6-8. http://analogue.r-forge.r-project.org

Simpson GL, Shilland EM, Winterbottom JM, Keay J (2005) Defining reference conditions for acidified waters using a modern analogue approach. Environ Pollut 137:119–133

Smol JP, Birks HJB, Lotter, AF, Juggins S (2012) Chapter 1: The march towards the quantitative analysis of palaeolimnological data. In: Birks HJB, Lotter AF, Juggins S, Smol JP (eds) Tracking environmental change using lake sediments. Volume 5: Data handling and numerical techniques. Springer, Dordrecht

Stevenson AC, Jones VJ, Battarbee RW (1990) The cause of peat erosion: a palaeolimnological approach. New Phytol 114:727–735

Stevenson AC, Juggins S, Birks HJB, Anderson DS, Anderson NJ, Battarbee RW, Berge F, Davis RB, Flower RJ, Haworth EY, Jones VJ, Kingston JC, Kreiser AM, Line JM, Munro MAR, Renberg I (1991) The Surface Waters Acidification Project Palaeolimnology Programme: modern diatom/lake-water chemistry data-set. ENSIS Publishing, London

Part II
Numerical Methods for the Analysis of Modern and Stratigraphical Palaeolimnological Data

Chapter 4
Introduction and Overview of Part II

H. John B. Birks

Abstract An overview of the major numerical methods for the analysis of modern surface-sample and stratigraphical data is given in this chapter. The methods are discussed in relation to data collection and assessment, data summarisation, data analysis, and data interpretation. The main methods and research approaches outlined include data storage and data-bases, exploratory data analysis, clustering and partitioning, classical and canonical ordination, partial ordinations, principal curves, classification and regression trees and related decision-tree methods, artificial neural networks and self-organising maps, estimating gradient lengths, estimating taxonomic richness, estimating species optima and tolerances, Procrustes rotation for comparing ordinations, and variation partitioning. Many of these techniques are also discussed in detail in some of the chapters in Parts III and IV.

Keywords Artificial neural networks • Clustering • Data analysis • Data assessment • Data-bases • Data interpretation • Data storage • Data summarisation • Diversity estimation • Error estimation • Exploratory data analysis • Gaussian logit regression • Gradient length • Hypothesis generation • Hypothesis testing • Identification • Ordination • Partitioning • Principal curves • Procrustes rotation • Regression • Richness estimation • Self-organising maps • Species optima and tolerance estimation • Turnover estimation

---

H.J.B. Birks (✉)
Department of Biology and Bjerknes Centre for Climate Research, University of Bergen, PO Box 7803, Bergen N-5020, Norway

Environmental Change Research Centre, University College London, Pearson Building, Gower Street, London WC1E 6BT, UK

School of Geography and the Environment, University of Oxford, Oxford OX1 3QY, UK
e-mail: john.birks@bio.uib.no

H.J.B. Birks et al. (eds.), *Tracking Environmental Change Using Lake Sediments*, Developments in Paleoenvironmental Research 5, DOI 10.1007/978-94-007-2745-8_4,

Introduction

Part II consists of five chapters, in addition to this introductory chapter, that describe statistical and numerical methods that are equally useful in the quantitative analysis of modern 'surface-sample' palaeolimnological data (counts of organisms in surface-sediments and associated environmental data) and of down-core stratigraphical data (counts of organisms at different depths in a sediment-sequence). As explained by Smol et al. (2012: Chap. 1), palaeolimnologists today devote almost as much time in collecting and analysing modern surface-sample data so as to develop modern organism-environment calibration or 'training' data-sets (see Smol 2008) as they do in analysing lake sediments and their contained fossils, geochemistry, and physical properties.

All the numerical procedures presented in this Part can be applied to the quantitative analysis of both modern and stratigraphical data, whereas the techniques discussed in Part III (e.g., age-depth modelling, rate-of-change analysis, quantitative environmental reconstructions, modern analogue analysis) are only applicable to stratigraphical data. It should, however, be emphasised that approaches such as modern analogue techniques and calibration functions for environmental reconstruction require *both* modern and stratigraphical palaeolimnological data.

Numerical analyses are useful tools at many stages in a palaeolimnological investigation (Birks 2010, 2012a: Chap. 2 Table 4.1). During *data collection* they can help in the identification of critical fossil remains. In *data assessment*, statistical techniques are essential in estimating the inherent errors associated with different laboratory procedures and with the resulting different counts of palaeolimnological variables such as diatoms, chironomids, and cladocerans. As palaeolimnological data-sets increase in both number and size, the efficient storage of these data-sets requires the establishment of relational data-bases. Prior to any multivariate analysis of palaeolimnological data, exploratory data analysis (EDA) and graphical display of data are essential to summarise the data, to identify potential unusual observations or outliers, and to consider the need for data transformations. The next stage is *data summarisation* and there is a range of numerical techniques that are useful in detecting and summarising major patterns of variation in modern and stratigraphical data and in generating hypotheses about the underlying processes that may have influenced the observed patterns. The third general stage is *data analysis* where particular numerical characteristics are estimated from palaeolimnological data such as taxonomic diversity, the amount of compositional change along particular gradients, and the responses of species to environmental variables. The last stage in numerical analysis is *data interpretation* where attempts are made to test competing hypotheses about underlying causative factors such as water-depth, water-chemistry, substrate type, etc. in determining the composition of modern assemblages of diatoms, chironomids, etc. and the relative abundance of taxa within these assemblages.

The aim of this introductory overview chapter is to put the following five chapters into the context of data collection and data assessment, data summarisation, data analysis, and data interpretation (Table 4.1) and to outline the sorts of research

Table 4.1 Overview of numerical methods that are widely used in the analysis of both modern and stratigraphical palaeolimnological data

Aim	Numerical methods	Relevant chapters
Data collection;	Discriminant analysis	2
fossil identification;	Classification and regression trees (CARTs)	2, 9
data storage	Relational data-bases	4
Data assessment;	Exploratory data analysis	5
error estimation	Laboratory and analytical uncertainties	6
Data summarisation	Clustering, partitioning	7
	Ordination, classical and canonical	8
	CARTs and related decision trees	2, 9
	Self-organising maps	9
	Principal curves and surfaces	9
Data analysis	Estimation of compositional turnover	4, 8
	Changes in taxonomic richness	4
	Estimation of species optima and tolerances	2, 14
	Comparison of clusterings and ordinations	4, 7, 8
	Species-environment relationships	2, 8, 9, 14, 15
Data interpretation	Interpretation of modern assemblages in relation to external causative variables using canonical ordination, variance partitioning, simple discriminants, or CARTs and related techniques	2, 7, 8, 9, 14, 15, 19
	Hypothesis testing about modern assemblages and possible causative factors using canonical ordination and variance partitioning	8, 14, 19

questions that palaeolimnologists try to answer using appropriate numerical methods. In addition, brief accounts are given of important numerical approaches or techniques in the analysis of modern and stratigraphical palaeolimnological data that are not covered as separate chapters because the approaches or methods are not sufficiently large to warrant individual chapters.

Data Collection and Data Assessment

Numerical techniques can help in the identification of biological remains in surface sediments and in sediment cores and hence in basic data collection (Table 4.1). Statistical techniques are essential in the assessment of the uncertainties in many laboratory procedures and in the counting of biological remains in surface and core sediment samples, namely assessment of analytical errors (Table 4.1). Well-designed data-bases are essential tools not only for the researchers who have collected and assembled the data, but also for the archiving of palaeolimnological data for future generations of scientists. Data-sets from the real world are inevitably noisy and highly variable and may contain seemingly unusual observations or outliers. Such outliers may be a result of analytical or coding errors or they may result from unusual limnological characteristics or taphonomic processes. Numerical techniques of exploratory data analysis can help summarise large data-sets and identify potential outliers (Table 4.1). They can also guide the researcher about the need for appropriate data transformations prior to further numerical analyses.

Identification

Careful and reliable identification of all the biological remains of interest preserved in surface or core sediment samples is the first essential step in any biologically based palaeolimnological study (see the various chapters in Smol et al. 2001a, b about the range of different organisms studied in palaeolimnology).

Different approaches to the identification of fossil remains preserved in lake sediments are outlined in Chap. 2 of this volume. Some of these involve numerical procedures to help with identification, classification, and assignment (e.g., classification and regression trees (Simpson and Birks 2012: Chap. 9) and linear two-group discriminant analysis and multiple discriminant analysis (= canonical variates analysis) (Birks 2012a: Chap. 2; Legendre and Birks 2012b: Chap. 8)).

Error Estimation

All counts of biological remains present in a surface or core sediment sample are, hopefully, unbiased sample counts of the total numbers of the remains preserved in the sample of interest (Birks and Gordon 1985). As in all sampling, there are statistical uncertainties associated with any sample count (Maher et al. 2012: Chap. 6). The larger the count, the smaller the uncertainties become. As larger counts require more time to obtain, there is a trade-off between time and level of uncertainty. It is therefore important to estimate the uncertainty associated with all counts, determinations, and measurements. Maher et al. (2012: Chap. 6) provide a comprehensive guide to error estimation for palaeolimnological variables, including loss-on-ignition, varve counts, chemical determinands, age estimates, etc. (see Last and Smol 2001a, b for accounts of the main physical and chemical variables studied in palaeolimnology).

Data Storage and Data-Bases

Palaeolimnological data are multivariate and often contain many samples and many variables. Data from one or two fossil cores or from one or two modern calibration-sets can be stored in a spreadsheet form, such as EXCEL® or TILIA.

However, as more variables are studied on these cores or as data from different sites or different calibration-sets are assembled together, the investigator quickly reaches the maximum size and effectiveness of spreadsheets, and other approaches to data storage are essential. Rather than waiting for the stage when spreadsheets cease to be useful, it is good research design to develop a relational data-base (e.g., in ACCESS®) from the outset, thereby ensuring effective storage and manipulation of the basic data. A multi-proxy relational data-base (e.g., Juggins 1996) ensures compatibility and consistency between different data types and provides a rapid and effective means of bringing together, comparing, and cross-correlating different palaeolimnological records within cores, between cores, and between sites. A relational data-base provides archival and research tables of, for example, basic site, core, and surface-sample data, physical and chemical variables, biological data, chronological information, age-depth model results, correlations, environmental reconstructions, etc. A well-designed data-base allows rapid retrieval of data and provides the basis for subsequent data manipulations and output for further analysis, such as meta-analyses of large data-sets to explore competing hypotheses (e.g., Smol et al. 2005; Rühland et al. 2008; Battarbee et al. 2011; Mitchell 2011; Rose et al. 2011; Stomp et al. 2011).

A data-base consists of tables and fields and makes the distinction between primary data and meta-data and between archival and research data. Meta-data are associated data such as site location and description, dates of coring, coring device(s), investigators, sample depths, thicknesses, and dates, geochronological data, age-depth models, lithology, and publications (Michener et al. 1997), whereas the primary archival data are the actual counts of different organisms, and the determinations of physical or chemical variables. It is essential to maintain the distinction between archival data and research data (Grimm et al. 2007). Archival primary data will never change except for the correction of any errors. These data include the basic counts and measurements and associated meta-data. Research data are derived data which may change but which are essential for the use of the data-base (e.g., age-depth models, environmental reconstructions). Maintaining a relational data-base that incorporates data from different investigators and laboratories requires much effort to ensure quality control, consistent taxonomy and nomenclature, and data accuracy (Grimm et al. 2007). Besides being a major scientific resource to the palaeolimnological community, such data-bases (e.g., European Diatom Database Initiative (EDDI) – http://craticula.ncl.ac.uk/Eddi/jsp/) fulfil the need to "make the most of palaeodata" (Anon 2001) and "make sure that the world's palaeodata do no get buried" (Alverson and Eakin 2001; Dittert et al. 2001). Zuur et al. (2007) discuss various aspects of data preparation prior to numerical analysis and Moe et al. (2008) discuss their experiences in compiling and analysing monitoring data from 5000 lakes in 20 European countries including phytoplankton, aquatic macrophyte, macroinvertebrate, fish, chemical, and site data. Hernández (2003) and Whitehorn and Marklyn (2001) provide excellent introductions to developing a relational data-base, whereas Michener and Brunt (2000) discuss in detail the various aspects in the storage and management of ecological data. Jones et al. (2001) discuss tools in managing a wide range of meta-data, and McPhillips et al. (2009) outline the wide range of tools now available for handling large and diverse data-sets including tools for analysis and for workflow. For those who persist in using spreadsheets for data storage and management, O'Beirne (2005) is essential reading.

Exploratory Data Analysis

Exploratory data analysis (EDA) (Juggins and Telford 2012: Chap. 5) is curiously given little or no attention in the quantitative analysis of palaeolimnological data. EDA involves summarising data-sets in terms of measures of location or 'typical value' such as means, medians, trimmed means, and geometric means, measures of dispersion such as range, quartiles, variance, standard deviation, coefficient of variation, and standard error of the mean, and measures of skewness and kurtosis. Such measures can guide the data analyst about questions of data transformations. Simple graphical tools like histograms, kernel density estimation plots, quantile-quantile, and box-whisker plots (also called box-plots) can all help in decisions about data transformation (Fox 2002, 2008; Borcard et al. 2011). For data consisting of two or three variables only, simple scatter plots and matrices of scatter plots are useful. For multivariate data, there are several graphical tools (Everitt 1978, 2005; Everitt and Dunn 2001; Everitt and Hothorn 2011) but with data-sets of 50–100 variables these graphical tools have limited value and methods of clustering, partitioning, and ordination (Borcard et al. 2011; Legendre and Birks 2012a; b: Chaps. 7 and 8) are generally more useful. EDA provides a means of identifying potential outlying or 'rogue' observations – observations that are, in some sense, inconsistent with the rest of the observations in the data-set. An observation can be an outlier for various numerical reasons such as one or more of the biological or environmental variables lie well outside their expected range. EDA provides powerful means of detecting outliers using measures of leverage (the potential for influence resulting from unusual values) and of influence (an observation or variable is influential if its deletion substantially changes the results) (Fox 2002, 2008). After detecting potential outliers, the palaeolimnologist then has the challenge of trying to ascertain why the observations or variables are outliers (Birks et al. 1990) – do the unusual values result from incorrect measurements, incorrect data entry, transcription or recording errors, or unusual site features? Graphical techniques and effective data display are key aspects of modern EDA (see Chambers et al. 1983; Tufte 1983; Hewitt 1992; Cleveland 1993, 1994; Borcard et al. 2011; Juggins and Telford 2012: Chap. 5). As Cleveland (1994) says "graphs allow us to explore data to see overall patterns and to see detailed behaviour; no other approach can compete in revealing the structure of data so thoroughly". Warton (2008) provides a detailed graphical analysis of a standard ecological data-set widely used in ordination analyses to show how this data-set has been misinterpreted in several influential methodological papers because the data were interpreted from ordination plots alone, with no consideration of EDA plots of the basic raw data.

Data Summarisation

The data summarisation stage (Table 4.1) overlaps with EDA (Juggins and Telford 2012: Chap. 5) but in data summarisation an attempt is made to detect clusters, groups, gradients, and patterns of variation when the data-sets are considered as multivariate data rather than as univariate or bivariate data-sets as discussed above. Data summarisation is a useful stage in that it can provide useful low-dimensional representations or groupings of observations that can provide an easy basis for description, discussion, hypothesis generation, and interpretation.

For some purposes it is useful to cluster or partition multivariate data-sets into a small number of groups of samples with similar biological or environmental characteristics (see Borcard et al. 2011; Legendre and Birks 2012a: Chap. 7). These purposes are outlined in Table 7.2 in Legendre and Birks (2012a: Chap. 7) and include detecting groups of samples with similar biological composition or with similar environmental variables, detecting indicator species for the groups, relating biologically based groups to environmental variables, and assessing similarities between fossil samples and groups of modern samples from known environmental variables (see Catalan et al. 2009 for a detailed palaeolimnological example). Clustering and partitioning procedures can impose one-dimensional (single environmental gradient or time) or two-dimensional (geographic co-ordinates) constraints (see Birks 2012c: Chap. 11; Legendre and Birks 2012a: Chap. 7) to detect spatially or temporally contiguous groups of samples with similar biological and/or environmental characteristics. Clustering and partitioning techniques are surprisingly little used in palaeolimnology, except in the partitioning of stratigraphical sequences into assemblage zones (see Birks 2012c: Chap. 11).

In contrast, ordination techniques, both classical or indirect gradient analysis and canonical, constrained, or direct gradient analysis (ter Braak and Prentice 1988; Borcard et al. 2011) are widely used in palaeolimnology (Birks 1998; Legendre and Birks 2012b: Chap. 8) to summarise patterns of variation in complex multivariate data, to provide convenient low-dimensional graphical representations of such data, and to detect relationships between modern biological assemblages and contemporary environmental variables (see Table 8.1 in Legendre and Birks 2012b: Chap. 8). The development of canonical correspondence analysis (CCA) (ter Braak 1986) provided palaeolimnologists with a powerful tool for exploring organism-environment relationships within modern calibration-sets, under the assumption that the organisms have unimodal responses to the underlying environmental gradients. Redundancy analysis (RDA) (= constrained or canonical principal component analysis) is the analogous technique when the responses can be assumed to be monotonic (see Birks 1995; Legendre and Birks 2012b: Chap. 8). In some research problems it may be useful to partial out, as 'covariables', the effects of 'nuisance' variables that are not of primary interest (e.g., sampling date, collecting situation) (ter Braak and Prentice 1988; Borcard et al. 2011; Legendre and Birks 2012b: Chap. 8). Spatial or temporal structuring within modern or fossil palaeolimnological data can be explored and quantified with newly developed ordination procedures such as principal coordinates of neighbour matrices (Borcard et al. 2011; Legendre and Birks 2012b: Chap. 8). Principal curves and surfaces (De'ath 1999; Simpson and Birks 2012: Chap. 9) provide additional powerful data-exploratory and summarisation tools based on ordination.

With ever-increasing computer power becoming available, a set of techniques generally known as classification and regression trees (CART – Breiman et al. 1984; Efron and Tibshirani 1991; Fielding 2007; Olden et al. 2008; Borcard et al. 2011; Birks 2012a: Chap. 2; Legendre and Birks 2012a: Chap. 7; Simpson and Birks 2012: Chap. 9) has been developed that combine data exploration and data interpretation. These techniques (De'ath and Fabricius 2000; De'ath 2002, 2007; Prasad et al. 2006; Cutler et al. 2007) have not yet been widely used in palaeolimnology, but judging from their applications in ecology, biogeography, marine and freshwater biology, palynology, and remote sensing, CARTs and their relatives such as bagging, boosted trees, random forests, and multivariate adaptive regression splines (Simpson and Birks 2012: Chap. 9) are likely to be powerful and robust means of summarising large heterogeneous palaeolimnological data-sets (e.g., Olden and Jackson 2002; D'heygere et al. 2003; Raymond et al. 2005; Pelánkova et al. 2008). De'ath's (2002) multivariate regression tree (MRT) approach is particularly attractive as it produces a clustering of multivariate biological data using a monothetic (single environmental variable) divisive approach (see Bjerring et al. 2009 and Davidson et al. 2010a, b for palaeolimnological applications). MRT is related to regression in that the explanation of the biological data involves explanatory or predictor variables. MRT is thus an alternative to RDA or CCA and provides a robust approach for the prediction of biological assemblages in samples where environmental data only are available (De'ath 2002; Borcard et al. 2011).

Artificial neural networks (Lek and Guégan 2000; Olden et al. 2008; Birks 2012a: Chap. 2: Simpson and Birks 2012: Chap. 9) and self-organising maps (Kohonen 2001; Simpson and Birks 2012: Chap. 9) provide alternative approaches to analyse large and heterogeneous data-sets for the purposes of data exploration and interpretation and statistical modelling. Examples involving palaeolimnology or related topics include Racca et al. (2001, 2003), Olden and Jackson (2002), and Weller et al. (2006).

CARTs and their relatives, artificial neural networks, and self-organising maps are part of what is called by Hastie et al. (2011) 'statistical learning'. Statistical-learning techniques permit the exploration and summary of huge data-sets and allow statistical inferences to be made without the usual concerns for mathematical tractability because traditional analytical approaches are replaced by specially designed algorithms. Other statistical-learning techniques discussed by Simpson and Birks (2012: Chap. 9) include Bayesian networks (= belief networks or Bayesian belief networks), genetic algorithms, principal curves and surfaces, and shrinkage methods and variable selection in statistical modelling involving ridge regression, the lasso, and the elastic net (Hastie et al. 2011).

Data Analysis

The term data analysis is used here to include specialised techniques that estimate particular numerical characteristics of palaeolimnological data. Examples include gradient lengths, richness and diversity, and species optima and tolerances for particular environmental variables.

Gradient Lengths and Compositional Turnover in Palaeolimnological Data

Ter Braak and Prentice (1988) suggested that the length of the first major gradient of variation in multivariate biological data, as estimated by detrended correspondence analysis (DCA) (Hill and Gauch 1980) (estimated as units of compositional turnover in standard deviation (SD) units of turnover) is a useful guide as to whether species responses are primarily monotonic (gradient length <2 SD) or primarily unimodal (gradient length >2 SD). Despite the potential limitations of DCA as a general purpose ordination method (see Borcard et al. 2011; Legendre and Birks 2012b: Chap. 8), it is a valuable means of estimating gradients lengths in biological data-sets. Ter Braak and Juggins (1993) and Birks (1995, 1998, 2007) extended this approach to estimate the amount of compositional turnover along specific environmental gradients (e.g., lake-water pH) or through time. Such estimates of turnover along an environmental gradient are a guide to the type of regression procedure that should be used in the development of calibration functions from modern calibration-sets for quantitative environmental reconstructions (Birks 1995, 1998; Birks et al. 2010; Juggins and Birks 2012: Chap. 14). Estimating compositional turnover over a specific time period at many sites (e.g., Smol et al. 2005; Hobbs et al. 2010) or in different groups of organisms over a specific time period at one site (e.g., Birks and Birks 2008) provides a means of quantifying and summarising the amount of biological change recorded in palaeolimnological data.

Estimating Richness from Palaeolimnological Data

There are increasing concerns about the conservation of biodiversity now and in the future. The palaeolimnological record can provide unique information about how species richness, an important component of biodiversity, has changed with time or how it changes with space today (Gregory-Eaves and Beisner 2011). As the number of taxa recorded from a sediment sample is a function not only of the richness but also of the sample size, it is necessary to standardise sample size. Rarefaction analysis (Birks and Line 1992) estimates how many taxa (e.g., diatom valves, chironomid head-capsules) would have been found if all the counts had been the same size.

$$E(T_n) = \sum_{i=1}^{T} 1 - \left[\frac{\binom{N - N_i}{n}}{\binom{N}{n}} \right] \tag{4.1}$$

$$= \sum_{i=1}^{T} 1 - \left[\frac{(N - N_i)!(N - n)!}{(N - N_i - n)!N!} \right] \tag{4.2}$$

where $E(T_n)$ is the expected number of taxa in a sample of n individual remains selected at random without replacement from a count of N remains containing T taxa, namely the estimated number of taxa that would have been found if only n remains had been counted. T is the number of taxa in the original count; N_i is the number of individual remains assigned to taxon i in the original count; and N is the total number of individual remains counted in the sample, where

$$N = \sum_{l=1}^{T} N_i$$

and n is the number of individual remains (count size) chosen for standardisation ($n \leq N$) in the rarefied sample, usually the smallest total count in the samples to be compared. The term

$$\binom{N}{n}$$

is the number of combinations of n remains that can be drawn from a count of N remains and equals $N!/n!(N-n)!$ where $N!$ is N factorial or $N(N-1)(N-2), \ldots,$ 1; and

$$\left(\frac{N - N_i}{n} \right)$$

is the number of combinations of n remains that can be drawn randomly from the count of N remains without drawing any remains of taxon i.

The right-hand term of Eq. 4.1 is the probability that a count of n remains will not include taxon i. The expected number of taxa in a random sample of n remains is therefore the sum of probabilities that each taxon will be included in the rarefied sample. If $(N - N_i - n)$ is negative in Eq. 4.2, this term is set to zero.

The variance ($\text{var}(T_n)$) of the expected number of taxa in a random sample of n individuals, when n is large, can also be estimated (Heck et al. 1975), and 95% confidence intervals for $E(T_n)$ can be calculated as $E(T_n) \pm 1.96$ times the square root of var (T_n).

There are several underlying assumptions about using rarefaction analysis in palaeolimnology. These are presented and discussed in detail by Tipper (1979) and Birks and Line (1992). Although the interpretation of rarefaction estimates of palynological richness as records of past diversity is complex and currently unresolved (Odgaard 1999, 2001, 2007; Peros and Gajewski 2008; Meltsov et al. 2011), the interpretation of rarefaction estimates of diatom richness as a record of past biodiversity may appear more straightforward (but see Smol 1981). Palaeolimnological examples of studying the response of diatom richness to productivity or climate include Rusak et al. (2004) and Anderson et al. (1996), respectively (see also Laird et al. 2010).

Rarefaction analysis can be implemented in PSIMPOLL, available from http://chrono.qub.ac.uk/psimpoll/psimpoll.html/. In addition there are old user-unfriendly MS-DOS programs that run under Windows® (RAREPOLL, RAREFORM, RARECEP) developed by H. John B. Birks and John Line available from http://chrono.qub.ac.uk/inqua/boutique.htm.

Besides simple taxon richness, there are several other richness and evenness indices and a plethora of diversity indices. Diversity indices incorporate both species richness and evenness into a single value (Peet 1974). The units of these indices differ greatly, making comparisons difficult, confusing, and even impossible. The series of diversity numbers presented by Hill (1973) are the easiest to interpret ecologically.

Hill's (1973) family of diversity numbers are

$$\mathrm{NA} = \sum_{l=1}^{T} (p_i)^{1/(i-\mathrm{A})} \tag{4.3}$$

where p_i is the proportion of the individual belonging to taxon i, T is the total number of taxa, A may be any real number, and

$$\sum_{l=1}^{T} p_i = 1$$

NA is an intrinsic diversity number for A $\geq$ 0. Hill shows that the 0th, 1st, and 2nd order of these diversity numbers (i.e., A $=$ 0, 1, 2 in Eq. 4.3) are three widely used measures of diversity, namely

A $= 0$ N0 $= T$ where T is the total number of taxa
A $= 1$ N1 $= e^{H'}$ where H' is the Shannon index (Shannon and Weaver 1949)
A $= 2$ N2 $= 1/\lambda$ where λ is Simpson's (1949) index.

Simpson's (1949) index is

$$\lambda = \sum_{i=1}^{T} {p_i}^2 \tag{4.4}$$

where p_i is the proportional abundance of the ith taxon given by

$$p_i = \frac{n_i}{N} \quad i = 1, 2, 3, ..., T$$

where n_i is the number of individuals of the ith taxon and N is the known total number of individuals for all T taxa in the population. Equation 4.4 applies only to finite assemblages where all the fossils have been counted, i.e., $n = N$ where n is the total number of individuals in the sample and N is the total number of individuals in the underlying population. As all palaeolimnological work deals with samples with infinite underlying populations where it is impossible to count all members, Simpson (1949) developed an unbiased estimator ($\hat{\lambda}$) for sampling from an infinite population. It is

$$\hat{\lambda} = \sum_{l=1}^{T} \frac{n_i(n_l - 1)}{n(n-1)}$$

These diversity numbers, N0, N1, and N2, which are all in units of number of taxa, measure what Hill (1973) calls the effective number of taxa present in a sample (often also referred to as the effective number of occurrences). This effective number of taxa is a measure of the degree to which proportional abundances are distributed among the taxa. N0 is the number of **all** taxa in the sample regardless of their relative abundance; N2 is the number of **very abundant** taxa in the sample; and N1 is the number of **abundant** taxa in the sample. In other words, the effective number of taxa is a measure of the number of taxa in the sample where each taxon is weighted by its abundance. van Dam et al. (1981), van Dam (1982), and ter Braak (1983) pioneered the use of Hill's (1973) effective number of taxa to detect changes in diversity as a result of recent acidification from diatom assemblages. More recently, palaeolimnologists have begun to examine diversity in modern assemblages of diatoms or chironomids (Weckström and Korhola 2001; Nyman et al. 2005; Telford et al. 2006; Engels and Cwynar 2011) in relation to environmental gradients, population processes, or biogeographical factors. N2 is also used in weighted averaging (Juggins and Birks 2012: Chap. 14) and in canonical correspondence analysis to derive unbiased estimates of taxon tolerances (ter Braak and Verdonschot 1995) where N2 is now the effective number of occurrences of taxon k

$$\text{N2} = \left\{ \sum_{i=1}^{n} \left(\frac{y_{ik}}{y_{+k}} \right)^2 \right\}^{-1}$$

and n is the number of samples, y_{ik} is the abundance of taxon k in sample i, and y_{+k} is the abundance total across all samples for taxon k. N2 is also being used as an index of species evenness (e.g., Laird et al. 2010), even though Hill (1973) derives only one evenness measure (the ratio of N2 to N1) and Smith and Wilson (1996) do not consider N2 as an evenness index.

A potentially important development will be partitioning diversity within modern and fossil palaeolimnological data into independent alpha, beta, and gamma components of diversity (e.g., Veech et al. 2002; Crist and Veech 2006; Jost 2007; Pélissier and Couteron 2007).

Estimating Species Optima and Tolerances Using Palaeolimnological Data

Estimating the optima and the tolerances of different taxa (see Fig. 2.1 in Birks 2012a: Chap. 2) for selected environmental variables from modern calibration data-sets is often the first stage in quantitative reconstructions of past environments from biological assemblages (see Juggins and Birks 2012: Chap. 14). Optima and tolerances can be estimated in various ways (ter Braak 1996). Gaussian logit regression (GLR) (ter Braak and Looman 1986; Birks 1995, 2012a: Chap. 2; Juggins and Birks 2012: Chap. 14) provides unbiased estimates of species optima and tolerances because these estimates from GLR are not heavily influenced by the distribution of samples along the environmental gradient of interest (ter Braak 1996; cf. Coudon and Gégout 2006). Because GLR involves maximum-likelihood estimation (Birks 2012a: Chap. 2), occasionally a taxon's response cannot be fitted (e.g., it has a minimum rather than an optimum). Not all responses can be modelled completely because the environmental gradient is of finite length, in which case sigmoidal linear increasing or decreasing responses are modelled (ter Braak and van Dam 1989; Birks et al. 1990). Occasionally a taxon may have no statistically significant relationship to the environmental variable. A major advantage of GLR is that a palaeolimnologist can discover which taxa have statistically significant unimodal or sigmoidal responses, or no significant responses to the environmental variable of interest, and the estimates of optima and tolerances obtained are useful parameters of the taxon's realised niche. The fitting of a hierarchical set of response models from the most complex asymmetric unimodal response, through a symmetric unimodal response and sigmoidal linear response, to a null response (Huisman et al. 1993; Oksanen and Minchin 2002) is a useful way of analysing and categorising the multitude of taxon responses within taxon-rich modern calibration data-sets (Birks 2012a: Chap. 2). GLR-estimated optima and tolerances provide a strong basis for deriving 'indicator values' for taxa for particular environmental variables that can be used in monitoring and ecological assessment studies (e.g., Peeters and Gardeniers 2002; Yuan 2004, 2007a, b).

Comparison of Clusterings and Ordinations of Palaeolimnological Data

When palaeolimnologists have both modern biological assemblage and modern environmental data-sets, or several modern biological data-sets for the same lakes but based on different organisms (e.g., diatoms, chironomids), it can be useful to compare groupings of the lakes based on the different sets of variables (e.g., Birks et al. 2004). Legendre and Birks (2012a: Chap. 7) discuss ways of comparing different groupings.

Similarly, when ordinations have been performed on different data-sets (e.g., lake-water chemistry, diatom assemblages, chironomid assemblages) from the same set of lakes, it is useful to compare the different ordinations. The questions are how similar are the ordinations and hence are the major underlying gradients the same in all the data-sets? Procrustes analysis (Gower 1975; Digby and Kempton 1987; Gower and Dijksterhuis 2004) is the standard way of comparing the results of two different ordinations applied to the same set of objects. It holds one set of sample coordinates fixed and finds the best fit of the second set of points to this using rotations, re-scalings, and translations (lateral movements). The effectiveness of the rotation is assessed by the m^2 statistic which is the residual sum-of-squares after the Procrustes operation has been applied. The approach has been generalised (Gower 1975) to allow the comparison of three or more ordinations. The basic idea is to find a consensus or centroid configuration so that the fit of an ordinary Procrustes rotation to this centroid over all ordination configurations is optimal. The idea is to minimise m^2 where m^2 is Σm_i^2 and where m_i^2 is the Procrustes statistic for each pair-wise comparison of ordinations. The statistical significance of m^2 values can be assessed by means of PROTEST, a Procrustes randomisation test (Peres-Nato and Jackson 2001). Palaeolimnological examples include Chen et al. (2010), Wang et al. (2010), and Wischnewski et al. (2011).

The curious name is derived from an inn-keeper of Greek mythology whom ensured that all his customers fitted perfectly into his bed by stretching them or chopping off their feet.

Procrustes rotations and associated PROTEST tests can be implemented in the vegan package (http://cran.r-project.org/ and http://vegan.r-forge.r-project.org/) (Oksanen et al. 2011).

Data Interpretation

In this chapter, discussion of data interpretation is restricted to the interpretation of possible causative factors in determining the composition and abundance of biological assemblages in modern calibration data-sets and in testing specific hypotheses relating to modern calibration data-sets. In the introduction to Part III (Birks 2012b: Chap. 10) various approaches to the quantitative interpretation of stratigraphical data are outlined.

Canonical or constrained ordination techniques (RDA, CCA, and their partial relatives: Borcard et al. 2011; Legendre and Birks 2012b: Chap. 8) with associated Monte Carlo permutation tests (Lepš and Šmilauer 2003) are the most used (and useful) tools for identifying and statistically testing important explanatory environmental variables (see ter Braak and Verdonschot 1995). Various aids such as variance inflation factors and forward selection and associated permutation tests (ter Braak and Šmilauer 2002) help the investigator to derive a minimal adequate model (Birks 2012a: Chap. 2) with the smallest number of significant predictor variables that explains, in a statistical sense, the biological data about as well as the full set of explanatory predictor variables. Oksanen et al. (2011) have developed what they describe as experimental 'unfounded and untested' statistics that resemble deviance and the Akaike Information Criterion (Godínez-Domínguez and Freire 2003) used in the fitting of regression models to help in model building and selection in CCA and RDA. This approach has proved useful in developing minimal adequate models for modern palaeolimnological data-sets (unpublished results).

Variation partitioning analysis (Borcard et al. 2011; Legendre and Birks 2012b: Chap. 8) can help in assessing the relative importance of different sets of environmental or other explanatory variables in explaining the variation in modern biological assemblages. Specific hypotheses about, for example, the importance of lake-water pH or total phosphorus in determining modern biological assemblages can be tested using CCA or RDA and associated Monte Carlo permutation tests (ter Braak and Šmilauer 2002; Lepš and Šmilauer 2003; Borcard et al. 2011).

Conclusions

The chapters in this Part explore many of the techniques outlined above in detail. These techniques plus the additional methods summarised briefly in this introduction and overview provide palaeolimnologists with powerful and robust tools for assessing, summarising, analysing, and interpreting modern palaeolimnological calibration data-sets. Some of these techniques are equally useful in the numerical analysis of stratigraphical data, as discussed in Part III.

Acknowledgements I am indebted to John Smol for useful comments on this chapter and to Cathy Jenks for her invaluable help in its preparation. This is publication A347 from the Bjerknes Centre for Climate Research.

References

Alverson K, Eakin MC (2001) Making sure that the world's palaeodata do not get burned. Nature 412:269

Anderson NJ, Odgaard BV, Segerström U, Renberg I (1996) Climate-lake interactions recorded in varved sediments from a Scottish Boreal Forest Lake. Global Chang Biol 2:399–405

Anon (2001) Make the most of palaeodata. Nature 411:1

Battarbee RW, Morley D, Bennion H, Simpson GL, Hughes M, Bauere V (2011) A palaeolimnological meta-database for assessing the ecological status of lakes. J Paleolimnol 45:405–414

Birks HJB (1995) Quantitative palaeoenvironmental reconstructions. In: Maddy D, Brew S (eds) Statistical modelling of Quaternary science data. Quaternary Research Association, Cambridge, pp 161–254

Birks HJB (1998) Numerical tools in palaeolimnology – progress, potentialities, and problems. J Paleolimnol 20:307–332
Birks HJB (2007) Estimating the amount of compositional change in late-Quaternary pollen-stratigraphical data. Veg Hist Archaeobot 16:197–202
Birks HJB (2010) Numerical methods for the analysis of diatom assemblage data. In: Smol JP, Stoermer EF (eds) The diatoms: application for the environmental and earth sciences, 2nd edn. Cambridge University Press, Cambridge, pp 23–54
Birks HJB (2012a) Chapter 2: Overview of numerical methods in palaeolimnology. In: Birks HJB, Lotter AF, Juggins S, Smol JP (eds) Tracking environmental change using lake sediments. Volume 5: Data handling and numerical techniques. Springer, Dordrecht
Birks HJB (2012b) Chapter 10: Introduction and overview of part III. In: Birks HJB, Lotter AF, Juggins S, Smol JP (eds) Tracking environmental change using lake sediments. Volume 5: Data handling and numerical techniques. Springer, Dordrecht
Birks HJB (2012c) Chapter 11: Stratigraphical data analysis. In: Birks HJB, Lotter AF, Juggins S, Smol JP (eds) Tracking environmental change using lake sediments. Volume 5: Data handling and numerical techniques. Springer, Dordrecht
Birks HJB, Birks HH (2008) Biological responses to rapid climate change at the Younger Dryas-Holocene transition at Kråkenes, western Norway. Holocene 18:19–30
Birks HJB, Gordon AD (1985) Numerical methods in Quaternary pollen analysis. Academic Press, London
Birks HJB, Line JM (1992) The use of rarefaction analysis for estimating palynological richness from Quaternary pollen-analytical data. Holocene 2:1–10
Birks HJB, Line JM, Juggins S, Stevenson AC, ter Braak CJF (1990) Diatoms and pH reconstruction. Philos Trans R Soc Lond B 327:263–278
Birks HJB, Monteith DT, Rose NL, Jones VJ, Peglar SM (2004) Recent environmental change and atmospheric contamination on Svalbard as recorded in lake sediments – modern limnology, vegetation, and pollen deposition. J Paleolimnol 31:411–431
Birks HJB, Heiri O, Seppä H, Bjune AE (2010) Strengths and weaknesses of quantitative climate reconstructions based on late-Quaternary biological proxies. Open Ecol J 3:68–110
Bjerring R, Becares E, Declerck S, Gross EM, Hansson L-A, Kairesalo T, Nykänen M, Halkiewicz A, Kornijów R, Conde-Porcuna JM, Seferlis M, Nõges T, Moss B, Amsinck SL, Odgaard BV, Jeppesen E (2009) Subfossil Cladocera in relation to contemporary environmental variables in 54 pan-European lakes. Freshw Biol 54:2401–2417
Borcard D, Gillet F, Legendre P (2011) Numerical ecology with R. Springer, New York
Breiman L, Friedman JH, Olshen RA, Stone CJ (1984) Classification and regression trees. Chapman & Hall, New York
Catalan J, Barbieri MG, Bartumeus F, Bitusik P, Botev I, Brancelj A, Cogalniceau D, Manca M, Marchetto A, Ognjanova-Rumenova N, Pla S, Rieradevall M, Sorvari S, Stefkova E, Stuchlik E, Ventura M (2009) Ecological thresholds in European alpine lakes. Freshw Biol 54:2494–2517
Chambers JM, Cleveland WS, Kleiner B, Tukey PA (1983) Graphical methods for data analysis. Wadsworth, Monterey
Chen G, Dalton C, Taylor D (2010) Cladocera as indicators of trophic state in Irish lakes. J Paleolimnol 44:465–481
Cleveland WS (1993) Visualizing data. AT&T Bell Laboratories, Murray Hill
Cleveland WS (1994) The elements of graphing data. AT&T Bell Laboratories, Murray Hill
Coudon C, Gégout J-C (2006) The derivation of species response curves with Gaussian logistic regression is sensitive to sampling intensity and curve characteristics. Ecol Model 199: 164–175
Crist TO, Veech JA (2006) Additive partitioning of rarefaction curves and species-area relationships: unifying α-, β- and γ-diversity with sample size and habitat area. Ecol Lett 9:923–932
Cutler DR, Edwards TC Jr, Beard KH et al. (2007) Random forests for classification in ecology. Ecology 88:2783–2792
Davidson TA, Sayer CD, Langdon PG, Burgess A, Jackson M (2010a) Inferring past zooplanktivorous fish and macrophyte density in a shallow lake: application of a new regression tree model. Freshw Biol 55:584–599

Davidson TA, Sayer CD, Perrow M, Bramm M, Jeppesen E (2010b) The simultaneous inference of zooplanktivorous fish and macrophyte density from sub-fossil Cladoceran assemblages: a multivariate regression tree approach. Freshw Biol 55:546–564
De'ath G (1999) Principal curves: a new technique for indirect and direct gradient analysis. Ecology 80:2237–2253
De'ath G (2002) Multivariate regression trees: a new technique for modelling species-environment relationships. Ecology 83:1105–1117
De'ath G (2007) Boosted trees for ecological modeling and prediction. Ecology 88:243–251
De'ath G, Fabricius KE (2000) Classification and regression trees: a powerful and yet simple technique for the analysis of complex ecological data. Ecology 81:3178–3192
D'heygere T, Goethals PLM, De Pauw N (2003) Use of genetic algorithms to select input variables in decision tree models for the prediction of benthic macroinvertebrates. Ecol Model 160: 291–300
Digby PGN, Kempton RA (1987) Multivariate analysis of ecological communities. Chapman & Hall, London
Dittert N, Dippenbroek M, Grobe H (2001) Scientific data must be made available to all. Nature 414:393
Efron B, Tibshirani R (1991) Statistical data analysis in the computer age. Science 253: 390–395
Engels S, Cwynar LC (2011) Changes in fossil Chironomid remains along a depth gradient: evidence for common faunal thresholds within lakes. Hydrobiologia 665:15–38
Everitt B (1978) Graphical techniques for multivariate data. Heinemann, London
Everitt B (2005) An R and S-PLUS® companion to multivariate analysis. Springer, London
Everitt BS, Dunn G (2001) Applied multivariate data analysis, 2nd edn. Arnold, London
Everitt BS, Hothorn T (2011) An introduction to applied multivariate analysis with R. Springer, New York
Fielding AH (2007) Cluster and classification techniques for the bioscences. Cambridge University Press, Cambridge
Fox J (2002) An R and S-PLUS® companion to applied regression. Sage, Thousand Oaks
Fox J (2008) Applied regression analysis and generalized linear models. Sage, Thousand Oaks
Godínez-Domínguez E, Freire J (2003) Information-theoretic approach for selection of spatial and temporal models of community organization. Mar Ecol Prog Ser 253:17–24
Gower JC (1975) Generalised Procrustes analysis. Psychometrika 40:33–51
Gower JC, Dijksterhuis GB (2004) Procrustes problems. Oxford University Press, Oxford
Gregory-Eaves I, Beisner BE (2011) Palaeolimnological insights for biodiversity science: an emerging field. Freshw Biol 56:2653–2661
Grimm EC, Keltner J, Cheddadi R, Hicks S, Lézineé A-M, Berrio JC, Wiliams JW (2007) Databases and their application. In: Elias SA (ed) Encyclopedia of Quaternary science. Volume 3. Elsevier, Oxford, pp 2521–2528
Hastie T, Tibshirani R, Friedman J (2011) The elements of statistical learning, 2nd edn. Springer, New York
Heck KL, van Belle G, Simberloff D (1975) Explicit calculation of the rarefaction diversity measurement and the determination of sufficient sample size. Ecology 56:1459–1461
Hernández MJ (2003) Database design for mere mortals. Addison Wesley, Boston
Hewitt CN (ed) (1992) Methods of environmental data analysis. Elsevier, London
Hill MO (1973) Diversity and evenness: a unifying notation and its consequences. Ecology 54: 427–432
Hill MO, Gauch HG (1980) Detrended correspondence analysis – an improved ordination technique. Vegetatio 42:47–58
Hobbs WO, Telford RJ, Birks HJB, Saros JE, Hazewinkel RRO, Perren BB, Saulnier-Talbot É, Wolfe AP (2010) Quantifying recent ecological changes in remote lakes of North America and Greenland using sediment diatom assemblages. PLoS One 5:e10026
Huisman J, Olff H, Fresco LFM (1993) A hierarchical set of models for species response analysis. J Veg Sci 4:37–46

Jones MB, Berkley C, Bojilova J, Schildhauer M (2001) Managing scientific metadata. IEEE Internet Comput 5:59–68

Jost L (2007) Partitioning diversity into independent alpha and beta components. Ecology 88:2427–2439

Juggins S (1996) The PALICLAS database. Mem Ist Ital Idrobiol 55:321–328

Juggins S, Birks HJB (2012) Chapter 14: Quantitative environmental reconstructions from biological data. In: Birks HJB, Lotter AF, Juggins S, Smol JP (eds) Tracking environmental change using lake sediments. Volume 5: Data handling and numerical techniques. Springer, Dordrecht

Juggins S, Telford RJ (2012) Chapter 5: Exploratory data analysis and data display. In: Birks HJB, Lotter AF, Juggins S, Smol JP (eds) Tracking environmental change using lake sediments. Volume 5: Data handling and numerical techniques. Springer, Dordrecht

Kohonen T (2001) Self-organising maps, 3rd edn. Springer, Berlin

Laird KR, Kingsbury MV, Cumming BF (2010) Diatom habitats, species diversity and water-depth inference models across surface-sediment transects in Worth Lake, northwest Ontario, Canada. J Paleolimnol 44:1009–1024

Last WM, Smol JP (eds) (2001a) Tracking environmental change using lake sediments. Volume 1: Basin analysis, coring, and chronological techniques. Kluwer Academic Publishers, Dordrecht

Last WM, Smol JP (eds) (2001b) Tracking environmental change using lake sediments. Volume 2: Physical and geochemical methods. Kluwer Academic Publishers, Dordrecht

Legendre P, Birks HJB (2012a) Chapter 7: Clustering and partitioning. In: Birks HJB, Lotter AF, Juggins S, Smol JP (eds) Tracking environmental change using lake sediments. Volume 5: Data handling and numerical techniques. Springer, Dordrecht

Legendre P, Birks HJB (2012b) Chapter 8: From classical to canonical ordination. In: Birks HJB, Lotter AF, Juggins S, Smol JP (eds) Tracking environmental change using lake sediments. Volume 5: Data handling and numerical techniques. Springer, Dordrecht

Lek S, Guégan J-F (eds) (2000) Artificial neuronal networks. Application to ecology and evolution. Springer, Berlin

Lepš J, Šmilauer P (2003) Multivariate analysis of ecological data using CANOCO. Cambridge University Press, Cambridge

Maher LJ, Heiri O, Lotter AF (2012) Chapter 6: Assessment of uncertainties associated with palaeolimnological laboratory methods and microfossil analysis. In: Birks HJB, Lotter AF, Juggins S, Smol JP (eds) Tracking environmental change using lake sediments. Volume 5: Data handling and numerical techniques. Springer, Dordrecht

McPhillips T, Bowers S, Zinn D, Ludäscher B (2009) Scientific workflow design for mere mortals. Futur Gen Comp Syst 25:541–551

Meltsov V, Poska A, Odgaard BV, Sammul M, Kull T (2011) Palynological richness and pollen sample evenness in relation to local floristic diversity in southern Estonia. Rev Palaeobot Palynol 166:344–351

Michener WK, Brunt JW (eds) (2000) Ecological data. Design, management and processing. Blackwell, Oxford

Michener WK, Brunt JW, Helly JJ, Kirchner TB, Stafford SG (1997) Non-geospatial meta-data for the ecological sciences. Ecol Appl 71:330–342

Mitchell FJG (2011) Exploring vegetation in the fourth dimension. Trends Ecol Evol 26: 45–52

Moe SJ, Dudley B, Ptacnik R (2008) REBECCA databases: experiences from compilation and analyses of monitoring data from 5,000 lakes in 20 European countries. Aquat Ecol 42: 183–201

Nyman M, Korhola A, Brooks SJ (2005) The distribution and diversity of Chironomidae (Insecta: Diptera) in western Finnish Lapland, with special emphasis on shallow lakes. Global Ecol Biogeogr 14:137–153

O'Beirne P (2005) Spreadsheet check and control: 47 key practices to detect and prevent errors. Systems Publishing, Wexford

Odgaard BV (1999) Fossil pollen as a record of past biodiversity. J Biogeogr 26:7–17

Odgaard BV (2001) Palaeoecological perspectives on pattern and process in plant diversity and distribution adjustments: a comment on recent developments. Divers Distrib 7:197–201
Odgaard BV (2007) Reconstructing past biodiversity development. In: Elias SA (ed) Encyclopedia of Quaternary science. Volume 3. Elsevier, Oxford, pp 2508–2514
Oksanen J, Minchin PR (2002) Continuum theory revisited: what shape are species responses along ecological gradients? Ecol Model 157:119–129
Oksanen J, Blanchet FG, Kindt R, Legendre P, O'Hara RB, Simpson GL, Solymos P, Stevens MHM, Wagner H (2011) vegan: community ecology package. R package version 1.17-8. http://CRAN.R-project.org/package=vegan
Olden JD, Jackson DA (2002) A comparison of statistical approaches for modelling fish species distributions. Freshw Biol 47:1976–1995
Olden JD, Lawler JJ, Poff NL (2008) Machine learning methods without tears: a primer for ecologists. Q Rev Biol 83:171–193
Peet RK (1974) The measurement of species diversity. Annu Rev Ecol Syst 5:285–307
Peeters ETHM, Gardeniers JJP (2002) Logistic regression as a tool for defining habitat requirements of two common gammarids. Freshw Biol 39:605–615
Pelánkova B, Kuneš P, Chytrý M, Jankovská V, Ermakov N, Svobodová-Svitavská H (2008) The relationships of modern pollen spectra to vegetation and climate along a steppe-forest-tundra transition in southern Siberia, explored by decision trees. Holocene 18:1259–1271
Pélissier R, Couteron P (2007) An operational, additive framework for species diversity partitioning and beta-diversity analysis. J Ecol 95:294–300
Peres-Nato PR, Jackson DA (2001) How well do multivariate data sets match? The robustness and flexibility of a Procrustean superimposition approach over the Mantel test. Oecologia 129: 169–178
Peros MC, Gajewski K (2008) Testing the reliability of pollen-based diversity estimates. J Paleolimnol 40:357–368
Prasad AM, Iverson LR, Liaw A (2006) Newer classification and regression tree techniques: bagging and random forests for ecological prediction. Ecosystems 9:181–199
Racca JMJ, Philibert A, Racca R, Prairie YT (2001) A comparison between diatom-based pH inference models using artificial neural networks (ANN), weighted averaging (WA) and weighted averaging partial least squares (WA-PLS) regressions. J Paleolimnol 26:411–422
Racca JMJ, Wild M, Birks HJB, Prairie YT (2003) Separating wheat from chaff: diatom taxon selection using an artificial neural network pruning algorithm. J Paleolimnol 29:123–133
Raymond B, Watts DJ, Burton H, Bonnice J (2005) Data mining and scientific data. Arct Antarct Alp Res 37:348–357
Rose NL, Morley D, Appleby PG, Battarbee RW, Alliksaar T, Guilizzoni P, Jeppesen E, Korhola A, Punning J-M (2011) Sediment accumulation rates in European lakes since AD 1850: trends, reference conditions and exceedence. J Paleolimnol 45:447–468
Rühland K, Paterson AM, Smol JP (2008) Hemispheric-scale patterns of climate-induced shifts in planktonic diatoms from North American and European lakes. Global Chang Biol 14: 2740–2754
Rusak JA, Leavitt PR, McGowan S et al. (2004) Millennial-scale relationships of diatom species richness and production in two prairie lakes. Limnol Oceanogr 49:1290–1299
Shannon CE, Weaver W (1949) The mathematical theory of communication. University of Illinois Press, Urbana
Simpson EH (1949) Measurement of diversity. Nature 163:686
Simpson GL, Birks HJB (2012) Chapter 9: Statistical learning in palaeolimnology. In: Birks HJB, Lotter AF, Juggins S, Smol JP (eds) Tracking environmental change using lake sediments. Volume 5: Data handling and numerical techniques. Springer, Dordrecht
Smith B, Wilson JB (1996) A consumer's guide to evenness indices. Oikos 76:70–82
Smol JP (1981) Problems associated with the use of "species diversity" in paleolimnological studies. Quaternary Res 15:209–212
Smol JP (2008) Pollution of lakes and rivers: a paleoenvironmental perspective, 2nd edn. Blackwell, Oxford

Smol JP, Birks HJB, Last WM (eds) (2001a) Tracking environmental change using lake sediments. Volume 3: Terrestrial, algal, and siliceous indicators. Kluwer Academic Publishers, Dordrecht

Smol JP, Birks HJB, Last WM (eds) (2001b) Tracking environmental change using lake sediments. Volume 4: Zoological indicators. Kluwer Academic Publishers, Dordrecht

Smol JP, Wolfe AP, Birks HJB et al. (2005) Climate-driven regime shifts in the biological communities of Arctic lakes. Proc Natl Acad Sci USA 102:4397–4402

Smol JP, Birks HJB, Lotter AF, Juggins S (2012) Chapter 1: The march towards the quantitative analysis of palaeolimnological data. In: Birks HJB, Lotter AF, Juggins S, Smol JP (eds) Tracking environmental change using lake sediments. Volume 5: Data handling and numerical techniques. Springer, Dordrecht

Stomp M, Huisman J, Mittelbach GG, Litchman E, Klausmeir CA (2011) Large-scale biodiversity patterns in freshwater phytoplankton. Ecology 92:2096–2107

Telford RJ, Vandvik V, Birks HJB (2006) Dispersal limitations matter for microbial morphospecies. Science 312:1015

ter Braak CJF (1983) Principal components biplots and alpha and beta diversity. Ecology 64: 454–462

ter Braak CJF (1986) Canonical correspondence analysis: a new eigenvector technique for multivariate direct gradient analysis. Ecology 67:1167–1179

ter Braak CJF (1996) Unimodal models to relate species to environment. DLO-Agricultural Maths Group, Wageningen

ter Braak CJF, Juggins S (1993) Weighted averaging partial least squares regression (WA-PLS): an improved method for reconstructing environmental variables from species assemblages. Hydrobiologia 269(270):485–502

ter Braak CJF, Looman CWN (1986) Weighted averaging, logit regression and the Gaussian response model. Vegetatio 65:3–11

ter Braak CJF, Prentice IC (1988) A theory of gradient analysis. Adv Ecol Res 18:271–317

ter Braak CJF, Šmilauer P (2002) CANOCO reference manual and CanoDraw for Windows user's guide: software for canonical community ordination (version 4.5). Microcomputer Power, Ithaca

ter Braak CJF, van Dam H (1989) Inferring pH from diatoms – a comparison of old and new calibration methods. Hydrobiologia 178:209–223

ter Braak CJF, Verdonschot PFM (1995) Canonical correspondence analysis and related multivariate methods in aquatic ecology. Aquat Sci 57:255–289

Tipper JC (1979) Rarefaction and rarefiction – the use and abuse of a method in paleoecology. Paleobiol 5:423–434

Tufte ER (1983) The visual display of quantitative information. Graphics Press, Cheshire

van Dam H (1982) On the use of measures of structure and diversity in applied diatom ecology. Beiheft 73 zur Nova Hedwigia 97–115

van Dam H, Suurmond G, ter Braak CJF (1981) Impact of acidification on diatoms and chemistry of Dutch moorland pools. Hydrobiologia 83:425–459

Veech JA, Summerville KS, Crist TO, Gering JC (2002) The additive partitioning of species diversity: recent revival of an old idea. Oikos 99:3–9

Wang Y, Liu X, Herzschuh U (2010) Asynchronous evolution of the Indian and East Asian Summer Monsoon indicated by Holocene moisture patterns in monsoonal central Asia. Earth Sci Rev 103:135–153

Warton DI (2008) Raw data graphing: an informative but under-utilized tool for the analysis of multivariate abundances. Austral Ecol 33:290–300

Weckström J, Korhola A (2001) Patterns in the distribution, composition, and diversity of diatom assemblages in relation to ecoclimatic factors in Arctic Lapland. J Biogeogr 28:31–45

Weller AF, Harris AJ, Ware JA (2006) Artificial neural networks as potential classification tools for dinoflagellate cyst images: a case using the self-organising map clustering algorithm. Rev Palaeobot Palynol 141:278–302

Whitehorn M, Marklyn B (2001) Inside relational databases. Springer, London

Wischnewski J, Mischke S, Wang Y, Herzschuh U (2011) Reconstructing climate variability on the northeastern Tibetan Plateau since the last Lateglacial – a multi-proxy, dual-site approach comparing terrestrial and aquatic signals. Quaternary Sci Rev 30:82–97

Yuan L (2004) Assigning macroinvertebrate tolerance classifications using generalised additive models. Freshw Biol 49:662–677

Yuan L (2007a) Using biological assemblage composition to infer the values of covarying environmental factors. Freshw Biol 52:1159–1175

Yuan L (2007b) Maximum likelihood method for predicting environmental conditions from assemblage composition: the R package bio.infer. J Stat Softw 22:1–20

Zuur AF, Ieno EN, Smith GM (2007) Analysing ecological data. Springer, New York

Chapter 5
Exploratory Data Analysis and Data Display

Steve Juggins and Richard J. Telford

Abstract Exploratory data analysis (EDA) is an essential first step in numerical or statistical analysis of palaeolimnological data. The main functions of EDA are exploration, analysis and model diagnosis, and presentation and communication. The main tools of EDA are graphical tools such as histograms, box-plots, scatter-plots, pie-charts, smoothers, co-plots, and scatter-plot matrices. EDA also considers questions of data transformation, outlier detection and treatment, and missing values. Quality of graphical presentation is also discussed and appropriate computer software is outlined. Careful and creative graphical EDA can be a great aid in data-exploration and in hypothesis-generation.

Keywords Bivariate data • Categorical data • Computer software • Data distribution • Graph drawing • Graphical display • Graphical tools • Hypothesis generation • LOESS smoother • Missing values • Model diagnostics • Multivariate data • Outliers • Time-series • Transformations • Univariate data

Introduction

Familiar statistical tests, such as the *t*-test, test a null hypothesis by taking some data, making assumptions about the properties of the data, and typically returning one number with an associated probability value. This is the so-called classical

S. Juggins (✉)
School of Geography, Politics & Sociology, Newcastle University,
Newcastle-upon-Tyne, NE1 7RU, UK
e-mail: stephen.juggins@ncl.ac.uk

R.J. Telford
Department of Biology and Bjerknes Centre for Climate Research, University of Bergen,
PO Box 7803, Bergen N-5020, Norway
e-mail: richard.telford@bio.uib.no

H.J.B. Birks et al. (eds.), *Tracking Environmental Change Using Lake Sediments*,
Developments in Paleoenvironmental Research 5, DOI 10.1007/978-94-007-2745-8_5,

approach to statistical inference. Exploratory data analysis (EDA) has a different philosophical approach, being concerned with hypothesis generation rather than with hypothesis testing. It is an essential first stage of any data-analytical or statistical work, even when there is already a hypothesis to test. Tukey (1977) advocated its widespread use, and since this work, increases in computer power and availability have permitted the development of many new techniques.

The primary goal of EDA is to maximise insight into a data-set and its underlying structure (NIST/SEMATECH 2006). It is based predominantly on graphical tools, relying on the brain's pattern-recognition ability to identify trends, relationships, and unusual features in the data. Specifically, EDA and graphical data display have three main functions in data analysis (Snee and Pfeifer 1983; Quinn and Keough 2002): (1) ***exploration***, including checking the data for unusual values (outliers), identifying relationships, assessing the need for transformation prior to other analyses, and suggesting the form and type of model to fit, (2) ***analysis and model diagnosis***, including checking that the data meet the assumptions of formal statistical tests and that the chosen model is a realistic fit to the data, and (3) ***presentation and communication of results*** using graphics to display complex numerical data.

Palaeolimnological data are often complex: they can contain many variables (i.e., highly multivariate) of mixed data types (e.g., continuous, counts, categorical, percentages, presence-absence), that frequently follow non-normal distributions, exhibit non-linear relationships between variables, and often contain missing values and outliers (see Birks 2012a: Chap. 2). Many of the techniques considered elsewhere in this volume can be considered as exploratory tools (e.g., clustering and partitioning (Legendre and Birks 2012a: Chap. 7), classical ordination (Legendre and Birks 2012b: Chap. 8), and constrained partitioning (Birks 2012b: Chap. 11)). Here we focus on some of the simpler graphical tools for exploring these properties, looking first at methods for visualising individual samples and the distribution of single variables, and then techniques for exploring relationships between variables. We illustrate these methods using limnological survey data containing physical, locational, and annual mean chemical data for 124 lakes in the UK (Bennion et al. 1996). The data-set was originally constructed by merging several smaller regional data-sets and combined with diatom data to develop a diatom-based calibration function for reconstructing epilimnetic phosphorus and contains a mix of continuous, categorical, and ordinal variables.

Exploring Univariate Distributions

Graphical Tools

A fundamental property of our sample data, and, by inference, of the population from which it came, is the shape of the distribution. For continuous variables a histogram provides a convenient graphical display of tabulated frequencies and

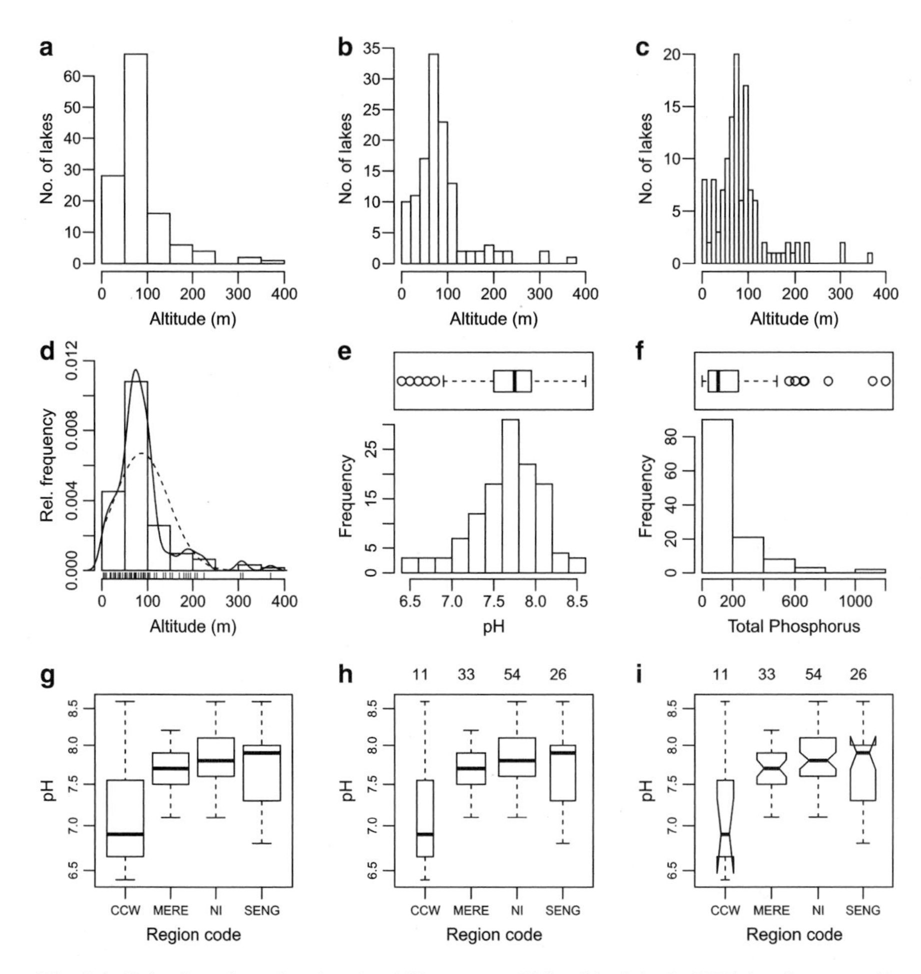

Fig. 5.1 Univariate plots, showing, (**a–c**) histograms of lake altitude in the UK lakes data-set, with the number of bins chosen by (**a**) Sturges' formula ($\log_2$ (n) + 1), the default in most packages, (**b**) Freedman and Diaconis's rule based on the inter-quartile range, and (**c**) manual choice of 30 bins (see Freedman and Diaconis 1981 for further discussion); (**d**) histogram of altitude with superimposed normal (*dotted*) and smoothed kernel-density distributions (*solid*) and rugs; (**e** and **f**) histograms and box-plots of pH and total phosphorus; (**g–i**) box-plots of pH by geographic region, with box width varied according to number of observation in each group (**h**) and notches added around the 95% confidence interval of the median (**i**)

allows an easy assessment of the distribution of a variable: the location, dispersion, and shape including possible bimodality. The number of classes or bins used in a histogram needs some consideration, especially for more complex distributions. If the bins are too few and broad, relevant details are obscured; conversely, too numerous and narrow bins start to capture random fluctuations in the data. Various rules for selecting the number of bins have been proposed and implemented (Fig. 5.1a–c).

Further insight into the shape of the distribution may be gained by superimposing a probability density function on the histogram (Silverman 1986). This may be derived from a formal distribution, for example a normal distribution based on the sample mean and variance, or determined by the data itself via non-parametric estimation. As with the choice of histogram bins, the choice of window width, or smoothing parameter, is important: too narrow and it will produce numerous artificial modes, too wide and it will miss important features (Fig. 5.1d). For lake altitude, the normal probability density curve is clearly a poor fit to the underlying data: in this case the non-parametric smoothing curve is a more faithful representation. A final enhancement to the basic histogram is to add a rug-plot to the x-axis in which ticks are drawn to represent the original data points. For large data-sets the rugs may be jittered to avoid overlap (see below).

Box-plots are an alternative and efficient way to examine the overall data distribution (Fig. 5.1e, f). The central box encloses the inter-quartile range, with the median marked by a horizontal line. Whiskers extend out to the extremes of the data, or 1.5 times the inter-quartile range, whichever is shorter. Data beyond this, which may be outliers, are marked individually by points. If the data distribution is symmetrical about the median, the median will be in the centre of the box, and the whiskers about the same length. Because box-plots are based on robust measures of central tendency and dispersion, they are resistant to extreme values. The box-plot for pH (Fig. 5.1e) reflects the essentially symmetrical nature of distribution with just a few unusually low values. The corresponding plot for total phosphorus clearly reflects the skewness observed in the histogram (Fig. 5.1f).

Box-plots are particularly useful for displaying a comparison of a single variable under different values of a categorical variable (Fig. 5.1g). A range of options can be used to enhance box-plots, for example the width of each box can be related to the size in each class and the number of observations in each class added to the plot (Fig. 5.1h), and the 95% confidence interval of the median can be represented by notches (Fig. 5.1g): if the notches do not overlap there is strong evidence that the medians are different.

Data Transformation

Most parametric statistical tests assume that the data have a normal distribution, and can give misleading or even invalid results if this assumption is violated. Unfortunately, palaeolimnological data often have a non-normal distribution, revealed using the methods described above and even if parametric testing is not the aim, highly skewed, non-normal distributions will often distort further exploratory analyses, and mask underlying patterns and trends. In such cases it may be necessary to transform the data to approximate a normal distribution. The choice of transformation is often difficult and may depend on the subsequent analysis. If regression is the aim and non-normal data are the response variables (e.g., species abundances), the data can be analysed without transformation using methods appropriate for the distribution, for example generalised linear models can be applied to species count data (which are Poisson distributed) (see Birks 2012a: Chap. 2). However, if the data are explanatory variables or the aim is further exploratory analysis using multivariate methods (Legendre and Birks 2012a, b: Chaps. 7 and 8), then some form of transformation will usually be required.

Log-normal distributions are common in palaeolimnology, for example in geochemical and biological analyses (Limpert et al. 2001). These highly right-skewed distributions can be normalised by taking logs. The log of zero is undefined so where there are zeros in the data use log(x + a), where a is an appropriate constant: 1.0 for count data or half the smallest value or half the detection limit for geochemical data. Less skewed data can be normalised with a square-root transformation. Cube roots and fourth roots are increasingly effective for biological count data containing many zeros and a few large values (Quinn and Keough 2002; Legendre and Birks 2012b: Chap. 8). Percentage and proportional data do not have a normal distribution but can be normalised with an arcsine transformation (but see Warton and Hui 2011 for a discussion of the limitations of this transformation).

Many of the above transformations are part of the Box-Cox family of transformations:

$$\frac{Y^{\lambda}-1}{\lambda} \quad when \;\; \lambda \neq 0 \quad \text{and} \quad \log(Y) \;\; when \;\; \lambda = 0$$

When $\lambda = 1$ we have no change to the distribution, $\lambda = 0.5$ gives a square-root transformation, $\lambda = 1.0$ gives a log transformation, and $\lambda = -1.0$ gives a reciprocal transformation. An automated procedure based on the Box-Cox family of transformations can be used to find the optimal transformation to normality by choosing fractional values of λ (Sokal and Rohlf 1995).

Figure 5.2 shows histograms and normal-probability plots for raw and variously transformed total phosphorus data from the UK lakes data-set. These plot ordered data against the corresponding quantiles of a normal distribution and provide a convenient graphical assessment of normality. If the sample has a normal distribution the points should fall on a straight line through the first and third quartile. Kinks and other departures from a straight line indicate skewness and/or multimodality, with long tails of the distribution shown by the relationship becoming steeper towards the ends, and short tails shown by the converse. The right-hand plots show the optimal Box-Cox transformation using $\lambda = 0.042$. This value is very close to 0 (i.e., log transformation) so in this example we would choose the latter for simplicity. The graphical methods shown in Fig. 5.2 are usually sufficient to suggest the most appropriate transformation: the optimal Box-Cox transformation may suggest fractional powers that have no direct environmental meaning, although it can help to confirm choice or clarify the transformation for difficult data. Different transformations can be chosen for each variable in a multivariate data-set, although unless there is strong evidence otherwise, we suggest using the same transformation for groups of variables derived from similar processes (e.g., groups of nutrient variables, or groups of anion and cation variables). This is especially important if the data are used as predictors in forward-selection regression-type analyses. Finally, data transformations discussed here should not be confused with data

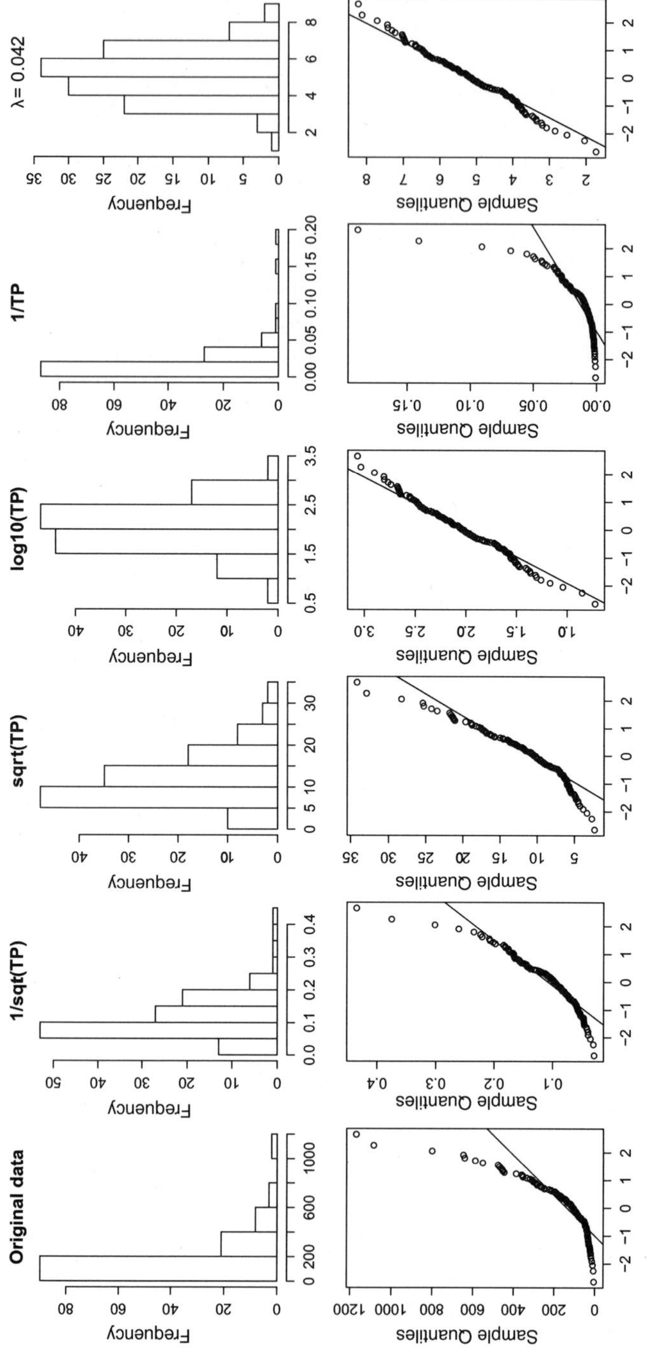

Fig. 5.2 Histograms and associated normal probability plots of total phosphorus under various data-transformations (see text for details)

standardisation to unit variance and zero mean, employed, for example, prior to a principal component analysis of environmental data (Legendre and Birks 2012b: Chap. 8). Standardisation changes the location and dispersion of the data, but not the shape of its distribution.

Graphical Techniques for Categorical (Nominal and Ordinal) Data

Categorical data contain observations that may be sorted into one of (usually) a small number of groups. For example, in the UK lakes data-set, region is a categorical variable, indicating which of the four regional data-sets a lake belongs to. Ordinal data are categorical but with an implied ordering between the categories. For example, in the UK lakes data-set, precise values of maximum depth are not available but each lake is classified into one of four depth classes.

Pie-charts are a familiar and popular way to present categorical data. However, they may not be optimal as it is difficult to judge differences in angles, and information about the sample size is lost (Fig. 5.3a). Pie-charts may be suitable if there are few categories, with values of a similar magnitude, and if the emphasis is on representing proportional, rather than absolute differences between categories. Pie-charts may be particularly useful in representing geographical differences in a categorical variable, especially if the diameter of the pies is scaled proportional to the count (Fig. 5.3b). Where information about the absolute differences between categories is to be displayed, bar-charts are generally more appropriate (Fig. 5.3c). Cross-classified data may be displayed using multiple bar-charts. Figure 5.3d shows the distribution of lake-depth classes by geographic region. Adjacent bars help visualise between-group differences in maximum depth, while stacked bars (not shown) make it easier to visualise differences in the total number of observations per category. Figure 5.3e shows the depth and geographical data displayed using a dot-plot (after Cleveland 1994): this contains the same information as Fig. 5.3d but is quite different visually.

Exploring Bivariate Relationships

Bivariate statistics are used when two variables have been measured on the same observations, and to quantify the relationship between these variables. The covariance is a numerical estimator of this relationship, calculated as the sum of the products of each centred variable divided by $n-1$, where n is the number of observations. Often the covariance is standardised by the variance of each variable to give Pearson's product-moment correlation coefficient (r). This unitless statistic varies between -1 and $+1$, for a perfect negative and positive relationship, respectively. Pearson's correlation coefficient is sensitive to outliers and non-linearity, and can be replaced,

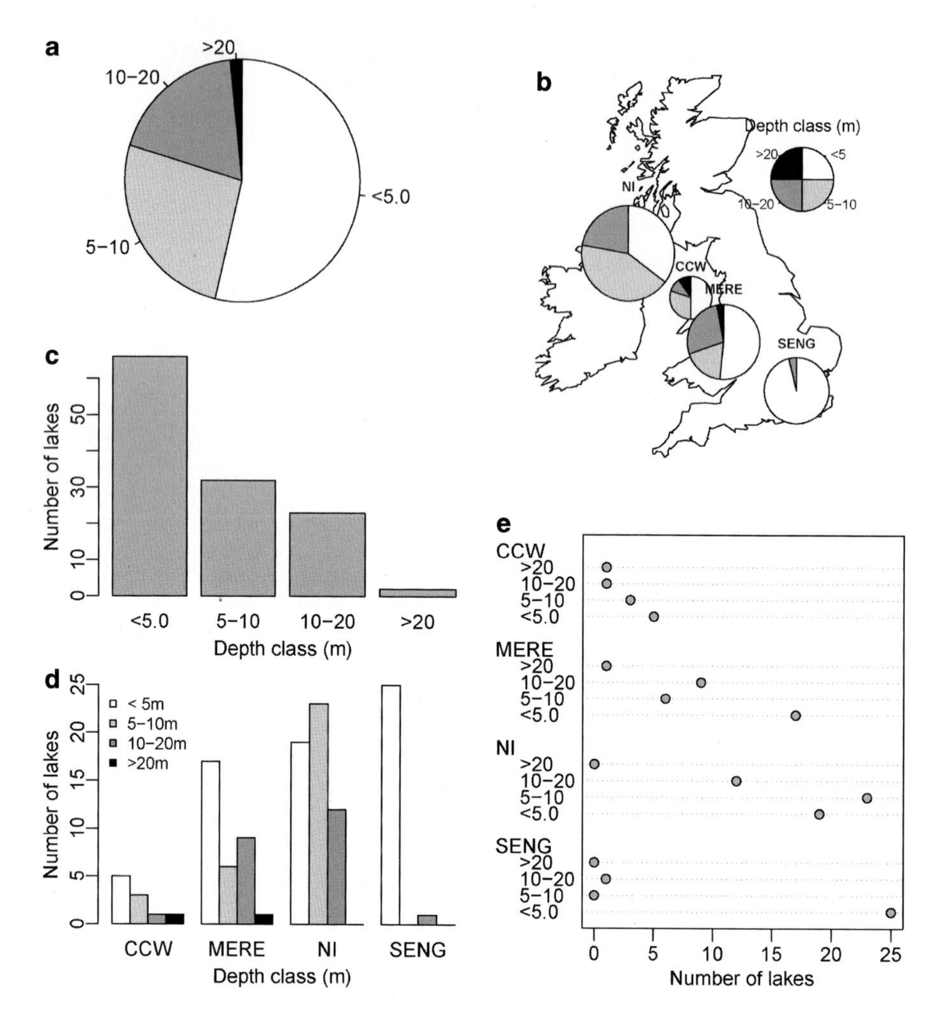

Fig. 5.3 Descriptive graphics for categorical and ordinal data, showing (**a** and **b**) pie-chart of maximum depth categories, (**b**) pie-charts showing maximum depth categories by region, (**c**) bar-chart of maximum depth categories, (**d**) bar-chart of maximum depth categories by region, and (**e**) dot-chart of maximum depth categories by region for the UK lakes data-set

with some loss of efficiency, with the non-parametric Kendall's rank coefficient tau or Spearman's rank coefficient rho. None of these statistics can identify curvilinear relationships, so it is vital to plot and examine the data first.

Scatter-plots are the main graphical method for exploring relationships between two variables (Tufte 1990, 2001). The independent variable is plotted on the x-axis or abscissa, and the dependent variable plotted on the y-axis or ordinate. Often there is no functional dependency between the two variables, and either can be plotted on the x-axis. If the data points are ordered by space or time they can be joined by lines. Figure 5.4a shows the relationships between chlorophyll-*a* and total phosphorus for

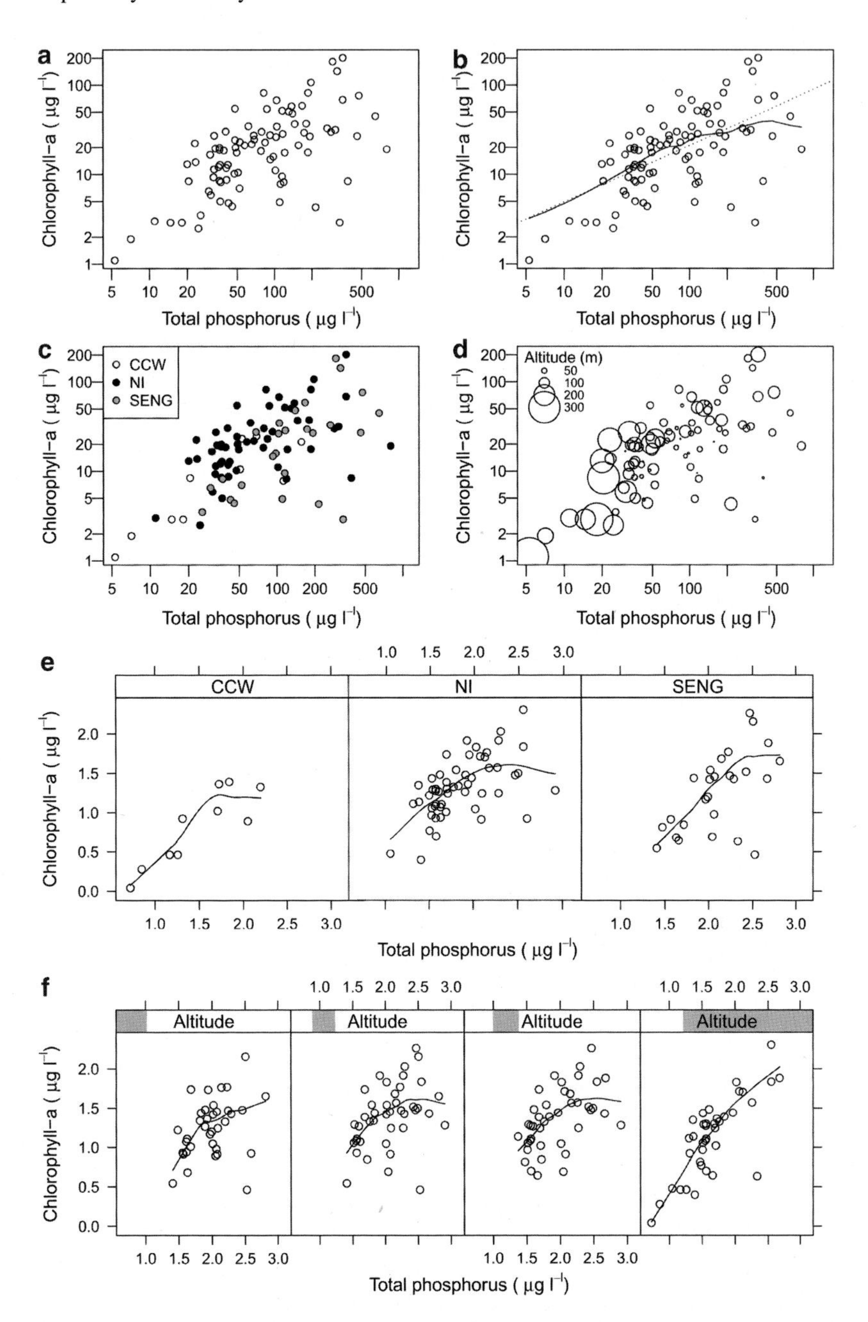

Fig. 5.4 Scatter-plot variants showing (**a**) relationship between chlorophyll-*a* and total phosphorus, (**b**) as (**a**) with linear regression (*dotted*) and LOESS smoother (*solid*) lines added, (**c**) as (**a**) with points coded by region, (**d**) as (**a**) with symbol size varied by lake altitude, (**e**) co-plot showing the relationship between chlorophyll-*a* and total phosphorus by region, and (**f**) as (**e**) co-plot showing the relationship between chlorophyll-*a* and total phosphorus by altitude (Data from the UK lakes data-set)

the UK lakes data-set. Both variables are plotted on log scales: initial inspection using histograms, box-plots, and normal probability plots show both variables to be strongly right-skewed. A log-transformation brings them to approximate normality and linearises the relationship between the two variables.

If the data are integers, or are heavily rounded, the same position on the graph may host multiple data points. This can obscure the relationship between the two variables. This can be solved by adding a small amount of random noise, known as jitter, to each point so they no longer overlap. Hexagonal binning is useful for very large data-sets where overplotting obscures patterns (Carr et al. 1987).

Sometimes it is useful to enhance scatter-plots by highlighting the general trends in the data. LOESS (locally weighted scatter-plot smoother: Cleveland 1979) is a non-parametric regression method that can be used to model the relationship between two variables, without having to specify the form of a global function (Birks 2012a: Chap. 2). At each point in the data-set, LOESS fits a low-order polynomial (typically linear or quadratic) to data in the neighbourhood, weighted by their distance from the point. The size of the neighbourhood is determined by setting the span. Large spans produce smooth curves, but may miss important features; small spans start to capture noise. Figure 5.4b shows a LOESS smoother added to the chlorophyll-*a*/total phosphorus plot, using the default span of 0.67, along with the fit from a least-squares linear regression. Notice how the smoother tracks the regression line for most of the data span but departs above 200 $\mu g\ L^{-1}$ total phosphorus, suggesting that total phosphorus ceases to become limiting above this value.

Information about a third variable influencing the relationship between the two variables of interest can be visualised in a number of ways. If the variable is categorical and there are not too many categories, the data points can be coded to represent the class of the third variable (Fig. 5.4c). Colour-coded plots are easiest to decode visually, but if colour is not available for publication, different symbols can be used, with categories coded using filled and open symbols being easiest to perceive (Cleveland 1994). Convex hulls, drawn around the outmost observations in each category, are an effective way to highlight distinct groups and a comparison of the original hull and a peeled hull, which includes all but the most extreme data is useful for identifying outliers (Ellison 1993). If the third variable is quantitative it can be displayed using a bubble-plot, where the size of the symbol is scaled according to the third variable (Fig. 5.4d).

Another way to visualise the influence of a third variable on a bivariate relationship is via a conditional scatter-plot, or co-plot (Cleveland 1993). These are a set of scatter-plots of the two variables, each drawn under a given range of values for the third (or even fourth) variable. Where the conditioning variable is categorical, a co-plot displays the relationship for each category (Fig. 5.4e). If the conditioning variable is quantitative, the co-plot displays the relationship for overlapping slices, or shingles, of the conditioning variable, with the shingle widths chosen to give equal spacing along the third variable or equal numbers of observations in each slice (Fig. 5.4f).

Multivariate Techniques

It is difficult to visualise data in more than two dimensions: static three-dimensional scatter plots displayed in two dimensions are difficult to interpret and rarely useful. Interactive, dynamic graphics can be useful for exploration, especially for large data-sets (Cook and Swayne 2007), but for most analyses and for presentation, multivariate exploratory data analysis relies on simple extensions to bivariate methods (Everitt and Dunn 2001; Everitt and Hothorn 2011), and dimension-reduction techniques of ordination (Legendre and Birks 2012b: Chap. 8) and clustering and partitioning (Legendre and Birks 2012a: Chap. 7). The graphical techniques described here should be undertaken before subsequent *numerical* analysis, to check that the assumptions of these methods are met.

The simplest multivariate technique is to draw a matrix of scatter-plots to show the pair-wise relationship between all the variables. If there are more than a dozen or so variables, these pair-plots become too small and crowded to be useful. Provided the normality of individual variables has already been assessed, and outliers dealt with, a correlation matrix, annotated with significance levels, can be a useful summary of the main patterns in the data. The basic scatter-plot matrix can be enhanced in a number of ways, for example by including histograms of each variable on the diagonal, and Pearson's product-moment correlation in the upper panel (Fig. 5.5). See Husson et al. (2010) and Everitt and Hothorn (2011) for additional graphical tools for multivariate data.

Stratigraphical diagrams are essentially a specific form of scatter-plot matrix, in which the variables of interest are plotted on the x-axis against depth or time on the y-axis. Such plots are extremely widely used in palaeoecology and many variants exist (Grimm 1988). They may be drawn as bars, lines, or filled silhouettes. Bars are recommended for short sequences with a small number of samples, and if used, the bar width may be varied to represent the depth or time interval sampled. If lines or filled silhouettes are used then the position of sample points should be indicated in at least one curve on the diagram.

Time-Series Data

Many palaeolimnological data are time- or temporal-series (Dutilleul et al. 2012: Chap. 16) and are best described and communicated using a line graph with time on the x-axis and the variable of interest on the y-axis (e.g., Fig. 5.6). Inspection of a simple time-series plot should reveal the presence of outliers and the need for data transformation, and if the series is non-stationary, that is if there are systematic changes in the mean value (step changes or trends) or in the variance.

Palaeolimnological time-series are inherently noisy and underlying trends or other patterns can often be obscured by short-term or high-frequency variation. It is very often useful to smooth the series to help visualise the underlying structure

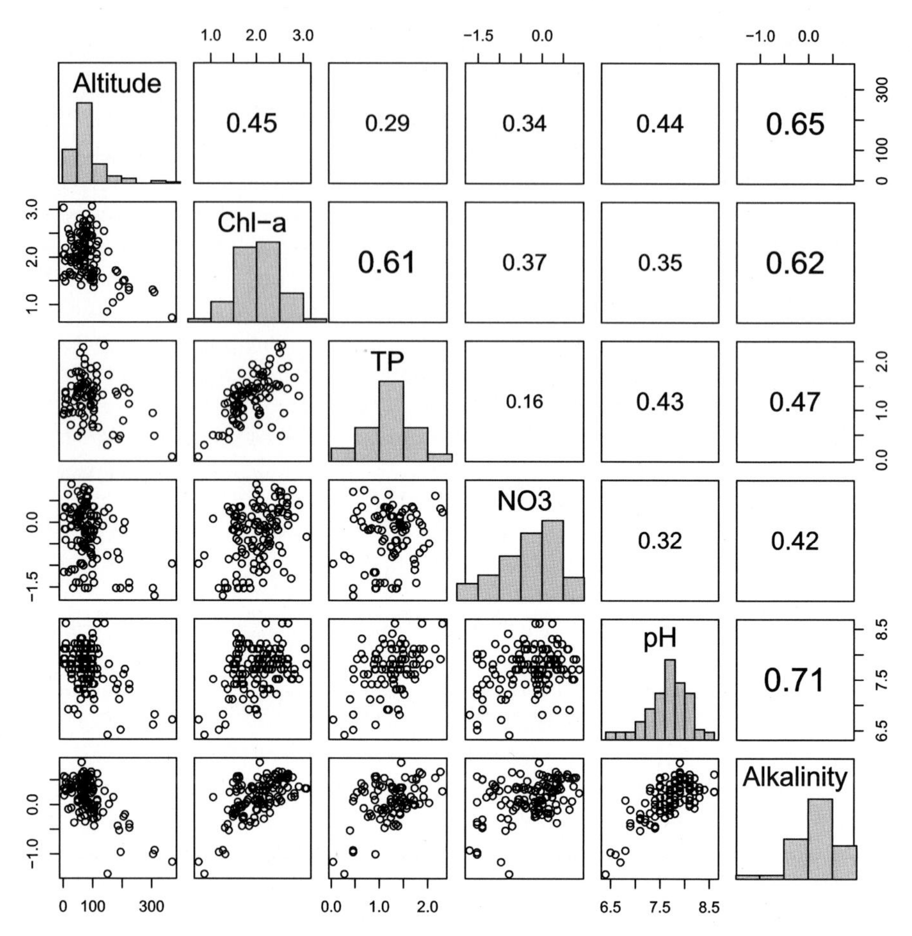

Fig. 5.5 Scatter-plot matrix showing scatter-plots of variable pairs (*bottom triangle*), histograms and Pearson product-moment correlations (*upper triangle*) (Data from the UK lakes data-set)

or broader-scale variation. The simplest form of smoothing is the centred moving average or running mean, in which the smoothed value is the average of a specified number (k) of values around each point in the original series. The degree of smoothing is controlled by varying k and because the running mean is centred, k is always an odd number, and the first and last few values of the smoothed series are indeterminate (Fig. 5.6a, b). A robust alternative to the running mean is the running median, which tends to give less smooth curves than the running mean but is less susceptible to any outliers in the series (Fig. 5.6c).

Smoothing a time-series is a form of filtering in which the high-frequency component of variation is reduced or filtered out. The moving average is the simplest form of low-pass filter and although intuitive and easily interpreted it is has a major disadvantage in that it does not damp all frequencies of the high-frequency variation equally and can, in some cases, actually accentuate noise in the filtered series. An

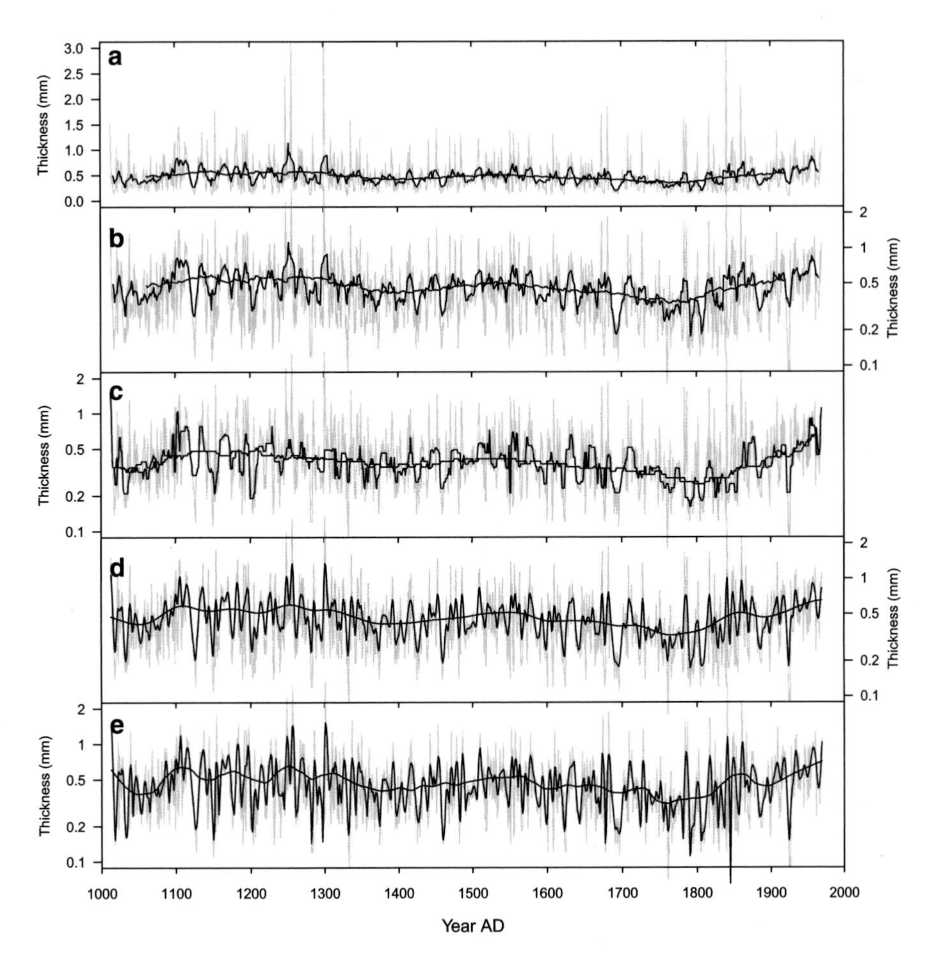

Fig. 5.6 Varve thickness for the last 1000 years measured in sediments from Lower Murray Lake, Canadian high arctic showing (**a**) varve thickness (*grey*) with 9-year and 99-year running averages (*black*), (**b**) as plot (**a**) but on $\log_{10}$ scale to stabilise variance and accentuate detail in short and long-term trends, (**c**) as plot (**b**) but showing 9- and 99-year running medians, (**d**) as plot (**b**) but showing 9- and 99-year Gaussian filtered data, and (**e**) as plot (**b**) but showing LOESS smoothers with 10 and 100 years spans (Redrawn in part from Besonen et al. 2008)

alternative that does not suffer from this problem is the Gaussian filter, in which the smoothed value is taken as the average of k points around the central value but with each point weighted according to the value of the appropriate Gaussian or normal density function (Janacek 2001). This filter tends to produce smoother curves than either the running mean or median (Fig. 5.6d). Finally, LOESS (see above) is often used to smooth time-series (Fig. 5.6e). LOESS tends to produce smoother results than the running mean or median and is also relatively robust to outliers. As with all smoothing the choice of span will be important: the two smoothers in Fig. 5.6 have been chosen to highlight approximate decadal and century-long patterns. However,

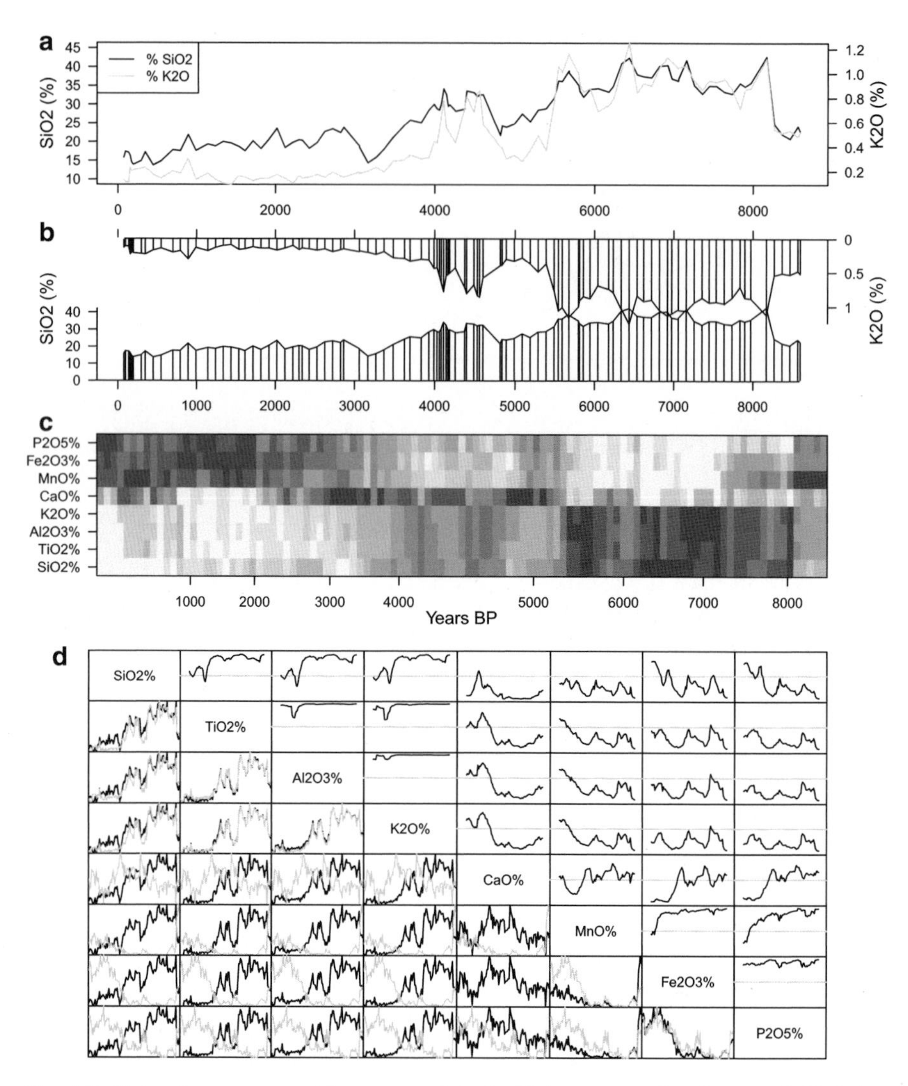

Fig. 5.7 Major element composition measured in the Holocene varved sediments of Elk Lake, Minnesota (From Dean 1993), showing (**a**) time-series line graphs of Si and K, (**b**) cave-plot of Si and K, (**c**) multivariate time-series plot (Peng 2008) of eight major elements (see text for details), and (**d**) pair-wise plots of eight major elements showing pair-wise time-series plots (*bottom left*; black = horizontal variable, grey = vertical variable) and moving window correlation coefficient (y-axis scaled from −1 to 1, grey line = 0). x-axis in all plots is scaled from 0 to 9,000 years BP

as we smooth to explore the data, the choice of an appropriate span is inevitably subjective (Birks 1998).

Another common aim of exploratory time-series analysis is the comparison of data from two or more series. Two series can be plotted on the same line graph and the y-axis scaled appropriately for each variable (Fig. 5.7a). The eye is drawn to common trends and peaks and troughs that are shared by the two series but it is sometimes difficult to get an overall impression of the correlation (or lack of correlation) between the two series. Figure 5.7b shows a cave-plot in which the two series in Fig. 5.7a are plotted to resemble the inside of a cave – with one series plotted as stalagmites from the bottom up and the other as stalactites from the top down (Becker et al. 1994). This plot takes no more space that the simple line graphs but depicts the trends, fluctuations, and individual values in each series and gives a better visualisation of the correlation between the series.

Comparison of multiple time-series is more challenging. Figure 5.7c shows an image plot of the variation in sedimentary major element composition with age. The values in each series have been converted into a small number of discrete classes (in this case 7) and each class plotted in a grey-scale, with lowest values of the series in white, mid values in grey, and high values in black (Peng 2008). Common patterns among the eight series plotted in Fig. 5.7c are clearly visualised (e.g., Si, Ti, Al, and K; Fe, Mn, and P; Ca) and can be further investigated using separate more detailed line graphs.

The correlation between two series can be quantified using Pearson's product-moment correlation coefficient, but with EDA we are often more interested in the pattern of correlation between two series and how this changes with time. Figure 5.7d shows a matrix of plots of the major element data with traditional time-series plots of pairs of variables shown in the lower left triangle and the value of the correlation coefficient calculated between the series along a moving window of 20 observations plotted in the upper right triangle (cf. Dean and Anderson 1974). This relatively simple plot shows a pair-wise comparison of the raw data-series together with a visualisation of the changing pattern of correlation between each series with time.

Outlier Detection and Treatment

An outlier is a value that is surprising given the rest of the data, being far away from the main cluster of values. They can have a large influence on many statistical techniques, and so need to be investigated. Outliers can be univariate, an extreme value on a single variable, or bivariate, having an unexpected value for one variable given its value for a second. A bivariate outlier need not be a univariate outlier. Multivariate outliers also occur, but are generally harder to identify.

There is no predetermined level at which a value should be considered an outlier, but values more than about two standard deviations away from the expected value should be investigated. There are a number of possible causes of outliers, and they should not be automatically deleted. They may be incorrect values, either as a result of analytical or transcription errors: these can be verified by re-checking notes, and, if possible, re-analysis. Alternatively, outliers may be correct values. If data have a normal distribution, more than 1 in a 100 values is expected to be more than 2.5 standard deviations away from the mean: deleting these values will artificially deflate the standard deviation. Other distributions, such as log-normal, have more extreme values: these apparent outliers are cured if an appropriate transformation is made. Outliers may also indicate that a second process is operating that produces values in a different range.

Outliers, and the reasons for deletion, should be clearly reported in any accompanying text. In general, only justifiably erroneous points should be removed: deleting outliers to improve a correlation or the r^2 value or the root mean squared error in a predictive model is bad practice and should be avoided.

Missing Values

Missing values are inevitable in any large sampling exercise, either because a variable was not possible to measure in some instances, or because outliers have been deleted. Most numerical techniques cannot handle missing values so they need to be removed. If data are missing for a large portion of the observations then one has little choice but to delete that variable. Similarly, if an observation is missing data for many variables you should delete that observation. However, deleting data is wasteful, and if only a few values are missing an alternative strategy is to substitute the missing value with either the variable's mean, or a value estimated using the correlation with other variables. Such imputed values are only estimates: this is not a problem if the aim is further data-exploration using ordination or clustering techniques but they are not real data and should be used with caution in statistical tests, as the number of degrees of freedom may be over-estimated, and in calibration-function development.

Values that are missing because they are less than the detection limit are known as censored. If the aim is further exploration and only a few values are censored they can be replaced by half the detection limit (McBride 2005). If many values are censored, more sophisticated methods based on statistical distribution fitting are available (Helsel 2005).

Graph Drawing

Production of publication quality figures always takes far longer than anticipated, and although a picture 'is worth a thousand words', they also take up almost as much space. Consideration should therefore first be given as to whether a figure is the most appropriate medium to display the data. If the data-set is small, it may be better to present it in a table or the text.

Figures should be as simple as is compatible with their purpose: everything on the figure should aid interpretation (Webb 2009). Anything that does not is 'chart junk', and acts to obscure the figure's message by swamping the reader with unnecessary information. For example, excessive use of grid-lines on a graph, or unnecessary shading patterns can be distracting. Particular ire is reserved for the use of three-dimensional graphs. At best, the extra complexity of these graphs requires more concentration from the reader, as the effect is distracting and makes it more difficult to compare values and relate them to the axes. At worst the perspective can be misleading.

Axis scales should be selected to present the data clearly. If data with different magnitudes are plotted on the same scale, variability in the smaller values may be obscured. This can be prevented by using a second y-axis, or perhaps by showing the data on a logarithmic scale. Changes in scale on an axis are potentially very misleading, and should be avoided, or marked clearly. Cumulative graphs are difficult to read, as the categories do not start from a common base. They should only be used if there are a few categories.

Complex figures may be easier to understand if they use colour to differentiate components of the graph. Fortunately, many journals now allow colour figures, though usually at some cost. If colour is used, care should be given to ensure that the figure will be understandable to people affected by colour blindness, or who print and photocopy in black and white. This can be achieved by avoiding pairs of easily confusable colours, especially red and green, using colours with a different tone, and, where possible, using redundant information, such as line or marker style.

Software

There are a number of statistical packages suitable for EDA, including SPSS (2006), Minitab (2008), S-PLUS® (2008), and R (R Development Core Team 2011). R is a powerful and flexible software for statistical computing and graphics, similar to S-PLUS® and available free of charge on a General Public License from http://www.r-project.org (see Murrel 2006; Petchey et al. 2009). Spreadsheets, such as Microsoft EXCEL®, have limited functionality, and are best reserved for data-entry. Stratigraphical plots can be drawn with C2 (Juggins 2007), TILIA (Grimm 1990), PSIMPOLL (Bennett 1994), and the rioja package in R (Juggins 2009).

Conclusions

Exploratory data analysis is an iterative process. It is essential at every step of the data-analysis cycle, from initial data-screening and data-exploration, through model diagnosis, to the presentation of complex data. Much of palaeolimnology is characterised by observational, or quasi-experimental data collection, rather than controlled experimentation. We therefore also stress graphical EDA as an important tool in hypothesis generation: as Everitt (2005) puts it, "Plotting data is a first step in trying to understand their message".

Acknowledgement This is publication number A385 from the Bjerknes Centre for Climate Research.

References

Becker RA, Clark LA, Lambert D (1994) Cave plots: a graphical technique for comparing time series. J Comput Graph Stat 3:277–283

Bennett KD (1994) PSIMPOLL version 2.23: a C program for analysing pollen data and plotting pollen diagrams. INQUA Commission for the Study of the Holocene: Working group on data-handling methods. Newsletter 11:4–6. http://chrono.qub.ac.uk/psimpoll/psimpoll.html

Bennion H, Juggins S, Anderson NJ (1996) Predicting epilimnetic phosphorus concentrations using an improved diatom-based transfer function, and its application to lake eutrophication management. Environ Sci Technol 30:2004–2007

Besonen MR, Partridge W, Bradley RS, Francus P, Stoner JS, Abbott MB (2008) A record of climate over the last millennium based on varved lake sediments from the Canadian High Arctic. Holocene 18:169–180

Birks HJB (1998) Numerical tools in palaeolimnology – progress, potentialities, and problems. J Paleolimnol 20:307–332

Birks HJB (2012a) Chapter 2: Overview of numerical methods in palaeolimnology. In: Birks HJB, Lotter AF, Juggins S, Smol JP (eds) Tracking environmental change using lake sediments. Volume 5: Data handling and numerical techniques. Springer, Dordrecht

Birks HJB (2012b) Chapter 11: Analysis of stratigraphical data. In: Birks HJB, Lotter AF, Juggins S, Smol JP (eds) Tracking environmental change using lake sediments. Volume 5: Data handling and numerical techniques. Springer, Dordrecht

Carr DB, Littlefield RJ, Nicholson WL, Littlefield JS (1987) Scatterplot matrix techniques for large *N*. J Am Stat Assoc 83:424–436

Cleveland WS (1979) Robust locally weighted regression and smoothing scatterplots. J Am Stat Assoc 74:829–836

Cleveland WS (1993) Visualizing data. Summit Press, New Jersey

Cleveland WS (1994) The elements of graphing data. Hobart Press, Monterey

Cook D, Swayne DF (2007) Interactive and dynamic graphics for data analysis. Springer, New York

Dean WE (1993) Physical properties, mineralogy, and geochemistry of Holocene varved sediments from Elk Lake, Minnesota. Geol Soc Am Spec Paper 276:135–157

Dean WE, Anderson RY (1974) Application of some correlation coefficient techniques to time-series analysis. Math Geol 6:363–372

Dutilleul P, Cumming BF, Lontoc-Roy M (2012) Chapter 16: Autocorrelogram and periodogram analyses of palaeolimnological temporal series from lakes in central and western North America to assess shifts in drought conditions. In: Birks HJB, Lotter AF, Juggins S, Smol JP (eds) Tracking environmental change using lake sediments. Volume 5: Data handling and numerical techniques. Springer, Dordrecht

Ellison AM (1993) Exploratory data analysis and graphic display. In: Scheiner SM, Gurevitch J (eds) Design and analysis of ecological experiments. Chapman & Hall, New York, pp 14–45

Everitt B (2005) An R and S-PLUS® companion to multivariate analysis. Springer, London

Everitt BS, Dunn G (2001) Applied multivariate data analysis, 2nd edn. Arnold, London

Everitt BS, Hothorn T (2011) An introduction to applied multivariate analysis using R. Springer, New York

Freedman D, Diaconis P (1981) On the histogram as a density estimator: *L_2* theory. Zeitschrift für Wahrscheinlichkeitstheorie und verwandte Gebiete 57:453–476

Grimm EC (1988) Data analysis and display. In: Huntley B, Webb T (eds) Vegetation history. Kluwer Academic Publishers, Dordrecht, pp 43–76

Grimm EC (1990) TILIA and TILIA GRAPH: PC spreadsheet and graphics software for pollen data. INQUA Commission for the Study of the Holocene: Working group on data-handling methods. Newsletter 4:5–7. http://www.ncdc.noaa.gov/paleo/tiliafaq.html, http://museum.state.il.us/pub/grimm/tilia/
Helsel DR (2005) Non-detects and data analysis: statistics for censored environmental data. Wiley, New York
Husson F, Lê S, Pagès J (2010) Exploratory multivariate analysis by example using R. CRC Press, Boca Raton
Janacek G (2001) Practical timer series. Arnold, London
Juggins S (2007) C2 version 1.5 user guide: software for ecological and palaeoecological data analysis and visualisation. Department of Geography, University of Newcastle. http://www.staff.ncl.ac.uk/stephen.juggins/software/C2home.htm
Juggins S (2009) rioja: analysis of Quaternary science data. R package version 0.5-6. http://cran.r-project.org/package=rioja
Legendre P, Birks HJB (2012a) Chapter 7: Ordination. In: Birks HJB, Lotter AF, Juggins S, Smol JP (eds) Tracking environmental change using lake sediments. Volume 5: Data handling and numerical techniques. Springer, Dordrecht
Legendre P, Birks HJB (2012b) Chapter 8: Clustering and partitioning. In: Birks HJB, Lotter AF, Juggins S, Smol JP (eds) Tracking environmental change using lake sediments. Volume 5: Data handling and numerical techniques. Springer, Dordrecht
Limpert E, Stahel WA, Abbt M (2001) Log-normal distributions across the sciences: keys and clues. Bioscience 51:341–352
McBride GB (2005) Using statistical methods for water quality management. Wiley, New York
Minitab (2008) MINITAB 15. Minitab, State College
Murrel P (2006) R graphics. Chapman & Hall, London
NIST/SEMATECH (2006) e-handbook of statistical methods. http://www.itl.nist.gov/div898/handbook/
Peng RD (2008) A method for visualizing multivariate time series data. J Stat Softw 25: Code Snippet 1–17
Petchey OL, Beckerman AP, Childs DZ (2009) Shock and awe by statistical software – why R? Bull Br Ecol Soc 40(4):55–58
Quinn G, Keough M (2002) Experimental design and data analysis for biologists. Cambridge University Press, Cambridge
R Development Core Team (2011) R: a language and environment for statistical computing. R Foundation for Statistical Computing, Vienna
Silverman BW (1986) Density estimation. Chapman & Hall, London
Snee RD, Pfeifer CG (1983) Graphical representation of data. In: Kotz S, Johnson NL (eds) Encyclopedia of statistical sciences. Wiley, New York, pp 488–511
Sokal RR, Rohlf FJ (1995) Biometry – the principles and practice of statistics in biological research. WH Freeman, New York
SPSS (2006) 15.0 for windows. SPSS Inc, Chicago
S-Plus (2008) S-PLUS® 8. Insightful Corporation, Seattle
Tufte ER (1990) Exploring information. Graphics Press, Cheshire
Tufte ER (2001) The visual display of quantitative information, 2nd edn. Graphics Press, Cheshire
Tukey J (1977) Exploratory data analysis. Addison-Wesley, Reading
Warton DU, Hui FKC (2011) The arcsine is asinine: the analysis of proportions in ecology. Ecology 92:3–10
Webb T (2009) Methods in statistical graphics. Bull Br Ecol Soc 40(4):53–54

Chapter 6
Assessment of Uncertainties Associated with Palaeolimnological Laboratory Methods and Microfossil Analysis

Louis J. Maher, Oliver Heiri, and André F. Lotter

Abstract Assessment of uncertainties associated with laboratory analytical methods and biostratigraphic analyses is an important but much neglected aspect of palaeolimnological research. Error assessment, method validation, and inter-laboratory or inter-method comparisons should play an important role in palaeolimnology, as they do in other natural sciences and in medical sciences. The chapter summarises the statistics in error-assessment trials developed in analytical chemistry to derive uncertainties associated with estimates of a single parameter such as loss-on-ignition. The statistical techniques for estimating errors for microfossil counts (e.g., pollen, diatoms) are presented for taxa expressed as percentages of an overall sum, for taxa expressed as ratios of selected types, and for taxa estimated as influx or accumulation rates. Assessments of counting errors in varve chronologies and of errors derived from multi-core palaeolimnological studies are also discussed. Availability of relevant computer software is summarised.

Keywords Accuracy • Confidence limits • Errors • Inter-laboratory comparisons • Microfossil counts • Multi-core studies • Precision • Random errors • Repeatability • Reproducibility • Systematic errors • Trueness • Varve counts

---

L.J. Maher (✉)
Geology and Geophysics, University of Wisconsin, 1215 West Dayton Street, Madison, WI 53706, USA
e-mail: maher@geology.wisc.edu

O. Heiri
Institute of Plant Sciences and Oeschger Centre for Climate Change Research, University of Bern, Altenbergrain 21, Bern CH-3013, Switzerland
e-mail: oliver.heiri@ips.unibe.ch

A.F. Lotter
Department of Biology, Laboratory of Palaeobotany and Palynology, University of Utrecht, Budapestlaan 4, Utrecht NL-3584 CD, The Netherlands
e-mail: a.f.lotter@uu.nl

H.J.B. Birks et al. (eds.), *Tracking Environmental Change Using Lake Sediments*, Developments in Paleoenvironmental Research 5, DOI 10.1007/978-94-007-2745-8_6,

Introduction

When interpreting sediment records the uncertainty associated with laboratory methods and analyses is an important consideration and will significantly affect the strength and validity of the conclusions that can be drawn from the data. As in other fields in the natural sciences, measurements should ideally be accompanied by an error estimate and analyses should be reproducible, whether re-assessed by the same analyst and equipment or by another team of researchers using identical methodology. Error assessment, method validation, and inter-laboratory or inter-method comparisons should therefore play an important role in all palaeolimnological research. These issues are of special importance in research projects involving a number of different laboratories and where conclusions are based on data-sets produced by different analysts and institutions.

Ideally, palaeolimnological measurements should be both accurate and precise. Precision and accuracy are treated as synonyms in most dictionaries, but they have distinct meanings in the natural sciences. *Precise* means a measurement is repeatable and reproducible. *Accurate* means a measurement is capable of correctly reflecting the size of the entity being measured. The distinction between the terms is often illustrated using the analogy of throwing darts at a target. The darts may miss the bull's eye. However, if they are scattered randomly around the bull's eye the throws may still be considered more accurate than darts clustering tightly together in, say, the lower right quadrant. In the first case we might consider the throws fairly accurate, but not very precise; in the second case the thrower is very precise but the throws are not very accurate. If all darts cluster in the bull's eye the pattern is both accurate and precise. In the field of analytical chemistry *trueness* is furthermore differentiated from accuracy (Fleming et al. 1996a). Trueness reflects whether the average of replicate measurements accurately reflects the true value of a measurand, whereas accuracy reflects whether an individual measurement is close to the true value of the measurand. It follows that for measurements to be accurate the applied method should be optimised with respect to both trueness and precision.

For error-assessment and method-evaluation trials, the error associated with a given methodology is commonly partitioned into two components to reflect the difference between trueness and precision. The *systematic error* (or bias) includes error components which in the course of repeated measurements remain constant or vary in a predictable way (Ellison et al. 2000). A method with a small or negligible systematic error will provide measurements which, on average, are an approximation of the true value of the measurand. The systematic error is commonly expressed as the difference between the mean of replicate measurements of the parameter of interest and an independent estimate of the 'true' value of this parameter based on a different methodology or measured by a different laboratory. The *random error* reflects the precision or scatter of a given analytical method (Walker and Lumley 1999; Ellison et al. 2000) and is usually expressed as a variance or standard deviation (see Birks 2012: Chap. 2). Conceptually, any analytical error associated with a given methodology, institution, or analyst can be partitioned into these two error components. It is useful to keep this partitioning in mind when designing and interpreting error assessment trials in order to delimit clearly which error component has been assessed within a given experiment.

A number of discrete steps is usually necessary to obtain a palaeolimnological record, including coring the lake sediments in the field, sediment subsampling, sample storage, sample preparation, sample measurement, and possibly statistical or numerical analysis of the results (Smol 2008). Error-assessment trials can consequently be designed to assess errors associated with these individual steps, errors associated with a sequence of steps, or errors associated with the full process from sediment coring to data interpretation. Error estimates can be based either on replicate single-method measurements of the same homogenised natural or artificial substance, replicate multi-method measurements, inter-laboratory comparisons, or multi-core assays. Alternatively, error estimates can be based on the statistical and numerical properties of the obtained data themselves. The latter approach has been widely used to produce error estimates for microfossil counts as the time-consuming nature of these analyses hampers extensive replicate counting experiments.

Error assessment, method validation, inter-laboratory comparisons, and quality control are of high relevance in analytical chemistry. A large number of peer-reviewed articles and textbooks dealing with these issues exist in the scientific literature (e.g., Williams 1996; Ellison 1998; Walker and Lumley 1999; Ellison et al. 2000) and most of the described methods are directly applicable to the geochemical methods used in palaeolimnology. We therefore restrict ourselves to providing a short summary of these error-assessment trials and a brief overview of the basic statistical methods commonly in use. Producing error estimates for microfossil counts is a problem more specific to palaeolimnology and the second part of this chapter provides a detailed review of this topic. The third and fourth parts of the chapter deal with assessing counting errors in varve chronologies and the use of multi-core palaeolimnological studies for error assessment, respectively.

Single Parameter Estimates

Experiments assessing the uncertainty of single parameter measurements (e.g., sediment chemistry, physical properties of sediments) can be designed to assess the trueness of a given analytical method, the precision, or both (Ellison 1998; Walker and Lumley 1999; Ellison et al. 2000). The precision of a given laboratory method is usually calculated as the variability of replicate or repeated measurements of a substance of high homogeneity with respect to the parameter being measured. The standard deviation of these replicate measurements will then be an estimate of the random error associated with the results and will provide information about the precision of the method. In this context, repeatability and reproducibility are usually separated. *Repeatability* is an estimate of the precision of replicate measurements performed on independent samples of the substance of interest by the same analyst, the same equipment, using the same conditions of use, at the same location, and over a short period of time (Fleming et al. 1996b). The repeatability standard deviation is commonly abbreviated as s_r and is considered equivalent to the within-laboratory standard deviation (s_w) (Walker and Lumley 1999; Ellison et al. 2000). *Reproducibility* reflects the precision of measurements performed on subsamples of the substance of interest by the same method but under changing conditions such as with different analysts and equipment, at a different location, or at a different time (Fleming et al. 1996b). The reproducibility standard deviation is usually abbreviated as s_R and can only be measured directly by inter-laboratory comparisons (Ellison et al. 2000).

The major complication in assessing the systematic error of a series of measurements is in obtaining a realistic estimate of the 'true' value of the measurand in the analysed substance. In analytical chemistry a given method is commonly evaluated against certified reference material (CRM) or against spiked samples (Walker and Lumley 1999). A chain of calibrations and inter-laboratory comparisons ensure the traceability of the measurements of a given laboratory or the properties of a given reference material back to primary national or international standards (see e.g., Walker and Lumley 1999 or Ellison et al. 2000 for more information on traceability, primary standards and CRMs). Namiesnik and Zygmunt (1999) and Boyle (2000, 2001) list a number of certified reference materials appropriate for lake-sediment studies and discuss some of the basic assumptions and problems associated with quality control in the chemical analysis of lake sediments. In practice it will often be difficult to find CRMs with properties similar to the sediments of interest. Alternatively, a given methodology can be assessed by comparing it with independent measurements of the parameter of interest using a different method considered to be more precise or which has been calibrated more rigorously (Ellison et al. 2000). For example, Beaudoin (2003) and Santisteban et al. (2004) compared loss-on-ignition (LOI) measurements with more sophisticated techniques of organic and inorganic carbon content determination to assess the trueness of the method. The silent assumption in these trials is that the wet oxidation/titration and carbon analyser methods used to evaluate LOI provide an accurate measurement of the organic and inorganic carbon content of the sediments. Spiked samples or samples of a pure substance with known chemical properties provide a further method of assessing the systematic error of a geochemical method. For example, Heiri et al. (2001) used samples of pure calcium carbonate to assess whether LOI is able to measure accurately carbonate content. Similarly, spiked samples where a known quantity of the measurand is added to a matrix substance can be used. Dean (1974) used this approach to evaluate the performance of LOI in estimating the carbonate content of sediments by mixing known quantities of calcium carbonate with anhydrite. However, considering the chemical complexity of most sediments, error trials with pure substances or spiked samples will only provide limited information about the suitability of a given methodology for palaeolimnological analyses. Finally, inter-laboratory comparisons provide a way of estimating the true value of a measurand (Walker and Lumley 1999). If a homogenised sediment sample is distributed to a range of laboratories which evaluate the accuracy of their chemical analyses on a regular basis (e.g., using CRMs), then the mean of the measurements of all involved laboratories can provide an estimate of the true value of the measurand in the substance of interest. It is important to define clearly which steps of analysis are being evaluated in error-assessment trials using inter-laboratory comparisons. For example, Heiri et al. (2001) distributed homogenised and freeze-dried sediments to a number of laboratories for LOI measurements. The results provided an assessment of the reproducibility of LOI performed on freeze-dried sediments. However, uncertainties associated with sample drying and sample storage were not taken into account in this study, even though sample preparation techniques and storage may be a significant source of variability in geochemical measurements (e.g., Tanner and Leong 1995; Lasorsa and Casas 1996; Dixon et al. 1997). The major drawback of inter-laboratory comparisons is that they are very time-intensive and sensitive to outliers (Walker and Lumley 1999).

Usually an experiment designed to detect a systematic error in a measurement method will involve replicate measurements of a material for which an estimate of the true value of the measurand exists. The apparent bias of the results is then defined as

$$\Delta = \bar{X} - \mu \tag{6.1}$$

where Δ indicates the apparent bias, $\bar{X}$ the mean of n measurements, and μ the estimate of the true value of the measurand (Walker and Lumley 1999). Any estimate of $\bar{X}$ will be associated with a random error and this will have to be taken into account when testing for the presence of a systematic error. Walker and Lumley (1999) suggest testing whether the apparent bias is within a laboratory's measurement precision to assess whether evidence exists for a systematic error. In order to calculate the laboratory's expected precision (σ), the between-laboratory component (s_b) of s_R needs to be known. This can be estimated as

$$s_b = \sqrt{s_R^2 - s_r^2} \tag{6.2}$$

if estimates of s_R based on inter-laboratory comparisons are available. Walker and Lumley (1999) discuss other ways of estimating the between-laboratory standard deviation of a method if this is not the case. The expected precision of a laboratory's measurement based on n replicates can then be estimated as

$$\sigma = \sqrt{s_b^2 + s_w^2/n} \tag{6.3}$$

where s_w is the standard deviation of the n replicate analyses. The apparent bias in a measurement can then be tested versus

$$-2\sigma < \bar{X} - \mu < 2\sigma \tag{6.4}$$

If this requirement is met, it can be concluded at an approximately 95% level of confidence that no evidence exists for the presence of bias (Walker and Lumley

1999). To provide reliable results, the estimate of both s_w and s_b should be based on a large number of replicate measurements (Walker and Lumley 1999).

Alternatively, a t-test can be used to assess whether the mean of a laboratory's replicate measurements is significantly different than the expected value of the measurand (Walker and Lumley 1999; Ellison et al. 2000) following the equation

$$t = \frac{\bar{X} - \mu}{s_{\bar{X}}} \tag{6.5}$$

where $s_{\bar{x}}$ refers to the standard deviation of the mean (equivalent to $\sqrt{s_w^2/n}$). Critical values of t ($t_{\alpha,\nu}$) for different significance values (α) and degrees of freedom ($\nu = n$-1) can be found in most textbooks on statistics (e.g., Zar 1999). If the absolute value of t exceeds the critical value of $t_{\alpha,\nu}$ it can be concluded that evidence for a systematic error exists at the significance level of α.

Obviously the statistical tests described above are highly dependent on an accurate estimate of the true value of the measurand (μ) in the analysed samples and they assume that any random error associated with μ is negligible. If the random error of the estimate of μ is of the same order of magnitude as the measurement error associated with a given method this will have to be taken into account when applying a statistical test. For comparing a laboratory's measurements with the expected value of a reference material, Walker and Lumley (1999) recommend expanding Eq. 6.3 to

$$\sigma = \sqrt{(s_b^2 + s_w^2/n + U_{RM}^2} \tag{6.6}$$

where U_{RM} is an estimate of the standard uncertainty of the reference material and then to apply the test outlined in Eq. 6.4.

The t-test can also be adapted to test whether a difference exists between two means (Zar 1999). The test then assumes that the two means have equal variances and t can be estimated as

$$t = \frac{\bar{X}_1 - \bar{X}_2}{\sqrt{s_p^2/n_1 + s_p^2/n_2}} \tag{6.7}$$

where $\bar{X}_1$ and $\bar{X}_2$ refer to the two means, and n_1 and n_2 to the number of replicates used to calculate the means. s_p^2 is calculated as $(SS_1 + SS_2)/(\nu_1 + \nu_2)$, where SS_1 and SS_2 refer to the sum-of-squares of the deviations of the means, and ν_1 and ν_2 to the degrees of freedom of the two samples. The critical value of t will have to be looked up for $\nu = \nu_1 + \nu_2 = n_1 + n_2 - 2$ degrees of freedom. The reader is referred to introductory textbooks on statistics (e.g., Zar 1999; Davis 2002; Blæsild and Granfeldt 2003; Borradaile 2003) for more information on the basic assumptions of t-tests and on the robustness of the test if these assumptions are violated. Examples of t-tests being used to test for a systematic error between different sediment chemistry measurements or between measurements and the certified value of reference materials include King et al. (1998), Tung and Tanner (2003), and Brunori et al. (2004).

The *t*-test is not suitable if more than three means are compared (Zar 1999), as will be the case in most inter-laboratory comparisons. In this case, analysis of variance (ANOVA) (see Birks 2012: Chap. 2) can be used to test the null hypothesis that mean values measured by a number of different laboratories, analysts, or methods are based on samples from statistical populations with the same unimodal distribution (i.e., with identical means and variances). Single-factor ANOVA (Sokal and Rohlf 1995) involves the calculation of the sum-of-squares and the degrees of freedom both within the groups forming the different means and among the groups. If the different samples are from identical statistical populations, the within-groups (or error) mean square (MS) and the among-groups (or simply groups) MS should both be an estimate of the variance of the populations and therefore approximately equal. This can be tested by calculating

$$F = \frac{\text{groups MS}}{\text{error MS}} \tag{6.8}$$

and comparing this value with the critical value of F for a given significance value, with the groups degrees of freedom and error degrees of freedom (available in statistical tables). If F is equal to or exceeds the critical value then the null hypothesis is rejected and it can be concluded that evidence exists that the different population means are not equal. Again, the reader is referred to textbooks on basic statistics (e.g., Zar 1999; Davis 2002; Blæsild and Granfeldt 2003; Borradaile 2003) for more information on the underlying assumptions associated with ANOVA and for more details about how F is calculated. ANOVA is widely used in inter-laboratory experiments to test for significant differences between the results of different laboratories. Examples of studies using single-factor ANOVA to test laboratory methods applied to sediments include Lasorsa and Casas (1996), Conley (1998), King et al. (1998), Sahuquillo et al. (1999), and Tung and Tanner (2003). In experiments which have been designed accordingly, ANOVA can also be expanded to test for the effects of two or more factors on population means simultaneously (Zar 1999). Somogyi et al. (1997) provide an example of this approach being used within an inter-method comparison for measuring the elemental composition of marsh sediments.

The null hypothesis in single-factor ANOVA is that all tested means are equal within the uncertainty of the measurement method. If the alternate hypothesis is accepted it can only be concluded that evidence exists that not all the means are equal. Single-factor ANOVA or multiple *t*-tests should not be used to assess whether all the means are different from each other or how many statistical differences there are (Zar 1999). Most introductory texts on statistics and commercially available statistical software packages present alternative tests such as the Tukey test or the Newman-Keuls test for this purpose (e.g., Zar 1999). Examples of these sorts of tests being applied to sediment samples include Tanner and Leong (1995: Tukey and Newman-Keuls tests) and King et al. (1998: Bonferroni-comparisons).

The tests described here are useful for comparisons of multiple measurements of a reference material with its certified value or for assessing whether significant differences exist in the measurements of multiple laboratories. For more comprehensive error-assessment trials, the trueness, precision, and sensitivity of a method will have to be assessed over a range of values of the measurand, as the bias and random error of chemical measurements will often vary with analyte concentrations (Ellison et al. 2000). This will commonly involve regression techniques, which are outlined in Birks (2012: Chap. 2).

Microfossil Counts

Microfossil counting tends to be time-consuming, and therefore the number of counts per sample will be one of the constraining parameters for the achievable sampling resolution. Ideally, enough microfossils should be counted to obtain stable percentage values. Commonly, a counting sum of between 300 and 500 microfossils provides stable percentage values in pollen and diatom analyses (Birks and Birks 1980; Battarbee et al. 2001), whereas concentration estimates stabilise at counts greater than 800 (Wolfe 1997). Assessing the uncertainty associated with the microfossil counting sum is of particular concern for proxies which can potentially occur at low abundances in sediments such as chironomids (Heiri and Lotter 2001; Larocque 2001; Quinlan and Smol 2001), testate ameobae (Payne and Mitchell 2009; Wall et al. 2010), cladocerans (Kurek et al. 2010), and mites (Solhøy 2001). Microfossil counts will usually be presented as percentages, ratios, concentrations, or accumulation rates. For the first three possibilities, estimates of the random error in the results attributable to count size can be calculated based on theoretical considerations as described in detail below. For accumulation rates, error estimates are more difficult to derive due to uncertainties associated with dating of the sediment sequence (Bennett 1994).

Percentages (Taxa as Proportions of an Overall Sum)

Mosimann (1965) provides a useful source of statistical methods for anyone working with microfossils. He presents clear descriptions of the relevant statistical tests, and solves example problems. While there were mainframe computers in 1965, programmable calculators and microcomputers still lay in the future, and some of Mosimann's most useful statistics seemed too cumbersome for general use. However, computer programs for solving these problems are nowadays available over the Internet and a number of these programs is discussed in the following pages. A separate paragraph describing the availability of these programs is given at the end of this section on microfossil counts.

In principle, the quantitative analysis of a microfossil taxon, say pine pollen in a pollen sample, can be considered to be a sequence of trials with two possible outcomes: the fossil belongs to pine or not (Mosimann 1965: p. 637). If in such a series of trials the outcome of any trial is independent of the outcome of any other, and if the probability, p, of pine occurring in any trial is constant, then the trials can be considered to be Bernoulli trials. Mosimann points out that even if the assumptions of Bernoulli trials are met, analysts cannot predict the distribution of their counts until they choose one of two possible sampling plans. If they choose to count inside the sum (e.g., for percentages or proportions), the counts will follow a binomial distribution. If they choose to count outside the sum (e.g., for ratios between different taxa or between a taxon and a marker taxon), the counts will follow a negative binomial distribution.

In the usual practice of counting inside the sum, analysts estimate the proportion (p) of a given taxon in their samples by calculating $\hat{p}$, a 'point estimate' of p, by dividing the number of fossils of the taxon of interest counted (x) by the sum of all the examined microfossils (n_p); that is, $\hat{p} = x/n_p$. For inside counts then, $\hat{p}$ and p may range from 0 to 1.0 (or to 100 if the analyst thinks in percentages) and the counts will follow a binomial distribution.

Tables for confidence limits of p based on $\hat{p}$ for counts inside the sum were developed by Mainland et al. (1956), and abridged versions appear in several statistics texts (e.g., Rohlf and Sokal 1969: Table W). Although these tables should be used when n_p is small, Mosimann (1965: p. 643) provides an equation that predicts approximate confidence limits of p when n_p is fairly large – say, 50 or more:

$$p_{(c\ \mathrm{limit})} = \frac{\hat{p} + Z^2/(2n_p) \pm Z\sqrt{\hat{p}(1-\hat{p})/n_p + Z^2/(4n_p^2)}}{1 + Z^2/n_p} \qquad (6.9)$$

where $p_{(c\ \mathrm{limit})}$ indicates the confidence limits for the proportion/percentage and Z represents the specified area under the normal curve to include in the confidence intervals (in standard deviation units). For 0.95 confidence limits, $Z = 1.960$; for 0.99 confidence limits, $Z = 2.576$; and if 0.995 limits are desired, $Z = 2.810$. Equation 6.9 relies on the fact that for high degrees of freedom (n_p - counts of several hundred microfossils) the binomial, normal, and Student's t distributions are essentially identical.

Using a computer to explore the relationship of x and n_p allows us to visualise the problem. For example, Fig. 6.1 shows how the confidence interval gets smaller as the total microfossil count increases. The limits are asymmetrically distributed around $\hat{p}$ for all values except $\hat{p} = 0.5$, and the asymmetry is most pronounced for extreme values of $\hat{p}$ and for low values of n_p. The confidence limits are skewed toward the upper limit for values of $\hat{p} < 0.5$ and toward the lower limit for $\hat{p} > 0.5$. The complement of p for a taxon X is, by definition, equal to q for the non-X category; that is to say, $q = 1 - p$. Figure 6.1 makes it clear that the lower limit of p is equal to one minus the upper limit of q, and the upper limit of p equals one minus the lower

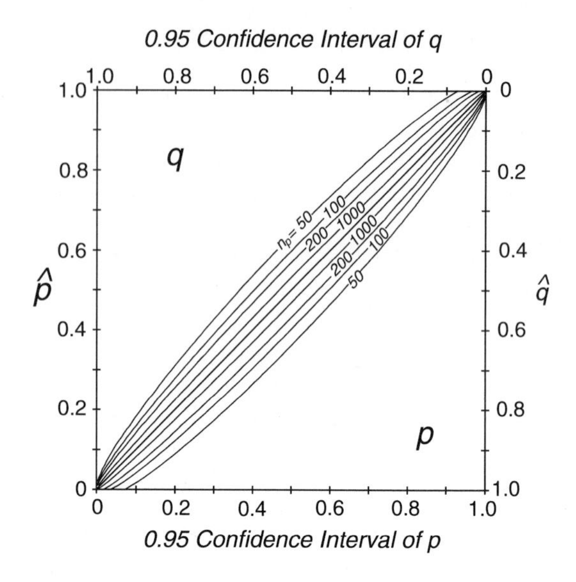

Fig. 6.1 Confidence limits for percentage data (counts inside a sum) based on Eq. 6.9; n_p is the count sum, p the proportion of the taxon of interest in the sample, $\hat{p}$ the estimated proportion based on the count sum n_p $(q = 1 - p, \hat{q} = \hat{p} - 1)$

limit of q. This relationship allows tables of binomial confidence limits to cover the whole range of p by listing only the values 0 to 0.5. Even for microfossil counts as low as 100 fossils, the maximum difference between Mosimann's Eq. 6.9 and binomial tables is 0.0068 scale unit, and the usual error is much less. Although the confidence interval appears to be widest at the $\hat{p}$ value of 0.5, its relative uncertainty (confidence interval divided by $\hat{p}$) is really at its lowest; the size of the integers in x and n_p are simultaneously at their maximum values.

Equation 6.9 can be used to predict how confidence intervals become narrower as n_p gets larger. It can be shown that as n_p increases, the confidence interval of $\hat{p}$ decreases as the reciprocal of the square root of n_p; to halve the confidence interval, n_p must be increased fourfold. As the confidence interval improves with higher counts (but at a continually decreasing rate) analysts must decide for themselves where the gain in precision is worth the extra effort in counting.

Treating Taxa as Ratios of Types (Counting Outside the Sum)

There are also cases where analysts need to estimate the true ratio (u) of one type of pollen (e.g., pine) to another type of pollen (e.g., non-pine) in a sample by calculating $\hat{u}$, a point estimate of u; that is, $\hat{u} = x/n_u$, where n_u indicates the number of microfossil counts not belonging to the taxon of interest (in our example the sum of non-pine pollen). For these counts outside the sum, u and $\hat{u}$ can range from 0 to arbitrarily large values, and the counts will follow a negative binomial distribution. Proportions (p) can be converted to ratios (u) by the relationship $u = p/(1 - p)$, and ratios can be converted to proportions by $p = u/(1 + u)$. Mosimann (1965) adapted

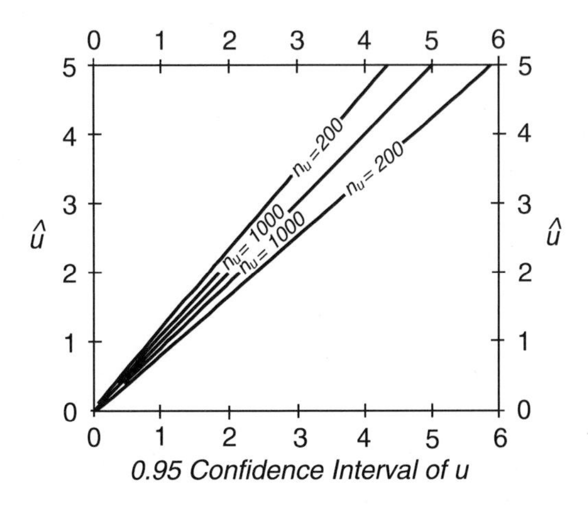

Fig. 6.2 Confidence limits for ratio data based on Eq. 6.10; n_u is the total counts excluding the taxon of interest, u the ratio between the taxon of interest and other fossils in the sample, and $\hat{u}$ the estimated ratio based on n_u counts

his equation for inside counts (binomial distribution) to the following equation for outside counts (negative binomial distribution):

$$u_{(c\ \mathrm{limit})} = \frac{\hat{u} + Z^2/(2n_u) \pm Z\sqrt{\hat{u}(1+\hat{u})/n_u + Z^2/(4n_u^2)}}{1 - (Z^2/n_u)} \tag{6.10}$$

where $u_{(c\ \mathrm{limit})}$ indicates the confidence limits for the ratio.

Figure 6.2 shows the negative binomial distributions for $n_u = 200$ and $n_u = 1{,}000$, for values of $\hat{u}$ from 0 to 5; this probably covers the size range for practical counts. Note that ratio confidence limits are always asymmetrically distributed around $\hat{u}$ with the upper limit further from $\hat{u}$ than is the lower limit. Equation 6.10 can also be used to demonstrate that as n_u increases, the confidence interval of $\hat{u}$ decreases as the reciprocal of the square root of n_u.

MOSLIMIT is a program for calculating confidence limits (0.95, 0.99, or 0.995) for counts both inside and outside the sum. The answers can be expressed either as decimal fractions or as percentages rounded to a specified number of places. The user enters the count for x and n and the upper limit, $\hat{p}$ (or $\hat{u}$), and the lower limit are displayed immediately. The executable program runs in a DOS window on computers operating under Microsoft Windows®.

Mosimann (1965: pp. 645–649, 662–666) provides examples of binomial and multinomial homogeneity tests based on the chi-square (χ^2) criterion for microfossil data expressed either as proportions or as ratios. These tests provide an objective means of judging whether microfossil counts for one or more taxa (k) from two or more samples (N) are sufficiently alike to have come from the same population. One generates the null hypothesis that the samples are from the same statistical population and then compares the computed χ^2 with critical values of χ^2 at $(k-1)(N-1)$ degrees of freedom. If the calculated χ^2 is greater than the critical value of α at the 0.05 or 0.01 level, then the null hypothesis is rejected; evidence exists that the samples do not come from the same population. The program

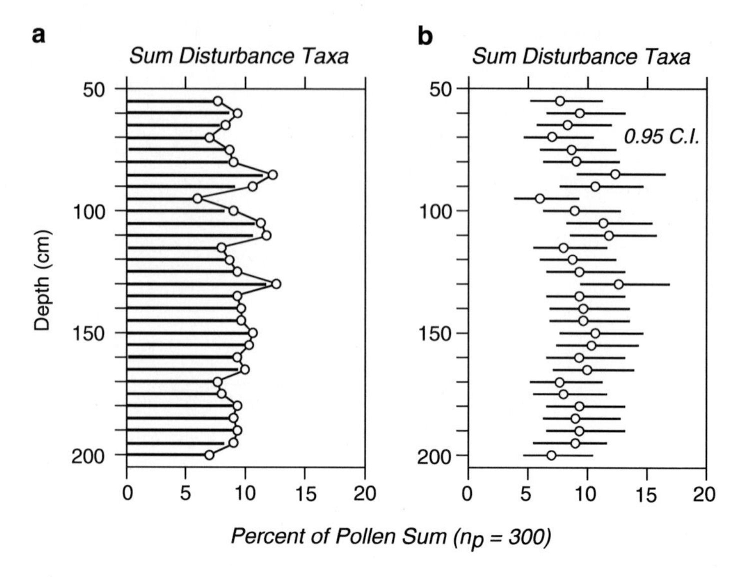

Fig. 6.3 Simulated pollen percentage diagram suggesting phases of settlement and land clearance plotted both without (**a**) and with 0.95 confidence limits (**b**)

MOSITEST allows one to make this test. Given the number of taxa and the number of samples, the program asks for the data in the required order, calculates the value of χ^2 and degrees of freedom, and states whether to accept or reject the null hypothesis. As with all such χ^2 tests, most of the taxon counts should be greater than, say, five microfossils. Taxa with low counts should be combined with others to exceed that limit.

MOSITEST offers a standard way for new researchers to check their microfossil counts against those of experienced colleagues. The apprentice counts several slides from different depths and zones in a core that the expert has previously studied. The results are run through MOSITEST to see if the null hypothesis can be accepted, and the training continues until it can. MOSITEST is also ideally suited for quality control among different laboratories, for example in diatom analysis (e.g., Munro et al. 1990).

A simple example may serve to illustrate how confidence intervals and χ^2 tests calculated by MOSITEST can be used to assess whether down-core changes in microfossil assemblages are significant beyond the variability that could be due to the total count sum of the sample. For example, assume palaeolimnologists are studying the Holocene history of settlement and land clearance. They take a core and the pollen analyst in the research group decides to amalgamate the pollen of plants that benefit from open-field conditions into a category labelled 'Disturbance Indicators'. Figure 6.3a shows that, in the 30 samples from the core, these taxa comprise a background abundance of from 5% to 10% of the 300 pollen grains in the counts from each level. The analyst notes that there are four levels (85, 110, 130, and

~150 cm) with peaks of disturbance indicators over 10%. Do these maxima reflect a real signal or are we simply dealing with noise in the data? Figure 6.3b shows the same data but with 0.95 confidence limits calculated with MOSLIMIT. Most of the 0.95 confidence intervals overlap. When we run the counts from the 30 levels through MOSITEST, the computed χ^2 is 24.91 with 29 degrees of freedom. The critical value of χ^2 for $\alpha = 0.05$ is 42.56; the null hypothesis that the 30 samples come from the same statistical population is clearly not rejected, and there is no reason to believe that any of the fluctuations are statistically significant. Analysts often report evidence of environmental change in their data. But apparent change is difficult to assess if we do not know the uncertainty inherent in the counts. To establish that the disturbance taxa really are more abundant in the four levels chosen, the pollen analyst would have to count far more than 300 pollen grains.

Treating Taxa as Numbers of Individuals Per Volume or Weight (Microfossil Concentrations)

Sediments that are well-suited for so-called 'absolute' techniques, such as wet lake muds, are homogeneous at the scale of the sample. It makes little sense to talk about how many microfossils reside in 1 cm^3 of gravel if a quartz pebble makes up most of the sample's volume, and it is difficult to measure accurately the volume of a fibrous peat. In the latter case using dry sediment weight might help solve the problem.

Percentage data are easy to calculate, but a taxon's percentage depends not only on itself, but on all the other taxa in the sample. Davis (1965) measured the number of pollen grains in a given volume or weight of sediment by suspending the processed material in a known volume of a volatile fluid (benzene; later, tertiary butyl alcohol). A similar method was described by Battarbee (1973) where a processed diatom suspension is left to settle on cover slips that are placed in an evaporation tray. These techniques allow one to estimate the number of microfossils in the original sediment.

In pollen and diatom analyses, as well as in the analysis of spherical carbonaceous particles (see e.g., Rose 1990), microfossil concentration is commonly estimated by 'spiking' each sample; that is, adding known numbers of an exotic marker to a measured volume or weight of sediment. The spike may be composed of a known quantity of tiny glass or polystyrene microspheres (Craig 1972), an aliquot drawn from a fluid with a measured concentration of microspheres or exotic pollen (Maher 1972; Battarbee and Kneen 1982), or of small tablets compressed from a mixture of exotic spores or pollen and dry powder (Stockmarr 1971, 1973). When markers are suspended in a fluid, one must make a number of assumptions: (1) the bulk material is always thoroughly mixed, (2) the carrier fluid neither evaporates over time nor affects the markers, and (3) the suspension parameters in the bulk container are the same as those pipetted into the sample. Tablets have the advantage that the overall mean and standard deviation of the tablets is set at the time of their fabrication, and will last as long as the tablets exist.

Stockmarr (1973) produced a large number of tablets of darkly stained *Lycopodium* spores, and determined the mean ($\bar{Y}$) and standard deviation (s) (e.g., 12,489 ± 491) of *Lycopodium* spores in the tablets. By adding multiple tablets, the standard deviation relative to the mean decreases. The expected number of spores in a combined group of N tablets is $N\bar{Y}$ and the standard deviation is $s\sqrt{N}$. If a single tablet is added to a sample in our example then the spike mean is 12,489 with standard deviation 491 and coefficient of variation 3.9%. If ten tablets are added the spike mean is 124,890 with a standard deviation 1553, and the group coefficient of variation falls to 1.25%. It can be shown that an analyst gets the best precision for the least amount of work if the indigenous microfossils in a sample are about twice as abundant as the markers (Maher 1981). A list of sources for marker tablets and for plastic microspheres can be found at the Indiana University Diatom Home Page (http://www.indiana.edu/~diatom/exopol.ann). Both of these marker types can be added to the raw sediment and processed along with its indigenous fossils.

The estimated concentration of microfossil X in a sample is generally based on the proportion:

$$\frac{\text{Markers in sample}}{\text{Markers counted}} = \frac{\text{Fossil } X \text{ in sample}}{\text{Fossil } X \text{ counted}} \tag{6.11}$$

Let M be the number of markers added to a vial containing the sediment sample, and let x be the number of microfossils of a taxon X counted while tallying n_m markers in a slide made from the material in the vial. Taxon X can be any taxon in the sample or groups of taxa (e.g., arboreal pollen, periphytic diatoms). Finally, let $R = x/n_m$ be the ratio of microfossils to markers in the count. An estimate of the number of microfossils of taxon X in the sample is then RM. If V is the volume of the sediment sample, the concentration of taxon X is RM/V (commonly expressed in microfossils cm^{-3}). The markers added to the sample will have a mean $\bar{M}$ and a standard deviation of S_m. Our methods of measuring the sediment volume will involve a mean and standard deviation ($\bar{V} \pm S_v$). Knowing the size of x and n_m, we can use Eq. 6.10 to obtain information about the distribution of the variable R (with n_u being equivalent to n_m). Maher (1981: Table 1, p. 179) developed the statistical basis for combining all these variables to estimate 0.95 or 0.99 confidence limits for concentration data. The procedure is involved, but the confidence limits are quickly solved with the computer program CONCENTR. The method makes certain assumptions about the variables' distribution. To meet these assumptions, the quotients $\bar{M}/s_m$ and $\bar{V}/s_v$ should both be greater than 11. Further, x should be greater than three, and the marker sum n should be greater than 99. These restrictions are not limiting for careful workers.

NULCONC, PAIRS, and COMBINE are three further programs useful for dealing with concentration data. Given a taxon's concentration confidence limits in two samples, NULCONC computes a null test for whether the data could have come from the same population. Given the raw data on the two samples' $\bar{M}/s_m$ and $\bar{V}/s_v$, and the size of their counts, PAIRS calculates the confidence limits for the two samples, a null test, and the confidence limits for the combined data. COMBINE takes a group of samples and calculates the group's mean and confidence limits, given either their raw data or their individual confidence limits.

As an example of how these programs might be used, assume that a series of samples were taken in contiguous 0.5 cm intervals and analysed for pollen or diatoms. Environmental factors (e.g., pollen-trap results, temperature, water chemistry) are available as annual means. After getting the sediment dating results (^{137}Cs, ^{210}Pb), the sediment accumulation is found to be 2 cm per year. To compare the two data-sets it will be necessary to combine the fossil samples in the sediment record; four samples being equivalent to a year. Calculating the new percentages is easy. The counts made on each of the year's four slides are analogous to counting four slides made from a single sample vial; we add all the grains together and compute the new percentages. The combined counts are larger, and the confidence intervals will thus get smaller. Because the spikes and sediment volumes may differ among the four samples, the microfossil counts for the concentration data should not simply be pooled. COMBINE makes it easy to combine the four concentration samples in each year, and the confidence intervals will also improve because of the larger counts (see also Maher 1981).

Treating Taxa as Numbers of Individuals Per Unit Surface Per Year (Accumulation Rates, Influx)

If a taxon's concentration as microfossils cm^{-3} was multiplied by the sedimentation rate S (in cm year^{-1}) the result would be an estimate of the taxon's accumulation rate or influx (in microfossils cm^{-2} year^{-1}) and, as with concentrations, we could chart the behaviour of a single taxon through time. Estimating the sedimentation rate S is relatively easy if the core is varved, but may otherwise be difficult (see Blaauw and Heegaard 2012: Chap. 12). In cases where the samples come from varved sediment, one need only count the number of varves per cm at the depth of the sample; its reciprocal would be the sedimentation rate in cm year^{-1}. Multiplying that rate by the sample's concentration limits would approximate the influx confidence limits for the sample.

It is harder to set confidence limits on samples whose sedimentation rates are estimated from a series of, say, ^{14}C dates from a few levels in a core. Maher (1981: pp. 188–190) concluded that an average sedimentation rate between two core depths does not provide hard evidence about the sedimentation rate in a sample taken somewhere in between. He suggested the term 'influx index' be used when the mean and limits of microfossil concentration are multiplied by the average sedimentation rate. Influx index limits would consider the uncertainties in sediment volume and count size, but not the uncertainties in the resolution of time – which may be large. Bennett (1994) and Blaauw and Heegaard (2012: Chap. 12) discuss possible ways of reducing this uncertainty (see also Telford et al. 2005).

Artificial Count Data to Assess the Errors Associated with Low Microfossil Counts

Instead of calculating confidence intervals for percentage abundances, proportions, or concentrations of individual microfossils in the manner described above, it may sometimes be more relevant for the interpretation of a sediment record to obtain information about the error associated with a parameter inferred from the entire fossil assemblage. Examples include environmental parameters reconstructed on the basis of fossil assemblages (e.g., temperature, lake nutrient concentrations, pH) or summary parameters describing the assemblage composition (e.g., richness, evenness, diversity, ratios between taxa with different ecological functions). For simple indices (e.g., the ratio between benthic and planktonic diatoms or cladocerans) the approach described above for ratios of different microfossil types may be appropriate. An alternative approach has been described by Heiri and Lotter (2001) and Quinlan and Smol (2001). Based on samples with a large number of analysed specimens these authors simulated low-count samples using sampling without replacement or sampling with replacement, respectively. Replicate low-count simulations then provided an estimate of the effects of low count sums on chironomid-inferred hypolimnetic oxygen (Quinlan and Smol 2001) or July air temperature (Heiri and Lotter 2001).

Inter-laboratory Comparisons

Inter-laboratory or inter-analyst comparisons have only rarely been conducted for microfossil analyses of lacustrine sediments. Munro et al. (1990) describe an inter-laboratory comparison involving four different diatom analysts. They emphasise the importance of harmonising nomenclature and establishing clear criteria for the identification of a taxon. Kreiser and Battarbee (1988) provide a more comprehensive discussion of this study and recommend ordination methods (see Legendre and Birks 2012: Chap. 8) for comparing the results of inter-analyst comparisons of fossil diatom samples. Wolfe (1997) presents the results of an inter-laboratory comparison where 15 different laboratories estimated the total diatom concentration in a uniform lake-sediment sample. He concludes that there is a clear need for additional efforts towards harmonisation of the techniques used for estimating total diatom concentrations (see also Mertens et al. 2009). Besse-Lototskaya et al. (2006) describe an attempt to assess the uncertainty in the use of periphytic diatom assemblages in river characterisation. By means of an extensive test involving ten researchers, it was concluded that the choice of site and substrate for sampling, inter-analyst differences in diatom taxonomy, and counting techniques were the primary sources of uncertainty. As a result of this quality control, clear protocols were developed for further sampling and analysis. Kelly et al. (2009) investigated questions of uncertainty in assessing the ecological status of lakes and rivers today using diatoms and developed a measure for estimating the confidence of a class and the risk of misclassification when samples are available from a site over a period of time.

Quantitative inference models using remains of aquatic organisms such as diatoms, cladocerans, or chironomids to reconstruct quantitatively climatological or limnological parameters play an increasingly important role in palaeolimnology (Smol 2008; Birks et al. 2010; Juggins and Birks 2012: Chap. 14). These models, usually based on microfossil counts by a single analyst, are commonly applied to records analysed by other investigators. An increasing effort to harmonise and evaluate the microfossil counts and identifications of different laboratories and analysts is therefore highly desirable. Heiri and Lotter (2010) assessed how taxonomic resolution affects chironomid-based temperature reconstructions. Inferences based on four different levels of taxonomic detail were compared and offsets between the reconstructions based on these taxonomic schemes were quantified. Since the chironomid assemblage data presented in this study had all been produced by the same analyst, the results presented by Heiri and Lotter (2010) only take into account the variability in inferred values due to using different levels of taxonomic detail. Uncertainties associated with variable identification skills and experience of analysts are not taken into account. Nevertheless, the approach described by Heiri and Lotter (2010) provides a means of estimating the expected variability between inferences produced by different laboratories if they are using different taxonomic frameworks and levels of taxonomic resolution when enumerating microfossil counts.

Software Availability

PALYHELP, a Microsoft Quickbasic® software package containing L.J. Maher's programs COMBINE, CONCENTR, MOSITEST, MOSLIMIT, NULCONC, and PAIRS is downloadable from the INQUA File Boutique mirror site (http://www.geology.wisc.edu/~maher/inqua.html) and the World Data Center for Palaeoclimatology (http://www.ngdc.noaa.gov/paleo/softlib/palyhelp.html). KD Bennett's PSIMPOLL, a program able to calculate confidence limits for fossil count data, is available at http://chrono.qub.ac.uk/psimpoll/psimpoll.html/.

Estimating Varve-Counting Errors

Few lakes are varved throughout their sedimentary record, whereas isolated annually laminated sediment sequences are often encountered and allow the construction of floating chronologies. In both cases, proof of the seasonal nature of deposition of the layers is essential. The layers should be carefully photographed when the cores are fresh and each picture must include labels tying them to the core logs. High-contrast prints and/or petrographic thin sections of epoxy-impregnated sediment can be optically scanned and/or separately counted by several individuals. The results may be used to assess counting errors which may differ markedly in various parts of the core. Lotter (1989), Zolitschka (1990, 2003), Bradbury and Dean (1993), Sprowl (1993), Lotter and Lemcke (1999), and Lamoureux (2001) discuss the possibilities and problems in counting varves. Hiatuses, coring artefacts, or problems with core correlation are often sources of errors in varve chronologies. Moreover, errors may also arise where low sediment-accumulation rates hamper the distinction of individual seasonal layers or make their identification ambiguous. Other sources of error result from sediment disturbances through chance events such as slumps, turbidites, tephras, or methane ebullition. To minimise these errors we strongly recommend that replicate varve counts be done in two or more cores from the same lake basin. The counts should be done by several analysts or by image analysis (e.g., Francus 2004). This would allow a simple root mean squared error of the multiple counts to be calculated. However, as this error statistic is rather optimistic, Lotter and Lemcke (1999) suggest calculating the range (i.e., maximum and minimum deviation from the mean varve counts). Root mean squared errors, standard deviations, 95% confidence limits, or ranges of replicate varve counts, if examined downcore, will allow an assessment of the quality of the varves. These statistics will remain more or less constant in well-developed varved sequences and increase disproportionally in critical parts of a varve chronology (e.g., Lamoureux 2001). Varve chronologies need to be cross-validated by independent dating methods (e.g., tephras of known age, ^{14}C, ^{137}Cs, ^{210}Pb).

Multi-core Studies

Multi-core studies provide the only way of quantifying the errors in palaeolimnological records associated with sediment coring and sediment heterogeneity. In principle, multi-core studies can be separated into two categories. Multi-core surface-sediment studies are commonly used to assess the within-lake variability of the measured parameter(s) and provide a snapshot of the quantitative distribution of the proxy of interest within the lake basin. Multiple down-core studies additionally incorporate variability associated with spatial differences in sedimentation rates, and provide information about changes in within-lake variability with time. Both approaches will also incorporate variability associated with the applied coring methods. In down-core studies correlation between the records and sediment dating will be an additional potential source of error. The studies can be designed to assess the variability of replicate cores taken in a specific location of the study lake (e.g., the deepest part of the lake basin), the comparability of littoral versus deep-water records, or to compare different sub-basins of a lake (e.g., Charles et al. 1991). Multi-core studies are of exceptional importance for determining whether quantitative inferences based on a single sediment record are representative for the whole lake basin (e.g., Charles et al. 1991; Anderson 1998; Heiri et al. 2003). Multi-core surface-sediment studies are available for a range of proxies including chironomids (e.g., Heiri et al. 2003; Heiri 2004; Eggermont et al. 2007; Holmes et al. 2009; Kurek and Cwynar 2009a, b; Engels and Cwynar 2011), diatoms and chrysophyte cysts (e.g., Charles et al. 1991; Lotter and Bigler 2000; Laird et al. 2011), pollen (e.g., Davis et al. 1971; Davis and Brubaker 1973), ostracods and molluscs (e.g., Stark 1976), cladocerans (e.g., Mueller 1964 in Frey 1988; Kattel et al. 2007), and a range of geochemical parameters (e.g., Kaushal and Binford 1999; Korsman et al. 1999; Shuman 2003). Examples of multiple down-core studies include Anderson (1990, 1998), Charles et al. (1991), Rose et al. (1999), and Shuman (2003). Replicate coring using different coring equipment can also be used to assess the reliability of different coring techniques (e.g., Lotter et al. 1997).

Conclusions

Palaeolimnological studies consist of a number of discrete steps ranging from the formulation of the hypothesis to be tested or the research question to be solved to obtaining the sediment cores in the field, analysing the samples in the laboratory, and interpreting the final results. The accuracy of the applied analytical methods and the uncertainty associated with a given experimental approach are clearly of paramount importance for solving the originally formulated research question during the final stages of a project. However, these issues are often neglected in palaeolimnological research, even though a broad palette of methods and approaches is available to quantify the uncertainties associated with the applied analytical techniques. For geochemical analyses, statistical procedures and quality control measures developed in the field of analytical chemistry can be adapted for assessing errors associated with a given method. For microfossil count data, statistical procedures and software have been developed which allow the calculation of confidence intervals for percentages, proportions, ratios, concentrations, and assemblage summary parameters, and which can be used to test for significant differences between individual samples. A number of detailed studies is available which address issues such as the assessment of counting errors in varve chronologies and the representativeness of analyses based on a single sediment core for reconstructing past lacustrine environments.

For some parameters commonly analysed in palaeolimnology, it remains difficult to assign uncertainty estimates and a number of sources of uncertainty remain difficult to quantify, either because appropriate statistical methods have not yet been developed or because the necessary error-assessment trials are extremely time-consuming. Examples are the calculation of uncertainty estimates for microfossil influx data in radiometrically dated sediment sequences and the assessment of the influence of different analysts on microfossil counts, respectively. Given the importance of uncertainty estimates for the interpretation of palaeolimnological records it is therefore highly desirable to (1) further develop statistical methods for quantifying error sources in palaeolimnological research, (2) produce additional case-studies quantifying these error sources, even if this necessitates extensive and labour-intensive inter-analyst and inter-laboratory comparisons, and (3) regularly report error estimates for palaeolimnological records irrespective of whether these are based on geochemical or microfossil analyses.

References

Anderson NJ (1990) Variability of diatom concentrations and accumulation rates in sediments of a small lake basin. Limnol Oceanogr 35:497–508

Anderson NJ (1998) Variability of diatom-inferred phosphorus profiles in a small lake basin and its implications for histories of lake eutrophication. J Paleolimnol 20:47–55

Battarbee RW (1973) A new method for estimating absolute microfossil numbers with special reference to diatoms. Limnol Oceanogr 18:647–653

Battarbee RW, Kneen MJ (1982) The use of electronically counted microspheres in absolute diatom analysis. Limnol Oceanogr 27:184–188

Battarbee RW, Jones VJ, Flower RJ, Cameron NG, Bennion H, Carvalho L, Juggins S (2001) Diatoms. In: Smol JP, Birks HJB, Last WM (eds) Tracking environmental change using lake sediments. Volume 3: Terrestrial, algal, and siliceous indicators. Kluwer Academic Publishers, Dordrecht, pp 155–202

Beaudoin A (2003) A comparison of two methods for estimating the organic content of sediments. J Paleolimnol 29:387–390

Bennett KD (1994) Confidence intervals for age estimates and deposition times in late-Quaternary sediment sequences. Holocene 4:337–348

Besse-Lototskaya A, Verdonschot PFM, Sinkeldam JA (2006) Uncertainty in diatom assessment: sampling, identification and counting variation. Hydrobiologia 566:247–260

Birks HJB (2012) Chapter 2: Overview of numerical approaches in palaeolimnology. In: Birks HJB, Lotter AF, Juggins S, Smol JP (eds) Tracking environmental change using lake sediments. Volume 5: Data handling and numerical techniques. Springer, Dordrecht

Birks HJB, Birks HH (1980) Quaternary palaeoecology. Arnold, London

Birks HJB, Heiri O, Seppä H, Bjune AE (2010) Strengths and weaknesses of quantitative climate reconstructions based on late-Quaternary biological proxies. Open Ecol J 3:68-110

Blaauw M, Heegaard E (2012) Chapter 12: Estimation of age-depth relationships. In: Birks HJB, Lotter AF, Juggins S, Smol JP (eds) Tracking environmental change using lake sediments. Volume 5: Data handling and numerical techniques. Springer, Dordrecht

Blæsild P, Granfeldt J (2003) Statistics with applications in biology and geology. Chapman & Hall/CRC, Boca Raton

Borradaile G (2003) Statistics of earth science data. Their distribution in time, space and orientation. Springer, Berlin

Boyle JF (2000) Rapid elemental analysis of sediment samples by isotopic source XRF. J Paleolimnol 23:213–221

Boyle JF (2001) Inorganic geochemical methods in paleolimnology. In: Last WM, Smol JP (eds) Tracking environmental change using lake sediments. Volume 2: Physical and geochemical methods. Kluwer Academic Publishers, Dordrecht, pp 83–141

Bradbury JP, Dean WE (eds) (1993) Elk Lake, Minnesota: evidence for rapid climatic change in the north-central United States. Geol Soc Am Spec Paper 276:1–336

Brunori C, Ipolyi I, Macaluso L, Morabito R (2004) Evaluation of an ultrasonic digestion procedure for total metal determination in sediment reference materials. Anal Chim Acta 510:101–107

Charles DF, Dixit SS, Cumming BF, Smol JP (1991) Variability in diatom and chrysophyte assemblages and inferred pH: paleolimnological studies of Big Moose Lake, New York, USA. J Paleolimnol 5:267–284

Conley DJ (1998) An interlaboratory comparison for the measurement of biogenic silica in sediments. Mar Chem 63:39–48
Craig AJ (1972) Pollen influx to laminated sediments: a pollen diagram from northeastern Minnesota. Ecology 53:46–57
Davis JC (2002) Statistics and data analysis in geology, 3rd edn. Wiley, New York
Davis MB (1965) A method for determination of absolute pollen frequency. In: Kummel B, Raup D (eds) Handbook of paleontological techniques. Freeman, San Francisco, pp 674–686
Davis MB, Brubaker LB (1973) Differential sedimentation of pollen grains in lakes. Limnol Oceanogr 18:635–646
Davis MB, Brubaker LB, Beiswenger JM (1971) Pollen grains in lake sediments: pollen percentages in surface sediments from southern Michigan. Quaternary Res 1:450–467
Dean WE (1974) Determination of carbonate and organic matter in calcareous sediments and sedimentary rocks by loss on ignition: comparison with other methods. J Sediment Petrol 44:242–248
Dixon EM, Gardner MJ, Hudson R (1997) The comparability of sample preparation techniques for the determination of metals in sediments. Chemosphere 35:2225–2236
Eggermont H, De Deyne P, Verschuren D (2007) Spatial variability of chironomid death assemblages in the surface sediments of a fluctuating tropical lake (Lake Naivasha, Kenya). J Paleolimnol 38:309–328
Ellison SLR (1998) ISO uncertainty and collaborative trial data. Accred Qual Assur 3:95–100
Ellison SLR, Rosslein M, Williams A (eds) (2000) Quantifying uncertainty in analytical measurements, 2nd edn. Eurachem/CITAC Guide 4, London
Engels S, Cwynar LC (2011) Changes in fossil chironomid remains along a depth gradient: evidence for common faunal thresholds within lakes. Hydrobiologia 665:15–38
Fleming J, Neidhart B, Albus H, Wegscheider W (1996a) Glossary of analytical terms (III). Accred Qual Assur 1:135
Fleming J, Neidhart B, Tausch C, Wegscheider W (1996b) Glossary of analytical terms (I). Accred Qual Assur 1:41–43
Francus P (ed) (2004) Image analysis, sediments, and paleoenvironments. Springer, Dordrecht
Frey DG (1988) Littoral and offshore communities of diatoms, cladocerans, and dipterous larvae, and their interpretation in paleolimnology. J Paleolimnol 1:179–191
Heiri O (2004) Within-lake variability of subfossil chironomid assemblages in shallow Norwegian lakes. J Paleolimnol 32:67–84
Heiri O, Lotter AF (2001) Effect of low count sums on quantitative environmental reconstructions: an example using subfossil chironomids. J Paleolimnol 26:43–350
Heiri O, Lotter AF (2010) How does taxonomic resolution affect chironomid-based temperature reconstruction? J Paleolimnol 44:589–601
Heiri O, Lotter AF, Lemcke G (2001) Loss on ignition as a method for estimating organic and carbonate content in sediments: reproducibility and comparability of results. J Paleolimnol 25:101–110
Heiri O, Birks HJB, Brooks SJ, Velle G, Willassen E (2003) Effects of within-lake variability of fossil assemblages on quantitative chironomid-inferred temperature reconstruction. Palaeogeogr Palaeoclim Palaeoecol 199:95–106
Holmes N, Langdon PG, Caseldine CJ (2009) Subfossil chironomid variability in surface sediment samples from Icelandic lakes: implications for the development and use of training sets. J Paleolimnol 42:281–295
Juggins S, Birks HJB (2012) Chapter 14: Quantitative environmental reconstructions from biological data. In: Birks HJB, Lotter AF, Juggins S, Smol JP (eds) Tracking environmental change using alke sediments. Volume 5: Data handling and numerical techniques. Springer, Dordrecht
Kattel GR, Battarbee RW, Mackay A, Birks HJB (2007) Are cladoceran fossils in lake sediment samples a biased reflection of the communities from which they are derived? J Paleolimnol 38:157–181

Kaushal S, Binford MW (1999) Relationship between C:N ratios of lake sediments, organic matter sources, and historical deforestation in Lake Pleasant, Massachusetts, USA. J Paleolimnol 22:439–442

Kelly M, Bennion H, Burgess A, Ellis J, Juggins S, Guthrie R, Jamieson J, Adriaenssens V, Yallop M (2009) Uncertainty in ecological status assessments of lakes and rivers using diatoms. Hydrobiologia 633:5–15

King P, Kennedy H, Newton PP, Jickells TD, Brand T, Calvert S, Cauwet G, Etcheber H, Head B, Khripounoff A, Manighetti B, Miquel JC (1998) Analysis of total and organic carbon and total nitrogen in settling oceanic particles and a marine sediment: an interlaboratory comparison. Mar Chem 60:203–216

Korsman T, Nilsson MB, Landgren K, Renberg I (1999) Spatial variability in surface sediment composition characterised by near-infrared (NIR) reflectance spectroscopy. J Paleolimnol 21:61–71

Kreiser A, Battarbee RW (1988) Analytical quality control (AQC) in diatom analysis. In: Miller U, Robertsson A-M (eds) Proceedings of Nordic diatomist meeting, Stockholm, June 10–12, 1987. University of Stockholm, Department of Quaternary Research Report, pp 41–44

Kurek J, Cwynar LC (2009a) Effects of within-lake gradients on the distribution of fossil chironomids from maar lakes in western Alaska: implications for environmental reconstructions. Hydrobiologia 623:37–52

Kurek J, Cwynar LC (2009b) The potential of site-specific and local chironomid-based inference models for reconstructing past lake levels. J Paleolimnol 42:37–50

Kurek J, Korosi JB, Jeziorski A, Smol JP (2010) Establishing reliable minimum count sizes for cladoceran subfossils sampled from lake sediments. J Paleolimnol 44:603–612

Laird KR, Kingsbury MV, Lewis CFM, Cumming BF (2011) Diatom-inferred depth models in 8 Canadian boreal lakes: inferred changes in the benthic:planktonic depth boundary and implications for assessment of past droughts. Quaternary Sci Rev 30:1201–1217

Lamoureux S (2001) Varve chronology techniques. In: Last WM, Smol JP (eds) Tracking environmental change using lake sediments. Volume 1: Basin analysis, coring, and chronological techniques. Kluwer, Dordrecht, pp 247–260

Larocque I (2001) How many chironomid head capsules are enough? A statistical approach to determine sample size for paleoclimatic reconstructions. Palaeogeogr Palaeoclim Palaeoecol 172:133–142

Lasorsa B, Casas A (1996) A comparison of sample handling and analytical methods for determination of acid volatile sulfides in sediment. Mar Chem 52:211–220

Legendre AF, Birks HJB (2012) Chapter 8: From classical to canonical ordination. In: Birks HJB, Lotter AF, Juggins S, Smol JP (eds) Tracking environmental change using lake sediments. Volume 5: Data handling and numerical techniques. Springer, Dordrecht

Lotter AF (1989) Evidence of annual layering in Holocene sediments of Soppensee, Switzerland. Aquat Sci 51:19–30

Lotter AF, Bigler C (2000) Do diatoms in the Swiss Alps reflect the length of ice cover? Aquat Sci 62:125–141

Lotter AF, Lemcke G (1999) Methods for preparing and counting biochemical varves. Boreas 28:243–252

Lotter AF, Merkt J, Sturm M (1997) Differential sedimentation *versus* coring artifacts: a comparison of two widely used piston-coring methods. J Paleolimnol 18:75–85

Maher LJ (1972) Nomograms for computing 0.95 confidence limits for pollen data. Rev Palaeobot Palynol 13:85–93

Maher LJ (1981) Statistics for microfossil concentration measurements employing samples spiked with marker grains. Rev Palaeobot Palynol 32:153–191

Mainland D, Herrera L, Sutcliffe MI (1956) Statistical tables for use with binomial samples, contingency tests, confidence limits, and sample size estimates. Department of Medical Statistics, New York University College of Medicine

Mertens KN, Verhoeven K, Verleye T, Louwye S et al. (2009) Determining the absolute abundance of dinoflagellate cysts in recent marine sediments: the *Lycopodium* marker-grain method put to the test. Rev Palaeobot Palynol 157:238–252
Mosimann JE (1965) Statistical methods for the pollen analyst: multinomial and negative multinomial techniques. In: Kummel B, Raup D (eds) Handbook of paleontological techniques. Freeman, San Francisco, pp 636–673
Mueller WP (1964) The distribution of cladoceran remains in surficial sediments from three northern Indiana lakes. Invest Indiana Lakes Streams 6:1–63
Munro MAR, Kreiser AM, Battarbee RW, Juggins S, Stevenson AC, Anderson DS, Anderson NJ, Berge F, Birks HJB, Davis RB, Flower RJ, Fritz SC, Haworth EY, Jones VJ, Kingston JC, Renberg I (1990) Diatom quality control and data handling. Philos Trans R Soc Lond B 327:257–261
Namiesnik J, Zygmunt B (1999) Role of reference materials in analysis of environmental pollutants. Sci Total Environ 228:243–257
Payne RJ, Mitchell EAD (2009) How many is enough? Determining optimal count tools for ecological and palaeoecological studies of testate amoebae. J Paleolimnol 42:483–495
Quinlan R, Smol JP (2001) Setting minimum head capsule abundance and taxa deletion criteria in chironomid-based inference models. J Paleolimnol 26:327–342
Rohlf FJ, Sokal RR (1969) Statistical tables. Freeman, San Francisco
Rose NL (1990) A method for the selective removal of inorganic ash particles from lake sediments. J Paleolimnol 4:61–67
Rose NL, Harlock S, Appleby PG (1999) Within-basin profile variability and cross-correlation of lake sediment cores using the spheroidal carbonaceous particle record. J Paleolimnol 21:85–96
Sahuquillo A, Lopez-Sanchez JF, Rubio R, Rauret G, Thomas RP, Davidson CM, Ure AM (1999) Use of a certified reference material for extractable trace metals to assess sources of uncertainty in the BCR three-stage sequential extraction procedure. Anal Chim Acta 382:317–327
Santisteban JI, Mediavilla R, López-Pamo E, Dabrio CJ, Ruiz Zapata MB, Gil García MJ, Castaño S, Martínez-Alfaro PE (2004) Loss on ignition: a qualitative or quantitative method for organic matter and carbonate content in sediments? J Paleolimnol 32:287–299
Shuman B (2003) Controls on loss-on-ignition variation in cores from two shallow lakes in the northeastern United States. J Paleolimnol 30:371–385
Smol JP (2008) Pollution of lakes and rivers. A paleoenvironmental perspective, 2nd edn. Blackwell, Oxford
Sokal RR, Rohlf FJ (1995) Biometry, 2nd edn. WH Freeman, New York
Solhøy T (2001) Orbatid mites. In: Smol JP, Birks HJB, Last WM (eds) Tracking environmental change using lake sediments. Volume 4: Zoological indicators. Kluwer, Dordrecht, pp 81–104
Somogyi A, Braun M, Posta J (1997) Comparison between X-ray fluorescence and inductively coupled plasma atomic emission spectrometry in the analysis of sediment samples. Spectrochim Acta Part B 52:2011–2017
Sprowl DR (1993) On the precision of the Elk Lake varve chronology. In: Bradbury JP, Dean WE (eds) Elk Lake, Minnesota: evidence for rapid climatic change in the north-central United States. Geol Soc Am Spec Paper 276:69–74
Stark DM (1976) Paleolimnology of Elk Lake, Itasca State Park, Northwestern Minnesota. Archiv für Hydrobiologie/Suppl 50:208–274
Stockmarr J (1971) Tablets with spores used in absolute pollen analysis. Pollen Spores 13:615–621
Stockmarr J (1973) Determination of spore concentration with an electronic particle counter. Geological Survey of Denmark Yearbook 1972, pp 87–89
Tanner PA, Leong LS (1995) The effects of different drying methods for marine sediment upon moisture content and metal determination. Mar Pollut Bull 31:325–329
Telford RJ, Heegaard E, Birks HJB (2005) All age-depth models are wrong: but how badly? Quaternary Sci Rev 23:1–5
Tung JWT, Tanner PA (2003) Instrumental determination of organic carbon in marine sediments. Mar Chem 80:161–170

Walker R, Lumley I (1999) Pitfalls in terminology and use of reference materials. Trends Anal Chem 18:594–616

Wall AAJ, Gilbert D, Magny M, Michell EAD (2010) Testate amoeba analysis of lake sediments: impact of filter size and total count on estimates of density, species richness and assemblage structure. J Paleolimnol 43:689–704

Williams A (1996) Measurement uncertainty in analytical chemistry. Accred Qual Assur 1:14–17

Wolfe AP (1997) On diatom concentrations in lake sediments: results from an inter-laboratory comparison and other tests performed on a uniform sample. J Paleolimnol 18:261–268

Zar JH (1999) Biostatistical analysis, 4th edn. Prentice-Hall, Upper Saddle River

Zolitschka B (1990) Spätquartäre jahreszeitlich geschichtete Seesedimente ausgewählter Eifelmaare. Paläolimnologische Untersuchungen als Beitrag zur spät- und postglazialen Klima- und Besiedlungsgeschichte. Documenta Naturae 60:1–226

Zolitschka B (2003) Dating based on freshwater- and marine-laminated sediments. In: Mackay AW, Battarbee RW, Birks HJB, Oldfield F (eds) Global change in the Holocene. Arnold, London, pp 92–106

Chapter 7
Clustering and Partitioning

Pierre Legendre and H. John B. Birks

Abstract Hierarchical clustering methods and partitioning techniques such as *K*-means partitioning and two-way indicator species analysis are useful tools for summarising group structure within large, complex, multivariate data-sets that are increasingly common in palaeolimnology. The incorporation of one- or two-dimensional constraints in the clustering algorithms provides means of exploring group structure in temporal, stratigraphical data and in geographical modern data, respectively. Indicator species analysis with its associated permutation tests is a simple and effective means of detecting statistically significant indicator species for any grouping of a set of objects. The newly developed approach of multivariate regression trees combines partitioning and data exploration with regression and data interpretation and modelling.

Keywords Agglomerative clustering • Constrained clustering • Indicator species • *K*-means partitioning • Multivariate regression trees • Partitioning • Two-way indicator species analysis

P. Legendre (✉)
Département de sciences biologiques, Université de Montréal, C.P. 6128, succursale Centre-ville, Montréal, QC, H3C 3J7, Canada
e-mail: pierre.legendre@umontreal.ca

H.J.B. Birks
Department of Biology and Bjerknes Centre for Climate Research, University of Bergen, PO Box 7803, Bergen N-5020, Norway

Environmental Change Research Centre, University College London, Pearson Building, Gower Street, London WC1E 6BT, UK

School of Geography and the Environment, University of Oxford, Oxford OX1 3QY, UK
e-mail: john.birks@bio.uib.no

H.J.B. Birks et al. (eds.), *Tracking Environmental Change Using Lake Sediments*, Developments in Paleoenvironmental Research 5, DOI 10.1007/978-94-007-2745-8_7,

Introduction

Hierarchical clustering methods were developed as heuristic (empirical) tools to produce tree-like arrangements of large numbers of observations. The original intention of the biologists who developed them was to obtain a tree-like representation of the data, in the hope that it would reflect the underlying pattern of evolution. Hierarchical clustering starts with the calculation of a similarity or dissimilarity (= distance) matrix using a coefficient which is appropriate to the data and research problem. The choice of an appropriate distance coefficient is discussed in Legendre and Birks (2012: Chap. 8). Here we will briefly describe the algorithms most commonly used for hierarchical clustering.

Partitioning methods were developed within a more rigorous statistical frame. K-means partitioning, in particular, attempts to find partitions that optimise the least-squares criterion, which is widely and successfully used in statistical modelling. Least-squares modelling methods include regression, analysis of variance, and canonical analysis (Birks 2012a: Chap. 2). Data may need to be transformed prior to K-means partitioning. This is the case, in particular, for assemblage data. Please refer to Table 8.2 of Legendre and Birks (2012: Chap. 8) for details of such transformations.

Palaeoecologists have long been interested in segmenting time series of data, such as sediment cores that represent depositional temporal-series. Several approaches have been proposed in the literature. They can all be seen as special cases of clustering or partitioning with constraints.

Artificial Example

Table 7.1 shows an artificial data dissimilarity and similarity matrix among five objects that will be used to illustrate various agglomerative clustering methods throughout this chapter. Clustering can be computed from either similarity (**S**) or dissimilarity (or distance) matrices (**D**); most software has preferences for either **S** or **D** matrices. Figure 7.1 shows the relationships among the five objects in the form of a two-dimensional principal coordinate analysis (PCoA) ordination diagram (see Legendre and Birks 2012: Chap. 8 for details of PCoA). PCoA axis 1 accounts for 65.9% of the variation of the data while axis 2 accounts for 27.6%, leaving a mere 6.5% for axis 3. So the five data points are very well represented in 2 dimensions. We will cluster these five objects using various methods. Two of the interpoint distances are especially distorted in the two-dimensional ordination: $D(1,2) = 1.389$ in the ordination instead of 1 in the original distance matrix (Table 7.1), and $D(4,5) = 1.159$ instead of 2. The other pair-wise distances in two dimensions are close to their original values (Table 7.1). We will see how the various clustering methods deal with these similarities or dissimilarities.

Table 7.1 Matrices showing the dissimilarity (or distance, **D**, on the left) and similarity (**S**, on the right, with $S_{ij} = 1 - D_{ij}/D_{max}$) relationships among five objects, numbered 1–5 (artificial data)

D	1	2	3	4	5	**S**	1	2	3	4	5
1	*0*					1	*1*				
2	1	*0*				2	0.8	*1*			
3	5	5	*0*			3	0.0	0.0	*1*		
4	5	4	3	*0*		4	0.0	0.2	0.4	*1*	
5	5	4	4	2	*0*	5	0.0	0.2	0.2	0.6	*1*

In each matrix, the upper-triangular portion (not shown) is symmetric to the lower-triangular. In the distance matrix (**D**), the main diagonal (italics) contains 0's, whereas it contains 1's in the similarity matrix (**S**)

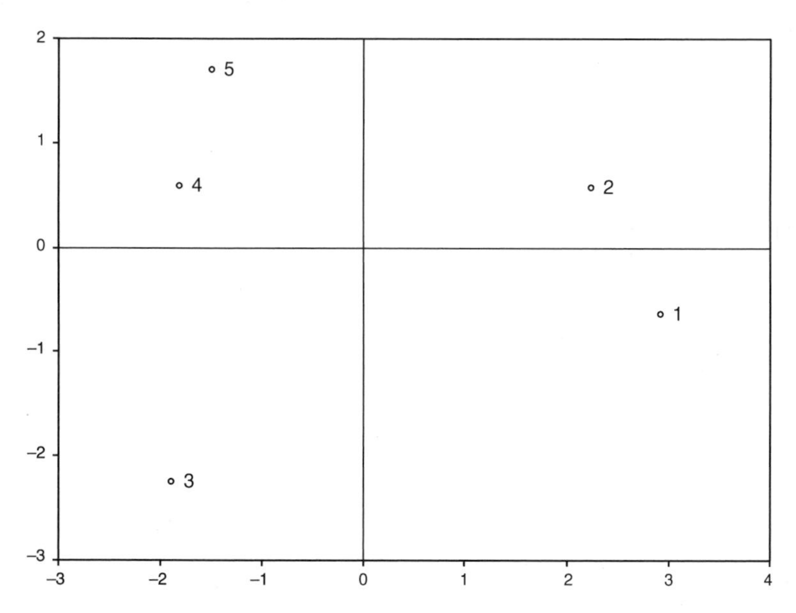

Fig. 7.1 Principal coordinate analysis (PCoA) ordination computed from the distance matrix of the artificial data (Table 7.1) showing the relationships among the five objects

Basic Concepts in Clustering

A cluster is a group of objects (observations, sites, samples, etc.) that are sufficiently similar to be recognised as members of the same group. Clustering results from an analysis of the similarities or dissimilarities among objects, calculated from the data of interest. The similarity and dissimilarity measures most commonly used by ecologists and palaeolimnologists are described in Legendre and Birks (2012:

Chap. 8). A partition, such as produced by the K-means partitioning method, is a set of non-overlapping clusters covering the whole collection of objects in the study; some clusters may be of size 1 (singletons). Hierarchical clustering produces a hierarchy of nested partitions of objects. Numerical clustering algorithms will always produce a partition or a hierarchical clustering, whatever the data. So, obtaining a partition or a hierarchical set of partitions does not demonstrate that there are real discontinuities in the data. Most hierarchical clustering methods are heuristic techniques, producing a solution to the problem but otherwise without any statistical justification. A few methods are based on statistical concepts such as sum-of-squares.

Clustering methods summarise data with an emphasis on pair-wise relationships. The most similar objects are placed in the same group, but the resulting dendrogram provides little information about among-group relationships. Ordination methods do the opposite: ordination diagrams depict the main trends in data but pair-wise distances may be distorted. For many descriptive purposes, it is often valuable to conduct both forms of analysis (e.g., Birks et al. 1975; Birks and Gordon 1985; Prentice 1986; Battarbee et al. 2011; Bennion and Simpson 2011).

The various potential uses of clustering and partitioning in palaeolimnology are summarised in Table 7.2. No attempt is made here to give a comprehensive review of palaeolimnological applications of clustering or partitioning. Emphasis is placed instead on basic concepts and on methods that have rarely been used but that have considerable potential in palaeolimnology.

Clustering with the constraint of spatial contiguity involves imposing that all members of a cluster be contiguous on the spatial map of the objects. Clustering with a one-dimensional contiguity constraint is often used on sediment cores to delineate sectors or zones where the core sections are fairly homogeneous in terms of their sediment texture, fossil composition, etc., and to identify transition zones (Birks and Gordon 1985; Birks 2012b: Chap. 11). Cores can be seen as one-dimensional geographic (or temporal) data series, so the concept of clustering with a contiguity constraint can be applied to them. Other forms of constraint can be applied to the data to be clustered through the use of canonical analysis (see Birks 2012a: Chap. 2 and Legendre and Birks 2012: Chap. 8) or multivariate regression trees (see Simpson and Birks 2012: Chap. 9).

The most simple form of clustering for multivariate data is to compute an ordination (principal component analysis (PCA), principal coordinate analysis (PCoA), correspondence analysis (CA) – see Legendre and Birks 2012: Chap. 8), draw the points in the space of ordination axes 1 and 2, and divide the points into boxes of equal sizes. This will produce a perfectly valid partition of the objects and it may be all one needs for some purposes, such as the basic summarisation of the data. In other cases, one prefers to delineate groups that are separated from other groups by gaps in multivariate space. The clustering methods briefly described in this chapter should then be used. For further details of clustering methods, see Legendre and Legendre (1998: Chap. 8), Borcard et al. (2011: Chap. 4), Everitt and Hothorn (2011: Chap. 6), Everitt et al. (2011), and Wehrens (2011).

Table 7.2 Palaeolimnological uses of clustering and partitioning techniques

Uses
Modern biological assemblages (e.g., diatoms, chironomids)
Detect groups of samples with similar biological composition – AHC, TWINSPAN, *K*-means
Detect groups of samples with similar biological composition along a single environmental gradient (e.g., altitude, pH) – CC1
Detect groups of geographically contiguous samples with similar biological composition – CC2
Detect indicator species for groups of samples – ISA, TWINSPAN
Comparison of groupings based on different groups of organisms – AHC, TWINSPAN, or *K*-means followed by CC
Detect groups of samples that can be overlain on an ordination of the same samples – AHC, TWINSPAN, *K*-means
Modern environmental data (e.g., lake-water chemistry)
Detect groups of lakes with similar environmental variables – AHC, *K*-means
Detect groups of geographically contiguous lakes with similar environmental variables – CC2
Comparison of groupings based on different types of environmental variables – AHC or *K*-means followed by CC
Detect groups of samples that can be overlain on an ordination of the same samples – AHC, *K*-means
Fossil biological assemblages (e.g., diatoms, chironomids)
Detect groups of samples (zones) with similar fossil composition – CC1, MRT
Detect indicator species for groups of samples (zones) with similar fossil composition – ISA
Comparison of groups of samples (zones) based on different fossil groups – CC1, MRT, CC
Detect groups of samples (zones) with similar fossil composition that can be overlain on an ordination of the same samples – CC1, MRT
Detect recurring groups of samples with similar fossil and/or sediment composition – AHC
Down-core non-biological data (e.g., geochemistry, magnetics)
Detect groups of samples with similar geochemical composition, magnetic properties, etc. – CC1, MRT
Comparison of clusterings of samples based on different set of variables – CC1, MRT followed by CC
Detect groups of samples (zones) with similar chemical composition or magnetic properties that can be overlain on an ordination of the same samples – CC1, MRT
Detect recurring groups of samples with similar sediment geochemical composition or magnetic properties – AHC
Modern and fossil biological assemblages (e.g., diatoms, chironomids)
Detect similarities between modern and fossil samples ('analogues') as an aid to interpreting fossil samples – AHC, TWINSPAN, *K*-means
Modern biological assemblages and modern environmental data (e.g., diatoms and lake-water chemistry)
Detect environmental variables that characterise clusters or partitions of modern samples – MRT, DA
Integrated clustering of sites on the basis of both biological and environmental data – CCC followed by *K*-means
Comparison of groups of samples of similar biological composition with groups of lakes based on environmental variables – AHC, TWINSPAN, or *K*-means followed by CC
Modern biological assemblages, modern environmental data, and fossil biological assemblages (e.g., diatoms and lake-water chemistry)
Detect groups of similar modern and fossil samples on basis of similar composition and then relate the groups of modern samples to modern environmental variables – TWINSPAN with fossil samples declared passive in WinTWINS, followed by DA using modern environmental data of TWINSPAN modern groups

(continued)

Table 7.2 (continued)

Fossil biological assemblages and palaeoenvironmental variables (e.g., diatoms, occurrences of volcanic tephras, stable-isotope data)
Relate fossil assemblage changes to palaeoenvironmental variables – MRT or CC1 of biological data followed by DA
Detect groups of samples with similar fossil composition ('zones') and palaeoenvironmental variables – CCC followed by CC1
Fossil biological assemblages from many sites
Detect groups of sites with similar fossil biological assemblages at a particular time – AHC, *K*-means, TWINSPAN
Detect groups of geographically contiguous sites with similar fossil biological assemblages at a particular time – CC2

AHC agglomerative hierarchical clustering, *TWINSPAN* two-way indicator species analysis, *K-means* *K*-means partitioning, *CC1* constrained clustering in one dimension, *MRT* multivariate regression tree, *CC2* constrained clustering in two dimensions, *ISA* indicator species analysis, *DA* simple discriminant analysis, *CC* clustering (or partitioning) comparison (Rand's index, etc.), *CCC* clustering constrained by canonical ordination results

Unconstrained Agglomerative Clustering Methods

Only the hierarchical clustering methods commonly found in statistical software will be described in this section. The most commonly used method is unweighted arithmetic average clustering (Rohlf 1963), also called UPGMA (for 'Unweighted Pair-Group Method using Arithmetic averages': Sneath and Sokal 1973) or 'group-average sorting' (Lance and Williams 1966, 1967). The algorithm proceeds by step-wise condensation of the similarity or dissimilarity matrix. Each step starts by the identification of the next pair that will cluster; this is the pair having the largest similarity or the smallest dissimilarity. This is followed by condensation of all the other measures of resemblance involving that pair, by the calculation of the arithmetic means of the similarities or dissimilarities.

The procedure is illustrated for similarities for the artificial data (Table 7.3). Objects 1 and 2 should cluster first because their similarity (0.8) is the highest. The similarity matrix is condensed by averaging the similarities of these two objects with all other objects in turn. Objects 4 and 5 should cluster during the second step because their similarity (0.6) is the highest in the condensed table. Again, the similarities of these two objects are averaged. During step 3, object 3 should cluster at $S = 0.3$ with the group (4,5) previously formed. In UPGMA, one has to weight the similarities by the number of objects involved when calculating the average similarity: $((1 \times 0.0) + (2 \times 0.1))/3 = 0.067$. This *weighted average* is actually equivalent to calculating the *simple (unweighted) mean* of the 6 similarities between objects 1 and 2 on the one hand and 3, 4, and 5 on the other, in the first panel of the table: $(0.0 + 0.0 + 0.0 + 0.0 + 0.2 + 0.2)/6 = 0.067$. In that sense, the method is 'unweighted'. The dendrogram representing the hierarchical clustering results is shown in Fig. 7.2.

Table 7.3 Step-wise condensation of the similarity matrix from Table 7.1 during unweighted pair-group method using arithmetic averages (UPGMA) agglomerative clustering

Objects	1	2	3	4	5
1	—				
2	***0.8***	—			
3	0.0	0.0	—		
4	0.0	0.2	0.4	—	
5	0.0	0.2	0.2	0.6	—
Cluster objects 1 and 2 at $S = 0.8$					
1–2		—			
3		0.0	—		
4		$(0 + 0.2)/2 = 0.1$	0.4	—	
5		$(0 + 0.2)/2 = 0.1$	0.2	***0.6***	—
Cluster objects 4 and 5 at $S = 0.6$					
1-2		—			
3		0.0	—		
4-5		$(0.1 + 0.1)/2 = 0.1$	$(0.4 + 0.2)/2 =$ ***0.3***	—	
Cluster object 3 with group (4,5) at $S = 0.3$					
1–2		—			
3–4–5		$(1\times0.0) + (2\times0.1)/3 =$ ***0.067***	—		
Cluster group (1,2) with group (3,4,5) at $S = 0.067$					

At each step, the highest similarity value is identified (italic boldface value in the previous step); then the values corresponding to the similarities of these two objects or groups, with all other objects or groups, are averaged in turn

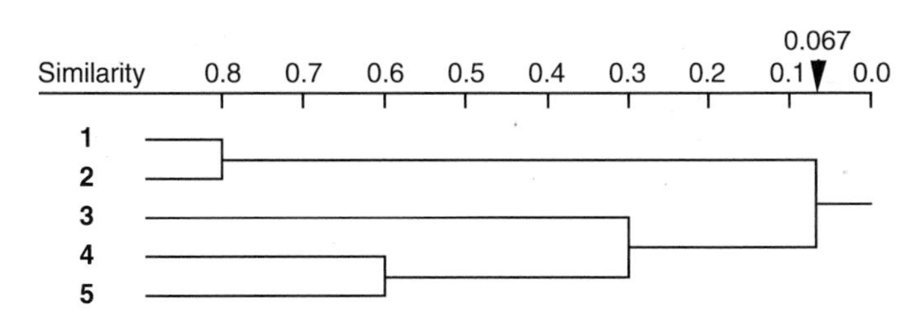

Fig. 7.2 Unweighted pair-group method using arithmetic averages (UPGMA) agglomerative clustering of the objects of the artificial example

Weighted arithmetic average clustering (Sokal and Michener 1958), also called WPGMA (for 'Weighted Pair-Group Method using Arithmetic averages': Sneath and Sokal 1973), only differs from UPGMA in the fact that *a simple, unweighted mean* is computed at each step of the similarity matrix condensation. This is equivalent to giving *different weights to the original similarities* (first panel of Table 7.3) when condensing the similarities, hence the name 'weighted'. For the data in our example, only the last fusion is affected; the similarity level of the last fusion is: $(0.0 + 0.1)/2 = 0.05$. Otherwise, the dendrogram is similar to that of Fig. 7.2.

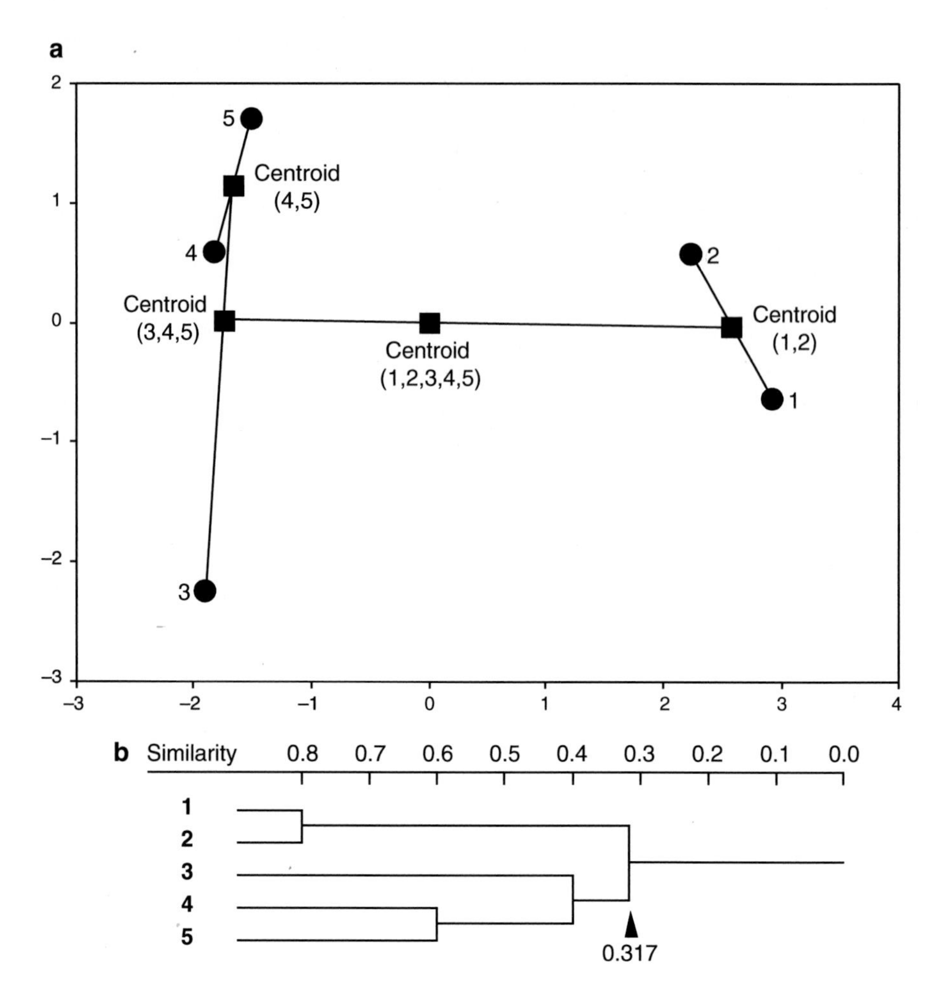

Fig. 7.3 (**a**) In principal coordinate space, each unweighted centroid (UPGMC) clustering step leads to the replacement of objects (*circles*) by their centroid (*squares*), computed as the centre of mass of all the objects members of a group. (**b**) The UPGMC dendrogram

Unweighted centroid clustering (Lance and Williams 1967; UPGMC in Sneath and Sokal 1973) proceeds from a different conceptual paradigm. Imagine the objects in multidimensional space: as in UPGMA, the first two objects to cluster are chosen as the pair having the largest similarity or smallest dissimilarity or distance. Instead of averaging their similarities to all other objects, the two clustered objects are replaced by their centroid, or centre of mass, in multivariate space. This is illustrated in Fig. 7.3a, a representation which is two- instead of three-dimensional. An UPGMC centroid is located at the centre of mass of all the objects that are members of a cluster.

Table 7.4 Matrices of normalised distances $[D_{ij}/D_{max}]$ and squared normalised distances $[(D_{ij}/D_{max})^2]$ used in Ward's agglomerative clustering

D_{ij}/D_{max}	1	2	3	4	5	$(D_{ij}/D_{max})^2$	1	2	3	4	5
1	*0*					1	*0*				
2	0.2	*0*				2	0.04	*0*			
3	1.0	1.0	*0*			3	1.00	1.00	*0*		
4	1.0	0.8	0.6	*0*		4	1.00	0.64	0.36	*0*	
5	1.0	0.8	0.8	0.4	*0*	5	1.00	0.64	0.64	0.16	*0*

Clustering step k	New cluster formed	Fusion D^2_k	Fusion D_k	ESS_k	$TESS_k$	R^2_k
0	Five separate objects	0	0	0	0	1
1	(1,2)	0.04	0.2	0.02	0.02	0.985
2	(4,5)	0.16	0.4	0.08	0.10	0.923
3	(3,4,5)	0.613	0.783	0.387	0.407	0.686
4	(1,2,3,4,5)	1.779	1.334	1.296	1.296	0

ESS error sum-of-squares, *TESS* total error sum-of-squares

In the weighted form of centroid clustering, called WPGMC (weighted centroid clustering: Gower 1967), a centroid is placed at the mid-point between the two objects of previously created centroids without regard for the number of objects in the cluster. Figure 7.3b shows the dendrogram corresponding to UPGMC of the five objects. The dendrogram for WPGMC only differs from that of Fig. 7.3b by the position of the last fusion level, which is at $S = 0.3$ instead of $S = 0.317$. The two forms of centroid clustering can lead to *reversals*. A reversal occurs when a later fusion occurs at a similarity value larger than that of the previous fusion. This phenomenon, which results from the geometric properties of centroid clustering, is explained in greater detail in Legendre and Legendre (1998: Sect. 8.6). Reversals are never large and can, most of the time, be interpreted as tri- or multi-furcations of the dendrogram represented by successive bifurcations.

Ward's (1963) minimum-variance clustering minimises, at each agglomerative step, the sum of squared distances to the group centroids. This criterion, called 'total error sum-of-squares' or TESS, is the same as used in analysis of variance and K-means partitioning. The example was calculated from a new distance matrix derived from Table 7.1 using the equation $\mathbf{D} = [D_{ij}/D_{max}]$. It is shown in Table 7.4 together with the matrix of squared distances $[(D_{ij}/D_{max})^2]$ which will be used in the calculations. Ward's agglomerative clustering can be understood and computed in two different ways.

First, it can be computed in the same way as UPGMA clustering, by successive fusions of values in the matrix of squared distances $[(D_{ij}/D_{max})^2]$. This is usually the strategy used in computer programs. The equation for the fusion of squared distances is given in textbooks describing cluster analysis, including Legendre and Legendre (1998: Eq. 8.10). Even though the cluster-fusion calculations are done using squared distances, it is useful to use the square roots of these fusion distances, as the scale for the dendrogram is then in the same units as the original distances (Table 7.1).

The second way of computing Ward's agglomerative clustering reflects the least-squares roots of the method. It is easier to understand but harder to compute; see Fig. 7.3. The first two objects forming the first cluster are objects 1 and 2. The fusion of objects 1 and 2 produces a cluster containing unexplained variation; its value is calculated as the sum of the squared distances of objects 1 and 2 to their centroid. It turns out that this value can be computed directly from the matrix of squared distances (Table 7.4, top right), using the equation for error sum-of-squares (ESS):

$$\mathrm{ESS}_k = (1/n_k) \sum D^2{}_{hi} \tag{7.1}$$

where the values $D^2{}_{hi}$ are the squared distances among the objects belonging to cluster k and n_k is the number of objects in that cluster. So for the first cluster, $\mathrm{ESS}_1 = 0.04/2 = 0.02$. Since this is the only cluster formed so far, the total sum-of-squares is also equal to that value: $\mathrm{TESS}_1 = 0.02$. To find the second cluster, the program has to search all possible fusions in turn and find the one that minimises TESS. As in UPGMA, the second cluster formed contains objects 4 and 5. The error sum-of-squares for that cluster is found using Eq. 7.1: $\mathrm{ESS}_2 = 0.16/2 = 0.08$. Since there are now two clusters, $\mathrm{TESS}_2 = 0.02 + 0.08 = 0.10$. The next cluster contains objects 3, 4, and 5. From Eq. 7.1, $\mathrm{ESS}_3 = (0.36 + 0.64 + 0.16)/3 = 0.38667$. There are still only two clusters and $\mathrm{TESS}_3 = 0.02 + 0.38667 = 0.40667$. The last fusion creates a cluster encompassing all five objects. ESS is found using Eq. 7.1: $\mathrm{ESS}_4 = 1.296$. This is also the total sum-of-squares for all objects in the study, $\mathrm{TESS}_{max} = \mathrm{TESS}_4 = 1.296$.

Depending on the computer program used, the results of Ward's agglomerative clustering may be presented using different scales (Fig. 7.4): different programs may use the fusion distance, the squared fusion distance, the total sum-of-squares error statistic TESS, the fraction of the variance (R^2) accounted for by the clusters formed at each partition level, etc. (see Grimm 1987). R^2 is computed as $(\mathrm{TESS}_{max} - \mathrm{TESS}_k)/\mathrm{TESS}_{max}$.[1]

Linkage clustering is a family of methods in which objects are assigned to clusters when a user-determined proportion (connectedness, *Co*) of the similarity links has been realised (Borcard et al. 2011). The similarities (Table 7.1 right) are first rewritten in order of decreasing values (or the distances, Table 7.1 left, in

[1] Using R software, Ward's agglomerative clustering is implemented by two functions, `agnes()` of package `cluster` and `hclust()` of package `stats`. Function `agnes()` with method = 'ward' implements Ward's (1963) minimum-variance criterion as described here. Function `hclust()` with method = 'ward' does all calculations on distances instead of squared distances, so the results differ. Users can obtain correct Ward clustering results by providing a matrix of squared distances to the `hclust()` function; then one has to modify the $height element of the output list to make it contain the square roots of the height values before calling the `plot()` function.

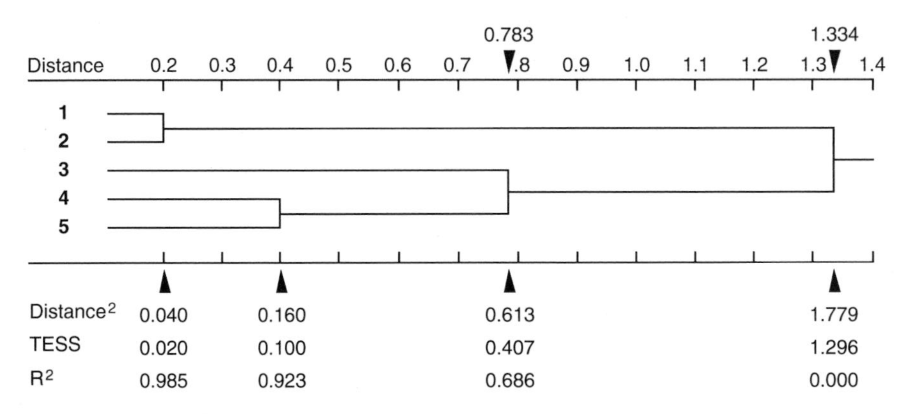

Fig. 7.4 Ward's agglomerative clustering of the objects of the artificial example. The dendrogram is drawn along a scale of distances (fusion D in Table 7.3). Alternative drawing scales are given underneath: distance^2, total error sum-of-squares (*TESS*), and R^2

order of increasing values). Clusters are formed as the program reads the list of ordered similarities. In *single linkage agglomerative clustering*, objects are placed in groups as soon as they have formed a single similarity link with at least one member of the group. For the example data, the highest similarity value is 0.8; it creates a link between objects 1 and 2 at level $S = 0.8$. The next pair, (4,5), is formed at $S = 0.6$. The next similarity value in the ordered list is 0.4; it attaches object 3 to the (4,5) cluster at $S = 0.4$. Finally, there are two similarity links, (2,4) and (2,5), formed at level $S = 0.2$. These links connect the previously-formed cluster (1,2) to the group (3,4,5) (Fig. 7.5a). In *complete linkage agglomerative clustering*, all possible similarity links must be formed before an object is admitted into a previously-formed cluster or two clusters can be fused. For the example (Fig. 7.5b), the pairs (1,2) and (4,5) are formed at the same levels as in single linkage since these clusters involve a single link. Incorporation of object 3 into cluster (4,5) must wait until the two possible similarity links (3,4) and (3,5) can be formed; this happens when the similarity level drops to $S = 0.2$ (Table 7.1 right). Likewise, fusion of the clusters (1,2) and (3,4,5) has to wait until the six similarity links (1,3), (1,4), (1,5), (2,3), (2,4), and (2,5) are formed; this only happens at $S = 0$. In *proportional-link linkage agglomerative clustering*, the connectedness level is set at any value between $Co = 0$ (single linkage) and $Co = 1$ (complete linkage). Figure 7.5c shows the dendrogram obtained with $Co = 0.5$. The pairs (1,2) and (4,5) are formed again at the same levels as in single linkage clustering since these clusters involve a single link. Incorporation of object 3 into cluster (4,5) must wait until 50% of the two possible similarity links (3,4) and (3,5) are formed; in other words, the cluster (3,4,5) is formed as soon as one of the two links is formed. Link (3,4) is formed at $S = 0.4$, so object 3 can cluster with objects 4 and 5 at that level. The fusion of cluster (1,2) with cluster (3,4,5) must wait until 50% of the six similarity links between the two clusters, or 3 links, are formed; this only happens at $S = 0$ (Table 7.1 right).

All the previously-described agglomerative clustering methods, including single and complete linkage but not proportional-link linkage with $0 < Co < 1$, can be

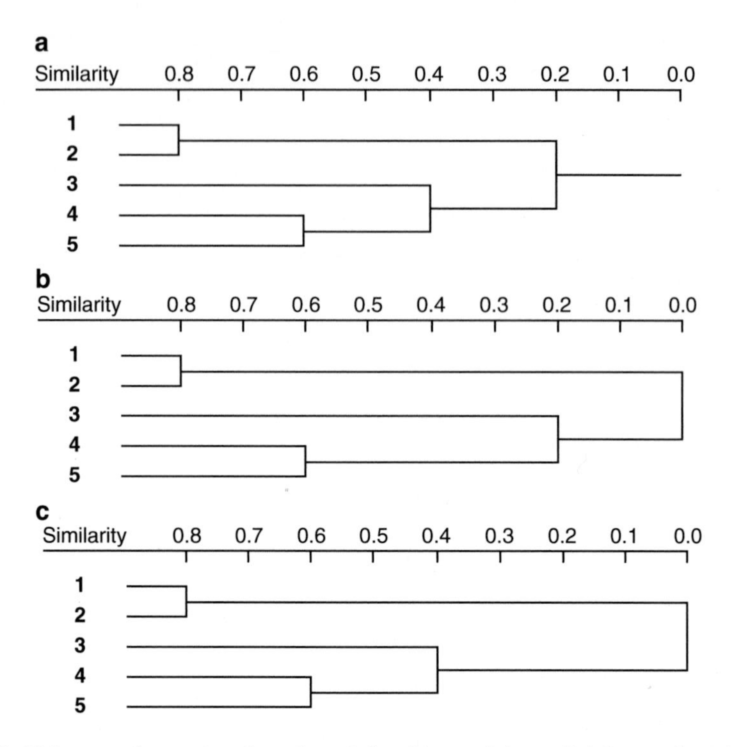

Fig. 7.5 Linkage agglomerative clustering of the objects of the artificial example. (**a**) Single linkage ($Co = 0$). (**b**) Complete linkage ($Co = 1$). (**c**) Proportional-link linkage ($Co = 0.5$). Co = connectedness

computed using an algorithm described by Lance and Williams (1966, 1967). Different methods are obtained by specifying different combinations of four parameters called α_i, α_j, β, and γ by these authors. The algorithm of Lance and Williams, which is described in more detail in textbooks on data analysis (including Legendre and Legendre 1998: Sect. 8.5 and Borcard et al. 2011: Chap. 4), is used in many computer packages that offer agglomerative clustering. The algorithm led Lance and Williams (1966, 1967) to propose a new family of methods called *flexible clustering*. In flexible clustering, $\alpha_i = \alpha_j = (1 - \beta)/2$ and $\gamma = 0$. Varying β in the range $-1 \le \beta < 1$ produces solutions with dense groups separated by long branches (when β is near -1), as in complete linkage, to loosely-chained objects as in single linkage clustering (when β is near $+1$). No reversals can occur in flexible clustering.

It is often useful to compare dendrograms to the original similarity or distance matrix in order to determine which, among several clustering methods, has preserved the original information best. To accomplish that, we need to turn the dendrograms into numbers. A cophenetic matrix is a similarity (or dissimilarity) matrix representing a dendrogram (Table 7.5). To construct it, one simply has to read, on the dendrogram, the S- or D-level where two objects become members of the same cluster and write that value into a blank matrix. Different measures of goodness-of-fit can be used to compare the original matrix to the cophenetic

Table 7.5 Cophenetic matrices for the unweighted pair-group method using arithmetic averages (UPGMA) (left) and Ward's dendrograms (right) representing the dendrograms shown in Figs. 7.2 and 7.4, respectively

UPGMA						Ward's					
S	1	2	3	4	5	**D**	1	2	3	4	5
1	*1*					1	*0*				
2	0.800	*1*				2	0.200	*0*			
3	0.067	0.067	*1*			3	1.334	1.334	*0*		
4	0.067	0.067	0.300	*1*		4	1.334	1.334	0.783	*0*	
5	0.067	0.067	0.300	0.600	*1*	5	1.334	1.334	0.783	0.400	*0*

similarities or dissimilarities. The most popular indices are the *matrix correlation* (also called *cophenetic correlation*), which is Pearson's linear correlation coefficient r computed between the values in the two half-matrices of similarities or dissimilarites, and the *Gower distance* which is the sum of the squared differences between the original and cophenetic values:

$$D_{\text{Gower}} = \sum \left(\text{original } S_{ij} - \text{cophenetic } S_{ij}\right)^2 \tag{7.2}$$

For the example data, the clustering method that best represents the original information, by these criteria, is UPGMA: the matrix correlation r is 0.946 (high values are better) while the Gower distance is 0.073 (small values are better).

A different problem is that of comparing classifications to one another. One can compute a consensus index for two classifications (reviewed in Rohlf 1982; Mickevich and Platnick 1989; Swofford 1991). Alternatively, one can compute a consensus tree using a choice of rules (strict consensus, majority rule, Adams consensus, etc.) summarised in Swofford (1991); another rule, called 'average consensus', was described by Lapointe and Cucumel (1997). Legendre and Lapointe (2004) also described a way of testing the congruence among dissimilarity matrices derived from data-sets containing different variables about the same objects. If they are congruent, the data-sets can be used jointly in statistical analysis. Borcard et al. (2011) discuss several useful graphical tools for displaying and evaluating the results from cluster analysis. They also outline methods for comparing clustering results with external environmental data and specific methods for identifying species associations in large biological data-sets.

K-Means Partitioning

The K-means problem was defined by MacQueen (1967) as that of partitioning a multivariate data-set (containing n objects and p variables) in Euclidean space into K non-overlapping groups in such a way as to minimise the sum (across the groups) of the within-group sums of squared Euclidean distances to the respective group centroids. The function to be minimised is TESS, the same function that is used in Ward's agglomerative clustering. K-means will produce a single partition of the objects, not a hierarchy. The number of groups, K, to be found is determined by the user of the method. If one asks for several values of K, the K partitions produced may not be nested.

The search for the partition that minimises TESS is done by an iterative algorithm, which begins with a starting configuration and tries to optimise it by modifying the group membership.

- A starting configuration is a preliminary partition of the objects into K groups given to the program. Depending on the program being used, one may have to provide the group membership for all objects, or the positions of the K cluster centroids in p-dimensional space. If a configuration is given as a hypothesis, one can use it as the starting point; the K-means algorithm will try to optimise this configuration, in the least-squares sense, by modifying the group membership if this results in a lower value for TESS. A second method is to restart the procedure several times, e.g., 50, 100, or 1000 times, using different random assignments of the objects to the K groups or random centroids as starting configurations. There are different ways of choosing random assignments of the objects or random centroids. A third method is to conduct agglomerative clustering, cut the dendrogram into K groups, find the positions of the group centroids in p-dimensional space, and use these as the starting configuration. Hand and Krzanowski (2005) advise against the use of this method, which has proved less efficient in simulations compared with random starts.
- Many different algorithms have been proposed to solve the K-means problem. K-means can even be computed from distance matrices. A simple alternating least-squares algorithm, used for instance in the SAS package, iterates between two steps: (1) compute the group centroids; they become the new group seeds, and (2) assign each object to the nearest seed. One may start with either step 1 or step 2, depending on whether the initial configuration is given as an assignment of objects to groups or a list of group centroids. Such an algorithm can easily cluster tens of thousands of objects.
- Note that K-means partitioning minimises the sum of squared Euclidean distances to the group centroids (TESS). The important expression is *Euclidean distance*. Many of the data tables studied by ecologists should not be directly analysed using Euclidean distances. The data need to be transformed first. This topic is discussed in detail in Legendre and Birks (2012: Chap. 8, refer to Table 8.2 of that chapter for a summary). Physical variables may need to be standardised or ranged to make them dimensionless, while assemblage composition data may need to be subjected to the chord, chi-square, or Hellinger transformation, prior to PCA, redundancy analysis (RDA), or K-means analysis.
- If one computed K-means partitioning for different values of K, how does one decide on the optimal number of groups? A large number of criteria have been proposed in the statistical literature to decide on the correct number of groups in cluster analysis. Fifteen or so of these criteria, including Calinski-Harabasz (see below), are available in the cclust package of the R computer language.

A simulation study by Milligan and Cooper (1985) compared 30 of these criteria. The best one turned out to be the Calinski and Harabasz (1974) criterion (C-H), which we will describe here. C-H is simply the F-statistic of multivariate analysis of variance and canonical analysis:

$$\text{C–H}_K = \left[R^2{}_K / (K - 1)\right] / \left[\left(1 - R^2{}_K\right) / (n - K)\right] \tag{7.3}$$

where $R^2{}_K = (\text{TESS}_{\max} - \text{TESS}(K))/\text{TESS}_{\max}$. $\text{TESS}_{\max}$ is the total sum of squared distances of all n objects to the overall centroid and TESS(K) is the sum of squared distances of the objects, divided into K groups, to their groups' own centroids. One is interested to find the number of groups, K, for which the Calinski-Harabasz criterion is maximum; this corresponds to the most compact set of groups in the least-squares sense. Even though C-H is constructed like an F-statistic, it cannot be tested for significance since there are no independent data, besides those that were used to obtain the partition, to test it. Another useful criterion, also found in the cclust package of the R language, is the Simple Structure Index (SSI: Dolnicar et al. 1999). It multiplicatively combines several elements which influence the interpretability of a partitioning solution. The best partition is indicated by the highest SSI value (see Borcard et al. 2011: Sect. 4.8 for further details).

The artificial example is too small for K-means partitioning. Notwithstanding, if we look for the best partition in two groups ($K = 2$), a distance K-means algorithm[2] finds a first group with objects (1,2) and a second group with (3,4,5). $R^2{}_{K=2} = 0.686$ (Table 7.4, step $k = 3$) so that $\text{C-H}_{K=2} = 6.561$.

Indices are available to compare different partitions of a set of objects. They can be used to compare partitions obtained with a given method, for example K-means partitions into 2–7 groups, or partitions across methods, for example a seven-group partition obtained by K-means to the partition obtained by UPGMA at the level where seven groups are found in the dendrogram. They can also be used to compare partitions obtained for different groups of organisms at the same sampling sites, for example fossil diatoms and pollen analysed at identical levels in sediment cores.

Consider all pairs of objects in turn. For each pair, determine if they are (or not) in the same group for partition 1, and likewise for partition 2. Create a 2×2 contingency table and place the results for that pair in one of the four cells of the table (Fig. 7.6). When all pairs have been analysed in turn, the frequencies a, b, c, and d can be assembled to compute the Rand index (1971), which is identical to the simple matching coefficient for binary data:

$$\text{Rand} = (a + d) / (a + b + c + d) \tag{7.4}$$

[2]A distance K-means algorithm had to be used here because the original data was a **D** matrix (Table 7.1). Turning **D** into a rectangular data matrix by PCoA, followed by K-means partitioning, would not have yielded the same exact value for C-H because PCoA of **D** produces negative eigenvalues (see Legendre and Birks 2012: Chap. 8 this volume for a discussion of negative eigenvalues).

Fig. 7.6 Contingency table comparing two partitions of a set of objects. *a*, *b*, *c*, and *d* are frequencies

First partition	Second partition: Same group	Second partition: Different groups
Same group	a	b
Different groups	c	d

The Rand index produces values between 0 (completely dissimilar) and 1 (completely similar partitions). Hubert and Arabie (1985) suggested a modified form of this coefficient. If the relationship between two partitions is comparable to that of partitions chosen at random, the corrected Rand index returns a value near 0, which can be slightly negative or positive; similar partitions have indices near 1. The modified Rand index is the most widely used coefficient to compare partitions. Birks and Gordon (1985) used the original Rand index to compare classifications of modern pollen assemblages from central Canada with the modern vegetation-landform types from which the pollen assemblages were collected in an attempt to establish how well modern pollen assemblages reflected modern vegetation types. Birks et al. (2004) compared independent classifications of modern diatom, chrysophyte cyst, and chironomid assemblages and of modern lake chemistry on Svalbard using Hill's information similarity statistic between classifications (Moss 1985). This index is related to Rand's index.

Example: The SWAP-UK Data

The UK Surface Waters Acidification Programme (SWAP-UK) data represent diatom assemblages comprising 234 taxa, from present-day surface samples from 90 lakes in England, Scotland, and Wales (Stevenson et al. 1991; Birks and Jones 2012: Chap. 3). The diatom counts were expressed as percentages relative to the total number of diatom valves in each surface sample. This means that the counts have been transformed into relative abundances, following Eq. 8.8 of Legendre and Birks (2012: Chap. 8), then multiplied by 100. They are thus ready for analysis using a method based on Euclidean distances. K-means partitioning was applied to the objects, with K values from two to ten groups. The partition that had the highest value of the Calinski-Harabasz criterion was $K = 5$ (C-H = 16.140); that partition is the best one in the least-squares sense. The five groups comprised 20, 38, 20, 8, and 4 lakes, respectively. The 90 diatom assemblages are represented in a scatter plot of pH and latitude of the lakes; the groups are represented by symbols as well as ellipses covering most of the points of each group (Fig. 7.7). The graph shows that

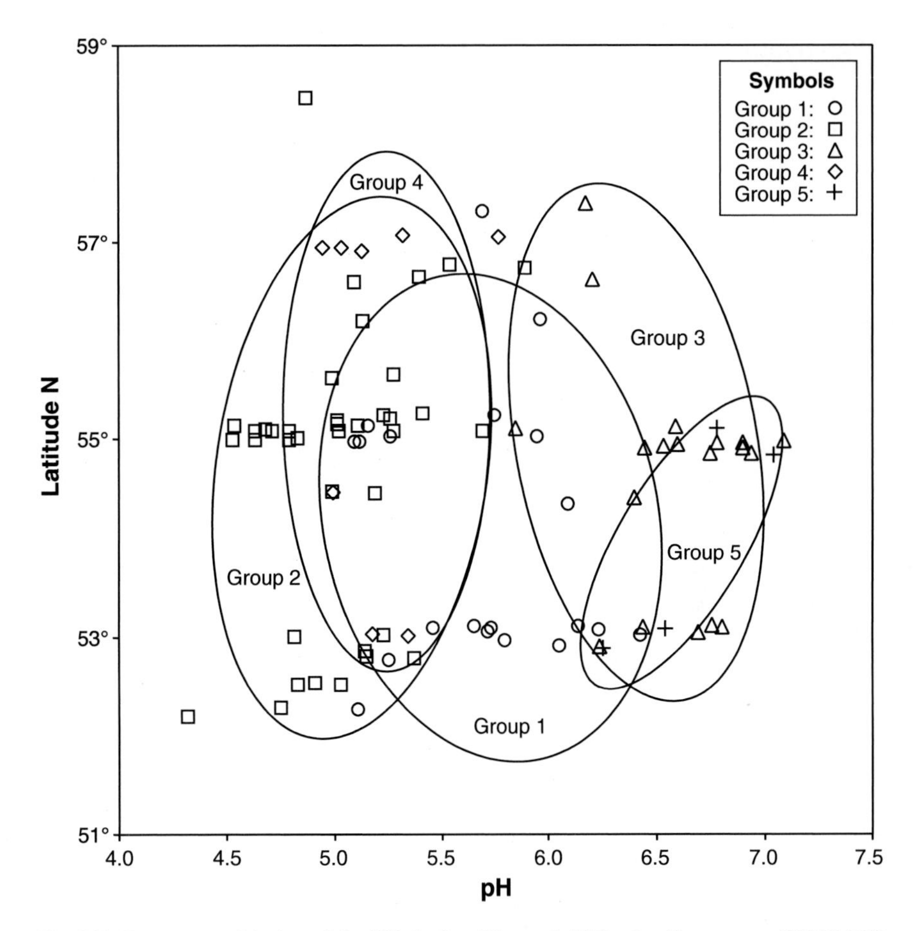

Fig. 7.7 *K*-means partitioning of the UK Surface Waters Acidification Programme (SWAP-UK) data (see Birks and Jones 2012: Chap. 3) into $K=5$ groups. The diatom assemblages are represented in a scatter diagram of pH and latitude. The ellipses summarise the extent of the five groups in the graph

the five groups of lakes are closely linked to lake-water pH but not to latitude. They are not related to longitude either. This strong pH relationship reflects the overriding influence of lake-water pH on modern diatom assemblages in temperate areas (Smol 2008; Battarbee et al. 2010). Palaeolimnological applications of *K*-means partitioning are surprisingly few. Catalan et al. (2009b) provide a detailed analysis of ecological thresholds in European alpine lakes based on *K*-means partitioning and indicator species analysis. Battarbee et al. (2011) and Bennion and Simpson (2011) provide examples involving recent changes in low alkalinity lakes and eutrophic lakes, respectively.

Constrained Clustering in One Dimension

Palaeolimnologists have always been interested in detecting discontinuities and segmenting stratigraphical data (sediment cores), an operation called zonation in Birks (2012b: Chap. 11). For univariate data, the operation can be conducted by eye on simple graphs, but for multivariate data like fossil assemblages, multivariate data analysis can be of help. One can, for instance, produce ordination diagrams from the multivariate data, using PCA or CA (see Legendre and Birks 2012: Chap. 8), and detect by eye the jumps in the positions of the data points. Palaeolimnologists more often use constrained clustering, a family of methods that was first proposed by Gordon and Birks (1972, 1974) who introduced a constraint of temporal or stratigraphical contiguity into a variety of clustering algorithms to analyse pollen stratigraphical data (see also Birks and Gordon 1985; Birks 1986). The *constraint of temporal contiguity* simply means that, when searching for the next pair of objects to cluster, one considers only the objects (or groups) that are adjacent to each other along the stratigraphical or temporal series.

Several examples of zonation using this type of algorithm are given in Birks (2012b: Chap. 11). One can use one of the stopping rules mentioned in the previous section, and in particular the Calinski-Harabasz (Eq. 7.3) and SSI criteria, to decide how many groups should be recognised in the stratigraphical series.

Another approach is to use multivariate regression tree (MRT) analysis, described in the last section of this chapter, to partition a multivariate data table representing a sediment core, for example, into homogeneous sections in the least-squares sense. A variable representing level numbers or ages in the core is used as the constraint. MRT finds groups of core levels with the minimum total error sum-of-squares (Borcard et al. 2011: Sect. 4.11.5; Simpson and Birks 2012: Chap. 9).

Example: The Round Loch of Glenhead (RLGH) Fossil Data

Another approach is the *chronological clustering* procedure of Legendre et al. (1985) who introduced a constraint of temporal contiguity into a proportional-link linkage agglomerative algorithm and used a permutation test as a stopping criterion to decide when the agglomeration of objects into clusters should be stopped. This method was applied to the RLGH3 fossil data, which consists of the counts of 139 diatom taxa observed in 101 levels of a Holocene sediment core from a small lake, The Round Loch of Glenhead, in Galloway, south-western Scotland (Jones et al. 1989; Birks and Jones 2012: Chap. 3). The data series covers the past 10,000 years. Level no. 1 is the top one (most recent) while no. 101 is at the bottom of the core (oldest). The diatom counts were expressed relative to the total number of diatom valves in each level of the core. This means that the counts have been transformed into relative abundances, following Eq. 8.8 in Legendre and Birks (2012: Chap. 8), where these data have also been analysed. There, principal coordinates of neighbour matrices (PCNM) analysis (Legendre and Birks 2012: Chap. 8) show that their temporal structure is complex.

Table 7.6 Results of indicator species analysis for the partition of The Round Loch of Glenhead (RLGH) core data into 12 groups by chronological clustering

Group no.	Membership (level no.)	Significant indicator diatoms (taxon names)
1	1–5	*Eunotia bactriana* (69), *Navicula pupula* (52), *Tabellaria quadriseptata* (46), *Suirella delicatissima* (44), *Cymbella aequalis* (39), *Navicula hoefleri* (37), *Tabellaria binalis* (34), *Eunotia exigua* (25), *Eunotia pectinalis* var. *minor* (16)
2	6–14	*Navicula cumbriensis* (42), *Eunotia tenella* (19)
3	15–17	*Neidium bisulcatum* (38), *Eunotia naegelii* (36), *Eunotia denticulata* (33), *Pinnularia microstauron* (26), *Peronia fibula* (26), *Suirella biseriata* (25), *Navicula leptostriata* (25), *Eunotia incisa* (16)
4	18–36	*Eunotia pectinalis* var. *minor* form *impressa* (26), *Achnanthes pseudoswazi* (25), *Tabellaria flocculosa* (19), *Achnanthes austriaca* (17), *Navicula mediocris* (16)
5	37–44	*Achnanthes umara* (56), *Navicula minima* (48), *Cymbella microcephala* (47), *Cyclotella kuetzingiana* agg. (38), *Navicula pupula* (38), *Nitzchia perminuta* (32), *Navicula arvensis* (30), *Nitzchia fonticola* (28), *Achnanthes flexella* (27), *Navicula minuscula* var. *muralis* (27), *Achnanthes minutissima* (25), *Navicula radiosa* var. *tenella* (22)
6	45–53	*Aulacoseira lirata* var. *lacustris* (48), *Eunotia diodon* (24)
7	54–62	*Navicula indifferens* (32)
8	63–65	*Eunotia vanheurckii* (32), *Frustulia rhomboides* var. *saxonica* (14), *Fragilaria elliptica* (43)
9	66–78	*Semiorbis hemicyclus* (38), *Brachysira* [sp.1] (28), *Aulacoseira perglabra* var. *floriniae* (24), *Fragilaria vaucheriae* (24)
10	79–90	*Navicula tenuicephala* (56), *Cymbella* [PIRLA sp.1] (37), *Eunotia iatriaensis* (31), *Navicula bremensis* (24)
11	91–94	*Navicula seminuloides* (57), *Aulacoseira distans* var. *tenella* (47), *Navicula seminulum* (42), *Navicula impexa* (35), *Aulacoseira* [cf. *distans distans*] (35), *Aulacoseira perglabra* (23), *Aulacoseira lirata* (26)
12	95–101	*Navicula hassiaca* (56), *Cymbella perpusilla* (50), *Stauroneis anceps* form *gracilis* (38), *Pinnularia subcapitata* var. *hilseana* (28), *Navicula angusta* (26), *Gomphonema acuminatum* var. *coronatum* (25), *Brachysira vitrea* (20), *Fragilaria virescens* var. *exigua* (19), *Achnanthes marginulata* (15)

The diatom taxa j with significant $IndVal_j$ values (significance level: 0.05) are listed for each group of the partition. Indicator values are given in parentheses. Group 1 is the most recent, group 12 the oldest of the 10,000-year fossil data series. Diatom nomenclature follows Stevenson et al. (1991)

For the present example, Euclidean distances were computed among the levels, then turned into similarities using the equation $S = 1 - D/D_{max}$. Chronological clustering (module *Chrono* of The R Package: Casgrain and Legendre 2004) produced 12 groups of contiguous sections, using $Co = 0.5$ in proportional-link linkage agglomeration and the significance level $\alpha = 0.01$ as the permutation clustering criterion (Table 7.6): levels 1–5, 6–14, 15–17, 18–36, 37–44, 45–53, 54–62, 63–65, 66–78, 79–90, 91–94, and 95–101. Clustering was repeated on the diatom data detrended against level numbers to remove the linear trend present in the data, as described in Legendre and Birks (2012: Chap. 8); the clustering results were identical. These 12 groups are almost entirely compatible with the two dendrograms shown in Fig. 11.1 in Birks (2012b: Chap. 11); the position of a single object (level no. 90) differs. The difference is due to the use of proportional-link linkage clustering with $Co = 0.5$ in this example, instead of CONISS (constrained incremental sum-of-squares (= Ward's) agglomerative clustering: Grimm 1987) or CONIIC (constrained incremental information clustering) in Birks (2012b: Chap. 11). This partition in Table 7.6 will serve as the basis for indicator species analysis (see below).

MRT analysis was also applied to the RLGH fossil core data. The constraint in the analysis was a single variable containing the sample numbers 1–101. Cross-validation results suggest that the best division of the core was into 12 groups, but the groups differed in part from those produced by chronological clustering: levels 1–12, 13–17, 18–36, 37–44, 45–53, 54–62, 63–66, 67–81, 82–90, 91–95, 96–99, and 100–101. Only five division points between groups were identical in the results of MRT and chronological clustering.

Constrained Clustering in Two Dimensions

Caseldine and Gordon (1978) extended the concept of temporal contiguity constraints to that of spatial contiguity constraints to analyse surface pollen spectra from three transects across a bog (see also Engels and Cwynar 2011). They showed that constraints can be applied to any data-set for which the graph-theory representation as a minimum spanning tree is such that removing any line joining pairs of adjacent samples divides the data into two connected groups (Gordon 1973). After this, time was ripe for the development of clustering procedures with the constraint of spatial contiguity, an idea that had been proposed by several other authors all at about the same time (e.g., Lebart 1978; Lefkovitch 1978, 1980; Monestiez 1978; Roche 1978; Perruchet 1981; Legendre and Legendre 1984).

The constraint generally consists of a set of geographical contiguity links describing the points that are close to each other on the map. Several types of planar connection networks can be used to connect neighbouring points: for regular grids, one can choose from among different types of connections named after the movements of chess pieces (rook, bishop, king); for irregularly-spaced points, a Delaunay triangulation (Fig. 7.9a shows an example), Gabriel graph, or relative neighbourhood graph can be used. These connection schemes are described in books on geographical statistics as well as in Legendre and Legendre (1998: Sect. 13.3). The connections between neighbouring objects are written as 1s in a spatial contiguity matrix (Fig. 7.8); 0s indicate non-neighbours.

In constrained clustering, the similarity (or dissimilarity) matrix is combined with the matrix of spatial contiguity by a Hadamard product which is the cell-by-cell product of two matrices. The cells corresponding to contiguous objects keep their similarity values whereas the cells that contain 0s in the contiguity matrix contribute

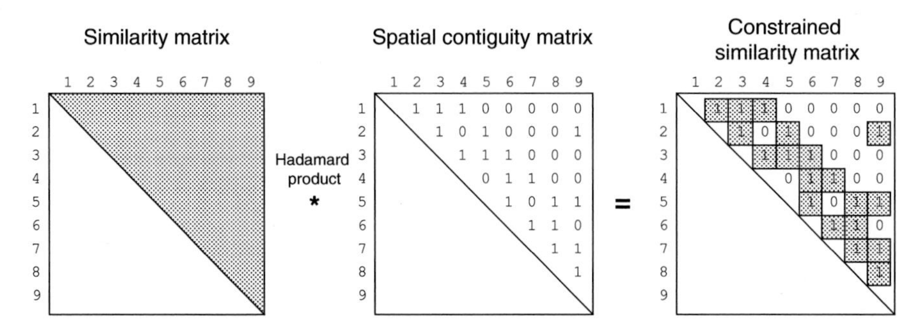

Fig. 7.8 The constrained similarity matrix used in constrained clustering is the Hadamard product (cell-by-cell product) of the similarity with the matrix of spatial contiguity. Only the upper triangular portion of each matrix is shown. Nine objects are used in this illustration

0s to the constrained similarity matrix. From that point on, a regular clustering algorithm is applied: the highest value found in the constrained similarity matrix designates the next pair to cluster and the values of similarity of these two objects with all the other objects in the study are condensed in the similarity matrix (left in Fig. 7.8), as in Table 7.3. The spatial contiguity matrix also has to be condensed: an object which is a neighbour of either of the two objects being clustered receives a 1 in the condensed spatial contiguity matrix. Figure 7.8 is a generalisation of constrained clustering in one dimension and applies to that case as well. The R package const.clust for constrained clustering in one (core data, time series) or two dimensions (maps), following the algorithm described in this paragraph, is available from http://www.bio.umontreal.ca/legendre/.

Example: The SWAP-UK Data

The SWAP-UK data used to illustrate K-means partitioning will now be clustered with the constraint of spatial contiguity. A Delaunay triangulation (Fig. 7.9a) was used to describe the neighbourhood relationships among lakes; the list of links was written to a file and passed to the constrained clustering program (module *Biogeo* in The R Package, Casgrain and Legendre 2004). Figure 7.9b is a map showing ten groups of lakes resulting from clustering with the constraint of spatial contiguity at level $S = 0.648$. Among the 90 lakes, 27 are not clustered at that level and do not appear in Fig. 7.9b. Contrary to the unconstrained clustering results (Fig. 7.7), the partition is now clearly related to latitude, with most of the Scottish lakes forming a single group (empty circles). The interpretation of these constrained clustering results is unclear ecologically. The potential influence of geography and associated components of bedrock geology, climate, and land-use at the scale of the UK on modern diatom assemblages has not, to date, been explored. The idea of regionalisation or groupings of lakes with similar biological, chemical, and ecosystem properties within and between geographical regions is a

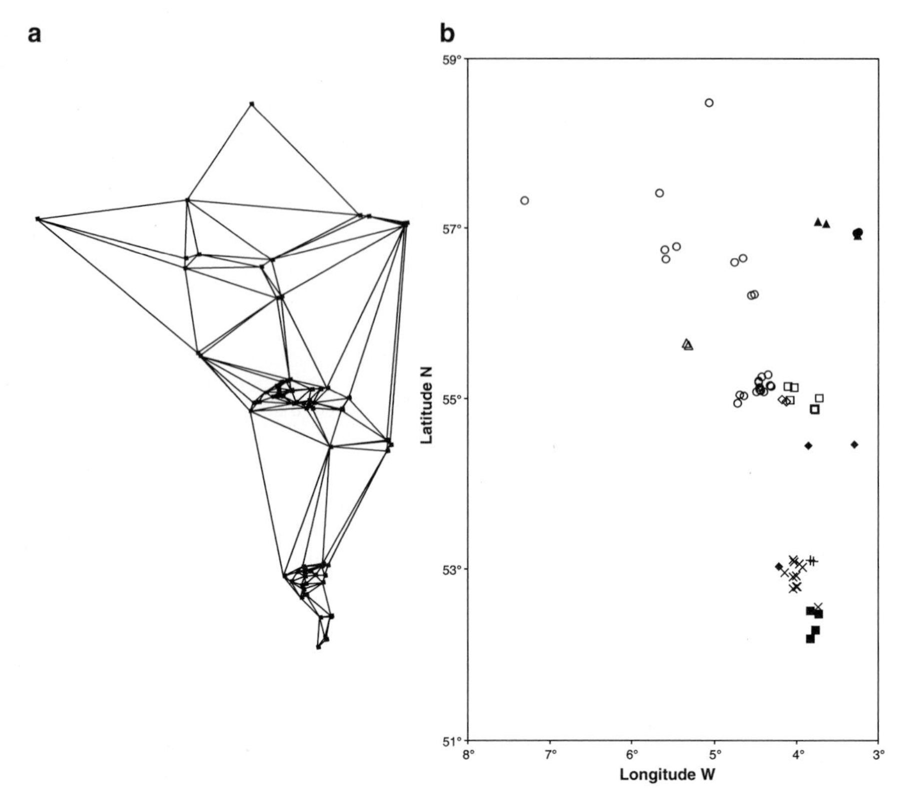

Fig. 7.9 The UK Surface Waters Acidification Programme (SWAP-UK) data set (see Birks and Jones 2012: Chap. 3). (**a**) Delaunay triangulation. The edges of the graph served as constraints in constrained clustering. (**b**) Results of spatially-constrained clustering for 63 lakes among 90: 10-group partition of the lakes (*symbols*). All members of a group are connected through adjacency links in the triangulation

topic of current research in applied freshwater science (e.g., Camero et al. 2009; Catalan et al. 2009a; Kernan et al. 2009) and is a research area where unconstrained and constrained clustering methods and a comparison of the resulting partitions could usefully be applied. An analysis, not of fresh waters but of Single Malt Scotch whiskies, showed that the organoleptic properties of these whiskies could be interpreted as reflecting the different geographical regions of Scotland (Lapointe and Legendre 1994). From an ecological viewpoint, there are strong theoretical reasons to hypothesise that broad-scale geographically-structured processes may be important in controlling the structure of ecological assemblages (Legendre 1993). Much work remains to be done on the analysis of modern assemblages of diatoms and other organisms widely studied in palaeolimnology in relation to the range of possible processes that may determine their composition and structure (e.g., Jones et al. 1993; Weckström and Korhola 2001).

Clustering Constrained by Canonical Analysis

A more general form of constraint can be provided by canonical analysis (redundancy analysis (RDA) or canonical correspondence analysis (CCA): see Legendre and Birks 2012: Chap. 8). The idea is to extract the portion of the information of the response or biological data matrix **Y** that can be explained by a table of explanatory or predictor variables **X** and apply clustering or partitioning analysis to that part of the information.

Typically, **Y** is a (fossil, recent) assemblage composition data table whereas **X** may contain environmental, spatial, or historical variables. Figure 8.2 in Legendre and Birks (2012: Chap. 8) shows that the first step of RDA is a series of multiple regressions. At the end of that step, which is also called multivariate linear regression (Finn 1974) and is available in some statistical packages under that name, the fitted values of the regressions are assembled in a table of fitted values $\hat{\mathbf{Y}}$. RDA is obtained by applying PCA to that table, producing a matrix **Z** of ordination scores in the space of the explanatory variables **X**, called "Sample scores which are linear combinations of environmental variables" in the output of the CANOCO program (ter Braak and Šmilauer 2002). Computing Euclidean distances on either of these matrices, $\hat{\mathbf{Y}}$ or **Z**, will produce the same matrix **D**. One can then apply cluster analysis to **D** or $\mathbf{S} = [1 - D_{ij}/D_{max}]$. An example of combining cluster analysis with canonical correspondence analysis in plant geography and ecology to derive an integrated biogeographical zonation is given by Carey et al. (1995).

Indicator Species Analysis

Indicator species represent a classical problem in ecology (Hill et al. 1975; Hill 1979). One may be interested to find indicator species for groups known *a priori*, for example pH classes or geographical regions, or for groups obtained by clustering. Dufrêne and Legendre (1997) developed an operational index to estimate the indicator value of each species. The indicator value of a species j in a group of sites k, $IndVal_{kj}$, is the product of the specificity A_{kj} and fidelity B_{kj} of the species to that group, multiplied by 100 to give percentages. *Specificity* estimates to what extent species j is found only in group k. *Fidelity* measures what proportion of the sites of group k species j is found in. The indicator value of species j is the largest value found for that species among all groups k of the partition under study:

$$IndVal_j = \max\left[IndVal_{kj}\right] \tag{7.5}$$

The index is maximum (100%) when individuals of species j are found at all sites belonging to a group k of the partition and in no other group. A permutation test, based on the random reallocation of sites to the various groups, is used to assess the statistical significance of $IndVal_j$. A significant $IndVal_j$

is attributed to the group j that has generated this value. The index can be calculated for a single partition or for all partitions of a hierarchical classification of sites. The IndVal program is distributed by M. Dufrêne on the site http://old.biodiversite.wallonie.be/outils/indval/. The method is also available in the package PC-ORD (MjM Software, P.O. Box 129, Gleneden Beach, Oregon 97388, USA: http://home.centurytel.net/~mjm/pcordwin.htm) and the R package labdsv (Borcard et al. 2011: Sect. 4.10.4). Catalan et al. (2009b) illustrate its use in their work on ecological thresholds in European alpine lakes and Penczak (2009) applied the INDVAL measure to clusterings derived from a self-organising map (see Simpson and Birks 2012: Chap. 9) based on an artificial neural network of fish assemblages. Other palaeolimnological examples include Battarbee et al. (2011) and Bennion and Simpson (2011). The INDVAL measure and approach have been extended (e.g. De Cáceres and Legendre 2009; De Cáceres et al. 2010; Podani and Csányi 2010). The new indices, developed by De Cáceres and Legendre (2009) and De Cáceres et al. (2010), are found in the R package indicspecies, available from http://sites.google.com/site/miqueldecaceres/software. Indicator species analysis is an extremely useful and robust procedure.

Example: The Round Loch of Glenhead (RLGH) Fossil Data

Indicator species analysis was conducted on the 12-group partition of the RLGH3 fossil data obtained by chronological clustering (see above), to identify the diatom taxa that were significantly related to the groups or zones. The diatom species with significant $IndVal_j$ values are listed in Table 7.6 for each group of the partition. The number of statistically significant indicator species varies from 1 (group 7) to 12 (group 5); there were 139 taxa in the study. These indicator species highlight and summarise the differences in diatom composition between the groups or zones. This approach deserves wide use in palaeolimnology because it provides a simple and effective means of identifying the biological features of each group or zone of levels (Birks 1993). It provides a more rigorous approach to detecting groups of species that characterise or are indicative of particular sediment sections than the early attempts by Janssen and Birks (1994a).

Two-Way Indicator Species Analysis

Two-way indicator species analysis (TWINSPAN) (Hill et al. 1975; Hill 1979) is a partitioning method that was specifically developed for the simultaneous grouping of objects and their attributes in large, heterogeneous ecological data-sets.

It has been widely used by plant community ecologists but, rather surprisingly, little used in palaeolimnology or palaeoecology (Grimm 1988). It is a polythetic divisive procedure. The division of the objects is constructed on the basis of a correspondence analysis (CA) of the data (see Legendre and Birks 2012: Chap. 8). Objects are divided into those on the negative (left) side and those on the positive (right) side on the basis of the object scores on the first CA axis. The division is at the centroid of the axis. This initial division into two groups is refined by a second CA ordination that gives greater weight to those attributes that are most associated with one side of the dichotomy. The algorithm used is complicated but the overall aim is to achieve a partitioning of the objects based on the attributes (usually species) typical of one part of the dichotomy, and hence a potential indicator of the group and its underlying ecological conditions. The process is continued for 4, 8, 16, etc. groups. The classification of the objects is followed by a classification of the attributes and the final structured table based on this two-way classification is constructed. Details of TWINSPAN, the underlying algorithm, and questions of data transformation are given by Hill (1979), Kent and Coker (1992), Lepš and Šmilauer (2003), and Fielding (2007). The computer program TWINSPAN has been modified, converted, and updated with a user-friendly interface to run under Microsoft Windows® (WinTWINS) by Petr Šmilauer and can be downloaded from http://www.canodraw.com. Despite its age and complex algorithm, TWINSPAN remains a very useful and robust technique for classifying very large heterogeneous data-sets containing may zero values ('absences'), keeping in mind that the method assumes the existence of a single, strong gradient dominating the data and that the divisions between neighbouring groups may not always be optimal (Belbin and McDonald 1993). A classification resulting from TWINSPAN can provide a useful starting configuration for *K*-means partitioning, particularly for large heterogeneous data-sets. Palaeolimnological applications of TWINSPAN include Brodersen and Lindegaard (1997), Brodersen and Anderson (2002), Bennion et al. (2004), Kernan et al. (2009), Engels and Cwynar (2011), and Mackay et al. (2011) (see Simpson and Hall 2012: Chap. 19).

Example: The SWAP-UK Data

A two-way indicator species analysis of the SWAP-UK data (90 objects by 234 taxa) was implemented using WinTWINS 2.3. Eight pseudospecies were used with cut-levels of 0%, 1%, 2%, 4%, 8%, 16%, 32%, and 64% (see Kent and Coker 1992 or Lepš and Šmilauer 2003 for an explanation of pseudospecies or conjoint coding). The classification into four groups and the associated indicator species are summarised in Fig. 7.10. The four groups of lakes differ in their pH values, just as the five groups in the *K*-means partitioning do, with group medians of pH 5.0, 5.3, 6.3, and 6.8.

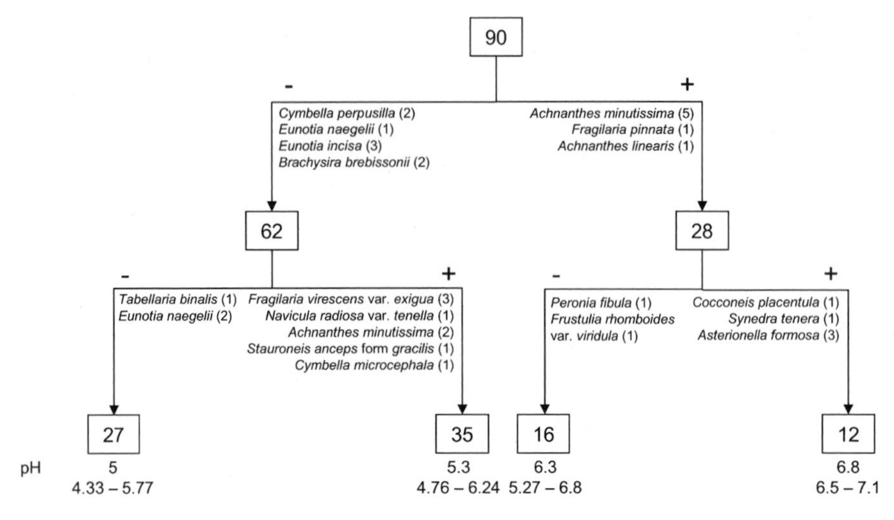

Fig. 7.10 Two-way indicator species analysis (TWINSPAN) of the UK Surface Waters Acidification Programme (SWAP-UK) data-set (90 lakes, 234 diatom species – see Birks and Jones 2012: Chap. 3). The numbers of lakes in the total data-set, the second-level (two groups), and the third-level (four groups) divisions are shown in *squares*. The indicator species for each division are shown, with their pseudospecies values in parentheses (1 = <1%, 2 = 1–2%, 3 = 2–4%, 4 = 4–8%, 5 = 8–16%). The median pH and range of pH values for the lakes in the four groups are also shown

Multivariate Regression Trees

Multivariate regression trees (MRT) produce a clustering of multivariate biological or 'response' data using a monothetic divisive approach, as explained below. The method combines data exploration and data interpretation (forecasting). MRT is related to regression in the sense that the explanation of the response data involves explanatory variables. It thus represents an alternative to multivariate explanatory methods such as RDA and CCA and belongs to the family of classification and regression trees discussed by Simpson and Birks (2012: Chap. 9). MRT is a least-squares method, but it does not use simple or multiple regression.

Monothetic divisive classification methods base each split on a single variable. For each branching point of the tree, MRT chooses one of the explanatory variables, and a splitting point along it, that maximises the separation of two daughter groups in the multivariate space of the response variables (e.g., species assemblages). Group separation, or homogeneity, is maximum when the total sum-of-squares error statistic, called TESS in Ward's clustering and K-means partitioning (see above), is minimised. MRT can be seen as a form of constrained clustering, the constraint being given by the environmental variables characterising each division step. The method was proposed by De'ath (2002) and Larsen and Speckman (2004) as an extension of univariate regression trees (Breiman et al. 1984; Simpson and Birks 2012: Chap. 9). De'ath also provided an R-language package, mvpart (De'ath 2007), implementing the method. Borcard et al. (2011: Sect. 4.11) discuss MRT and combine MRT with indicator species analysis (e.g. Amsinck et al. 2006). Besides its emphasis on interpretation and forecasting, MRT is well-suited for the analysis of unbalanced ecological data (groups of different sizes), data containing missing values, or explanatory variables that are not necessarily related to the species in a linear or unimodal way. Davidson et al. (2010a, b) present palaeolimnological examples of the use of MRT to relate cladoceran assemblage relationships to zooplanktivorous fish density and to submerged macrophyte abundance in shallow lakes in England and Denmark. Bjerring et al. (2009) use MRT to relate modern assemblages of cladocerans in 54 lakes along a latitudinal gradient (36–48°N) in Europe to a range of environmental variables. Other palaeolimnological examples include Amsinck et al. (2006), Pelánková et al. (2008), and Herzschuh and Birks (2010).

The result of MRT analysis is a hierarchical classification of the data represented by a tree, plus information about the explanatory (environmental) variables that best explain each split and the distribution of the response variables, which are often species in ecological applications, in each terminal group.

A somewhat similar approach is that of ter Braak (1986) implemented in the DOS program DISCRIM. This method starts with a hierarchical classification based on the species composition of the objects (in his case derived by two-way indicator species analysis, TWINSPAN: Hill 1979). It finds the environmental or other external predictor variables (expressed as presence/absence, qualitative or nominal variables, ranks, quantitative, etc.) that optimally predict the classification of the objects into two groups, four groups, etc. Simple discriminant functions are constructed in a very simple way (Hill 1977, 1979). A presence/absence environmental variable is a possible group indicator if its frequency of occurrence is higher in the group than in the alternative group. The *n* variables with the highest absolute difference in frequency of occurrence are included in the discriminant function, where *n* is the smallest integer that minimises the number of misclassifications. The great advantage of Hill's (1977, 1979) simple discriminant functions is their simplicity: the sign of a variable is given the same sign as that of the frequency difference, and the number of possible sets of indicator variables is restricted by ordering the variables on the basis of the absolute frequency difference. These restrictions avoid the need for optimisation by linear programming and facilitate ease of interpretation of the simple discriminants constructed (ter Braak 1986). Hill's simple discriminants make no assumptions about species–environmental relationships or about the underlying nature of the data. They simply consider if environmental variables differ in their frequencies between groups. They are close in concept to Gower's (1974) maximal predictive classification as the emphasis is on classification prediction. With quantitative environmental variables, linear discriminant analysis or multiple discriminant analysis (= canonical variates analysis) can be used *a posteriori* to discriminate between groups (Birks 2012a: Chap. 2). For nominal environmental variables, correspondence analysis could be applied to a $2 \times c$ table where the rows correspond to the two branches of the node and c is the total number of categories of the nominal variables. Linear discriminant functions may, however, be difficult to interpret and as one moves down the classification hierarchy the groups may contain so few objects that the coefficients of the discriminant functions cannot be reliably estimated, if at all. Using presence/absence data throughout for his simple discriminant functions, Hill (1977) circumvented these problems by proposing simple discriminant functions in which the coefficients can only take three values; −1 and +1 for variables that are characteristic of one group or another, and 0 for non-discriminating variables. Such functions are easy to interpret and both quantitative and nominal variables can easily be incorporated into this approach after careful recoding using disjoint or conjoint coding (ter Braak 1986). Janssen and Birks (1994b) applied this approach to stratigraphical pollen and plant macrofossil assemblages to detect predictor or indicator variables for constrained classifications of stratigraphical samples.

Example: The SWAP-UK Data

A MRT was computed for the SWAP-UK data (90 sites, 234 diatom taxa) using the `mvpart()` function (mvpart package) written by G. De'ath in the R computer language. Only three explanatory variables were available for these data: pH, latitude, and longitude of the lakes. Cross-validation is available in De'ath's function; it can be used to select the tree having the smallest predicted mean squared error, or some other tree slightly longer or shorter than that. The cross-validated relative error criterion indicated that the partition in two groups (57 and 33 lakes, respectively) explained by pH was statistically the best; the Calinski-Harabasz criterion led to the same conclusion. For illustrative purposes only, we present a more developed tree containing nine binary partitions and ten terminal groups (Fig. 7.11) along a scale representing the proportion of the species variation explained by each partition (R^2, as in Fig. 7.4). The most important explanatory variable is clearly pH; this was also the case in the unconstrained partitions shown in Figs. 7.7 and 7.10. The mean value of each taxon in each group is available in De'ath's function output; it is not presented for this example because the taxa are too numerous. This example is presented here simply to illustrate the potential value of MRT in analysing modern sets of biological and associated environmental data, which are now such an important part of palaeolimnological research (Smol 2008). Multivariate regression trees provide a very powerful means of exploring complex biological-environmental relationships (De'ath 2002) and are of obvious wide application in a range of palaeolimnological problems. Simpson and Birks (2012: Chap. 9) discuss the range of classification and regression trees and other decision trees of potential value in palaeolimnology.

Simple discriminants based on pH, latitude, and longitude coded as ranks and converted into +/− variables based on quartiles (ter Braak 1986) applied to the two-way indicator species analysis of the SWAP-UK data (Fig. 7.10) similarly show the over-riding importance of pH in discriminating between the four TWINSPAN groups, but with some influence of latitude apparent at the four-group level, in particular between the two left-hand groups in Fig. 7.10.

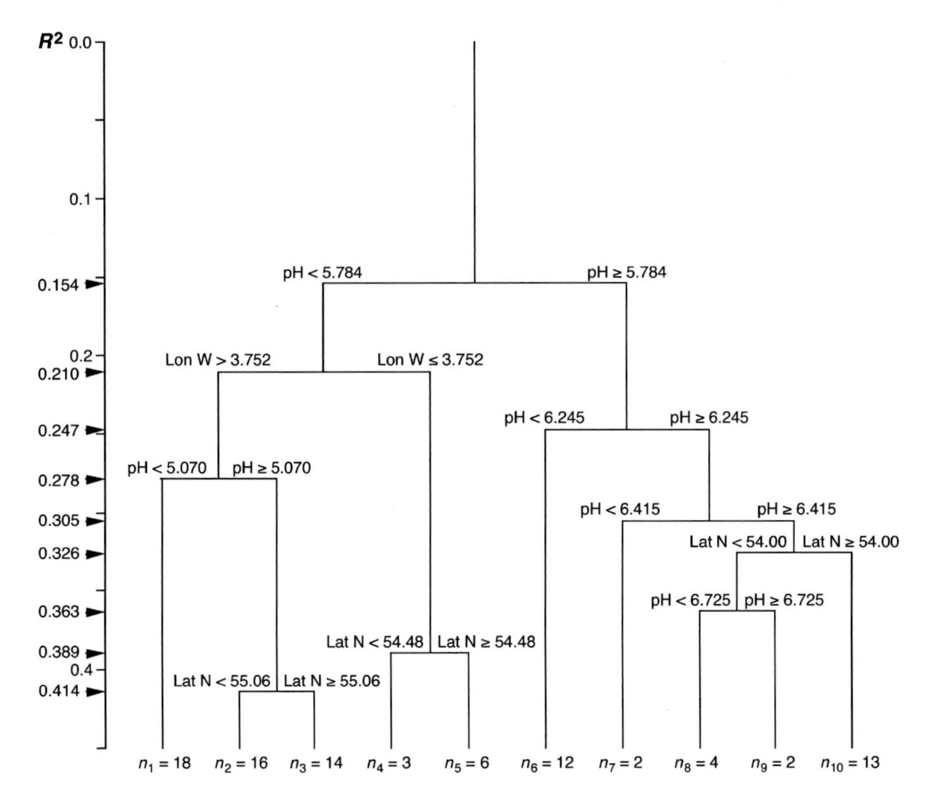

Fig. 7.11 Multivariate regression tree (MRT) for the UK Surface Waters Acidification Programme (SWAP-UK) data-set (90 lakes, 234 diatom taxa – see Birks and Jones 2012: Chap. 3) using three explanatory variables: pH, longitude (*Lon W*) and latitude (*Lat N*) of the lakes. The best partition is two groups. n_k: number of lakes in group k

Conclusions

Hierarchical clustering methods and partitioning techniques such as K-means partitioning and two-way indicator species analysis are useful tools for summarising group structure within large, complex, multivariate data-sets that are increasingly common in palaeolimnology. The incorporation of one- or two-dimensional constraints in the clustering algorithms provides means of exploring group structure in temporal, stratigraphical data and in geographical modern data, respectively. Indicator species analysis with its associated permutation tests is a simple and effective means of detecting statistically significant indicator species for any grouping of a set of objects. The newly developed approach of multivariate regression trees combines partitioning and data exploration with regression and data interpretation and modelling. Agglomerative clustering and K-means partitioning are available in most statistical software packages, as well as in the R statistical language (functions `hclust()` and `kmeans()`). Multivariate regression trees and indicator species analysis can be computed using functions of the mvpart, labdsv, and indicspecies packages in R (see Borcard et al. 2011).

Palaeolimnologists have largely concentrated on the use of classical ordination and canonical ordination methods to explore patterns in their data. Modern classification and partitioning techniques along with indicator species analysis and multivariate regression trees are attractive and useful complementary tools for exploring and summarising large, complex, palaeolimnological data-sets (see Table 7.2). They deserve wider use than they currently receive within the palaeolimnological research community.

Acknowledgements We are indebted to Keith Bennett, Steve Juggins, and Gavin Simpson for comments and discussion and to Cathy Jenks for her invaluable help in editing this chapter. This is publication number A209 from Bjerknes Centre for Climate Research.

References

Amsinck SL, Strzelczak A, Bjerring R, Landkildehus F, Lauridsen TL, Christoffersen K, Jeppesen E (2006) Lake depth rather than fish planktivory determines cladoceran community structure in Faroese lakes – evidence from contemproary data and sediments. Freshw Biol 51:2124–2142

Battarbee RW, Charles DF, Bigler C, Cumming BF, Renberg I (2010) Diatoms as indicators of surface-water acidity. In: Smol JP, Stoermer EF (eds) The diatoms: applications for the environmental and earth sciences, 2nd edn. Cambridge University Press, Cambridge

Battarbee RW, Simpson GL, Bennin H, Curtis C (2011) A reference typology of low alkalinity lakes in the UK based on pre-acidification diatom assemblages from lake sediment cores. J Paleolimnol 45:489–505

Belbin L, McDonald C (1993) Comparing three classification strategies for use in ecology. J Veg Sci 4:341–348

Bennion H, Simpson GL (2011) The use of diatom records to establish reference conditions for UK lakes subject to eutrophication. J Paleolimnol 45:469–488

Bennion H, Fluin J, Simpson GL (2004) Assessing eutrophication and reference conditions for Scottish freshwater lochs using subfossil diatoms. J Appl Ecol 41:124–138

Birks HJB (1986) Numerical zonation, comparison and correlation of Quaternary pollen–stratigraphical data. In: Berglund BE (ed) Handbook of Holocene palaeoecology and palaeohydrobiology. Wiley, Chichester, pp 743–774

Birks HJB (1993) Quaternary palaeoecology and vegetation science – current contributions and possible future developments. Rev Palaeobot Palynol 79:153–177

Birks HJB (2012a) Chapter 2: Overview of numerical methods in palaeolimnology. In: Birks HJB, Lotter AF, Juggins S, Smol JP (eds) Tracking environmental change using lake sediments. Volume 5: Data handling and numerical techniques. Springer, Dordrecht

Birks HJB (2012b) Chapter 11: Analysis of stratigraphical data. In: Birks HJB, Lotter AF, Juggins S, Smol JP (eds) Tracking environmental change using lake sediments. Volume 5: Data handling and numerical techniques. Springer, Dordrecht

Birks HJB, Gordon AD (1985) Numerical methods in Quaternary pollen analysis. Academic Press, London

Birks HJB, Jones, VJ (2012) Chapter 3: Data-sets. In: Birks HJB, Lotter AF, Juggins S, Smol JP (eds) Tracking environmental change using lake sediments. Volume 5: Data handling and numerical techniques. Springer, Dordrecht

Birks HJB, Webb T, Berti AA (1975) Numerical analysis of pollen samples from central Canada: a comparison of methods. Rev Palaeobot Palynol 20:133–169

Birks HJB, Monteith DT, Rose NL, Jones VJ, Peglar SM (2004) Recent environmental change and atmospheric contamination on Svalbard as recorded in lake sediments – modern limnology, vegetation, and pollen deposition. J Paleolimnol 31:411–431

Bjerring R, Becares E, Declerck S et al. (2009) Subfossil Cladocera in relation to contemporary environmental variables in 54 pan-European lakes. Freshw Biol 54:2401–2417

Borcard D, Gillet F, Legendre P (2011) Numerical ecology with R. Springer, New York

Breiman L, Friedman JH, Olshen RA, Stone CJ (1984) Classification and regression trees. Chapman & Hall, New York

Brodersen KP, Anderson NJ (2002) Distribution of chironomids (Diptera) in low Arctic West Greenland lakes: trophic conditions, temperature and environmental reconstruction. Freshw Biol 47:1137–1157

Brodersen KP, Lindegaard C (1997) Significance of subfossil remains in classification of shallow lakes. Hydrobiologia 342(343):125–132

Calinski T, Harabasz J (1974) A dendrite method for cluster analysis. Commun Stat 3:1–27

Camero L, Rogora M, Mosello R et al. (2009) Regionalisation of chemical variability in European mountain lakes. Freshw Biol 54:2452–2469

Carey PD, Preston CD, Hill MO, Usher MB, Wright SM (1995) An environmentally defined biogreographical zonation of Scotland designed to reflect species distributions. J Ecol 83:833–845

Caseldine CJ, Gordon AD (1978) Numerical analysis of surface pollen spectra from Bankhead Moss, Fife. New Phytol 80:435–453

Casgrain P, Legendre P (2004) The R package for multivariate and spatial analysis, version 4.0 (development release 7) – user's manual. Département de sciences biologiques, Université de Montréal. 148 pp. Manual and program available from http://www.bio.umontreal.ca/casgrain/en/labo/R/index.html

Catalan J, Curtis CJ, Kernan M (2009a) Remote European mountain lake ecosystems: regionalisation and ecological status. Freshw Biol 54:2419–2432

Catalan J, Barbieri MG, Bartumeus F et al. (2009b) Ecological thresholds in European alpine lakes. Freshw Biol 54:2494–2517

Davidson T, Sayer CD, Perrow M, Bramm M, Jeppesen E (2010a) The simultaneous inference of zooplanktivorous fish and macrophyte density from sub-fossil cladoceran assemblages: a multivariate regression tree approach. Freshw Biol 55:546–564

Davidson T, Sayer CD, Langdon PG, Burgess A, Jackson M (2010b) Inferring past zooplanktivorous fish and macrophyte density in a shallow lake: application of a new regression tree model. Freshw Biol 55:584–599

De'ath G (2002) Multivariate regression trees: a new technique for modeling species-environment relationships. Ecology 83:1105–1117

De'ath G (2007) mvpart: multivariate partitioning. R package version 1.2–6. URL http://cran.r-project.org/

De Cáceres M, Legendre P (2009) Associations between species and groups of sites: indices and statistical inference. Ecology 90:3566–3574

De Cáceres M, Legendre P, Moretti M (2010) Improving indicator species analysis by combining groups of sites. Oikos 119:1674–1684

Dolnicar S, Grabler K, Mazanec JA (1999) A tale of three cities: perceptual charting for analyzing destination images. In: Woodside A (ed) Consumer psychology of tourism, hospitality and leisure. CAB International, New York, pp 39–62

Dufrêne M, Legendre P (1997) Species assemblages and indicator species: the need for a flexible asymmetrical approach. Ecol Monogr 67:345–366

Engels S, Cwynar LC (2011) Changes in fossil chironomid remains along a depth gradient: evidence for common faunal thresholds within lakes. Hydrobiologia 665:15–38

Everitt BS, Hothorn T (2011) An introduction to applied multivariate analysis using R. Springer, New York

Everitt BS, Landan S, Leese M, Stahl D (2011) Cluster analysis, 5th edn. Wiley, Chichester

Fielding AH (2007) Cluster and classification techniques for the biosciences. Cambridge University Press, Cambridge

Finn JD (1974) A general model for multivariate analysis. Holt, Rinehart & Winston, New York

Gordon AD (1973) Classification in the presence of constraints. Biometrics 29:821–827

Gordon AD, Birks HJB (1972) Numerical methods in Quaternary palaeoecology. I. Zonation of pollen diagrams. New Phytol 71:961–979

Gordon AD, Birks HJB (1974) Numerical methods in Quaternary palaeoecology. II. Comparison of pollen diagrams. New Phytol 73:221–249

Gower JC (1967) A comparison of some methods of cluster analysis. Biometrics 23:623–637

Gower JC (1974) Maximal predictive classification. Biometrics 30:643–654

Grimm EC (1987) CONISS: a FORTRAN 77 program for stratigraphically constrained cluster analysis by the method of incremental sum-of-squares. Comput Geosci 13:13–35

Grimm EC (1988) Data analysis and display. In: Huntley B, Webb T III (eds) Vegetation history. Kluwer Academic Publishers, Dordrecht, pp 43–76

Hand DJ, Krzanowski WJ (2005) Optimising *k*-means clustering results with standard software packages. Comput Stat Data Anal 49:969–973

Herzschuh U, Birks HJB (2010) Evaluating the indicator value of Tibetan pollen taxa for modern vegetation and climate. Rev Palaeobot Palynol 160:197–208

Hill MO (1977) Use of simple discriminant functions to classify quantitative phytosociological data. In: Diday E, Lebart L, Pagés JP, Tomassone R (eds) First international symposium on data analysis and informatics. Volume 1: Institut de Recherche d'Informatique et D'Automatique, Le Chesnay, pp 181–189

Hill MO (1979) TWINSPAN – a FORTRAN program for arranging multivariate data in an ordered two-way table by classification of individuals and attributes. Cornell University, Ithaca

Hill MO, Bunce RGH, Shaw MW (1975) Indicator species analysis, a divisive polythetic method of classification and its application to a survey of native pinewoods of Scotland. J Ecol 63: 597–613

Hubert LJ, Arabie P (1985) Comparing partitions. J Classification 2:193–218

Janssen CR, Birks HJB (1994a) Recurrent groups of pollen types in time. Rev Palaeobot Palynol 79:153–177

Janssen CR, Birks HJB (1994b) Examples of recurrent groups of pollen and macrofossils in space and time delimited by simple discriminant functions. Diss Bot 234:337–351

Jones VJ, Stevenson AC, Battarbee RW (1989) Acidification of lakes in Galloway, southwest Scotland: a diatom and pollen study of the post-glacial history of The Round Loch of Glenhead. J Ecol 77:1–23

Jones VJ, Juggins S, Ellis-Evans JC (1993) The relationship between water chemistry and surface sediment diatom assemblages in maritime Antarctic lakes. Antarc Sci 5:339–348

Kent M, Coker PD (1992) Vegetation description and analysis – a practical approach. Wiley, Chichester

Kernan M, Ventura M, Bitušík P, Brancelj A, Clarke G, Velle G, Raddum GG, Stuchlík E, Catalan J (2009) Regionalisation of remote European mountain lake ecosystems according to their biota: environmental versus geographical patterns. Freshw Biol 54:2470–2493

Lance GN, Williams WT (1966) A generalized sorting strategy for computer classifications. Nature 212:218

Lance GN, Williams WT (1967) A general theory of classificatory sorting strategies. I. Hierarchical systems. Comput J 9:373–380

Lapointe F-J, Cucumel G (1997) The average consensus procedure: combination of weighted trees containing identical or overlapping sets of taxa. Syst Biol 46:306–312

Lapointe F-J, Legendre P (1994) A classification of pure malt Scotch whiskies. Appl Stat 43: 237–257

Larsen DR, Speckman PL (2004) Multivariate regression trees for analysis of abundance data. Biometrics 60:543–549

Lebart L (1978) Programme d'agrégation avec containtes (C. A. H. contiguïté). C Anal Données 3:275–287

Lefkovitch LP (1978) Cluster generation and grouping using mathematical programming. Math Biosci 41:91–110
Lefkovitch LP (1980) Conditional clustering. Biometrics 36:43–58
Legendre P (1993) Spatial autocorrelation: trouble or new paradigm? Ecology 74:1659–1673
Legendre P, Birks HJB (2012) Chapter 8: From classical to canonical ordination. In: Birks HJB, Lotter AF, Juggins S, Smol JP (eds) Tracking environmental change using lake sediments. Volume 5: Data handling and numerical techniques. Springer, Dordrecht
Legendre P, Lapointe F-J (2004) Assessing the congruence among distance matrices: single malt Scotch whiskies revisited. Aust N Z J Stat 46:615–629
Legendre P, Legendre L (1998) Numerical ecology, 2nd English edn. Elsevier, Amsterdam
Legendre P, Legendre V (1984) Postglacial dispersal of freshwater fishes in the Québec peninsula. Can J Fish Aquat Sci 41:1781–1802
Legendre P, Dallot S, Legendre L (1985) Succession of species within a community: chronological clustering, with applications to marine and freshwater zooplankton. Am Nat 125:257–288
Lepš J, Šmilauer P (2003) Multivariate analysis of ecological data using CANOCO. Cambridge University Press, Cambridge
Mackay AW, Davidson TA, Wolski P, Woodward S, Mazebedi R, Masamba WRL, Todd M (2011) Diatom sensitivity to hydrological and nutrient variability in a subtropical, flood-pulse wetland. Ecohydrology. doi:10.1002/eco.242
MacQueen J (1967) Some methods for classification and analysis of multivariate observations. In: Le Cam LM, Neyman J (eds) Proceedings of the fifth Berkeley symposium on mathematical statistics and probability. Volume 1: University of California Press, Berkeley, pp 281–297
Mickevich MF, Platnick NI (1989) On the information content of classifications. Cladistics 5: 33–47
Milligan GW, Cooper MC (1985) An examination of procedures for determining the number of clusters in a data set. Psychometrika 50:159–179
Monestiez P (1978) Méthodes de classification automatique sous contraintes spatiales. In: Legay JM, Tomassone R (eds) Biométrie et écologie. Inst nat Rech agronomique, Jouy-en-Josas, pp 367–379
Moss D (1985) An initial classification of 10-km squares in Great Britain from a land characteristic data bank. Appl Geogr 5:131–150
Pelánková B, Kunes P, Chytry M, Jankovská V, Ermakov N, Svobodová-Svitavaská H (2008) The relationships of modern pollen spectra to vegetation and climate along a steppe-forest-tundra transition in southern Siberia, explored by decision trees. Holocene 18:1259–1271
Penczak T (2009) Fish assemblage compositions after implementation of the IndVal method on the Narew River system. Ecol Model 220:419–423
Perruchet C (1981) Classification sous contrainte de contiguïté continue. In: Classification automatique et perception par ordinateur. Séminaires de l'Institut national de Recherche en Informatique et en Automatique (C 118), Rocquencourt, pp 71–92
Podani J, Csányi B (2010) Detecting indicator species: some extensions of the IndVal measure. Ecol Indic 10:1119–1124
Prentice IC (1986) Multivariate methods for data analysis. In: Berglund BE (ed) Handbook of Holocene palaeoecology and palaeohydrology. Wiley, Chichester, pp 775–797
Rand WM (1971) Objective criteria for the evaluation of clustering methods. J Am Stat Assoc 66:846–850
Roche C (1978) Exemple de classification hiérarchique avec contrainte de contiguïté. Le partage d'Aix-en-Provence en quartiers homogènes. C Anal Données 3:289–305
Rohlf FJ (1963) Classification of Aedes by numerical taxonomic methods (Diptera: Culicidae). Ann Entomol Soc Am 56:798–804
Rohlf FJ (1982) Consensus indices for comparing classifications. Math Biosci 59:131–144
Simpson GL, Birks HJB (2012) Chapter 9: Statistical learning in palaeolimnology. In: Birks HJB, Lotter AF, Juggins S, Smol JP (eds) Tracking environmental change using lake sediments. Volume 5: Data handling and numerical techniques. Springer, Dordrecht

Simpson GL, Hall RI (2012) Chapter 19: Human impacts – applications of numerical methods to evaluate surface-water acidification and eutrophication. In: Birks HJB, Lotter AF, Juggins S, Smol JP (eds) Tracking environmental change using lake sediments. Volume 5: Data handling and numerical techniques. Springer, Dordrecht

Smol JP (2008) Pollution of lakes and rivers: a paleoenvironmental perspective, 2nd edn. Blackwell, Oxford

Sneath PHA, Sokal RR (1973) Numerical taxonomy – the principles and practice of numerical classification. Freeman, San Francisco

Sokal RR, Michener CD (1958) A statistical method for evaluating systematic relationships. Univ Kans Sci Bull 38:1409–1438

Stevenson AC, Juggins S, Birks HJB, Anderson DS, Anderson NJ, Battarbee RW, Berge F, Davis RB, Flower RJ, Haworth EY, Jones VJ, Kingston JC, Kreiser AM, Line JM, Munro MAR, Renberg I (1991) The Surface Waters Acidification Project Palaeolimnology Programme: modern diatom/lake-water chemistry data-set. Ensis Publishing, London

Swofford DL (1991) When are phylogeny estimates from molecular and morphological data incongruent? In: Miyamoto MM, Cracraft J (eds) Phylogenetic analysis of DNA sequences. Oxford University Press, Oxford, pp 295–333

ter Braak CJF (1986) Interpreting a hierarchical classification with simple discriminant functions: an ecological example. In: Diday E et al. (eds) Data analysis and informatics 4. North Holland, Amsterdam, pp 11–21

ter Braak CJF, Šmilauer P (2002) CANOCO reference manual and CanoDraw for Windows user's guide: software for canonical community ordination (version 4.5). Microcomputer Power, Ithaca

Ward JH (1963) Hierarchical grouping to optimize an objective function. J Am Stat Assoc 58: 236–244

Weckström J, Korhola A (2001) Patterns in the distribution, composition, and diversity of diatom assemblages in relation to ecoclimatic factors in Arctic Lapland. J Biogeogr 28:31–45

Wehrens R (2011) Chemometrics with R. Springer, New York

Chapter 8
From Classical to Canonical Ordination

Pierre Legendre and H. John B. Birks

Abstract The simple or classical ordination methods mostly used by palaeoecologists and palaeolimnologists are *principal component analysis* (PCA) and *correspondence analysis* (CA), and, more rarely, *principal coordinate analysis* (PCoA) and *non-metric multidimensional scaling* (NMDS). These methods are reviewed in a geometric framework. They mostly differ by the types of distances among objects that they allow users to preserve during ordination. Canonical ordination methods are generalisations of the simple ordination techniques; the ordination is constrained to represent the part of the variation of a table of response variables (e.g., species abundances) that is maximally related to a set of explanatory variables (e.g., environmental variables). *Canonical redundancy analysis* (RDA) is the constrained form of PCA whereas *canonical correspondence analysis* (CCA) is the constrained form of CA. Canonical ordination methods have also been proposed that look for polynomial relationships between the dependent and explanatory variables. Tests of statistical significance using permutation tests can be obtained in canonical ordination, just as in multiple regression. Canonical ordination serves as the basis for variation partitioning, an analytical procedure widely used by palaeolimnologists.

P. Legendre (✉)
Département de sciences biologiques, Université de Montréal, C.P. 6128, succursale Centre-ville, Montréal, QC, H3C 3J7, Canada
e-mail: pierre.legendre@umontreal.ca

H.J.B. Birks
Department of Biology and Bjerknes Centre for Climate Research, University of Bergen, PO Box 7803, Bergen N-5020, Norway

Environmental Change Research Centre, University College London, Pearson Building, Gower Street, London WC1E 6BT, UK

School of Geography and the Environment, University of Oxford, Oxford OX1 3QY, UK
e-mail: john.birks@bio.uib.no

H.J.B. Birks et al. (eds.), *Tracking Environmental Change Using Lake Sediments*, Developments in Paleoenvironmental Research 5, DOI 10.1007/978-94-007-2745-8_8,

Keywords Canonical correspondence analysis • Correspondence analysis • Non-metric multidimensional scaling • Ordination • PCNM analysis • Principal component analysis • Principal coordinate analysis • Principal coordinates of neighbour matrices • Redundancy analysis • Variation partitioning

Introduction

To ordinate is to arrange objects in some order (Goodall 1954). Ordination procedures are well-known to ecologists who wish to represent and summarise their observations along one, two, or a few axes. The most simple case is the ordination of sites along a single variable representing an environmental gradient (e.g., lake-water pH), or a sampling variable such as depth along a sediment core or along the estimated ages of levels in a sediment core. Ordination diagrams are simply scatter-plots of the objects (e.g., core levels) on two or sometimes three axes according to the values taken by the objects along the variables comprising the axes.

When the data are multivariate, the problem is either to choose two pertinent variables for plotting the observations, or to construct synthetic variables that represent, in some optimal mathematical way, the set of variables under study; these synthetic variables may then be used as the major axes for the ordination. The data matrix subjected to analysis may contain a set of environmental variables, or the multi-species composition of the assemblage under study. In such cases, we will say that we are performing an ordination in a space of reduced dimensionality, or an ordination in reduced space, since the original data-set has many more dimensions (variables) than the ordination graph we want to produce.

This chapter describes the choices that have to be made in order to obtain a meaningful and useful ordination diagram. It will also show how the methods of canonical ordination, which are widely used to relate species to environmental data in palaeolimnology, are extensions within the framework of regression modelling of two classical ordination methods. Some forms of ordination analysis, classical or canonical, are widely used by palaeolimnologists as tools in the handling, summarisation, and interpretation of palaeolimnological data, either modern assemblages or core fossil assemblages (Smol 2008). The various types of use of ordination analysis in palaeolimnology are summarised in Table 8.1. No attempt is made here to provide a comprehensive review of palaeolimnological applications of ordination methods. Emphasis is placed instead on basic concepts and the critical methodological questions that arise in the use of ordination methods in palaeolimnology. Birks (2008, 2010) provides a short overview of the range of ordination methods currently available and of the general use and value of ordination techniques in ecology and palaeoecology. Borcard et al. (2011) discuss classical (unconstrained) and canonical (constrained) ordinations and their implementation with R.

Table 8.1 Palaeolimnological uses of ordination analysis

Modern biological assemblages (e.g., diatoms, chironomids)
Estimate the amount of compositional change or turnover – DCA
Summarise graphically the major patterns of variation – PCA, tb-PCA, CA, DCA, more rarely PCoA or NMDS
Display results of clustering or partitioning of data in a few dimensions – PCA, tb-PCA, CA, DCA, more rarely PCoA or NMDS
Modern environmental data (e.g., lake-water chemistry)
Summarise graphically the major patterns of variation – PCA, more rarely PCoA or NMDS
Display results of clustering or partitioning of data in a few dimensions – PCA, more rarely PCoA or NMDS
Fossil biological assemblages (e.g., diatoms, chironomids)
Estimate the amount of compositional change or turnover – DCA or its canonical relative DCCA with object age or depth as the sole constraining variable
Summarise graphically the major patterns of variation – PCA, tb-PCA, CA, DCA, more rarely PCoA or NMDS
Summarise stratigraphically the major patterns of variation – plot PCA, tb-PCA, CA, or DCA ordination axis object scores (e.g., axes 1–3) stratigraphically
Modelling temporal structure – RDA, tb-RDA, db-RDA, or CCA with PCNM temporal constraints
Down-core non-biological data (e.g., geochemistry, magnetics)
Summarise graphically the major patterns of variation – PCA
Summarise stratigraphically the major patterns of variation – plot PCA ordination axis object scores stratigraphically
Modelling temporal structure – RDA with PCNM temporal constraints
Modern and fossil biological assemblages (e.g., diatoms, chironomids)
Display similarities and dissimilarities between modern and fossil assemblages – PCA, tb-PCA, CA, DCA, more rarely PCoA or NMDS with either modern or fossil analysed passively or analysed together
Modern biological assemblages and modern environmental data (e.g., diatoms and lake-water chemistry)
Estimate the amount of compositional change or turnover along individual environmental gradients – DCCA
Summarise graphically the major patterns of biological variation explained by the environmental variables – RDA, tb-RDA, db-RDA, or CCA
Summarise graphically the major patterns of biological variation remaining after the partialling of other environmental variables – partial RDA, partial tb-RDA, partial db-RDA, or partial CCA
Assessment of statistical significance of single or combined environmental variables as predictors of the biological variation – RDA, tb-RDA, db-RDA, or CCA with Monte Carlo permutation tests
Development of 'minimal adequate model' of environmental variables that explain statistically the biological variation almost as well as the full set of environmental variables – RDA, tb-RDA, db-RDA, or CCA with variable selection (e.g., forward selection)
Partitioning biological variation among two or more sets of explanatory variables – RDA and partial RDA, tb-RDA and partial tb-RDA, db-RDA and partial db-RDA, CCA and partial CCA
Modelling spatial structure – RDA, tb-RDA, db-RDA, or CCA with PCNM spatial constraints

(continued)

Table 8.1 (continued)

Modern biological assemblages, modern environmental data, and fossil biological assemblages (e.g., diatoms and lake-water chemistry)
Display similarities and dissimilarities between modern and fossil assemblages in relation to modern environmental gradients – RDA, tb-RDA, db-RDA, or CCA with the fossil assemblages analysed passively
Fossil biological assemblages and palaeoenvironmental variables (e.g., diatoms, occurrences of volcanic tephras)
Test hypotheses of biological responses to particular environmental variables – RDA and partial RDA, tb-RDA and partial tb-RDA, db-RDA and partial db-RDA, CCA and partial CCA with Monte Carlo permutation tests
Modelling temporal structure – PCNM
Fossil biological assemblages from many sites
Summarise graphically the major patterns of variation – PCA, tb-PCA, CA, DCA, more rarely PCoA or NMDS
Modelling spatial structure – RDA, tb-RDA, db-RDA, or CCA with PCNM spatial constraints

CA correspondence analysis, *CCA* canonical correspondence analysis, *db-RDA* distance-based canonical redundancy analysis, *DCA* detrended correspondence analysis, *DCCA* detrended canonical correspondence analysis, *NMDS* non-metric multidimensional scaling, *PCA* principal component analysis, *PCNM* principal coordinates of a neighbour matrix, *PCoA* principal coordinate analysis, *tb-PCA* transformation-based principal component analysis, *tb-RDA* transformation-based canonical redundancy analysis

Basic Concepts in Simple Ordination

The simple ordination methods mostly used by (palaeo)ecologists and (palaeo) limnologists are principal component analysis (PCA), correspondence analysis (CA) and its relative, detrended correspondence analysis (DCA), principal coordinate analysis (PCoA), and non-metric multidimensional scaling (NMDS) (Prentice 1980, 1986). These methods will be reviewed here in a geometric framework. They mostly differ in the types of distances among objects that they attempt to preserve in the ordination.

Simple ordination is used in palaeolimnology to address two main types of questions. (1) In a study of sediment cores, ordination is used to identify the main gradients in the species assemblage data, which are multivariate by nature, and to interpret these gradients using species loadings on the ordination axes (see Birks 2012b: Chap. 11). Ordinations are also used as graphical templates to draw groups of sampling units obtained by clustering, as well as trajectories of the multivariate species data through time to estimate the magnitude and rates of change in species assemblage composition (Birks and Gordon 1985; Jacobson and Grimm 1986; Birks 1992, 2012b: Chap. 11). (2) Ordination of modern objects from various locations is also used as a basis on which fossil objects can be projected as passive objects for comparison between modern and fossil assemblages (Lamb 1984; Birks and Gordon 1985; Birks 1992, 2012b: Chap. 11).

Starting with a data-set, several choices have to be made before obtaining an ordination (Table 8.2). These choices will be described in some detail because a

Table 8.2 Questions that must be addressed prior to ordination analysis

Transform physical data
Univariate distributions are not symmetrical ⇒ Apply skewness-reduction transformation
Variables are not in the same physical units ⇒ Apply standardisation or ranging
Multistate qualitative variables ⇒ In some cases, transform them to dummy variables
Transform biological composition data (species presence-absence or abundance)
Reduce asymmetry of distributions ⇒ Apply square root or $\log(y + c)$ transformation
Make biological composition data suitable for Euclidean-based ordination methods ⇒ Use the chord, chi-square, or Hellinger transformation
Choose an appropriate distance function
Popular similarity or distance functions are:
Physical binary data: simple matching coefficient
Species presence-absence data: Jaccard, Sørensen, and Ochiai coefficients. The transformation $D = \sqrt{1 - S}$ ensures a fully Euclidean representation in principal coordinate analysis
Quantitative physical data: Euclidean distance on standardised or ranged variables
Physical data of mixed precision levels (quantitative, qualitative, binary): Gower similarity
Species abundance data: the chord, chi-square, Hellinger coefficients, as well as Clark's coefficient of divergence, are Euclidean. The Steinhaus similarity (equivalent to the Odum/Bray-Curtis distance) and Whittaker's index of association may not be Euclidean

good understanding of their implications is likely to produce more informative and useful ordination diagrams. Users of ordination methods should not let themselves be guided blindly by the implicit choices that are inherent to some methods or computer programs. The critical decisions to be made are the following:

- Do the data (environmental or assemblage data) need to be transformed prior to ordination analysis?
- Which distance measure should be preserved by the ordination method?
- Should a metric or non-metric ordination method be used?
- How many axes are required?

These decisions will now be discussed in some detail.

Transformation of Physical Data

Physical, chemical, or geological variables are often used as explanatory variables in palaeolimnological studies. They may also be used directly to obtain ordinations of the objects or sites on the basis of these variables (Table 8.1). Three problems may require pre-processing of the data before ordination: (a) if the distributions of the data along the variables are not symmetrical, skewness may need to be reduced; (b) if the variables are not all expressed in the same physical units, they need to be transformed to eliminate their physical units; (c) multistate qualitative variables (e.g., rare, common, abundant) may require, in some cases, transformation into dummy variables prior to ordination. Possible solutions to these three problems are as follows.

1. An ordination in which some of the points are clumped in a big mass while other points are stretched across the diagram is not very useful or informative. It is better to have the points scattered in a fairly homogeneous fashion across the diagram, with perhaps some clumping in the centre of the diagram, or in some areas of higher density if the data are clumped; the latter case may suggest that a cluster analysis might produce a more interesting and useful multivariate description of the data (see Legendre and Birks 2012: Chap. 7).

   The data should be initially examined using univariate methods, such as computing skewness statistics, or drawing frequency histograms (see Juggins and Telford 2012: Chap. 5). Depending on the type of asymmetry found, various transformations can be applied, such as square root, double square root, or log transformation. General methods, such as the Box-Cox transformations, are available to find automatically the most efficient normalising transformation; see Sokal and Rohlf (1995), Legendre and Legendre (1998, 2012), and Juggins and Telford 2012: Chap. 5. These are often referred to as *normalising transformations* because removing the asymmetry is an important step towards obtaining normally-distributed data. We emphasise, however, that the objective prior to ordination is not to obtain a multinormal distribution of the data, but simply to reduce the asymmetry of the distributions. Tests of normality may be useful to screen the data and identify the variables whose distributions should be examined more closely in order to find, if possible, a skewness-reducing transformation (see Juggins and Telford 2012: Chap. 5).

   Scientists often worry about transforming variables. Is it permissible? The original physical unit in which an environmental variable is measured imposes a scale to the data that is as unlikely to be related to the response of the species to this variable as any other scale that we may impose by applying a nonlinear transformation to the data. In order to relate a physical variable to the response of the species, physiological studies would be required to determine what the most appropriate transformation is. So, short of having such information available to them, users of ordination methods are left with statistical criteria only, such as skewness of the distributions, to decide on the transformation of physical variables.
2. In most cases, physical variables are not expressed in the same physical units; some may be in cm, others in $\mu g\,L^{-1}$, in °C, or in pH units. Such variables need to be transformed to eliminate the physical dimensions before being used together to produce an ordination. Note that log-transformed data are dimensionless because logarithms are exponents of a base and exponents are dimensionless. There are two main methods for eliminating physical dimensions: standardisation and ranging. They both eliminate the physical units by dividing the original data by a value possessing the same physical units.

   - Standardise variable **y** to **z**:

$$z_i = \frac{y_i - \bar{y}}{s_y} \tag{8.1}$$

where y_i is the original value of variable **y** for object i, $\bar{y}$ is the mean value of **y**, and s_y is the estimated standard deviation of **y**. z_i is the standardised value of variable **y** for object i. Variable standardisation is available in the `decostand()` function of the veganR-language package (method = 'standardize').

- Relative-scale variables: range variable **y** to **y**' using equation

$$y'_i = y_i / y_{max} \tag{8.2}$$

where y'_i is the ranged value of **y** for object i and y_{max} is the maximum value of **y** in the whole data table. This form of ranging is used for relative-scale variables, where 'zero' means the absence of the characteristic of interest. This transformation is available in the `decostand()` function of the vegan R-language package (method = 'max').

- Interval-scale variables: range variable **y** to **y**' using equation

$$y'_i = \frac{y_i / y_{min}}{y_{max} - y_{min}} \tag{8.3}$$

where y'_i is the ranged value of **y** for object i, whereas y_{min} and y_{max} are, respectively, the minimum and maximum values of **y** in the whole data table. This form of ranging is used for interval-scale variables, in which the value 'zero' is chosen arbitrarily and whose range may include negative values. Temperatures in °C are an example of an interval-scale variable. This transformation is available in the `decostand()` function of the vegan R-language package (method = 'range').

Variables may also be standardised in order to bring their variances to unity. It is preferable to apply skewness-reducing transformations before standardising the data. If the opposite is done, standardisation would produce negative values which are incompatible with square root, log, or Box-Cox transformations. Ranging, which brings all values of a variable into the interval [0,1], may be used before or after applying a skewness-reducing transformation.

3. Multistate qualitative variables may be handled in different ways. If the ordination is to be obtained through a method requiring the prior calculation of a distance matrix (PCoA, NMDS), resemblance coefficients are available that are capable of handling mixtures of quantitative and qualitative variables, as discussed in the section below on Choice of an Appropriate Distance Function and in Simpson (2012: Chap. 15). If, on the other hand, the ordination is to be obtained through a method that will implicitly preserve the Euclidean distance among objects (PCA, redundancy analysis (RDA)), the qualitative data must be transformed in some way prior to being subjected to the ordination method because a qualitative variable is not a *metric* or *measurement* variable; in other words, the distance between states 1 and 3 of a qualitative variable is not twice as large as the distance between states 1 and 2. Variables from which Euclidean

distances are calculated must be metric (or quantitative). The transformation can be done in one of two ways:

- A qualitative variable possessing *p* states can be recoded into *p* binary (0–1) variables, called *dummy variables*, using one dummy variable for each state of the qualitative variable. The coding method is described in Legendre and Legendre (1998, 2012: Sect. 1.5.7). Dummy variables can be used in PCA or RDA only if the program provides a possibility for weighting the variables. Indeed, if the variables are standardised or ranged prior to the ordination, a qualitative variable recoded into *p* dummy variables occupies *p* dimensions in the full-dimensional representation of the data. Each dummy variable should be downweighted to have a weight of 1/*p* in the analysis while the other quantitative variables have a weight of 1. The program CANOCO (ter Braak 1988a; ter Braak and Šmilauer 2002) offers the possibility of specifying weights for variables in PCA or RDA.
- Redundancy analysis (RDA) or canonical correspondence analysis (CCA) can be used to find a transformation of a qualitative multistate variable into a quantitative variable which is optimal with respect to a table of assemblage composition data (Legendre and Legendre 1998: p. 597). This is done as follows. Recode the qualitative variable into dummy variables as in the previous paragraph. Remove one of the dummy variables because, with all *p* dummy variables, the variance-covariance matrix of the dummy variables is singular and cannot be inverted; this is an obligatory step for explanatory variables in multiple regression and canonical analysis. For RDA or CCA, use the table of species composition data as the response matrix and the table of dummy variables as the explanatory matrix. If the first canonical ordination axis explains most of the canonical variance, it can be used in further analyses as a quantitative representation of the original qualitative variable. [Note: in the program CANOCO, the last of a set of dummy variables is automatically removed from the calculations. In the same way, the last state of a 'factor' variable is removed from the calculations in the `rda()` and `cca()` functions of the vegan R-language package, but the centroids of all states are drawn in the biplot.]

Transformation of Assemblage Composition Data

Assemblage composition data (species abundances) for short gradients, which contain relatively few zeros, can be ordinated by PCA or RDA: in that case the Euclidean distance is a meaningful measure of the ecological distance among the observations. These variables may, however, have asymmetric distributions because species tend to have exponential growth when conditions are favourable. This well-known fact has been embedded in the theory of species-abundance models; see He and Legendre (1996, 2002) for a synthetic view of these models. To reduce the asymmetry of the distributions, the species abundance variable **y** may be transformed to **y**' by taking the square root or the fourth root (which is equivalent to taking the square root twice), or by using a log transformation:

$$y' = y^{0.5} \quad \text{or} \quad y' = y^{0.25} \quad \text{or} \quad y' = \log(y + c) \tag{8.4}$$

where y is the species abundance and c is a constant. Usually, $c = 1$ in species log transformations, so that an abundance $y = 0$ is transformed into $y' = \log(0 + 1) = 0$ for any logarithmic base. Michael Palmer (http://www.okstate.edu/artsci/botany/ordinate/) does not recommend this transformation for absolute biomass data because it gives different values depending on the mass units (e.g., g or kg) used to record biomass. Another transformation that reduces the asymmetry of heavily skewed abundance data is the one proposed by Anderson et al. (2006). The abundance data y_{ij} are transformed to an exponential scale that makes allowance for zeros: $y'_{ij} = \log_{10}(y_{ij}) + 1$ when $y_{ij} > 0$ or $y'_{ij} = 0$ when $y_{ij} = 0$. Hence, for $y_{ij} = \{0, 1, 10, 100, 1000\}$, the transformed values are $\{0, 1, 2, 3, 4\}$. This transformation is available in the `decostand()` function of the vegan package (method = 'log').

Community composition data sampled along long ecological gradients typically contain many zero values because species are known to have generally unimodal responses along environmental gradients (ter Braak and Prentice 1988). The proportion of zeros is greater when the sampling has crossed a long environmental gradient. This is because species have optimal niche conditions, where they are found in greater abundances along environmental variables (see Juggins and Birks 2012: Chap. 14). The optimum for a species along an environmental variable corresponds to the centre of its theoretical Hutchinsonian niche along that factor. These propositions are discussed in most texts of community ecology and, in particular, in Whittaker (1967) and ter Braak (1987a). Because ordination methods use a distance function as their metric to position the objects with respect to one another in ordination space, it is important to make sure that the chosen distance is meaningful for the objects under study. Choosing an appropriate distance measure means trying to model the relationships among the sites appropriately for the assemblage composition data at hand. The choice of a distance measure is an ecological, not a statistical decision.

An example presented in Legendre and Legendre (1998: p. 278, 2012: Sect. 7.4.1) shows that the Euclidean distance function may produce misleading results when applied to assemblage composition data. Alternative (dis)similarity functions described in the next section, which were specifically designed for assemblage composition data, do not have this drawback. In some cases, distance measures that are appropriate for assemblage composition data can be obtained by a two-step procedure: first, transform the species abundance data in some appropriate way, as described below; second, compute the Euclidean distance among the sites using the transformed data (Fig. 8.1). This also means that assemblage composition data transformed in these ways can be directly used to compute ordinations by the Euclidean-based methods of PCA or RDA; this approach is called transformation-based PCA (tb-PCA) or transformation-based RDA (tb-RDA). The transformed data matrices

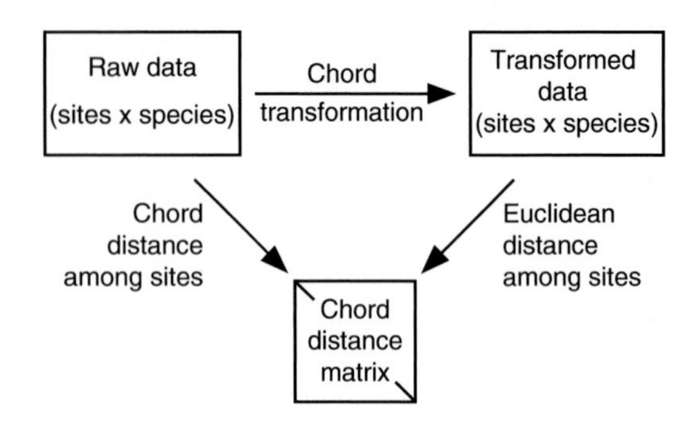

Fig. 8.1 The role of the data transformations as a way of obtaining a given distance function. The example uses the chord distance (Modified from Legendre and Gallagher 2001)

can also be used in *K*-means partitioning, which is another Euclidean-based method (see Legendre and Birks 2012: Chap. 7). Legendre and Gallagher (2001) have shown that the following transformations can be used in that context (some of these transformations have been in use in community ecology and palaeoecology for a long time, e.g., by Noy-Meir et al. (1975) and by Prentice (1980)).

1. Transform the species abundances from each object (sampling unit) into a vector of length 1, using the equation:

$$y'_{ij} = y_{ij} \Bigg/ \sqrt{\sum_{j=1}^{p} y_{ij}^2} \tag{8.5}$$

   where y_{ij} is the abundance of species *j* in object *i*. This equation, called the 'chord transformation' in Legendre and Gallagher (2001), is one of the transformations available in the program CANOCO (Centring and standardisation for 'samples': *Standardise by norm*) and in the `decostand()` function of the vegan R-language package (method = 'normalize'). If we compute the Euclidean distance

$$D_{\text{Euclidean}}\left(\mathbf{x}'_1, \mathbf{x}'_2\right) = \sqrt{\sum_{j=1}^{p} \left(y'_{1j} - y'_{2j}\right)^2} \tag{8.6}$$

   between two rows ($\mathbf{x}'_1$, $\mathbf{x}'_2$) of the transformed data table, the resulting value is identical to the chord distance (Eq. 8.18) that could be computed between the rows of the original (untransformed) species abundance data table (Fig. 8.1). The interest of this transformation is that the chord distance, proposed by Orlóci (1967) and Cavalli-Sforza and Edwards (1967), is one of the distances recommended for species abundance data. Its value is maximum and equal to $\sqrt{2}$ when two objects have no species in common. As a consequence, after the chord transformation, the assemblage composition data are suitable for PCA or RDA which are methods preserving the Euclidean distance among the objects.
2. In the same vein, if the data $[y_{ij}]$ are subjected to the 'chi-square distance transformation' as follows:

$$y'_{ij} = \sqrt{y_{++}} \frac{y_{ij}}{y_{i+}\sqrt{y_{+j}}} \tag{8.7}$$

where y_{i+} is the sum of the row (object) values, y_{+j} is the sum of the column (species) values, and y_{++} is the sum of values of the whole data table, then Euclidean distances computed among the rows of the transformed data table $[y'_{ij}]$ are equal to chi-square distances (Eq. 8.19) among the rows of the original, untransformed data-table. The chi-square distance, preserved in correspondence analysis, is another distance often applied to species abundance data. Its advantage or disadvantage, depending upon the circumstances, is that it gives higher weight to the rare than to the common species. The chi-square distance transformation is available in the `decostand()` function of the vegan R-language package (method = 'chi.square').

3. The data can be transformed into profiles of relative species abundances through the equation:

$$y'_{ij} = \frac{y_{ij}}{y_{i+}} \tag{8.8}$$

which is a widespread method of data standardisation, prior to analysis, especially when the sampling units are not all of the same size as is commonly the case in palaeolimnology. Data transformed in that way are called *compositional data*. In palaeolimnology and community ecology, the species assemblage is considered to represent the response of the community to environmental, historical, or other types of forcing; the variation of any single species has no clear interpretation. Compositional data are used because ecologists and palaeoecologists believe that the vectors of relative proportions of species can lead to meaningful interpretations. Many fossil or recent assemblage data-sets are presented as profiles of relative abundances, for example, in palynology and palaeolimnology, or as percentages if the values y'_{ij} are multiplied by 100. Computing Euclidean distances among rows (objects) of a data-table transformed in this way produces 'distances among species profiles' (Eq. 8.20). The transformation to profiles of relative abundances is available in the `decostand()` function of the vegan R-language package (method = 'total'). Statistical criteria investigated by Legendre and Gallagher (2001) show that this is not the best transformation; the Hellinger transformation (next paragraph) is preferable. Log-ratio analysis has been proposed as a way of analysing compositional data (Aitchison 1986). This method is, however, only appropriate for data that do not contain many zeros (ter Braak and Šmilauer 2002).

4. A modification of the species profile transformation is the Hellinger transformation:

$$y'_{ij} = \sqrt{\frac{y_{ij}}{y_{i+}}} \tag{8.9}$$

Computing Euclidean distances among rows (objects) of a data table transformed in this way produces a matrix of Hellinger distances among sites (Eq. 8.21). The Hellinger distance, described in more detail below, is a measure recommended for clustering or ordination of species abundance data (Prentice 1980; Rao 1995). It has good statistical properties as assessed by the criteria investigated by Legendre and Gallagher (2001). The Hellinger transformation is available in the `decostand()` function of the vegan R-language package (method = 'hellinger').

Before using these transformations, one may apply a square root or log transformation to the species abundances in order to reduce the asymmetry of the species distributions (Table 8.2). The transformations described above can also be applied to presence-absence data. The chord and Hellinger transformations appear to be the best for general use. The chi-square distance transformation is interesting when one wants to give more weight to the rare species; this is the case when the rare species are considered to be good indicators of special ecological conditions. We will come back to the use of these transformations in later sections. Prior to these transformations, any of the standardisations investigated by Noy-Meir et al. (1975), Prentice (1980), and Faith et al. (1987) may also be used if the study justifies it: species adjusted to equal maximum abundances or equal standard deviations, sites standardised to equal totals, or both.

Choice of an Appropriate Distance Function

Most statistical and numerical analyses assume some form of distance relationship among the observations. Univariate and multivariate analyses of variance and covariance, for instance, assume that the Euclidean distance is the appropriate way of describing the relationships among objects; likewise for methods of multivariate analysis such as K-means partitioning and PCA (see Legendre and Birks 2012: Chap. 7). It is the responsibility of the scientist doing the analyses either to make sure that this assumption is met by the data, or to model explicitly relationships of other forms among the objects by computing particular distance functions and using them in appropriate methods of data analysis.

Many similarity or distance functions have been used by ecologists; they are reviewed by Legendre and Legendre (1998, 2012: Chap. 7), Borcard et al. (2011: Chap. 3) and other authors. We will only mention here those that are most commonly used in the ecological, palaeoecological, and palaeolimnological literature.

1. The Euclidean distance (Eq. 8.6) is certainly the most widely used coefficient to analyse tables of physical descriptors, although it is not always the most appropriate. This is the coefficient preserved by PCA and RDA among the rows of the data matrix (objects), so that if the Euclidean distance is considered appropriate to the data, these methods can be applied directly to the data matrix, perhaps after one of the transformations described in the two previous sections, to obtain a meaningful ordination.

2. For physical or chemical data, an alternative to the Euclidean distance is to compute the Gower (1971) coefficient of similarity, followed by a transformation of the similarities to distances. The Gower coefficient is particularly important when one is analysing a table containing a mixture of quantitative and qualitative variables. In this coefficient, the overall similarity is the mean of the similarities computed for each descriptor *j* (see Simpson 2012: Chap. 15). Each descriptor is treated according to its own type. The partial similarity (s_j) between objects $\mathbf{x}_1$ and $\boldsymbol{x}_2$ for a quantitative descriptor *j* is computed as follows:

$$s_j(\mathbf{x}_1, \mathbf{x}_2) = 1 - \frac{|y_{1j} - y_{2j}|}{R_j} \tag{8.10}$$

where R_j is the range of the values of descriptor *j* across all objects in the study. The partial similarity s_j is a value between 0 (completely dissimilar) and 1 (completely similar). For a qualitative variable *j*, $s_j = 1$ if objects $\mathbf{x}_1$ and $\mathbf{x}_2$ have the same state of the variable and $s_j = 0$ if they do not. The Gower similarity between $\mathbf{x}_1$ and $\mathbf{x}_2$ is obtained from the equation:

$$S(\mathbf{x}_1, \mathbf{x}_2) = \sum_{j=1}^{p} s_j(\mathbf{x}_1, \mathbf{x}_2) / p \tag{8.11}$$

where *p* is the number of variables. The variables may receive different weights in this coefficient; see Legendre and Legendre (1998, 2012) for details. See also the note at the end of this section about implementations in R.

For presence-absence of physical descriptors, one may use the simple matching coefficient:

$$S(\mathbf{x}_1, \mathbf{x}_2) = \frac{a + d}{a + b + c + d} = \frac{a + d}{p} \tag{8.12}$$

where *a* is the number of descriptors for which the two objects are coded 1, *d* is the number of descriptors for which the two objects are coded 0, whereas *b* and *c* are the numbers of descriptors for which the two objects are coded differently. *p* is the total number of physical descriptors in the table.

There are different ways of transforming similarities (*S*) into distances (*D*). The most commonly used equations are:

$$D(\mathbf{x}_1, \mathbf{x}_2) = 1 - S(\mathbf{x}_1, \mathbf{x}_2) \tag{8.13}$$

and

$$D(\mathbf{x}_1, \mathbf{x}_2) = \sqrt{1 - S(\mathbf{x}_1, \mathbf{x}_2)} \tag{8.14}$$

For the coefficients described in Eqs. 8.11, 8.12, 8.15, 8.16 and 8.17, Eq. 8.14 is preferable for transformation prior to principal coordinate ordination because the distances so obtained produce a fully Euclidean representation of the objects in the ordination space, except possibly in the presence of missing values; Eq. 8.13 does not guarantee such a representation (Legendre and Legendre 1998, 2012: Table 7.2). The concept of Euclidean representation of a distance matrix is explained below in the section on Euclidean or Cartesian Space, Euclidean Representation. Equation 8.14 is used for transformation of all binary coefficients computed by the `dist.binary()` function of the ade4 R-language package.

3. For species presence-absence data,

   1. the Jaccard coefficient:

$$S(\mathbf{x}_1, \mathbf{x}_2) = \frac{a}{a + b + c} \tag{8.15}$$

   2. and the Sørensen coefficient of similarity:

$$S(\mathbf{x}_1, \mathbf{x}_2) = \frac{2a}{2a + b + c} \tag{8.16}$$

   are widely used. In these coefficients, a is the number of species that the two objects have in common, b is the number of species found at site or sample 1 but not at site or sample 2, and c is the number of species found at site or sample 2 but not at site or sample 1.

   3. The Ochiai (1957) coefficient:

$$S(\mathbf{x}_1, \mathbf{x}_2) = \frac{a}{\sqrt{(a + b)(a + c)}} \tag{8.17}$$

   deserves closer attention on the part of palaeoecologists since it is monotonically related to the binary form of the widely used chord and Hellinger distances described below (Eqs. 8.18 and 8.21).

For principal coordinate ordination analysis, the three similarity coefficients described above (Eqs. 8.15, 8.16, and 8.17) can be transformed into Euclidean-embeddable distances using the transformation $D(\mathbf{x}_1,\mathbf{x}_2) = \sqrt{(1 - S(\mathbf{x}_1, \mathbf{x}_2))}$ (Eq. 8.14). After these transformations, these distances will not produce negative eigenvalues in principal coordinate analysis and will thus be entirely represented in Euclidean space.

An interesting similarity coefficient among sites, applicable to presence-absence data, has been proposed by the palaeontologists Raup and Crick (1979): the coefficient is the probability of the data under the hypothesis of no association between objects. The number of species in common in two sites, a, is tested for significance under the null hypothesis H_0 that there is no association between sites $\mathbf{x}_1$ and $\mathbf{x}_2$ because each site in a region (or each level in a core) receives a random subset of species from the regional pool (or the whole sediment core). The association between objects, estimated by a, is tested using permutations. The probability (p) that the data conform to the null hypothesis is used as a measure of distance, or $(1-p)$ as a measure of similarity. The permutation procedure of Raup and Crick (1979) was re-described by Vellend (2004). Legendre and Legendre (2012: coefficient S_{27}) describe two different permutational procedures that can be used to test the significance of the number of species in common between two sites (i.e., the statistic a). These procedures correspond to different null hypotheses. Birks (1985) discusses the application of this and other probabilistic similarity measures in palaeoecology.

4. Several coefficients have been described by ecologists for the analysis of quantitative assemblage composition data. The property that these coefficients share is that the absence of any number of species from the two objects under comparison does not change the value of the coefficient. This property avoids producing high similarities, or small distances, between objects from which many species are absent. The Euclidean distance function, in particular, is not appropriate for assemblage composition data obtained from long environmental gradients because the data table then contains many zeros, and the objects that have many zeros in common have small Euclidean distance values; this is considered to be an inappropriate answer in most ecological and palaeoecological problems. This question is discussed at length in many texts of quantitative community ecology. The coefficients most widely used by ecologists for species abundance data tables are:

   1. The chord distance, occasionally called the *cosine-θ distance*:

$$D(\mathbf{x}_1,\mathbf{x}_2) = \sqrt{\sum_{j=1}^{p}\left(\frac{y_{1j}}{\sqrt{\sum_{j=1}^{p} y_{1j}^2}} - \frac{y_{2j}}{\sqrt{\sum_{j=1}^{p} y_{2j}^2}}\right)^2} = \sqrt{2\left(1 - \frac{\sum_{j=1}^{p} y_{1j}y_{2j}}{\sqrt{\sum_{j=1}^{p} y_{1j}^2 \sum_{j=1}^{p} y_{2j}^2}}\right)} \tag{8.18}$$

   which consists of subjecting the species data to the chord transformation (Eq. 8.5) followed by calculation of the Euclidean distance (Eq. 8.6). The chord distance is closely related to the Hellinger distance (Eq. 8.21).

   2. The chi-square distance:

$$D(\mathbf{x}_1,\mathbf{x}_2) = \sqrt{y_{++}}\sqrt{\sum_{j=1}^{p}\frac{1}{y_{+j}}\left(\frac{y_{1j}}{y_{1+}} - \frac{y_{2j}}{y_{2+}}\right)^2} \tag{8.19}$$

where y_{i+} is the sum of the frequencies in row i, y_{+j} is the sum of the frequencies in column j, and y_{++} is the sum of all frequencies in the data table. It is equivalent to subjecting the species data to the chi-square distance transformation (Eq. 8.7) followed by calculation of the Euclidean distance (Eq. 8.6).

3. The distance between species profiles:

$$D(\mathbf{x}_1, \mathbf{x}_2) = \sqrt{\sum_{j=1}^{p} \left(\frac{y_{1j}}{y_{1+}} - \frac{y_{2j}}{y_{2+}} \right)^2} \tag{8.20}$$

is equivalent to subjecting the species data to the transformation to profiles of relative abundances (Eq. 8.8) followed by calculation of the Euclidean distance (Eq. 8.6).

4. The Hellinger distance (Rao 1995):

$$D(\mathbf{x}_1, \mathbf{x}_2) = \sqrt{\sum_{j=1}^{p} \left[\sqrt{\frac{y_{1j}}{y_{1+}}} - \sqrt{\frac{y_{2j}}{y_{2+}}} \right]^2} \tag{8.21}$$

It is equivalent to subjecting the species data to the Hellinger transformation (Eq. 8.9) followed by calculation of the Euclidean distance (Eq. 8.6). This equation is occasionally called the *chord distance* (Prentice 1980) described in Eq. 8.18, because the Hellinger distance is the chord distance computed on square-root transformed frequencies. In the Hellinger distance, the relative species abundances ('compositional data', used directly in Eq. 8.20) are square-root transformed in order to lower the importance of the most abundant species, which may grow exponentially when they encounter favourable conditions. This coefficient thus increases the importance given to the less abundant species (Prentice 1980). The chord (Eq. 8.18) and Hellinger (Eq. 8.21) functions produce distances in the range $[0, \sqrt{2}]$. For presence-absence data, they are both equal to

$$\sqrt{2}\sqrt{1 - \frac{a}{\sqrt{(a+b)(a+c)}}}$$

where

$$\frac{a}{\sqrt{(a+b)(a+c)}}$$

is the Ochiai (1957) similarity coefficient for binary data described in Eq. 8.17.

5. A coefficient first described by Steinhaus (in Motyka 1947) and rediscovered by other authors, such as Odum (1950) and Bray and Curtis (1957), is called the percentage difference (Odum 1950):

$$D(\mathbf{x}_1, \mathbf{x}_2) = \frac{\sum_{j=1}^{p} |y_{1j} - y_{2j}|}{\sum_{j=1}^{p} (y_{1j} + y_{2j})} \tag{8.22}$$

This coefficient has excellent descriptive properties for community composition data (Hajdu 1981; Gower and Legendre 1986). Taking the square root of this distance will avoid negative eigenvalues and complex principal axes in principal coordinate analysis. A particular form of this coefficient, for data transformed into percentages by sites (y'_{ij} of Eq. 8.8 multiplied by 100), has been described by Renkonen (1938). When presence-absence data are used in Eq. 8.22, the resulting coefficient is the one-complement of the Sørensen coefficient of similarity (Eq. 8.16) computed over the same data (i.e., $D_{(\text{eq. }8.22)} = 1 - S_{(\text{eq. }8.16)}$).

6. Whittaker's (1952) index of association is:

$$D(\mathbf{x}_1, \mathbf{x}_2) = \frac{1}{2} \sum_{j=1}^{p} \left| \frac{y_{1j}}{y_{1+}} - \frac{y_{2j}}{y_{2+}} \right| \tag{8.23}$$

7. Clark's (1952) coefficient of divergence:

$$D(\mathbf{x}_1, \mathbf{x}_2) = \sqrt{\frac{1}{p} \sum_{j=1}^{p} \left(\frac{y_{1j} - y_{2j}}{y_{1j} + y_{2j}} \right)^2} \tag{8.24}$$

is a form of the Canberra metric (Lance and Williams 1967) rescaled to the [0, 1] range.

Most of the distances described in this section can be computed using the R-language functions `dist()` (stats package), `vegdist()` (vegan), `dist.binary()` (ade4), `gowdis()` (FD) and `daisy()` (cluster); see footnote of Table 8.3 for references. This statement calls for some remarks. (1) These packages do not all produce the same results for the binary Jaccard coefficient: `dist()` and `vegdist()` use the transformation $D = (1 - S)$ (Eq. 8.13) whereas `dist.binary()` uses $D = \sqrt{1 - S}$ (Eq. 8.14) to transform similarities into distances. The latter guarantees that a fully Euclidean representation, without negative eigenvalues and complex eigenvectors, will result from principal coordinate analysis. (2) The chord, chi-square and Hellinger distances are not obtained directly but after two calculation steps:

Table 8.3 Computer programs for ordination. The list makes no pretence at being exhaustive

Simple ordination

CANOCO: PCA, CA

PC-ORD: PCA, CA, NMDS

PrCoord: PCoA (available dissimilarity measures: 7)

WynKyst: NMDS (available dissimilarity measures: 7)

R-language functions for PCA and CA: `rda()`, `cca()` (vegan package); `dudi.pca()`, `dudi.coa()` (ade4 package); `pca()` (labdsv package)

R-language dissimilarity functions: 10 binary measures in `dist.binary()` of ade4, 6 dissimilarity measures in `dist()` of stats, 13 in `vegdist()` of vegan, 3 in `daisy()` of cluster, and `gowdis()` in FD

R-language functions for PCoA: `dudi.pco()` (ade4 package); `pcoa()` (ape package); `cmdscale()` (stats package) and its wrappers `cmds.diss()` (mvpart package), `pco()` (labdsv package), and `capscale()` (vegan package).

R-language functions for NMDS: `isoMDS()` (MASS package) and its wrappers `nmds()` and `bestnmds()` of labdsv, and `metaMDS()` of vegan. Dissimilarity measures available in the R language: see previous entry

SYN-TAX: PCA, CA, PCoA, NMDS (available dissimilarity measures: 39)

Canonical ordination

CANOCO: linear RDA and CCA; partial RDA and CCA

PC-ORD: linear CCA

Polynomial RDACCA: linear and polynomial RDA, linear and polynomial CCA

R-language functions: `rda()` and `cca()` (vegan package): linear RDA and CCA; partial RDA and CCA

R-language function for variation partitioning: `varpart()` (vegan package) partitions the variation of a response table **Y** with respect to two, three, or four explanatory tables **X**, using partial RDA

SYN-TAX: linear RDA and CCA

R-language package cocorresp: co-correspondence analysis

Biplots and triplots

CanoDraw

PC-ORD

SYN-TAX

R language: `plot.cca()` (vegan package) produces PCA and CA biplots as well as RDA and CCA triplots

CANOCO, CanoDraw, and PrCoord (for Windows): available as a bundle from Microcomputer Power http://www.microcomputerpower.com. PC-ORD (for Windows): available from MjM Software http://home.centurytel.net/~mjm/pcordwin.htm. R language (for Windows, Linux, and MacOS X): freely downloadable from the Comprehensive R Archive Network (CRAN) http://cran.r-project.org/. Packages ade4 (Chessel et al. 2004, Dray et al. 2007), labdsv (Roberts 2007), cluster (Maechler et al. 2005), cocorresp (Simpson 2009), stats (R Development Core Team 2011), ape (Paradis et al. 2010), FD (Laliberté and Shipley 2010), and vegan (Oksanen et al. 2011). SYN-TAX (for Windows and Macintosh): available from Scientia Publishing http://ramet.elte.hu/~scientia/ and Exeter Software http://www.exetersoftware.com. Polynomial RDACCA (for Windows and Macintosh): freely downloadable from P. Legendre's Web page http://www.bio.umontreal.ca/legendre/indexEn.html.

transformation of the data (Eqs. 8.5, 8.7 and 8.9) followed by calculation of the Euclidean distance (Eq. 8.6). (3) Several functions implement the Gower distance: `vegdist()` (vegan), `daisy()` (cluster), and `gowdis()` (FD);

see footnote of Table 8.3 for references. The latter is the only function that can handle missing values and variables of all precision levels, including multistate qualitative variables ('factors' in R), and allows users to give different weights to the variables involved in a calculation.

Euclidean or Cartesian Space, Euclidean Representation

A *Cartesian space*, named after René Descartes (1596–1650), French mathematician and philosopher, is a space with a Cartesian system of coordinates. It is also called a *Euclidean space* because the distances among points are measured by Eq. 8.6 in that space. The multidimensional ordination spaces of PCA, CA, PCoA, NMDS, etc., are Cartesian or Euclidean spaces; hence the distances among points embedded in these spaces are measured by the Euclidean distance formula. A few dimensions that represent a good deal of the variance of the data will be chosen from these multidimensional spaces to create a reduced-space ordination.

A distance function is said to have the Euclidean property, or (in short) to be Euclidean, if it always produces distance matrices that are fully embeddable in a Euclidean space. The test, available in the R-language package ade4 (function `is.euclid()`), is that a principal coordinate analysis (PCoA) of a Euclidean distance matrix produces no negative eigenvalues. This is not always the case in ordination. Some distance functions are not Euclidean, meaning that the distances in the matrix cannot be fully represented in a Euclidean ordination space. A principal coordinate analysis of the distance matrices produced by these coefficients may generate negative eigenvalues; these eigenvalues indicate the non-Euclidean nature of the distance matrix (Gower 1982). They measure the amount of variance that needs to be added to the distance matrix to obtain a full Euclidean representation. To be a metric is a necessary but not a sufficient condition for a distance coefficient to be Euclidean. Many of the commonly-used similarity coefficients are not Euclidean when transformed into distances using Eq. 8.13. The transformation described by Eq. 8.14 often solves the problem, however. For instance, the similarity coefficients of Gower, simple matching, Jaccard, Sørensen, Ochiai, and Steinhaus, described above, all become Euclidean when transformed into distances using Eq. 8.14 (Gower and Legendre 1986; Legendre and Legendre 1998, 2012: Table 7.2).

If the analysis is carried out to produce a PCoA ordination in a few (usually 2 or 3) dimensions, negative eigenvalues do not matter as long as their absolute values are not large when compared to the positive eigenvalues of the axes used for the reduced-space ordination. If the analysis requires that all coordinates be kept, as will be the case when testing multivariate hypotheses using the db-RDA method (see the subsection below on Linear RDA), negative eigenvalues should either be avoided or corrected for. They can be avoided by selecting a distance coefficient that is known to be Euclidean. When a non-Euclidean coefficient is used (for example, the Steinhaus/Odum/Bray-Curtis coefficient of Eq. 8.22), there are ways of correcting for negative eigenvalues in PCoA to obtain a fully Euclidean solution; see Legendre and Legendre (1998, 2012) for details. These corrections are available in some PCoA computer programs, including function `pcoa()` of the **ape** R-language package.

Metric or Non-metric Ordination?

Metric ordinations are obtained by the methods of principal component analysis (PCA), correspondence analysis (CA), and principal coordinate analysis (PCoA). These methods all proceed by eigenvalue decomposition. The eigenvalues measure the amount of variation of the observations along the ordination axes. The distances in the full-dimensional ordination space are projected onto the space of reduced dimensionality (usually two dimensions) chosen for ordination. Non-metric ordinations are obtained by non-metric multidimensional scaling (NMDS) which is not an eigenvalue method. This method only approximately preserves the rank-order of the original distances in the reduced ordination space.

PCA is the method of choice to preserve Euclidean distances among objects, and CA when the chi-square distance is to be preserved. For other forms of distance, users have to choose between PCoA (also called metric scaling) and NMDS. PCoA is the preferred method (1) when one wishes to preserve the original distances in full-dimensional space, (2) when many (or all) ordination axes are sought, or (3) when the data-set is fairly large. NMDS may be preferred when the user wants to represent as much as possible of the distance relationships among objects in a few dimensions, at the cost of preserving only the rank-order of the distances and not the distances themselves.

The size of the data-sets is also of importance. PCA and CA can easily be computed on very large data-sets (tens or hundreds of thousand objects) as long as the number of variables is small (up to a few hundred), because the eigenvalue decomposition is done on the covariance matrix, which is of size p, the number of variables in the data-set.

For tables containing assemblage composition data, three paths can be followed: (1) one can transform the data using one of the transformations described by Eqs. 8.5, 8.7, 8.8, or 8.9, and produce the ordination by PCA (tb-PCA approach), or (2) compute a distance matrix using Eqs. 8.15, 8.16, 8.17, 8.18, 8.19, 8.20, 8.21, 8.22, 8.23 and 8.24, followed by PCoA or NMDS. For large data-sets of intermediate sizes (up to a few thousand objects), PCoA will produce the ordination solution faster than NMDS. For very large data-sets, PCA should be used. (3) For data-sets of any size, one can produce the ordination using CA if the chi-square distance is appropriate.

An alternative and biologically useful approach to deciding between ordinations based on PCA (Euclidean distance) of untransformed data and CA (chi-square distance) of multivariate species assemblage data is that emphasised by ter Braak (1987a) and ter Braak and Prentice (1988), namely the underlying species response model that is assumed when fitting either PCA or CA and extracting synthetic latent variables that are then used as the major ordination axes. PCA assumes an underlying linear response model, whereas CA assumes an underlying unimodal response model between the variables and the unknown but to be determined latent variables or ordination axes. The question is thus how to know whether a linear-based or a unimodal-based ordination is appropriate for a given data-set. The detrended relative of CA, detrended correspondence analysis (DCA: Hill and Gauch 1980; ter Braak 1987a), is a heuristic modification of CA designed to minimise two of the disadvantages of CA, namely the so-called arch-effect and the so-called edge-effect (ter Braak and Prentice 1988). As a result of the non-linear rescaling of the axes that removes the edge-effect, the object scores are scaled and standardised in a particular way. The lengths of the resulting ordination axes are given by the range of object scores and are expressed in 'standard deviation units' (SD) or units of compositional turnover. The tolerance or amplitude of the species' curves along the rescaled DCA axes are close to 1; each curve will therefore rise and fall over about 4 SD (ter Braak and Prentice 1988). Objects that differ by 4 SD can be expected to have no species in common. A preliminary DCA of an assemblage data-set, with detrending by segments and non-linear rescaling, provides an estimate of the underlying gradient length. If the gradient length is less than about 2.5 SD, the assemblage variation is within a relatively narrow range, and the linear approach of PCA is appropriate. If the gradient length is 3 or more SD, the assemblage variation is over a larger range, and the unimodal-based approach of CA is appropriate (ter Braak and Prentice 1988). Transformation-based PCA (tb-PCA) is also appropriate in that case.

How Many Axes Are Required?

In most instances, ordination analysis is carried out to obtain an ordination in two, sometimes three, dimensions. The ordination is then used to illustrate the variability of the data along the ordination axes and attribute it to the variables that are most highly correlated with those axes. Simple interpretation of the variability in the ordination diagram can be obtained by projecting interpretative variables in the ordination plane, or by representing other properties of the data (for instance, the groups produced by cluster or partitioning analysis (Legendre and Birks 2012: Chap. 7)), or some other grouping of the objects known *a priori* (for example, the type of lake, or the nature of the sediment) (see Lepš and Šmilauer 2003).

There are instances where ordination analysis is carried out as a pre-treatment, or transformation, of the original data, before carrying out some further analysis. For example, one may wish to preserve the Steinhaus/Odum/Bray-Curtis distance in a canonical redundancy analysis (RDA) or K-means partitioning (see Legendre and Birks 2012: Chap. 7). To achieve that, one may compute the distance matrix using Eq. 8.22 (or its square root) and carry out a PCoA of that matrix. One then keeps all eigenvectors from this analysis (after perhaps a correction for negative eigenvalues) and uses that matrix of eigenvectors as input to redundancy analysis (RDA) or K-means partitioning. This is an example of distance-based RDA (db-RDA) described in more detail in the subsection on Linear RDA.

Tests of significance for individual eigenvalues are available for PCA; see the review papers of Burt (1952) and Jackson (1993). They are not often useful because, in most instances, ecologists do not have a strong null hypothesis to test; they rather use PCA for an exploratory representation of their data. Also, the parametric tests of significance assume normality of all descriptors, which is certainly a drawback for palaeolimnological data. Users most often rely on criteria that help them determine how many axes represent 'important' variation with respect to the original data table. The two best criteria at the moment are the simple broken-stick model proposed by Frontier (1976) as well as the bootstrapped eigenvalue method proposed by Jackson (1993).

Simple Ordination Methods: PCA, CA, PCoA, NMDS

The simple ordination methods mostly used by palaeoecologists and palaeolimnologists (Table 8.1) are the following.

1. Principal component analysis (PCA) is the oldest (Hotelling 1933) and best-known of all ordination methods. Consider a group of data points in multidimensional space, placed at Euclidean distances (Eq. 8.6) of one another. Imagine a lamp behind the cloud of points, and the shadows of the points projected onto a white wall. The geometric problem consists of rotating the points in such a way that the shadows have as much variance as possible on the wall. The mathematics of eigenvalues and eigenvectors, which is part of matrix algebra, is the way to find the rotation that maximises the variance of the projection in any number of dimensions. The variables are first transformed if required (Table 8.2), then centred by column, forming matrix **Y**. One computes the dispersion (or variance-covariance) matrix **S** among the variables, followed by the eigenvalues (λ_j) and eigenvectors of **S**. The eigenvectors are assembled in matrix **U**. The principal components, which provide the coordinates of the points on the successive ordination axes, are the columns of matrix $\mathbf{F} = \mathbf{YU}$. The eigenvalues measure the variance of the points along the ordination axes (the columns of matrix **F**). The first principal component has the highest eigenvalue λ_1, hence the largest variance; and so on for the following components, with the constraint that all components are orthogonal and uncorrelated to one another.

A scatter diagram with respect to the first two ordination axes, using the coordinates in the first two columns of matrix **F**, accounts for an amount of variance equal to $\lambda_1 + \lambda_2$. The distances among points in two dimensions are projections of their original, full-dimensional Euclidean distances. The contributions of the variables to the ordination diagram can be assessed by drawing them using the loadings found in matrix **U**. For two dimensions again, the first two columns of matrix **U** provide the coordinates of the end-points of vectors (arrows) representing the successive variables. A graph presenting the variables (as arrows) on top of the dispersion of the points, as described above, is called a *distance biplot*. Another type of biplot, called a *correlation biplot*, can also be produced by many PCA programs; the correlations among variables are represented by the angles of projection of the variables, in two dimensions, after rescaling the eigenvectors to the square root of their respective eigenvalues (ter Braak 1994; Lepš and Šmilauer 2003). The objects projected onto these modified axes are not at Euclidean distances but are at Mahalanobis distances to one another. Supplementary or passive objects and variables can be projected onto a PCA ordination diagram. This option is available in most of the programs offering a PCA procedure listed in Table 8.3. The mathematics behind such projections is described in Legendre and Legendre (1998, 2012: Sect. 9.1.9) and ter Braak and Šmilauer (2002).

The approach of fitting fossil objects as supplementary objects onto a PCA ordination of modern data has been used by palaeoecologists (e.g., Lamb 1984) as an aid in detecting similarities between modern and fossil assemblages. It is important, however, to calculate the residual distances when adding additional supplementary objects into any low-dimensional ordination, as new objects may appear to be positioned close to other objects on the first few axes and yet be located some distance from these other objects when further dimensions are considered (Birks and Gordon 1985). Gower (1968) discusses the calculation and interpretation of the residual distances from the true position of the added points to the fitted plane giving the best two-dimensional representation of the objects.

Alternatively, one may perform a PCA of fossil assemblage data and add modern objects into the ordination (e.g., Ritchie 1977), or perform a PCA of fossil and modern assemblage data combined (MacDonald and Ritchie 1986). Prentice (1980) and Birks and Gordon (1985) discuss the advantages and disadvantages of fitting objects, modern or fossil, into low-dimensional PCA representations.

The most common application of PCA in palaeolimnology is to produce biplot diagrams of the objects (sites, lakes, core subunits, etc.) with respect to physical or chemical variables (e.g., Jones et al. 1993) or assemblage composition data (after appropriate transformation: Table 8.2) (e.g., Birks and Peglar 1979). Another useful representation of PCA results of core assemblages is to plot the object scores on the first few principal components in stratigraphical order for each axis (e.g., Birks and Berglund 1979; Birks 1987; Lotter and Birks 2003; Wall et al. 2010; Wang et al. 2010; Birks 2012b: Chap. 11), thereby providing a summarisation of the major patterns of variation in the stratigraphical data in two or three axes. PCA can also be used to detect outliers in data, which may correspond to legitimate outliers, or to erroneous data. PCA may be used to identify groups of variables that are highly correlated and, thus, form bundles of arrows in the ordination diagram; look, in particular, for variables that are highly but negatively correlated: their arrows are opposite in the diagram (e.g., Gordon 1982; MacDonald and Ritchie 1986). Another application is to simplify data-sets containing many highly collinear variables; the PCA axes that account for, say, 95% of the total variance form a simplified set of variables and allow discarding of the remaining 5%, which can be regarded as noise (Gauch 1982; Lotter et al. 1992).

2. Correspondence analysis (CA) is a form of PCA that preserves the chi-square distance (Eq. 8.19) among the objects or variables. CA is appropriate for frequency data, and in particular for species presence-absence or abundance data, subject to the caveat that the chi-square distance gives high weights to rare species. There are several ways of presenting CA (Hill 1974). We will look at it here as the eigenanalysis of a table of components of chi-square. The assemblage composition data matrix **Y** is transformed into a matrix of components of chi-square $\mathbf{Q} = [q_{ij}]$ where

$$q_{ij} = \left[\frac{O_{ij} - E_{ij}}{\sqrt{E_{ij}}} \right] / \sqrt{y_{++}} \tag{8.25}$$

The part inside the square parentheses is easily recognised as the component of the chi-square statistic computed in each cell of a frequency (or contingency) table; they are obtained from the observed (O_{ij}) and the expected values (E_{ij}) of cell ij of the table. These components can be added to produce the Pearson chi-square statistic used to test the hypothesis of absence of relationship between the rows and columns of a contingency table. Here, the components of chi-square are divided by a constant, the square root of the sum of values in the whole table (y_{++}), which turns them into the values $[q_{ij}]$ of the transformed data table **Q**. From this point, one can compute a cross-product matrix (the covariance matrix computed in PCA is also a cross-product matrix, but it is computed here without further centring since centring is part of Eq. 8.25), and from it the eigenvalues and eigenvectors are extracted. An alternative approach is to carry out singular value decomposition of the matrix **Q**, as explained in Legendre and Legendre (1998: Sect. 9.4, 2012: Sect. 9.2). The eigenvalues measure the amount of inertia accounted for by each ordination axis. Matrices are obtained that contain the positions of the objects (rows) and species (columns) along the successive axes of the ordination space. Two types of scaling can be used for biplots: one can (1) preserve the chi-square distances among objects (rows), the objects being at the centroids of the species (columns); or (2) preserve the chi-square distance among the variables (columns), the variables being at the centroids of the objects (rows) (ter Braak and Verdonschot 1995). The most common application of CA in palaeolimnology is to produce biplot diagrams of species and objects or other sampling units (e.g., Jones and Birks 2004). As in PCA, supplementary objects and variables can be projected onto a CA ordination diagram (e.g., Jones and Birks 2004). This option is available, for instance, in the program CANOCO. In R, functions to that effect are also available in vegan and ade4. vegan: `predict.rda()` and `predict.cca()` for adding new points to PCA, RDA, CA and CCA, and `envfit()` for adding supplementary variables to all of the above (`envfit()` does weighted fitting in CCA so that it is consistent with the original). ade4: `suprow()` to add supplementary objects and `supcol()` to add supplementary variables to PCA and CA plots.

Usually, ecologists who see the organisms they are sampling consider rare species as potential indicators of rare environmental conditions, whereas those who have to sample blindly or use traps are more wary of the interpretation of rare species. In animal ecology, a single presence of a species at a site may be due to a species that does not belong to the site but was travelling between two other favourable sites. In palynology, likewise, pollen may be brought by far transport from distant sites. In aquatic ecology, rare species may appear in spurious ways in sampling units from sites where they are found at low abundance. Because of their influence on the chi-square distance (Eq. 8.19), one should pay special attention to rare species in CA. One must understand that rare species affect the ordination of objects very little, but these species will be represented by points located far from the origin. Users of CA who are worried about the interpretation of rare species often decide to remove, not the species that have low abundance, but those that occur in the data-set very rarely. One may try removing first the species that occur only once in the data-set, then those that occur once or twice, and so on, repeating the analysis every time. One can remove the rarest species up to the point where the first few eigenvalues, expressed as percentages of the inertia (= variation) *in the original data-set*, are little affected by the removal. This approach has been suggested by Daniel Borcard, Université de Montréal.

Palaeolimnologists often use the detrended relative of CA, detrended correspondence analysis (DCA), as a preliminary tool in establishing the extent of compositional turnover in modern calibration data-sets as a guide as to whether to use calibration procedures that assume linear or unimodal responses of species to environmental gradients (Birks 1995). Detrending by segments is an arbitrary method for which no theoretical justification has been offered, while the assumptions behind the nonlinear rescaling procedure have not been fully substantiated (Wartenberg et al. 1987, but see ter Braak 1985). Jackson and Somers (1991) showed that DCA ordinations of sites greatly varied with the number of segments one arbitrarily decides to use, so that the ecological interpretation of the results may vary widely. In simulation studies conducted on artificial data representing unimodal species responses to environmental gradients in one or two dimensions, DCA did not perform particularly well in recovering complex gradients (Kenkel and Orlóci 1986; Minchin 1987). For these reasons, detrended correspondence analysis (DCA) should generally be avoided for the production of ordination plots when a detailed interpretation of the object relative positions is sought. Palaeolimnologists (e.g., Birks et al. 2000: Birks and Birks 2001; Bradshaw et al. 2005) have plotted the object scores on the first DCA axis in stratigraphical order for different palaeolimnological variables (e.g., diatoms, chironomids) as a means of comparing the major trends and compositional turnover between different proxies within the same stratigraphical sequence (see Birks 2012b: Chap. 11).

3. In principal coordinate analysis (PCoA), the objective is to obtain an ordination, in any number of dimensions, representing as much as possible of the variation of the data while preserving the distance that has explicitly been computed. The algebra used to find a solution to the geometric problem proceeds directly from a pre-computed square, symmetric distance matrix **D**. The first step is to transform

the distances d_{hi} of **D** into values $a_{hi} = -0.5\,d_{hi}^2$, then to centre the resulting matrix **A** to produce a third matrix $\Delta = [\delta_{hi}]$ using the equation:

$$\delta_{hi} = a_{hi} - \bar{a}_h - \bar{a}_i + \bar{a} \tag{8.26}$$

where $\bar{a}_h$ and $\bar{a}_i$ are the means of row h and column i corresponding to element a_{hi}, whereas $\bar{a}$ is the mean of all a_{hi} values in the matrix. Eigenvalue decomposition is applied to matrix Δ, producing eigenvalues and eigenvectors. When the eigenvectors are normalised to the square root of their respective eigenvalues, they directly provide the coordinates of the objects on the given ordination axis. The eigenvalues give the variance (not divided by degrees of freedom) of the objects along that axis. If some eigenvalues are negative and all ordination axes are needed for subsequent analyses, corrections can be applied to the distance matrix; this was mentioned in the section on Euclidean or Cartesian Space, Euclidean Representation.

A simple example may help explain PCoA. From an object-by-variable data matrix **Y**, compute matrix **D** of Euclidean distances among the objects. Run PCA using matrix **Y** and PCoA using matrix **D**. The eigenvalues of the PCoA of matrix **D** are proportional to the PCA eigenvalues computed for matrix **Y** (they differ by the factor $(n-1)$), while the eigenvectors of the PCoA of **D** are identical to matrix **F** of the PCA of **Y**. Normally, one would not compute PCoA on a matrix of Euclidean distances since PCA is a faster method to obtain an ordination of the objects in **Y** that preserves the Euclidean distance among the objects. This was presented here simply as a way of understanding the relationship between PCA and PCoA in the Euclidean distance case. The real interest of PCoA is to obtain an ordination of the objects from some other form of distance matrix more appropriate to the data at hand—for example, a Steinhaus/Odum/Bray-Curtis distance matrix in the case of assemblage composition data. Surprisingly, PCoA has rarely been used in palaeoecology (e.g., Birks 1977; Engels and Cwynar 2011) in contrast to the extensive use of PCA, CA, and DCA.

4. Non-metric ordinations are obtained by non-metric multidimensional scaling (NMDS); several variants of this method have been proposed (Prentice 1977, 1980). The distances in the low-dimensional space are not rigid projections of the original distances in full-dimensional space. In NMDS, the user sets the dimensionality of the space in which the ordination is to be produced; the solution sought is usually two-dimensional. The computer program proceeds by successive iterations, trying to preserve in the ordination the rank-order of the original distances. Different formulae, called Stress (formula 8.1 or 8.2), Sstress, or Strain, may be used to measure the goodness-of-fit of the solution in reduced space. Non-metric ordinations are rarely used in palaeoecology. Early applications include Birks (1973), Gordon and Birks (1974), and Prentice (1978), whereas more recent applications include Brodersen et al. (1998, 2001), Simpson et al. (2005), Soininen and Weckström (2009), Tremblay et al. (2010), Wiklund et al. (2010), Allen et al. (2011), and Wischnewski et al. (2011). Overall, there seem to be few theoretical advantages in using NMDS in palaeoecology (Prentice 1980).

Introduction to Canonical Ordination

The methods of canonical ordination are generalisations of simple ordination methods; the ordination is forced or constrained to represent the part of the variation in a table of response variables (e.g., species abundances) that is maximally related to a set of explanatory variables (e.g., environmental variables). Canonical redundancy analysis (RDA) is the constrained form of PCA whereas canonical correspondence analysis (CCA) is the constrained form of CA. Canonical ordination is a hybrid between regression and ordination, as will be described below. The classical forms of RDA and CCA use multiple linear regression between the variables in the two data tables. Canonical ordination methods have also been described that look for polynomial relationships between the dependent (response) and explanatory (predictor) variables. Tests of statistical significance of the relationship between the species and environmental data can be performed in canonical ordination, just as in multiple regression.

Canonical ordination methods are widely used in palaeolimnological studies. The Birks et al. (1998) bibliography on the use of canonical analysis in ecology for the period 1986–1996 contained 804 titles, 96 of which are in the fields of palaeobotany, palaeoecology, and palaeolimnology. Applications of these methods in palaeoecology (Table 8.1) try to establish links between species assemblages and environmental factors, or use canonical analysis as a first step in calibration studies to guide the selection of significant environmental variables that may be estimated by biological assemblages (ter Braak and Juggins 1993; Birks 1995) (see Juggins and Birks 2012: Chap. 14). Palaeolimnologists also try to estimate how much of the assemblage variation can be attributed to different groups of environmental factors, such as sediment types, geology, climatic factors, geography, topography, land-use, etc. (e.g., Lotter et al. 1997; Simpson and Hall 2012: Chap. 19).

Canonical Ordination Methods

The types of canonical ordination methods that palaeoecologists are mostly interested in are redundancy analysis (RDA) and canonical correspondence analysis (CCA). They are asymmetric forms of analysis, combining regression and ordination. These analyses focus on a clearly identified table of response variables (containing, very often, assemblage composition data), which is related to a table of explanatory variables (e.g., environmental variables). Other forms of canonical analysis are available in the major statistical packages: canonical correlation analysis (CCorA) and canonical variates analysis (CVA), also called multiple discriminant analysis (see ter Braak 1987a). These forms will not be discussed in this chapter because they do not treat the community composition (or other quantitative data) as a response data table matrix; they are briefly outlined in Birks (2012a: Chap. 2). Other more general approaches to the linking of two or more ecological data tables are co-inertia analysis (Dolédec and Chessel 1994; Dray et al. 2003) and multiple factor analysis (Escofier and Pagès 1994); they allow the analysis of a wide range of different data tables (Dray et al. 2003), with no constraints on the number of species and environmental variables in relation to the number of objects or on the role of the different tables as response and predictor variables. All these methods of canonical analysis are described and illustrated in Chap. 6 of Borcard et al. (2011) and in Chap. 11 of Legendre and Legendre (2012).

In the asymmetric forms of canonical analysis, after regressing the **Y** variables (responses) on **X** (explanatory variables), an ordination is computed on the regression fitted values. The preliminary questions that have to be resolved before ordination (Table 8.2) will also have to be answered about the data in **Y** prior to canonical ordination: the choice of transformations for the physical or species data, and of an appropriate distance measure among objects. The table of explanatory variables, called **X**, contains the independent (or constraining) variables used in the regression part of the analysis. The decisions normally made prior to or during regression will have to be considered prior to canonical analysis: transformation of the regressors; coding of multi-state qualitative variables into dummy (binary or orthogonal) variables; coding the factors of experiments into (orthogonal) dummy variables; and choice of a linear or polynomial regression model. We do not have to worry about (multi)normality of the residuals since the tests of significance in canonical analysis are carried out by Monte Carlo permutation tests (see Legendre and Legendre 1998, 2012; Lepš and Šmilauer 2003; Birks 2012a: Chap. 2; Lotter and Anderson 2012: Chap. 18).

Linear RDA

Canonical redundancy analysis (RDA) combines two steps: linear regression and PCA. The analysis is schematically described in Fig. 8.2. (1) Each variable (column) of **Y** is regressed on **X**, which contains the explanatory variables. The fitted values of the multiple regressions are assembled in matrix $\hat{\mathbf{Y}}$, whereas the residuals are placed in the columns of matrix $\mathbf{Y}_{\mathbf{res}}$. $\hat{\mathbf{Y}}$ thus contains that part of **Y** that is explained by linear models of **X**, whereas $\mathbf{Y}_{\mathbf{res}}$ contains that part of **Y** that is linearly independent of (or orthogonal to) **X**. At this point, the matrices $\hat{\mathbf{Y}}$ and $\mathbf{Y}_{\mathbf{res}}$ have the same number of columns as **Y**. (2) The matrix of fitted values $\hat{\mathbf{Y}}$ usually contains (much) less information, measured by its total variance, than **Y**. A PCA of $\hat{\mathbf{Y}}$ is computed to reduce its dimensionality, producing eigenvalues (that are now called canonical eigenvalues), a matrix of eigenvectors **U** (now called canonical eigenvectors, which will be used as the matrix of response variable scores for the biplot), and a matrix **Z** of principal components, obtained in the same way as matrix **F** of the principal components in PCA, which contains the sampling unit scores for the ordination biplot; for details, see the description of PCA in the previous section on Simple Ordination Methods. In some applications, ecologists prefer to use, for biplots, the sampling unit scores obtained by the operation $\mathbf{F} = \mathbf{YU}$ (upper-right in Fig. 8.2).

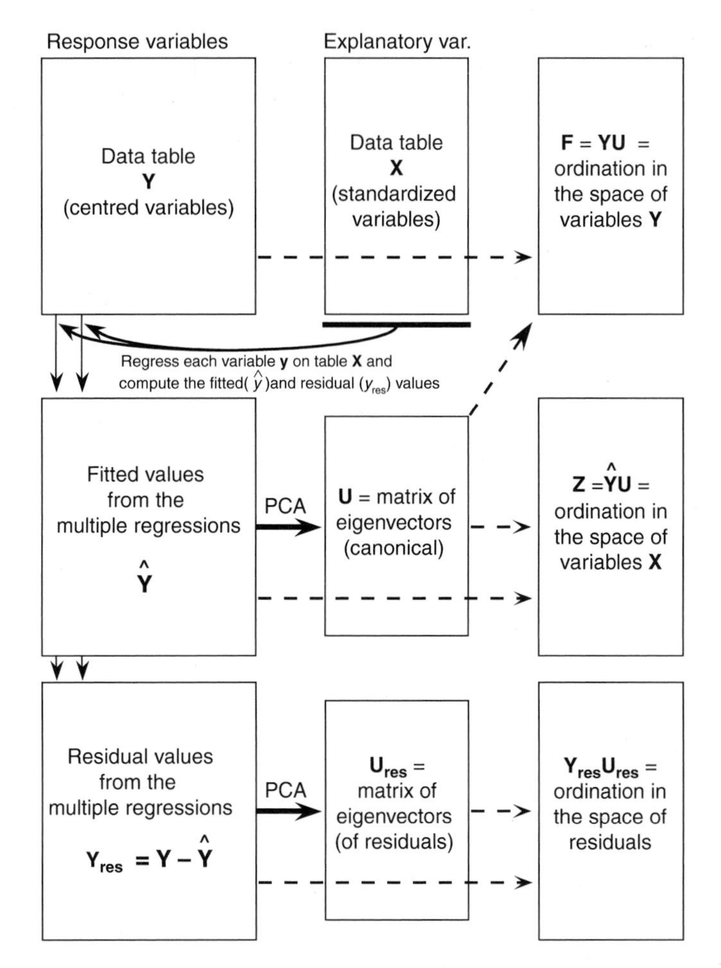

Fig. 8.2 Redundancy analysis involves two steps: regression which produces fitted values $\hat{\mathbf{Y}}$ and residuals $\mathbf{Y}_{\mathbf{res}}$, followed by principal component analysis (PCA) of the matrix of fitted values. PCA of the matrix of residuals may also be of interest (Modified from Legendre and Legendre 1998). *Var.* variables

These scores are not the direct result of the PCA of the fitted values $\hat{\mathbf{Y}}$; they are based on the original data **Y**, which contain the fitted values plus the residuals (noise). These sampling unit scores (column vectors of matrix **F**) are not orthogonal to each other. They differ from the vectors of matrix **Z**, which are orthogonal as in any PCA. (3) In some applications, the effect of the explanatory variables on **Y** is already well documented and understood; for instance, the effect of water depth on aquatic macroinvertebrates. RDA can be used to go beyond what is already known, by examining the residuals of the regression, found in matrix $\mathbf{Y}_{\mathbf{res}}$. In those cases, one is interested in obtaining an ordination of the matrix of residual variation: a PCA is performed on matrix $\mathbf{Y}_{\mathbf{res}}$, as shown in the lower part of Fig. 8.2. This analysis is called a partial PCA.

Scalings in RDA biplots follow the same rules as in PCA: one may be primarily interested in an ordination preserving the Euclidean distances among sampling unit fitted values (distance biplot), or in illustrating the correlations among the columns of $\hat{\mathbf{Y}}$ (correlation biplot) (ter Braak 1994). The explanatory environmental variables can also be represented in the ordination diagrams, which become triplots, by using their correlations with the canonical ordination axes. The correlation coefficients must be slightly modified to account for the stretching of the canonical ordination axes; the biplot scores of environmental variables are obtained by multiplying the correlation coefficients by $(\lambda_k/\text{total variance in } \mathbf{Y})^{0.5}$. States of binary or multistate qualitative variables can be usefully represented in triplots by the centroids (mean coordinates) of the sampling units that possess the given state (ter Braak 1994).

The number of canonical axes is limited by either the number of variables in **Y** or the number of variables in **X**. Example 1: if **Y** contains a single variable, regressing it on **X** produces a single vector of fitted values and, hence, a single canonical axis. Example 2: if **X** contains a single column, regressing **Y** (which contains *p* columns) on **X** will produce a matrix $\hat{\mathbf{Y}}$ with *p* columns, but since they are the result of regression on the same explanatory variable, matrix $\hat{\mathbf{Y}}$ is actually one-dimensional. So, the PCA will come up with a single non-zero eigenvalue that will contain all the variance of **Y** explained by **X**. The analysis of a matrix $\mathbf{Y}(n \text{ x } p)$ by a matrix $\mathbf{X}(n \text{ x } m)$ produces at most $(n-1)$, p, or m canonical axes, whichever is the smallest.

Like PCA, RDA can be tricked into preserving some distance that is appropriate to assemblage composition data, instead of the Euclidean distance (Fig. 8.3). Figure 8.3b shows that assemblage composition data can be transformed using Eqs. 8.5 or 8.7, 8.8, 8.9 (transformation-based RDA, or tb-RDA, approach). RDA computed on data transformed by these equations will actually preserve the chord, chi-square, profile, or Hellinger distance among sites. One can also directly compute one of the distance functions appropriate for assemblage composition data (Eqs. 8.15, 8.16, 8.17, 8.18, 8.19, 8.20, 8.21, 8.22, 8.23 and 8.24), carry out a principal coordinate analysis of the distance matrix, and use all the PCoA eigenvectors as input to RDA (Fig. 8.3c). This is the distance-based RDA approach (db-RDA) advocated by Legendre and Anderson (1999).

Partial RDA offers a way of controlling for the effect of a third data-set, called the matrix of covariables **W**. Computationally, the analysis first calculates the residuals $\mathbf{Y}_{\mathbf{res|w}}$ of the response variables **Y** on **W** and the residuals $\mathbf{X}_{\mathbf{res|w}}$ of the explanatory variables **X** on **W**; then an RDA of $\mathbf{Y}_{\mathbf{res|w}}$ on $\mathbf{X}_{\mathbf{res|w}}$ is computed; see details in Legendre and Legendre (2012). This is quite different from a PCA of $\mathbf{Y}_{\mathbf{res}}$ mentioned at the end of the introductory paragraph of the present section. Partial RDA is a generalisation of partial linear regression to multivariate data, for example species assemblages. It is used in many different situations, including the following: (1) controlling for the effect of **W** (e.g., geographic positions) in tests of the relationship between **Y** (e.g., modern biological assemblages) and **X** (e.g., modern environmental data) (Peres-Neto and Legendre 2010); (2) determining the partial, singular effect of an explanatory vari-

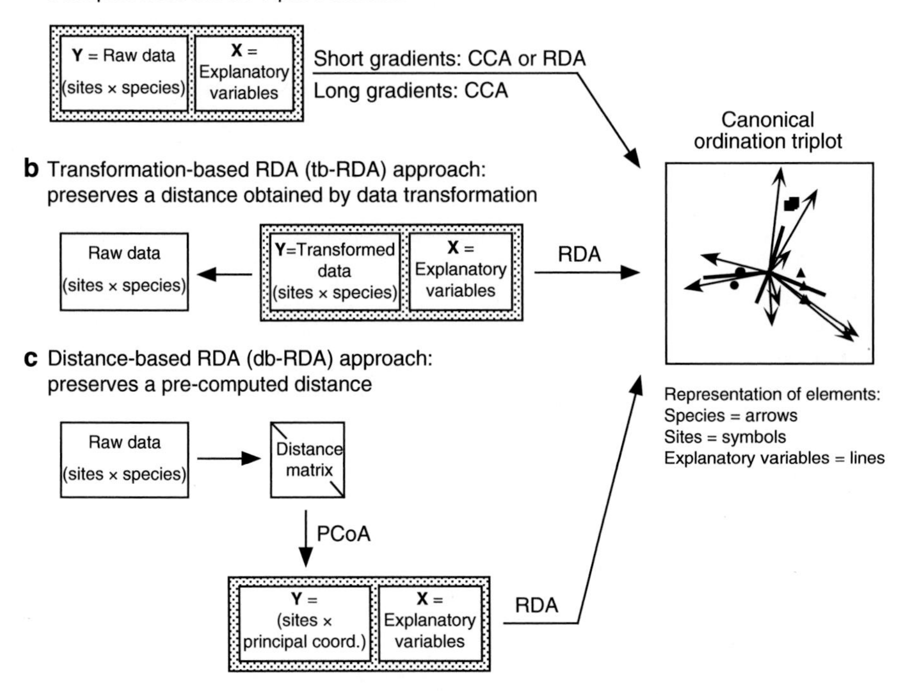

Fig. 8.3 Comparison of (**a**) classical redundancy analysis (RDA) and canonical correspondence analysis (CCA) to (**b**, **c**) alternative approaches forcing canonical analyses to preserve other distances adapted to assemblage composition data (Modified from Legendre and Gallagher 2001)

able of interest (e.g., environmental), and testing its significance, while controlling for the effect of all the other explanatory variables in the study; (3) partial RDA is used to test the significance of single factors and interaction terms in two-way or multi-way experimental designs where species assemblages are the response variable (see Testing hypotheses in (multi-)factorial experiments below); (4) partial RDA is also used to test the significance of individual fractions in variation partitioning (see Spatial or Temporal Analysis Through Variation Partitioning below). For details of these applications, see Legendre and Legendre (2012: Sect. 11.1.10).

In terms of algorithms, RDA and CCA can be obtained either by global regression and PCA, as described here, or by the iterative algorithm described by ter Braak (1987a) and used in the CANOCO program. In large analyses, the global algorithm produces more precise results when many canonical ordination axes are to be extracted and used in further analyses; the iterative algorithm is computationally faster when one is only interested in obtaining the first few (4–8) canonical axes.

Linear CCA

Canonical correspondence analysis (CCA) (ter Braak 1986, 1987b) only differs from RDA in two aspects. First, it is the matrix **Q** of CA (see Simple ordinations methods above) that is used as the response data matrix, instead of the data matrix **Y**. This ensures that the chi-square distance is preserved by CCA among the rows of the response data table; the assumption of unimodal species responses is made as in CA. Second, the regression step is carried out using weights p_{i+}, where p_{i+} is the sum of frequencies in row i (y_{i+}) divided by the grand total (y_{++}) of all frequencies in the table. Using these weights is tantamount to repeating each row of the response and explanatory data tables y_{i+} times before computing the regressions. Scalings for biplots or triplots are the same as in CA (see ter Braak and Verdonschot 1995). Just as one can compute a partial RDA, it is possible to perform a partial CCA (ter Braak 1988b; ter Braak and Prentice 1988). Odgaard (1994) provides an illustrative application of partial CA in palaeoecology and Bradshaw et al. (2005) provide a detailed application of partial CCA in palaeolimnology.

Fossil assemblages can be positioned as supplementary or passive objects in a CCA or RDA of modern biological assemblages, in relation to modern environmental variables, to provide a projection of fossil samples (from an unknown past environment) into modern 'environment–species–object' space (e.g., Birks et al. 1990a; Allott et al. 1992; Juggins and Birks 2012: Chap. 14; Simpson and Hall 2012: Chap. 19).

There have been many applications of RDA and CCA and their partial forms in palaeoecology and palaeolimnology in either a descriptive mode to display modern species–object–environment relationships (e.g., Birks et al. 1990a) or in an analytical, hypothesis-testing mode. Illustrative examples of the latter approach include Lotter and Birks (1993), Renberg et al. (1993), Birks and Lotter (1994), Korsman et al. (1994), Anderson et al. (1995), Korsman and Segerström (1998), Odgaard and Rasmussen (2000), and Bradshaw et al. (2005) (see Lotter and Anderson 2012: Chap. 18).

Other Forms of Asymmetric Canonical Analyses

There is no special reason why nature should linearly relate changes in community composition to changes in environmental variables. While they know that the assumption of linearity is often unrealistic, users of RDA and CCA sometimes use the linear forms of these methods simply because more appropriate models are not available. Makarenkov and Legendre (2002) proposed a nonlinear form of RDA and CCA, based on polynomial regression, to do away with the assumption of linearity in modelling the relationships between tables of response and explanatory variables. Their algorithm includes a step-wise procedure for selection of the best combination of linear and quadratic terms of the explanatory variables.

Palaeolimnologists may want to relate two types of assemblages, for example predators and preys. For a top-down model, the predators would form the dataset explaining the variation of the prey, and the opposite for a bottom-up model. To relate two communities, ter Braak and Schaffers (2004) proposed a model of co-correspondence analysis (see also Schaffers et al. 2008). An alternative method is to transform the two community data tables using one of the transformations described in the section above on Transformation of Assemblage Composition Data, as proposed by Pinel-Alloul et al. (1995), and analyse the two tables using RDA. As noted by ter Braak and Schaffers (2004), one should not use forward selection of the species in the explanatory table during this type of analysis. The R-package cocorresp (Simpson 2009) implements co-correspondence analysis.

Spatial or Temporal Analysis Through Variation Partitioning

Variation partitioning is an approach to the analysis of a response variable or data table, using two or more explanatory variables or data tables. For simple response variables, the analysis is carried out using partial linear regression; see Legendre and Legendre (1998, 2012: Sect. 10.3.5). Partial canonical analysis, which is available in CANOCO and vegan, allows ecologists to partition the variation of a response data table among two explanatory tables, using RDA or CCA.

In the original proposal (Borcard et al. 1992), the proportion of variation of a response variable or data table accounted for by a table of explanatory variables was estimated using the ordinary coefficient of determination (R^2). It has long been known that R^2 is a biased estimator of the proportion of explained variation. Ohtani (2000, for regression) and Peres-Neto et al. (2006, for canonical analysis) have shown that the adjusted coefficient of determination R_a^2 (Ezekiel 1930),

$$R_a^2 = 1 - (1 - R^2)\left(\frac{n-1}{n-m-1}\right) \tag{8.27}$$

is unbiased, where n is the number of observations and m is the number of explanatory variables. Peres-Neto et al. (2006) have also shown how to compute the fractions of variation described in the next paragraph using R_a^2. The R-language function `varpart()` available in the vegan package allows users to partition the variation of a response data table $\mathbf{Y}$ among 2, 3, or 4 tables of explanatory variables $\mathbf{X}_1$ to $\mathbf{X}_4$.

The variation-partitioning approach was first advocated by Borcard et al. (1992) in the context of spatial analysis in which a species composition response table $\mathbf{Y}$ is partitioned between a matrix of environmental variables and one describing the spatial relationships among the sampling sites. The variation in $\mathbf{Y}$ is partitioned into four fractions, three of which can be interpreted separately or in combinations (Fig. 8.4): [a] is the non-spatially-structured component of the variation of $\mathbf{Y}$ explained by the environmental variables, [b] is the spatially-structured component

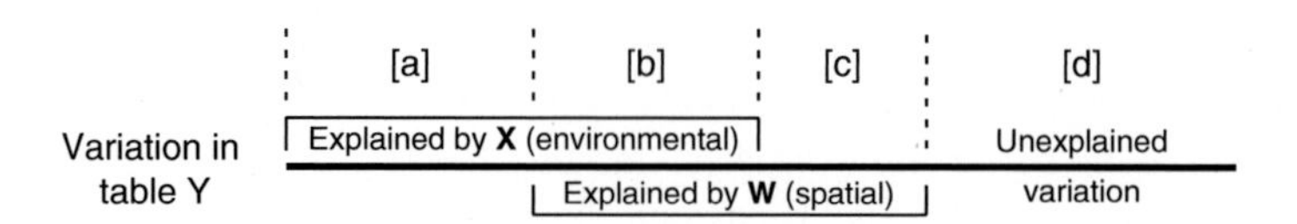

Fig. 8.4 Partitioning the variation of a response data table **Y** with respect to a table **X** of environmental variables and a table **W** of spatial variables. The length of the thick horizontal line represents the total variation in **Y** (Modified from Borcard et al. 1992 and Legendre 1993)

explained by the environmental variables, [c] is the amount of spatially-structured variation of **Y** not explained by the environmental variables used in the analysis, and [d] is the unexplained (residual) variation.

In Borcard et al. (1992) and Borcard and Legendre (1994), as well as in many applications published since 1992, the spatial relationships were represented in the analysis by a polynomial function of the geographical coordinates of the sampling sites. A new form of spatial partitioning, based on principal coordinates of neighbour matrices (PCNM), has been proposed by Borcard and Legendre (2002). In PCNM analysis, the polynomial function of the geographic coordinates of the sites of Borcard et al. (1992) is replaced by a set of spatial eigenfunctions, the PCNMs, corresponding to a spectral decomposition of the spatial relationships among the sites. PCNM analysis allows the modelling of spatial or temporal relationships at all spatial scales that can be perceived by the sampling design. Borcard et al. (2004) and Legendre and Borcard (2006) present several applications to the analysis of multivariate spatial patterns. Telford and Birks (2005) have also applied PCNM analysis to explore the spatial structures within core-tops of foraminiferal assemblages in the Atlantic.

Dray et al. (2006) examined the link between PCNM analysis and spatial autocorrelation structure functions, and generalised the method to different types of spatial weightings. The generalised eigenfunctions are called Moran's Eigenvector Maps (MEM). In the MEM framework, PCNM is called distance-based MEM (dbMEM) (see Legendre and Legendre 2012: Chap. 14). Griffith and Peres-Neto (2006) unified Dray's MEM spatial eigenfunctions with Griffith's (2000) spatial eigenfunctions. Blanchet et al. (2008) developed Asymmetric Eigenvector Maps (AEM) to model species spatial distributions generated by hypothesised directional physical processes such as migrations in river networks and currents in water bodies.

Several R-language functions are available to compute PCNM and MEM spatial eigenfunctions: `pcnm()` in vegan, `pcnm()` in spacemakeR and `PCNM()` and `quickPCNM()` in the PCNM package; the last two packages are available at http://r-forge.r-project.org/R/?group_id=195. Several applications of spatial eigenfunction analysis to ecological data in R are presented by Borcard et al. (2011). A stand-alone program called SpaceMaker2 (Borcard and Legendre 2004) is also available at http://www.bio.umontreal.ca/legendre/indexEn.html to compute PCNM eigenfunctions.

Modelling Temporal Structure in Sediment Cores [and Environmental Structure in Modern Assemblages]

Variation partitioning of stratigraphical palaeolimnological data has been used in various studies to partition variation in a biostratigraphical sequence (e.g., diatoms) into components explained by the occurrence of volcanic ash or other potential perturbations, by climatic changes, and by natural temporal shifts (e.g., Lotter and Birks 1993, 1997, 2003; Birks and Lotter 1994; Lotter et al. 1995; Barker et al. 2000; Eastwood et al. 2002). It has also been used to partition variation in modern biological assemblages (e.g., diatoms) in relation to a range of explanatory variables such as lake-water chemistry, climate, geography, etc. (e.g., Gasse et al. 1995; Jones and Juggins 1995; Pienitz et al. 1995; Lotter et al. 1997, 1998; Potopova and Charles 2002; Kernan et al. 2009; Simpson and Hall 2012: Chap. 19) and in fossil assemblages in relation to spatial and temporal variables (e.g., Ammann et al. 1993). Variation partitioning is being increasingly applied as a hypothesis-testing approach in palaeolimnology, to quantify the proportion of total variation in assemblage composition over time explicable by various environmental variables. For example, Hall et al. (1999) used high-resolution diatom core data and 100 years of historical data to quantify the effects of climate, agriculture, and urbanisation on diatom assemblages in lakes in the northern Great Plains (Saskatchewan). They showed that human impact was the major determinant of biotic change. Quinlan et al. (2002) obtained similar results for the same area using fossil chironomid assemblages. The use of variation partitioning requires careful project design to exploit 'natural experiments' (e.g., factorial designs) and to test critical hypotheses. Other detailed palaeolimnological applications of variation partitioning to test specific hypotheses include Vinebrooke et al. (1998) and Leavitt et al. (1999). Birks (1998) reviewed the use of variation partitioning as a means of testing hypotheses in palaeolimnology (see also Lotter and Anderson 2012: Chap. 18).

We now present an example to illustrate the use of canonical ordination as a form of spatial or time-series analysis for multivariate ecological response data. The Round Loch of Glenhead (RLGH) fossil data consist of the counts of 139 Holocene diatom taxa observed in 101 levels of a sediment core from a small lake in Galloway, south-western Scotland (Jones et al. 1989; see Birks and Jones 2012: Chap. 3 of this volume). The data-series covers the past 10,000 years. Level no. 1 is the top (most recent), no. 101 is the bottom of the core (oldest). The diatom counts were expressed as proportions relative to the total number of cells in each section of the core. This means that the counts had been transformed into profiles of relative abundances following Eq. 8.8. Polynomial trend-surface and PCNM analyses will be used to detect structures in the multivariate diatom data within the core.

RDA of the multivariate diatom data from the RLGH core against level numbers showed that the core data contained a highly significant linear gradient ($R^2 = 0.190$, $R_a^2 = 0.182$, $p = 0.001$ after 999 random permutations; Fig. 8.5a). We then analysed the response data against a 3rd-order polynomial of the core level numbers (1–101): the three monomials contributed significantly to the explanation of the diatom

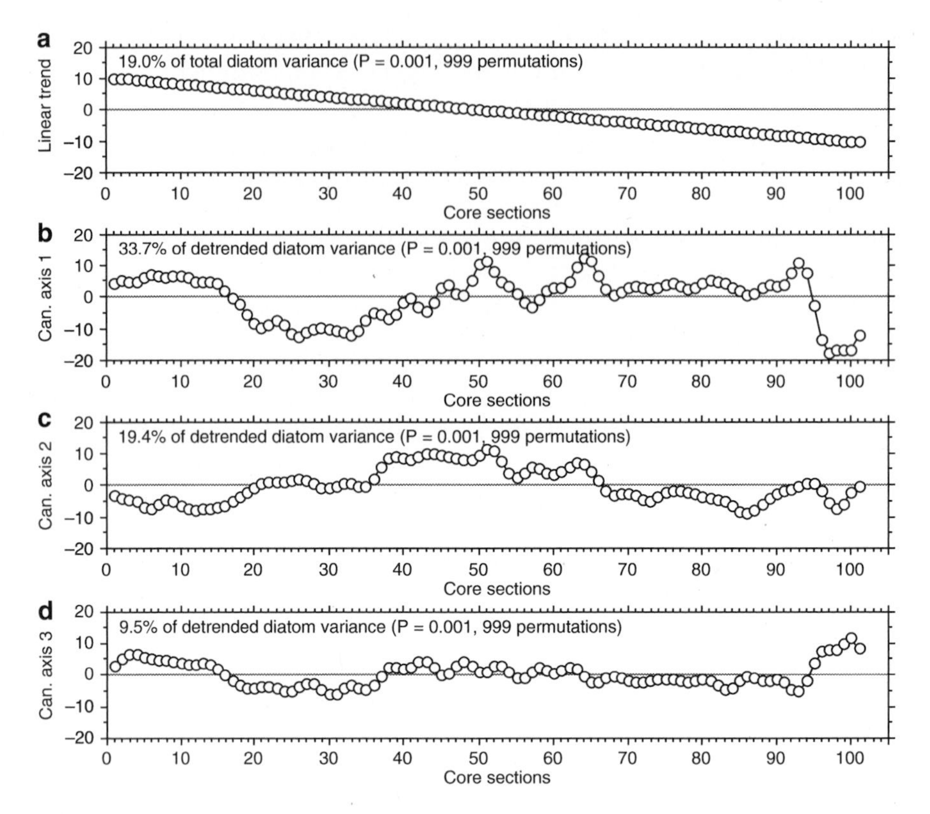

Fig. 8.5 The linear gradient (**a**) and first three canonical axes of the principal coordinates of neighbour matrices (PCNM) model (**b–d**), as a function of the core level or section numbers in The Round Loch of Glenhead diatom stratigraphical data. *P* probability

data, producing a model (not shown) with high explanatory power ($R^2 = 0.460$, $R_a^2 = 0.443$, $p = 0.001$). All monomials of a 5th-order polynomial also contributed significantly to the explanation of the diatom data, producing a model with an even higher coefficient of determination ($R^2 = 0.567$, $R_a^2 = 0.545$, $p = 0.001$). Since the data seemed to be structured in an intricate series of scales, we turned to PCNM analysis to extract submodels corresponding to the different temporal scales present in the data.

The diatom data were regressed on level numbers to extract the linear gradient, as recommended by Borcard et al. (2004). PCNM analysis was then conducted on the detrended data, namely the residuals of these regressions. Sixty-eight PCNM variables were created using the `PCNM()` function of the PCNM R-language package (last paragraph of the previous section); these variables, which have the form of sine waves of decreasing periods, represent variation at the various scales that can be identified in the series of 101 core levels. The first 50 PCNM variables, which had Moran's *I* coefficients larger than the expected value of *I* and thus modelled positive correlation, were retained for canonical analysis. They were subjected to forward selection against the detrended diatom data, using the `forward.sel()` function of the packfor package available at http://r-forge.r-project.org/R/?group_id=195; forward selection of explanatory variables in RDA is also available in the program CANOCO, version 4.5. Thirty PCNM variables were selected at the $\alpha = 0.05$ significance level (Monte Carlo permutation tests, 999 permutations). The selected PCNM variables were numbers 1–20, 28, 30, 32, 33, 35, 37, 38, 41, 42, and 45. Canonical redundancy analysis of the detrended diatom data by this subset of 30 PCNM variables explained $R_a^2 = 70.1\%$ of the variance in the detrended data. The RDA produced nine significant canonical axes; three of them, which accounted for more than 5% of the detrended species variation, are displayed in Fig. 8.5b–d. The diatom taxa contributing in an important way to the variation along these axes vary depending on the axis. Six species were highly positively correlated ($r > 0.6$) to the core level numbers (linear trend): *Tabellaria quadriseptata* (TA004A), *Navicula hoefleri* (NA167A), *Navicula cumbrensis* (NA158A), *Peronia fibula* (PE002A), *Eunotia denticulata* (EU015A), and *Eunotia naegelii* (EU048A); these species are found in the sections on the positive side of the linear trend (Fig. 8.5a). Two taxa were highly negatively correlated ($r < -0.5$) to the same trend: *Brachysira brebissonii* (BR006A), *Cymbella* [PIRLA sp. 1] (CM9995). Two species were highly positively correlated to canonical axis 1 ($r > 0.5$, Fig. 8.5b): *Aulacoseira perglabra* (AU010A), *Eunotia incisa* (EU047A); four taxa were highly negatively correlated ($r < -0.5$) to the same wave form: *Brachysira vitrea* (BR001A), *Achnanthes minutissima* (AC013A), *Tabellaria flocculosa* (TA001A), *Cymbellla perpusilla* (CM010A). And so on (Fig. 8.5c, d). Each canonical axis displays structures representing a mixture of stratigraphical and temporal scales. This information could also be displayed in the form of biplots of the species together with the trend or with the PCNM variables.

Another useful way to describe the structure of the multivariate diatom data along the core is to separate the PCNM variables into an arbitrary number of groups, made of contiguous PCNMs, and examine the resulting submodels. We chose to divide them into three submodels. The broad-scale submodel contains PCNMs numbers 1–10 as explanatory variables; it explains $R_a^2 = 47.7\%$ of the detrended diatom variation. Canonical axes 1–3 of this fraction are significant and explain more than 5% of the detrended diatom variation (Fig. 8.6a–c, $p = 0.001$). The taxa that are positively correlated with axis 1 ($r > 0.6$) are *Navicula krasskei* (NA044A) and *Aulacoseira perglabra* (AU010A); these species are found in the sections on the positive side of the wave form (Fig. 8.6a). Other species that are highly negatively correlated with that axis ($r < -0.6$) are *Achnanthes linearis* (AC002A), *Tabellaria flocculosa* (TA001A), and *Eunotia iatriaensis* (EU019A); they are present in the sections found on the negative side of the wave form (Fig. 8.6a). The medium-scale submodel uses PCNMs numbers 11–20 as explanatory variables; it explains $R_a^2 = 9.1\%$ of the detrended diatom variation. Only canonical axis 1 of that submodel is significant and explains more than 5% of the detrended diatom variation (Fig. 8.6d). The fine-scale submodel uses PCNMs numbers 28, 30, 32, 33, 35, 37, 38, 41, 42, and 45 as explanatory variables. Taken alone, this submodel does not

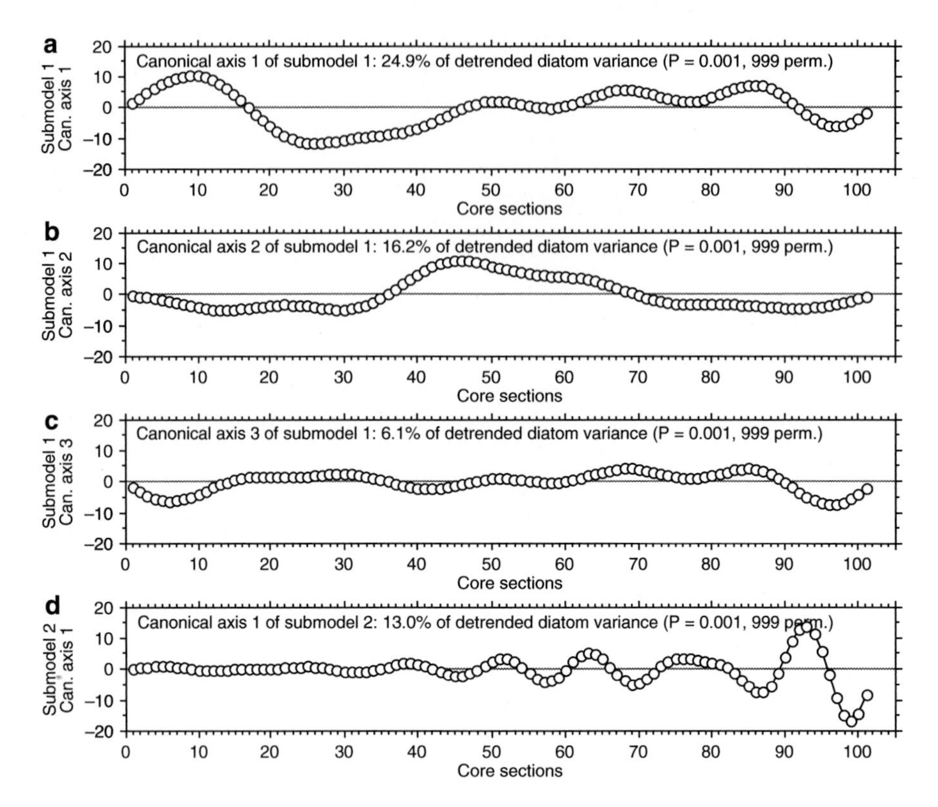

Fig. 8.6 Significant canonical axes of the first two principal coordinates of neighbour matrices (PCNM) submodels, as a function of the core level or section numbers in The Round Loch of Glenhead diatom stratigraphical data. *P* probability

explain a significant portion of the diatom variation ($p = 0.916$, 999 permutations); the PCNM variables it contains were significant in the global model of 30 PCNMs, after the broad- and medium-scale PCNMs had been selected. None of its canonical axes is significant. We conclude that the core is mainly structured by processes operating at broad ($5 \times 10^3 - 10^4$ year) and medium ($5 \times 10^2 - 10^3$ year) scales. This PCNM example is presented simply to illustrate the potential of PCNM analysis in palaeolimnology. A more detailed analysis would naturally consider the estimated age of each level (see Blaauw and Heegaard 2012: Chap. 12) in the sediment core, rather than simply level numbers.

Testing Hypotheses in (Multi-) Factorial Experiments

RDA and CCA provide ways of testing hypotheses about multivariate data, as in analysis of variance (ANOVA). Assemblage composition data can be used as the response table in RDA provided that they are transformed in appropriate ways, as shown in Fig. 8.3b, c. Examples of the use of RDA to test ANOVA-like hypotheses are found in Sabatier et al. (1989), ter Braak and Wiertz (1994), Verdonschot and ter Braak (1994), Anderson and Legendre (1999), and Hooper et al. (2002). The principle of this analysis is the following: multiple regression can be used to calculate any ANOVA model, provided that the factors are coded in appropriate ways in the matrix of predictors **X**. Since RDA and CCA are simply regression followed by PCA, they can be used in the same way as regression to carry out analysis of variance. The PCA portion of the procedure is only needed to illustrate the ANOVA results using bi- or triplots, as in Hooper et al. (2002); it is not needed nor computed for the test of significance of the canonical relationship. RDA and CCA use Monte Carlo permutation tests to assess the significance of the relationship between the response matrix **Y** (or **Q**) and the factor coded into matrix **X**.

Here are examples of such potential hypotheses in palaeolimnology. For sediment cores: in time-series, are there differences between time periods of interest? In the comparison of cores: are there differences among cores, which can be related to sampling regions? (In the latter example, one can control for the time pairing of core subunits by coding them into a matrix of covariables.) In such analyses, the factors (or ANOVA classification criteria) must be coded using dummy variables. A set of ordinary (binary 0–1) dummy variables will do the job when analysing a single factor. For two or more factors and their interactions, the factors must be coded using Helmert contrasts, also called orthogonal dummy variables. A method of coding such factors is described in Appendix C of Legendre and Anderson (1999) and in Legendre and Legendre (2012: Sect. 1.5.7). In R, factors can be automatically coded into Helmert contrasts by the function `model.matrix()` in the R package stats using an appropriate contrast type specification.

The use of RDA and CCA to test hypotheses in palaeolimnology is discussed in detail by Lotter and Anderson (2012: Chap. 18). Birks (1996, 1998, 2010) reviewed hypothesis testing in palaeolimnology both directly through rigorous project design and through site selection (e.g., Birks et al. 1990b) and indirectly through RDA or CCA.

Software

A list of programs and packages available for simple and canonical ordination of ecological and palaeoecological data is presented in Table 8.3. The list of functions available, especially in the R language (R Development Core Team 2011), is rapidly increasing.

Most general-purpose statistical programs contain procedures for principal component analysis (PCA). Very few allow, however, the direct drawing of biplots of species and objects, and many do not even compute the coordinates of the species and objects necessary to construct distance or correlation biplots. PCA and biplots are available in CANOCO (biplots: CanoDraw), in PC-ORD, in SYN-TAX, and in the `rda()` function of vegan, the `dudi.pca()` function of ade4, and the `pca()` function of labdsv (R-language packages).

Correspondence analysis (CA) is offered in few general-purpose statistical packages. In R, palaeoecologists will find it in the same packages as PCA. Principal coordinate analysis (PCoA) is available in the PrCoord program distributed with CANOCO, in functions of the R language (`cmdscale()` and its wrappers `capscale()`, `pco()` and `cmds.diss()`, in `pcoa()`, and in `dudi.pco()`; see Table 8.3 for references), and in SYN-TAX. Non-metric multidimensional scaling (NMDS) is found in PC-ORD, in function `metaMDS()` of the R language, and in isoMDS and its wrappers `nmds()` and `bestnmds()`, and in SYN-TAX. NMDS is also found in some general-purpose statistical packages; they offer, however, a poor choice of dissimilarity functions.

CANOCO and the `rda()` and `cca()` functions of the vegan R-language package are widely used for unconstrained or constrained ordination analysis. Other programs and packages allow the computation of some forms of canonical analysis: the PC-ORD and SYN-TAX packages, and the program Polynomial_RDACCA of Makarenkov and Legendre (2002). CANOCO contains many interesting features for palaeoecologists, not shared by most other canonical analysis packages (Rejmánek and Klinger 2003), such as a procedure for selecting the environmental variables of **X** that contribute significantly to modelling **Y**; selection of explanatory variables is also available in the R language: functions `ordistep()` and `ordiR2step()` (vegan), as well as `forward.sel()` (packfor on http://r-forge.r-project.org/R/?group_id=195). CANOCO also offers tests of significance for individual canonical eigenvalues (also in vegan), partial canonical analysis (also in vegan), and permutation methods especially designed for time series and blocked experimental designs.

Acknowledgements We are indebted to Gavin Simpson for comments and to Cathy Jenks for her invaluable help in the editing of this chapter. This is publication A210 for the Bjerknes Centre for Climate Research.

References

Aitchison J (1986) The statistical analysis of compositional data. Chapman & Hall, London

Allen MR, van Dyke JN, Cáceres CE (2011) Metacommunity assembly and sorting in newly formed lake communities. Ecology 92:269–275

Allott TEH, Harriman R, Battarbee RW (1992) Reversibility of lake acidification at The Round Loch of Glenhead, Galloway, Scotland. Environ Pollut 77:219–225

Ammann B, Birks HJB, Drescher-Schneider R, Juggins S, Lang G, Lotter A (1993) Patterns of variation in Late-glacial pollen stratigraphy along a northwest-southeast transect through Switzerland – a numerical analysis. Quaternary Sci Rev 12:277–286

Anderson MJ, Legendre P (1999) An empirical comparison of permutation methods for tests of partial regression coefficients in a linear model. J Stat Comput Simul 62:271–303

Anderson MJ, Ellingsen KE, McArdle BH (2006) Multivariate dispersion as a measure of beta diversity. Ecol Lett 9:683–693

Anderson NJ, Renberg I, Segerström U (1995) Diatom production responses to the development of early agriculture in a boreal forest lake-catchment (Kassjön, northern Sweden). J Ecol 83:809–822

Barker P, Telford RJ, Merdaci O, Williamson D, Taieb M, Vincens A, Gibert E (2000) The sensitivity of a Tanzanian crater lake to catastrophic tephra input and four millennia of climate change. Holocene 10:303–310

Birks HH, Birks HJB (2001) Recent ecosystem dynamics in nine North African lakes in the CASSARINA project. Aquat Ecol 35:461–478

Birks HH, Battarbee RW, Birks HJB (2000) The development of the aquatic ecosystem at Kråkenes Lake, western Norway, during the late-glacial and early-Holocene – a synthesis. J Paleolimnol 23:91–114

Birks HJB (1973) Modern pollen rain studies in some Arctic and alpine environments. In: Birks HJB, West RG (eds) Quaternary plant ecology. Blackwell Scientific Publications, Oxford, pp 143–168

Birks HJB (1977) Modern pollen rain and vegetation of the St Elias Mountains, Yukon Territory. Can J Bot 55:2367–2382

Birks HJB (1985) Recent and possible mathematical developments in quantitative palaeoecology. Palaeogeogr Palaeoclim Palaeoecol 50:107–147

Birks HJB (1987) Multivariate analysis of stratigraphic data in geology: a review. Chemometr Intell Lab Syst 2:109–126

Birks HJB (1992) Some reflections on the application of numerical methods in Quaternary palaeoecology. Publ Karelian Inst 102:7–20

Birks HJB (1995) Quantitative palaeoenvironmental reconstructions. In: Maddy D, Brew S (eds) Statistical modelling of Quaternary science data. Volume 5: Technical Guide. Quaternary Research Association, Cambridge, pp 161–254

Birks HJB (1996) Achievements, developments, and future challenges in quantitative Quaternary palaeoecology. INQUA Commission for the Study of the Holocene: Sub-commission on data-handling methods. Newsletter 14:1–8

Birks HJB (1998) Numerical tools in palaeolimnology – progress, potentialities, and problems. J Paleolimnol 20:307–332

Birks HJB (2008) Ordination – an ever-expanding tool-kit for ecologists? Bull Brit Ecol Soc 39:31–33

Birks HJB (2010) Numerical methods for the analysis of diatom assemblage data. In: Smol JP, Stoermer EF (eds) The diatoms: applications for the environmental and earth sciences, 2nd edn. Cambridge University Press, Cambridge, pp 23–54

Birks HJB (2012a) Chapter 2: Overview of numerical methods in palaeolimnology. In: Birks HJB, Lotter AF, Juggins S, Smol JP (eds) Tracking environmental change using lake sediments. Volume 5: Data handling and numerical techniques. Springer, Dordrecht

Birks HJB (2012b) Chapter 11: Stratigraphical data analysis. In: Birks HJB, Lotter AF, Juggins S, Smol JP (eds) Tracking environmental change using lake sediments. Volume 5: Data handling and numerical techniques. Springer, Dordrecht

Birks HJB, Berglund BE (1979) Holocene pollen stratigraphy of southern Sweden: a reappraisal using numerical methods. Boreas 8:257–279

Birks HJB, Gordon AD (1985) Numerical methods in Quaternary pollen analysis. Academic Press, London

Birks HJB, Jones VJ (2012) Chapter 3: Data-sets. In: Birks HJB, Lotter AF, Juggins S, Smol JP (eds) Tracking environmental change using lake sediments. Volume 5: Data handling and numerical techniques. Springer, Dordrecht

Birks HJB, Lotter AF (1994) The impact of the Laacher See Volcano (11000 yr. BP) on terrestrial vegetation and diatoms. J Paleolimnol 11:313–322

Birks HJB, Peglar SM (1979) Interglacial pollen spectra at Fugla Ness, Shetland. New Phytol 68:777–796

Birks HJB, Juggins S, Line JM (1990a) Lake surface-water chemistry reconstructions from palaeolimnological data. In: Mason BJ (ed) The Surface Waters Acidification Programme. Cambridge University Press, Cambridge, pp 301–313
Birks HJB, Berge F, Boyle JF, Cumming BF (1990b) A palaeoecological test of the land-use hypothesis for recent lake acidification in south-west Norway using hill-top lakes. J Paleolimnol 4:69–85
Birks HJB, Austin HA, Indrevær NE, Peglar SM, Rygh C (1998) An annotated bibliography of canonical correspondence analysis and related constrained ordination methods 1986–1996. Bergen, Norway. Available from http://www.bio.umontreal.ca/Casgrain/cca_bib/index.html
Blaauw M, Heegaard E (2012) Chapter 12: Estimation of age-depth relationships. In: Birks HJB, Lotter AF, Juggins S, Smol JP (eds) Tracking environmental change using lake sediments. Volume 5: Data handling and numerical techniques. Springer, Dordrecht
Blanchet FG, Legendre P, Borcard D (2008) Modelling directional spatial processes in ecological data. Ecol Model 215:325–336
Borcard D, Legendre P (1994) Environmental control and spatial structure in ecological communities: an example using oribatid mites (Acari, Oribatei). Environ Ecol Stat 1:37–53
Borcard D, Legendre P (2002) All-scale spatial analysis of ecological data by means of principal coordinates of neighbour matrices. Ecol Model 153:51–68
Borcard D, Legendre P (2004) SpaceMaker2 – user's guide. Département de sciences biologiques, Université de Montréal. Program and user's guide. Available from http://www.bio.umontreal.ca/legendre/indexEn.html
Borcard D, Legendre P, Drapeau P (1992) Partialling out the spatial component of ecological variation. Ecology 73:1045–1055
Borcard D, Legendre P, Avois-Jacquet C, Tuomisto H (2004) Dissecting the spatial structure of ecological data at multiple scales. Ecology 85:1826–1832
Borcard D, Gillet F, Legendre P (2011) Numerical ecology with R. Springer, New York
Bradshaw EG, Rasmussen P, Odgaard BV (2005) Mid- to late-Holocene land-use change and lake development at Dallund Sø, Denmark: synthesis of multiproxy data, linking land and lake. Holocene 15:1152–1162
Bray JR, Curtis JT (1957) An ordination of the upland forest communities of southern Wisconsin. Ecol Monogr 27:325–349
Brodersen KP, Whiteside MC, Lindegaard C (1998) Reconstruction of trophic state in Danish lakes using subfossil chydorid (Cladocera) assemblages. Can J Fish Aquat Sci 55:1093–1103
Brodersen KP, Odgaard BV, Vestergaard O, Anderson NJ (2001) Chironomid stratigraphy in the shallow and eutrophic Lake Søbygaard, Denmark: chironomid–macrophyte occurrence. Freshw Biol 46:253–267
Burt C (1952) Tests of significance in factor analysis. Brit J Psychol Stat Sect 5:109–133
Cavalli-Sforza LL, Edwards AWF (1967) Phylogenetic analysis: models and estimation procedures. Evolution 21:550–570
Chessel D, Dufour AB, Thioulouse J (2004) The ade4 package – I: one-table methods. R News 4:5–10
Clark PJ (1952) An extension of the coefficient of divergence for use with multiple characters. Copeia 2:61–64
Dolédec S, Chessel D (1994) Co-inertia analysis: an alternative method for studying species-environment relationships. Freshw Biol 31:277–294
Dray S, Chessel D, Thioulouse J (2003) Co-inertia analysis and the linking of ecological data tables. Ecology 84:3078–3089
Dray S, Legendre P, Peres-Neto PR (2006) Spatial modelling: a comprehensive framework for principal coordinate analysis of neighbour matrices (PCNM). Ecol Model 196:483–493
Dray S, Dufour AB, Chessel D (2007) The ade4 package – II: two-table and K-table methods. R News 7:47–52
Eastwood WJ, Tibby J, Roberts N, Birks HJB, Lamb HF (2002) The environmental impact of the Minoan eruption of Santorini (Thera): statistical analysis of palaeoecological data from Göhlisar, southwest Turkey. Holocene 12:431–444

Engels S, Cwynar LC (2011) Changes in fossil chironomid remains along a depth gradient: evidence for common faunal thresholds within lakes. Hydrobiologia 665:15–38
Escofier B, Pagès J (1994) Multiple factor analysis (AFMULT package). Comput Stat Data Anal 18:121–140
Ezekiel M (1930) Methods of correlation analysis. Wiley, New York
Faith DP, Minchin PR, Belbin L (1987) Compositional dissimilarity as a robust measure of ecological distance. Vegetatio 69:57–68
Frontier S (1976) Étude de la décroissance des valeurs propres dans une analyse en composantes principales: comparaison avec le modèle du bâton brisé. J Exp Mar Biol Ecol 25:67–75
Gasse F, Juggins S, Ben Khelifa L (1995) Diatom-based transfer functions for inferring past hydrochemical characteristics of African lakes. Palaeogeogr Palaeoclim Palaeoecol 117:31–54
Gauch HG (1982) Noise reduction by eigenvector ordination. Ecology 63:1643–1649
Goodall DW (1954) Objective methods for the classification of vegetation. III. An essay on the use of factor analysis. Aust J Bot 2:304–324
Gordon AD (1982) Numerical methods in Quaternary palynology V. Simultaneous graphical representation of the levels and taxa in a pollen diagram. Rev Palaeobot Palynol 37:155–183
Gordon AD, Birks HJB (1974) Numerical methods in Quaternary palaeoecology. II. Comparisons of pollen diagrams. New Phytol 73:221–249
Gower JC (1968) Adding a point to vector diagrams in multivariate analysis. Biometrika 55:582–585
Gower JC (1971) A general coefficient of similarity and some of its properties. Biometrics 27:857–871
Gower JC (1982) Euclidean distance geometry. Math Sci 7:1–14
Gower JC, Legendre P (1986) Metric and Euclidean properties of dissimilarity coefficients. J Class 3:5–48
Griffith DA (2000) A linear regression solution to the spatial autocorrelation problem. J Geogr Syst 2:141–156
Griffith DA, Peres-Neto PR (2006) Spatial modeling in ecology: the flexibility of eigenfunction spatial analysis. Ecology 87:2603–2613
Hajdu LJ (1981) Graphical comparison of resemblance measures in phytosociology. Vegetatio 48:47–59
Hall RI, Leavitt PR, Quinlan R, Dixit AS, Smol JP (1999) Effects of agriculture, urbanization, and climate on water quality in the northern Great Plains. Limnol Oceanogr 44:739–756
He F, Legendre P (1996) On species-area relations. Am Nat 148:719–737
He F, Legendre P (2002) Species diversity patterns derived from species-area models. Ecology 83:1185–1198
Hill MO (1974) Correspondence analysis: a neglected multivariate method. Appl Stat 23:340–354
Hill MO, Gauch HG (1980) Detrended correspondence analysis – an improved ordination technique. Vegetatio 42:47–58
Hooper E, Condit R, Legendre P (2002) Responses of 20 native tree species to reforestation strategies for abandoned farmland in Panama. Ecol Appl 12:1626–1641
Hotelling H (1933) Analysis of a complex of statistical variables into principal components. J Educ Psychol 24(417–441):498–520
Jackson DA (1993) Stopping rules in principal components analysis: a comparison of heuristical and statistical approaches. Ecology 74:2204–2214
Jackson DA, Somers KM (1991) Putting things in order: the ups and downs of detrended correspondence analysis. Am Nat 137:704–712
Jacobson GL, Grimm EC (1986) A numerical analysis of Holocene forest and prairie vegetation in Central Minnesota. Ecology 67:958–966
Jones VJ, Birks HJB (2004) Lake-sediment records of recent environmental change on Svalbard: results of diatom analysis. J Paleolimnol 31:445–466

Jones VJ, Juggins S (1995) The construction of a diatom-based chlorophyll *a* transfer function and its application at three lakes on Signy Island (maritime Antarctic) subject to differing degrees of nutrient enrichment. Freshw Biol 34:433–445

Jones VJ, Stevenson AC, Battarbee RW (1989) Acidification of lakes in Galloway, southwest Scotland: a diatom and pollen study of the post-glacial history of The Round Loch of Glenhead. J Ecol 77:1–23

Jones VJ, Juggins S, Ellis-Evans JC (1993) The relationship between water chemistry and surface sediment diatom assemblages in maritime Antarctic lakes. Antarct Sci 5:339–348

Juggins S, Birks HJB (2012) Chapter 14: Quantitative environmental reconstructions from biological data. In: Birks HJB, Lotter AF, Juggins S, Smol JP (eds) Tracking environmental change using lake sediments. Volume 5: Data handling and numerical techniques. Springer, Dordrecht

Juggins S, Telford RJ (2012) Chapter 5: Exploratory data analysis and data display. In: Birks HJB, Lotter AF, Juggins S, Smol JP (eds) Tracking environmental change using lake sediments. Volume 5: Data handling and numerical techniques. Springer, Dordrecht

Kenkel NC, Orlóci L (1986) Applying metric and non-metric multidimensional scaling to ecological studies: some new results. Ecology 67:919–928

Kernan M, Ventura M, Bitušík P, Brancelj A, Clarke G, Velle G, Raddum GG, Stuchlík E, Catalan J (2009) Regionalisation of remote European mountain lake ecosystems according to their biota: environmental versus geographical patterns. Freshw Biol 54:2470–2493

Korsman T, Segerström U (1998) Forest fire and lake-water acidity in a northern Swedish boreal area: Holocene changes in lake-water quality at Makkassjön. J Ecol 86:113–124

Korsman T, Renberg I, Anderson NJ (1994) A palaeolimnological test of the influence of Norway spruce (*Picea abies*) immigration on lake-water acidity. Holocene 4:132–140

Laliberté E, Shipley B (2010) FD: measuring functional diversity from multiple traits, and other tools for functional ecology. R package version 1.0-7. http://cran.r-project.org/web/packages/FD/index.html

Lamb HF (1984) Modern pollen spectra from Labrador and their use in reconstructing Holocene vegetational history. J Ecol 72:37–59

Lance GN, Williams WT (1967) A general theory of classificatory sorting strategies. I. Hierarchical systems. Comput J 9:373–380

Leavitt PR, Findlay DL, Hall RI, Smol JP (1999) Algal responses to dissolved organic carbon loss and pH decline during whole-lake acidification: evidence from palaeolimnology. Limnol Oceanogr 44:757–773

Legendre P (1993) Spatial autocorrelation: trouble or new paradigm? Ecology 74:1659–1673

Legendre P, Anderson MJ (1999) Distance-based redundancy analysis: testing multispecies responses in multifactorial ecological experiments. Ecol Monogr 69:1–24

Legendre P, Birks HJB (2012) Chapter 7: Clustering and partitioning. In: Birks HJB, Lotter AF, Juggins S, Smol JP (eds) Tracking environmental change using lake sediments. Volume 5: Data handling and numerical techniques. Springer, Dordrecht

Legendre P, Borcard D (2006) Chapter 19: Quelles sont les échelles spatiales importantes dans un écosystème? In: Droesbeke J-J, Lejeune M, Saporta G (eds) Analyse Statistique de Données Spatiales. Éditions TECHNIP, Paris

Legendre P, Gallagher E (2001) Ecologically meaningful transformations for ordination of species data. Oecologia 129:271–280

Legendre P, Legendre L (1998) Numerical ecology, 2nd English edn. Elsevier, Amsterdam

Legendre P, Legendre L (2012) Numerical ecology, 3rd English edn. Elsevier, Amsterdam

Lepš J, Šmilauer P (2003) Multivariate analysis of ecological data using CANOCO. Cambridge University Press, Cambridge

Lotter AF, Anderson NJ (2012) Chapter 18: Limnological response to environmental changes at inter-annual to decadal time-scales. In: Birks HJB, Lotter AF, Juggins S, Smol JP (eds) Tracking environmental change using lake sediments. Volume 5: Data handling and numerical techniques. Springer, Dordrecht

Lotter AF, Birks HJB (1993) The impact of the Laacher See Tephra on terrestrial and aquatic ecosystems in the Black Forest southern Germany. J Quaternary Sci 8:263–276

Lotter AF, Birks HJB (1997) The separation of the influence of nutrients and climate on the varve time-series of Baldegersee, Switzerland. Aquat Sci 59:362–375

Lotter AF, Birks HJB (2003) The Holocene palaeolimnology of Sägistalsee and its environmental history – a synthesis. J Paleolimnol 30:333–342

Lotter AF, Eicher U, Birks HJB, Sigenthaler U (1992) Late-glacial climatic oscillations as recorded in Swiss lake sediments. J Quaternary Sci 7:187–204

Lotter AF, Birks HJB, Zolitschka B (1995) Late-glacial pollen and diatom changes in response to two different environmental perturbations: volcanic eruption and Younger Dryas cooling. J Paleolimnol 14:23–47

Lotter AF, Birks HJB, Hofmann W, Marchetto A (1997) Modern diatom, Cladocera, chironomid, and chrysophytes cyst assemblages as quantitative indicators for the reconstruction of past environmental conditions in the Alps. I. Climate. J Paleolimnol 18:395–420

Lotter AF, Birks HJB, Hofmann W, Marchetto A (1998) Modern diatom, Cladocera, chironomid, and chrysophytes cyst assemblages as quantitative indicators for the reconstruction of past environmental conditions in the Alps. II. Nutrients. J Paleolimnol 18:443–463

MacDonald GM, Ritchie JC (1986) Modern pollen spectra from the western interior of Canada and the interpretation of late Quaternary vegetation development. New Phytol 103:245–268

Maechler M, Rousseeuw P, Struyf A, Hubert M (2005) Cluster analysis basics and extensions. R package version 1.12.1. http://cran.r-project.org/web/packages/cluster/index.html

Makarenkov V, Legendre P (2002) Nonlinear redundancy analysis and canonical correspondence analysis based on polynomial regression. Ecology 83:1146–1161

Minchin PR (1987) An evaluation of the relative robustness of techniques for ecological ordination. Vegetatio 69:89–107

Motyka J (1947) O zadaniach i metodach badan geobotanicznych. Sur les buts et les méthodes des recherches géobotaniques. Annales Universitatis Mariae Curie-Sklodowska (Lublin, Polonia), Sectio C, Supplementum I

Noy-Meir I, Walker D, Williams WT (1975) Data transformations in ecological ordination II. On the meaning of data standardization. J Ecol 63:779–800

Ochiai A (1957) Zoogeographic studies on the soleoid fishes found in Japan and its neighbouring regions. Bull Japan Soc Sci Fish 22:526–530

Odgaard BV (1994) The Holocene vegetation history of northern West Jutland, Denmark. Op Bot 123:1–71

Odgaard BV, Rasmussen P (2000) Origin and temporal development of macro-scale vegetation patterns in the cultural landscape of Denmark. J Ecol 88:733–748

Odum EP (1950) Bird populations of the highlands (North Carolina) plateau in relation to plant succession and avian invasion. Ecology 31:587–605

Ohtani K (2000) Bootstrapping R^2 and adjusted R^2 in regression analysis. Econ Model 17:473–483

Oksanen J, Blanchet B, Kindt R, Legendre P, O'Hara B, Simpson GL, Solymos P, Stevens MHH, Wagner H (2011) vegan: community ecology package. R package version 1.17–0. http://cran.r-project.org/web/packages/vegan/index.html

Orlóci L (1967) An agglomerative method for classification of plant communities. J Ecol 55:193–205

Paradis E, Bolker B, Claude J, Cuong HS, Desper R, Durand B, Dutheil J, Gascuel O, Jobb G, Heibl C, Lawson D, Lefort V, Legendre O, Lemon J, Noel Y, Nylander J, Opgen-Rhein R, Strimmer K, de Vienne D (2010) ape: analyses of phylogenetics and evolution. R package version 2.5. http://cran.r-project.org/web/packages/ape/index.html

Peres-Neto PR, Legendre P (2010) Estimating and controlling for spatial structure in the study of ecological communities. Global Ecol Biogeogr 19:174–184

Peres-Neto PR, Legendre P, Dray S, Borcard D (2006) Variation partitioning of species data matrices: estimation and comparison of fractions. Ecology 87:2614–2625

Pienitz R, Smol JP, Birks HJB (1995) Assessment of freshwater diatoms as quantitative indicators of past climatic change in the Yukon and Northwest Territories, Canada. J Paleolimnol 13: 21–49
Pinel-Alloul B, Niyonsenga T, Legendre P (1995) Spatial and environmental components of freshwater zooplankton structure. Écoscience 2:1–19
Potapova MG, Charles DF (2002) Benthic diatoms in USA rivers; distributions along spatial and environmental gradients. J Biogeogr 29:167–187
Prentice IC (1977) Non-metric ordination methods in ecology. J Ecol 65:85–94
Prentice IC (1978) Modern pollen spectra from lake sediments in Finland and Finnmark, North Norway. Boreas 7:131–153
Prentice IC (1980) Multidimensional scaling as a research tool in Quaternary palynology: a review of theory and methods. Rev Palaeobot Palynol 31:71–104
Prentice IC (1986) Multivariate methods for data analysis. In: Berglund BE (ed) Handbook of Holocene palaeoecology and palaeohydrology. Wiley, Chichester, pp 775–797
Quinlan R, Leavitt PR, Dixit AS, Hall RI, Smol JP (2002) Landscape effects of climate, agriculture, and urbanization on benthic invertebrate communities of Canadian prairie lakes. Limnol Oceanogr 47:378–391
R Development Core Team (2011) R: a language and environment for statistical computing. R Foundation for Statistical Computing, Vienna, Austria. http://www.R-project.org
Rao CR (1995) A review of canonical coordinates and an alternative to correspondence analysis using Hellinger distance. Qüestiió (Quaderns d'Estadística i Investigació Operativa) 19:23–63
Raup DM, Crick RE (1979) Measurement of faunal similarity in paleontology. J Paleontol 53:1213–1227
Rejmánek M, Klinger R (2003) CANOCO 4.5 and some comparisons with PC-ORD and SYN-TAX. Bull Ecol Soc Am 84:69–74
Renberg I, Korsman T, Birks HJB (1993) Prehistoric increases in the pH of acid-sensitive Swedish lakes caused by land-use changes. Nature 362:824–826
Renkonen O (1938) Statisch-ökologische Untersuchungen über die terrestiche Kaferwelt der finnischen Bruchmoore. Ann Zool Soc Bot Fenn Vanamo 6:1–231
Ritchie JC (1977) The modern and late Quaternary vegetation of the Campbell-Dolomite Uplands, near Inuvik, N.W.T., Canada. Ecol Monogr 47:401–423
Roberts DW (2007) labdsv: ordination and multivariate analysis for ecology. R package version 1.3-1. http://cran.r-project.org/web/packages/labdsv/index.html, http://ecology.msu.montana.edu/labdsv/R
Sabatier R, Lebreton J-D, Chessel D (1989) Principal component analysis with instrumental variables as a tool for modelling composition data. In: Coppi R, Bolasso S (eds) Multiway data analysis. Elsevier, Amsterdam, pp 341–352
Schaffers AP, Raemakers IP, Sýkora KV, ter Braak CJF (2008) Arthropod assemblages are best predicted by plant species composition. Ecology 89:782–794
Simpson GL (2009) cocorresp: co-correspondence analysis ordination methods. R package version 0.1-9. http://cran.r-project.org/web/packages/cocorresp/index.html
Simpson GL (2012) Chapter 15: Analogue methods in palaeolimnology. In: Birks HJB, Lotter AF, Juggins S, Smol JP (eds) Tracking environmental change using lake sediments. Volume 5: Data handling and numerical techniques. Springer, Dordrecht
Simpson GL, Hall RI (2012) Chapter 19: Human impacts – applications of numerical methods to evaluate surface-water acidification and eutrophication. In: Birks HJB, Lotter AF, Juggins S, Smol JP (eds) Tracking environmental change using lake sediments. Volume 5: Data handling and numerical techniques. Springer, Dordrecht
Simpson GL, Shilland EM, Winterbottom JM, Keay J (2005) Defining reference conditions for acidified waters using a modern analogue approach. Environ Pollut 137:119–133
Smol JP (2008) Pollution of lakes and rivers: a paleoenvironmental perspective, 2nd edn. Blackwell Publishing, Oxford
Soininen J, Weckström J (2009) Diatom community structure along environmental and spatial gradients in lakes and streams. Archiv Hydrobiol 174:205–213

Sokal RR, Rohlf FJ (1995) Biometry – the principle and practice of statistics in biological research, 3rd edn. Freeman, New York

Telford RJ, Birks HJB (2005) The secret assumption of transfer functions: problems with spatial autocorrelation in evaluating model performance. Quaternary Sci Rev 24:2173–2179

ter Braak CJF (1985) Correspondence analysis of incidence and abundance data: properties in terms of a unimodal response model. Biometrics 41:859–873

ter Braak CJF (1986) Canonical correspondence analysis: a new eigenvector technique for multivariate direct gradient analysis. Ecology 67:1167–1179

ter Braak CJF (1987a) Ordination. In: Jongman RHG, ter Braak CJF, van Tongeren OFR (eds) Data analysis in community and landscape ecology. Pudoc, Wageningen, The Netherlands. Reissued in 1995 by Cambridge University Press, Cambridge, pp 91–173

ter Braak CJF (1987b) The analysis of vegetation-environment relationships by canonical correspondence analysis. Vegetatio 69:69–77

ter Braak CJF (1988a) CANOCO – an extension of DECORANA to analyze species-environment relationships. Vegetatio 75:159–160

ter Braak CJF (1988b) Partial canonical correspondence analysis. In: Bock HH (ed) Classification and related methods of data analysis. North-Holland, Amsterdam, pp 551–558

ter Braak CJF (1994) Canonical community ordination. Part I: Basic theory and linear methods. Écoscience 1:127–140

ter Braak CJF, Juggins S (1993) Weighted averaging partial least squares regression (WA-PLS): an improved method for reconstructing environmental variables from species assemblages. Hydrobiologia 269:485–502

ter Braak CJF, Prentice IC (1988) A theory of gradient analysis. Adv Ecol Res 18:271–317

ter Braak CJF, Schaffers AP (2004) Co-correspondence analysis: a new ordination method to relate two community compositions. Ecology 85:834–846

ter Braak CJF, Šmilauer P (2002) CANOCO reference manual and CanoDraw for Windows user's guide – software for canonical community ordination (version 4.5). Microcomputer Power, Ithaca

ter Braak CJF, Verdonschot PFM (1995) Canonical correspondence analysis and related multivariate methods in aquatic ecology. Aquat Sci 57:255–289

ter Braak CJF, Wiertz J (1994) On the statistical analysis of vegetation change: a wetland affected by water extraction and soil acidification. J Veg Sci 5:361–372

Tremblay V, Larocque-Tobler I, Sirois P (2010) Historical variability of subfossil chironomids (Diptera: Chironomidae) in three lakes impacted by natural and anthropogenic disturbances. J Paleolimnol 44:483–495

Vellend M (2004) Parallel effects of land-use history on species diversity and genetic diversity of forest herbs. Ecology 85:3043–3055

Verdonschot PFM, ter Braak CJF (1994) An experimental manipulation of oligochaete communities in mesocosms treated with chlorpyrifos or nutrient additions: multivariate analyses with Monte Carlo permutation tests. Hydrobiologia 278:251–266

Vinebrooke RD, Hall RI, Leavitt PR, Cumming BF (1998) Fossil pigments as indicators of phototrophic response to salinity and climatic change in lakes of Western Canada. Can J Fish Aquat Sci 55:668–681

Wall AAJ, Magny M, Mitchell EAD, Vannière B, Gilbert D (2010) Response of testate amoeba assemblages to environmental and climatic changes during the Lateglacial-Holocene transition at Lake Lautrey (Jura Mountains, Eastern France). J Quaternary Sci 25:945–956

Wang Y, Liu X, Herzschuh U (2010) Asynchronous evolution of the Indian and East Asian Summer Monsoon indicated by Holocene moisture patterns in monsoonal Central Asia. Earth Sci Rev 103:135–153

Wartenberg D, Ferson S, Rohlf FJ (1987) Putting things in order: a critique of detrended correspondence analysis. Am Nat 129:434–448

Whittaker RH (1952) A study of summer foliage insect communities in the Great Smoky Mountains. Ecol Monogr 22:1–44

Whittaker RH (1967) Gradient analysis of vegetation. Biol Rev 42:207–264

Wiklund JA, Bozinovski N, Hall RI, Wolfe BB (2010) Epiphytic diatoms as flood indicators. J Paleolimnol 44:25–42

Wischnewski J, Mischke S, Wang Y, Herzschuh U (2011) Reconstructing climate variability on the northeastern Tibetan Plateau since the last Lateglacial – a multi-proxy, dual-site approach comparing terrestrial and aquatic signals. Quaternary Sci Rev 30:82–97

Chapter 9
Statistical Learning in Palaeolimnology

Gavin L. Simpson and H. John B. Birks

Abstract This chapter considers a range of numerical techniques that lie outside the familiar statistical methods of linear regression, analysis of variance, and generalised linear models or data-analytical techniques such as ordination, clustering, and partitioning. The techniques outlined have developed as a result of the spectacular increase in computing power since the 1980s. The methods make fewer distributional assumptions than classical statistical methods and can be applied to more complicated estimators and to huge data-sets. They are part of the ever-increasing array of 'statistical learning' techniques (*sensu* Hastie et al. (2011). The elements of statistical learning, 2nd edn. Springer, New York) that try to make sense of the data at hand, to detect major patterns and trends, to understand 'what the data say', and thus to learn from the data.

A range of tree-based and network-based techniques are presented. These are classification and regression trees, multivariate regression trees, bagged trees, random forests, boosted trees, multivariate adaptive regression splines, artificial neural networks, self-organising maps, Bayesian networks, and genetic algorithms. Principal curves and surfaces are also discussed as they relate to unsupervised self-organising maps. The chapter concludes with a discussion of current developments in shrinkage methods and variable selection in statistical modelling that can help in model selection and can minimise collinearity problems. These include principal components regression, ridge regression, the lasso, and the elastic net.

G.L. Simpson (✉) • H.J.B. Birks
Environmental Change Research Centre, University College London, Pearson Building, Gower Street, London WC1E 6BT, UK

H.J.B. Birks
Department of Biology and Bjerknes Centre for Climate Research, University of Bergen, PO Box 7803, Bergen N-5020, Norway

School of Geography and the Environment, University of Oxford, Oxford OX1 3QY, UK
e-mail: john.birks@bio.uib.no

H.J.B. Birks et al. (eds.), *Tracking Environmental Change Using Lake Sediments*, Developments in Paleoenvironmental Research 5, DOI 10.1007/978-94-007-2745-8_9,

Keywords Artificial neural networks • Bagging trees • Bayesian belief networks • Bayesian decision networks • Bayesian networks • Boosted trees • Classification trees • Data-mining • Decision trees • Genetic algorithms • Genetic programmes • Model selection • Multivariate adaptive regression splines • Multivariate regression trees • Principal curves and surfaces • Random forests • Regression trees • Ridge regression • Self-organising maps • Shrinkage • Statistical learning • Supervised learning • The elastic net • The lasso • Unsupervised learning

Introduction

This chapter considers a range of numerical techniques that lie outside the familiar statistical methods of linear regression, analysis of variance, and maximum-likelihood estimation or data-analytical techniques such as ordination or clustering. The techniques outlined here have developed as a result of the spectacular increase in computational power since the 1980s. They make fewer distributional assumptions than classical statistical methods and can be applied to more complicated estimators and to huge data-sets (Efron and Tibshirani 1991; Raymond et al. 2005; Witten and Frank 2005; Hastie et al. 2011). They allow the exploration and summary of vast data-sets and permit valid statistical inferences to be made without the usual concerns for mathematical tractability (Efron and Tibshirani 1991) because traditional analytical approaches are replaced by specially designed computer algorithms (Hastie et al. 2011).

Many of the techniques discussed in this chapter are part of the ever-increasing battery of techniques that are available for what Hastie et al. (2011) call ‘statistical learning’. In this, the aim of the numerical analysis is to make sense of the relevant data, to detect major patterns and trends, to understand ‘what the data say’, and thus to learn from the data (Hastie et al. 2011). Statistical learning includes prediction, inference, and data-mining (Hastie et al. 2011). Data-mining (Ramakrishnan and Grama 2001; Witten and Frank 2005) usually involves very large data-sets with many objects and many variables. In conventional statistical analyses, the formulation of the hypotheses to be tested usually follows the observation of the phenomena of interest and associated data collection. In statistical learning and data-mining, observations on the numerical properties of previously collected data can also stimulate hypothesis generation (Raymond et al. 2005). Hypotheses generated in this manner can be tested using existing independent data (so-called test-data) or where these are inadequate, by further observations and data-collection. Data-mining within statistical learning is, like exploratory data analysis (Juggins and Telford 2012: Chap. 5), clustering and partitioning (Legendre and Birks 2012a: Chap. 7), and classical ordination (Legendre and Birks 2012b: Chap. 8), a data-driven hypothesis-generation tool as well as a data-summarisation technique. Classical statistical techniques such as regression (Birks 2012a: Chap. 2; Blaauw and Heegaard 2012: Chap. 12), temporal-series analysis (Dutilleul et al. 2012: Chap. 16), and canonical ordination (Legendre and Birks 2012b: Chap. 8; Lotter and Anderson 2012: Chap. 18) are model-based hypothesis-testing techniques. Statistical learning and data-mining can thus play a critical role, not only in data-analysis but also in the design of future data-collection and research projects.

Statistical learning from large data-sets has provided major theoretical and computational challenges and has led to a major revolution in the statistical sciences (Efron and Tibshirani 1991; Hastie et al. 2011). As a result of this revolution, statistical learning tends now to use the language of machine learning of inputs which are measured or preset (Hastie et al. 2011). These have some influence on one of more outputs. In conventional statistical terminology, inputs are usually called predictors or independent exploratory variables, whereas outputs are called responses or dependent variables. In palaeolimnology, the outputs are usually quantitative variables (e.g., lake-water pH), qualitative (categorical 1/0) variables, (e.g., lake type), or ordered categorical variables (e.g., low, medium, high water-depth). The inputs can also vary in measurement type and are usually quantitative variables. In a typical palaeolimnological study, we have an outcome measurement, usually quantitative (e.g., lake-water pH) or categorical (e.g., fish present/absent) that we want to predict on a set of features (e.g., modern diatom assemblages). We have a training-set of data in which we observe the outcome and feature measurements for a set of objects (e.g., lakes). Using this training-set, we construct a prediction model or learner that will enable us to predict or infer the outcome for new unseen objects with their feature measurements (e.g., fossil diatom assemblages). A good learner is one that accurately predicts such an outcome. The distinction in output type has resulted in the prediction tasks being called regression when predicting quantitative outputs and classification when predicting qualitative outputs (Hastie et al. 2011).

Statistical learning can be roughly grouped into supervised or unsupervised learning. In supervised learning, the aim is to predict the value of an output measure based on a number of input measures. It is called supervised because the presence of the outcome measure(s) can guide the learning process. In unsupervised learning, there is no outcome measure, only input features. The aim is not to predict but to describe how the data are organised or clustered and to discern the associations and patterns among a set of input measures. Table 9.1 summarises the major data-analytical techniques used in palaeolimnology that are discussed by Birks (2012a: Chap. 2), Legendre and Birks (2012a, b: Chaps. 7 and 8), Blaauw and Heegaard (2012: Chap. 12), Juggins and Birks (2012: Chap. 14), Simpson (2012: Chap. 15), and Lotter and Anderson (2012: Chap. 18) in terms of supervised and unsupervised statistical learning.

This chapter outlines several tree-based and network-based data-analytical techniques that permit data-mining and statistical learning from large data-sets (over 500–1000 samples and variables) so as to detect the major patterns of variation within such data-sets, to predict responses to future environmental change, and to summarise the data as simple groups. These techniques are listed in Table 9.2 in relation to whether they are supervised or unsupervised statistical-learning techniques.

Table 9.1 Summary of the major analytical techniques used in palaeolimnology in terms of supervised and unsupervised statistical learning

	Type of statistical learning	
Numerical technique	Unsupervised	Supervised
Clustering (Chap. 7)	+	
K-means partitioning (Chap. 7)	+	
Ordination (e.g. PCA) (Chap. 8)	+	
Canonical ordination (Chaps. 8 and 18)		+
Weighted averaging regression and calibration (Chap. 14)		+
Weighted averaging partial least squares (Chap. 14)		+
Modern analogue technique (Chap. 15)		+
Discriminant analysis (Chap. 2)		+
Regression analysis (Chaps. 2 and 12)		+

Table 9.2 Summary of statistical machine-learning techniques in terms of supervised and unsupervised learning

	Type of statistical learning	
Machine-learning technique	Unsupervised	Supervised
Classification trees		+
Regression trees		+
Multivariate regression trees		+
Bagging trees		+
Boosted trees		+
Random forests	+	+
Multivariate adaptive regression splines		+
Artificial neural networks		+
Self-organising maps (SOMs)	+	
X-Y-fused SOMs, Bi-directional Kohonen networks, and super-organised maps		+
Bayesian belief networks		+
Bayesian decision networks		+
Genetic algorithms		+
Principal curves and surfaces	+	+
Shrinkage methods (ridge regression, the lasso, the elastic net)		+

Classification and Regression Trees

Dichotomous identification keys are common in fields such as biology, medicine, and ecology, where decisions as to the identification of individual specimens or the presence of disease are reduced to a set of simple, hierarchical rules that lead the user through the decision-making process. An example that will be familiar to many readers is the numerous plant identification keys used by field botanists. Computer-generated versions of these keys were first discussed in the social sciences arising from the need to cope with complex data and scientific questions resulting from questionnaire responses leading to the Automatic Interaction Detection programme of Morgan and Sonquist (1963). Around the same time, similar tree-based methodologies were being developed independently in the machine-learning field (e.g., Quinlan 1993). The seminal work of Breiman et al. (1984) brought the main ideas and concepts behind tree-based models into the statistical arena. De'ath and Fabricius (2000) and Vayssieres et al. (2000) introduced classification and regression trees to the ecological literature. Fielding (2007) provides a simple introduction to tree-based modelling procedures in biology. Witten and Frank (2005) discuss classification and regression trees in the context of data-mining large, heterogeneous data-sets.

The general idea behind tree-based modelling is to identify a set of decision rules that best predicts (i) the 'identities' of a categorical response variable (a classification tree), or (ii) a continuous response variable (a regression tree). By 'best predicts', we mean minimises a loss function such as least-squares errors

$$D_N = \sum_{i=1}^{n} (y_i - \hat{y}_N) \tag{9.1}$$

where D_N is the deviance (impurity) of node N, y_i refers to the ith observation in node N and $\hat{y}_N$ is the mean of y_i in node N. The total deviance (impurity) of a tree (D) consisting of N nodes is the sum of the deviances of the individual N nodes

$$D = \sum_{i=1}^{N} D_i \tag{9.2}$$

Building trees using the recursive binary partitioning method is by far the most commonly used technique. At each stage of fitting a tree, the algorithm identifies a split that best separates the observations in the current node into two groups; hence the binary part of the algorithm's name. The recursive partitioning aspect refers to the fact that each node is in turn split into two child nodes, and those child nodes are subsequently split, and so on in a recursive fashion (see Legendre and Birks 2012a: Chap. 7). We have glossed over many of the details of model fitting in the above description of recursive partitioning. We now expand on the detail of how trees are fitted to data.

The recursive partitioning algorithm starts with all the available data arranged in a single group or node (see also Legendre and Birks (2012a: Chap. 7) and Birks (2012b: Chap. 11) for other partitioning techniques that use this type of recursive algorithm (TWINSPAN, binary partitioning)). The data are a single matrix of n observations on m variables. The response variable y is also known; if y is a categorical variable (e.g., species presence/absence, or different species of pollen or diatom) then a classification tree will be fitted, whereas, if y is a continuous variable (e.g., lake-water pH or temperature) a regression tree is fitted. Each of the

m predictor variables is taken in turn and all possible locations for a split within the variable are assessed in terms of its ability to predict the response. For binary predictor variables, there is a single possible split (0 or 1). Categorical variables present a greater number of potential splits. An unordered categorical variable (e.g., red, green, blue) with number of levels (categories) L has $2(L - 1) - 1$ potential splits, whilst an ordered categorical variable (e.g., dry < moist < wet < very wet) conveys $L - 1$ potential splits. For continuous variables, imagine the observations lain out on a scale in ascending order of values of the variable. A split may be located between any pair of adjacent values. If there are U unique values, then each continuous variable conveys U – 1 potential splits. At each stage in the algorithm all of these potential split locations need to be evaluated to determine how well making each split predicts the response. Once the variable and split location that best predict the response have been identified, the data are separated into two groups on the basis of the split and the algorithm proceeds to split each of the two child groups (or nodes) in turn, using the same procedure as outlined above. Splitting continues until no nodes can be further subdivided or until some stopping criteria have been met, usually the latter. At this point fitting is complete and a full tree has been fitted to the data.

An important question remains; how do we quantify which split location best predicts the response? Splits are chosen on the basis of how much they reduce node impurity. For regression trees, the residual sums-of-squares (RSS, Eq. 9.1) about the child-node means or residual sums of absolute deviations (RSAD) from the child-node medians are used to measure node impurity, although the latter (RSAD) is of lesser utility with ecological data (De'ath and Fabricius 2000). Several alternative measures of node impurity (D_N) are commonly used in classification trees, including

(i) deviance

$$D_N = -2\sum_k n_{Nk}\log(p_{Nk}) \tag{9.3.1}$$

(ii) entropy

$$D_N = -2\sum_k p_{Nk}\log(p_{Nk}) \tag{9.3.2}$$

and
(iii) the Gini index

$$D_N = 1 - \sum_k p_{Nk}^2 \tag{9.3.3}$$

where D_N is the node impurity, n_{Nk} is the number of observations of class k in the Nth node, and p_{Nk} is the proportion of observations in the Nth node that are of type k. The overall node impurity evaluated for all possible splits is the sum of the impurities of the two groups formed by the split.

A final problem we face is how big a tree to grow? Above, we mentioned that the algorithm will continue until either it cannot split any node further (i.e., all nodes have zero impurity) or some stopping criteria are reached (e.g., fewer than five observations in a node). Such an approach will produce a large, complex tree that will tend to over-fit the observed data. Such a tree is unlikely to generalise well and will tend to produce poor out-of-sample predictions. A small tree, on the other hand, will be unlikely to capture important features in the response. Tree-size is a tuning parameter that controls the complexity of the fitted tree-model. The optimal tree-size can be determined from the data using a procedure known as cost-complexity pruning. The cost-complexity of a tree, CC, is computed as $CC = T_{impurity} + \alpha(T_{complexity})$, where $T_{impurity}$ is the impurity of the current tree over all terminal nodes, $T_{complexity}$ is the number of terminal leaves, and α a real number >0. α is the tuning parameter we aim to minimise in cost-complexity pruning, and represents the trade-off between tree-size and goodness-of-fit. Small values of α result in larger trees, whilst large values of α lead to smaller trees. Starting with the full tree, a search is made to identify the terminal node that results in the lowest CC for a given value of α. As the penalty α on tree complexity is increased, the tree that minimises CC will become smaller and smaller until the penalty is so great that a tree with a single node (i.e., the original data) has the lowest CC. This search produces a sequence of progressively smaller trees with associated CC. The solution now is to choose a value of α that is optimal in some sense. κ-fold cross-validation (Birks 2012a: Chap. 2; Juggins and Birks 2012: Chap. 14) is used to choose the value of α that has the minimal root mean squared error (RMSE). An alternative strategy is to select the smallest tree that lies within 1 standard error of the RMSE of the best tree.

Once the final tree is fitted, identified, and pruned, the data used to train the tree are passed down the branches to produce the fitted values for the response. In a regression tree, the predicted value is the mean of the observed values of the response in the terminal node that an observation ends up in. All the observations that are in the same terminal node therefore get the same fitted value. We say that regression trees fit a piece-wise constant model in the terminal nodes of the tree. The fitted values for classification trees are determined using a different procedure; the majority vote. The classes of all the observations in the same terminal node provide votes as to the fitted class for that node. The class that receives the highest number of votes is then the predicted class for all observations in that node.

Palaeolimnological data often contain missing data where, for one reason or another, a particular measurement on one or more samples is not available (Birks 2012a: Chap. 2; Juggins and Birks 2012: Chap. 14; Juggins and Telford 2012: Chap. 5). Deleting missing data reduces the number of samples available for analysis and may also introduce bias into the model if there is a systematic reason for the 'missingness' (Nakagawa and Freckleton 2008). Trees can handle missing data in the predictor variables in a number of ways. The first is to propagate a sample as far down the tree as possible until the variable used to split a node is one for which the data are missing. At that point we assign a fitted value as the average or majority vote of all the samples that pass through that particular node in the tree. The rationale for this is that we have sufficient information to make a partial prediction for a sample with missing data, but we are unable to provide a final prediction because of the missing data.

An alternative strategy is to use *surrogate splits* to decide how to propagate a sample with missing data further down a fitted tree. During the exhaustive search for split locations, a record is made of which alternative split locations provide a similar binary split of the data in the current node to that of the best split. Surrogate splits are those splits that provide the division of the samples in a node that most closely resembles the division made by using the best split location. When a sample with missing data is passed down a tree during prediction, the sample proceeds until it reaches a node where data on the splitting variable is missing. At this point, the best surrogate split is used to attempt to assign the sample to one of the two child nodes. If the variable used in the best surrogate split is also missing, the next best surrogate split is used, and so on until all available surrogate splits have been examined. If it is not possible to assign the sample to one of the two child nodes, then the sample is left in the current node and its predicted value is taken as the average or majority vote of samples passing through the node as previously described.

Surrogate splits are those that produce a similar binary division of a set of samples to that of the best split for a given node. There may also be split variables that reduce node impurity almost as much as the best split but do so using a different predictor variable and result in a different binary partition of a node. Such splits are known as *alternative splits*. Replacing the best split with an alternative split might lead to the fitting of a very different tree simply because of the legacy of having chosen one predictor over another early on in the tree-building process. Examination of the alternative splits can help provide a fuller description of the system under study by highlighting alternative models that explain the training data to a similar degree as the fitted tree.

High temperature combustion of coal and oil produces, amongst other pollutants and emissions, spheroidal carbonaceous particles (SCPs) (Rose 2001). Rose et al. (1994) studied the surface chemistry of a range of SCPs produced by burning coal, oil, and oil-shale fuels, and used linear discriminant analysis to identify linear combinations of surface chemistry variables that best distinguished between particles of the different fuel sources (see Birks 2012a: Chap. 2). To illustrate tree-based models, we re-analyse these data using a classification tree. The data consist of 6000 particles (3000 coal, 1000 oil, and 2000 oil-shale). A full classification tree was fitted using the rpart package (Therneau and Atkinson 2011) for the R statistical language and environment (R Core Development Team 2011). Apparent and ten-fold cross-validation (CV) relative error rates for trees of various size up to the full tree are shown in Fig. 9.1. The tendency for trees to over-fit the training data is illustrated nicely as the apparent relative error rate continues decreasing as

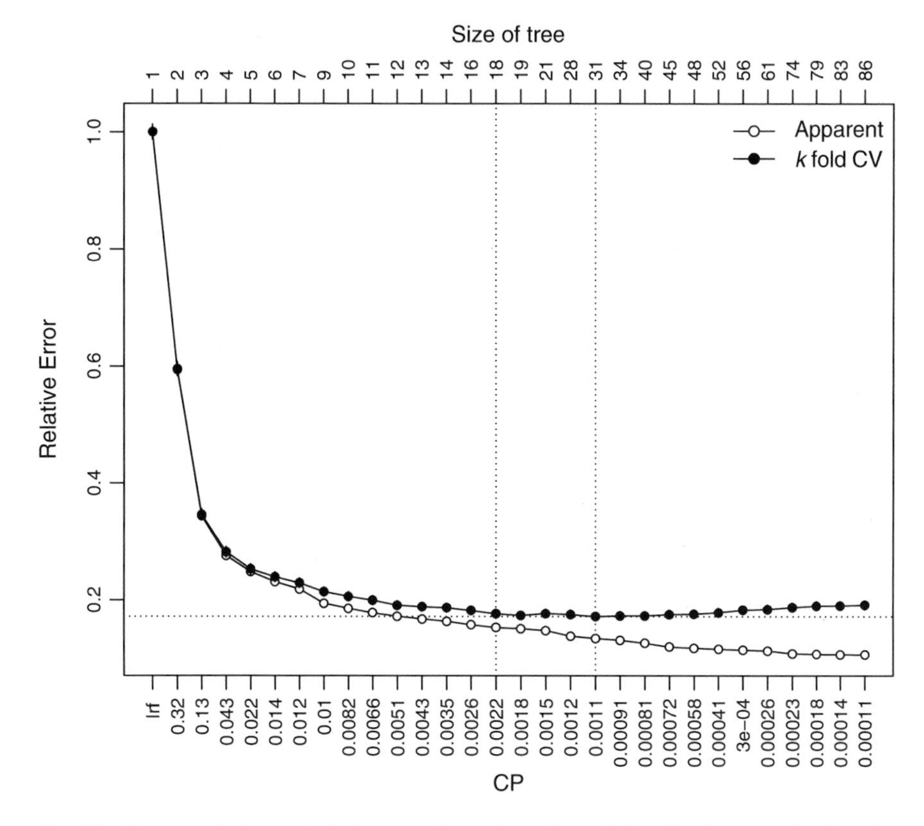

Fig. 9.1 Cost complexity and relative error for various sizes of classification trees fitted to the three-fuel spheroidal carbonaceous particle (SCP) example data. Apparent (*open circles*) and ten-fold cross-validated (CV; *filled circles*) relative error to the simplest tree (size one) are shown. The tree with the smallest CV relative error has 31 leaves, whilst the smallest tree within one standard error of the best tree has 18 leaves

the tree is grown and becomes more complex, whilst the ten-fold CV error rate stabilises after the tree contains 18 nodes or leaves and increases once the size of the tree exceeds 31 nodes. The values on the x-axis of Fig. 9.1 are the values of the cost-complexity parameter to which one must prune in order to achieve a tree of the indicated size. The best sized tree is one consisting of 31 nodes, with a CV relative error of 0.172 (CV standard error 0.007), and is indicated by the right-most vertical line. The smallest tree within one standard error of this best tree, is a model with 18 nodes and a CV relative error of 0.177 (CV standard error 0.007), and is indicated by the left-most vertical line.

Trees between sizes 18 and 48 all do a similar job, but we must guard against over-fitting the training data and producing a model that does not generalise well, so we select a tree size using the one standard-error rule and retain the tree with 18 nodes. This tree is shown in Fig. 9.2. The first split is based on Ca, with SCPs having

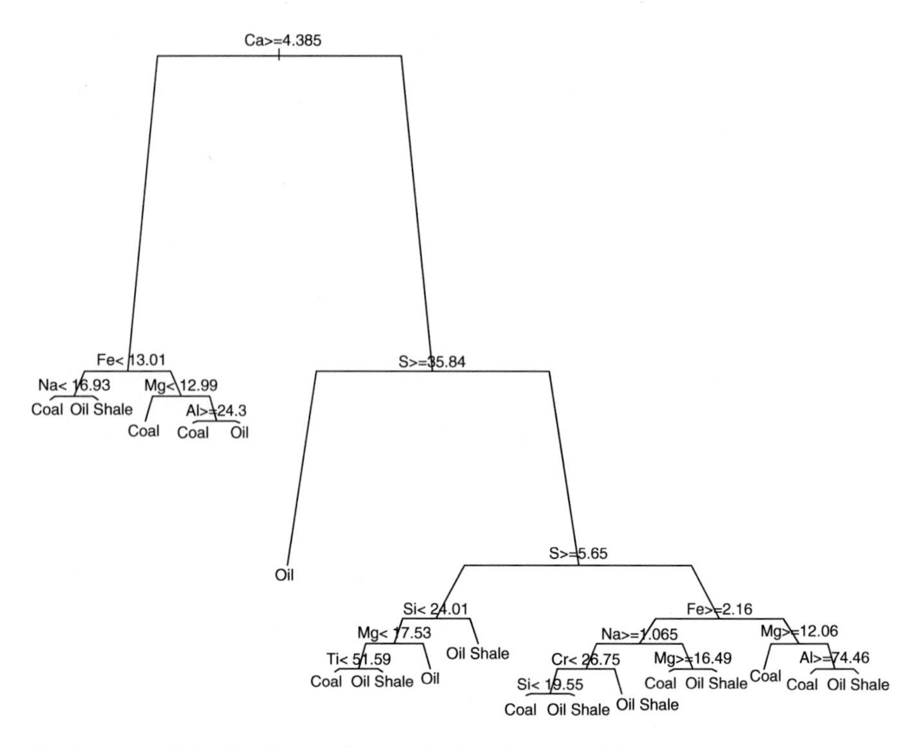

Fig. 9.2 Pruned classification tree fitted to the three-fuel spheroidal carbonaceous particle (SCP) example data. The predicted fuel types for each terminal node are shown, as are the split variables and thresholds that define the prediction rules

low amounts of Ca passing into the right-hand branch of the tree and those particles with Ca ≥ 4.385 passing into the left-hand branch. The right-hand branch is further split on the basis of S, with particles having ≥ 35.84 (and Ca < 4.385) classified as being produced by oil-fired power stations. By convention, the tree is plotted in such a way that the heights of the stems between nodes indicate the degree of importance attached to a split in terms of decreased node impurity. The first split on Ca and the split on S in the right-hand branch of the tree are clearly the most important rules for predicting SCP fuel type. The remaining splits are largely a fine tuning of these two main rules. The tree correctly classifies 5680 of the particles in the training data, giving an apparent error rate of 0.0533. Table 9.3 contains a summary of the predictions from the classification tree in the form of a confusion matrix. Individual error rates for the three fuel-types are also shown. Using ten-fold cross-validation to provide a more reliable estimate of model performance yields an error rate of 0.1 for the classification tree.

Of the machine-learning techniques described in this chapter, with the exception of artificial neural networks, trees are the most widely used method in palaeoecology and palaeolimnology, being employed in a variety of ways. Lindblah et al. (2002) used a classification tree to classify *Picea* pollen grains from three different species;

Table 9.3 Confusion matrix of predicted fuel type for the three-fuel classification tree

	Coal	Oil	Oil-shale	Error rate
Coal	2871	49	118	0.055
Oil	16	938	11	0.028
Oil-shale	113	13	1817	0.063

The rows in the table are the predicted fuel types for the 6000 spheroidal carbonaceous particles (SCPs) based on the majority vote rule. The columns are the known fuel-types. The individual fuel-type error rates of the classification tree are also shown. The overall error rate is 0.053

P. glauca, *P. mariana*, and *P. rubens* in eastern North America. Seven morphological measurements were made on approximately 170 grains of each species' pollen, and were used as predictor variables in the classification tree. An overall classification tree was fitted to assign grains to one of the three species, as well as individual species-specific binary classifications which aimed to predict whether a grain belonged to one of the three pollen taxa or not. Lindblah et al. (2003) used this approach to assign *Picea* pollen grains from a sediment core to one of the three species in late-glacial and Holocene sediments at a number of sites in New England, USA. Barton et al. (2011) employed a similar approach, using a classification tree to differentiate between pollen of red pine (*Pinus resinosa*) and jack pine (*Pinus banksiana*) in eastern North America. The habitat characteristics of sites where terrestrial snails, typical of full-glacial conditions in southern Siberia, are found have been described using a classification tree (Horsak et al. 2010). Other palaeoecological examples include Pelánková et al. (2008). CARTs are widely used in forestry (e.g., Baker 1993; Iverson and Prasad 1998, 2001; Iverson et al. 1999), ecology (e.g., Olden and Jackson 2002; Caley and Kuhnert 2006; Spadavecchia et al. 2008; Keith et al. 2010), biogeography (e.g., Franklin 1998, 2010), species-environment modelling (e.g., Iverson et al. 1999; Cairns 2001; Miller and Franklin 2002; Thuiller et al. 2003; Bourg et al. 2005; Kallimanis et al. 2007; Aho et al. 2011), limnology (e.g., Rejwan et al. 1999; Pyšek et al. 2010), hydrology (e.g., Carlisle et al. 2011), conservation biology (e.g., Ploner and Brandenburg 2003; Chytrý et al. 2008; Pakeman and Torvell 2008; Hejda et al. 2009), analysis of satellite data (e.g., Michaelson et al. 1994; DeFries et al. 2010), and landscape ecology (Scull et al. 2005).

Trees, whilst being inherently simple and interpretable, have a major drawback: the fitted model has high variance. A small change in the data can often lead to large changes in the form of the fitted tree, where a very different series of splits is identified. This makes trees somewhat difficult to interpret reliably; you might get a very different answer if you collected a different sample of data to fit the model. This is the downside of such a simple model structure. Solutions to this problem exist, and they all involve fitting many different trees to the data and averaging the predictions from each tree in some way. Collectively, these approaches are ensemble methods and include bagging, boosting, and random forests. We will discuss each of these techniques in later sections of this chapter.

Multivariate Regression Trees

The trees described in the previous section are univariate, dealing with a single response variable. Their extension to the multivariate response case is reasonably trivial (De'ath 2002; Larsen and Speckman 2004) yet the resulting technique is surprisingly versatile and is a useful counterpart to constrained ordination techniques such as redundancy analysis (RDA) and canonical correspondence analysis (CCA) (De'ath 2002; Legendre and Birks 2012a, b: Chaps. 7 and 8). Typically we have a response matrix of observations on m species for n sites. In addition, observations on p predictor variables (e.g., lake-water chemistry, climate-related variables) for the same n sites are available. In multivariate regression trees (MRT), the aim is to find a set of simple rules from the p predictor variables that best explains variation in the multivariate species-response matrix. Whilst MRT is closely related to constrained ordination, it can also be instructive to view MRT as a constrained clustering technique, where we partition the n observations in k groups or clusters on the basis of similar species composition *and* environment (Legendre and Birks 2012a: Chap. 7).

Regression trees use the concept of sum of squared errors as their measure of node impurity. This is inherently univariate, but can be extended to the multivariate case by considering sum of squared errors about the multivariate mean (centroid) of the observations in each node (De'ath 2002). In geometric terms, this amounts to being simply the sum of squared Euclidean distances of sites about the node centroid. In all other respects, the fitting and pruning of multivariate trees is the same as for univariate regression trees. However, the interpretation of multivariate trees requires additional techniques owing to the more complex nature of the response variable being modelled.

The Euclidean distance is often not suitable for use with ecological data as it focuses on absolute values, does not ignore or downweight double zeros, and imposes a linear framework on the analysis (Legendre and Birks 2012b: Chap. 8). MRTs can be adapted to work with any dissimilarity coefficient via direct decomposition of a supplied dissimilarity matrix to derive within-node sum of squared distances between node members. De'ath (2002) calls this method distance-based MRTs (db-MRTs). Note that in db-MRTs the within-node sum-of-squares are not computed with respect to the node centroid but instead with respect to pairs of samples. Minimising the sum of all pair-wise squared distances between samples within nodes is equivalent to computing the within-node sum-of-squares where the response data are species abundances. The response data in a db-MRT are a dissimilarity matrix computed using a chosen dissimilarity or distance coefficient (see Legendre and Birks 2012b: Chap. 8). As such, the raw data are not available during fitting to enable computation of the node centroids. Therefore, db-MRT uses the sum of pair-wise within-node distances as the measure of node impurity.

Univariate trees describe the mean response and a single tree-diagram can be used to convey in a simple fashion a large amount of information about the fitted model and the mean response. In MRTs, the mean response is multivariate, being the mean abundance of each species for the set of samples defined by the tree nodes. A biplot is a natural means for displaying the mean response. De'ath (2002) suggests that principal component analysis (PCA) (Legendre and Birks 2012b: Chap. 8) be used as the base plot, with PCA being performed on the fitted values of the response (the mean abundance for each species in each of the MRT terminal nodes). The observed data are also shown on the biplot. The samples themselves can thus be located in the biplot about their node centroid. Species loadings can be added to the biplot either as simple PCA loadings (species scores), in which case they are represented as biplot arrows, or as a weighted mean of the node means, in which case the species are represented as points in ordination space. The branching tree structure can also be drawn on the biplot to aid visualisation.

Earlier, we mentioned that MRTs can be viewed as a constrained form of cluster analysis. From the description of the technique we have provided, it should be clear that MRTs find k groups of samples that have the lowest within-group dispersion for the kth partition. If the constraints or predictor variables were not involved in the analysis then MRTs would be a way of fitting a minimum variance-cluster analysis (Legendre and Birks 2012a: Chap. 7). However, because the constraints are included in a MRT analysis, the identification of the group structure in the data is supervised, with groups being formed by partitioning the response variables on the basis of thresholds in the constraints. Chronological or constrained clustering and partitioning have a long tradition in palaeoecology and several numerical approaches to the problem of zoning stratigraphical data have been suggested (e.g., Gordon and Birks 1972, 1974; Gordon 1973; Birks 2012b: Chap. 11; Legendre and Birks 2012a: Chap. 7). One proposed solution to the problem is the binary divisive procedure using the sum-of-squares criterion (SPLITLSQ) method of Gordon and Birks (1972) which fits a sequence of b boundaries to the stratigraphical diagram, where $b \in \{1, 2, \ldots, n-1\}$. The boundaries are placed to minimise the within-group sums-of-squares of the groups formed by the boundaries. The process is sequential or hierarchical; first the entire stratigraphical sequence is split into two groups by the placement of a boundary that most reduces within-group sums of squares. Subsequently, one of the groups formed by positioning the first boundary is split by the placement of a second boundary, and so on until b boundaries have been positioned. The SPLITLSQ approach is exactly equivalent to the MRT when the Euclidean distance is used (see Legendre and Birks 2012b: Chap. 8). The utility of the MRT as a means of zoning stratigraphical diagrams is that the cross-validation procedure provides a simple way to assess the number of groups into which the sequence should be split.

To illustrate MRTs and to emphasise the constrained clustering nature of the technique, we turn to the classic Abernethy Forest pollen data of Birks and Mathewes (1978) (see Birks and Gordon 1985 for details). We fit a MRT to the pollen percentage data without transformation. A plot of the apparent and cross-validated relative error as a function of the cost-complexity parameter (or tree-size) for the MRT-fit to the Abernethy Forest data is shown in Fig. 9.3. Of the tree-sizes considered, the minimum cross-validated relative error is achieved by a tree with

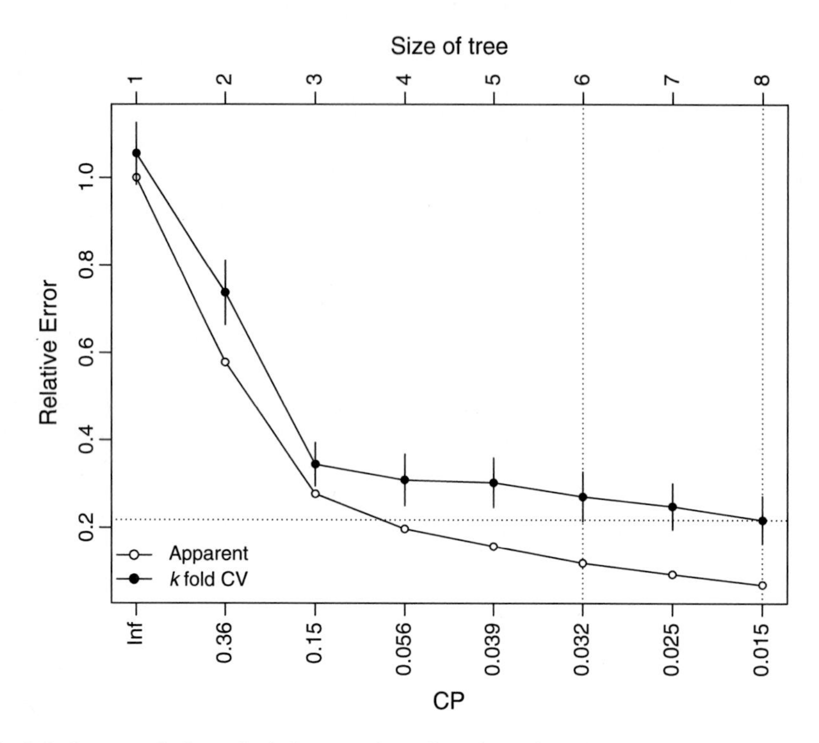

Fig. 9.3 Cost complexity and relative error for various sizes of multivariate regression trees fitted to the late-glacial and early-Holocene Abernethy Forest pollen sequence. Apparent (*open circles*) and ten-fold cross-validated (CV; *filled circles*) relative error to the simplest tree (size one) are shown. The tree with the smallest CV relative error has 8 leaves, whilst the smallest tree within one standard error of the best tree has 6 leaves

eight terminal nodes (seven splits), whilst the one standard-error rule would select the six-node sized tree. We select the latter and show the pruned, fitted MRT in Fig. 9.4. The first split is located at 7226 radiocarbon years BP and the second at 9540 BP. These two splits account for much larger proportions of the variance in the pollen data than the subsequent splits, as shown by the heights of the bars below the splits. The bar charts located at the terminal nodes in Fig. 9.4 provide a representation of the mean abundance for each pollen type over the set of samples located in each terminal node. A better representation of the mean response is given by the tree biplot (Fig. 9.5). The first split separates the samples dominated by *Pinus*, *Quercus*, and *Ulmus* pollen from the other samples, and is aligned with the first principal component (PC) axis. The second PC axis separates a group of samples characterised by *Juniperus*, *Corylus*, and *Betula* pollen.

MRTs have proved a relatively popular machine-learning technique in the palaeoenvironmental sciences. Davidson et al. (2010a) employed MRT to infer simultaneously the densities of zooplanktivorous fish and aquatic macrophytes from cladoceran species composition. The MRT was applied to a training-set of 39 lakes,

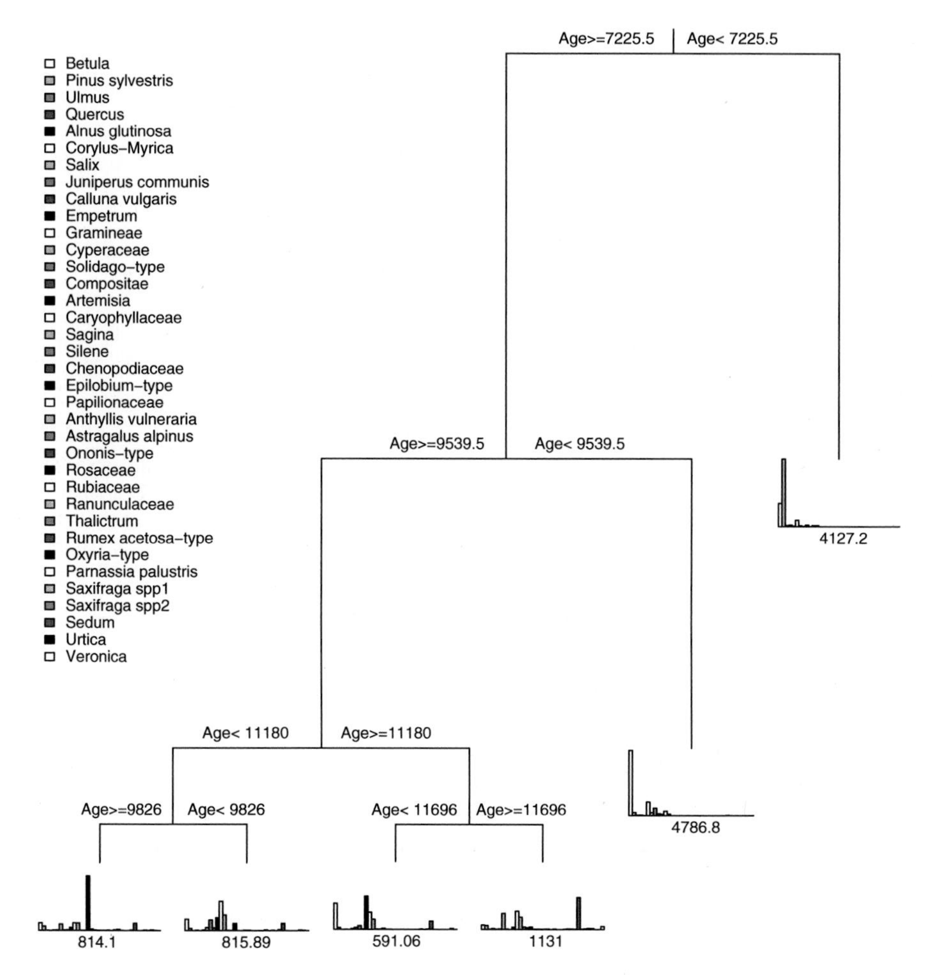

Fig. 9.4 Pruned multivariate regression tree (MRT) fitted to the late-glacial and early-Holocene Abernethy Forest pollen sequence. The major stratigraphic zones in the pollen stratigraphy are identified by the MRT. The bar charts in the terminal nodes describe the abundance of the individual species in each zone. The numbers beneath the bar charts are the within-zone sums of squares

using the cladoceran taxa as response variables and 14 environmental variables as predictors. The resulting pruned MRT had six clusters of samples resulting from splits on zooplanktivorous fish density (ZF) and plant volume infestation (PVI) and explained 67% of the variance in the species data. Davidson et al. (2010b) then applied their MRT model in conjunction with redundancy analysis (Legendre and Birks 2012b: Chap. 8) to cladoceran assemblages from a sediment core from Felbrigg Lake to investigate past changes in fish abundance and macrophyte abundance. Herzschuh and Birks (2010) used MRT in their investigation of the indicator value of Tibetan pollen and spore taxa in relation to modern vegetation

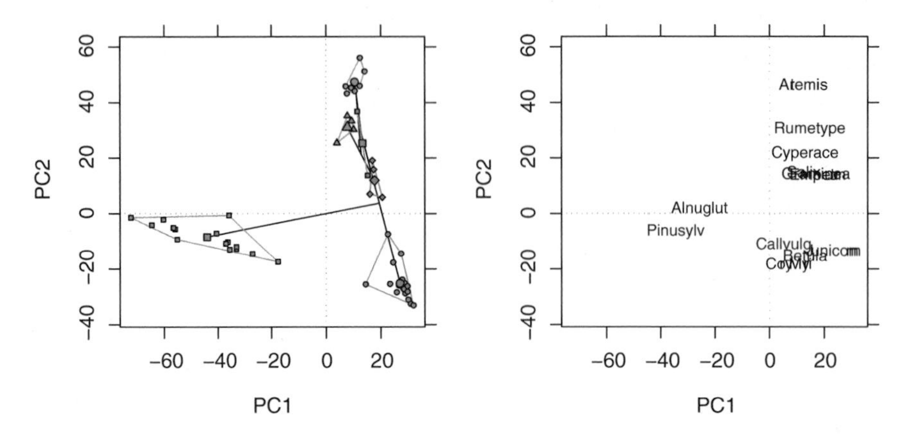

Fig. 9.5 Principal component analysis (PCA) display of the multivariate regression tree (MRT) fitted to the late-glacial and early-Holocene Abernethy Forest pollen sequence (*left*). The terminal nodes of the MRT are shown by large *open circles*, joined by line segments that represent the hierarchy. The samples within each node are differentiated by symbol shape and colour. Species scores (*right*) for the most common taxa in the Abernethy data-set are positioned using weighted averages instead of weighted sums

and climate. Their analysis showed that annual precipitation was the most important climatic variable in grouping the pollen counts in modern assemblages, with a value of ~390 mm precipitation identified as a critical threshold. Temperature was identified as then playing a role in separating the two groups of pollen assemblages resulting from the 'low' and 'high' precipitation split. The resulting MRT produced four pollen groupings associated with four climate types: dry and warm, dry and cool, wet and warm, and wet and cool. Other palaeolimnological examples include Amsinck et al. (2006) and Bjerring et al. (2009). Surprisingly, MRTs do not appear to have been widely used in ecology or biogeography except in a recent biogeographical study by Chapman and Purse (2011).

Other Types of Tree-Based Machine-Learning Methods (Bagging, Boosted Trees, Random Forests, Multivariate Adaptive Regression Splines)

Earlier, we mentioned the instability problem of single-tree based models, which can be viewed as sampling uncertainty in the model outputs. If we were to take a new sample of observations and fit a model to those and use it to predict for a test-set of observations, we would get a different set of predictions for the test-set samples. If this process were repeated many times for each observation in the test-set, a posterior distribution of predicted values would be produced. The mean of each of these posterior distributions can be used as the predictions for the test-set samples, and in addition, the standard error of the mean or the upper and lower 2.5th

quantiles can be used to form uncertainty estimates on the predictions. In general, however, taking multiple samples of a population is not feasible. Instead, we can use the training-set observations themselves to derive the posterior distributions using bootstrap re-sampling (see Birks 2012a: Chap. 2; Juggins and Birks 2012: Chap. 14; Simpson 2012: Chap. 15). Such approaches are often termed ensemble or committee methods.

This general description applies neatly to bagging and random forests, but less so to the technique of boosting and not at all to multivariate adaptive regression splines (MARS: Friedman 1991). Boosting employs many trees in a manner similar to bagging and random forests, but each additional tree focuses on the hard-to-predict observations in the training-set, thereby learning different features in the data (Schapire 1990; Freund 1995; Friedman et al. 2000; Friedman 2001; Hastie et al. 2011). MARS, on the other hand, relaxes the piece-wise constant models fitted in the nodes of regression trees to allow piece-wise linear functions and in doing so discards the hierarchical nature of the simple tree structure (Friedman 1991). Whilst the switch to piece-wise linear functions is not that dramatic in itself, MARS employs these piece-wise linear functions in a flexible way combining several such functions to fit regression models capable of identifying complex, non-linear relationships between predictor variables and the response (Friedman 1991). Prasad et al. (2006) provide a comprehensive comparison of these newer tree techniques.

Bagging

Bagging, short for bootstrap aggregating, is a general method, proposed by Breiman (1996), for producing ensembles for any type of model, though it has typically been applied to tree-based models. In palaeolimnology, when we perform bootstrapping (Efron and Tibshirani 1993) to estimate calibration-function errors and provide sample-specific errors (Birks et al. 1990; Juggins and Birks 2012: Chap. 14; Simpson 2012: Chap. 15), we are using bagging. The idea is quite simple and draws upon the power of Efron's (1979) bootstrap to produce a set or ensemble of models that replicate the uncertainty in the model arising from sampling variation.

In bagging, a large number of models, b, is produced from a single training-set by drawing a bootstrap sample from the training-set with which to fit each model. Recall that a bootstrap sample is drawn from the training-set with replacement, and that, on average, approximately two thirds of the training-set samples will be included in the bootstrap sample. The remaining samples not selected for the bootstrap sample are set to one side and are known as the out-of-bag (OOB) samples. A tree model without pruning (or any other model) is fitted to this bootstrap sample. The fitted tree is used to generate predictions for the OOB samples, which are recorded, as are the fitted values for the in-bag samples. This procedure is repeated b times to produce a set of b trees. The sets of fitted values for each training-set sample are averaged to give the bagged estimates of the fitted values. In the case of a regression tree the mean is used to average the fitted values, whilst the majority

Table 9.4 Confusion matrix of predicted fuel type for the bagged three-fuel classification tree (number of trees = 500)

	Coal	Oil	Oil-shale	Error rate
Coal	2794	50	116	0.056
Oil	18	930	6	0.025
Oil-shale	188	13	1878	0.100

The rows in the table are the predicted fuel types for the 6000 spheroidal carbonaceous particles (SCPs) based on the majority vote rule over the ensemble of trees. The columns are the known fuel-types. The individual fuel-type error rates of the bagged classification tree are also shown. The overall error rate is 0.066

vote rule is used for classification trees, where each of the b bagged trees supplies a vote as to the fitted class for each observation, and the class with the largest number of votes is selected as the fitted class for that observation. Alternatively, posterior class probabilities can be produced for each observation from the set of bagged classification trees (though not using the relative proportions of votes for each class) and the class with the highest posterior probability is taken as the predicted class. The same procedures are used to provide bagged predictions for new observations not included in the training-set.

Table 9.4 shows the confusion matrix for a bagged classification tree model applied to the three fuel-type SCP data analysed earlier. Error rates for the three fuel-types are also shown. These statistics were computed using the OOB samples and are honest, reliable estimates of the true error rates as opposed to those for the single classification tree we produced earlier. The overall error rate for the bagged model is 0.066 (6.6%), a modest improvement over the single classification tree (k-fold cross-validation error = 0.1). Table 9.4 contains a summary of the predictions from the bagged classification tree. The predictions for the Coal and Oil classes are very similar to the apparent predictions from the classification tree (Table 9.3). The main difference between the bagged tree and the single tree is in their abilities to discriminate between coal- and oil-shale-derived particles, with the single tree being somewhat over-optimistic in its ability to discriminate these two fuel-types. The bagged tree gives a more honest appraisal of its ability to discriminate; the error rate for the oil-shale class is similar to the overall k-fold CV error rate of the classification tree.

Model error for bagged regression trees can be expressed as RMSE

$$\text{RMSE} = \sqrt{\sum_{i=1}^{n} (\hat{y}_i - y_i) / n} \qquad (9.4)$$

using the fitted values, but this is an apparent error statistic and is not reflective of the real expected error. Instead, we can compute the equation above for each observation using only the OOB predictions. The OOB predictions are for the samples not used to fit a given tree. As such they provide an independent estimate of the model error when faced with new observations. A similar quantity can be computed for classification trees and is known as the error rate (number of misclassifications / number of OOB observations). Again, only the OOB samples should be used in generating the error rate of the model to achieve an honest error estimate.

How does bagging help with the tree instability problem? Individual trees are unstable and hence have high variance. Model uncertainty is a combination of bias (model error or mean squared error: MSE) and variance (the variation of model estimates about the mean). Bagging improves over single tree models because averaging over b trees reduces the variance whilst leaving the bias component unchanged, hence the overall model uncertainty is reduced. This does not hold for classification trees, however, where squared loss is not appropriate and 0–1 loss is used instead, as bias and variance are not additive in such cases (Hastie et al. 2011). Bagging a good classification model can make that model better but bagging a bad classification model can make the model worse (Hastie et al. 2011).

The improved performance of bagged trees comes at a cost; the bagged model loses the simple interpretation that is a key feature of a single regression tree or classification tree. There are now b trees to interpret and it is difficult, though not impossible, to interrogate the set of trees to determine the relative importance of predictors. We discuss this in the following section on the related technique of random forests.

Random Forests

With bagged trees, we noted that reduction in model uncertainty is achieved through variance reduction because averaging over many trees retains the same bias as that of a single tree. Each of the b trees is statistically identically distributed, but not necessarily independent because the trees have been fitted to similar data-sets. The degree of pair-wise correlation between the b trees influences the variance of the trees and hence the uncertainty in the model; the larger the pair-wise correlation, the larger the variance. One way to improve upon bagging is to reduce the correlation between the b trees. Random forests (Breiman 2001) is a technique that aims to do just that. Prasad et al. (2006) and Cutler et al. (2007) provide accessible introductions to random forests from an ecological view-point, whilst Chap. 15 of Hastie et al. (2011) provides an authoritative discussion of the method.

The key difference between bagging as described above and random forests is that random forests introduces an additional source of stochasticity into the model-building process (Breiman 2001), which has the effect of de-correlating the set of trees in the ensemble of trees or the forest (Hastie et al. 2011). The tree-growing algorithm, as we saw earlier, employs an exhaustive search over the set of available explanatory variables to find the optimal split criterion to partition a tree node into two new child nodes. In standard trees and bagging, the entire set of explanatory variables is included in this search for splits. In random forests, however, the set of explanatory variables made available to determine each split is a randomly determined, usually small, subset of the available variables. As a result, each tree in the forest is grown using a bootstrap sample, just as with bagging, and each split in each and every tree is chosen from a random subset of the available predictors.

The number of explanatory variables chosen at random for each split search is one of two tuning parameters in random forests that needs to be chosen by the user. The number of explanatory variables used is referred to as m and is usually small. For classification forests, the recommended value is $\lfloor\sqrt{p}\rfloor$, and $\lfloor p/3\rfloor$ is suggested for regression forests, where the brackets represent the floor (rounding down to the nearest integer value), and p is the number of explanatory variables (Hastie et al. 2011). The recommended minimum node size, the size in number of observations beyond which the tree growing algorithm will stop splitting a node, is one and five for classification and regression forests, respectively (Hastie et al. 2011). This has the effect of growing large trees to each bootstrap sample with the result that each individual tree has low bias.

The trees are not pruned as part of the random-forest algorithm; the intention is to grow trees until the stopping criteria are met so that each tree in the forest has a low bias. Each of the individual trees is therefore over-fitted to the bootstrap sample used to create it, but averaging over the forest of trees effectively nullifies this over-fitting. It is often claimed that random forests do not over-fit. This is not true, however, and, whilst the details of why this is the case are beyond the scope of this chapter, it is worth noting that as the number of fully grown trees in the forest becomes large, the average of the set of trees can result in too complex a model and consequently suffer from increased variance. Section 15.3.4 of Hastie et al. (2011) explains this phenomenon, but goes on to state that using fully grown trees tends not to increase the variance too much and as such we can simplify our model building by not having to select an appropriate tree depth via cross-validation.

Random forests suffer from the same problem of interpretation as bagged trees owing to the large number of trees grown in the forest. Several mechanisms have been developed to allow a greater level of interpretation for random forests. We will discuss two main techniques: (i) variable importance measures and (ii) proximity measurements.

The importance of individual predictors is easy to identify with a single tree as the relative heights of the branches between nodes represent this, and alternative and surrogate splits can be used to form an idea of which variables are important at predicting the response and which are not. With the many trees of the bagged or random forest ensemble this is not easy to do by hand, but is something that the computer can easily do as it is performing the exhaustive search to identify splits. Two measures of variable importance are commonly used: (i) the total decrease in node impurity averaged over all trees and (ii) a measure of the mean decrease in the model's ability to predict the OOB samples before and after permuting the values of each predictor variable in turn (Prasad et al. 2006). Recall that node impurity can be measured using several different functions. In random forests, the first variable importance measure is computed by summing the total decrease in node impurity for each tree achieved by splitting on a variable and averaging by the number of trees. Variables that are important will be those that make the largest reductions in node impurity. The accuracy importance measure is generated by recording the prediction error for the OOB samples for each tree, and then repeating the exercise after randomly permuting the values of each predictor variable. The difference between the recorded prediction error and that achieved after permutation is averaged over the set of trees. Important variables are those that lead to a large increase in prediction error when randomly permuted. The mean decrease in node impurity measure tends to be the most useful of the two approaches because there is often a stronger demarcation between important and non-important variables compared with the decrease in accuracy measure, which tends to decline steadily from important to non-important predictors.

A novel feature of random forests is that the technique can produce a proximity matrix that records the dissimilarity between observations in the training-set. The dissimilarity between a pair of observations is based on the proportion of times the pair is found in the same terminal node over the set of trees in the model. Samples that are always found in the same terminal node will have zero dissimilarity and likewise those that are never found in the same node will have dissimilarity of 1. This matrix can be treated as any other dissimilarity matrix and ordinated using principal coordinate analysis (see Legendre and Birks 2012b: Chap. 8) or non-metric multidimensional scaling (see Legendre and Birks 2012b: Chap. 8) or clustered using hierarchical clustering or *K*-means partitioning (see Legendre and Birks 2012a: Chap. 7).

We continue the three-fuel SCP example by analysing the data using random forests. Five hundred trees were grown using the recommended settings for classification forests; minimum node size of five, $m = \lfloor\sqrt{21}\rfloor = 4$. Figure 9.6 shows the error rate for the OOB samples of the random-forest model as additional trees are added to the forest. The overall OOB error rate and that of each of the three fuel-types is shown. Error rates drop quickly as additional trees are added to the model, and stabilise after 100–200 trees have been grown. Table 9.5 shows the confusion matrix and error rates for the individual fuel-types for the random-forest model. The overall error rate is 6.6%. Figure 9.7 shows the variable importance measures for the overall model, with Ca and S, and, to a lesser extent, Si, having the largest decrease in node impurity as measured by the Gini coefficient. A similar result is indicated by the decrease in the accuracy measure, although it is more difficult to identify clear winners using this index. These same variables are also important for predicting the individual fuel-types, where Fe and Mg also appear as important indicators for the Oil and Oil-shale fuel-types (Fig. 9.8).

Random forests, whilst having recently been used in ecology as a method for broad-scale prediction of species presence/absence or ecological niche modelling (Iverson and Prasad 2001; Benito Garzón et al. 2006, 2008; Lawler et al. 2006; Rehfeldt et al. 2006; Cutler et al. 2007; Peters et al. 2007; Brunelle et al. 2008; Iverson et al. 2008; Williams et al. 2009; Chapman 2010; Chapman et al. 2010; Franklin 2010; Dobrowski et al. 2011; Vincenzi et al. 2011), have been little used in palaeoecology, which is surprising given the accuracy, simplicity, and speed of the method relative to other machine-learning techniques. Brunelle et al. (2008) use

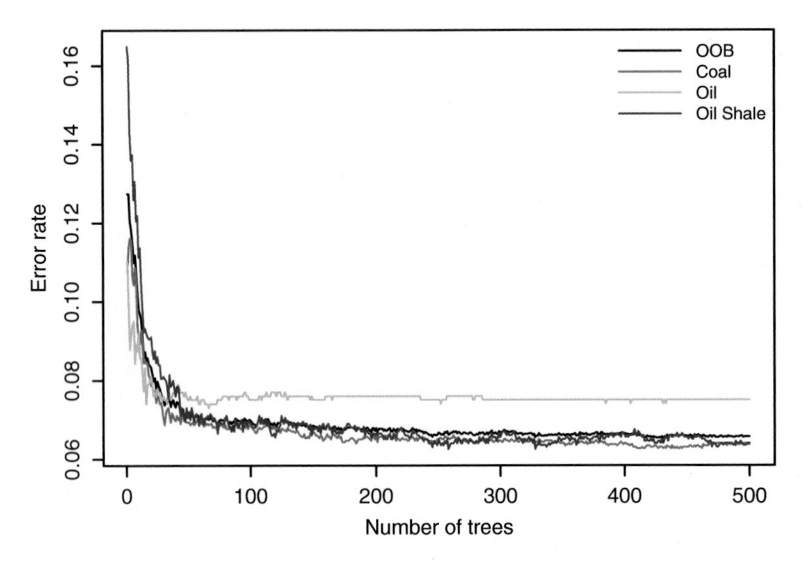

Fig. 9.6 Error rate for the classification random forest fitted to the three-fuel spheroidal carbonaceous particle (SCP) example data as trees are added to the ensemble. The *black line* is the overall error rate for the random forest model. The remaining lines are the error rates for the individual fuel types. The error rates are determined from the out-of-bag (OOB) samples for each tree

Table 9.5 Confusion matrix of predicted fuel type for the three-fuel random forest model (number of trees = 500)

	Coal	Oil	Oil-shale	Error rate
Coal	2809	8	183	0.064
Oil	56	925	19	0.075
Oil-shale	128	0	1872	0.064

The rows in the table are the predicted fuel types for the 6000 spheroidal carbonaceous particles (SCPs) based on the majority vote rule over the ensemble of trees. The columns are the known fuel-types. The individual fuel-type error rates of the random forest classifier are also shown. The overall error rate is 0.066

random forests to investigate the climatic variables associated with the presence, absence, or co-occurrence of lodgepole and whitebark pine in the Holocene, whilst Benito Garzón et al. (2007) employ random forests to predict tree species distribution on the Iberian Peninsula using climate data for the last glacial maximum and for the mid-Holocene. Other palaeoecological examples include Goring et al. (2010) and Roberts and Hamann (2012). Random forests are widely used in genomic and bioinformatical data-analysis (e.g., Cutler and Stevens 2006; van Dijk et al. 2008) and epidemiology (e.g., Furlanello et al. 2003).

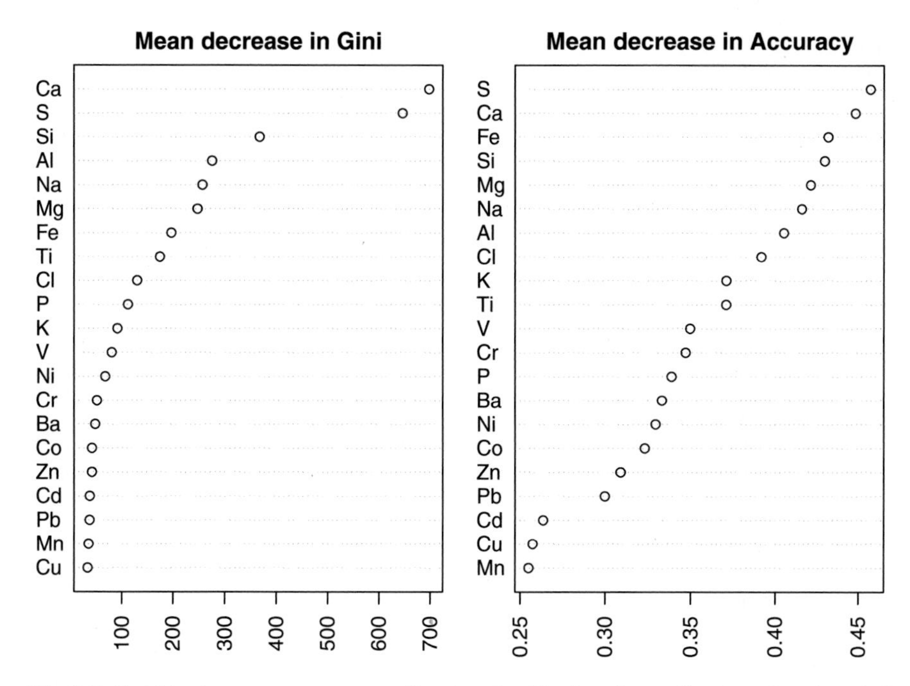

Fig. 9.7 Variable importance measures for the classification forest fitted to the three-fuel spheroidal carbonaceous particle (SCP) example data, showing the mean decrease in the Gini index when each variable is not included in the model (*left*) and the mean decrease in accuracy measure (*right*). See main text for details

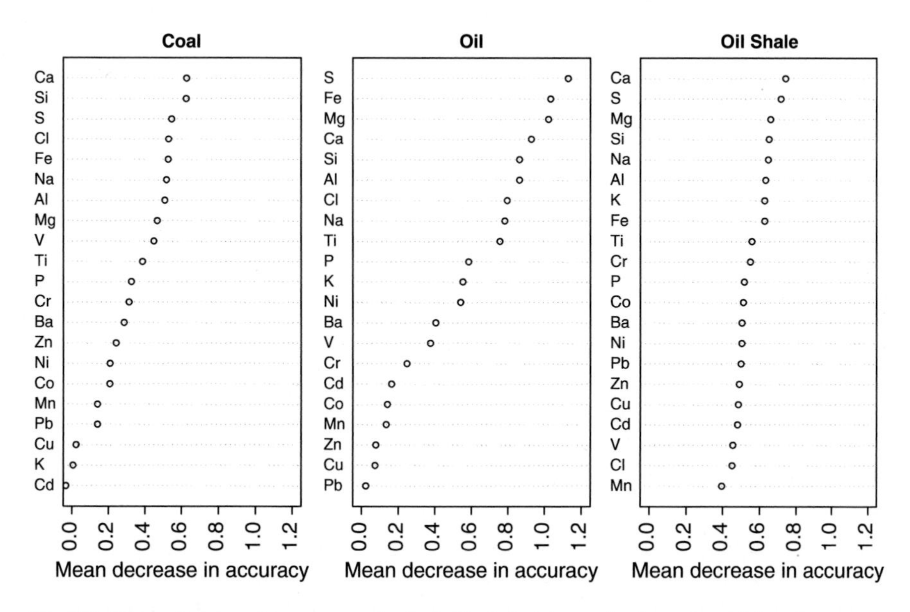

Fig. 9.8 Mean decrease in accuracy for individual fuel types in the spheroidal carbonaceous particle (SCP) example data, determined from the fitted classification forest

Boosting

In the discussion of bagging and random forests, we saw that modelling involves a trade-off between the bias and the variance of the fitted model. Bagging and random forests attempt to reduce the variance of a fitted model through the use of an ensemble of trees in place of the single best tree. These techniques do not reduce the bias of the fitted model. Boosting, a loosely related technique, uses an ensemble of models (in our case trees) to reduce both the bias *and* the variance of a fitted model. Boosting is an incredibly powerful technique that today relates to a whole family of approaches. Here we restrict our discussion to gradient boosting, which also goes by the name multiple additive regression trees (MART), and its variant stochastic gradient boosting. Hastie et al. (2011) contains a lengthy discussion of boosting and is essential reading for anyone attempting to use the technique for modelling data. Elith et al. (2008) is a user friendly, ecologically-related introduction to both the theory and practice of fitting boosting models (see also Witten and Frank 2005; De'ath 2007).

As with bagging and random forests, boosting begins from the realisation that it is easier to identify and average many rough predictors than it is to find one, all encompassing, accurate, single model. The key difference with boosting is that it is sequential; additional models are added to the ensemble with the explicit aim of trying to improve the fit to those observations that are poorly modelled by the previous trees already included in the model. With bagging and random forests each new tree is fitted to a bootstrap sample of the training data with no recourse to how well any of the previous trees did in fitting observations. As such, bagging and random forests do not improve the bias in the fitted model: they just attempt to reduce the variance. Boosting, in contrast, aims to reduce the bias in the fitted model by focussing on the observations in the training data that are difficult to model, or are poorly modelled, by the preceding set of trees. In the terminology of Hastie et al. (2011), boosting is a forward, stage-wise procedure.

Our discussion proceeds from the point of view of regression; this includes models for discrete responses such as logistic or multinomial regression thus encompassing classification models (Birks 2012a: Chap. 2). We have already mentioned loss functions, a function or measure, such as the deviance, that represents the loss in predictive power due to a sub-optimal model (Elith et al. 2008). Boosting is an iterative computational technique for minimising a loss function by adding a new tree to the model that at each stage in the iteration provides the largest reduction in loss. Such a technique is said to descend the gradient of the loss function, something known as functional gradient descent. For boosted regression trees, the algorithm starts by fitting a tree of a known size to the training data. This model, by definition, provides the largest reduction in the loss function. In subsequent iterations, a tree is fitted to the *residuals* of the previously fitted trees, which maximally reduces the loss function. As such, subsequent trees are fitted to the variation that remains unexplained after considering the previous set of trees. Each subsequent tree added to the ensemble has as its focus those poorly modelled observations that are not well fitted by the combination of previous trees, and as such can have quite different structures incorporating different variables and splits into the tree. Boosting is a stage-wise procedure because the preceding trees in the ensemble are not altered during the current iteration, which contrasts with step-wise procedures where the entire model is updated at each iteration (step-wise regression procedures, for example). Elith et al. (2008) summarise the boosted ensemble as a "linear combination of many trees... that can be thought of as a regression model where each term is a tree."

A further important aspect of boosting is the concept of regularisation. The logical conclusion of the boosting algorithm if no restriction on the learning rate was imposed is that the sequence of trees could be added until the training-set samples were perfectly explained and the model was hopelessly over-fitted to the data. In the standard regression setting, the number of terms in the model is often constrained by dropping out covariates (variables) or functions thereof, via a set of step-wise selection and elimination steps. A better, alternative approach is to fit a model with many terms and then down-weight the contributions of each term using shrinkage, as is done in ridge regression (Hoerl and Kennard 1970) or the lasso (Tibshirani 1996) (see below). With ridge regression or the lasso, the shrinkage that is applied is global, acting on the full model. In boosting, shrinkage is applied incrementally to each new tree as it is added to the ensemble and is controlled via the learning rate, lr, which, together with the number of trees in the ensemble, tr, and tree complexity, tc (the size of the individual trees), form the set of parameters optimised over by boosted trees.

Stochasticity was introduced into bagging and random forests through the use of bootstrap samples, where it introduces randomness that can improve the accuracy and speed of model fitting and help to reduce over-fitting (Friedman 2002) at the expense of increasing the variance of the fitted values. In boosting, stochasticity is introduced through randomly sampling a fraction, f, of the training samples at each iteration. This fraction is used to fit each tree. f lies between 0 and 1 and is usually set to 0.5 indicating that 50% of the training observations are randomly selected to fit each tree. In contrast to bagging and random forests, the sampling is done without replacement.

Recent work (Elith et al. 2008) on boosting has shown that it works best when learning is slow and the resulting model includes a large (>1,000) number of trees. This requires a low learning rate, say $lr = 0.001$. We still need a way of being alerted to over-fitting the model so as to guide how many trees should be retained in the ensemble. If using stochastic boosting, each tree has available a set of OOB samples with which we can evaluate the out-of-sample predictive performance for the set of trees up to and including the current tree. A plot of this predictive performance as new trees are added to the ensemble can be used to guide as to when to stop adding new trees to the ensemble. If stochastic boosting is not being used, other methods are required to guide selection of the number of trees. An independent test-set can also be employed, if available, in place of the OOB samples. Alternatively, k-fold cross-validation (CV) can be used if computational time and storage are not issues, and there is evidence that this procedure performs best for a wide range of test data-sets (Ridgeway 2007). In k-fold cross-validation, the training data are divided into k subsets of (approximately) equal size. A boosting model is fitted to the k-1 subsets and the subset left out is used as an independent test-set. A large boosting model is fitted and the prediction error for the left-out subset is recorded as the number of trees in the model increases. This process is repeated until each of the k subsets has been left out of the model-building process, and the average CV error is computed for a give number of trees. We take as the number of trees to retain in the model as that number of trees with lowest CV error.

Tree complexity, tc, is a tuning parameter in boosting; it affects the learning rate required to yield a large ensemble of trees, and also determines the types of interactions that can be fitted by the final model. Earlier, we saw how trees were able to account flexibly for interactions between predictor variables by allowing additional splits within the separate nodes of the tree, namely the interaction that only affects the predicted values for the set of samples in the node that is subsequently split by a further predictor. The more complex the individual trees in the boosted model are, the more quickly the model will learn to predict the response and hence will require fewer trees to be grown before over-fitting, for a fixed learning rate. The complexity of the individual trees should ideally be chosen to reflect the true interaction order in the training data. However, this is often unknown and selection via an independent test-set or optimisation-set will be required.

To illustrate the fitting of boosted regression trees we demonstrate their use in a calibration setting using the European Diatom Database Initiative (EDDI) combined pH-diatom training-set. The combined pH data-set contains diatom counts and associated lake-water pH measurements for 622 lakes throughout Europe with a pH gradient of 4.32–8.40. As an independent test-set, we applied a stratified random sampling strategy to select a set of 100 samples from across the entire pH gradient by breaking the pH gradient into ten sections of equal pH interval and subsequently choosing ten samples at random from within each section of the gradient. The remaining 522 samples formed the training-set to which a boosted regression-tree model is fitted using the gbm package (Ridgeway 2010) for the R statistical software. The squared error loss-function was employed during fitting and we explored various learning rates of 0.1, 0.01, 0.001, and 0.0001 and tree complexities of 2, 5, 10, and 20 to identify the best set of learning parameters to predict lake-water pH from the diatom percentage abundance data. Preliminary exploration suggested that a large number of trees was required before error rates stabilised at their minimum and that a modest degree of tree complexity is required to optimise model fit, so we fitted models containing 20,000 trees. Throughout, we assessed model fit using five-fold cross-validation on-line during model fitting.

Figure 9.9a shows the value of the loss-function as trees are added to the model for a variety of learning rates. A tree complexity of 10 was used to build the models. The two fastest learning rates (0.1 and 0.01) converge quickly to their respective minima and then slowly start to over-fit, as shown by the increasing CV squared error loss. Conversely, the model fitted using the smallest learning rate is slow to fit the features of the training data-set and has still to converge to a minimum squared error loss when 20,000 trees had been fitted. The best fitting of all the models shown

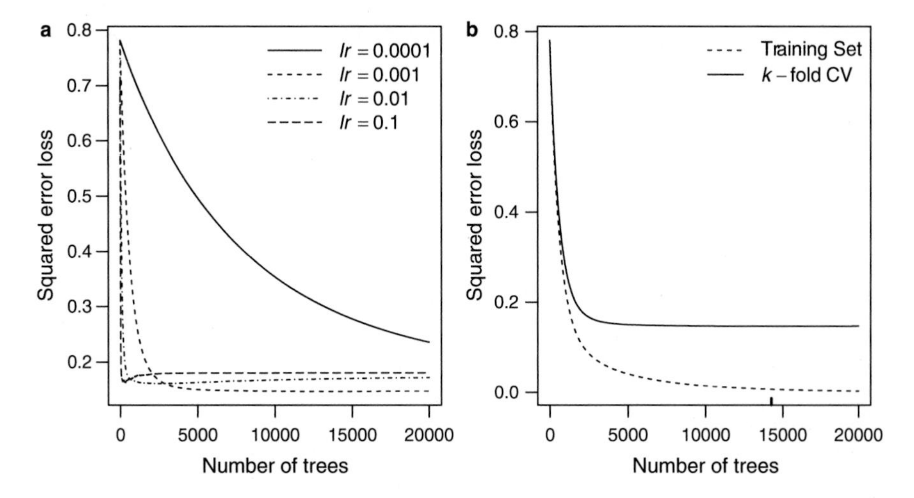

Fig. 9.9 Relationship between squared error loss, number of boosted trees, and learning rate (*lr*) for a boosted regression tree fitted to the European Diatom Database Initiative (EDDI) calibration set predicting lake-water pH from diatom species composition. (**a**) *k*-fold cross-validated and apparent squared error loss for the tuned boosted regression tree fitted to the EDDI data (**b**). The apparent squared error loss is derived using the training data to test the model and continues to decline as additional trees are added to the ensemble, indicating over-fitting. The thick tick mark on the x-axis of panel (**b**) is drawn at the optimal number of trees (14,255)

is the one with a learning rate of 0.001, which reaches a minimum squared error loss after approximately 14,000 trees. Figure 9.9b shows the CV squared error loss for this model alongside the training-set based estimate or error. We can clearly see that the boosted-tree model over-fits the training data converging towards an error of 0 given sufficient trees. This illustrates the need to evaluate model fit using a cross-validation technique, such as *k*-fold CV, or via a hold-out test-set that has not taken part in any of the model building.

The learning rate is only one of the parameters of a boosted regression tree for which optimal values must be sought. Tree complexity, *tc*, controls the size of the individual trees: the more complex the trees, the higher the degree of flexible interactions that can be represented in the model. Models that employ more complex trees also learn more quickly than models using simpler trees. This is illustrated in Fig. 9.10, which shows the effect of tree complexity on the squared error loss as trees are fitted for several values of complexity and for two learning rates ($lr = 0.001$ and 0.0001). The effect of tree complexity on the speed of learning is easier to see in the plot for the slowest learning rate (right hand panel of Fig. 9.10). The simplest trees, using tree complexities of 2 and 5, respectively, converge relatively slowly compared to the boosted trees using trees of complexity 10 or 20. Of the latter two, there is little to choose between the loss functions once tree complexity reaches a value of 10. Figure 9.10 combines the three parameters that users of boosted trees need to set to control the fitting of the model, and illustrates the key feature of

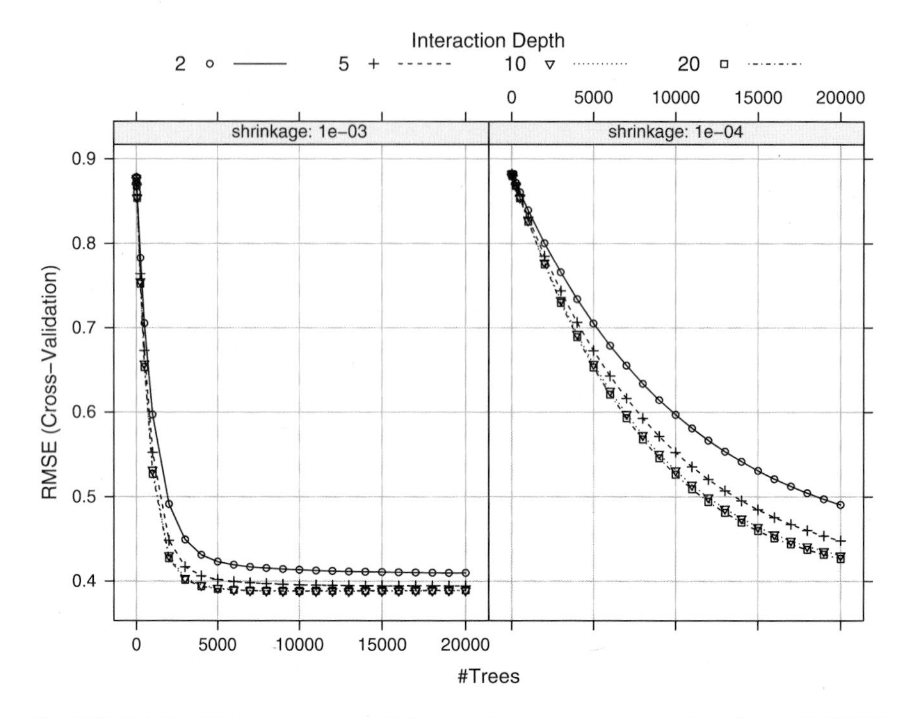

Fig. 9.10 Relationship between cross-validated root mean squared error of prediction (RMSEP) and number of boosted trees for a range of tree interaction depths and two learning rates ($lr = 0.001$, *left*; $lr = 0.0001$, *right*), for the European Diatom Database Initiative (EDDI) diatom-pH boosted regression tree model

requiring a sufficiently slow learning rate to allow averaging over a large number of trees, whilst using trees of sufficient complexity to capture the degree of interaction between predictors in the training data.

We can assess the quality of the boosted-tree calibration model by using the best fitting model ($lr = 0.001$, $tc = 10$, $nt = 13{,}000$). This model was chosen as the one giving the lowest five-fold CV error over a grid of tuning parameters. The RMSEP of the boosted tree for the test-set is 0.463 pH units. On the basis of Fig. 9.9a, one might consider using the model with $lr = 0.01$, $tc = 10$, and $nt = 2500$ instead of the best model as it has a similar, though slightly higher, squared error loss than the best model identified. Using this model gives a RMSEP for the test-set samples of 0.533, which is substantially higher than the best model. For comparison, we fitted weighted averaging (WA) calibration models (Juggins and Birks 2012: Chap. 14) to the EDDI training data using inverse and classical deshrinking and then applied each of these models to the held-out test-set. RMSEP for the WA models was 0.467 and 0.439 using inverse and classical deshrinking, respectively. There is little to choose between these models, with WA with classical deshrinking having the lowest hold-out sample RMSEP. It is always surprising how well the simple heuristic WA performs on such a complex problem of predicting lake-water pH from hundreds of diatom species. In this example, one of the state-of-the-art machine-learning methods is unable to beat WA in a real-world problem!

Weighted averaging, whilst being very simplistic and powerful, is not a very transparent modelling technique as we do not have any useful variable importance measures that we can use to interrogate a WA calibration model. Bagged trees and random forests employ various variable importance measures to indicate to the user which predictors are important in modelling the response. In boosted trees, Friedman (2001) proposed to use the relative improvement in the model by splitting on a particular variable, as used in single tree models, as a variable importance measure in a boosted tree model but to average this relative importance over all trees produced by the boosting procedure. Figure 9.11 shows a needle plot of the 20 most important predictor variables (diatom species) for the boosted pH calibration model fitted to the EDDI data-set. The most important taxon is *Eunotia incisa* (EU047A), an acid-tolerant diatom, whilst *Achnanthes minutissima* agg. (AC048A) is a diatom that tends to be found in circum-neutral waters. The suite of taxa shown in Fig. 9.11 are often identified as indicator species for waters of different pH, so it is encouraging that the boosted model has identified these taxa as the most important in predicting lake-water pH (see Legendre and Birks 2012a: Chap. 7). Ecological examples of boosted regression trees are given by Elith et al. (2008), De'ath and Fabricius (2010), and Dobrowski et al. (2011).

Multivariate Adaptive Regression Splines

Multivariate adaptive regression splines (MARS) (Friedman 1991; Friedman and Meulman 2003; Leathwick et al. 2005) are an attempt to overcome two perceived problems of the single regression tree. The hierarchical nature of the tree imposes a severe restriction on the types of model that can be handled by such models. A change made early on in growing the tree is very difficult to undo with later splits, even if it would make sense to change the earlier split criteria in light of subsequent splits. Furthermore, as regression trees (as described above) fit piece-wise constant models in the leaves of the tree, they have difficulties fitting smooth functions; instead, the response is approximated via a combination of step functions determined by the split criteria. MARS does away with the hierarchical nature of the tree and uses piece-wise linear basis functions, combined in an elegant and flexible manner, to approximate smooth relationships between the responses and the predictors.

MARS proceeds by forming sets of reflected pairs of simple, piece-wise linear basis functions. These functions are defined by a single knot location, and take the value 0 on one side of the knot, and a linear function on the opposite side. Each such basis function has a reflected partner, where the 0-value region and the linear-function region are reversed. Figure 9.12 shows an example of a reflected pair of basis functions for variable or covariate x, with a single knot located at $t = 0.5$.

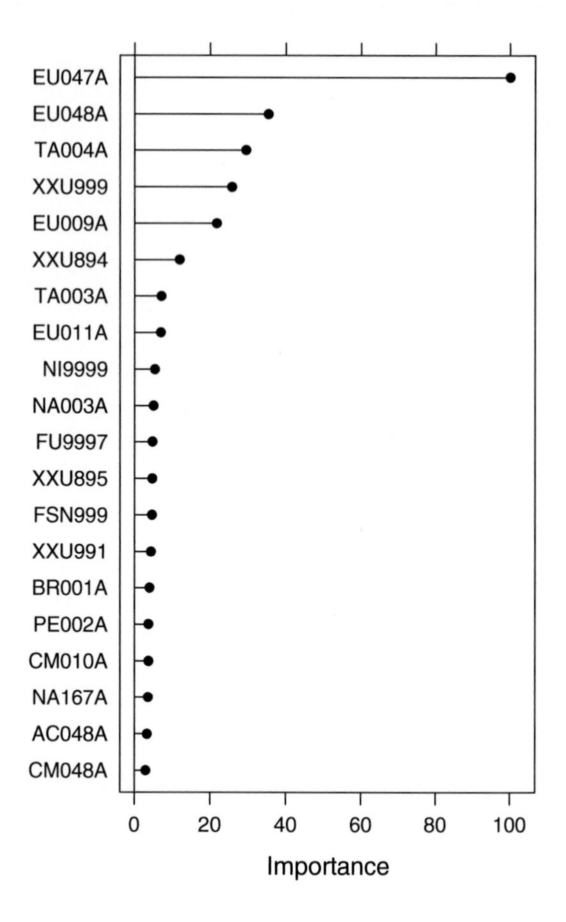

Fig. 9.11 Relative variable importance measure for the 20 most important diatom taxa in the European Diatom Database Initiative (EDDI) diatom training-set

The solid line is denoted $(x - t)_+$, with the subscript + indicating that we take the positive part of the function only, with the negative part set to 0. As a result, the basis function illustrated by the solid line in Fig. 9.12 is zero until the knot location ($x = 0.5$) is exceeded. The reflected partner has the form $(t - x)_+$, and is illustrated by the dashed line in Fig. 9.12. For each quantitative covariate in the training data, a reflected pair of basis functions is formed by setting each t to be a unique value taken by that covariate. Qualitative covariates are handled by forming all possible binary partitions of the levels of a categorical covariate to form two groups. A pair of piece-wise constant functions are formed for each binary partition, which act as indicator functions for the two groups formed by the binary partition, and are treated like any other reflected pair of basis functions during fitting.

Model building in MARS is similar to step-wise linear regression except the entire set of basis functions is used as input variables and not the covariates themselves. The algorithm begins with a constant term, the intercept, and performs an exhaustive search over the set of basis functions to identify the pair that minimises the model residual sum-of-squares. That pair and their least-squares coefficients are added to the model-set of basis functions and the algorithm continues. Technically the

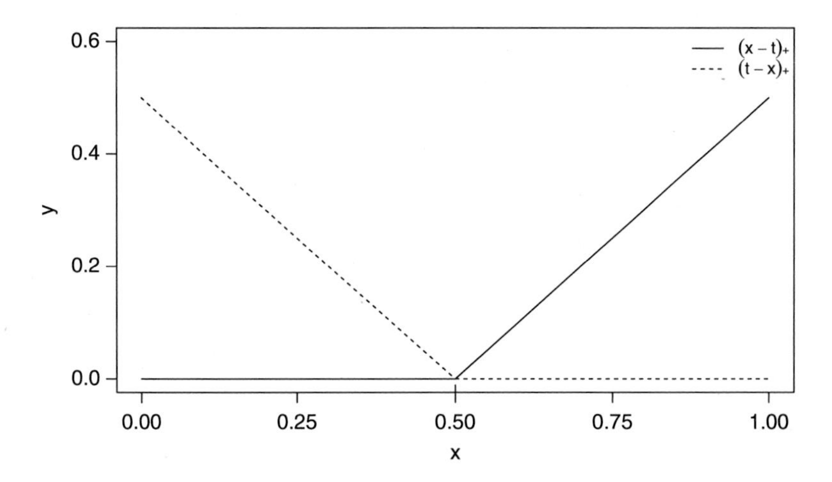

Fig. 9.12 Examples of a reflected pair of basis functions used in multivariate adaptive regression splines. The basis functions are shown for the interval $(0,1)$ with a knot located at $t = 0.5$. See main text for details

algorithm finds the pair that, when multiplied by a term *already* included in the model, results in the lowest residual sum-of-squares, but as the only term in the model at the first iteration is the constant term, this amounts to finding the pair of basis functions that affords the largest improvement in fit. At the second and subsequent steps of the algorithm, the products of each existing model-term with the set of paired basis functions are considered and the basis function that results in the largest decrease in residual sum-of-squares is added to the model along with its partner basis function and their least-squares coefficients. The process continues until some stopping criteria are met; for example, the improvement in residual sum-of-squares falls below a threshold or a pre-specified number of model terms is reached. An additional constraint is that a single basis function pair may only be involved in a single product term in the model. Because products of basis functions are considered, interactions between covariates are handled naturally by the MARS model. The degree of interactions allowed is controlled by the user; if the degree is set to 1, an additive model in the basis functions is fitted.

At the end of this forward stage of model building a large model of basis functions has been produced that will tend to strongly over-fit the training data. A backwards elimination process is used to remove sequentially from the model the term that causes the smallest increase in the residual sum-of-squares. At each stage of the forward model-building phase, we added a basis function *and* its partner to the model during each iteration. The backwards elimination stage will tend to remove one of the pair of basis functions unless both contribute substantially to the model-fit (Hastie et al. 2011). A generalised cross-validation (GCV) procedure is used to determine the optimal model-size as ordinary cross-validation is too computationally expensive to apply to MARS for model-building purposes. The size of a MARS model is not simply the number of model terms included within it; a penalty must be paid for selecting the knots for each term. The effective degrees of freedom (EDF) used by a MARS model is given by $\mathrm{EDF} = \lambda + c((\lambda - 1)/2)$, where λ is the number of terms in the model, and c is a penalty term on the number of knots $((\lambda - 1)/2)$ and is usually equal to 3, or 2 in the case of an additive MARS model where no interactions are allowed. The EDF term is part of the GCV criterion that is minimised during the backwards elimination phase.

MARS was originally derived using least squares to estimate the coefficients for each basis function included in the model. The technique is not restricted to fitting models via least squares, however. The scope of MARS can be expanded by estimating the basis function coefficients via a generalised linear model (GLM), which allows the error distribution to be one of the exponential family of distributions (see Birks 2012a: Chap. 2).

We illustrate MARS via a data-set of ozone measurements from the Los Angeles Basin, USA, collected in 1976. A number of predictor variables are available; *inter alia*, air temperature, humidity, wind speed, and inversion base height and temperature. The aim is to predict the ozone concentration as a function of the available predictor variables. The variance in ozone concentrations increases with the mean concentration and as negative concentrations are not possible, a sensible fitting procedure for MARS is to estimate the coefficients of the model terms via a gamma GLM and the inverse link function (Birks 2012a: Chap. 2). Only first-order interaction terms were considered during fitting. The MARS model was fitted using the R package earth (Milbarrow 2011). A MARS model comprising ten terms, including the intercept and seven predictor variables, was selected using the GCV procedure. Four model terms involve the main effects of air temperature (two terms), pressure gradient[1] (DPG), and visibility. The remaining terms involve interactions between variables. A summary of the model terms and the estimated coefficients is shown in Table 9.6, whilst Fig. 9.13 displays the model terms graphically. The upper row of Fig. 9.13 shows the main effect terms. A single knot location was selected for air temperature at 58°F, with terms in the model for observations above and below this knot. Both air-temperature terms have different coefficients as illustrated by the differences in slopes of the fitted piece-wise functions. Note that the terms are non-linear on the scale of the response due to fitting the model via a gamma GLM.

Variable importance measures are also available to aid in interpreting the model fit, and are shown in Fig. 9.14 for the ozone example. The 'number of subsets' measurement indicates how many models, during the backward elimination stage, included the indicated term. The residual sum-of-squares (RSS) measure indicates the reduction in RSS when a term is included in one of the model subsets considered. The decrease in RSS is summed over all the model subsets in which a term is involved and is expressed relative to the largest summed decrease in RSS (which is notionally given the value 100) to aid interpretation. The GCV measure is

[1]Pressure gradient between Los Angeles airport (LAX) and Daggert in mm Hg.

Table 9.6 MARS model terms and their coefficients

Term	$\hat{\beta}$
Intercept	0.0802293
h(temp-58)	−0.0007115
h(58-temp)	0.0058480
h(2-dpg)	0.0018528
h(200-vis)	−0.0002000
h(wind-7) × h(1069-ibh)	0.0000681
h(55-humidity) × h(temp-58)	0.0000196
h(humidity-44) × h(ibh-1069)	0.0000005
h(temp-58) × h(dpg-54)	0.0000435
h(258-ibt) × h(200-vis)	0.0000010

The h() terms refer to basis functions, the numeric value inside the parentheses is the knot location for the piece-wise linear function, and the name inside the parentheses is the variable associated with the basis function *temp* Air Temperature (°F), *dpg* pressure gradient (mm Hg) from LAX airport to Daggert, *vis* visibility in miles, *wind* wind speed in MPH, *ibh* temperature inversion base height (feet), *humidity* percent humidity at LAX airport, *ibt* inversion base temperature (°F)

computed in the same manner as the RSS measure but involves summing the GCV criterion instead of RSS. A variable might increase the GCV score during fitting, indicating that it makes the model worse. As such, it is possible for the GCV importance measure to become negative. For the ozone model, air temperature is clearly the most influential variable, with the remaining variables included in the model all being of similar importance. The model explains approximately 79% of the variance in the response (76% by the comparable GCV measure). Ecological and biogeographical applications of MARS are relatively few and include Moisen and Frescino (2002), Leathwick et al. (2005, 2006), Prasad et al. (2006), Elith and Leathwick (2007), Balshi et al. (2009), and Franklin (2010).

Artificial Neural Networks and Self-organising Maps

Artificial neural networks (ANNs) and self-organising maps (SOMs) were developed for applications in artificial-intelligence research and are often conflated into a general machine-learning technique that is based on the way biological nervous systems process information or generate self-organising behaviour. However, despite these similarities, ANN and SOM are best considered from very different vantage points. There are also a large number of variations that fall under the ANN or SOM banner – too many to consider here. Instead we focus on the techniques most frequently used in ecological and limnological research (Lek and Guégan 1999).

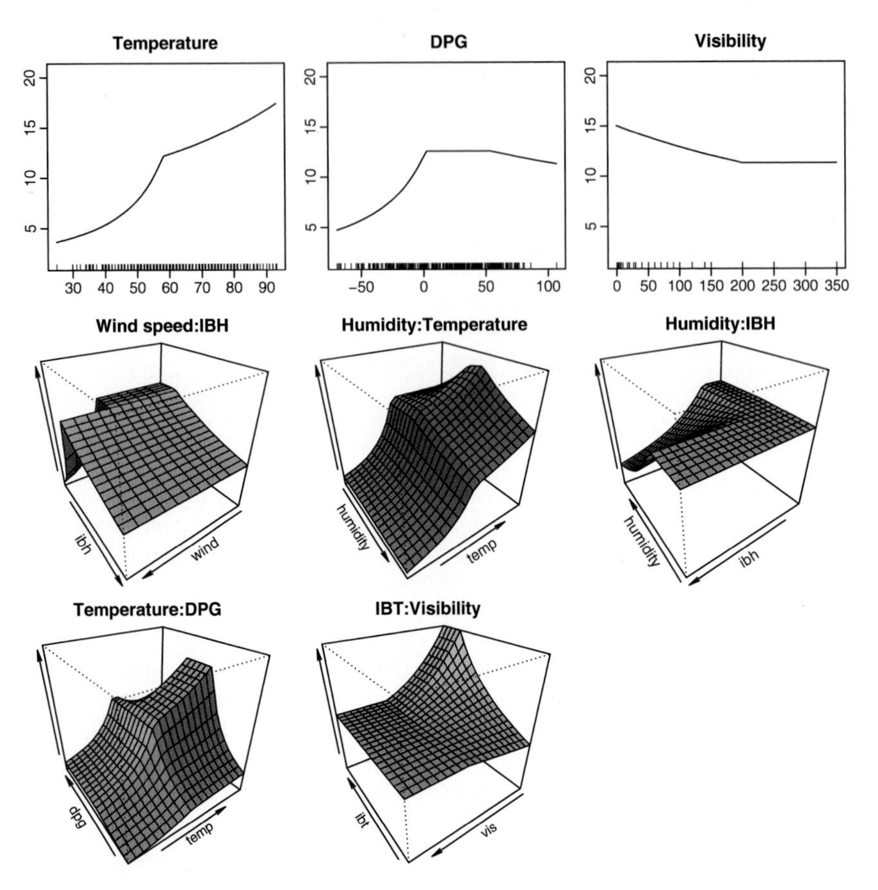

Fig. 9.13 Partial response plots for the multivariate adaptive regression spline (MARS) model fitted to the ozone concentration data from the Los Angeles basin

Artificial Neural Networks

An artificial neural network is a particularly flexible, non-linear modelling technique that is based on the way neurons in human brains are thought to be organised (Chatfield 1993; Warner and Misra 1996; Witten and Frank 2005; Ripley 2008). The term ANN today encompasses a large number of different yet related modelling techniques (Haykin 1999). The most widely used form of ANN is the forward-feed neural network, which is sometimes known as a back-propagation network or multi-layer perceptron. The general form of a forward-feed ANN is shown in Fig. 9.15. Configurations for both regression and classification settings are shown. The main feature of a forward-feed ANN is the arrangement of 'neurons' or units into a series of layers. The input layer contains m units, one per predictor variable in the training data-set, whilst the output layer contains units for the response variable or

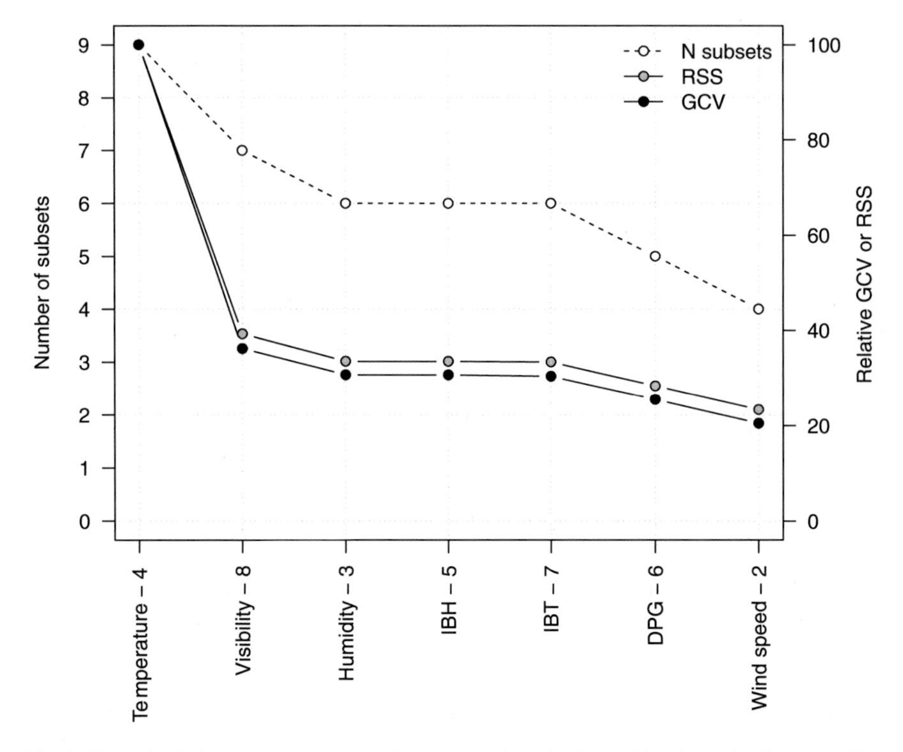

Fig. 9.14 Variable importance measures for the covariates in the multivariate adaptive regression spline (MARS) model fitted to the ozone concentration data from the Los Angeles basin. See main text for details of the various measures. *RSS* residual sum-of-squares, *GCV* generalised cross-validation

variables. In the univariate regression setting, there is a single unit in the output layer (Fig. 9.15a). In a classification setting, where the response takes one of k possible classes, there are k units in the output layer, one per class. The predicted class in a classification ANN is the largest value taken by $\mathbf{Y}_k$ for each input. Between the input and output layers a hidden layer of one or more units is positioned. Units in the input layer each have a connection to each unit in the hidden layer, which in turn have a connection to every unit in the output layer. The number of units in the hidden layer is a tuning parameter to be determined by the user. Additional hidden layers may be accommodated in the model, though these do not necessarily improve model fit. In addition, bias units may be connected to each unit in the hidden and output layers, and play the same role as the constant term in regression analysis.

Each unit in the network receives an input signal, which in the case of the input layer is an observation on m predictor variables, and outputs a transformation of the input signal. Where a unit receives multiple inputs, a transformed sum of these inputs is outputted. Bias units output a value of $+1$. The connections between units are represented as lines in Fig. 9.15 and each is associated with a weight. The output

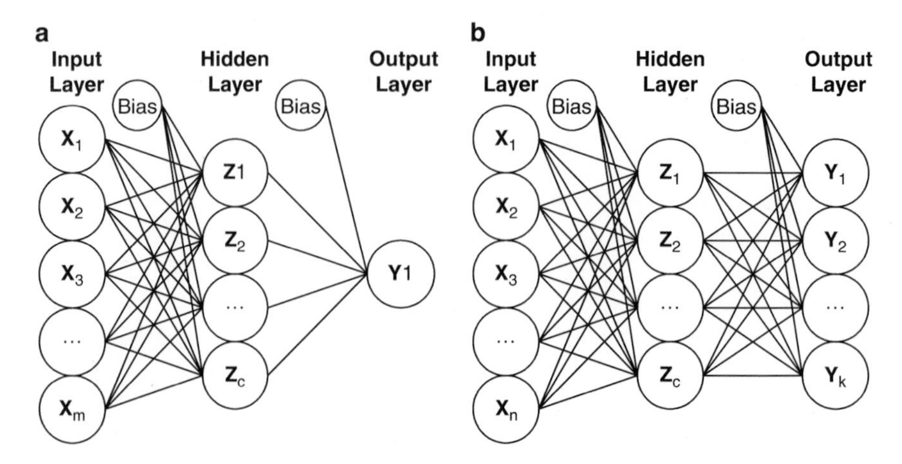

Fig. 9.15 Structure of a forward-feed, back-propagation neural network in a regression (**a**) and a classification (**b**) setting. A single hidden layer (Z_c) is shown. The lines connecting the layers of the network carry weights that are estimated from the data during fitting to minimise the loss of the final model. It is clear that the response is modelled as a function of a series of linear combinations (Z_c) of the input data

signal from an individual unit is multiplied by the connection weight and passed on to the next layer in the network along the connection. The weights are the model coefficients and optimal values for these are sought that best fit the response data provided to the network during training.

We said that the inputs to a unit are transformed. The identity function (Birks 2012a: Chap. 2) is generally used for the input layer, as a result the input data for the ith sample are passed on to the hidden layer units untransformed. The hidden layer generally employs a non-linear transformation, typically a sigmoid function of the form

$$\sigma(sv) = 1/(1 + e^{-v}) \tag{9.5}$$

where v is the sum of the inputs to the unit and s is a parameter that controls the activation rate. Figure 9.16 shows the sigmoid function for various activation rates. As s becomes large, the function takes the form of a hard activation or threshold once a particular value of the inputs is reached. The origin can be shifted from 0 to ν_0 by replacing the terms in the parentheses on the left hand side of Eq. 9.5 with $s(\nu - \nu_0)$. If an identity function is used in place of the sigmoid, the entire model becomes a simple linear regression. For the output layer, an identity function is used for regression models, whilst the softmax function, which produces positive outputs that sum to one, is used for classification.

The connection weights are estimated using gradient descent, known as back-propagation in the ANN field. For regression ANNs, sum-of-squares error is used to estimate the lack-of-fit for the current set of weights, whilst cross-entropy is

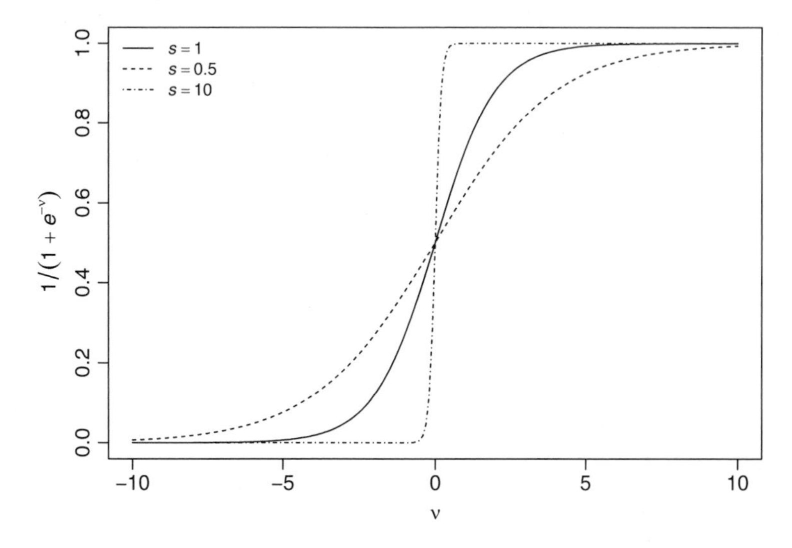

Fig. 9.16 Sigmoid function used to non-linearly transform inputs to hidden layer units in an artificial neural network (ANN) shown using a variety of activation rates s. See main text for further details

generally used in classification. The weights are sequentially updated to improve the fit of the model and each pass over the data is termed a training epoch. Generally, a large number of training epochs is performed to optimise the weights and thus improve the accuracy of the model. The set of weights that provides the global minimum of the model error is likely over-fitted to the training data. To alleviate over-fitting, training is often stopped early, before the global minimum is reached (Hastie et al. 2011). A validation data-set is useful in determining the appropriate stopping point, where the prediction error for the validation samples begins to increase. An alternative procedure, called weight-decay, provides a more explicit regularisation of the model weights, and is analogous to that used in ridge regression (see below). Details of the weight-decay procedure are given in Sect. 11.5.2 of Hastie et al. (2011).

It is instructive to consider what the units in the hidden layer represent; they are linear combinations of the input variables with the loading (or weighting) of each input variable in each unit $\mathbf{Z}_c$ given by the connection weight of the relevant unit in the input layer. We can then think of the forward-feed ANN as a general linear model in the linear combinations $\mathbf{Z}_c$ of the inputs (Hastie et al. 2011). A key feature of the forward-feed ANN is that the connection weights that define the linear combinations $\mathbf{Z}_c$ are learnt from the data during training. In other words, a set of optimal linear combinations of the inputs are sought to best predict the response.

ANNs are often considered black-box prediction tools (Olden and Jackson 2002) owing to how ANNs learn patterns from the data and encode this information in the connection weights, which makes it more difficult to extract and interpret than more simple, parametric techniques. To some extent this is a valid criticism;

however the connection weights are available for inspection along with the linear combinations of the inputs reconstructed ($\mathbf{Z}_c$) from these. Several methods for inspecting ANN model structure have been proposed, including the connection weighting approach of Olden et al. (2004) to derive a measure of variable importance, sensitivity analyses (Lek et al. 1996a), and various pruning algorithms (Bishop 1995, 2007; Despagne and Massart 1998; Gevrey et al. 2003). An example of a pruning algorithm applied in a palaeoecological context is the skeletonisation procedure of Racca et al. (2003), which for the Surface Waters Acidification Programme (SWAP) diatom-pH training-set allowed the removal of 85% of the taxa from the training data without drastically affecting model performance. This pruning also improved the robustness of the resulting calibration (Racca et al. 2003) (see Juggins and Birks 2012: Chap. 14).

Several factors can affect optimisation in ANNs which ultimately can determine the quality of the resulting model. We have already mentioned the potential for over-fitting the training data. In addition, the number of hidden layers and units within those layers needs to be decided. In general a single hidden layer will be sufficient, but additional layers can speed up model fitting. The number of units in the hidden layer controls the flexibility of functions of the input data that can be described by the model. Too many hidden units and the model may over-fit the data quickly, whilst too few units will unnecessarily restrict the very flexibility that ANNs afford. The optimal number of units in the hidden layer can be determined analytically (Bishop 1995, 2007; Ripley 2008) but in practice, treating the number of units as a tuning parameter to be optimised using *k*-fold cross-validation is generally used. Özesmi et al. (2006) reviewed other aspects of ANN assessment.

ANNs, when compared to the majority of the machine-learning tools described in this chapter, have been used relatively frequently to model palaeoecological data, particularly as a means of implementing calibration models (Borggaard and Thodberg 1992; Næs et al. 1993; Wehrens 2011). At one time ANNs were becoming a popular means of producing palaeoenvironmental reconstructions as they were seen as highly competitive when compared to modern analogue technique (MAT), weighted averaging (WA), and weighted-averaging partial least squares (WAPLS) because the calibration functions produced using ANNs had comparatively low root mean squared errors of prediction (RMSEP). Malmgren and Nordlund (1997) compared ANNs with Imbrie and Kipp factor analysis (IKFA), MAT, and soft independent modelling of class analogy (SIMCA) on a data-set of planktonic foraminifera and achieved substantially lower RMSEP than the other techniques. Racca et al. (2001) compared ANN, WA, and WAPLS calibration models for a data-set of diatom counts from 76 lakes in the Quebec region of Ontario, Canada. In this study, ANNs gave modest improvements in RMSEP over WA and WAPLS. Other palaeoecological applications of ANNs include Peyron et al. (1998, 2000, 2005), Tarasov et al. (1999a, b), Malmgren et al. (2001), Grieger (2002), Nyberg et al. (2002), Racca et al. (2004), Barrows and Juggins (2005), and Kucera et al. (2005). Limnological, environmental, biogeographical, and ecological examples are numerous, as reviewed by Lek and Guégan (2000). Illustrative examples include Lek et al. (1996a, b), Recknagel et al. (1997), Guégan et al. (1998), Lindstrom et al.

(1998), Brosse et al. (1999), Manel et al. (1999a, b), Spitz and Lek (1999), Olden (2000), Belgrano et al. (2001), Cairns (2001), Černá and Chytrý (2005), Steiner et al. (2008), and Chapman and Purse (2011).

The popularity of ANNs in palaeoecology has waned recently following the discovery that many published ANN-derived calibration functions may have greatly under-estimated model RMSEP by failing to account for spatial autocorrelation in the training data (Birks 2012a: Chap. 2). The autocorrelation problem can be accounted for using appropriate cross-validation techniques, such as the h-block approach of Burman et al. (1994) as used by Telford and Birks (2009). Typically, when one accounts for the dependence structure in the input data, the performance of ANNs is comparable to or worse than the best fits produced using WA and WAPLS.

Self-organising Maps

The self-organising map (SOM; also known as a self-organising feature map) is a relatively popular machine-learning tool for mapping and clustering high-dimensional data (Wehrens 2011), which has been used in a wide variety of ecological, environmental, and biological contexts (see e.g., Chon 2011 for a recent overview, and Giraudel and Lek 2001 for a comparison of SOMs and standard ordination techniques used in palaeoecology). The SOM is superficially similar to an artificial neural network, but this analogy only gets one so far and it is simpler to consider SOMs as a constrained version of the K-means clustering or partitioning method (Legendre and Birks 2012a: Chap. 7). As we will see, SOMs are also similar to principal curves and surfaces (see below and Hastie et al. 2011) and can be likened to a non-linear form of principal component analysis (PCA).

In a SOM, p prototypes are arranged in a rectangular or hexagonal grid of units of pre-defined dimension (number of rows and columns). The number of prototypes, p, is usually small relative to the dimensionality (number of variables or species) of the input data. A prototype is assigned to each grid unit. The SOM algorithm forces each of the samples in the input data to map onto one of the grid units during an iterative learning process. The goal of the SOM is to preserve the similarities between samples such that similar samples map on to the same or neighbouring units in the grid, whilst dissimilar samples are mapped on to non-neighbouring units.

At the start of the algorithm, the p prototypes are initialised via a random sample of p observations from the input data. Alternatively, the first two principal components of the input data can be taken and a regular grid of points on the principal component plane used as the prototypes (Hastie et al. 2011). Regardless of how the prototypes are initialised, each is characterised by a codebook vector that describes the typical pattern for the unit to which it has been assigned. If the prototypes are initialised using a random sample from the input data, then the codebook vector for an individual prototype will be the values of the species abundances, for example, in the sample assigned to that prototype. The aim of the SOM algorithm is to update these codebook vectors so that the input data are best described by the small number of prototypes.

During training, samples from the input data are presented to the grid of units in random order. The distance between the species abundances in the presented sample and the codebook vectors for each of the units is determined, usually via the Euclidean distance, but other distance measures can be used. The unit whose codebook vector is closest, i.e., most similar, to the presented sample is identified as the winning unit. The winning unit is then made more similar to the presented sample by updating its codebook vector. Geometrically, we can visualise this update as moving the unit in the m-dimensional space towards the location of the presented sample. By how much the codebook vector of the winning unit is updated (moved towards the presented sample) is governed by the learning rate, α, which is typically a small value of the order of 0.05. The learning rate is gradually decreased to 0 during learning to allow the SOM to converge.

Earlier, we noted that the SOM can be considered a constrained form of K-means clustering or partitioning: the constraint is spatial and arises because neighbouring units in the grid are required to have similar codebook vectors. To achieve this, not only is the winning unit updated to become more similar to the presented sample, but those units that neighbour the winning unit are also updated in the same way. Which units are considered neighbours of the winning unit is determined via another tuning parameter, r, which can be thought of as the distance within which a grid unit is said to be a neighbour of the winning unit. This distance, r, is topological, i.e., it is the distance between units on the grid, not the distance between the units in the m-dimensional space defined by the input data. The value of r, and hence the size of the neighbourhood around the winning unit, is also decreased during training; the implication is that as learning progresses, eventually only the winning units are updated. The SOM algorithm proceeds until an *a priori*-defined number of learning iterations, known as epochs, has been performed. The standard reference work for SOM is Kohonen (2001) where further details of the learning algorithm can be found.

As described above, SOM is an unsupervised technique, learning features of the data from the data themselves. However, the simplicity of the SOM algorithm allows scope for significant adaptation. One such adaptation allows SOMs to be used in a supervised fashion. If additional, dependent variables are available then these can be modelled alongside the independent or predictor variables. Such a supervised SOM then allows for predictions of the dependent variable to be made at new values of the predictor variables. One simple means of achieving this is to take an indirect approach and fit the SOM without regard to the dependent (response) variable(s) of interest and then take as the predicted value for each sample in the input the mean of the values of the response for all the samples that end up in the same grid unit as the sample of interest. This approach is very much in the spirit of the indirect ordination approach (Legendre and Birks 2012b: Chap. 8), but cannot be considered truly supervised.

Kohonen (2001) considered a supervised form of SOM and suggested building the map on the concatenation of the response variables (**Y**) and the predictor variables (**X**). In practice however, it may be difficult to find a scaling of **X** and **Y** such that both contribute similarly in the analysis. Indeed, if one of **X** or **Y** contains many more variables than the other, it will dominate the distance computations when identifying the winning unit. Melssen et al. (2006) introduce two variations of supervised SOMs that have wide applicability as general techniques for analysing palaeoenvironmental data: (i) the X-Y Fused Kohonen Network (XYF) and (ii) the Bi-directional Kohonen Network (BDK). Both approaches make use of two grids of prototypes, the first providing a mapping of **X**, the second a mapping of **Y**, into low dimensions. The networks are trained in the same manner as described for the unsupervised SOM, but differ in how the two mappings are combined to identify the winning unit during each learning epoch.

XYF networks operate on a fused distance, where the total distance between each observation and the prototypes is a weighted sum of the scaled distance between each observation and the prototypes on the individual maps. The winning unit is the one that has the lowest weighted sum distance to the observation. The relative weighting is given by α, taking values between 0 and 1, with the distances on the **X** map weighted by $\alpha(t)$ and the distances on the **Y** map weighted by $1 - \alpha(t)$. The distances between observations and prototypes on the individual maps are normalised by the maximum distance on each map so that the maximal distance on each map is 1. This scaling allows for very different magnitudes of distances on the maps, such as might arise when computing distances where **X** and **Y** are measured in different units or where different dissimilarity coefficients are used for the different maps. This latter point is particularly useful when applying the supervised SOM in a classification setting where the distance used for the response **Y** should consider group membership (0, 1). In such cases, the Jaccard distance (Legendre and Birks 2012b: Chap. 8; often called the Tanimoto distance in the chemometrics literature where the XYF and BDK methods were developed) is generally used. The t in $\alpha(t)$ indexes the learning epoch, allowing α to be decreased linearly during learning. Initially, this results in the determination of the winning unit being dominated by distances to prototypes on the **X** map. As learning proceeds, α is slowly decreased such that at the end of learning, distances to prototypes on both the **X** and **Y** maps contribute equally. It should be noted that a single epoch entails presenting, at random, each observation in the training-set to the maps.

The BDK network is similar to that described for the XYF network, but differs in that the two maps are considered individually during separate passes over the data. First, in the forward pass, the winning unit on the **X** map is identified as a weighted sum of distances on the two maps, as described above, and updated in the usual SOM manner. A reverse pass over the data is then performed, where the winning unit in the **Y** map is determined, again via a weighted sum of distances on the two maps, but this time using the **X** map updated in the forward pass. Learning proceeds in this alternating manner until convergence or an upper limit of epochs is reached. In practice there is generally little difference between the networks learned via the XFY or BDK methods (Melssen et al. 2006).

The XYF supervised SOM can be generalised to any number of maps, where the winning unit is identified as a weighted sum of distances over i maps, each map weighted by α_i, where $\Sigma\alpha_i = 1$, and the distances on each map scaled so the maximal distance is 1. Such a network is known as a super-organised SOM.

One problem with supervised SOMs as presented above is that in a regression setting, the number of possible fitted (or predicted) values of the response **Y** is limited by the number of units in the grid used. The fitted values for each observation are the mean of the response over the set of observations that map to the same unit. The predicted value for new observations is likewise the mean of the response for the training samples mapped to the same unit as each new observation. This is the same problem as identified for regression trees; in the terminology introduced there, a piece-wise constant model is fitted in the units of the trained supervised SOM. Melssen et al. (2007) combine the BDK or XYF networks with partial least squares regression (PLS) (Martens and Næes 1989; Wehrens 2011: Juggins and Birks 2012: Chap. 14) to overcome this deficiency in supervised SOMs.

We illustrate the utility and applicability of SOMs for palaeoecological data analysis using the SWAP-138 diatom calibration data-set, using the R package kohonen (Wehrens and Buydens 2007). Figure 9.17 shows output from a SOM fitted to the standardised, log-transformed (except pH, and conductivity was excluded from this analysis) water-chemistry for the 138-lake training-set. Figure 9.17a shows how the mean distance to the winning unit (per epoch) improves as the network is trained. The SOM appears to have converged after approximately 60 iterations. There is a clear conductivity signal in the data that is captured by the SOM (Fig. 9.17b), with units to the left of the map identified by high values of various ions and high pH and alkalinity. The upper right section of the map is characterised by dilute, low pH waters, whilst very low pH waters with high aluminium concentrations are located in the lower right area of the map. High total organic carbon (TOC) concentrations are found towards the lower left. The average distance of observations to the unit onto which they map is a measure of the quality of the mapping achieved by the SOM, and is shown in Fig. 9.17c for the SWAP water-chemistry SOM. There are few units with high mean distances, which suggests that the low-dimensional mapping closely fits the data. Figure 9.17d shows which unit each of the 138 SWAP sites maps onto and the number of samples within each unit. Given the small numbers of observations within some of the map units, it might be prudent to sequentially refit the SOM with reduced grid sizes until the degree of fit drops appreciably.

A supervised SOM can be fitted to the SWAP-138 diatom and lake-water chemistry data to investigate relationships between chemistry and diatom species composition. Here we use the square-root transformed species data as the response data, **Y**, and the standardised water chemistry data in the predictor role, **X**. Only diatom taxa that were present in at least 5 samples at 2% abundance or greater were included in the analysis. Both maps converged after approximately 60 epochs (Fig. 9.18a) and achieved similar levels of fit. The codebook vectors for the **X** map (chemistry: Fig. 9.18b) are very similar to those produced by the unsupervised SOM (Fig. 9.17b), indicating the strong influence on diatom species composition exerted by the water chemistry. In general, the supervised SOM **X** map is a reflected, about the vertical, version of the unsupervised SOM; higher ionic strength waters are found to the right and the more acid sites to the left. The high aluminium, low pH units are now located to the upper left, with the low pH and low aluminium units to the lower left.

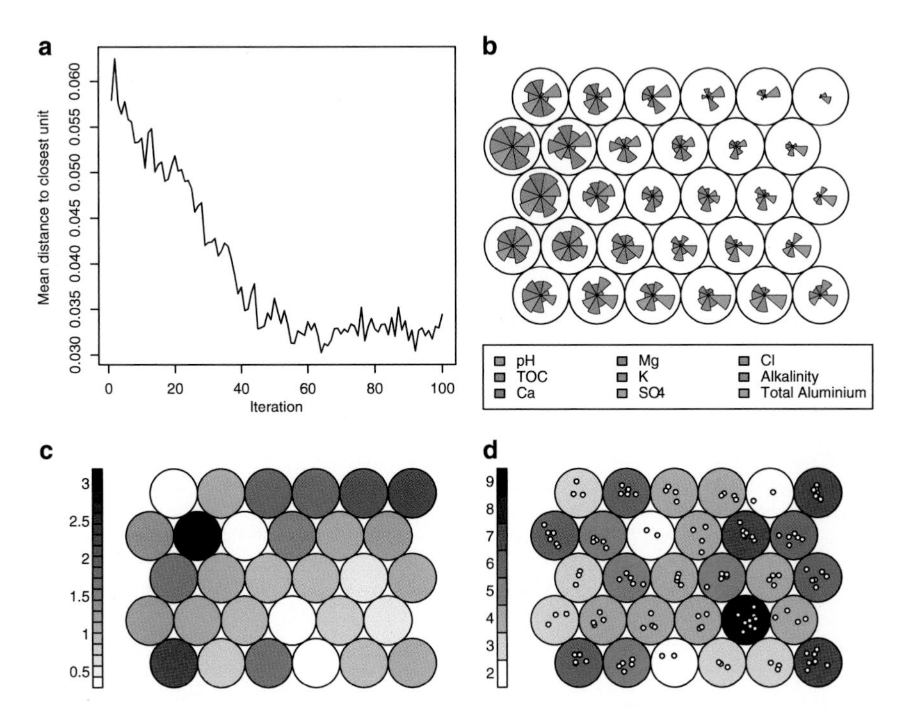

Fig. 9.17 Graphical summary of the self-organising map (SOM) fitted to the Surface Waters Acidification Programme (SWAP) water chemistry data-set: (**a**) shows how the mean distance to the closest map unit falls steadily as the SOM is trained and learns the features of the data, stabilising after 60 iterations or training epochs. The codebook vectors for the trained SOM map units are shown in (**b**) where each segment represents one of the nine water chemistry determinands and the radius of each segment represents the 'abundance' of the determinand (*large radii* indicate large values and *small radii* indicate small values). The degree of heterogeneity in the water chemistry of samples within each map unit is shown in panel (**c**) with higher values indicating units with samples of more heterogeneous chemistry. The number of samples in the SWAP training-set mapped to each unit in the SOM grid is shown in (**d**); the background shading refers to the number of samples and each map unit on the panel contains that number of samples (*circles*) plotted using a small amount of jitter

Due to the large number of taxa, the codebook vectors for the **Y** map are best visualised on a per taxon basis. Figure 9.19 shows the XYF SOM-predicted abundances (back-transformed) for four taxa with differing water chemistry responses. *Achnanthes minutissima* is restricted to the high pH, high alkalinity units to the right of the map. Predicted abundances for *Brachysira brebissonii* are positive for many units indicating the wide tolerance of this taxon, however it is most abundant in the slightly more-acidic units on the map. *Tabellaria binalis*, an acid-loving species, is found most abundantly in the very acid, high aluminium map units towards the upper left of the map, whilst *Eunotia incisa*, an acid-tolerant species common in nutrient-poor, acid waters, is most abundant in a range of the low pH units but particularly in those with lower aluminium concentrations.

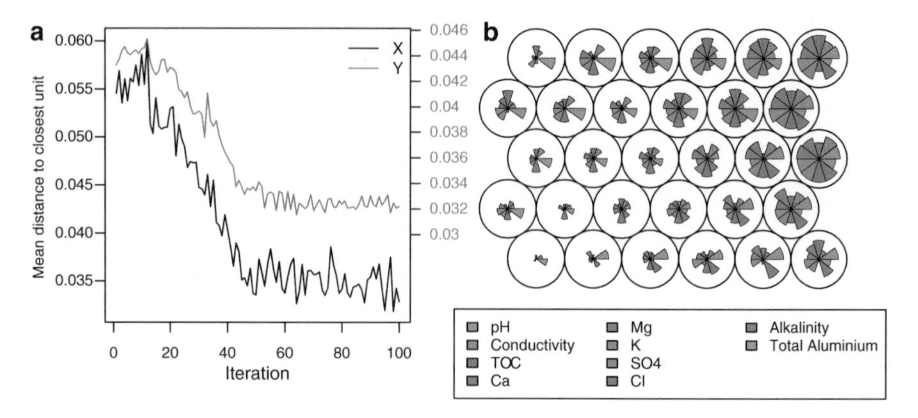

Fig. 9.18 Graphical summary of the X-Y fused Kohonen network self-organising map (XYF-SOM) fitted to the Surface Waters Acidification Programme (SWAP) diatom training-set. The square-root transformed diatom data were used as the response map Y with the water chemistry data used as predictor map X. (**a**) Shows how the mean distance to the closest unit for both X and Y maps decreases steadily as the XYF-SOM is trained, apparently converging after 50 iterations. The codebook vectors for the X map (water chemistry) are shown in (**b**). See Fig. 9.17 for details on interpretation

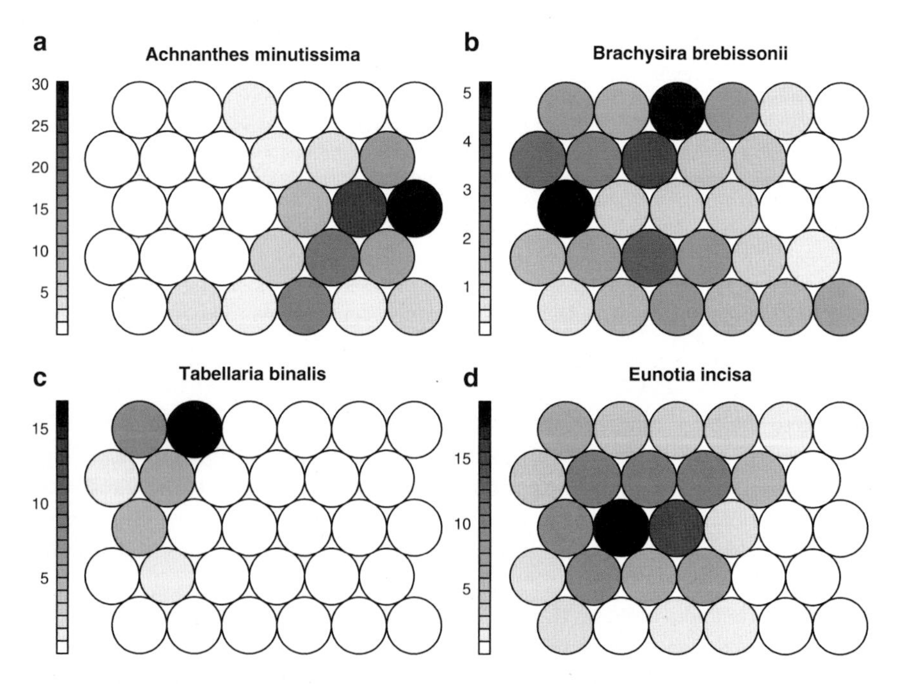

Fig. 9.19 Predicted percentage abundance for four diatom taxa using a X-Y fused Kohonen network self-organising map (XYF-SOM) fitted to the Surface Waters Acidification Programme (SWAP) training-set data

A supervised SOM can also be used as a multivariate calibration tool; here the species data play the role of the predictor variables (**X** map), whilst the variable(s) of interest to be predicted are now used in the response role (**Y** map). Here we build a supervised SOM to predict lake-water pH from the diatom data, using the same data as for the previous example except in reverse roles. We also only include pH as the sole **Y** map variable, although, where appropriate, two or more response variables may be included in a calibration SOM. The fitted model has an apparent root mean squared error (RMSE) of 0.215 pH units when assessed using the training-set data. Further analysis of the fitted codebook vectors of the species (**X** map) can be performed, to identify those taxa most influential for predicting pH and also the species composition of the SOM map unit. We use the fitted XYF SOM to predict lake-water pH values for the Holocene core from The Round Loch of Glenhead (Birks and Jones 2012: Chap. 3). Only those taxa used to fit the XYF SOM were selected from the fossil data. The pH reconstruction is shown in the upper panel of Fig. 9.20, whilst the pH codebook vector is shown for each map unit in the lower panel with the fossil samples projected on to the map. Whilst the general form of the reconstruction is similar to previously published reconstructions (e.g., Birks et al. 1990) and the recent acidification period is captured by the reconstruction, a major deficiency in the reconstruction is immediately apparent; the predicted values for the core samples only take on one of nine possible values. This is due to the predicted pH for each fossil sample being the fitted pH value from the map unit onto which each fossil sample is projected. As the fossil samples project onto only nine map units, only nine possible values can be predicted for the reconstruction. This deficiency is addressed by Melssen et al. (2007) by combining supervised SOMs with PLS. Although we will not consider this technique further here, the general idea is that a BDK SOM is trained on the input data and the similarities between the objects and the codebook vectors of the trained SOM are computed to form a similarity matrix. The elements of this matrix are weighted by a kernel function to form a so-called kernel matrix. The columns of this kernel matrix are then used as predictor variables in a PLS model to predict the response (Melssen et al. 2007). In this way, the information contained in the trained SOM is used to predict the response, but continuous predictions can now be produced because of the use of PLS. Examples of the use of SOMs in limnology and palaeoecology include Malmgren and Winter (1999), Céréghino et al. (2001), Holmqvist (2005), and Weller et al. (2006).

Bayesian Networks

Bayesian networks (also known as belief networks or Bayesian belief networks) are a powerful modelling technique that describes a means by which reasoning in the face of uncertainty about a particular outcome can be performed (Witten and Frank 2005; Bishop 2007; Jensen and Nielsen 2007; Ripley 2008). A Bayesian network can be viewed as a graphical description of the system under study, where

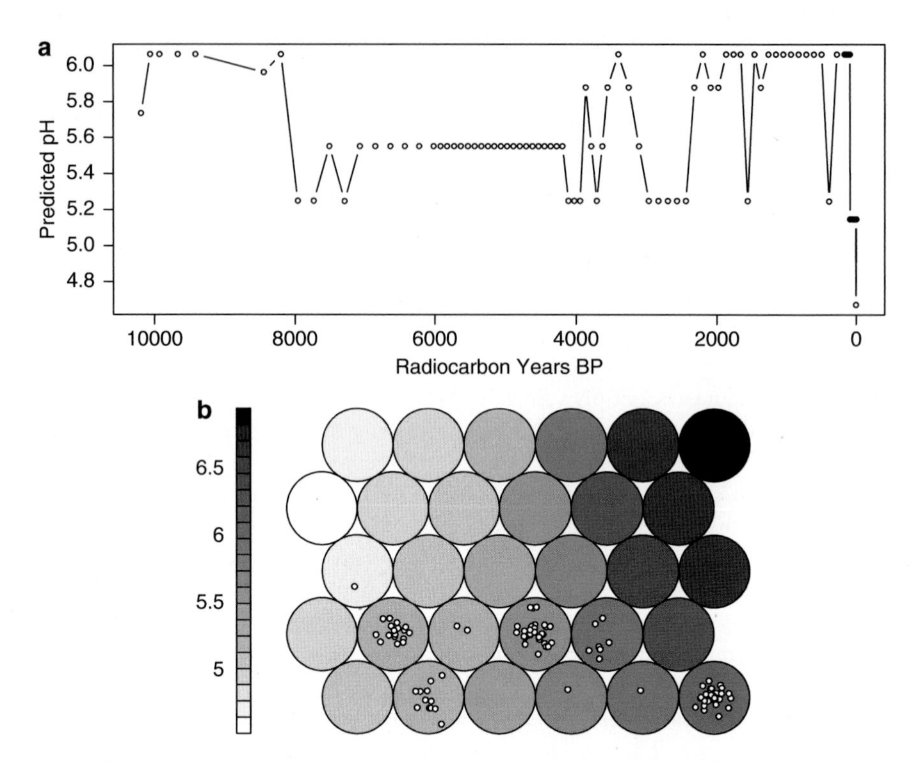

Fig. 9.20 Graphical summary of a X-Y fused Kohonen network self-organising map (XYF-SOM) fitted to the Surface Waters Acidification Programme (SWAP) training-set in calibration mode, with lake-water pH used as the response data Y and square-root transformed diatom abundance data used as prediction data X, and applied to a Holocene diatom sequence from The Round Loch of Glenhead, Scotland, UK. (**a**) Reconstructed lake-water pH history for the loch. The predicted pH for each map unit is shown in (**b**) with The Round Loch of Glenhead sediment core samples mapped on it

key features of the system are represented by nodes that are linked together in some fashion so that the cause-and-effect relationships between the nodes are described. Bayesian networks are more formally known as directed acyclic graphs (DAGs), where the nodes represent random variables and the linkages between nodes represent the conditional dependencies between the joined nodes. The graph is acyclic, meaning that there are no loops or feedbacks in the network structure, and is directed because the relationships between nodes have stated directions; A causes B (Ripley 2008).

Consider a simple system with two nodes, A and B, which are the nodes in the network. A and B are linked by a directional arrow from A to B indicating that A influences B. In this network, A is the parent of B, and B is the child of A. A has no parents and thus is also known as a root node, and plays the role of an input variable in the network. A node that does not have any children is known as a leaf node and plays the role of an output variable. Each node in the network is associated with a set of states, that may be discrete or continuous, which represent the set of possible conditions that the node may take. A conditional probability table is produced for each node, which states the probability with which a node will take each of its states conditional upon the states (or values) of the parent nodes. As such, root nodes are not initialised with conditional probability tables and instead are provided unconditional probabilities: the probability that the input variable (root node) is in a particular state. Conditional independence is a key property of Bayesian networks: two events X and Y given a third event Z are said to be conditionally independent if, given knowledge about the state of Z, knowledge of X conveys no information about the state of Y or vice versa. Independent and interactive (conditional) effects of variables on the modelled response (output nodes) can be examined. Bayesian networks also assume the Markov property, namely that the conditional probability tables can be completed only by considering the immediate parents of a particular node. If we know the probabilities of the states for the parents of a particular node, given the conditional probability table for that node, the probabilities for the child nodes can be computed using Bayes Theorem

$$P\left(y|x\right) = \frac{P\left(x|y\right)P(y)}{P(x)} \tag{9.6}$$

where $P(y)$ is the prior probability of the child node, $P(x \mid y)$ is the likelihood or the conditional probability of x given y, $P(x)$ is the probability of the parent node and is a normalising constant in the equation, and $P(y \mid x)$ is the posterior probability of the child node given the state of the parent x. The posterior probability $P(y \mid x)$ is the probability of a particular state of the child node conditional upon the probabilities of the states of the parent. The prior probabilities and the conditional probability tables for the nodes may be specified using expert judgement and knowledge of the system under study or learned from the training data via one of several Bayesian learning algorithms.

Bayesian networks can be operated bottom-up or top-down. Consider again our system with two nodes, A and B. In bottom-up mode, we might observe a particular state for B, thus setting the probability for that state to 1, and then propagate this information back up the network to A to determine the most likely state of A, given that we have observed the state of B. Conversely, we might be interested in determining the effect on B of altering the state of A, therefore we set the probability for one of the A states to 1 and then propagate this information down the network to see the most likely response of B to the state of A.

As an example, consider a study relating nutrient loadings, through trophic levels, to provide an estimate of water quality (Castelletti and Soncini-Sessa 2007a). Nitrogen and phosphorus loadings influence the trophic level of a water body, stimulating primary production when nutrient levels are elevated, and thus the trophic level is an influence on the perceived water quality. The network associated with this hypothetical system/problem is shown in Fig. 9.21. In this simplified illustration, each of the nodes is characterised by two states; low and high. Table 9.7

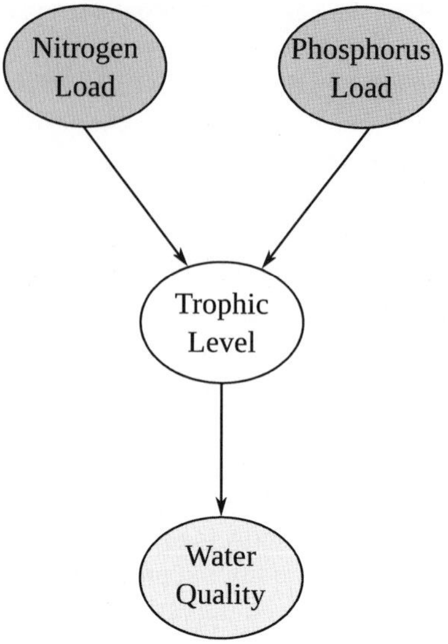

Nitrogen Load
Phosphorus Load
Trophic Level
Water Quality

Fig. 9.21 Example of a Bayesian Network discussed in the text, showing the directional relationship of the effects of nutrient loadings on trophic level and consequently upon water quality. Input/root nodes are shown in dark grey, whilst leaf/output nodes are shown in light grey (Modified from Castelletti and Soncini-Sessa 2007a)

Table 9.7 Conditional probability tables for the Trophic Level (a) and Water Quality (b) nodes in Fig. 9.21

(a)					
Nitrogen loading		L		H	
Phosphorus loading		L	H	L	H
Trophic level	**L**	1.0	0.3	0.5	0.0
	H	0.0	0.7	0.5	1.0
(b)					
Trophic level		L	H		
Water quality	**L**	0.0	0.8		
	H	1.0	0.2		

L Low, *H* High

shows the conditional probability tables for the trophic level and water quality nodes for this illustrative example. If the prior beliefs of the states for the phosphorus and nitrogen loading nodes are set to the values shown in the left-hand section of Table 9.8, the posterior probabilities computed using the conditional probability tables (Table 9.7) of the trophic level and water quality states would be those shown in the right-hand section of Table 9.8. If our prior beliefs about the probabilities of the nutrient-loading states were to change or be updated, then the conditional probabilities of the states for trophic levels and water quality would likewise be updated in light of the new prior beliefs.

Bayesian networks can be used to inform the decision-making process via the inclusion of a decision node into the network (Korb and Nicholson 2004; Bishop 2007). Returning to our simple two-node network example (A and B), we could turn this network into a Bayesian decision network (BDN) by assigning a decision parent node to A. This decision node might also be associated with a cost function describing the cost of enacting the decision. The decision node describes the states

Table 9.8 Prior beliefs for the states of nitrogen and phosphorus loading, which when combined with the conditional probability tables in Table 9.7, yield the posterior probabilities for the states of trophic level and water quality. Arrows show the directional relationships of the effects of the nutrient loadings on trophic level and hence water quality (see Fig. 9.21)

	Nitrogen loading	Phosphorus loading		Trophic level		Water quality
L	0.1	0.3	→	0.1	→	0.7
H	0.9	0.7		0.9		0.3

L Low, *H* High

of possible management actions, for example restoration strategies or water-quality limits or standards, whilst the cost function describes the cost of enacting a particular restoration strategy or setting a particular water-quality standard. The output node in our example, B, is linked to a utility node, which describes the desirability (utility) of particular states of the outcome node. Node A now needs to be assigned a conditional probability table to describe the probabilities of the states of A conditional upon the different states of the decision node. The utility output from the network is the sum of the individual utilities of the output state in node B, weighted by the probabilities of each of the output states. Management decisions can then be based on selecting the intervention that maximises the output utility of the network relative to the cost of intervention. As with the simpler Bayesian networks, the prior and conditional probabilities of the BDN nodes can be set *a priori* using expert judgement or learned from available training data or a combination of the above; probabilities for decision nodes and utility values for outcome states are set by the user.

Bayesian networks have seen little use in palaeoecology, but have had some limited use in conservation management in freshwater ecology. Stewart-Koster et al. (2010), for example, use Bayesian networks to investigate the cost effectiveness of flow and catchment restoration for impacted river ecosystem, the output of which would be used to guide investments in different types of restoration. Other examples include the use of Bayesian networks in water-resource management (Castelletti and Soncini-Sessa 2007b; Allan et al. 2011), the evaluation of management alternatives on fish and wildlife population viability (Marcot et al. 2001), and the effects of land-management practices on salmonids in the Columbia River basin (Rieman et al. 2001), whilst Newton et al. (2006, 2007), Aalders (2008), Kragt et al. (2009), Murphy et al. (2010), and Ticehurst et al. (2011) employ Bayesian networks in vegetation conservation and management. Pourret et al. (2008) present a wide range of case studies from many disciplines that have found Bayesian networks useful.

Genetic Algorithms

Genetic algorithms are one of a number of stochastic optimisation tools that fall under the heading of evolutionary computing. Numerical optimisation is a general catch-all term for algorithms that given a cost (or loss) function aim to find a globally optimal solution to a modelling problem, for example a set of model coefficients that minimises the lack of fit of a model to a set of training samples. Numerical optimisation techniques that use derivatives of the loss function proceed towards an optimal solution in an iterative fashion but which may not, however, converge to a globally optimal solution, instead they find a locally optimal solution. This is akin to always walking downhill to find the lowest point in a landscape; eventually you will not be able to proceed further because to do so would involve moving uphill. A much lower valley just over a small rise from the one you are currently in would be out of reach if you could only walk downhill. Evolutionary computing introduces ideas from natural selection and evolution to add elements of stochasticity to the optimisation search in an attempt to avoid becoming trapped in sub-optimal local solutions.

Of the various evolutionary computing techniques, genetic algorithms have been most frequently used in ecology, especially the Genetic Algorithm for Rule-set Prediction (GARP) procedure, which has seen extensive use in modelling spatial distributions of species (Anderson et al. 2003; Jeschke and Strayer 2008; Franklin 2010). Here we describe genetic algorithms in a general sense, and then we briefly discuss genetic programmes and GARP.

Genetic algorithms consider a population of solutions to a modelling problem rather than a single solution (D'heygere et al. 2003). Each of the solutions is described by a string of numbers, each number representing a gene and the set of numbers an individual chromosome in the terminology of genetic algorithms. The strings represent terms in the model. If we consider a simple least-squares regression, then we could use a string of length m zeroes and ones indicating which of the m predictor variables is in the model (Wehrens 2011). Alternatively, we could just record the index of the variables included in the model, where the string of values would be of length M (the number of variables in the model, its complexity) and the individual values in the string would be in the set $(1, 2, \ldots, m)$ (Wehrens 2011). The size of the population of chromosomes (the number of solutions) considered by the genetic algorithm needs to be set by the user; with too small a population the algorithm will take a long time to reach a solution, whilst too large a population entails fitting many models to evaluate each of the chromosomes in every generation. The initial population of chromosomes is generally seeded by assigning a small random selection of the available predictor variables to each of the C chromosomes.

Offspring solutions (chromosomes) are produced via a sexual reproduction procedure whereby genes from two parent solutions are mixed. The fitness of the offspring determines which of them persist to produce offspring of their own, with fitness being defined using a loss function, such as least-squares error. Offspring with low fitness have a low probability of reproducing, whilst the fittest offspring have the highest chance of reproducing. This process of sexual selection is repeated a large number of times with the result that subsequent generations will tend to consist of better solutions to the modelling problem. The sexual reproduction step consists of two random processes termed crossover or sharing of parents' genes, and mutation. These processes are random and as such are not influenced by the fitness of individual parents. Sexual reproduction mixes the genes from two parents in a random fashion to produce an offspring that contains a combination of the genes from the two parents. Mutation introduces a stochastic component to the genetic algorithm, and allows predictor variables not selected in the initialisation of the population of chromosomes a chance to enter the genetic code of the population. Mutation is a low-probability event; say 0.01 indicating that one time in a hundred a mutation will take place during reproduction. Mutations can involve the addition of a new variable to the chromosome, the removal of an existing variable, or both addition and removal of variables. Mutation allows the genetic diversity of the population to be maintained.

Each iteration of a genetic algorithm produces a new generation of offspring by sexual reproduction of the fittest members of the current population. The candidates for reproduction are chosen at random from those models that reach a minimum fitness threshold. The selection of two candidates for reproduction may be done at random from within this set of fittest chromosomes or at random with the probability of selection weighted by the fitness of each chromosome. The latter gives greater weight to the best of the best solutions in the current population.

The genetic algorithm is run for a large number of iterations (generations) and the fittest solution at the end of the evolutionary sequence is taken as the solution to the modelling problem. It is possible that the population of solutions will converge to the same, identical solution before the stated number of generations has been produced. Likewise, there is no guarantee of convergence to the best solution in the stated number of iterations. As such, it is important that the evolutionary process is monitored during iteration, say by recording the fitness of the best solution and the median fitness over the population of solutions for each generation (Wehrens 2011). If the fitness of the best solution is still rising and not reached an asymptote by the end of the generations then it is unlikely that the algorithm has converged.

Genetic algorithms are a general purpose optimisation tool, and as such they require far more user interaction than many of the other statistical machine-learning methods described in this chapter. The size of the population of solutions, the minimum and maximum number of variables included in a single solution, the number of iterations or generations to evolve, the mutation rate, the fitness threshold required to select candidates for sexual reproduction, and the loss function all need to be specified by the user. The flexibility of the genetic algorithm thus comes with a price. However, the algorithm can be applied to a wide range of problems, simply by introducing a new loss function that is most appropriate to the modelling problem to hand. The loss function can be any statistical modelling function, such as least-squares, linear discriminants, principal components regression, or partial least squares, for example, and as such a wide range of problems can be tackled. Genetic algorithms can also be slow to converge to an optimal solution, especially when faced with a complex modelling problem consisting of many observations and predictor variables.

Genetic programmes are related to genetic algorithms, but now each chromosome in the population is a computer program that uses combinations of simple arithmetic rules (using $+$, $-$, $\times$, etc.) and mathematical functions or operators. The various rules and functions are combined into a syntax tree to combine numeric values with mathematical operators and functions that form a solution to a problem. Reproduction now takes the form of randomly swapping sub-trees in the syntax trees of two parents to produce new offspring that include aspects of both parents' genetic programme. Mutation is performed by randomly selecting a sub-tree in the syntax tree of an individual and replacing that sub-tree with a randomly generated sub-tree. Which programmes are allowed to reproduce is controlled by a fitness criterion in the same way as described for genetic algorithms. The key difference between a genetic algorithm and a genetic programme is that genetic algorithms optimise an *a priori* specified model by evolving solutions to the modelling problem (regression coefficients for example) that give the best fit of the model to the training data, whereas genetic programmes aim to find an optimal solution to an unspecified modelling problem by combining simple mathematical steps to best fit or explain the training data.

GARP (Stockwell and Noble 1992; Stockwell and Peters 1999) is a genetic algorithm where the genes do not represent inclusion or exclusion of particular predictor variables, but instead are simple rules that are very much akin to the rules produced by the tree models we described earlier. In GARP, each of the rules follows a similar form: *if* 'something' is true, *then* 'this' follows, where 'something' is a simple rule and 'this' is a predicted value say. For example, a rule might be *if* pH is less than Y and aluminium is greater than X, *then* the abundance of the diatom *Tabellaria binalis* is Z%. The set of possible rules using combinations of predictor variables is impossibly large for most problems for an exhaustive search to be made. Instead, genetic algorithms are used to evolve the rules into a set of optimal combinations that best predict the response. The algorithm starts by identifying all rules consisting of a single predictor; at this point, the algorithm is very much similar to the exhaustive search used in tree models to identify the first split. A predefined number, *r*, of these rules is then chosen as the initial set of rules upon which the genetic algorithm will operate. The *r* best rules are chosen as the initial set. Each of several predefined operators is then applied to the initial set of rules to evolve a new generation of rules. These operators include a random operator which creates a rule with a random number of conditions (*if* 'something's) and values (*then* 'this's), a mutation operation which randomly changes the values used in a condition, and a concatenation operation which combines two randomly chosen rules from the existing set. Having applied these operators to the current set of rules, the rules are ordered in terms of fitness, and the least fit rules are discarded. The remaining set of rules then undergo another round of operator application to evolve a new generation of rules and the least fit rules again are discarded. This process is repeated a large number of times in order to evolve a set of rules that best predicts the response. GARP is most useful in situations where the user has little reliable background knowledge to guide model choice and in situations where rules are sought in noisy, high dimensional, discontinuous data with many local optima. However, GARP is considered computer intensive relative to the many of the statistical machine-learning tools described here.

Genetic algorithms and programmes and GARP are very flexible, general optimisation tools. However, they are not well suited to all problems. More-specific statistical machine-learning tools, such as regression or classification trees and related methods will tend to perform as well or better than the evolutionary computing approaches for general regression or classification problems (D'heygere et al. 2003; Olden et al. 2008), and as we have seen, bagging, random forests, and boosting can all improve upon single tree models by combining information from several weak learners. In addition, Elith et al. (2006) and Lawler et al. (2006) both observed that GARP tended to over-fit species distributions compared with other modelling techniques. As such, and given the availability of powerful alternative techniques plus the additional effort required by the user to use evolutionary computing techniques, we cannot recommend their use over the other statistical machine-learning techniques described earlier. GARP is, however, widely used in species-climate modelling in biogeography and climate-change predictive biology (e.g., Elith and Burgman 2002; Stockwell and Peterson 2002; Pearson et al. 2006; Tsaor et al. 2007; Jeshcke and Strayer 2008).

Principal Curves and Surfaces

Principal component analysis (PCA) (Jolliffe 2002; Legendre and Birks 2012b: Chap. 8) is used in a large number of fields as a means of dimension reduction by expressing on the first few principal components orthogonal linear combinations of the input data that explain the data best in a statistical sense. These first few principal component axes are often used as synthesisers of the patterns of change found in stratigraphical data for example (Birks 2012b: Chap. 11). PCA is also the basis of the linear, multivariate calibration technique principal components regression (Juggins and Birks 2012: Chap. 14), where the input data are reduced to $p \ll m$ components, which are then used in a multiple regression to predict the known response variable. In the high-dimensional space of the input data, the principal components represent lines, planes, or manifolds (where manifold is the generic term for these surfaces in m dimensions). These principal components are inherently linear, and where data do not follow linear patterns, PCA may be sub-optimal at capturing this non-linear variation. This is why correspondence analysis, principal coordinates, and non-metric multidimensional scaling (Legendre and Birks 2012b: Chap. 8) are popular in ecology where the input data are assumed to be inherently non-linear.

SOMs can be viewed as a non-linear two-dimensional manifold, one that is best fitted to the data in m dimensions. One of the options for choosing the starting points of a SOM grid is to select points on the two-dimensional principal component plane, which are then bent towards the data to improve the quality of fit. A number of other techniques have been developed in the last 20 years or so that generalise the problem of fitting non-linear manifolds in high dimensions. Here we discuss one particular technique – that of principal curves and surfaces.

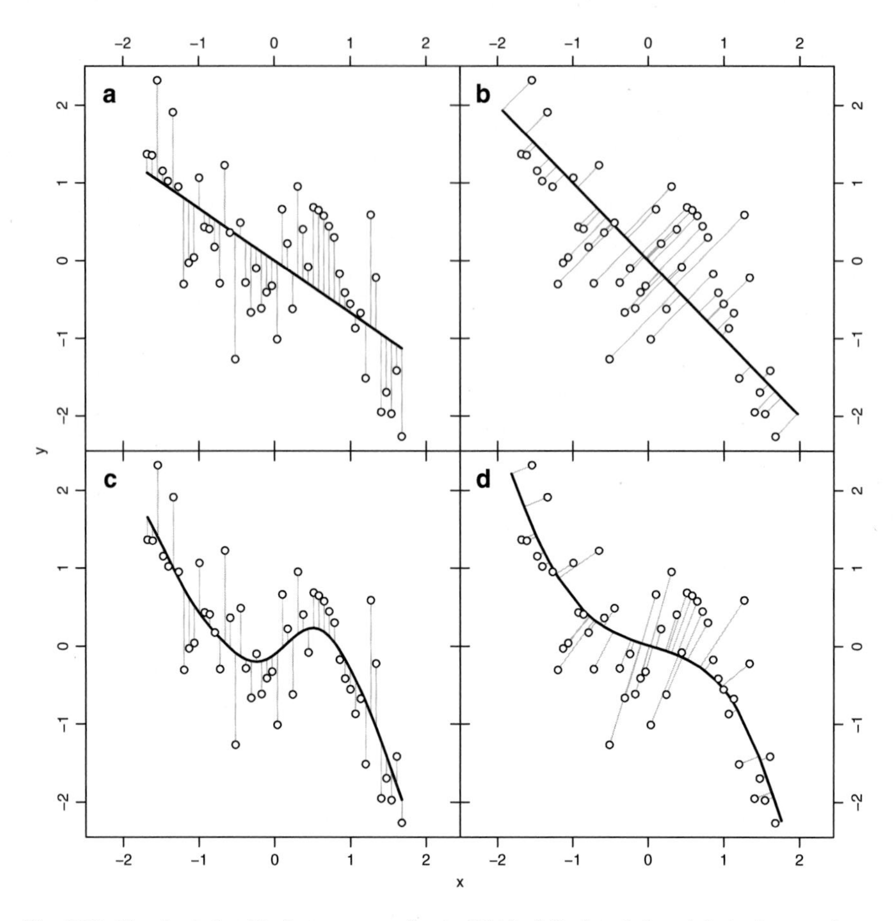

Fig. 9.22 Fitted relationship between x and y (*solid black line*) and the minimised errors (*grey line segments*) for least-squares regression (**a**), principal component analysis (**b**), cubic smoothing spline (**c**), and a principal curve (**d**). Where relevant, y is treated as the response variable and x as the predictor variable

Principal curves (PCs: Hastie and Stuetzle 1989) are a generalisation of the first principal component line, being a smooth, one-dimensional curve fitted through the input data in m dimensions such that the curve fits the data best, i.e., the distances of the samples to the PC are in some sense minimised (Hastie et al. 2011). In least-squares regression, the model lack-of-fit is computed as the sum of squared distances between the fitted values and the observations for the response variable. These errors are shown as vertical lines in Fig. 9.22a for the function

$$y = -0.9x + 2x^2 + -1.4x^3 + \varepsilon \quad \varepsilon \sim N(\mu = 0, \sigma = 0.05) \tag{9.7}$$

In PCA, the first principal component is fitted such that it minimises the lack-of-fit in terms of both the 'response' variable and the 'predictor' variable. These errors are shown in Fig. 9.22b for the function in Eq. 9.7 and are the orthogonal distances of the observations to the principal component line. We can generalise the simple least-squares regression to a smooth function of the covariates (= variables) using smoothing splines (or, for example, in a generalised additive model; Birks 2012a: Chap. 2). A smoothing spline fit to the data generated from Eq. 9.7 is shown in Fig. 9.22c. As with the least-squares regression, the lack-of-fit is measured in terms of the sum of squared distances in the response between the fitted values and the observations. Principal curves generalise the first principal component line by combining the orthogonal errors aspect of PCA with the concept of a smooth function of the covariates. A PC fitted to the data generated from Eq. 9.7 is shown in Fig. 9.22d with the errors shown as orthogonal distances between the observations and the points on the PC onto which they project. The degree of smoothness of the fitted PC is constrained by a penalty term, just as with smoothing splines (Birks 2012a: Chap. 2), and the optimal degree of smoothing is identified using a generalised cross-validation (GCV) procedure. The point on the PC to which an observation projects is the point on the curve that is closest to the observation in m dimensions.

Principal curves are fitted to data using a two-stage iterative algorithm. Initially, a starting point for each observation is determined, usually from the sample scores on the first principal component or correspondence analysis axis. These starting points define a smooth curve in the data. The first stage of the algorithm then proceeds by projecting each point in m dimensions onto a point on the initial curve to which they are closest. The distances of the projection points along the curve from one arbitrarily selected end are determined. This is known as the projection step. In the second stage of the algorithm, the local averaging step, the curve is bent towards the data such that the sum of orthogonal distances between the projection points and the observed data are reduced. This local averaging is achieved by fitting a smoothing spline to each species' abundance using distance along the curve as the single predictor variable. The fitted values of these individual smoothing splines combine to describe a new smooth curve that more closely fits the data. At this point, a self-consistency check is performed such that if the new curve is sufficiently close to the previous curve, convergence is declared to have been reached and the algorithm terminates. If the new curve is not sufficiently similar to the previous curve, the projection and local averaging steps are iterated until convergence, each time bending the curve closer to the data.

The algorithm used to fit a PC is remarkably simple, yet several choices need to be made by the user that can affect the quality of the fitted curve and ultimately the interpretation of the fitted curve. The first choice is the selection of suitable starting points for the algorithm. A logical starting point is the first principal component line, however De'ath (1999) found that better results were achieved using the first correspondence analysis (CA) axis. The second choice involves the fitting of smooth functions to the individual species during the local averaging step. Above we used the general term *smoothing splines* to describe the functions used.

Here we use a cubic smoothing spline (Birks 2012a: Chap. 2) for the example, but LOESS or kernel smoothers may also be used, as could generalised additive models (GAMs). GAMs (Birks 2012a: Chap. 2) are particularly useful when the individual species responses are not thought to be normally distributed; for example, for count abundances, a Poisson GAM may provide a better fit to each species. Whichever type of smoother is used, it is effectively a plug-in component used by the algorithm to perform the local averaging.

Having chosen a type of smoother, the degree of smoothness for the fitted PC needs to be determined. De'ath (1999) suggests that an initial smoother is fitted to each species in the data using GCV to determine, separately for each species, the degree of smoothness required for each curve. The median degree of smoothness (span or degrees of freedom) over the set of fitted smoothers is then chosen for the degree of smoothness used to fit the PC. Alternatively, the complexity of the individual smoothers fitted during the local averaging step can be allowed to vary between the different species, with GCV used to select an appropriate degree of smoothness for each species during each of the averaging steps (GL Simpson unpublished). This allows the individual smoothers to adapt to the varying degrees of response along the PC exhibited by each species; some species will respond linearly along the curve whilst others will show unimodal or skew-unimodal responses, and it seems overly restrictive to impose the same degree of smoothing to each species in such situations.

It is essential that the algorithm is monitored during fitting and that the resulting PC is explored to identify lack-of-fit. Choosing good starting locations can help with over-fitting, but overly complex, over-fitted PCs are most easily identified via examination of the final smoothers for each species, which tend to show complex fitted responses along the curve. The PC can be visualised by projecting it into a PCA of the input data. De'ath (1999) contains further advice on fitting, evaluating, and interpreting PCs.

One use of PCs is in summarising patterns of species compositional change in a stratigraphical sequence. PCA, CA, and DCA axes one and two scores are often used in palaeoecological studies to illustrate where the major changes in species composition occur (Birks 2012b: Chap. 11). Given the additional flexibility of a PC, it is likely to explain similar, or even greater, amounts of temporal compositional change in a single variable (distance along the PC) than that explained by two or more ordination axes. We illustrate the use of PCs in this setting by describing temporal compositional change in a sequence of pollen counts from Abernethy Forest for the period 12,150–5515 radiocarbon years BP (Birks and Mathewes 1978).

As the starting curve we used sample scores on the first CA axis, and fitted the PC to the data using cubic smoothing splines allowing the complexity of the individual smoothers used in the local averaging step to vary between pollen taxa, using GCV to choose the optimal degree of smoothing for each taxon. A penalty term of 1.4 was used to increase the cost of degrees of freedom in the GCV calculations. The PC converged after six iterations and is shown in Fig. 9.23, as projected onto a PCA of the pollen data. The configuration of the samples in PCA

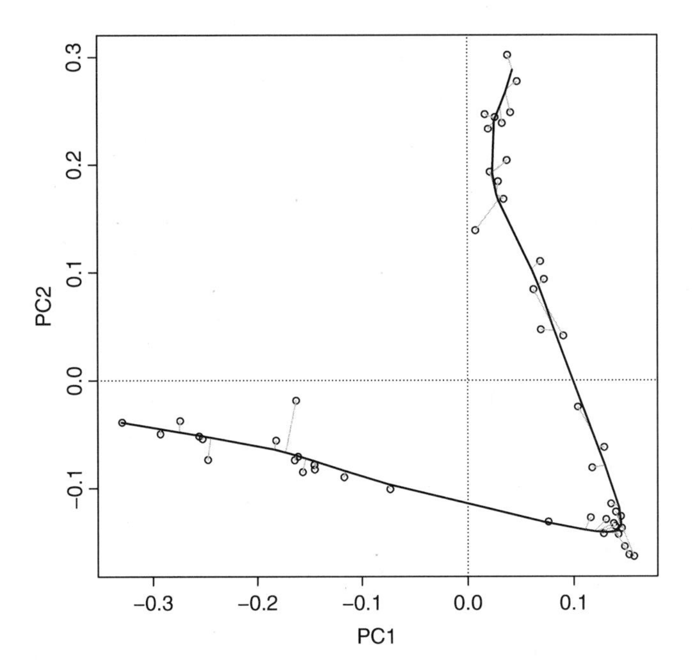

Fig. 9.23 Principal component analysis (PCA) plot of the Abernethy Forest late-glacial and early-Holocene pollen data with the fitted principal curve superimposed (*thick line*). The *thin, grey lines* join each observation with the point on the principal curve on to which they project, and are the distances minimised during fitting. *PC* principal component

space shows a marked horseshoe-like shape that is commonly encountered when a single, dominant gradient is projected onto 2 dimensions. The fitted PC is shown by the thick curved line in Fig. 9.21 with the orthogonal errors represented by thin segments drawn between the sample points and the curve. The PC explains 95.8% of the variation in the Abernethy Forest pollen sequence, compared with 46.5% and 30.9% for the first principal component axis and the first correspondence analysis axis, respectively. The PC accounts for substantially more of the variation in species composition than two PCA or CA axes (80.2% and 52.3%, respectively), which might conventionally be used. Figure 9.24a shows the distance along the PC between adjacent samples in the sequence expressed as a rate of change per 1000 years, clearly illustrating four periods of substantial compositional change in the pollen taxa. The actual distances along the PC are shown in Fig. 9.22b, alongside similar measures for the first PCA and CA axis scores. The total gradient described by each method has been normalised to the range (0,1) to allow a direct comparison between the three methods. Although the changes in PCA and CA axis 1 scores appear more marked, exhibiting apparently greater variation during periods of change, the PC adequately captures these periods of change but also places them within the context of overall compositional change as ~96% of the variation in the pollen taxa is described by the PC.

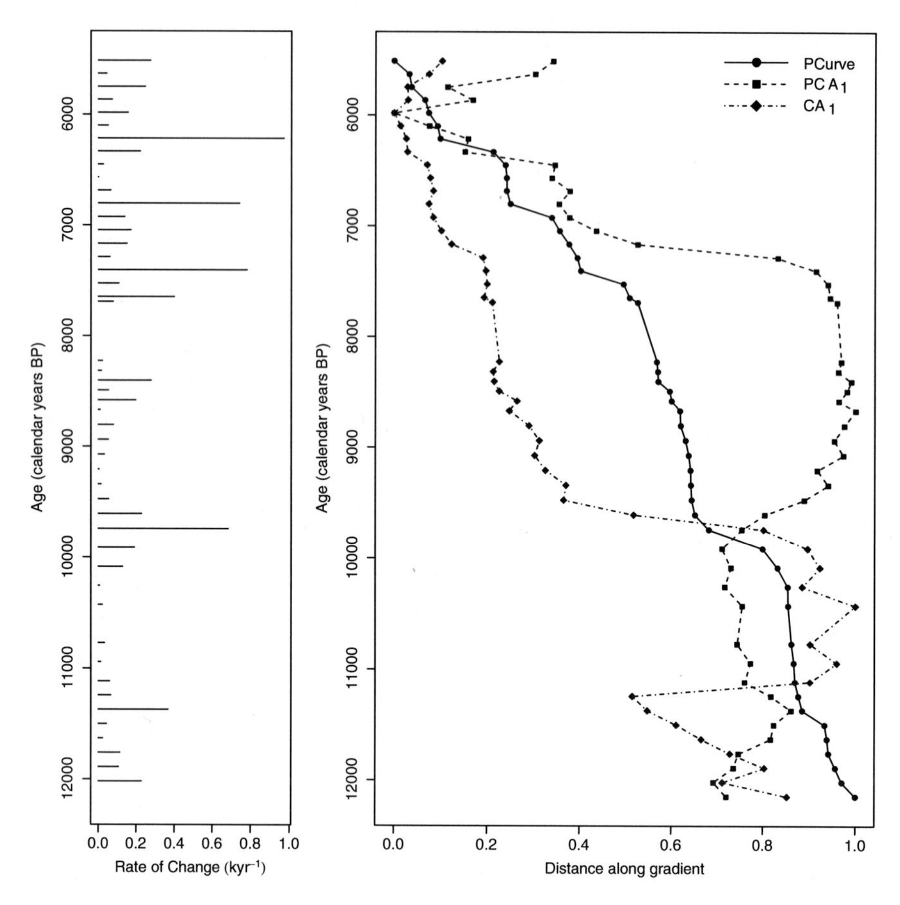

Fig. 9.24 (*left*) Distance along the principal curve expressed as a rate of change per kyr between samples for the Abernethy Forest pollen data-set. Several periods of rapid compositional change are detected. (*right*) Distance along the gradient expressed as a proportion of the total gradient for the fitted principal curve and the first ordination axes respectively of a principal component analysis (PCA) and a correspondence analysis (CA) fitted to the Abernethy Forest data

Figure 9.25 shows cubic smoothing splines fitted to the nine most abundant pollen taxa in the Abernethy Forest sequence. Each smoothing spline models the proportional abundance of the taxon as a function of the distance along the PC (expressed in temporal units). The degrees of freedom associated with each smoothing spline was taken from the smoother fitted to each taxon during the final local averaging step at convergence. As expected, given the amount of variation explained, the PC clearly captures the dynamics present in the pollen data and further illustrates that the data represent a single gradient of successive temporal change in pollen composition.

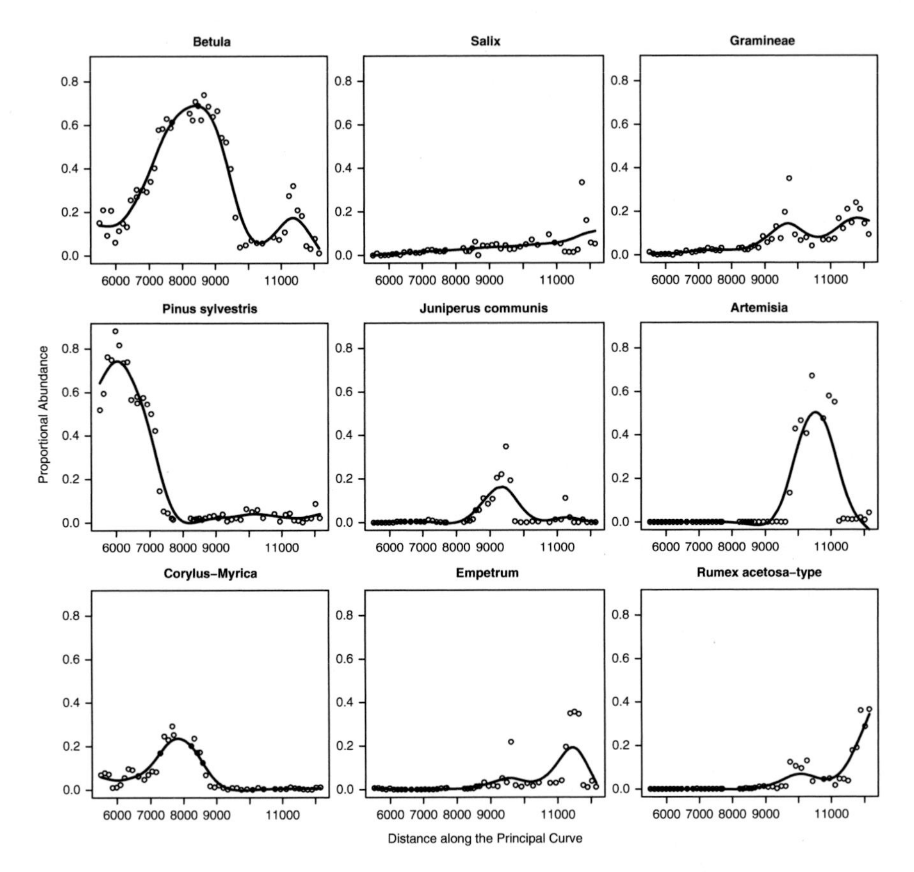

Fig. 9.25 Fitted response curves for the nine most abundant pollen taxa in the Abernethy Forest data as estimated using a principal curve. *Open circles* are the observed proportional abundance and the *solid line* is the optimised smoother from the final iteration of the principal curve. The distance along the principal curve is expressed here in radiocarbon years BP

When combined with the rate-of-change along the curve, the PC approach is far better at describing compositional change than either PCA or CA. This is particularly apparent when the stratigraphical data are best described by a single dominant, though not necessarily long, gradient. The PC degrades to the first principal component solution when all taxa are described by 1 degree-of-freedom linear functions; as a result the method can perform no worse than PCA and can, in the right circumstances, perform substantially better.

Principal curves can be generalised to principal surfaces, analogous to a plane described by the first two principal components. The algorithm described above is adapted in this case to use two-dimensional smoothers for the individual species and the projection points on the curve now become projection points on the principal surface. Higher dimensional principal surfaces can, in theory, be fitted but their use is infrequent owing not least to problems in visualising such curves and in performing the smoothing in multiple dimensions. An unsupervised SOM is very similar to a two-dimensional principal surface, although motivated from a very different view point. Both principal surfaces and SOMs fit a manifold that is progressively warped towards the response data in order to achieve a closer fit to the data points. Geological examples of PCs include Banfield and Raftery (1992) and medical examples include Jacob et al. (1997).

Shrinkage Methods and Variable Selection

A fundamental problem in the statistical analysis of a data-set is in finding a minimal set of model terms or parameters that fit the data well (Murtaugh 2009; Birks 2012a: Chap. 2). By removing terms or parameters from the model that do not improve the fit of the model to the data we aim to produce a more easily interpretable model that is not over-fitted to the training data. The assumption that there is a single 'best' model is, in general, wrong. A more likely situation is that there will be a number of candidate models that all do a similar job in terms of explaining the training data without being over-fitted to them. Without further external criteria it may be wrong to assume that the 'best' of the candidate models is the one that describes the relationship between predictors and response for the population from which the sample of data used to fit the model was collected.

The information theoretic approach advocated by a number of authors (Burnham and Anderson 2002; Whittingham et al. 2006) proceeds by ranking candidate models in terms of the Akaike Information Criterion (AIC) and combining the terms in the various models by averaging over the set of models, and weighting each model in proportion to a likelihood function that describes the probability that each model is the best model in terms of AIC if the training data were collected again under the same circumstances (Whittingham et al. 2006). Often, AIC is used to select between nested models and the model averaging step skipped, to identify the 'best' model. In such cases, selection via AIC (or Bayesian Information Criterion (BIC), etc.) suffers from the same problems as forward-selection or backward-elimination and step-wise selection procedures, in particular, selection bias in the estimates of the model parameters. Anderson (2008) provides a gentle introduction to model-based inference.

Forward-selection and backward-elimination techniques are routinely used in ecology and palaeolimnology to prune models of unimportant terms. Starting from a model containing only an intercept term, forward selection proceeds by adding to the model that predictor variable that affords the largest reduction in model residual sum-of-squares (RSS). The procedure continues by identifying the predictor that provides the largest reduction in RSS conditional upon the previously selected terms included in the model. When the reduction in RSS afforded by inclusion of an additional predictor in the model is insignificant (usually assessed using an F-ratio test between models including and excluding the predictor, or an information statistic such as AIC), selection stops. Backward elimination operates in a similar manner, except in reverse, starting with a model containing all the available predictor variables. The predictor whose removal from the current model would result in the smallest increase in RSS is eliminated from the model if doing so does not result in a significantly worse model. Backward elimination proceeds until either all predictors are removed from the model or no terms can be removed from the model without significantly affecting the fit to the response. An important difference is that forward selection can be performed on a model fitted to any data-set consisting of two or more predictors, whereas backward selection can only be performed on data-sets where there are $n-1$ predictors.

Step-wise selection combines both forward selection and backward elimination; at each step in the selection procedure, all single-term additions or deletions are considered and the change that results in the most parsimonious model is made subject to the condition that the added term significantly improves, or the deleted term does not significantly harm, the model fit. An alternative approach to step-wise selection is best-subsets selection, in which models using all possible combinations of predictor variables are generated and the best model of a given size, or the best model over all subsets, is selected from the set of models. The feasibility of this exhaustive search depends on the number of available predictor variables and becomes computationally difficult when only a modest number are available. The branch and bound algorithm (Miller 2002), however, allows an exhaustive search to be performed in a feasible amount of time.

There are several problems with the sequential selection and best-subsets methods, most notably (i) selection bias in the estimates of the model parameters, (ii) increased variability of the selected model, and (iii) bias in the standard errors of model parameters and its effect on the interpretation of p-values. Selection bias arises because the selection techniques described above amount to the imposition of a hard threshold on the size of the model coefficients; the estimate for a coefficient is either zero when the term is not included in the model, or some value $\hat{\beta}_i$ when included in the model. An extreme example, adapted from Whittingham et al. (2006), is shown in Fig. 9.26, where 5000 data-sets of size 10 were drawn from the model

$$y_i = 1 + 0.8x_i + \varepsilon_i \tag{9.8}$$

where x_i are the values $\{1, 2, \ldots, 10\}$ and ε_i are model errors consisting of independent Gaussian random variables with mean 0 and σ_i equal to 1. The subscripts i index the 10 observations in each data-set. In the above model, the coefficient is known ($\beta = 0.8$). Given values for x_i and y_i, we can fit a linear regression to estimate β for each of the 5000 data-sets. The distribution of the estimates for β is shown in the upper panel of Fig. 9.26 with the known value superimposed. If we set the estimates of β to zero for models where the estimate is not statistically different from 0 at the $\alpha = 0.95$ level (i.e., with a p-value >0.05) and retain those estimates that are statistically significant (i.e., those with a p-value ≤ 0.05), a process which amounts to selecting whether to include the term in the model or not, we observe

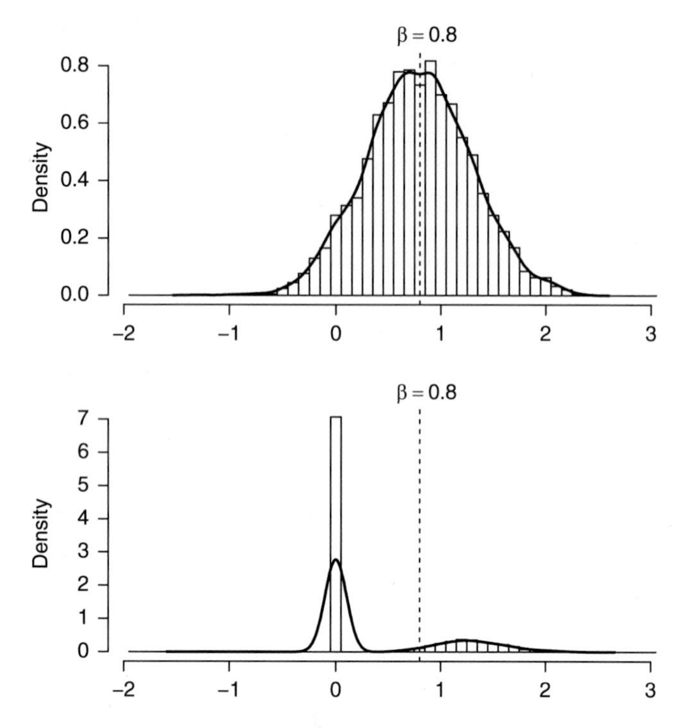

Fig. 9.26 An illustration of selection bias of regression coefficients. The *upper panel* shows the distribution of estimates of a single regression coefficient from models fitted to random samples from a model with known coefficient $\beta = 0.8$. The estimates from 5000 random draws are centred about the known value of β. If we retain the estimates of β from the 5000 random draws that are significant at the $\alpha = 0.95$ (95%) level and set the insignificant coefficients to 0, equivalent to a hard selection threshold, we observe the distribution shown in the lower panel, which contains coefficient estimates that are very different from the known value of β

the distribution of the estimates of β for the 5000 models shown in the lower panel of Fig. 9.26. Note that the retained values are all substantially different from the known population value of β; they are biased low when the term is not selected or biased high when the term is retained. No such bias occurs in the set of unselected parameter estimates (Fig. 9.26); it is the act of selection that introduces the bias and arises because the term is either included in the model or not. This bias occurs whether terms are selected using *p*-values or via some other statistic, such as AIC.

Models resulting from forward selection and/or backward elimination are prone to increased variance, and hence, ultimately higher model error (Mundry and Nunn 2009). The argument behind this statement is the same as that used to explain the instability of single tree-based models (see above). Small changes in the sample data may lead to a different variable entering the model in the early stages of selection, especially if there are two or more predictors that have similar predictive ability as in the case of collinear predictors. The resultant model may be over-fitted to the training sample and generalise poorly when making predictions for other observations from the population. Such models are said to have high variability; the uncertainty in the predicted values is large.

An often overlooked issue with model selection is that the standard errors of the estimated coefficients in a selected model are biased and too small, suggesting apparent precision in their estimation; their construction knows nothing of the previous, often convoluted, selection process that led to the selected model. Consequently, test statistics and their p-values are too optimistic and the possibility of making a Type I error is increased. It is not clear how this bias can be corrected for in a practical sense (Hastie et al. 2011). This problem affects best-subsets selection as well as forward selection/backward elimination.

Model selection often results in models that contain too many parameters unless steps are taking during selection to manage the entry of variables to the model. Consider the situation where a p-value threshold of 0.05 is used to decide whether to include a variable in a model at each stage of a forward-selection procedure. Each of the tests performed to decide whether to include the predictor or not is subject to a Type I error-rate of 0.05, and as such the final model has a much larger Type I error-rate. A correction to the p-value used in each test may be made, to guard against this inflated Type I error-rate. For example, a Bonferroni-type correction can be made of p/t, where p is the user-selected p-value threshold (0.05 in the above discussion) and t is the number of tests conducted thus far. In deciding whether to include the first predictor variable, using 0.05 as the threshold for inclusion, the variable is included if it achieves a p-value of $0.05/1 = 0.05$ or lower. For the second variable to enter the model it must achieve a p-value of $0.05/2 = 0.025$ or lower to be selected, and so on for the subsequent rounds of selection. Using BIC instead of AIC to decide on inclusion or elimination penalises more-complex models to a stronger degree and thus may help to guard against selecting overly complex models.

Correlated predictors, as well as making model selection more difficult, cause additional problems in estimating model coefficients; they are poorly determined and have high variance (large standard errors). Consider two correlated predictors; a large positive value as the estimate for the model coefficient for one of the predictors can be counteracted by a large negative coefficient for the other predictor (Hastie et al. 2011). If the interest in fitting the model is to interpret the coefficients to shed light on ecological or environmental mechanisms, spurious inflation of effects due to multicollinearity, if undetected, may lead to erroneous statements about the mechanisms under study.

There are a number of approaches that can be applied to help with model selection and collinearity problems. These approaches are known as shrinkage methods. Two shrinkage techniques familiar to palaeolimnologists are principal components regression (PCR) and partial least squares (PLS) (Martens and Næes 1989; Birks 1995; Næs et al. 2002; Juggins and Birks 2012: Chap. 14). In both approaches, the aim is to identify a small number of orthogonal (uncorrelated) components that explain maximal amounts of variance in the predictors (PCR) or maximal amounts of the covariance between the response and predictors (PLS). Predictors that exhibit low variance (PCR) or are unrelated to the response (PLS) will have low weights in the components retained for modelling; in a sense, the coefficients for these variables have been shrunk from their least-squares estimates (Hastie et al. 2011). PCR and PLS are also useful simplification techniques in situations where there are many more predictor variables than observations, as in chemometrics (Wehrens 2011). However, these techniques suffer in terms of model interpretation; the regression coefficients no longer apply to individual predictors but to linear combinations of the predictors. If the aim of modelling is prediction, and not explanation, then the aim of selecting a minimal adequate model is to achieve lower prediction error, and PCR or PLS are useful techniques.

PCR and PLS impose a size constraint on the coefficients of predictors in the model by retaining a small number of orthogonal components as predictors in the model. Information on those variables that are useful in predicting the response or have high variance is retained, whilst those variables unrelated to the response or have low variance are discarded – their coefficients are effectively, or close to, 0 (Hastie et al. 2011). A number of other techniques have been proposed that also impose size restrictions on model coefficients, namely ridge regression (Hoerl and Kennard 1970; Copas 1983, Hastie et al. 2011), the lasso (Tibshirani 1996; Hastie et al. 2011), and a technique known as the elastic net which combines ridge-like and lasso-like constraints (Zou and Hastie 2005; Hastie et al. 2011).

Ridge regression was proposed as a means to handle collinearity in the set of available predictors. Earlier we saw that two correlated variables may have large coefficients but of opposite sign. Imposing a constraint on the size of the model coefficients helps to alleviate this problem. Ridge regression imposes a quadratic constraint on the size of the coefficients, but can also be seen to shrink components of the predictors that have low variance, in other words, that explain low amounts of the variance in the set of predictors available (Hastie et al. 2011). Ridge regression coefficients β_{ridge} are chosen to minimise a penalised RSS criterion.

$$\beta_{\text{ridge}} = \underset{\beta}{\arg\min} \left\{ \sum_{i=1}^{n} \left(y_i - \beta_0 - \sum_{j=1}^{p} x_{ij}\beta_j \right)^2 + \lambda \sum_{j=1}^{p} \beta_j^2 \right\} \tag{9.9}$$

The first term in the braces is the RSS and the second term is the quadratic penalty imposed on the ridge coefficients. Equivalently, in ridge regression, the estimated coefficients minimise the RSS subject to the constraint that $\sum_{j=1}^{p} \beta_j^2 \leq t$ where t is a threshold limiting the size of the coefficients. There is a one-to-one relationship between λ and t; as λ is increased, indicating greater penalty, t is reduced, indicating a lower threshold on the size of the coefficients (Hastie et al. 2011). Software used to fit ridge regression solves the penalised RSS criterion for a range of values of either λ or t and cross-validation is used to identify the value of λ or t that has the lowest prediction error. Note that the model intercept (β_0) is not included in the penalty and that the predictor variables are standardised to zero mean and unit variance before estimation of the ridge coefficients. Where $\lambda = 0$, the ridge coefficients are equivalent to the usual least-squares estimates of the model coefficients.

It is important to note that ridge regression does not perform variable selection; all available predictor variables remain in the model, it is just their coefficients that are shrunk away from the least-squares estimates. The lasso (Tibshirani 1996) is related to ridge regression but can also perform variable selection because it employs a different penalty on the coefficients to that of the ridge penalty. The lasso (least absolute shrinkage and selection operator) imposes a restriction on the size of the absolute values of the coefficients instead of a restriction on the squared values of the coefficients used in ridge regression. The lasso finds coefficients β_{lasso} that minimise the following penalised RSS criterion

$$\beta_{\text{lasso}} = \arg\min_{\beta} \left\{ \frac{1}{2} \sum_{i=1}^{n} \left(y_i - \beta_0 - \sum_{j=1}^{p} x_{ij} \beta_j \right)^2 + \lambda \sum_{j=1}^{p} |\beta_j| \right\} \tag{9.10}$$

which is equivalent to minimising the RSS subject to the constraint that $\sum_{j=1}^{p} |\beta_j| \leq t$ (Hastie et al. 2011). This penalty allows variables whose coefficients are shrunk to zero to be removed from the model. As before, cross-validation is used to identify the value of λ or t with the lowest prediction error. It can be shown that ridge regression shrinks all coefficients proportionally, and the lasso shrinks each coefficient by a constant factor λ and truncates at zero (e.g., a positive coefficient that would otherwise go negative when shrunk by the factor λ is removed from the model). The lasso is a general technique and has been successfully applied to generalised linear models (Tibshirani 1996) and is used as a form of shrinkage in boosted trees (De'ath 2007). A fast computer algorithm, least angle regression (LARS) was developed by Efron et al. (2004) that can compute the entire lasso path from no predictors in the model to the full least-squares solution for the same computational cost as the least-squares solution. Park and Hastie (2007) have developed similar path algorithms for the lasso in a GLM setting.

Ridge regression shrinks the coefficients of correlated predictors and the lasso selects predictors via shrinkage. Ideally, these two characteristics would be combined into a single technique that handles correlated predictors and could perform model selection. This is exactly what the elastic-net penalty does, via a weighted combination of ridge-like and lasso-like penalties to form the elastic-net penalty

$$\lambda \sum_{j=1}^{p} \left(\alpha \beta_j^2 + (1 - \alpha) \left| \beta_j \right| \right) \tag{9.11}$$

where α controls the relative weighting of ridge-like and lasso-like penalties (Zou and Hastie 2005). Where there are correlated predictors, the elastic net will tend to shrink the coefficients for those predictors rather than necessarily dropping one of the predictors giving full weight in the model to the other predictor, which is how the lasso operates with collinear variables. Friedman et al. (2010) demonstrate an efficient path algorithm for fitting the elastic net regularisation path for GLMs.

Figure 9.27 shows ridge regression (Fig. 9.27a), lasso (Fig. 9.27b), and elastic net (Fig. 9.27c) regularisation paths for the ozone data considered in the MARS example earlier. The models were fitted to the log-transformed ozone concentration because gamma GLMs are not supported in the glmnet R package (version 1.6: Friedman et al. 2010) used here. We consider only the main effects of the nine predictor variables, and for the elastic net we use $\alpha = 0.5$, indicating equal amounts of ridge-like and lasso-like penalties. The left-hand panels of each figure show the regularisation path with the full least-squares solutions on the right of these plots; the y-axis represents the values of the coefficients for each predictor, whilst the lines on the plots describe how the values of the coefficients vary from total shrinkage to the their least-squares values. The right-hand panels show *k*-fold cross-validated mean squared error (MSE) for each regularisation path, here expressed on the $\log(\lambda)$ scale. The numbers on the top of each plot indicate the complexity of the models along the regularisation path or as a function of $\log(\lambda)$. For ridge regression, we note that all nine predictor variables remain in the model throughout the path, whereas for the lasso and elastic-net paths predictors are selected out of the model as an increasing amount of regularisation is applied.

An interesting feature of the ridge-regression path is the coefficient value for wind speed, which is negative in the least-squares solution but becomes positive after a small amount of shrinkage, before being shrunk back to zero as a stronger penalty is applied to the size of the coefficients. The coefficient value for wind speed does not show this pattern in either the lasso or the elastic-net regularisation paths because of the property that both these penalties share, whereby coefficients are truncated at zero and not allowed to change their sign. The elastic-net regularisation path is intermediate between those of the ridge and lasso, although it is most similar to the lasso path. The effect of the lower lasso-like penalty in the elastic-net path for the ozone model is for predictor variables to persist in the model until a higher overall penalty is applied than under the lasso path. However, whilst the nine predictors persist in the path for longer, the ridge part of the penalty is shrinking the size of the coefficients.

The right-hand panels in Fig. 9.27 indicate the optimal degree of shrinkage by identifying the value of λ that affords the lowest CV MSE (the left vertical line) or that is within one standard error of the minimum (the right vertical line). On these plots, model complexity *increases* from left to right. The optimal amount of shrinkage indicates that nine, five, and seven predictors should be included in the model for the ridge regression, lasso, and elastic-net penalties, respectively. Temperature is the most important variable in predicting the log ozone concentration, followed by humidity. At larger penalties in the lasso and elastic-net paths, pressure gradient replaces humidity as the second predictor, after temperature, to be selected in the model. We do not interpret these models further here.

This is an area of considerable research activity, much of which is of direct relevance to ecologists and palaeolimnologists but whose importance is poorly known (e.g., Dahlgren 2010). For example, ter Braak (2009) has developed a new

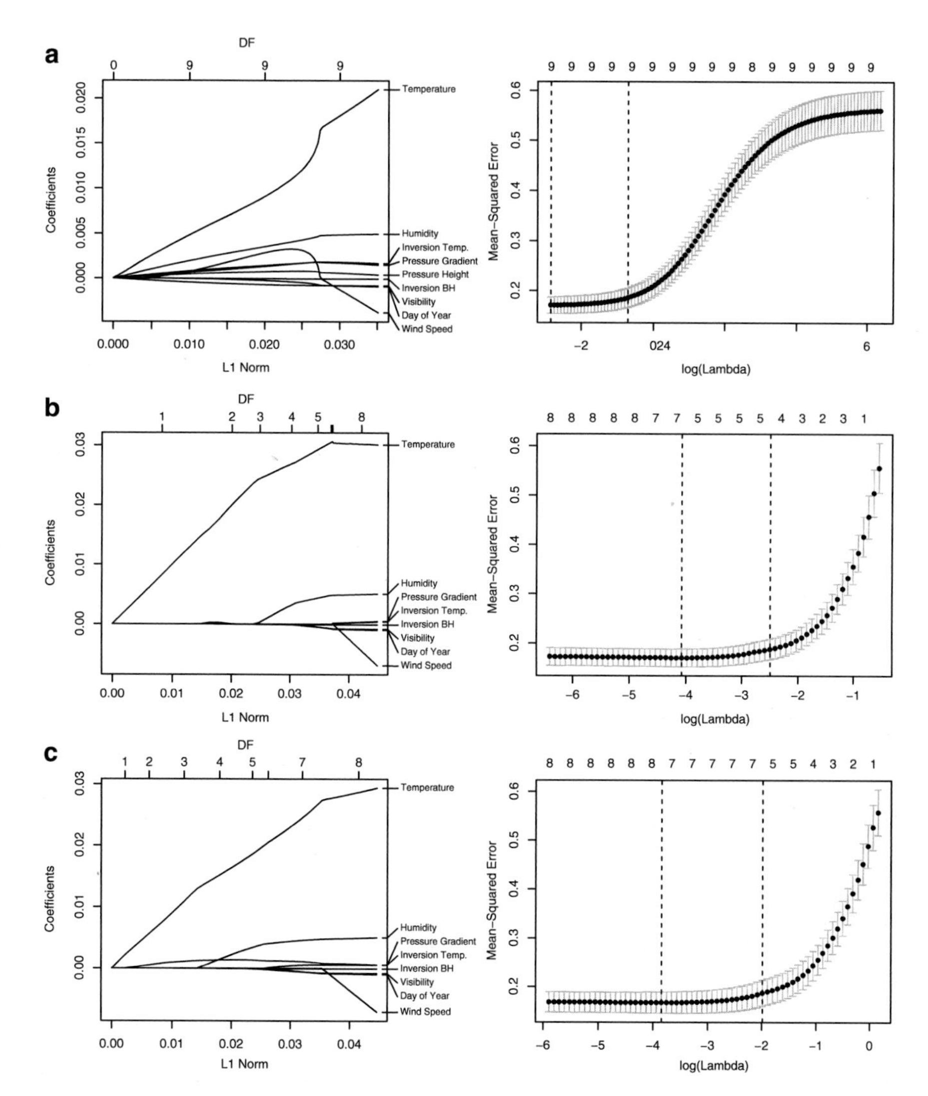

Fig. 9.27 Illustration of three shrinkage methods fitted to the ozone concentration data; (**a**) ridge regression, (**b**) the lasso, (**c**) the elastic net with $\alpha = 0.5$. The *left-hand panels* show the estimates of the regression coefficients for the entire regularisation path estimated, with the least complex model to the left. Estimates of the degrees of freedom associated with various values of the penalty are show on the upper axis of each panel. The *right-hand panels* show k-fold cross-validated model error for increasing (*left to right*) penalty. *Error bars* show the range of model errors across the k folds for each value of the penalty. The best model, with lowest mean squared error is highlighted by the *left-most dashed vertical line* in each panel, whilst the simplest model within one standard error of the best model is shown by the *right-most vertical line*. The values on the upper axis of each panel indicate the number of covariates included in the model for the value of the penalty

regression method, regularisation of smart contrasts and sums (ROSCAS), that outperforms the lasso, elastic net, ridge regression, and PLS when there are groups of predictors with each group representing an independent feature that influences the response and when the groups differ in size.

Discussion and Conclusions

This chapter has described several statistical machine-learning techniques, which can be loosely categorised into supervised and unsupervised learning techniques. The discussion for individual methods was intentionally brief, with the aim of introducing palaeolimnologists to the key features of the main machine-learning methods and illustrating their use. The references cited in each section should provide access to additional sources of information on each technique, and wherever possible we have referred to relevant palaeoecological or ecological papers.

A recurring theme in this chapter has been the reduction of bias, variance, or both in order to identify a model that has low prediction error. Given a model, $y = f(x) + \varepsilon$, that relates a response y to covariate x, we define the prediction error of a model as the expected difference between the true, unknown value of the response (y_0) and the predicted value for the response from the model, $\hat{f}(x)$. This prediction error can be decomposed into three components; (i) bias^2, (ii) variance, and (iii) ε, the irreducible error present even if we knew the true $f(x)$. We are unable to do anything about ε, so we must concern ourselves with trying to reduce bias, variance, or both in order to reduce prediction error. The bias^2 and variance together yield the mean squared error of the model (MSE).

To understand what each of these components is, consider a simple regression model fitted to a response y and covariate x. The relationship is quadratic and we have five observations. A simple least-squares model using one degree of freedom fitted to the data will yield predictions that follow a straight line. This model is very simple, but the straight line does not fit the data well; the model under-fits the data. Such a model will have high bias; over large parts of the observed data, the model systematically fails to capture the true relationship between x and y. Alternatively, we could fit a high degree polynomial that interpolates the training data perfectly, thus having zero bias. This is a more complex model but it over-fits the training data and is unlikely to generalise well to new observations for which we want to predict y. Such a model has high variance; each coefficient in the model has a high degree of uncertainty because we have used all the data to fit a large number of coefficients. In between these extremes is a model that has higher bias than the over-fitted model and lower bias than the simple model and the opposite features for model variance. Figure 9.28 illustrates this bias–variance tradeoff.

Several methods that we have introduced focus on reducing the variance part of MSE, such as bagged trees, random forests, and model averaging in an information theoretic framework. Shrinkage methods, introduced towards the end of the chapter,

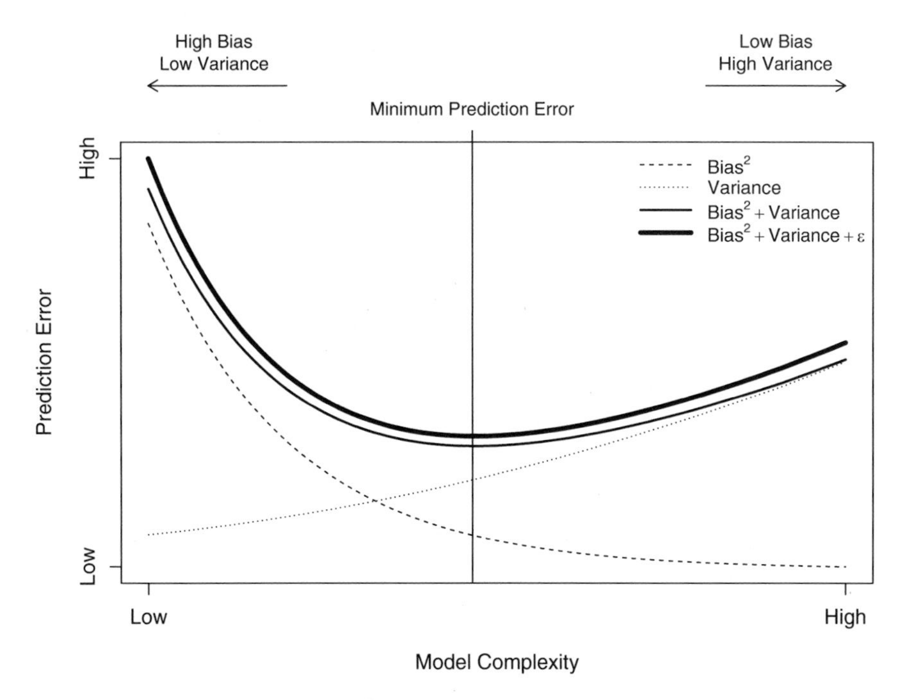

Fig. 9.28 Illustration of the bias–variance trade-off. At low complexity, models under-fit the observations and consequently have high bias which dominates the prediction error. At high complexity, models over-fit the data and as a result have low bias but high variance and the variance component dominates prediction error. Often the aim in statistical machine-learning is to fit a model that has minimal prediction error. Identifying such a model will require trading off bias against variance to achieve an overall lower prediction error. Bias^2 + Variance = MSE (mean squared error). ε is the irreducible error that is present in the model even if one knew the true relationship between the response and the predictors rather than having to estimate it

sacrifice a small increase in model bias (the estimates of regression coefficients using the methods are biased) for a larger reduction in model variance by shrinking coefficient estimates to zero. Of the methods discussed, only boosting has the potential to reduce both the bias and the variance of the fitted model. Bias is reduced by focussing on those observations that are poorly fitted by previous trees in the ensemble, whilst variance is reduced by averaging predictions over a large ensemble of trees.

Understanding the bias–variance trade-off is key to the successful use of statistical machine-learning where the focus is on producing a model for prediction that has the lowest possible prediction error given the available training data.

One feature of all of the techniques discussed is that they use the power of modern computers to learn aspects of the training data that allows the model to make accurate predictions. How well one of these algorithms or methods performs tends to be evaluated on the basis of its ability to predict the response variable on an independent test-set of samples. However, many, if not the majority of the techniques we describe do now have a thorough statistical underpinning (Hastie et al. 2011). This is especially so for the tree-based methods and boosting in particular.

What we have not been able to do here is illustrate *how* to go about fitting these sorts of models to data. Clearly, the availability of suitable software environments and code that implements these modern machine-learning methods is a prerequisite. All of the detailed examples have been performed by the authors with the R statistical software (version 2.13.1 patched r56332: R Core Development Team 2011) using a variety of add-on packages available on the Comprehensive R Archive Network (CRAN). A series of R scripts are available from the book website which replicate the examples used in this chapter and demonstrate how to use R and the add-on packages to fit the various models. We have used R because it is free and open source, and because of the availability of high-quality packages that implement all the machine-learning methods we have discussed. Other computational statistical software packages, such as MATLAB®, should also be able to fit most if not all the methods described here.

The technical and practical learning curves are far steeper for software such as R and the statistical approaches we discuss than the usual suspects of ordination, clustering, and calibration most commonly employed by palaeolimnologists. Machine-learning methods tend to place far higher demands on the user to get the best out of the techniques. One might reasonably ask if this additional effort is worthwhile? Ecological and palaeoecological data are inevitably noisy, complex, and high-dimensional. The sorts of machine-learning tools we have introduced here were specifically designed to handle such data and are likely to perform as well if not better than the traditional techniques most commonly used in the palaeolimnological realm. Furthermore, if all one knows is how to use CANOCO or C2 there will be a tendency to view all problems as ordination, calibration, or something else that cannot be handled. This situation is succinctly described as Maslow's Hammer; "it is tempting, if the only tool you have is a hammer, to treat every problem as if it were a nail" (Maslow 1966: p.15).

This chapter aims to provide an introduction to the statistical machine-learning techniques that have been shown to perform well in a variety of settings. We hope that it will suitably arm palaeolimnologists with the rudimentary knowledge required to know when to put down the hammer and view a particular problem as something other than a nail.

Acknowledgements We are indebted to Richard Telford, Steve Juggins, and John Smol for helpful comments and/or discussion. Whilst writing this chapter, GLS was supported by the European Union Seventh Framework Programme projects REFRESH (Contract N. 244121) and BioFresh (Contract No. 226874), and by the UK Natural Environment Research Council (grant NE/G020027/1). We are particularly grateful to Cathy Jenks for her editorial help. This is publication A359 from the Bjerknes Centre for Climate Research.

References

Aalders I (2008) Modeling land-use decision behavior with Bayesian belief networks. Ecol Soc 13:16

Aho K, Weaver T, Regele S (2011) Identification and siting of native vegetation types on disturbed land: demonstration of statistical methods. Appl Veg Sci 14:277–290

Allan JD, Yuan LL, Black P, Stockton T, Davies PE, Magierowski RH, Read SM (2011) Investigating the relationships between environmental stressors and stream conditions using Bayesian belief networks. Freshw Biol. doi:10.1111/j.1365-2427.2011.02683.x

Amsinck SL, Strzelczak A, Bjerring R, Landkildehus F, Lauridsen TL, Christoffersen K, Jeppesen E (2006) Lake depth rather than fish planktivory determines cladoceran community structure in Faroese lakes – evidence from contemporary data and sediments. Freshw Biol 51:2124–2142

Anderson DR (2008) Model based inference in the life sciences: a primer on evidence. Springer, New York

Anderson RP, Lew D, Peterson AT (2003) Evaluating predictive models of species' distributions: criteria for selecting optimal models. Ecol Model 162:211–232

Baker FA (1993) Classification and regression tree analysis for assessing hazard of pine mortality caused by *Heterobasidion annosum*. Plant Dis 77:136–139

Balshi MS, McGuire AD, Duffy P, Flannigan M, Walsh J, Melillo J (2009) Assessing the response of area burned to changing climate in western boreal North America using a Multivariate Adaptive Regression Splines (MARS) approach. Global Change Biol 15:578–600

Banfield JD, Raftery AE (1992) Ice floe identification in satellite images using mathematical morphology and clustering about principal curves. J Am Stat Assoc 87:7–16

Barrows TT, Juggins S (2005) Sea-surface temperatures around the Australian margin and Indian Ocean during the Last Glacial Maximum. Quaternary Sci Rev 24:1017–1047

Barton AM, Nurse AM, Michaud K, Hardy SW (2011) Use of CART analysis to differentiate pollen of red pine *(Pinus resinosa)* and jack pine (*P. banksiana*) in New England. Quaternary Res 75:18–23

Belgrano A, Malmgren BA, Lindahl O (2001) Application of artificial neural networks (ANN) to primary production time-series data. J Plankton Res 23:651–658

Benito Garzón M, Blazek R, Neteler M, Sánchez de Dios R, Sainz Ollero H, Furlanello C (2006) Predicting habitat suitability with machine learning models: the potential area of *Pinus sylvestris* L. in the Iberian Peninsula. Ecol Model 197:383–393

Benito Garzón M, Sánchez de Dios R, Sainz Ollero H (2007) Predictive modelling of tree species distributions on the Iberian Peninsula during the Last Glacial Maximum and Mid-Holocene. Ecography 30:120–134

Benito Garzón M, Sánchez de Dios R, Sainz Ollero H (2008) Effects of climate change on the distribution of Iberian tree species. Appl Veg Sci 11:169–178

Birks HH, Mathewes RW (1978) Studies in the vegetational history of Scotland. V. Late Devensian and early Flandrian pollen and macrofossil stratigraphy at Abernethy Forest, Inverness-shire. New Phytol 80:455–484

Birks HJB (1995) Quantitative palaeoenvironmental reconstructions. In: Maddy D, Brew J (eds) Statistical modelling of Quaternary science data. Volume 5: Technical guide. Quaternary Research Association, Cambridge, pp 161–254

Birks HJB (2012a) Chapter 2: Overview of numerical methods in palaeolimnology. In: Birks HJB, Lotter AF, Juggins S, Smol JP (eds) 2012. Tracking environmental change using lake sediments. Volume 5: Data handling and numerical techniques. Springer, Dordrecht

Birks HJB (2012b) Chapter 11: Stratigraphical data analysis. In: Birks HJB, Lotter AF, Juggins S, Smol JP (eds) Tracking environmental change using lake sediments. Volume 5: Data handling and numerical techniques. Springer, Dordrecht

Birks HJB, Gordon AD (1985) Numerical methods in Quaternary pollen analysis. Academic Press, London

Birks HJB, Jones VJ (2012) Chapter 3: Data-sets. In: Birks HJB, Lotter AF, Juggins S, Smol JP (eds) Tracking environmental change using lake sediments. Volume 5: Data handling and numerical techniques. Springer, Dordrecht

Birks HJB, Line JM, Juggins S, Stevenson AC, ter Braak CJF (1990) Diatoms and pH reconstruction. Philos Trans R Soc Lond B 327:263–278

Bishop CM (1995) Neural networks for pattern recognition. Clarendon, Oxford

Bishop CM (2007) Pattern recognition and machine learning. Springer, Dordrecht

Bjerring R, Becares E, Declerck S et al. (2009) Subfossil Cladocera in relation to contemporary environmental variables in 54 pan-European lakes. Freshw Biol 54:2401–2417

Blaauw M, Heegaard E (2012) Chapter 12: Estimation of age-depth relationships. In: Birks HJB, Lotter AF, Juggins S, Smol JP (eds) Tracking environmental change using lake sediments. Volume 5: Data handling and numerical techniques. Springer, Dordrecht

Borggaard C, Thodberg HH (1992) Optimal minimal neural interpretation of spectra. Anal Chem 64:545–551

Bourg NA, McShea WJ, Gill DE (2005) Putting a CART before the search: successful habitat prediction for a rare forest herb. Ecology 86:2793–2804

Breiman L (1996) Bagging predictors. Mach Learn 24:123–140

Breiman L (2001) Random forests. Mach Learn 45:5–32

Breiman L, Friedman JH, Olshen RA, Stone CJ (1984) Classification and regression trees. Wadsworth, Belmont

Brosse S, Guégan J-F, Tourenq J-N, Lek S (1999) The use of artificial neural networks to assess fish abundance and spatial occupancy in the littoral zone of a mesotrophic lake. Ecol Model 120:299–311

Brunelle A, Rehfeldt GE, Bentz B, Munson AS (2008) Holocene records of *Dendroctonus* bark beetles in high elevation pine forests of Idaho and Montana, USA. Forest Ecol Manage 255:836–846

Burman P, Chow E, Nolan D (1994) A cross-validatory method for dependent data. Biometrika 81:351–358

Burnham KP, Anderson DR (2002) Model selection and multimodel inference: a practical information-theoretic approach, 2nd edn. Springer, New York

Cairns DM (2001) A comparison of methods for predicting vegetation type. Plant Ecol 156:3–18

Caley P, Kuhnert PM (2006) Application and evaluation of classification trees for screening unwanted plants. Austral Ecol 31:647–655

Carlisle DM, Wolock DM, Meador MR (2011) Alteration of streamflow magnitudes and potential ecological consequences: a multiregional assessment. Front Ecol Environ 9:264–270

Castelletti A, Soncini-Sessa R (2007a) Bayesian networks and participatory modelling in water resource management. Environ Model Softw 22:1075–1088

Castelletti A, Soncini-Sessa R (2007b) Coupling real-time and control and socio-economic issues in participatory river basin planning. Environ Model Softw 22:1114–1128

Céréghino R, Giraudel JL, Compin A (2001) Spatial analysis of stream invertebrates distribution in the Adour-Garonne drainage basin (France), using Kohonen self-organizing maps. Ecol Model 146:167–180

Černá L, Chytrý M (2005) Supervised classification of plant communities with artificial neural networks. J Veg Sci 16:407–414

Chapman DS (2010) Weak climatic associations among British plant distributions. Global Ecol Biogeogr 19:831–841

Chapman DS, Purse BV (2011) Community versus single-species distribution models for British plants. J Biogeogr 38:1524–1535

Chapman DS, Bonn A, Kunin WE, Cornell SJ (2010) Random forest characterization of upland vegetation and management burning from aerial imagery. J Biogeogr 37:37–46

Chatfield C (1993) Neural networks: forecasting breakthrough or passing fad? Int J Forecast 9:1–3

Chon T-S (2011) Self-organising maps applied to ecological sciences. Ecol Inform 6:50–61

Chytrý M, Jarošik V, Pyšek P, Hájek O, Knollová I, Tichý L, Danihelka J (2008) Separating habitat invasibility by alien plants from the actual level of invasion. Ecology 89:1541–1553

Copas JB (1983) Regression, prediction and shrinkage. J R Stat Soc Ser B 45:311–354
Cutler A, Stevens JR (2006) Random forests for microarrays. Methods Enzymol 411:422–432
Cutler DR, Edwards TC, Beard KH, Cutler A, Hess KT, Gibson J, Lawler JJ (2007) Random forests for classification in ecology. Ecology 88:2783–2792
Dahlgren JP (2010) Alternative regression methods are not considered in Murtaugh (2009) or by ecologists in general. Ecol Lett 13:E7–E9
Davidson TA, Sayer CD, Perrow M, Bramm M, Jeppesen E (2010a) The simultaneous inference of zooplanktivorous fish and macrophyte density from sub-fossil cladoceran assemblages: a multivariate regression tree approach. Freshw Biol 55:546–564
Davidson TA, Sayer CD, Langdon PG, Burgess A, Jackson MJ (2010b) Inferring past zooplanktivorous fish and macrophyte density in a shallow lake: application of a new regression tree model. Freshw Biol 55:584–599
De'ath G (1999) Principal curves: a new technique for indirect and direct gradient analysis. Ecology 80:2237–2253
De'ath G (2002) Multivariate regression trees: a new technique for modeling species-environment relationships. Ecology 83:1108–1117
De'ath G (2007) Boosted trees for ecological modeling and prediction. Ecology 88:243–251
De'ath G, Fabricius KE (2000) Classification and regression trees: a powerful yet simple technique for ecological data analysis. Ecology 81:3178–3192
De'ath G, Fabricius KE (2010) Water quality as a regional driver of coral biodiversity and macroalgae on the Great Barrier Reef. Ecol Appl 20:840–850
DeFries RS, Rudel T, Uriarte M, Hansen M (2010) Deforestation driven by urban population growth and agricultural trade in the twenty-first century. Nat Geosci 3:178–181
Despagne F, Massart D-L (1998) Variable selection for neural networks in multivariate calibration. Chemometr Intell Lab Syst 40:145–163
D'heygere T, Goethals PLM, de Pauw N (2003) Use of genetic algorithms to select input variables in decision tree models for the prediction of benthic macroinvertebrates. Ecol Model 160:291–300
Dobrowski SZ, Thorne JH, Greenberg JA, Safford HD, Mynsberge AR, Crimins SM, Swanson AK (2011) Modeling plant ranges over 75 years of climate change in California, USA: temporal transferability and species traits. Ecol Monogr 81:241–257
Dutilleul P, Cumming BF, Lontoc-Roy M (2012) Chapter 16: Autocorrelogram and periodogram analyses of palaeolimnological temporal series from lakes in central and western North America to assess shifts in drought conditions. In: Birks HJB, Lotter AF, Juggins S, Smol JP (eds) Tracking environmental change using lake sediments. Volume 5: Data handling and numerical techniques. Springer, Dordrecht
Efron B (1979) Bootstrap methods: another look at the jackknife. Ann Stat 7:1–26
Efron B, Tibshirani R (1991) Statistical data analysis in the computer age. Science 253:390–395
Efron B, Tibshirani R (1993) An introduction to the bootstrap. Chapman & Hall, London
Efron B, Hastie T, Johnstone I, Tibshirani R (2004) Least angle regression. Ann Stat 32:407–499
Elith J, Burgman M (2002) Predictions and their validation: rare plants in the Central Highlands, Victoria, Australia. In: Scott JM, Heglund P, Morrison ML, Raven PH (eds) Predicting species occurrences: issues of accuracy and scale. Island Press, Washington, DC
Elith J, Leathwick JR (2007) Predicting species distributions from museum and herbarium records using multiresponse models fitted with multivariate adaptive regression splines. Divers Distrib 13:265–275
Elith J, Graham CH, Anderson RP, Dudík M, Ferrier S, Guisan A et al. (2006) Novel methods improve prediction of species' distributions from occurrence data. Ecography 29:129–151
Elith J, Leathwick JR, Hastie T (2008) A working guide to boosted regression trees. J Anim Ecol 77:802–813
Fielding AH (2007) Cluster and classification techniques for the biosciences. Cambridge University Press, Cambridge
Franklin J (1998) Predicting the distribution of shrub species in southern California from climate and terrain-derived variables. J Veg Sci 9:733–748

Franklin J (2010) Mapping species distributions — spatial inference and prediction. Cambridge University Press, Cambridge
Freund Y (1995) Boosting a weak learning algorithm by majority. Inf Comput 121:256–285
Friedman JH (1991) Multivariate adaptive regression splines. Ann Stat 19:1–67
Friedman JH (2001) Greedy function approximation: a gradient boosting machine. Ann Stat 29:1189–1232
Friedman JH (2002) Stochastic gradient boosting. Comput Stat Data Anal 38:367–378
Friedman G, Meulman JJ (2003) Multivariate adaptive regression trees with application in epidemiology. Stat Med 22:1365–1381
Friedman JH, Hastie T, Tibshirani R (2000) Additive logistic regression: a statistical view of boosting. Ann Stat 28:337–407
Friedman J, Hastie T, Tibshirani R (2010) Regularization paths for generalized linear models via coordinate descent. J Stat Software 33:1–22
Furlanello C, Neteler M, Merler S, Menegon S, Fontanari S, Donini A, Rizzoli A, Chemini C (2003) GIS and the random forests predictor: integration in R for tick-borne disease risk. In: Hornik K, Leitch F, Zeileis A (eds) Proceedings of the third international workshop on distributed statistical computings, Vienna, Austria. pp 1–11
Gevrey M, Dimopoulos I, Lek S (2003) Review and comparison of methods to study the contribution of variables in artificial neural network models. Ecol Model 160:249–264
Giraudel JL, Lek S (2001) A comparison of self-organising map algorithm and some conventional statistical methods for ecological community ordination. Ecol Model 146:329–339
Gordon AD (1973) Classifications in the presence of constraints. Biometrics 29:821–827
Gordon AD, Birks HJB (1972) Numerical methods in Quaternary palaeoecology. I. Zonation of pollen diagrams. New Phytol 71:961–979
Gordon AD, Birks HJB (1974) Numerical methods in Quaternary palaeoecology. II. Comparison of pollen diagrams. New Phytol 73:221–249
Goring S, Lacourse T, Pellatt MG, Walker IR, Matthewes RW (2010) Are pollen-based climate models improved by combining surface samples from soil and lacustrine substrates? Rev Palaeobot Palynol 162:203–212
Grieger B (2002) Interpolating paleovegetation data with an artificial neural network approach. Global Planet Change 34:199–208
Guégan J-F, Lek S, Oberdorff T (1998) Energy availability and habitat heterogeneity predict global riverine fish diversity. Nature 391:382–384
Hastie T, Stuetzle W (1989) Principal curves. J Am Stat Assoc 84:502–516
Hastie T, Tibshirani R, Friedman J (2011) The elements of statistical learning, 2nd edn. Springer, New York
Haykin S (1999) Neural networks, 2nd edn. Prentice-Hall, Upper Saddle River
Hejda M, Pyšek P, Jarošik V (2009) Impact of invasive plants on the species richness, diversity and composition of invaded communities. J Ecol 97:393–403
Herzschuh U, Birks HJB (2010) Evaluating the indicator value of Tibetan pollen taxa for modern vegetation and climate. Rev Palaeobot Palynol 160:197–208
Hoerl AE, Kennard R (1970) Ridge regression: biased estimation for nonorthogonal problems. Technometrics 12:55–67
Holmqvist BH (2005) Classification of large pollen datasets using neural networks with application to mapping and modelling pollen data. LUNDQUA report 39, Lund University
Horsak M, Chytrý M, Pokryszko BM, Danihelka J, Ermakov N, Hajek M, Hajkova P, Kintrova K, Koci M, Kubesova S, Lustyk P, Otypkova Z, Pelánková B, Valachovic M (2010) Habitats of relict terrestrial snails in southern Siberia: lessons for the reconstruction of palaeoenvironments of full-glacial Europe. J Biogeogr 37:1450–1462
Iverson LR, Prasad AM (1998) Predicting abundance of 80 tree species following climate change in the eastern United States. Ecol Mongr 68:465–485
Iverson LR, Prasad AM (2001) Potential changes in tree species richness and forest community types following climate change. Ecosystems 4:186–199

Iverson LR, Prasad AM, Schwartz MW (1999) Modeling potential future individual tree-species distributions in the eastern United States under a climate change scenario: a case study with *Pinus virginiana*. Ecol Model 115:77–93

Iverson LR, Prasad AM, Matthews SN, Peters M (2008) Estimating potential habitat for 134 eastern US tree species under six climate scenarios. Forest Ecol Manage 254:390–406

Jacob G, Marriott FHC, Robbins PA (1997) Fitting curves to human respiratory data. Appl Stat 46:235–243

Jensen FV, Nielsen TD (2007) Bayesian networks and decision graphs, 2nd edn. Springer, New York

Jeschke JM, Strayer DL (2008) Usefulness of bioclimatic models for studying climate change and invasive species. Ann NY Acad Sci 1134:1–24

Jolliffe IT (2002) Principal component analysis, 2nd edn. Springer, New York

Juggins S, Birks HJB (2012) Chapter 14: Quantitative environmental reconstructions from biological data. In: Birks HJB, Lotter AF, Juggins S, Smol JP (eds) Tracking environmental change using lake sediments. Volume 5: Data handling and numerical techniques. Springer, Dordrecht

Juggins S, Telford RJ (2012) Chapter 5: Exploratory data analysis and data display. In: Birks HJB, Lotter AF, Juggins S, Smol JP (eds) Tracking environmental change using lake sediments. Volume 5: Data handling and numerical techniques. Springer, Dordrecht

Kallimanis AS, Ragia V, Sgardelis SP, Pantis JD (2007) Using regression trees to predict alpha diversity based upon geographical and habitat characteristics. Biodivers Conserv 16:3863–3876

Keith RP, Veblen TT, Schoennagel TL, Sherriff RL (2010) Understory vegetation indicates historic fire regimes in ponderosa pine-dominated ecosystems in the Colorado Front Range. J Veg Sci 21:488–499

Kohonen T (2001) Self-organising maps, 3rd edn. Springer, Berlin

Korb KB, Nicholson AE (2004) Bayesian artificial intelligence. Chapman & Hall, Boca Raton

Kragt ME, Newham LTH, Jakeman AJ (2009) A Bayesian network approach to integrating economic and biophysical modelling. In: Anderssen RS, Braddock RD, Newham LTH (eds) 18th world IMACS congress and MODSIM09 international congress on modelling and simulation. Modelling and Simulation Society of Australia and New Zealand and International Association for Mathematics and Computers in Simulation, Cairns, Australia. pp 2377–2383

Kucera M, Weinelt M, Kiefer T, Pflaumann U, Hayes A, Chen MT, Mix AC, Barrows TT, Cortijo E, Duprat J, Juggins S, Waelbroeck C (2005) Reconstruction of sea-surface temperatures from assemblages of planktonic foraminifera: multi-technique approach based on geographically constrained calibration data sets and its application to glacial Atlantic and Pacific Oceans. Quaternary Sci Rev 24:951–998

Larsen DR, Speckman PL (2004) Multivariate regression trees for analysis of abundance data. Biometrics 60:543–549

Lawler JJ, White D, Neilson RP, Blaustein AR (2006) Predicting climate-induced range shifts: model differences and model reliability. Global Change Biol 12:1568–1584

Leathwick JR, Rowe D, Richardson J, Elith J, Hastie T (2005) Using multivariate adaptive regression splines to predict the distributions of New Zealand's freshwater diadromous fish. Freshw Biol 50:2034–2052

Leathwick JR, Elith J, Hastie T (2006) Comparative performance of generalized additive models and multivariate adaptive regression splines for statistical modelling of species distributions. Ecol Model 199:188–196

Legendre P, Birks HJB (2012a) Chapter 7: Clustering and partitioning. In: Birks HJB, Lotter AF, Juggins S, Smol JP (eds) Tracking environmental change using lake sediments. Volume 5: Data handling and numerical techniqlues. Springer, Dordrecht

Legendre P, Birks HJB (2012a) Chapter 8: From classical to canonical ordination. In: Birks HJB, Lotter AF, Juggins S, Smol JP (eds) Tracking environmental change using lake sediments. Volume 5: Data handling and numerical techniques. Springer, Dordrecht

Lek S, Guégan JF (1999) Artificial neural networks as a tool in ecological modelling, an introduction. Ecol Model 120:65–73

Lek S, Guégan J-F (2000) Artificial neuronal networks: application to ecology and evolution. Springer, Berlin

Lek S, Delacoste M, Baran P, Dimopoulos I, Lauga J, Aulagnier S (1996a) Application of neural networks to modelling nonlinear relationships in ecology. Ecol Model 90:39–52

Lek S, Dimopoulos I, Fabre A (1996b) Predicting phosphorus concentration and phosphorus load from watershed characteristics using backpropagation neural networks. Acta Oecol 17:43–53

Lindblah M, O'Connor R, Jacobson GL Jr (2002) Morphometric analysis of pollen grains for palaeoecological studies: classification of *Picea* from eastern North America. Am J Bot 89:1459–1467

Lindblah M, Jacobson GL Jr, Schauffler M (2003) The postglacial history of three *Picea* species in New England, USA. Quaternary Res 59:61–69

Lindström J, Kokko H, Ranta E, Lindén H (1998) Predicting population fluctuations with artificial neural networks. Wildl Biol 4:47–53

Lotter AF, Anderson NJ (2012) Chapter 18: Limnological responses to environmental changes at inter-annual to decadal time-scales. In: Birks HJB, Lotter AF, Juggins S, Smol JP (eds) Tracking environmental change using lake sediments. Volume 5: Data handling and numerical techniques. Springer, Dordrecht

Malmgren BA, Nordlund U (1997) Application of artificial neural networks to paleoceanographic data. Palaeogeogr Palaeoclim Palaeoecol 136:359–373

Malmgren BA, Winter A (1999) Climate zonation in Puerto Rico based on principal component analysis and an artificial neural network. J Climate 12:977–985

Malmgren BA, Kucera M, Nyberg J, Waelbroeck C (2001) Comparison of statistical and artificial neural network techniques for estimating past sea surface temperatures from planktonic foraminfer census data. Paleoceanography 16:520–530

Manel S, Dias JM, Buckton ST, Ormerord SJ (1999a) Alternative methods for predicting species distribution: an illustration with Himalayan river birds. J Appl Ecol 36:734–747

Manel S, Dias JM, Ormerord SJ (1999b) Comparing discriminant analysis, neural networks and logistic regression for predicting species distributions: a case study with a Himalayan river bird. Ecol Model 120:337–347

Marcot BG, Holthausen RS, Raphael MG, Rowland MG, Wisdom MJ (2001) Using Bayesian belief networks to evaluate fish and wildlife population viability under land management alternatives from an environmental impact statement. Forest Ecol Manage 153:29–42

Martens H, Næs T (1989) Multivariate calibration. Wiley, Chichester

Maslow AH (1966) The psychology of science: a reconnaissance. Harper & Row, New York

Melssen W, Wehrens R, Buydens L (2006) Supervised Kohonen networks for classification problems. Chemometr Intell Lab Syst 83:99–113

Melssen W, Bulent U, Buydens L (2007) SOMPLS: a supervised self-organising map-partial least squares algorithm for multivariate regression problems. Chemometr Intell Lab Syst 86:102–120

Michaelson J, Schimel DS, Friedl MA, Davis FW, Dubayah RC (1994) Regression tree analysis of satellite and terrain data to guide vegetation sampling and surveys. J Veg Sci 5:673–686

Milbarrow S (2011) earth. R package version 3.2-0. http://cran.r-project.org/packages=earth

Miller AJ (2002) Subset selection in regression, 2nd edn. Chapman & Hall/CRC, Boca Raton

Miller J, Franklin J (2002) Modeling the distribution of four vegetation alliances using generalized linear models and classification trees with spatial dependence. Ecol Model 157:227–247

Moisen GG, Frescino TS (2002) Comparing five modelling techniques for predicting forest characteristics. Ecol Model 157:209–225

Morgan JN, Sonquist JA (1963) Problems in the analysis of survey data, and a proposal. J Am Stat Assoc 58:415–434

Mundry R, Nunn CL (2009) Stepwise model fitting and statistical inference: turning noise into signal pollution. Am Nat 173:119–123

Murphy B, Jansen C, Murray J, de Barro P (2010) Risk analysis on the Australian release of *Aedes aegypti* (L.) (Diptera: Culicidae) containing *Wolbachia*. CSIRO, Canberra

Murtaugh PA (2009) Performance of several variable-selection methods applied to real ecological data. Ecol Lett 12:1061–1068

Nakagawa S, Freckleton RP (2008) Missing inaction: the danger of ignoring missing data. Trends Ecol Evol 23:592–596

Newton AC, Marshall E, Schreckenberg K, Golicher D, te Velde DW, Edouard F, Arancibia E (2006) Use of a Bayesian belief network to predict the impacts of commercializing non-timber forest products on livelihoods. Ecol Soc 11:24

Newton AC, Stewart GB, Diaz A, Golicher D, Pullin AS (2007) Bayesian belief networks as a tool for evidence-based conservation management. J Nat Conserv 15:144–160

Nyberg H, Malmgren BA, Kuijpers A, Winter A (2002) A centennial-scale variability of tropical North Atlantic surface hydrology during the late Holocene. Palaeogeogr Palaeoclim Palaeoecol 183:25–41

Næs T, Kvaal K, Isaksson T, Miller C (1993) Artificial neural networks in multivariate calibration. J NIR Spectrosc 1:1–11

Næs T, Isaksson T, Fearn T, Davies T (2002) A user-friendly guide to multivariate calibration and classification. NIR Publications, Chichester

Olden JD (2000) An artificial neural network approach for studying phytoplankton succession. Hydrobiologia 436:131–143

Olden JD, Jackson DA (2002) Illuminating the 'black box': a randomization approach for understanding variable contributions in artificial neural networks. Ecol Model 154:135–150

Olden JD, Joy MK, Death RG (2004) An accurate comparison on methods for quantifying variable importance in artificial neural networks using simulated data. Ecol Model 178:389–397

Olden JD, Lawler JJ, Poff NL (2008) Machine learning methods without tears: a paper for ecologists. Quaternary Rev Biol 83:171–193

Ôzesmi SL, Tan CO, Özesmi U (2006) Methodological issues in building, training, and testing artificial neural networks in ecological applications. Ecol Model 195:83–93

Pakeman RJ, Torvell L (2008) Identifying suitable restoration sites for a scarce subarctic willow (*Salix arbuscula*) using different information sources and methods. Plant Ecol Divers 1:105–114

Park MY, Hastie T (2007) *l*1-Regularization path algorithm for generalised linear models. J R Stat Soc Ser B 69:659–677

Pearson RG, Thuiller W, Araújo MB, Martinez-Meyer E, Brotons L, McClean C, Miles L, Segurado P, DawsonTP Lees DC (2006) Model-based uncertainty in species range prediction. J Biogeogr 33:1704–1711

Pelánková B, Kuneš P, Chytrý M, Jankovská V, Ermakov N, Svobodová-Svitavaská H (2008) The relationships of modern pollen spectra to vegetation and climate along a steppe-forest-tundra transition in southern Siberia, explored by decision trees. Holocene 18:1259–1271

Peters J, De Baets B, Verhoest NEC, Samson R, Degroeve S, de Becker P, Huybrechts W (2007) Random forests as a tool for predictive ecohydrological modelling. Ecol Model 207:304–318

Peyron O, Guiot J, Cheddadi R, Tarasov P, Reille M, de Beaulieu J-L, Bottema S, Andrieu V (1998) Climatic reconstruction of Europe for 18,000 yr BP from pollen data. Quaternary Res 49:183–196

Peyron O, Jolly D, Bonnefille R, Vincens A, Guiot J (2000) Climate of East Africa 6000 ^{14}C yr BP as inferred from pollen data. Quaternary Res 54:90–101

Peyron O, Bégeot C, Brewer S, Heiri O, Magny M, Millet L, Ruffaldi P, van Campo E, Yu G (2005) Lateglacial climatic changes in eastern France (Lake Lautrey) from pollen, lake-levels, and chironomids. Quaternary Res 64:197–211

Ploner A, Brandenburg C (2003) Modelling visitor attendance levels subject to day of the week and weather: a comparison between linear regression models and regression trees. J Nat Conserv 11:297–308

Pourret O, Naïm P, Marcot B (eds) (2008) Bayesian networks. A practical guide to applications. Wiley, Chichester

Prasad AM, Iverson LR, Liaw A (2006) Newer classification and regression tree techniques: bagging and random forests for ecological prediction. Ecosystems 9:181–199

Pysek P, Bacher S, Chytrý M, Jarosik V, Wild J, Celesti-Grapow L, Gassó N, Kenis M, Lambdon PW, Nentwig W, Pergl J, Roques A, Sádlo J, Solarz W, Vilà M, Hiulme PE (2010) Contrasting patterns in the invasions of European terrestrial and freshwater habitats by alien plants, insects and vertebrates. Global Ecol Biogeogr 19:317–331

Quinlan J (1993) C4.5: programs for machine learning. Morgan Kaufman, San Mateo

R Development Core Team (2011) R: a language and environment for statistical computing. R foundation for statistical computing. Vienna, Austria. http://www.r-project.org

Racca JMJ, Philibert A, Racca R, Prairie YT (2001) A comparison between diatom-pH-inference models using artificial neural networks (ANN), weighted averaging (WA) and weighted averaging partial least square (WA-PLS) regressions. J Paleolimnol 26:411–422

Racca JMJ, Wild M, Birks HJB, Prairie YT (2003) Separating wheat from chaff: diatom taxon selection using an artificial neural network pruning algorithm. J Paleolimnol 29:123–133

Racca JMJ, Gregory-Eaves I, Pienitz R, Prairie YT (2004) Tailoring palaeolimnological diatom-based transfer functions. Can J Fish Aquat Sci 61:2440–2454

Ramakrishnan N, Grama A (2001) Mining scientific data. Adv Comput 55:119–169

Raymond B, Watts DJ, Burton H, Bonnice J (2005) Data mining and scientific data. Arct Antarct Alp Res 37:348–357

Recknagel F, French M, Harkonen P, Yabunaka K-I (1997) Artificial neural network approach for modelling and prediction of algal blooms. Ecol Model 96:11–28

Rehfeldt GE, Crookston NL, Warwell MV, Evans JS (2006) Empirical analyses of plant-climate relationships for the western United States. Int J Plant Sci 167:1123–1150

Rejwan C, Collins NC, Brunner LJ, Shuter BJ, Ridgway MS (1999) Tree regression analysis on the nesting habitat of smallmouth bass. Ecology 80:341–348

Ridgeway G (2007) Generalized boosted models: a guide to the gbm package. http://cran.r-project.org/web/packages/gbm/vignettes/gbm.pdf. Accessed 20 July 2011

Ridgeway G (2010) gbm. R package version 1.6-3.1. http://cran.r-project.org/web/packages/gbm/

Rieman B, Peterson JT, Clayton J, Howell P, Thurow R, Thompson W, Lee D (2001) Evaluation of potential effects of federal land management alternatives on trends of salmonids and their habitats in the interior Columbia River basin. Forest Ecol Manage 153:43–62

Ripley BD (2008) Pattern recognition and neural networks. Cambridge University Press, Cambridge

Roberts DR, Hamann A (2012) Predicting potential climate change impacts with bioclimate envelope models: a palaeoecological perspective. Global Ecol Biogeogr 21:121–133

Rose NL (2001) Fly-ash particles. In: Last WM, Smol JP (eds) Tracking environmental change using lake sediments. Volume 2: Physical and geochemical methods. Kluwer Academic Publishers, Dordrecht, pp 319–349

Rose NL, Juggins S, Watt J, Battarbee RW (1994) Fuel-type characterization of spheroidal carbonaceous particles using surface chemistry. Ambio 23:296–299

Schapire RE (1990) The strength of weak learnability. Mach Learn 5:197–227

Scull P, Franklin J, Chadwick OA (2005) The application of classification tree analysis to soil type prediction in a desert landscape. Ecol Model 181:1–15

Simpson GL (2012) Chapter 15: Modern analogue techniques. In: Birks HJB, Lotter AF, Juggins S, Smol JP (eds) Tracking environmental change using lake sediments. Volume 5: Data handling and numerical techniques. Springer, Dordrecht

Spadavecchia L, Williams M, Bell R, Stoy PC, Huntley B, van Wijk MT (2008) Topographic controls on the leaf area index and plant functional type of a tundra ecosystem. J Ecol 96:1238–1251

Spitz F, Lek S (1999) Environmental impact prediction using neural network modelling. An example in wildlife damage. J Appl Ecol 36:317–326

Steiner D, Pauling A, Nussbaumer SU, Nesje A, Luterbacher J, Wanner H, Zumbühl HJ (2008) Sensitivity of European glaciers to precipitation and temperature – two case studies. Clim Change 90:413–441

Stewart-Koster B, Bunn SE, Mackay SJ, Poff NL, Naiman RJ, Lake PS (2010) The use of Bayesian networks to guide investments in flow and catchment restoration for impaired river ecosystems. Freshw Biol 55:243–260
Stockwell DRB, Noble IR (1992) Induction of sets of rules from animal distribution data: a robust and informative method of data analysis. Math Comput Sims 33:385–390
Stockwell DRB, Peters D (1999) The GARP modelling system: problems and solutions to automated spatial prediction. Int J Geogr Info Sci 13:143–158
Stockwell DRB, Peterson AT (2002) Effects of sample size on accuracy of species distribution models. Ecol Model 148:1–13
Tarasov P, Peyron O, Guiot J, Brewer S, Volkova VS, Bezusko LG, Dorofeyuk NI, Kvavadze EV, Osipova IM, Panova NK (1999a) Late glacial maximum climate of the former Soviet Union and Mongolia reconstructed from pollen and plant macrofossil data. Clim Dyn 15:227–240
Tarasov P, Guiot J, Cheddadi R, Andreev AA, Bezusko LG, Blyakharchuk TA, Dorofeyuk NI, Filimonova LV, Volkova VS, Zernitskayo VP (1999b) Climate in northern Eurasia 6000 years ago reconstructed from pollen data. Earth Planet Sci Lett 171:635–645
Telford RJ, Birks HJB (2009) Design and evaluation of transfer functions in spatially structured environments. Quaternary Sci Rev 28:1309–1316
ter Braak CJF (2009) Regression by L_1 regularization of smart contrasts and sums (ROSCAS) beats PLS and elastic net in latent variable model. J Chemometr 23:217–228
Therneau TM, Atkinson B [R port by Ripley B] (2011) rpart: recursive partitioning. R package version 3.1-50. http://cran.r-project.org/package/rpart
Thuiller W, Araújo MB, Lavorel S (2003) Generalized models vs, classification tree analysis: predicting spatial distributions of plant species at different scales. J Veg Sci 14:669–680
Tibshirani R (1996) Regression shrinkage and selection via the lasso. J R Stat Soc Ser B 58:267–288
Ticehurst JL, Curtis A, Merritt WS (2011) Using Bayesian networks to complement conventional analyses to explore landholder management of native vegetation. Environ Model Softw 26:52–65
Tsaor A, Allouche O, Steinitz O, Rotem D, Kadmon R (2007) A comparative evaluation of presence-only methods for modelling species distribution. Divers Distrib 13:397–405
van Dijk ADJ, ter Braak CJF, Immink RG, Angenent GC, van Ham RCHJ (2008) Predicting and understanding transcription factor interactions based on sequence level determinants of combinatorial control. Bioinformatics 24:26–33
Vayssieres MP, Plant RE, Allen-Diaz BH (2000) Classification trees: an alternative non-parametric approach for predicting species distributions. J Veg Sci 11:679–694
Vincenzi S, Zucchetta M, Franzoi P, Pellizzato M, Pranovi F, de Leo GA, Torricelli P (2011) Application of a random forest algorithm to predict spatial distribution of the potential yield of *Ruditapes philippinarum* in the Venice lagoon, Italy. Ecol Model 222:1471–1478
Warner B, Misra M (1996) Understanding neural networks as statistical tools. Am Stat 50:284–293
Wehrens R (2011) Chemometrics with R: multivariate analysis in the natural sciences and life sciences. Springer, New York
Wehrens R, Buydens LMC (2007) Self- and super-organising maps in R: the Kohonen package. J Stat Softw 21:1–19
Weller AF, Harris AJ, Ware JA (2006) Artificial neural networks as potential classification tools for dinoflagellate cyst images: a case using the self-organizing map clustering algorithm. Rev Palaeobot Palynol 141:287–302
Whittingham MJ, Stephens PA, Bradbury RB, Freckleton RP (2006) Why do we still use step-wise modelling in ecology and behaviour? J Anim Ecol 75:1182–1189
Williams JN, Seo C, Thorne J, Nelson JK, Erwin S, O'Brien JM, Schwartz MW (2009) Using species distribution models to predict new occurrences for rare plants. Divers Distrib 15:565–576
Witten IH, Frank E (2005) Data mining: practical machine learning tools and techniques. Morgan Kaufmann/Elsevier, Amsterdam
Zou H, Hastie T (2005) Regularization and variable selection via the elastic net. J R Stat Soc Ser B 67:301–320

Part III
Numerical Methods for the Analysis of Stratigraphical Palaeolimnological Data

Chapter 10
Introduction and Overview of Part III

H. John B. Birks

Abstract This chapter attempts to put the other six chapters in Part III into the context of the four main phases of a palaeolimnological investigation. These are data collection and data assessment, data summarisation, data analysis, and data interpretation. The relevant numerical techniques described in the chapters in this Part and in Part II are mentioned along with other techniques such as difference diagrams, population modelling, changes in taxonomic richness through time, basic time-series analysis, and the use of SiZer and its relatives that are not discussed elsewhere in the book.

Keywords Autoregressive models • β-diversity • BSiZer • Calibration functions • Clustering • Data analysis • Data assessment • Data interpretation • Data summarisation • Difference diagrams • Error estimation • Exponential population models • Hypothesis testing • Identification • LOESS • Logistic population models • Ordination • Partitioning • Rate-of-change analysis • REDFIT • Richness estimation • Runs tests • SiNos • SiZer • Spectral analysis • Time-series analysis • Transfer functions • Wavelet analysis

H.J.B. Birks (✉)
Department of Biology and Bjerknes Centre for Climate Research, University of Bergen, PO Box 7803, Bergen N-5020, Norway

Environmental Change Research Centre, University College London, Pearson Building, Gower Street, London WC1E 6BT, UK

School of Geography and the Environment, University of Oxford, Oxford OX1 3QY, UK
e-mail: john.birks@bio.uib.no

H.J.B. Birks et al. (eds.), *Tracking Environmental Change Using Lake Sediments*, Developments in Paleoenvironmental Research 5, DOI 10.1007/978-94-007-2745-8_10,

Introduction

Stratigraphical palaeolimnological data are counts or measurements of different biological (e.g., diatoms, chironomids, ostracods), chemical (e.g., inorganic chemistry, organic compounds), physical (e.g., stable-isotope records, magnetic properties, loss-on-ignition, bulk density), lithostratigraphical (e.g., grain size, sediment type, volcanic tephras), or chronological (e.g., ^{210}Pb dates, ^{14}C dates) variables in several sediment samples at known depths in one or more sedimentary sequences (see Last and Smol 2001a, b; Smol et al. 2001a, b; Smol 2008; Pienitz et al. 2009 for reviews of the range of palaeolimnological variables currently studied). The most common data type in biologically-based palaeolimnology consists of counts of many (typically 50–500) different types of diatoms, chrysophytes, chironomids, etc. in a large number of samples (~50–500) from a sedimentary sequence. Such data are highly multivariate containing many variables (e.g., diatom taxa) and many objects (e.g., sediment samples) (see Birks 2012a: Chap. 2). In diatom analysis, the number of variables (diatom taxa) is usually much larger than the number of samples, whereas in chironomid, cladoceran, and ostracod analysis, the number of variables is usually less than the number of samples, as in pollen analysis (Birks 2007a). Stratigraphical palaeolimnological data can thus be 'fat and short' (diatom data) or 'thin and long' (chironomid, pollen, ostracod data). In contrast to the majority of Quaternary pollen-stratigraphical studies, palaeolimnological studies are often of a fine-resolution nature with a temporal resolution of 1–10 years. Data from such studies are often inherently noisy, highly variable, large, as well as being taxonomically diverse (Birks 1998). Usually the samples are not equally spaced in time, with the exception of samples from well-studied annual varved deposits (see Lotter and Anderson 2012: Chap. 18). Stratigraphical data often become noisier and noisier as the temporal resolution becomes finer and finer (e.g., Green 1983; Green and Dolman 1988; Peglar 1993), with the result that the inherent year-to-year variation may mask any underlying long-term trends. Such fine-resolution data provide a major challenge for numerical analysis (Birks 1998).

Numerical methods are useful tools at many stages in stratigraphical studies in palaeolimnology (Birks 2010; Table 10.1). During data collection they can be potentially useful in assisting in the identification of certain fossil types. In data assessment, statistical techniques are essential in estimating the inherent errors associated with the resulting counts or measurements. A range of numerical techniques are available to detect and summarise major underlying patterns in stratigraphical data. For single stratigraphical sequences, the numerical delimitation of assemblage zones can be a useful first step in data summarisation. Other numerical techniques such as ordination procedures can summarise temporal trends in one or several data-sets. Numerical methods are essential for quantitative data analysis in terms of quantifying rates of changes, estimating taxonomic richness, modelling population changes, detecting temporal patterns such as trends and periodicities, and quantitatively reconstructing past environmental conditions. The final stage of data interpretation can be greatly aided by numerical techniques in

Table 10.1 Overview of widely used numerical methods for the analysis of stratigraphical palaeolimnological data

Aim	Numerical methods	Relevant chapters
Data assessment and error estimation	Exploratory data analysis	5
	Laboratory and analytical uncertainties	6
	Age-depth model uncertainties	12
Data summarisation		
Single stratigraphical sequence	Clustering, partitioning, zonation	7, 9, 11, 16
	Ordination	8, 11
Two or more sequences	Core correlation by sequence-slotting	13
	Combined ordinations	8, 11
	Canonical ordination statistical modelling	8, 10, 11, 14, 18
Data analysis		
One or more sequences	Rate-of-change analysis	11
	Population modelling	10, 21
	Quantifying recent biotic change	7, 8, 11, 14, 15
	Changes in taxonomic richness	4
	Variation partitioning	8, 14, 18, 19, 20
	Detection of 'signal' at the expense of 'noise' – LOESS smoothers	2, 5, 11, 19
	Detection of 'significant' trends – SiZer, SiNos, BSiZer	10
	Temporal-domain temporal-series analysis	16
	Frequency-domain temporal-series analysis	16
	Quantitative palaeoenvironmental reconstructions	14, 15, 19, 20
Data interpretation	Community and assemblage reconstructions	7, 8, 15, 19
	Hypothesis testing about causative factor using canonical ordination and/or variation partitioning	8, 18, 19, 20

the reconstruction of past communities and in the testing of competing hypotheses about underlying causative factors such as climate change, pollution, and other human activities in determining patterns in stratigraphical palaeolimnological data.

Many of the numerical methods discussed here and in Part II are also widely used in Quaternary pollen analysis (see Birks and Gordon 1985; Birks 1987, 1992, 1998, 2007a; Bennett 2001; Bennett and Willis 2001; Birks and Seppä 2004).

The aim of this introductory overview chapter is to place the chapters in this Part and in other Parts of this volume in the context of stratigraphical data analysis and of the phases of data collection and data assessment, data summarisation, data analysis, and data interpretation (Table 10.1). In addition, brief accounts are given, where appropriate, of other quantitative techniques that can be useful in the numerical analysis of stratigraphical palaeolimnological data but that are not sufficiently important as to justify chapters in their own right.

Data Collection and Data Assessment

Numerical and statistical methods can assist in the identification of fossil taxa and hence in the basic collection of stratigraphical data (Birks 2012a: Chap. 2). Statistical techniques are essential in the assessment of the statistical uncertainties associated with microfossil counting and with establishing age-depth relationships, namely error estimation (Table 10.1).

Identification

Identification of fossil remains in stratigraphical sequences follows the procedures outlined by Birks (Birks 2012a: Chap. 2). Sometimes problems of differential preservation of fossils can arise and numerical procedures cannot really help with these problems – there is simply no substitute for careful microscopy and the experience of the analyst. Brodersen (2008) discusses the effects of misidentification of fossil chironomid head capsules in terms of Type-I (to see differences that in fact do not exist) and Type-II (not recognise differences that actually exist) errors on resulting palaeolimnological interpretations. The decline of taxonomic expertise world-wide is causing what Bortolus (2008) calls "error cascades in the biological sciences". This decline has serious impacts on palaeolimnology as reliable identifications are the basis of sound palaeolimnology and reliable identifications depend on sound taxonomy and meticulous documentation (Dolan 2011). The development of easy-to-use electronic iconographs and monographs (e.g., Kelly and Telford 2007) is an important contribution in this time of declining basic taxonomy. Heiri and Lotter (2010) (see also Maher et al. 2012: Chap. 6) illustrate the influence of different taxonomic resolutions on the robustness of chironomid-based temperature reconstructions.

Data Assessment and Error Estimation

All counts of fossil remains in sediment samples are, hopefully, an unbiased sample count of the total amount of fossils preserved in the sediment sample of interest. Several of the statistical approaches presented by Maher et al. (2012: Chap. 6) for error estimation are specifically designed for estimating the uncertainty associated with microfossil counts. Similarly the range of techniques within exploratory data analysis (Juggins and Telford 2012: Chap. 5) involves graphical summaries such as box-and-whisker plots, scatter-plots, stratigraphical-plots, and cave-plots are essential tools in data assessment (Table 10.1).

An important pre-requisite for almost all palaeolimnological studies is a reliable and robust chronology for the sequence(s) under study. Chronologies are usually based on ^{210}Pb or ^{14}C age determinations or, more rarely, are based on annual varves or volcanic ash layers. Blaauw and Heegaard (2012: Chap. 12) discuss the steps required in establishing age-depth relationships for sediment sequences, including the calibration of radiocarbon ages into calendar ages, the estimation of age-depth models and their associated uncertainties, and the selection of which model to use. Age-depth modelling is an area of active research at present and important advances can be expected in the future, particularly within a Bayesian framework (e.g., Blockley et al. 2004, 2007, 2008; Blaauw and Christen 2005; Wohlfarth et al. 2006; Bronk Ramsey 2007, 2009; Buck and Bard 2007; Haslett and Parnell 2008).

Data Summarisation

Data summarisation involves detecting the major patterns of variation within a single stratigraphical data-set and the patterns of similarity and difference between two or more stratigraphical sequences (Table 10.1).

Single Stratigraphical Data-Sets

A useful first step in the numerical analysis of a multivariate palaeolimnological data-set is to summarise the data as a few groups (zones) or in a few dimensions (Birks 2012c: Chap. 11). These are selected to fulfil predefined mathematical criteria such that the groups maximise the between-group variation relative to the within-group variation or the selected dimensions and ordination axes capture as much of the total variation in the data with the constraint that all the ordination axes are uncorrelated to each other. However, standard techniques for clustering and partitioning (e.g., agglomerative hierarchical clustering, K-means partitioning, two-way indicator species analysis – see Borcard et al. 2011; Legendre and Birks 2012a: Chap. 7) and ordination or dimension-reduction techniques (e.g., principal component analysis, correspondence analysis – see Borcard et al. 2011; Legendre and Birks 2012b: Chap. 8) do not take account of the stratigraphical nature of the data. One may lose important information when these methods are applied to stratigraphical data, with the result that objects with similar assemblages may be clustered together even though the objects are far apart stratigraphically, hence the need for clustering, partitioning, and ordination methods that can take specific account of the stratigraphical ordering of the data (see Legendre and Birks 2012a, b: Chaps. 7 and 8; Birks 2012c: Chap. 11). In certain specific studies, cluster analysis without stratigraphical constraints can be useful when the groupings within the clusters are plotted stratigraphically to detect patterns of recurrence in fossil and/or sedimentary variables (e.g., Grimm et al. 2006, 2011).

Stratigraphical sequences are most commonly summarised as a series of assemblage zones. Numerical zonation is implemented by stratigraphically constrained clustering or partitioning algorithms (Birks and Gordon 1985; Legendre and Birks 2012a: Chap. 7; Birks 2012c: Chap. 11) followed by a comparison of the variation of the zones with the variation expected under the broken-stick model (Bennett 1996; Birks 2012c: Chap. 11) to provide a basis for deciding how many zones should be selected and defined. Of the several constrained zonation techniques currently available, optimal partitioning using a sum-of-squares criterion is consistently the most robust and hence most useful numerical zonation procedure.

Ordination of stratigraphical data, without constraints of sample depth or age, by, for example, principal component analysis or correspondence analysis (Borcard et al. 2011; Legendre and Birks 2012b: Chap. 8) can be valuable in identifying the major patterns of variation, the nature of change, and any trends in the sequence (Birks 1987). Plotting the ordination results (sample scores) in a stratigraphical context can provide a useful summary of the major patterns and trends and allows the detection of abrupt changes within the sequence which may be obscured in zonation with its primary concern on partitioning the stratigraphical sequence into 'homogeneous' units or zones. Careful examination of the taxon (variable) scores or loadings on the ordination axes can also reveal which taxa are most influential to the sample scores for a given ordination axis, thereby providing an ecological interpretation of the observed patterns of stratigraphical variation.

With fossil assemblages that contain many taxa ($\sim$50 or more), it is important to plot their relative frequencies in simple stratigraphical-plots in a way that displays the major patterns of variation within the data as a whole. A very simple but effective way is to calculate the weighted average or 'optimum' of each taxon for age or depth, and to reorder the taxa in order of the optima, with taxa having high optima for age or depth being plotted first at the bottom left of the stratigraphical-plot and with taxa having low optima being plotted last at the top right of the plot (Janssen and Birks 1994). Alternatively taxa can be ordered on the basis of their modern weighted average optima for a particular environmental variable (e.g., total phosphorus – see Juggins and Birks 2012: Chap. 14).

Two or More Stratigraphical Sequences

When two or more palaeolimnological variables (e.g., diatoms, chironomids) have been studied in the same stratigraphical sequence, numerical zonations based on each set of variables (Birks 2012c: Chap. 11) and a comparison of the resulting partitions can help to identify common and unique changes in the different variables (Birks and Gordon 1985). Separate ordinations of the different data-sets (Legendre and Birks 2012b: Chap. 8) can help to summarise the major patterns within each data-set, and these patterns can be compared using, for example, oscillation logs (Birks 1987). In cases where one data-set can be regarded as representing 'response' variables (e.g., diatoms) and another data-set can be regarded as reflecting potential 'predictor' or explanatory variables (e.g., stable-isotope ratios, sediment geochemistry), constrained ordinations (e.g., redundancy analysis) and associated Monte Carlo permutation tests (see Borcard et al. 2011; Birks 2012a: Chap. 2; Legendre and Birks 2012b: Chap. 8; Lotter and Anderson 2012: Chap. 18) can be used to assess the statistical relationship between the two data-sets. This approach involves specific statistical testing of hypotheses and is strictly part of data interpretation discussed below (Table 10.1).

Correlating two or more stratigraphical sequences is often necessary for a variety of reasons (see Thompson et al. 2012: Chap. 13). It may be necessary to correlate several cores taken within the same area of a lake with a master dated core, to correlate cores taken within different parts of a lake, or to correlate stratigraphical sequences from different sites. Core correlation can be made on the basis of variables such as loss-on-ignition, or biostratigraphy and can incorporate external constraints (e.g., volcanic tephra layers) in the core correlation using a sequence-slotting algorithm (see Thompson et al. 2012: Chap. 13).

For detailed within-lake studies with multiple cores from shallow and deep water, a palaeolimnologist might be interested in the spatial variations in different taxa within the lake through time. After the various cores have been correlated by sequence-slotting (Thompson et al. 2012: Chap. 13), difference diagrams (Birks and Gordon 1985) can be constructed. The differences in fossil composition ($\Delta\, y_{ik}$) at the same time between two cores are calculated as

$$\Delta_1 y_{ik} \equiv y_{1ik} - y_{2ik}$$

where y_{1ik} and y_{2ik} denote the accumulation rates or relative percentages of taxon k in the sample corresponding to time i at cores 1 and 2, respectively. Alternatively the difference can be calculated (Jacobson 1979) as

$$\Delta_2 y_{ik} \equiv \log\left(y_{1ik} / y_{2ik}\right)$$

Both $\Delta_1 y_{ik}$ and $\Delta_2 y_{ik}$ can be plotted stratigraphically (for varying i) for each taxon k.

Difference diagrams are useful in highlighting differences between cores within a lake or between sequences from different sites. They require a reliable and independent time control to ensure that similar times are being compared. An important point is that the observed data are subject to statistical counting errors, as discussed by Maher et al. (2012: Chap. 6). As one is examining the differences between pairs of counts, each of which is subject to counting errors, the observed difference may have a relatively large standard error. A useful precaution would therefore be to plot not only the observed difference but also its associated standard error.

Data Analysis

Data analysis is used here to include specialised techniques that estimate particular numerical characteristics from stratigraphical data. Examples include rates-of-change, population changes and expansion and contraction rates, taxonomic richness, temporal-series analysis, quantification of recent change, and inferred past environment (Table 10.1).

Rate-of-Change Analysis

Rate-of-change analysis (Grimm and Jacobson 1992; Birks 2012c: Chap. 11) estimates the amount of compositional change per unit time in stratigraphical data. It is estimated by calculating a multivariate dissimilarity (e.g., chord distance (=Hellinger distance)) (Legendre and Birks 2012b: Chap. 8) between stratigraphically adjacent samples and by dividing the dissimilarity by the estimated age interval between the sample pairs. An alternative approach is to interpolate the stratigraphical data to constant time intervals and calculate the dissimilarity. They can, if required, be smoothed prior to interpolation. Rate-of-change analysis is critically dependent on a reliable chronology for the sequence (Birks 1998). As radiocarbon years do not equal calendar years (Blaauw and Heegaard 2012: Chap. 12), a carefully calibrated timescale or an independent absolute chronology (e.g., from annually laminated sediments: Lotter et al. 1992) is essential for reliable rate-of-change estimation (Birks 2012c: Chap. 11).

Population Analysis

Biostratigraphical data, when expressed as accumulation rates (individuals per unit area per unit time), can be viewed as temporal records of past populations within the lake under study. Splitting of individual taxon sequences (Birks 2012c: Chap. 11) (=sequence splitting of Birks and Gordon 1985) divides the accumulation rates of individual taxa into units of presence or absence, and when the taxon is present, into units of uniform mean and variance. It focuses on the individualistic behaviour of taxa and on similarities and differences in the patterns of change of taxa within one or more sequences (Birks and Gordon 1985).

Patterns in accumulation rates of individual taxa can be modelled using logistic or exponential population-growth models and population doubling-times can be estimated for individual taxa as a means of comparing population rates of change between taxa, sites, or geographical regions (e.g., Bennett 1983, 1986, 1988; Walker and Chen 1987; Magri 1989; MacDonald 1993; Fuller 1998).

If y denotes the population size (estimated as numbers cm^{-2} year^{-1}), the exponential and logistic models can be derived from simple assumptions about the rate of increase of y with time dy/dt. In the exponential model, it is assumed that dy/dt is proportional to y, namely

$$\frac{dy}{dt} = ry$$

where r is the intrinsic rate of population growth per unit time. This can be integrated to give

$$\log y = rt + a \text{ or } y = \exp(rt + a) \tag{10.1}$$

where a is a constant of integration. Bennett (1983) plotted log y against t (in radiocarbon years) for several taxa, estimating r from the slope of the best-fitting least-squares regression line. Population doubling-times can be estimated as log $2/r$ where r is estimated from Eq. 10.1.

We would not expect y to be able to increase exponentially without limit; an upper carrying capacity K for the environment can be incorporated into the logistic model, in which

$$\frac{dy}{dt} = by\,(K - y) \tag{10.2}$$

where b is a constant. This can be integrated, to give

$$\log\left[(K - y)/y\right] = c - bKt \tag{10.3}$$

where c is a constant of integration. If an estimate of K is available, we can obtain estimates of b and c by plotting $\log[(K-y)y]$ against t. However, we are assuming that K and b remain constant through time.

The same approach of fitting exponential or logistic models to phases of declining taxon accumulation rates can be used to estimate rates of decline and population halving-times (e.g., Peglar 1993).

Population modelling (Birks 2012d: Chap. 21) has considerable potential in palaeolimnology, especially in fine-resolution studies and in assessing rates of population change of alien, invasive species in lakes (Smol 2008).

Stratigraphical Changes in Taxonomic Richness

Changes in taxonomic richness and diversity with time can be explored using rarefaction analysis to estimate the number of taxa (N0: Hill 1973) that would have been found if all the fossil counts had been the same size and by estimating

Hill's (1973) N1 and N2 diversity measures. Details are given in Birks (2012b: Chap. 4). Palaeolimnological examples include Solovieva et al. (2005), Velle et al. (2005, 2010), Birks and Birks (2008), Bjerring et al. (2008), and Wall et al. (2010).

Differences in compositional turnover (= β-diversity: Whittaker 1972; or differentiation diversity: Jurasinski et al. 2009) at different sites for a given time period can be estimated by means of detrended canonical correspondence analysis (ter Braak 1986; Birks 2007b). The different amounts of turnover can be mapped or plotted in relation to location or present-day climate or other environmental variables (e.g., Smol et al. 2005; Hobbs et al. 2010). Alternatively, differences in compositional turnover at different times or between different types of organisms within one sequence can be estimated using the same approach (Birks and Birks 2008).

Temporal-Series Analysis

Palaeolimnological stratigraphical data represent temporal series of the changing relative frequencies or accumulation rates of different fossils (e.g., diatoms), chemical variables, or inferred environmental variables (e.g., lake-water pH) at selected times in the past (see Dutilleul et al. 2012: Chap. 16). In the great majority of standard time-series analytical methods, the term 'time-series' is reserved for partial realisations of a discrete-time stochastic process, namely that the observations are made repeatedly on the same random variable at equal spacings in time (Diggle 1990; Dutilleul et al. 2012: Chap. 16). There are clearly difficulties in meeting this requirement in many palaeolimnological studies unless the observations are from annually varved sediments (e.g., Anderson 1992; Young 1997; Young et al. 1998, 2000; Bradbury et al. 2002; Dean 2002; Dean et al. 2002; Fagel et al. 2008; Brauer et al. 2009). Palaeolimnological temporal-series may show long-term trends, short-term variations, cyclical variations, phases of values well above or well below the long-term means or trends, or irregular or random variation (cf. Blaauw et al. 2010).

Prior to any statistical analysis, it can be useful to perform exploratory data analysis (EDA) (Juggins and Telford 2012: Chap. 5) on the individual variables (e.g., individual taxa) within the temporal series, such as simple tests for trends within each variable that estimate the statistical significance of any trends by randomisation tests (e.g., Manly 2007). Simple tests for trend include 'runs' tests and regression models. If the abundances of a chironomid taxon in a stratigraphical sequence are seemingly randomly distributed with time, then the length of 'runs' (series of samples all successively larger (or less) than the previous sample) may still show a pattern. The lengths of observed runs (how many runs of a single sample, runs of two samples, runs of three samples, etc.) can be counted and the resulting distribution compared with the runs null-distribution for the same samples repeatedly randomised (e.g., 999 times). The significance of the observed 'runs' statistic can then be compared with the randomisation distribution to decide if the sequence of runs for the taxon is significantly different from random expectation (Manly 2007). Parametric or non-parametric correlation coefficients (Pearson's product–moment coefficient, Spearman's non-parametric rank correlation coefficient rho, and Kendall's non-parametric correlation coefficient tau) can be calculated to detect if there are trends in the values for an individual taxon (e.g., Gaulthier 2001). The observed values can be compared with the randomisation distribution based on the repeated randomisation of the taxon values (Manly 2007). There are also randomisation tests for autocorrelation (serial correlation) and periodicity but the results from such tests when applied to irregularly spaced series can be problematic (see Manly 2007).

Other useful EDA approaches for temporal-series include non-parametric regression techniques such as locally weighted regression scatter plot smoothing (LOESS – see Efron and Tibshirani 1991; Birks 1998, 2012a: Chap. 2; Juggins and Telford 2012: Chap. 5). These are useful graphical tools for highlighting the 'signal' or major patterns in individual temporal-series. A LOESS curve is a non-parametric regression estimate because it does not assume a particular parametric form (e.g., quadratic) for the entire regression (Cleveland 1979). It is conceptually similar to 'running means' except that it takes into account the uneven spacing of the independent time variable. In LOESS fitting, the degree of smoothing or span (α) can be varied and lies between 0 and 1. As α increases, the fitted curve becomes smoother. Choosing α requires careful judgement for each temporal-series. The goal is generally to make α as large as possible and thus to make the fitted curve as smooth as possible without distorting the underlying pattern in the data (Birks 2012a: Chap. 2; Juggins and Telford 2012: Chap. 5). Residual plots can help identify appropriate values of α (see Birks 1998 for further discussion of the use of LOESS smoothers and related techniques in palaeolimnology and Seppä et al. 2009a for the use of LOESS smoothers with stratigraphical data).

An alternative approach to LOESS smoothing that combines graphical display, hypothesis-testing, and temporal-series analysis is provided by the SiZer (**Si**gnificance of **Zer**o crossing of the derivative) (Chaudhuri and Marron 1999; Holmström and Erästö 2002; Sonderegger et al. 2009) and the BSiZer (Erästö and Holmström 2005) and related SiNos (**Si**gnificant **No**n-**s**tationarities: Godtliebsen et al. 2003) procedures. The SiZer approach was introduced to palaeolimnology by Korhola et al. (2000). The SiZer approach asks which features in the smoothers are real and which may be sampling artefacts. It finds trends and curves within a temporal-series that are statistically significant. It uses a whole family of smooth curves fitted to the temporal-series, each based on various smoothing window sizes, and provides information about the underlying curve at different levels of detail. The features detected typically depend on the level of detail for which the temporal-series is considered.

At each point in state space (time and bandwidth), confidence intervals for the derivatives of the smoothers are calculated, thereby allowing an assessment of which observed features are statistically significant, i.e., what may be 'signal' and what may be 'noise'. The results are presented as coloured SiZer maps that are a function of location and scale. The amount of smoothing is controlled by parameter h and for each value of $\log(h)$, the effective smoothing window is described by a horizontal space between two dash-dotted curves. The optimal smoother, based on numerical criteria (Ruppert et al. 1995), is conventionally drawn as a white line. Red on a SiZer map indicates that the curve is significantly increasing, blue that the curve is significantly decreasing, purple that no conclusions can be made about the slope, and grey areas indicate that the available data are too sparse at that smoothing level for any conclusions to be drawn about statistical significance. SiZer is a very robust and useful technique and it warrants further use in palaeolimnology (e.g., Weckström et al. 2006).

There is also BSiZer (Erästö and Holmström 2006), a Bayesian extension of SiZer (Erästö and Holmström 2005). This allows the inclusion in the model of the errors in the dependent variable (e.g., sample-specific errors of prediction in an environmental reconstruction – see Juggins and Birks 2012: Chap. 14) and sample dating uncertainties (see Blaauw and Heegaard 2012: Chap. 12). Erästö and Holmström (2005, 2006) present palaeolimnological applications of BSiZer.

SiNos (Godtliebsen et al. 2003) handles temporal-series where there is stochastic dependence between data points whereas SiZer assumes that the data are independent, or at least it can suggest spurious details to be significant at small sizes. SiNos looks simultaneously for significant changes in the mean, variance, and first-lag autocorrelation of the observed temporal-series when the null hypothesis suggests that the process is stationary. In general, SiZer typically detects too many features for dependent temporal data, but is superior to SiNos with independent data. SiNos can detect other types of stationarities (e.g., changes in the first-lag autocorrelation) in a temporal-series that SiZer cannot do. Both SiZer and SiNos are useful tools that can help to detect 'signal' in palaeolimnological temporal-series and to test if particular changes are statistically significant or not.

Dutilleul et al. (2012: Chap. 16) discuss the two main approaches to time-series analysis with constant inter-sample time intervals. There is the time-domain approach that is based on the concept of autocorrelation, namely the correlation between samples in the same sequence that are k time intervals apart. The autocorrelation coefficient is a measure of the similarity between samples separated by different time intervals and it is usually plotted as a correlogram to assess the autocorrelation structure and the behaviour (e.g., periodicities) in the values of the fossil type of interest over time. Temporal-series of two different variables can be compared by the cross-correlation coefficient to detect patterns of temporal variation and relationships between variables (e.g., Legendre and Legendre 1998; Tinner et al. 1999; Davis 2002; Seppä et al. 2009b). Comparison of two temporal-series is complicated by the inherent autocorrelation within the series. In such cases regression models between temporal-series require specialised approaches such as generalised least squares (Venables and Ripley 2002; Shumway and Stoffer 2006; Cryer and Chan 2008). Willis et al. (2007) present an example of such modelling in a palaeoecological context based on annually laminated sediments.

The second approach involves the frequency domain. It focuses on bands of frequency or wave-length over which the variance of a time-series is concentrated (Dutilleul 1995). It estimates the proportion of the variance attributable to each of a continuous range of frequencies. The power spectrum of a time-series can help detect periodicities within the data. The main tools are spectral density functions, cross-spectral functions, and periodograms. See Birks (1998) for various palaeolimnological examples of Fourier (amplitude) and power (variance) spectral analysis and Willis et al. (1999) for an example of spectral density and cross-spectral functions applied to a palaeoecological example involving annually laminated sediments.

As Dutilleul et al. (2012: Chap. 16) emphasise, conventional time-series analysis makes stringent assumptions of the data, namely that the inter-sample intervals are consistent throughout the time-series and that the data are stationary and thus there are no trends in mean or variance in the time-series. In the absence of equally-spaced samples in time, the usual procedure is to interpolate samples to equal time intervals using one of several interpolation procedures. This is equivalent to low-pass filtering and may result in an under-estimation of the high-frequency components in the spectrum. Thus the estimated spectrum of an interpolated time-series becomes too 'red' compared to the time spectrum (Schulz and Mudelsee 2002). Stationarity is usually achieved by some detrending or differencing procedure (Diggle 1990).

Techniques for spectral analysis and cross-spectral analysis for unevenly-spaced time-series involve the fitting of sine- and cosine-functions by a least-squares criterion (Schulz and Stattegger 1997) (see Dutilleul et al. 2012: Chap. 16 for a critique of this approach). Spectra of palaeolimnological and palaeoclimatic time-series frequently show a continuous decrease of spectral amplitude with increasing frequency, so-called 'red noise' (Schulz and Mudelsee 2002). The conventional technique for assessing whether the variability recorded in a time-series is consistent with red noise is to estimate the first-order autoregressive (AR1) parameter for the time-series of interest. An autoregressive model postulates that observation x_i at time t_i is a linear function of previous values in the series, and an AR1 model is of the form

$$x_i = \emptyset x_{t-1} + a_t$$

where a_i is a random disturbance and $\emptyset$ is a parameter of the model. For evenly spaced time-series, estimation of AR1 is relatively straightforward. Schulz and Mudelsee (2002) have developed the program REDFIT to perform spectral analysis directly on temporally unevenly spaced temporal data. It uses a Monte Carlo simulation approach to test if peaks in the time-series spectrum are significant with respect to the red-noise background from a theoretical AR1 process. Nyberg et al. (2001), Brown et al. (2005), and Allen et al. (2007) present applications of REDFIT to Holocene palaeoecological data.

A limitation of conventional power-spectral analysis is that it provides an integrated estimate of variance for the entire time-series. This can be overcome by wavelet power-spectral analysis (Torrence and Compo 1999) that identifies the dominant frequencies in different variables and displays how these frequencies have varied through the time-series. Palaeoecological examples include Bradbury et al. (2002), Dean (2002), Dean et al. (2002), Brown et al. (2005), and Fagel et al. (2008). Witt and Schumann (2005) have developed Foster's (1996) approach for deriving wavelets for unevenly sampled time-series to explore Holocene climate variability at millennial scales recorded in Greenland ice-cores (see also Prasad et al. 2009).

Dutilleul et al. (2012: Chap. 16) demonstrate the use of two robust procedures, namely auto-correlograms using distance classes and a novel frequency-domain technique, multi-frequential periodogram analysis, to analyse four different palaeolimnological time-series, all of which consist of unequally spaced observations. These methods have considerable potential in palaeolimnology. Hammer (2007) presents a related approach based on the Mantel correlogram, the Mantel periodogram, to conduct spectral analysis of Plio-Pleistocene fossil data.

Quantifying Recent Change

Many palaeolimnological studies address questions relating to 'before and after' changes or reference conditions at, for example, AD 1850. A range of numerical techniques such as classical ordination (Legendre and Birks 2012b: Chap. 8), dissimilarity analysis between 'top' and 'bottom' samples (Legendre and Birks 2012b: Chap. 8; Simpson 2012: Chap. 15; Simpson and Hall 2012: Chap. 19), indicator-species analysis (Dufrêne and Legendre 1997; Legendre and Birks 2012a: Chap. 7), and quantitative reconstructions of changes in lake-water chemistry such as total phosphorus (Juggins and Birks 2012: Chap. 14) provide robust means for quantifying recent change in, for example, diatom, chrysophyte cyst, or cladoceran assemblages in the last 100–150 years.

Quantitative Palaeoenvironmental Reconstructions

Since the pioneering work by Imbrie and Kipp (1971) on the quantitative reconstruction of past sea-surface temperatures and salinity from planktonic foraminiferal assemblages preserved in deep-ocean sediment cores, several numerical approaches have been developed in palaeolimnology to derive so-called modern calibration or 'transfer' functions that model the relationship between modern biological assemblages (e.g., diatoms, chironomids) and modern environment (e.g., lake-water pH). These modern calibration functions are then used to transform fossil assemblages into quantitative estimates of the past environmental variable. The various approaches are reviewed by Juggins and Birks (2012: Chap. 14) (see also Birks 1995, 1998, 2010; Smol 2008; Birks et al. 2010). The commonest approaches in palaeolimnology involve two-way weighted averaging or its relative weighted-averaging partial least squares, or the more formal classical approach of Gaussian logit regression and maximum likelihood calibration (see Juggins and Birks 2012: Chap. 14). When several calibration functions appear equally appropriate (e.g., Simpson and Hall 2012: Chap. 19), a consensus reconstruction can be derived by fitting a robust smoother (e.g., LOESS smoother) (Birks 2012a: Chap. 2; Juggins and Telford 2012: Chap. 5) through the reconstructed values derived from the different calibration-function models. Similar regression and calibration techniques, especially partial least squares (PLS), can be used to infer total organic carbon, total inorganic carbon, lake-water pH, nitrogen, phosphorus, and biogenic silica of lake sediments from the results of Fourier transformed infrared spectroscopy (e.g., Rosén and Persson 2006; Rosén et al. 2010) or of near-infrared spectroscopy (e.g., Rosén et al. 2000, 2001; Rosén 2005; Rosén and Hammarlund 2007; Kokfelt et al. 2009; Reuss et al. 2010; Cunningham et al. 2011).

Data Interpretation

Stratigraphical palaeolimnological data are a complex reflection of the biota living within the lake and its immediate surroundings. Interpretation of these data can be in terms of past communities or assemblages or in terms of possible factors that have influenced changes in the lake biota and the lake environment and hence in the biostratigraphy (Table 10.1).

Community and Assemblage Reconstruction

The reconstruction of past aquatic communities can be attempted by finding fossil remains of 'indicator species' characteristic of particular communities today, or by comparing quantitatively fossil assemblages with modern assemblages in surface-sediments from lakes of known biota and environment today, so-called analogue matching (AM) (Simpson 2007; Simpson 2012: Chap. 15; Simpson and Hall 2012: Chap. 19). The basic idea of AM is simple – compare the fossil assemblage with all modern assemblages using an appropriate dissimilarity measure (e.g., chord distance (= Hellinger coefficient) – see Simpson 2007; Legendre and Birks 2012b: Chap. 8), find the modern assemblage(s) most similar to the fossil assemblage, assign the modern lake or assemblage type to the closest analogue(s) to the fossil assemblage, and repeat for all fossil assemblages. Problems arise in defining a critical threshold to decide whether the modern and fossil assemblages are sufficiently similar to indicate that they could be derived from the same lake type (Simpson 2007, 2012: Chap. 15). Fossil assemblages may, for a variety of reasons, have no close modern analogues (Jackson and Williams 2004). In such cases, AM will fail as a tool for interpreting fossil assemblages in terms of past limnological assemblages or lake type. Attempts have been made to put AM on a more rigorous numerical basis using receiver operating characteristic analysis (Gavin et al. 2003; Simpson 2007, 2012: Chap. 15), Monte Carlo simulation (Lytle and Wahl 2005), and logistic regression (Simpson 2012: Chap. 15).

A less formal approach to analogue analysis is to ordinate or cluster together modern samples and the fossil stratigraphical data (Birks and Gordon 1985) and to examine which modern and fossil samples are positioned near each other in ordinations or are grouped together in clusterings or partitionings (see Legendre and Birks 2012a, b: Chaps. 7 and 8). Alternatively the modern data can be ordinated (e.g., principal component analysis, principal coordinates analysis – see Legendre and Birks 2012b: Chap. 8) and the modern lake types displayed on the first few ordination axes. The fossil samples can be positioned as 'passive' or 'supplementary' objects onto the plane formed by the modern ordination axes (see Legendre and Birks 2012b: Chap. 8). Alternatively, fossil samples can be the basis of the primary ordination and modern samples added as 'supplementary' samples. In either approach the similarities between the modern and fossil samples can be assessed visually as an aid in interpreting the fossil samples in relation to modern lake type or limnological conditions. Simpson and Hall (2012: Chap. 19) illustrate a more complex approach of relating fossil samples to modern environmental conditions. This involves positioning fossil samples onto the plane formed by the first two axes of a canonical correspondence analysis (CCA) of modern assemblages and environmental variables (e.g., lake-water pH) from a set of lakes (see Legendre and Birks 2012b: Chap. 8). The fossil samples are positioned on the basis of similarities in their composition with the modern assemblages. Their position on the plane of CCA axes 1 and 2 illustrates not only their similarities in composition with the modern samples but also the passive relationship with the modern environmental variables (see also Birks et al. 1990; Allott et al. 1992).

Causative Factors

Many factors can influence lakes and their biota, and hence fossil assemblages – climatic changes, soil development, lake-water chemistry, human impact, pollution, disturbance, species interactions, and complex interactions between these factors. It is a major challenge in palaeolimnology to test competing hypotheses about underlying causative factors (see Lotter and Anderson 2012: Chap. 18; Simpson and Hall 2012: Chap. 19). Statistical techniques such as constrained ordination (e.g., reduced rank multivariate regression (= redundancy analysis), canonical correspondence analysis), variation partitioning analysis, and Monte Carlo permutation tests (see Birks 2012a: Chap. 2; Legendre and Birks 2012b: Chap. 8) can be used to assess the statistical relationships between biostratigraphical data ('response' variables) and external forcing factors ('predictor' variables), in an attempt to assess the relative impacts of, for example, climatic change and volcanic tephra deposition on diatom assemblages (Lotter and Birks 1993; Birks 2010). Although statistical techniques exist with appropriate permutation tests that can take account of the time-ordered nature of palaeolimnological stratigraphical data (ter Braak and Šmilauer 2002), the major problem is deriving, for meaningful statistical analysis, independent data that reflect potential forcing factors (Lotter and Birks 2003).

This remains a major challenge in the quantitative and statistical interpretation of palaeolimnological data (Birks 2012d: Chap. 21).

Conclusions

This introductory and overview chapter and the chapters in Parts II and III discuss the wide range of numerical and statistical techniques currently available to palaeolimnologists. These techniques can help in the collection and assessment, summarisation, analysis, and interpretation of stratigraphical palaeolimnological data. Some of the techniques are more robust than others and some techniques make stronger assumptions than others, especially in the general area of time-series analysis. Techniques for analysing temporal data are rapidly developing and within a few years palaeolimnologists may have a set of robust and well-tried techniques to detect trends, periodicities, and correlations within their stratigraphical data-sets.

Acknowledgements I am indebted to John Smol and Gavin Simpson for useful comments and to Cathy Jenks for her invaluable help in the preparation of this chapter. This is publication A348 from the Bjerknes Centre for Climate Research.

References

Allen JRM, Long AJ, Ottley CJ, Pearson DG, Huntley B (2007) Holocene climate variability in northernmost Europe. Quaternary Sci Rev 26:1432–1453
Allott TEH, Harriman R, Battarbee RW (1992) Reversibility of lake acidification at the Round Loch of Glenhead, Galloway, Scotland. Environ Pollut 77:219–225
Anderson RY (1992) Possible connection between surface winds, solar activity and the Earth's magnetic field. Nature 358:51–53
Bennett KD (1983) Postglacial population expansion of forest trees in Norfolk, UK. Nature 303:164–167
Bennett KD (1986) The rate of spread and population increase of forest trees during the postglacial. Philos Trans R Soc Lond B 314:523–531
Bennett KD (1988) Holocene geographic spread and population expansion of *Fagus grandifolia* in Ontario, Canada. J Ecol 76:547–557
Bennett KD (1996) Determination of the number of zones in a biostratigraphical sequence. New Phytol 132:155–170
Bennett KD (2001) Data-handling methods for Quaternary microfossils. Uppsala and Belfast. www.chrono.qub.ac.uk/datah/. Accessed 4 Jan 2011
Bennett KD, Willis KJ (2001) Pollen. In: Smol JP, Birks HJB, Last WM (eds) Tracking environmental change using lake sediments. Volume 3: Terrestrial, algal, and siliceous indicators. Kluwer, Dordrecht, pp 5–32
Birks HJB (1987) Multivariate analysis of stratigraphic data in geology: a review. Chemometr Intell Lab Syst 2:109–126
Birks HJB (1992) Some reflections on the application of numerical methods in Quaternary palaeoecology. Publications of Karelian Institute, University of Joensuu 102, pp 7–20

Birks HJB (1995) Quantitative palaeoenvironmental reconstructions. In: Maddy D, Brew S (eds) Statistical modelling of Quaternary science data. Quaternary Research Association, Cambridge, pp 161–254

Birks HJB (1998) Numerical tools in palaeolimnology – progress, potentialities, and problems. J Paleolimnol 20:307–332

Birks HJB (2007a) Numerical analysis methods. In: Elias SA (ed) Encyclopedia of Quaternary science. Volume 3. Elsevier, Oxford, pp 2515–2521

Birks HJB (2007b) Estimating the amount of compositional change in late-Quaternary pollen-stratigraphical data. Veg Hist Archaeobot 16:197–202

Birks HJB (2010) Numerical methods for the analysis of diatom assemblage data. In: Smol JP, Stoermer EF (eds) The diatoms: application for the environmental and earth sciences, 2nd edn. Cambridge University Press, Cambridge, pp 23–54

Birks HJB (2012a) Chapter 2: Overview of numerical methods in palaeolimnology. In: Birks HJB, Lotter AF, Juggins S, Smol JP (eds) Tracking environmental change using lake sediments. Volume 5: Data handling and numerical techniques. Springer, Dordrecht

Birks HJB (2012b) Chapter 4: Introduction and overview of Part II. In: Birks HJB, Lotter AF, Juggins S, Smol JP (eds) Tracking environmental change using lake sediments. Volume 5: Data handling and numerical techniques. Springer, Dordrecht

Birks HJB (2012c) Chapter 11: Stratigraphical data analysis. In: Birks HJB, Lotter AF, Juggins S, Smol JP (eds) Tracking environmental change using lake sediments. Volume 5: Data handling and numerical techniques. Springer, Dordrecht

Birks HJB (2012d) Chapter 21: Conclusions and future challenges. In: Birks HJB, Lotter AF, Juggins S, Smol JP (eds) Tracking environmental change using lake sediments. Volume 5: Data handling and numerical techniques. Springer, Dordrecht

Birks HJB, Birks HH (2008) Biological responses to rapid climate change at the Younger Dryas-Holocene transition at Kråkenes, Western Norway. Holocene 18:19–30

Birks HJB, Gordon AD (1985) Numerical methods in Quaternary pollen analysis. Academic Press, London

Birks HJB, Seppä H (2004) Pollen-based reconstructions of late-Quaternary climate in Europe – progress, problems, and pitfalls. Acta Palaeobot 44:317–334

Birks HJB, Juggins S, Line JM (1990) Lake surface-water chemistry reconstructions from palaeolimnological data. In: Mason BJ (ed) The surface waters acidification programme. Cambridge University Press, Cambridge, pp 301–313

Birks HJB, Heiri O, Seppä H, Bjune AE (2010) Strengths and weaknesses of quantitative climate reconstructions based on late-Quaternary biological proxies. Open Ecol J 3:68–110

Bjerring R, Bradshaw EG, Amsinck SL, Johansson LS, Odgaard BV, Nielsen AB, Jeppesen E (2008) Inferring recent changes in the ecological state of 21 Danish candidate reference lakes (EU Water Framework Directive) using palaeolimnology. J Appl Ecol 45:1566–1575

Blaauw M, Christen JA (2005) Radiocarbon peat chronologies and environmental change. Appl Stat 54:805–816

Blaauw M, Heegaard E (2012) Chapter 12: Estimation of age-depth relationships. In: Birks HJB, Lotter AF, Juggins S, Smol JP (eds) Tracking environmental change using lake sediments. Volume 5: Data handling and numerical techniques. Springer, Dordrecht

Blaauw M, Bennett KD, Christen JA (2010) Random walk simulations of fossil pollen data. Holocene 20:645–649

Blockley SPE, Lowe JJ, Walker MJC et al. (2004) Bayesian analysis of radiocarbon chronologies: examples from the European late-glacial. J Quaternary Sci 19:159–175

Blockley SPE, Blaauw M, Ramsey CB, van der Plicht J (2007) Building and testing age models for radiocarbon dates in Lateglacial and early Holocene sediments. Quaternary Sci Rev 26:1915–1926

Blockley SPE, Ramsey CB, Lane CS, Lotter AF (2008) Improved age modelling approaches as exemplified by the revised chronology for the Central European varved lake Soppensee. Quaternary Sci Rev 27:61–71

Borcard D, Gillet F, Legendre P (2011) Numerical ecology with R. Springer, New York

Bortolus A (2008) Error cascades in the biological sciences: the unwarranted consequences of using bad taxonomy in ecology. Ambio 37:114–115

Bradbury JP, Cumming B, Laird K (2002) A 1500-year record of climatic and environmental change in Elk Lake, Minnesota III: measures of past primary productivity. J Paleolimnol 27:321–334

Brauer A, Dulski P, Mangili C, Mingram J, Liu J (2009) The potential of varves in high-resolution paleolimnological studies. PAGES News 17:96–98

Brodersen KP (2008) Book review SJ Brooks, PG Langdon and O Heiri. The identification and use of Palearctic Chironomidae larvae in palaeoecology. J Paleolimnol 40:751–752

Bronk Ramsey C (2007) Deposition models for chronological records. Quaternary Sci Rev 27:42–60

Bronk Ramsey C (2009) Dealing with outliers and offsets in radiocarbon dating. Radiocarbon 51:1023–1045

Brown KJ, Clark JS, Grimm EC et al. (2005) Fire cycles in North American interior grasslands and their relation to prairie drought. Proc Natl Acad Sci USA 102:8865–8870

Buck CE, Bard E (2007) A calendar chronology for Pleistocene mammoth and horse extinction in North America based on Bayesian radiocarbon calibration. Quaternary Sci Rev 26:2031–2035

Chaudhuri P, Marron JS (1999) SiZer for exploration of structures in curves. J Am Stat Assoc 94:807–823

Cleveland WS (1979) Robust locally weighted regression and smoothing scatterplots. J Am Stat Assoc 74:829–836

Cryer JD, Chan K-S (2008) Time series analysis with applications in R, 2nd edn. Springer, New York

Cunningham L, Bishop K, Mettävainio E, Rosén P (2011) Paleoecological evidence of major declines in total organic carbon concentrations since the nineteenth century in four nemoboreal lakes. J Paleolimnol 45:507–518

Davis JC (2002) Statistics and data analysis in geology, 3rd edn. Wiley, New York

Dean W (2002) A 1500-year record of climatic and environmental change in Elk Lake, Clearwater County, Minnesota II: geochemistry, mineralogy, and stable isotopes. J Paleolimnol 27:301–319

Dean W, Anderson R, Bradbury JP, Anderson D (2002) A 1500-year record of climatic and environmental change in Elk Lake, Clearwater County, Minnesota I: varve thickness and gray-scale density. J Paleolimnol 27:287–299

Diggle PJ (1990) Time-series – a biostatistical introduction. Clarendon, Oxford

Dolan JR (2011) The legacy of the last cruise of the Carnegie: a lesson in the value of dusty old taxonomic monographs. J Plankton Res 33:1317–1324

Dufrêne M, Legendre P (1997) Species assemblages and indicator species: the need for a flexible asymmetrical approach. Ecol Monogr 67:345–366

Dutilleul P (1995) Rhythms and autocorrelation analysis. Biol Rhythm Res 26:173–193

Dutilleul P, Cumming BF, Lontoc-Roy M (2012) Chapter 16: Autocorrelogram and periodogram analyses of palaeolimnological temporal series from lakes in central and western North America to assess shifts in drought conditions. In: Birks HJB, Lotter AF, Juggins S, Smol JP (eds) Tracking environmental change using lake sediments. Volume 5: Data handling and numerical techniques. Springer, Dordrecht

Efron B, Tibshirani R (1991) Statistical data analysis in the computer age. Science 253:390–395

Erästö P, Holmström L (2005) Bayesian multiscale smoothing for making inferences about features in scatter plots. J Comput Graph Stat 14:569–589

Erästö P, Holmström L (2006) Selection of prior distributions and multiscale analysis in Bayesian temperature reconstructions based on fossil assemblages. J Paleolimnol 36:69–80

Fagel N, Boës X, Loutre MF (2008) Climate oscillations evidenced by spectral analysis of southern Chilean lacustrine sediments: the assessment of ENSO over the last 600 years. J Paleolimnol 39:253–266

Foster G (1996) Wavelets for period analysis of unevenly spaced time series. Astronom J 112:1709–1729

Fuller JL (1998) Ecological impact of the mid-Holocene hemlock decline in southern Ontario, Canada. Ecology 79:2337–2351
Gaulthier TD (2001) Detecting trends using Spearman's rank correlation coefficient. Environ Forensics 2:359–362
Gavin DG, Oswald WW, Wahl ER, Williams JW (2003) A statistical approach to evaluating distance metrics and analog assignments for pollen records. Quaternary Res 60:356–367
Godtliebsen F, Olsen LR, Winter J-G (2003) Recent developments in statistical time series analysis: examples of use in climate research. Geophys Res Lett 30. doi:10.1029/2003GLO17229
Green DG (1983) The ecological interpretation of fine resolution pollen records. New Phytol 94:459–477
Green DG, Dolman GS (1988) Fine resolution pollen analysis. J Biogeogr 15:685–701
Grimm EC, Jacobson GL Jr (1992) Fossil-pollen evidence for abrupt climate changes during the past 18000 yrs in eastern North America. Clim Dyn 6:179–184
Grimm EC, Watts WA, Jacobson GL, Hansen BCS, Almquist HR, Dieffenbacher-Krall AC (2006) Evidence for warm wet Heinrich events in Florida. Quaternary Sci Rev 25:2197–2211
Grimm EC, Donovan JJ, Brown KJ (2011) A high-resolution record of climatic variability and landscape response from Kettle Lake, northern Great Plains, North America. Quaternary Sci Rev 30:2626–2650
Hammer Ø (2007) Spectral analysis of a Plio-Pleistocene multispecies time series using the Mantel periodogram. Palaeogeogr Palaeoclim Palaeoecol 243:373–377
Haslett J, Parnell A (2008) A simple monotone process with application to radiocarbon-dated depth chronologies. Appl Stat 57:399–418
Heiri O, Lotter AF (2010) How does taxonomic resolution affect chironomid-based temperature reconstruction? J Paleolimnol 44:589–601
Hill MO (1973) Diversity and evenness: a unifying notation and its consequences. Ecology 54:427–432
Hobbs WO, Telford RJ, Birks HJB, Saros JE, Hazewinkel RRO, Perren BB, Saulnier-Talbot É, Wolfe AP (2010) Quantifying recent ecological changes in remote lakes of North America and Greenland using sediment diatom assemblages. PLoS One 5:e10026. doi:10.1371/journal.pone.0010026
Holmström L, Erästö P (2002) Making inferences about past environmental change using smoothing in multiple time scales. Comput Stat Data Anal 41:289–309
Imbrie J, Kipp NG (1971) A new micropaleontological method for quantitative paleoclimatology: application to a late Pleistocene Caribbean core. In: Turekian KK (ed) The late Cenozoic glacial ages. Yale University Press, New Haven, pp 71–181
Jackson ST, Williams JW (2004) Modern analogs in Quaternary paleoecology: here today, gone yesterday, gone tomorrow? Ann Rev Earth Planet Sci 32:495–537
Jacobson GL (1979) The palaeoecology of white pine (*Pinus strobus*) in Minnesota. J Ecol 67:697–726
Janssen CR, Birks HJB (1994) Recurrent groups of pollen types in time. Rev Palaeobot Palynol 82:165–173
Juggins S, Birks HJB (2012) Chapter 14: Quantitative environmental reconstructions from biological data. In: Birks HJB, Lotter AF, Juggins S, Smol JP (eds) Tracking environmental change using lake sediments. Volume 5: Data handling and numerical techniques. Springer, Dordrecht
Juggins S, Telford RJ (2012) Chapter 5: Exploratory data analysis and data display. In: Birks HJB, Lotter AF, Juggins S, Smol JP (eds) Tracking environmental change using lake sediments. Volume 5: Data handling and numerical techniques. Springer, Dordrecht
Jurasinski G, Retzer V, Beierkuhnlein C (2009) Inventory, differentiation, and proportional diversity: a consistent terminology for quantifying species diversity. Oecologia 159:15–26
Kelly M, Telford RJ (2007) Common freshwater diatoms of Britain and Ireland: an interactive identification key. Environment Agency, UK
Kokfelt U, Rosén P, Schoning S, Christensen TR, Förster J, Karlsson J, Reuss N, Rundgren M, Callaghan TV, Jonasson C, Hammarlund D (2009) Ecosystem responses to increased

precipitation and permafrost decay in subarctic Sweden inferred from peat and lake sediments. Glob Change Biol 15:1652–1663

Korhola A, Weckström J, Holmström L, Erästö P (2000) A quantitative Holocene climatic record from diatoms in Northern Fennoscandia. Quaternary Sci Res 54:284–294

Last WM, Smol JP (eds) (2001a) Tracking environmental change using lake sediments. Volume 1: Basin analysis, coring, and chronological techniques. Kluwer Academic Publishers, Dordrecht

Last WM, Smol JP (eds) (2001b) Tracking environmental change using lake sediments. Volume 2: Physical and geochemical methods. Kluwer Academic Publishers, Dordrecht

Legendre P, Birks HJB (2012a) Chapter 7: Clustering and partitioning. In: Birks HJB, Lotter AF, Juggins S, Smol JP (eds) Tracking environmental change using lake sediments. Volume 5: Data handling and numerical techniques. Springer, Dordrecht

Legendre P, Birks HJB (2012b) Chapter 8: From classical to canonical ordination. In: Birks HJB, Lotter AF, Juggins S, Smol JP (eds) Tracking environmental change using lake sediments. Volume 5: Data handling and numerical techniques. Springer, Dordrecht

Legendre P, Legendre L (1998) Numerical ecology, 2nd English edn. Elsevier, Amsterdam

Lotter AF, Anderson NJ (2012) Chapter 18: Limnological responses to environmental changes at inter-annual to decadal time-scales. In: Birks HJB, Lotter AF, Juggins S, Smol JP (eds) Tracking environmental change using lake sediments. Volume 5: Data handling and numerical techniques. Springer, Dordrecht

Lotter AF, Birks HJB (1993) The impact of the Laacher See Tephra on terrestrial and aquatic ecosystems in the Black Forest, Southern Germany. J Quaternary Sci 8:263–276

Lotter AF, Birks HJB (2003) The Holocene palaeolimnology of Sägistalsee and its environmental history – a synthesis. J Paleolimnol 30:333–342

Lotter AF, Ammann B, Sturm M (1992) Rates of change and chronological problems during the late-glacial period. Clim Dyn 6:233–239

Lytle DE, Wahl ER (2005) Palaeoenvironmental reconstructions using modern analogue techniques: the effects of sample size and decision rules. Holocene 15:554–566

MacDonald GM (1993) Fossil pollen analysis and the reconstruction of plant invasions. Adv Ecol Res 24:67–110

Magri D (1989) Interpreting long-term exponential growth of plant populations in a 250,000-year pollen record from Valle di Castiglione (Roma). New Phytol 112:123–128

Maher LJ, Heiri O, Lotter AF (2012) Chapter 6: Assessment of uncertainties associated with palaeolimnological laboratory methods and microfossil analysis. In: Birks HJB, Lotter AF, Juggins S, Smol JP (eds) Tracking environmental change using lake sediments. Volume 5: Data handling and numerical techniques. Springer, Dordrecht

Manly BFJ (2007) Randomization, bootstrap and Monte Carlo methods in biology, 3rd edn. Chapman & Hall/CRC, Boca Raton

Nyberg J, Kuijpers A, Malmgren BA, Kunzendorf H (2001) Late Holocene changes in precipitation and hydrography recorded in marine sediments from the northeastern Caribbean Sea. Quaternary Res 56:87–102

Peglar SM (1993) The mid-Holocene *Ulmus* decline at Diss Mere, Norfolk, UK: a year-by-year pollen stratigraphy from annual laminations. Holocene 3:1–13

Pienitz R, Lotter AF, Newman L, Kiefer T (eds) (2009) Advances in palaeolimnology. PAGES News 17:89–136

Prasad S, Witt A, Kienel W, Dulski P, Bauer E, Yancheva G (2009) The 8.2 ka event: evidence for seasonal differences and the rate of climate change in western Europe. Global Planet Change 67:218–226

Reuss NS, Hammarlund D, Rundgren M, Segerström U, Erikson L, Rosén P (2010) Lake ecosystem responses to Holocene climate change at the subarctic tree-line in northern Sweden. Ecosystems 13:393–409

Rosén P (2005) Total organic carbon (TOC) of lake water during the Holocene inferred from lake sediments and near infrared spectroscopy (NIRS) in eight lakes from northern Sweden. Biogeochemistry 76:503–516

Rosén P, Hammarlund D (2007) Effects of climate, fire and vegetation development on Holocene changes in total organic carbon concentrations in three boreal forest lakes in northern Sweden. Biogeosciences 4:975–984

Rosén P, Persson P (2006) Fourier-transformed infrared spectroscopy (FTIRS), a new method to infer past changes in tree-line position and TOC using lake sediment. J Paleolimnol 35:913–923

Rosén P, Dåbakk E, Renberg I, Nilsson M, Hall RI (2000) Near-infrared spectroscopy (NIRS): a new tool for inferring past climatic changes from lake sediments. Holocene 10:161–166

Rosén P, Segerström U, Erikson L, Renberg I, Birks HJB (2001) Holocene climatic change reconstructed from diatoms, chironomids, pollen and near-infrared spectroscopy at an alpine lake (Sjuodljavre) in northern Sweden. Holocene 11:551–562

Rosén P, Vogel H, Cunningham L, Reuss N, Conley DJ, Persson P (2010) Fourier transform infrared spectroscopy, a new method for rapid determination of total organic and inorganic carbon and biogenic silica concentration in lake sediments. J Paleolimnol 43:247–259

Ruppert D, Sheather SJ, Wand MP (1995) An effective bandwidth selector for local least squares regression. J Am Stat Assoc 90:1257–1270

Schulz M, Mudelsee M (2002) REDFIT: estimating red-noise spectra directly from unevenly spaced paleoclimatic time series. Comput Geosci 28:421–426

Schulz M, Stattegger K (1997) SPECTRUM: Spectral analysis of unevenly spaced paleoclimatic time series. Comput Geosci 23:929–945

Seppä H, Alenius T, Muukkonen P, Giesecke T, Miller PA, Ojala AEK (2009a) Calibrated pollen accumulation rates as a basis for quantitative tree biomass reconstructions. Holocene 19:209–220

Seppä H, Alenius T, Bradshaw RHW, Giesecke T, Heikkilä M, Muukkonen P (2009b) Invasion of Norway spruce (*Picea abies*) and the rise of the boreal ecosystem in Fennoscandia. J Ecol 97:629–640

Shumway RH, Stoffer DS (2006) Time series analysis and its applications, 2nd edn. Springer, New York

Simpson GL (2007) Analogue methods in palaeoecology: using the analogue package. J Stat Software 22:1–29

Simpson GL (2012) Chapter 15: Modern analogue techniques. In: Birks HJB, Lotter AF, Juggins S, Smol JP (eds) Tracking environmental change using lake sediments. Volume 5: Data handling and numerical techniques. Springer, Dordrecht

Simpson GL, Hall RI (2012) Chapter 19: Applications of numerical methods to evaluate surface-water acidification and eutrophication. In: Birks HJB, Lotter AF, Juggins S, Smol JP (eds) Tracking environmental change using lake sediments. Volume 5: Data handling and numerical techniques. Springer, Dordrecht

Smol JP (2008) Pollution of lakes and rivers – a paleoenvironmental perspective, 2nd edn. Blackwell, Oxford

Smol JP, Birks HJB, Last WM (eds) (2001a) Tracking environmental change using lake sediments. Volume 3: Terrestrial, algal, and siliceous indicators. Kluwer Academic Publishers, Dordrecht

Smol JP, Birks HJB, Last WM (eds) (2001b) Tracking environmental change using lake sediments. Volume 4: Zoological indicators. Kluwer Academic Publishers, Dordrecht

Smol JP, Wolfe AP, Birks HJB et al. (2005) Climate-driven regime shifts in the biological communities of Arctic lakes. Proc Natl Acad Sci USA 102:4397–4402

Solovieva N, Jones VJ, Nazarova L, Brooks SJ, Birks HJB, Grytnes J-A, Appleby P, Kauppila T, Kondratenok B, Renberg I, Ponomarev V (2005) Palaeolimnological evidence for recent climatic change in lakes from the northern Urals, arctic Russia. J Paleolimnol 33:463–482

Sonderegger DL, Wang H, Clements WH, Noon BR (2009) Using SiZer to detect thresholds in ecological data. Front Ecol Environ 7:190–195

ter Braak CJF (1986) Canonical correspondence analysis: a new eigenvector technique for multivariate direct gradient analysis. Ecology 67:1167–1179

ter Braak CJF, Šmilauer P (2002) CANOCO Reference manual and CanoDraw for Windows user's guide: software for canonical community ordination (version 4.5). Microcomputer Power, Ithaca, New York

Thompson R, Clark RM, Boulton GS (2012) Chapter 13: Core correlation. In: Birks HJB, Lotter AF, Juggins S, Smol JP (eds) Tracking environmental change using lake sediments. Volume 5: Data handling and numerical techniques. Springer, Dordrecht

Tinner W, Hubschmid P, Wehrli M, Ammann B, Conedera M (1999) Long-term forest fire ecology and dynamics in southern Switzerland. J Ecol 87:273–289

Torrence C, Compo GP (1999) A practical guide to wavelet analysis. Bull Am Meteorol Soc 79:61–78

Velle G, Larsen J, Eide W, Peglar SM, Birks HJB (2005) Holocene environmental history and climate of Råtåsjøen, a low-alpine lake in south-central Norway. J Paleolimnol 33:129–153

Velle G, Bjune AE, Larsen J, Birks HJB (2010) Holocene climate and environmental history of Brurskardstjørni, a lake in the catchment of Øvre Heimdalsvatn, south-central Norway. Hydrobiologia 642:13–34

Venables WN, Ripley BD (2002) Modern applied statistics with S, 4th edn. Springer, New York

Walker D, Chen Y (1987) Palynological light on tropical rainforest dynamics. Quaternary Sci Rev 6:77–92

Wall AAJ, Magny M, Mitchell EAD, Vannière B, Gilbert D (2010) Response of testate amoeba assemblages to environmental and climatic changes during the Lateglacial-Holocene transition at Lake Lautrey (Jura Mountains, Eastern France). J Quaternary Sci 25:945–956

Weckström J, Korhola A, Erästö P, Holmström L (2006) Temperature patterns over the past eight centuries in northern Fennoscandia inferred from sedimentary diatoms. Quaternary Res 66:78–86

Willis KJ, Kleczkowski A, Crowhurst SJ (1999) 124,000-year periodicity in terrestrial vegetation during the late Pliocene epoch. Nature 397:685–688

Willis KJ, Kleczkowski A, New M, Whittaker RJ (2007) Testing the impact of climate variability on European plant diversity: 320 000 years of water–energy dynamics and its long-term influence on plant taxonomic richness. Ecol Lett 10:673–679

Witt A, Schumann AY (2005) Holocene climate variability on millennial scales recorded in Greenland ice cores. Nonlinear Process Geophys 12:345–352

Whittaker RH (1972) Evolution and measurement of species diversity. Taxon 21:213–251

Wohlfarth B, Blaauw M, Davies SM et al. (2006) Constraining the age of late Glacial and early Holocene pollen zones and tephra horizons in southern Sweden with Bayesian probability methods. J Quaternary Sci 21:321–334

Young R (1997) Time-series analysis and time-basis reconstructions in palaeoecology. PhD thesis, University of Amsterdam

Young R, Walanus A, Lingeman R, Ran ETH, van Geel B, Goslar T, Ralska-Jasiewiczowa M (1998) Spectral analysis of pollen influxes from varved sediments of Lake Gościąź, Poland. In: Ralska-Jasiewiczowa M, Goslar T, Madeyska T, Starkel L (eds) Lake Gościąź, central Poland. A monographic study part 1. Szafer Institute of Botany, Krakow, pp 232–239

Young R, Walanus A, Goslar T (2000) Autocorrelation analysis in search of short-term patterns in varve data from sediments of Lake Gościąź, Poland. Boreas 29:251–260

Chapter 11
Analysis of Stratigraphical Data

H. John B. Birks

Abstract This chapter discusses numerical methods that can aid in the summarisation and analysis of patterns in complex multivariate stratigraphical palaeolimnological data. These methods are constrained cluster analysis and constrained partitioning for zonation, partitioning of individual stratigraphical sequences for exploring patterns within and between taxa, ordination techniques for summarising trends within stratigraphical data, rate-of-change analysis for estimating rates of biotic change through time, and dissimilarity analysis and ordination techniques for quantifying the magnitude of recent change. Availability of appropriate computer software is outlined and possible future developments are discussed.

Keywords Broken-stick model • Canonical ordination • Cluster analysis • Constrained clustering • Ordination • Partial ordination • Partitioning • Quantifying recent change • Rate-of-change analysis • Splitting of individual sequences • Statistical modelling • Zonation

Introduction

Palaeolimnological data are most commonly multivariate and quantitative. Typically, they consist of estimates of the abundance of variables (e.g., diatoms, chironomids, pollen), expressed in various ways, in a large number of samples.

H.J.B. Birks (✉)
Department of Biology and Bjerknes Centre for Climate Research, University of Bergen, PO Box 7803, Bergen N-5020, Norway

Environmental Change Research Centre, University College London, Pearson Building, Gower Street, London WC1E 6BT, UK

School of Geography and the Environment, University of Oxford, Oxford OX1 3QY, UK
e-mail: john.birks@bio.uib.no

H.J.B. Birks et al. (eds.), *Tracking Environmental Change Using Lake Sediments*, Developments in Paleoenvironmental Research 5, DOI 10.1007/978-94-007-2745-8_11,

Samples may have various relationships to each other, notably spatial ('surface samples') or temporal (stratigraphical) relationships (see Birks 2012a: Chap. 2). This chapter is concerned with the analysis of temporally-related stratigraphical samples. Such samples have a known order of deposition which is used to constrain the numerical analyses, providing the basis for presenting data in stratigraphical diagrams, dividing the data into zones (whether numerical or not), and for interpreting the data as reflecting temporal changes in past systems (Birks 1986, 1998, 2007a, 2010).

Analysis of samples in a stratigraphical sequence has traditionally been carried out subjectively ('by eye'). The use of numerical techniques has been dependent on the availability of computers, and the development of numerical techniques has closely paralleled computer development, beginning with their initial use for zonation (Gordon and Birks 1972). Some of the techniques can be applied by hand or with simple calculating aids (such as spreadsheets), but the quantity of data present in typical palaeolimnological data-sets precludes much numerical analysis without access to a computer. The most important advantages of numerical techniques are that the results are repeatable and are less subjective than traditional approaches (Birks and Gordon 1985). Different workers can obtain the same results from the same data by techniques that are clear and explicit. They enable the recognition of patterns and trends in data separately from the interpretation of the data (Birks 2007a) (something that is difficult to do in a repeatable way 'by hand' or 'by eye'). Nevertheless, the adoption of numerical techniques for the basic analysis of stratigraphical data in palaeolimnology has been slow and patchy (Birks 1986, 1998).

The aim of this chapter is to describe the main numerical techniques in use for the analysis of stratigraphical data for the purposes of data summarisation and description (as distinct from the reconstruction of past environments, ecosystems, or climates): zonation, independent splitting, summarising stratigraphical patterns, quantifying recent changes, and rate-of-change analysis. Additional techniques for the analysis of stratigraphical data are described elsewhere in this volume (e.g., canonical ordination, Legendre and Birks 2012b: Chap. 8; core correlation, Thompson et al. 2012: Chap. 13; environmental reconstructions, Juggins and Birks 2012: Chap. 14; analogue analysis, Simpson 2012: Chap. 15; temporal-series analysis, Dutilleul et al. 2012: Chap. 16). The techniques described are illustrated with the Holocene diatom data from The Round Loch of Glenhead (RLGH) (core RLGH3) (Jones et al. 1989; Birks and Jones 2012: Chap. 3), using series 1 radiocarbon-age data from Harkness et al. (1997). The age-depth model used here is a 3-term polynomial, based on the radiocarbon dates SRR 2811–SRR 2821, excluding SRR 2814, and including a value of 0 ± 50 for the surface. Development of this and other models, including radiocarbon-date calibration, is discussed further by Blaauw and Heegaard (2012: Chap. 12). It differs from the age-depth model used by Jones et al. (1989), especially in the basal portion of the sequence. It is used here solely for illustration, and should not be taken as a replacement for the earlier published age-depth model.

All the numerical analyses in this chapter were carried out by Keith Bennett using PSIMPOLL v. 4.00 (or above). The program runs on all operating systems with an ANSI C compiler (which includes DOS, Windows®, Apple®, and all Unix systems), and is available at URL http://chrono.qub.ac.uk/psimpoll/psimpoll.html/. Other relevant programs available on the internet are listed in the appropriate sections.

Zonation

In order to facilitate the description and correlation of microfossil data, it is useful to divide stratigraphical data (e.g., chironomids, pollen, diatoms, cladocerans) into 'zones'. There are many different types of zones (see Hedberg 1976), of which the most useful for Quaternary palaeolimnology is the 'assemblage zone'. These may be defined as bodies of sediment that are characterised by distinctive natural assemblages of pollen, diatoms, etc. In Quaternary palynology, this approach was pioneered by Cushing (1967), and formed the background to the development of numerical zonation schemes by Gordon and Birks (1972). Traditionally, sequences have been split into zones 'by eye', but the process can now be done easily by numerical procedures, which are not only fast and repeatable, but also reduce considerably the element of subjectivity. Zonation of biostratigraphical data may be carried out by divisive splitting or partitioning techniques, or by agglomerative techniques of stratigraphically constrained cluster analysis (see Legendre and Birks 2012a: Chap. 7). The basic principles were established by Gordon and Birks (1972) and Birks and Gordon (1985), and remain largely unchanged. Bennett (1996) discusses methods for establishing how many zones should be defined.

Techniques

Splitting or divisive techniques successively divide a stratigraphical sequence into smaller and smaller segments. They divide the data-set in such a way that the sum of the variation of the resulting portions is minimised. Variation may be measured in a number of ways. Two in common use are sums-of-squares and information content. The way that the data-set is split may be either 'binary' or 'optimal'. The binary approach splits the data-set into successively smaller groups by splitting existing zones. Results for any given number of zones are thus an extension of results for all results with fewer zones. As Legendre and Birks (2012a: Chap. 7) and Simpson and Birks (2012: Chap. 9) discuss, these binary splitting procedures of Gordon and Birks (1972) are an early implementation of De'ath's (2002) multivariate regression trees where a vector of sample depths or ages is used as the sole explanatory predictor variable (Borcard et al. 2011: Sect. 4.11.5).

The optimal partitioning approach starts afresh for each successive number of splits, and thus there will not necessarily be any correspondence between the results for division into different numbers of zones. The principles are discussed fully in Gordon (1982) and Birks and Gordon (1985). Optimal splitting can be demanding of computer time, but this is rapidly becoming irrelevant as computer processor speeds improve. In principle and in practice the optimal approach is more satisfactory than binary splitting, and it is now used routinely (e.g., Bennett et al. 1992; Birks and Birks 2008).

Agglomerative techniques cluster samples into successively larger groups. They are based on cluster analysis (Gordon 1999; Legendre and Birks 2012a: Chap. 7), with the constraint that clusters must be based on agglomeration of stratigraphically adjacent samples (Gordon and Birks 1972; Birks and Gordon 1985). Two techniques have been employed. Both begin with a dissimilarity matrix of all pairwise combinations of samples, and search for the most similar, stratigraphically adjacent pair of samples. This pair is then combined to form a cluster. The two methods differ in how the new cluster is then treated. With the first, termed constrained single-link analysis (or CONSLINK), the dissimilarity between the cluster and a sample outside the cluster is considered to be the smallest of the dissimilarities between the sample and any of the samples within the cluster (see Legendre and Birks 2012a: Chap. 7). With the second technique, termed constrained incremental sum-of-squares cluster analysis (or CONISS) and described in detail by Grimm (1987), a statistic termed 'dispersion', or 'sum-of-squares' is calculated for each cluster, and recalculated as clusters are merged (see Legendre and Birks 2012a: Chap. 7). The matrix is searched for the two stratigraphically adjacent clusters whose merger gives the least increase in total dispersion. In both cases, agglomeration proceeds until the whole data-set is combined into a single cluster. The measure of dissimilarity used by Grimm (1987) in his program CONISS is squared Euclidian distances (see Legendre and Birks 2012b: Chap. 8), calculated from untransformed or transformed (standardised, square-root, or normalised – see Prentice 1980 for details) data but other distances are possible. Padmore (1987) described a program called SHEFFPOLL that carries out agglomerative zonation using either total within group sum-of-squares (equivalent to squared Euclidian distance, but using a less-efficient algorithm) or information content. There seems to be no reason, in principle, for preferring one method over the other. In PSIMPOLL 4.00, CONISS is available, based on the program by Grimm (1987), and a second method, called CONIIC (constrained incremental information clustering), based on information content as a dissimilarity measure, is also available.

An additional agglomerative method is the variable-barriers approach (Gordon and Birks 1974; Birks and Gordon 1985; Birks 1986). This approach attempts to distinguish groups of samples with transitional assemblage composition from groups of samples with broadly similar composition. It thus tries to delimit two types of zones; phases of relatively 'stable' composition and phases of abrupt or systematic changes in composition (Birks and Gordon 1985).

Instead of placing boundaries between pairs of adjacent levels as in the divisive procedures, barriers of any height between 0 and 1 are positioned, subject to the constraint that the sum of the barrier heights equals h, a value pre-set by the investigator. The local mean for sample i is influenced by all levels that can be reached from i by 'jumping over' barriers of total height less than 1. A measure of the total variation in the fossil values from the local mean is then minimised by means of a computational iterative procedure, thereby ensuring that levels with dissimilar fossil composition are separated by high barriers. Levels belonging to phases of rapid transition will appear between adjacent high barriers. Transitional levels appear as late additions to the clusters formed early in the analysis. The variable-barriers approach is particularly useful in the analysis of high-resolution data where stratigraphical changes may be gradual. It has proved particularly useful in partitioning late-glacial biostratigraphies (e.g., Birks and Mathewes 1978; Birks 1981).

Determining the Number of Zones

A key difficulty in all numerical zonations is the determination of the number of zones that can be reasonably recognised in a sequence. Birks (1986) suggested that binary divisive analyses should proceed 'until little further reduction in variance occurs', leaving the judgement entirely to the analyst. Two methods, randomised data-sets and the 'broken-stick model', can be used to help determine how many zones should be recognised numerically in a sequence (Bennett 1996).

Randomised data-sets. If data-sets consisted of samples that were drawn from the same, perfectly uniform, body of sediment, any differences would be due to stochastic (random) variation. One way of assessing the success of a zonation analysis is to examine the change in residual variation as new zones are established and to compare this with the way that the residual variation would change if samples in the data-set were distributed along the stratigraphical sequence at random. A successful zone is then one which produces a greater fall in total variation than would be obtained from a zone in the randomised data-set. Randomised data-sets are generated by shuffling the samples of the existing data-set. Two random integers (i, j) are obtained, and the ith and j^{th} samples in the data-set swapped. This should be repeated many times (at least 10 $\times$ *number of samples*) to achieve a thorough shuffling.

This approach is similar to that of Legendre et al. (1985), who clustered samples along a chronological sequence until the samples within each cluster behaved as if they resulted from random sampling (see Legendre and Birks 2012a: Chap. 7). The aim in both cases is to separate variation that results from structure in the data-set from variation which results from stochastic processes. The approach demonstrates that much of the variation in biostratigraphical data-sets is essentially stochastic and that there is a point beyond which further subdivision of the variation is worthless. It also gives some information on the number of potentially useful zones.

Broken-stick model. The total variation in a data-set can be considered as a stick of unit length with $n-1$ markers positioned on it at random. The lengths of the n resulting segments are the proportion of total variation that would be due to each level of zonation if the sequence consisted of samples with no stratigraphical structure. Thus, if the reduction in variation for a particular zone exceeds the proportion expected from this model, the zone concerned accounts for more variation than would have been expected if the data-set consisted of samples arranged at random and consequently may be considered 'significant'. The idea is that each successive 'split' accounts for part of the total variation in the data-set and this part is tested against the portion expected from the model. The equation for calculating the values is:

$$Pr = \frac{1}{n}\sum_{i=k}^{n}\frac{1}{i} \quad (11.1)$$

where Pr is the expected proportion for the kth component out of n. It can be readily obtained on a calculator (for small n) or a spreadsheet, and results compared with the appropriate measure of variation from the output of a numerical zonation method. A similar application of this model (termed 'broken-stick' by MacArthur 1957) has been adopted to interpret the eigenvalues of principal component analysis (Legendre and Legendre 1983; Jackson 1993; see Legendre and Birks 2012b: Chap. 8).

This procedure is readily applied to binary divisive analyses. Each successive split accounts for a portion of residual variation, until the data-set is split completely, and the residual variation is nil. However, there is no hierarchy of successive splits with optimal divisive procedures, since each level of zonation is begun anew. If an optimal splitting procedure into n zones ($n - 1$ divisions) happened to be the same as a binary split into the same number of zones, then there would be the same residual variation. An optimal splitting will always result in a drop in variation that is at least as great as would be obtained from binary splitting. Therefore, if the reduction in variation after an optimal split into n zones is compared with the broken-stick model for $n - 1$ divisions, it is less likely to exceed the broken-stick proportions than the binary split into the same number of zones. Comparison of an optimal-split result with a broken-stick model can therefore help recognise levels of splitting that are significant (those that have reductions in variation greater than those expected from the broken-stick model) but cannot show that a level of splitting is not significant.

CONISS and CONIIC are agglomerative procedures. Combination of samples into clusters increases a measure of variation termed 'dispersion' (Grimm 1987). Each increase corresponds to the variation associated with each cluster (potentially a zone). The analysis thus generates a total dispersion (after all clusters are combined), and the values associated with each stage which sum to give the total dispersion. This is all that is needed to apply the broken-stick model: the increase of dispersion associated with the final stage as a proportion of the total dispersion is equivalent to the first break of the 'stick', the increase due to the penultimate stage is equivalent to the second break, and so on.

The broken-stick model provides a precise recommendation for the number of potentially significant and useful zones. For a data-set with structure, division into n

zones accounts for a higher proportion of the total variation than expected from the model until all the variation that is due to the structure has been accounted for. The number of significant zones is simply the last value for n that accounts for a portion of the variation greater than the model. Birks (1998) discusses the use of the broken-stick model in palaeolimnology and Lotter (1998) provides an elegant example applied to high-resolution diatom-stratigraphical data.

Software

There are many implementations of the different numerical zonation techniques.

1. PSIMPOLL provides six methods: binary splitting by least-squares and information content, optimal splitting by least-squares and information content, and constrained agglomeration by CONISS and CONIIC. PSIMPOLL also fully implements the procedures described by Bennett (1996) to determine the number of zones that can be considered significant and useful. It is available from http://chrono.qub.ac.uk/psimpoll/psimpoll.html/
2. The R package rioja (Juggins 2009) provides constrained hierarchical clustering by CONISS and CONSLINK and comparison of the clustering results with the broken-stick model (Bennett 1996). It is available from http://www.staff.ncl.ac.uk/staff/stephen.juggins/
3. TILIA provides CONISS. This can be used as a stratigraphically-constrained cluster analysis for numerical zonation or, as an unconstrained analysis, for clustering surface-samples. For availability of TILIA and TGView, see http://www.ncdc.noaa.gov/paleo/tiliafaq.html
4. PolPal provides constrained single-link cluster analysis as a procedure called CONSLINK. A square-root transformation is used to enhance the representation of taxa with low percentages. See Walanus and Nalepka (1997) for details.
5. Zone is an MS-DOS program that also runs under Windows® developed by Steve Juggins that combines CONSLINK, CONISS, binary splitting by least-squares and information content, optimal splitting by least-squares, and the variable barriers approach (see Birks and Gordon 1985). Zone is available from Steve Juggins' home-page http://www.staff.ncl.ac.uk/staff/stephen.juggins/software/ZoneHome.htm.

Example of Use

To illustrate use and results, the main zonation techniques have been applied to the RLGH diatom data-set, and summary results are shown in Fig. 11.1. In each case, the number of zones has been determined using the broken-stick criterion.

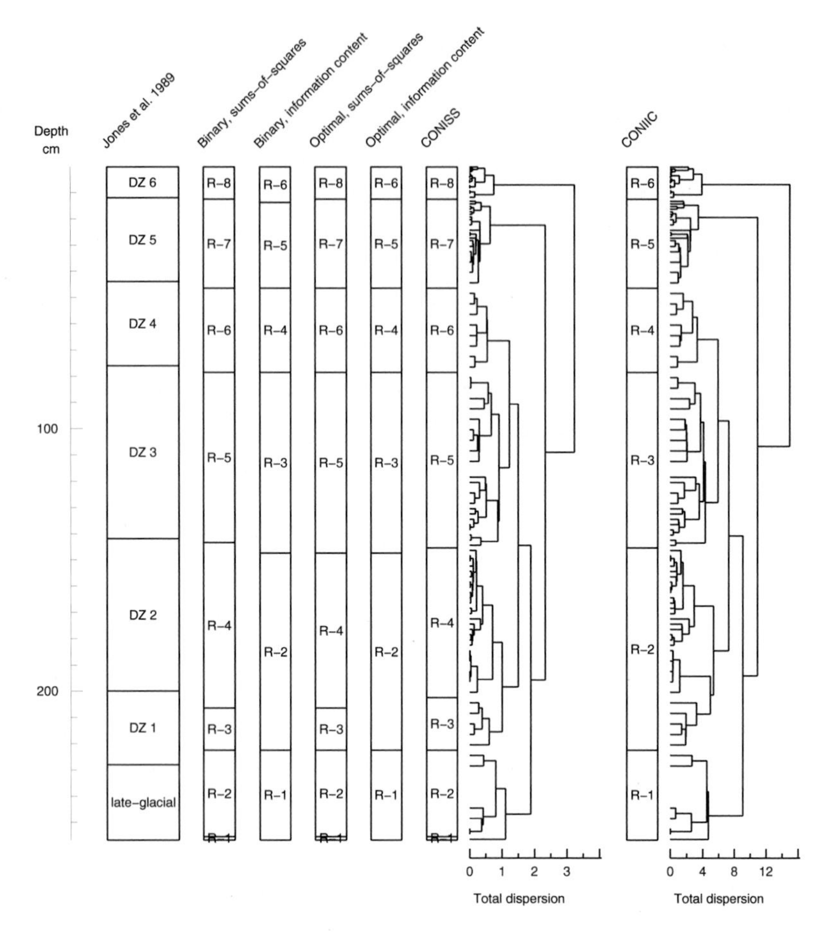

Fig. 11.1 Results of numerical zonation of The Round Loch of Glenhead (RLGH) diatom data, compared with the original zonation of Jones et al. (1989). The 'binary' and 'optimal' methods are splitting or divisive techniques, shown as columns. CONISS and CONIIC are agglomerative techniques, shown as columns and dendrograms illustrating the relationships between adjacent samples

Results from the six numerical techniques are similar to those defined subjectively by Jones et al. (1989), but differ in detail. Among the results from the numerical methods, the greatest similarity is found amongst the sums-of-squares techniques versus the information-content procedures. It appears that the choice of dissimilarity measure may be more important than the choice of agglomerative or splitting technique. The dendrograms from both CONISS and CONIIC show clearly the distinctiveness of the uppermost part of the sequence. The sums-of-squares methods identify more zones (8 *versus* 6). Of these, the lowest is based on only one sample, and should perhaps not be used in a final zonation scheme. The third zone (R-3) is not recognised by the information-content methods, but a similar zone was defined by Jones et al. (1989). It is probably questionable whether this zone can be justified in terms of the overall distribution of variation in the data-set: it might be best treated as a subzone within a larger zone (R-3 plus R-4).

Splitting of Individual Stratigraphical Sequences

Walker and Wilson (1978) argued that individual curves for pollen types should be considered independently of each other, especially when the data are expressed as pollen accumulation rates, and are therefore statistically independent. They presented an approach, with FORTRAN programs written by Y Pittelkow (Walker and Pittelkow 1981), for making independent splits of the records for individual taxa in a sequence, so-called sequence splitting (Birks and Gordon 1985). The approach involves first identifying portions of the stratigraphical record that are effectively non-zero sequences, distinguishing them from other portions of effectively zero sequences. This may be termed presence-absence splitting. The approach then looks at the non-zero sequences of each individual taxon and splits them quantitatively with a maximum-likelihood estimation method (Walker and Wilson 1978). Pittelkow's programs implemented the approach in two stages. In PSIMPOLL, these are combined, making presence-absence splits first, and then looking for quantitative splits within non-zero sequences. The detection of either sort of split is a statistical matter, depending on a level of significance that is a function of the number of samples (Table 2 in Walker and Wilson 1978) and there may be no splits of either sort for any given taxon. Splits are ignored if they would create 'zones' of less than 4 samples. Walker and Pittelkow (1981) present results using the method from three late-Quaternary palynological sequences. Walker and Wilson (1978) statistically fitted curves by regression to portions of the record of particular taxa, identified from presence-absence and quantitative splitting. However, this was not included in Pittelkow's programs, and has not been implemented in PSIMPOLL. Instead, means and standard deviations are given for each identified segment of the record for each taxon.

This method is analysed in more detail by Birks and Gordon (1985), using an example pollen data-set from Abernethy Forest, Scotland (Birks and Mathewes 1978), and comparing this with analyses from other sites. Splits may be concentrated at certain periods of time, suggesting linked behaviour of the taxa concerned, or may be spread more diffusely along a sequence. Periods of linked splits suggest periods when strong environmental control (such as climate change) brought about changes in the abundances of many taxa within the same interval. One such linked phase occurs at the beginning of the Holocene in the Abernethy Forest data-set. Birks and Line (1994) developed the method further with an analysis of pollen accumulation rates from three late-glacial sequences and ten Holocene sequences (including Abernethy Forest), all from Scotland, by independent splitting. They applied a statistical test that was able to show that most of the splits identified are clumped, mainly as a result of changes in sediment accumulation rates (caused partly by what may be artefacts in pollen accumulation rate data), and partly from clustering of responses to individual environmental changes.

Green (1982) applied independent splitting in conjunction with time-series procedures such as cross-correlograms to fine-resolution pollen-stratigraphical and charcoal data from a lake in Nova Scotia. The results provided insights into the temporal frequency of tree population change and into questions of long-term stability and dynamics and rapid change following forest fires.

Independent splitting was developed for so-called 'absolute' data, namely concentrations or accumulation rates (Birks 2012a: Chap. 2). One of the biggest limitations of sequence splitting may be the large uncertainties associated with accumulation rates of, first, sediments, and then of the fossils themselves (Bennett 1994). The log-ratio transformation of Aitchison (1986) should, in theory, make it possible to use the technique with percentage or proportional data (Birks and Line 1994), but so far there have been no applications of that. No uses of independent splitting of palaeolimnological data (e.g., diatom or chironomid accumulation rates) have been published, as far as I know, so a preliminary application using the RLGH diatom data is presented below. I believe that the approach has considerable potential in analysing fine-resolution palaeolimnological data, especially from annually laminated sediments (e.g., Lotter 1998).

Independent splitting of stratigraphical sequences (Walker and Wilson 1978) is a precursor to regression trees (Gavin Simpson, personal communication) where a quantitative response variable (in this case a stratigraphical sequence for taxon A) is repeatedly split so that at each partition the sequence is divided into two mutually exclusive groups, each of which is as homogenous as possible. A vector of sample depths or ages is used as the sole explanatory predictor variable. The splitting is then applied to each group separately until some stopping rule is applied (De'ath and Fabricus 2000; Simpson and Birks 2012: Chap. 9).

Example of Use

Selected results using independent splitting on the RLGH diatom concentration data are shown in Fig. 11.2. Presence-absence splits delimit sections of the curves where the taxon's status changes from continuously present to (more-or-less) absent and quantitative splits separate sections of the curves where abundance and variances change. One of the most striking features is that boundaries are very variable between taxa, indicating independent behaviour in response to forcing factors for change (or independent timing of such responses). Independent splitting is most useful for purposes that involve interpreting individualistic behaviour of different taxa in palaeoecological data-sets (Birks and Gordon 1985; Birks 1986, 1992, 2007a).

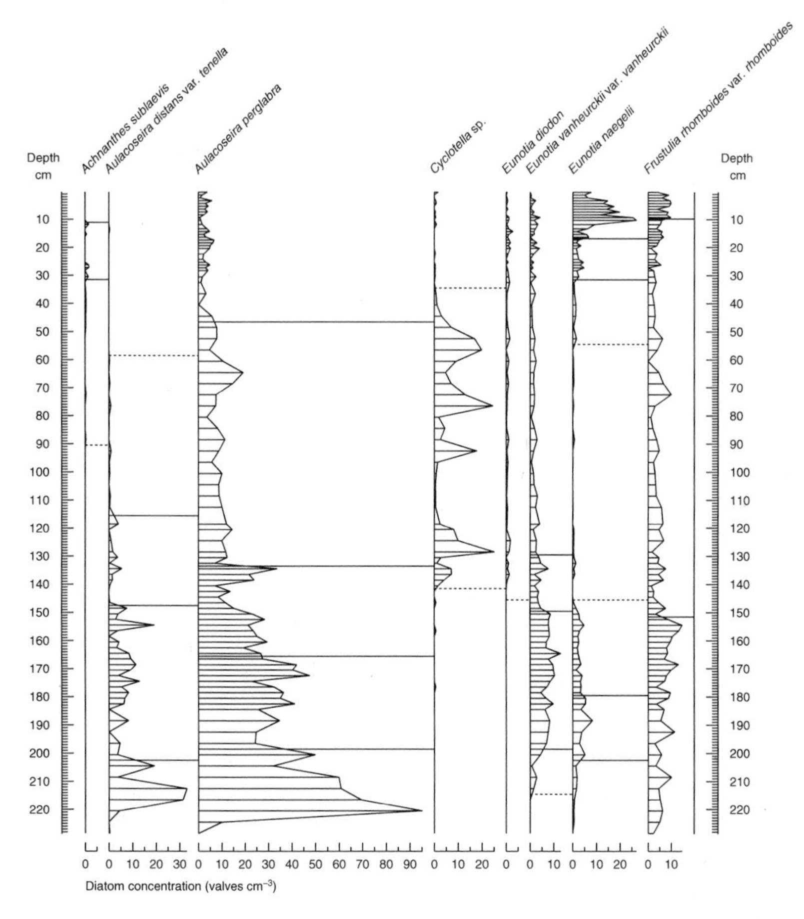

Fig. 11.2 Selected results of independent splitting of The Round Loch of Glenhead (RLGH) diatom concentration data (8 taxa only), using the methods of Walker and Wilson (1978). *Dashed lines* define 'presence—absence' splits, and *solid lines* define 'quantitative' splits

Summarising Stratigraphical Patterns Using Ordination Results

An additional useful step in the numerical analysis of a multivariate palaeolimnological data-set is to summarise the data in a few dimensions, gradients of variation, or ordination axes (Prentice 1980, 1986; Birks 1992, 1998, 2007a). Such axes are selected to fulfil predefined mathematical criteria so that the selected dimensions and axes capture as much of the total variation in the total data as possible under stated mathematical assumptions with the constraint that all the axes are uncorrelated to each other. The results of ordination procedures, such as principal component analysis (PCA), correspondence analysis (CA), detrended correspondence analysis (DCA), principal coordinate analysis (PCoA), and non-metric multidimensional scaling (NMDS) (see Legendre and Birks 2012b: Chap. 8), can be presented as low-dimensional plots of sample points on, for example, PCA axes 1 and 2. The results can also be presented as stratigraphical plots of sample scores on each of the first few major ordination axes (those with eigenvalues greater than expected under the broken-stick model – see Jackson 1993; Legendre and Birks 2012b: Chap. 8). Such stratigraphical plots are 'composite curves' based on many taxa that contribute to the major axes of variation or underlying latent structure in the total data. They provide a useful visual summary of the major patterns of variation within the data (see Lotter and Birks 2003; Haberle and Bennett 2004 for examples). Changes in the sample scores on the first few ordination axes can be used to delimit zones or the results of numerical zonations can be superimposed onto a two- or three-dimensional ordination plot (Birks 1986, 1998). This is particularly useful when there are gradual and often complex transitions from one assemblage to another and in assessing and interpreting results of different zonation techniques (e.g., Birks and Berglund 1979; Gordon 1982). With very large data-sets, it may be useful rather than plotting the scores for all samples on scatter plots of the first ordination axes, to plot the mean or median sample scores and the range of the sample scores for the samples within each zone (Birks and Berglund 1979) to avoid over-crowded ordination scatter plots.

Summarising palaeolimnological data-sets as the first few ordination axes provides a useful way of comparing and detecting common trends and identifying differences in two or more stratigraphical sequences from different sites (e.g., Haberle and Bennett 2004) or of comparing two or more stratigraphical sequences (e.g., diatoms, chironomids, geochemistry) from the same site (e.g., Birks 1987; Ammann et al. 2000; Birks et al. 2000; Lotter and Birks 2003; Massaferro et al. 2005; Haberle et al. 2006; Wall et al. 2010).

One important feature of ordination techniques like PCA, CA, DCA, and PCoA is their ability to concentrate 'signal' within the total data into the first few significant axes and to relegate 'noise' within the data to later non-significant axes (Gauch 1982). This is a useful feature when trying to test statistically using regression modelling the relationship between, for example, diatom assemblages containing several hundred taxa as response variables and stable-isotope data (e.g., ^{18}O, ^{13}C, ^{15}N) consisting of only a few variables acting as predictor variables. Reducing the diatom data to a few major axes of variation allows the use of these axes only as 'composite response variables' in regression modelling with isotope data as predictors. Lotter et al. (1992a) adopted this approach to test statistically the relationships between late-glacial pollen stratigraphies (reduced to the first few DCA axes) as responses and stable oxygen-isotope stratigraphies as predictors using canonical ordination (see Lotter and Anderson 2012: Chap. 18 for further details of statistical testing of hypotheses involving palaeolimnological data).

When using classical ordination methods (Legendre and Birks 2012b: Chap. 8), such as PCA, CA, or DCA, potentially important information is lost as these methods do not take account of the stratigraphical ordering of the samples. Canonical techniques such as redundancy analysis (RDA (= constrained PCA)), canonical correspondence analysis (CCA (= constrained CA)), and detrended CCA (DCCA

(= constrained DCA)) (see Legendre and Birks 2012b: Chap. 8) can incorporate the external constraint of sample age or depth to detect the major patterns of variation, the nature of any stratigraphical changes, and any trends in the sequence when the stratigraphical data are constrained by order, depth, or age (Birks 1992, 1998).

For some research questions, it may be useful not to impose the stratigraphical constraint and to use PCA, CA, DCA, etc., and simply detect the major patterns of variation irrespective of the ordering of the samples (e.g., Ammann et al. 2000; Birks et al. 2000; Birks and Birks 2001; Lotter and Birks 2003). In other research problems, it may be important to partial out, as covariables, the effects of 'nuisance' variables that are not of primary interest (ter Braak and Prentice 1988). Partialling out depth or age in partial PCA, CA, DCA, RDA, CCA, or DCCA can help to reduce temporal autocorrelation and to allow the analysis of ecological features within the stratigraphical sequences when the effects of stratigraphy have been allowed for statistically. See Odgaard (1994) for the use of partial CA to interpret stratigraphical patterns in terms of differences in the shade tolerances of taxa in Holocene pollen sequences and Bradshaw et al. (2005) for the use of partial CCA to model the role of external catchment factors on Holocene diatom assemblages in a shallow lake in Denmark.

One specialised use of DCCA is to estimate the amount of compositional change or 'turnover' in stratigraphical sequences (Birks 2007b). The idea is to constrain the data by sample age, depth, or order as the sole external variable and to do a DCCA with detrending-by-segments, non-linear rescaling, and Hill's scaling in standard-deviation units of turnover. The amount of turnover within a sequence in different time periods can be estimated and compared (e.g., Birks 2007b; Birks and Birks 2008). Turnover within several sequences for the same time period can also be estimated and compared (e.g., Smol et al. 2005; Birks 2007b; Hobbs et al. 2010). Turnover in different ecological systems but for the same time duration (e.g., biostratigraphical sequences and in primary ecological successions on chronosequences such as glacial forefields) can be estimated and compared (e.g., Birks and Birks 2008).

PCA and CA are implemented in PSIMPOLL (http://chrono.qub.ac.uk/psimpoll/psimpoll.html). PCA, CA, DCA, RDA, CCA, and DCA are implemented in CANOCO version 4.5 (ter Braak and Šmilauer 2002), along with additional software for PCoA and NMDS (see www.microcomputerpower.com, www.canoco.com, and www.canodraw.com for details). Other software for ordination methods is discussed by Legendre and Birks (2012b: Chap. 8) and Borcard et al. (2011).

Summarising Palaeoecological Patterns Using Cluster Analysis

In certain situations, it can be useful to use cluster analysis (e.g., incremental sum-of-squares (CONISS: Grimm 1987) or Ward's (1963) method) without any stratigraphical constraints to derive clusters or samples with similar biotic or geochemical composition, and/or magnetic properties. The sample composition of the individual clusters can then be plotted stratigraphically to detect temporal patterns of recurrence in the composition of the samples. Grimm et al. (2006) adopted this procedure to detect recurrence of fossil assemblages in a 60,000 year lake-sediment record from Florida. Grimm et al. (2011) have used the same approach to detect recurring clusters of samples with similar mineralogical and pollen composition in a 600+ sample sediment-sequence from Kettle Lake (North Dakota) covering the last 13,000 years.

Quantifying Recent Change

Answering broad-scale research questions such as 'how many lakes in eastern North America show evidence for recent surface-water acidification?' or 'what were the reference conditions of low alkalinity lakes in the UK at AD 1850?' is too labour-intensive to be addressed by detailed centimetre-by-centimetre stratigraphical analysis (Smol 2008). Instead a simpler sampling design is often used, the so-called 'top-bottom' or 'before-after' sampling strategy (Smol 2008), where the 'top' surface-sediment sample reflects current biota and environmental conditions ('after') whereas the 'bottom' sample at ~25–30 cm depth is assumed to represent pre-impact or 'before' conditions (e.g., Cumming et al. 1992; Dixit et al. 1992; Enache et al. 2011).

This 'before-after' approach was used in eastern and central North America to compare contemporary pollen assemblages in lakes with pollen assemblages deposited just before or at European settlement marked by the recent rise of *Ambrosia*-type pollen (e.g., McAndrews 1966; Webb 1973). Spatial patterns in the modern and pre-settlement pollen assemblages were compared by principal component analysis (see Legendre and Birks 2012b: Chap. 8) and the mapping of the sample scores on the first two PCA axes for the two time periods (Webb 1973) St. Jacques et al. (2008a) developed a pollen-climate calibration-function (Juggins and Birks 2012: Chap. 14) for Minnesota (USA) using a data-set of pre-settlement AD 1870 pollen assemblages from 133 lakes in Minnesota and adjacent States. This calibration function was then used to reconstruct temperature and effective moisture for the last 900 years from pollen assemblages in varved sediments at Lake Mina, west-central Minnesota (St. Jacques et al. 2008b). This study showed significant bias in the climatic reconstructions based on modern pollen assemblages and less bias when the pre-settlement assemblages are used (St. Jacques et al. 2008a, b).

In defining reference conditions at AD 1850 as part of the European Union Water Framework Directive (Bennion et al. 2004), diatom analyses have been performed on ^{210}Pb-dated sediment cores and ordination methods and dissimilarity measures (e.g., squared chord (Hellinger) distance – see Legendre and Birks 2012b: Chap. 8) have been used to quantify the magnitude of diatom-assemblage change since AD 1850. This general approach has been extensively used to define reference conditions in north-west European lakes. In some studies, the ordination methods used are principal component analysis (e.g., Battarbee et al. 2011; Bennion et al. 2011a), detrended correspondence analysis (e.g., Bennion et al. 2004), correspondence analysis (e.g., Bjerring et al. 2008), or non-metric multidimensional scaling (e.g., Bennion and Simpson 2011). Problems of defining critical thresholds to identify biotic change using squared chord distance (Simpson 2012: Chap. 15) as a measure of recent assemblage change arise (e.g., Simpson et al. 2005; Bjerring et al. 2008; Bennion et al. 2011a, b). Indicator species analysis and associated permutation tests (Dufrêne and Legendre 1997; Legendre and Birks 2012a: Chap. 7) have been shown to be useful tools in identifying taxa indicative of recent biotic change (e.g., Bjerring et al. 2008; Battarbee et al. 2011; Bennion and Simpson 2011). In addition, quantitative reconstructions of recent changes in lake-water chemistry using calibration functions (Juggins and Birks 2012: Chap. 14) provide a measure of the magnitude of change in, for example, lake-water pH or total phosphorus (e.g., Bennion et al. 2004, 2011b; Battarbee et al. 2011).

Rate-of-Change Analysis

Biostratigraphical sequences record changes with time: how rapidly do these changes take place? Chronologies give a measure of the passage of time: a measure of change or difference is also required.

Rate-of-change analysis was introduced by Jacobson et al. (1987) and Jacobson and Grimm (1988), and further explored by Lotter et al. (1992b) and Bennett and Humphry (1995). Two approaches have been used with pollen data, based on change in ordination units and on change measured by a dissimilarity coefficient.

Measuring change by ordination units is described by Jacobson and Grimm (1986). Their method involves:

1. Smooth the data
2. Interpolate to constant time intervals
3. Carry out an ordination (e.g., detrended correspondence analysis (DCA); see Legendre and Birks (2012b: Chap. 8))
4. Calculate the change in DCA units of compositional change or turnover between the scores for adjacent samples.

The effect of doing this is seen in Fig. 11.3. The RLGH diatom percentage data were smoothed with a 5-term smoothing function (fitted as a cubic spline (see Birks 2012a: Chap. 2) on the values for a particular sample, and its age) and then interpolated to 250-year time intervals. A DCA was then carried out. The first three axes were all significant according to a comparison with the broken-stick model, and so the difference between values for samples scores in three dimensions was calculated. These are the values plotted in Fig. 11.3. The main features of this diagram are a peak in values at about 8500 BP, and then higher values from about 4500 BP to the top of the sequence.

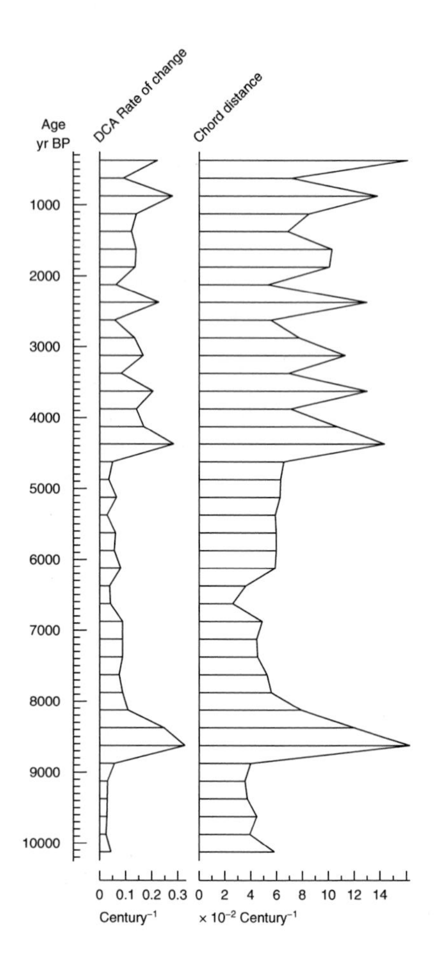

Fig. 11.3 Rate-of-change results from The Round Loch of Glenhead (RLGH) diatom data. Detrended correspondence analysis (DCA) rate-of-change values are smoothed and interpolated to 250 year intervals, following the method of Jacobson and Grimm (1986). Chord distance values are units of dissimilarity per century, smoothed and interpolated to 250 year intervals, following the method of Jacobson et al. (1987) and Jacobson and Grimm (1988)

It is also possible to use the dissimilarity measure approach. There are many ways of calculating dissimilarity (see e.g., Prentice 1980; Legendre and Birks 2012b: Chap. 8), and the measure chosen can be applied directly to the data, or after a transformation. One approach is to smooth the sequence first, interpolate to constant time intervals, and then to calculate the dissimilarity measures and divide by the time interval (Jacobson et al. 1987; Jacobson and Grimm 1988; Birks 1997). The effect of this is shown in Fig. 11.3. The general pattern of change is almost identical to the results from the DCA ordination approach also shown in Fig. 11.3. Laird et al. (1998) obtained rates of change by interpolating first, then smoothing the data, before finally calculating the dissimilarity. Interpolation before smoothing should help remove bias that might result from uneven sampling over time.

Another approach is to estimate the chord distance dissimilarity (Prentice 1980) (= Hellinger distance in Legendre and Birks 2012b: Chap. 8) directly between any pair of samples (Lotter et al. 1992b; Bennett and Humphry 1995), and then divide by the age interval between the pair of samples. This is simpler, with much less tinkering with the data. However, dissimilarity measures are non-linear (in a sequence of three samples, the dissimilarity between samples 1 and 3 will not be the same as the sum of the dissimilarities between 1 and 2 and between 2 and 3). This should not be a serious problem where the time difference between samples is approximately constant, or changing smoothly and slowly.

Most attempts to calculate rates of change hitherto have used radiocarbon years for the time scale (Birks 1998). It would clearly be vastly more satisfactory to base rate-of-change calculations on calendar years, obtained from varves (Lotter et al. 1992b) or from calibration of radiocarbon ages determinations (e.g., Birks et al. 2000; Birks and Birks 2008), in order to avoid inconsistencies arising from the uneven length of the radiocarbon 'year' (see Blaauw and Heegaard 2012: Chap. 12).

Data transformations before analysis may be appropriate in order to stabilise the variance in each data-set. Birks et al. (2000) analysed a data-set of several palaeolimnological variables after first transforming concentration data (macrofossils, mites, Trichoptera) to $\log(y+1)$, and all percentage data to their square roots. They then interpolated the data-sets to constant time intervals without smoothing and carried out the analysis using a chi-square distance measure (Prentice 1980). Time-ordered Monte Carlo permutation tests (ter Braak and Šmilauer 2002) were used to identify rates of change greater than would be expected by chance.

The method of rate-of-change analysis involves more choices in terms of approach and technique (e.g., choice of dissimilarity measure, interpolation, smoothing) than any of the other types of analysis discussed in this chapter. It is important to experiment with the range of possibilities in order to discern which features of the results are due to structure in the data-set and which may be artefacts of a particular numerical approach (Birks 1998). However, it is desirable to use constant methods across all variables of interest, even if these are not optimal for any particular variable in order to facilitate comparability (Birks et al. 2000).

Example of Use

Results from the RLGH data-set are shown in Fig. 11.4. Regardless of dissimilarity measure, there is a clear rise in rate-of-change values from about 600 BP (31 cm depth), and these changes dominate the figure. These changes are absent from the diagrams in Fig. 11.3 as a consequence of the smoothing and interpolation. Other perceptible changes occur, however, notably at around 8400 BP and an increase to generally higher values after about 6000 BP.

This particular data-set has clear changes within the last few centuries that are not detected by the smoothed and interpolated data-set, regardless of whether this is then analysed by a dissimilarity measure or by ordination. The changes in the last few centuries are at too fine scale for the particular interpolation used. On the other hand, longer term changes are well displayed. Calculating rates of change directly from the percentage values emphasises the rapidity of change in the last few centuries and displays it well in the context of changes over the whole Holocene (see Birks 1997).

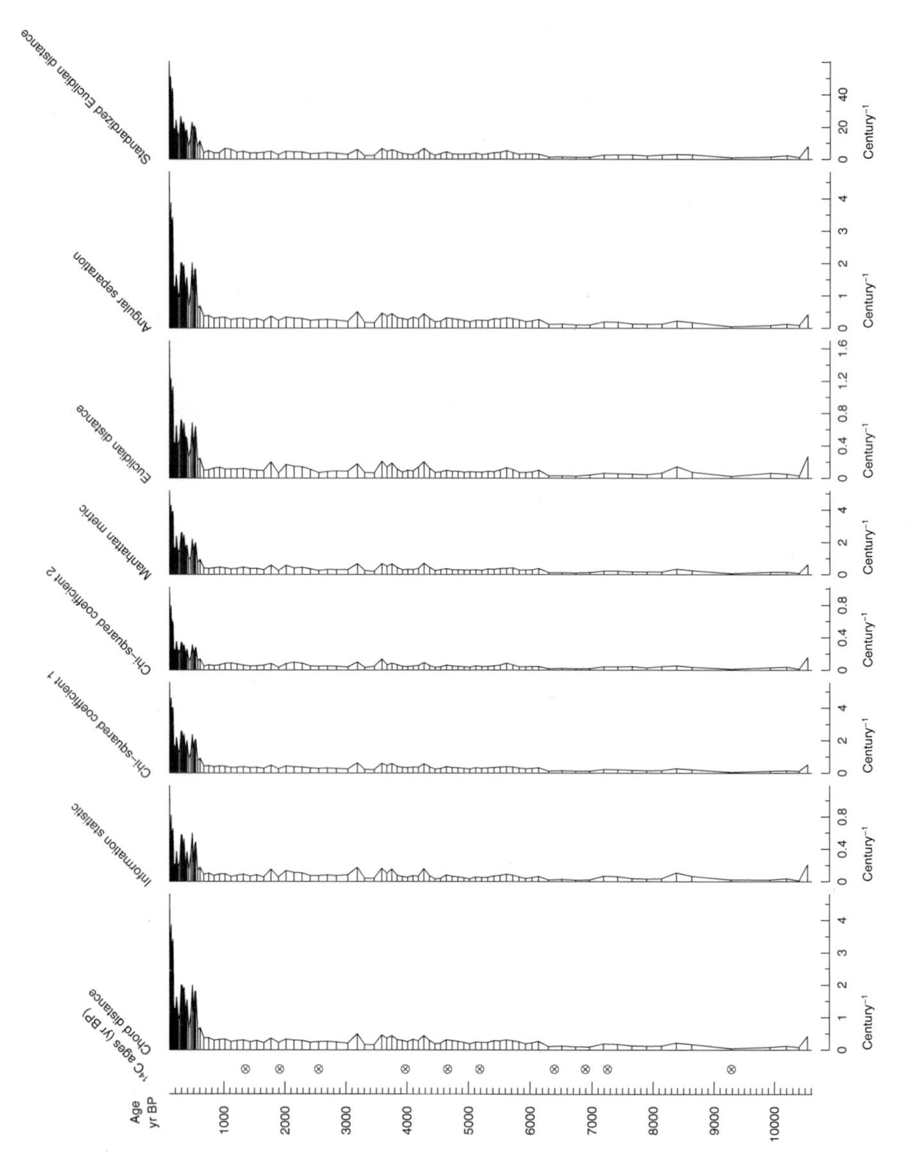

Fig. 11.4 Comparison of rate-of-change results using eight dissimilarity measures (units of dissimilarity per century) with The Round Loch of Glenhead (RLGH) diatom data, following the method of Bennett and Humphry (1995)

Future Developments

The analysis of stratigraphical data has developed surprisingly little over recent decades (Birks 1998). Most of the principles of zonation, for example, were established in the 1970s (see above). What has changed, dramatically, is the improvement in computer performance. Gordon and Birks (1972) wrote "The search for the best positions for three markers would be beyond the computing facilities available to most scientists" (p. 969). Today, optimal zonation can be performed on any desktop, for any number of markers. Unfortunately, the availability of more computer power has had relatively little impact on the choice of zonation methods routinely used. CONISS is probably the commonest used by a considerable margin, chiefly because it is included with TILIA, the most commonly used software for drawing microfossil diagrams. That is not to say that there is anything wrong with CONISS, but one feels that it may be used uncritically too often, without any examination of alternatives or the assumptions behind incremental sum-of-squares clustering (Prentice 1986; Kovach 1989; Baxter 1994; Gordon 1999).

The lesson from CONISS/TILIA is probably that new developments in the analysis of stratigraphical data are only going to come from the availability of software that makes it easy to run the analyses on the types of data that are being collected in palaeolimnology, probably in combination with other analyses. Researchers are generally not receptive to using multiple analytical tools on data, but they will use the tools available simply and easily along with other, more conventional, tools (such as the ability to plot diagrams). The statistical techniques for the analysis of stratigraphical data available are already some way ahead of those in common use in palaeolimnology (see Birks 1998, 2012b: Chap. 21 for a review of additional numerical tools for stratigraphical data-analysis in palaeolimnology and Juggins 2009 for details of his R package for the analysis of palaeolimnological data).

Conclusions

Palaeolimnologists have been analysing their stratigraphical data for many decades and using computer-based numerical techniques for the last 35 years or so. A range of techniques are available for zonation of whole data-sets or of individual taxa, for summarising patterns within and between sequences, and analysing rates of change along sequences. The use of these techniques in practice is limited, and varies between research areas depending upon tradition and software availability. There is considerable scope for expanding the use of numerical techniques for stratigraphical data analysis. The main limiting factor is probably now the availability of convenient software to facilitate this. The availability of useful numerical and statistical methods is far ahead of their practical use in palaeolimnology.

Acknowledgements I am extremely grateful to Keith Bennett for generously providing an extensive first draft of this chapter. I also thank Hilary Birks, Eric Grimm, Steve Juggins, Pierre Legendre, Andy Lotter, Gavin Simpson, John Smol, and Kathy Willis for their thoughtful and helpful comments and advice, and Cathy Jenks for her invaluable help in preparing this chapter. This is publication A211 from the Bjerknes Centre for Climate Research.

References

Aitchison J (1986) The statistical analysis of compositional data. Chapman & Hall, London

Ammann B, Birks HJB, Brooks SJ, Eicher U, von Grafenstein U, Hofmann W, Lemdahl G, Schwander J, Tobolski K, Wick L (2000) Quantification of biotic responses to rapid climatic changes around the Younger Dryas – a synthesis. Palaeogeogr Palaeoclim Palaeoecol 159:313–347

Battarbee RW, Simpson GL, Bennin H, Curtis C (2011) A reference typology of low alkalinity lakes in the UK based on pre-acidification diatom assemblages from lake sediment cores. J Paleolimnol 45:489–505

Baxter MJ (1994) Exploratory multivariate analysis in archaeology. Edinburgh University Press, Edinburgh

Bennett KD (1994) Confidence intervals for age estimates and deposition times in late-Quaternary sediment sequences. Holocene 4:337–348

Bennett KD (1996) Determination of the number of zones in a biostratigraphical sequence. New Phytol 132:155–170

Bennett KD, Humphry RW (1995) Analysis of late-glacial and Holocene rates of vegetational change at two sites in the British Isles. Rev Palaeobot Palynol 85:263–287

Bennett KD, Boreham S, Sharp MJ, Switsur VR (1992) Holocene history of environment, vegetation and human settlement on Catta Ness, Lunnasting, Shetland. J Ecol 80:241–273

Bennion H, Simpson GL (2011) The use of diatom records to establish reference conditions for UK lakes subject to eutrophication. J Paleolimnol 45:469–488

Bennion H, Fluin J, Simpson GL (2004) Assessing eutrophication and reference conditions for Scottish freshwater lochs using subfossil diatoms. J Appl Ecol 41:124–138

Bennion H, Simpson GL, Anderson NJ, Clarke G, Dong X, Hobæk A, Guilizzoni P, Marchetto A, Sayer CD, Thies H, Tolotti M (2011a) Defining ecological and chemical reference conditions and restoration targets for nine European lakes. J Paleolimnol 45:415–431

Bennion H, Battarbee RW, Sayer CD, Simpson GL, Davidson TA (2011b) Defining reference conditions and restoration targets for lake ecosystems using palaeolimnology: a synthesis. J Paleolimnol 45:533–544

Birks HH, Birks HJB (2001) Recent ecosystem dynamics in nine north African lakes in the CASSARINA project. Aquat Ecol 35:461–478

Birks HH, Mathewes RW (1978) Studies in the vegetational history of Scotland. V. Late Devensian and early Flandrian pollen and macrofossil stratigraphy at Abernethy Forest, Inverness-shire. New Phytol 80:455–484

Birks HH, Battarbee RW, Birks HJB (2000) The development of the aquatic ecosystem at Kråkenes Lake, western Norway, during the late glacial and early Holocene – a synthesis. J Paleolimnol 23:91–114

Birks HJB (1981) Late Wisconsin vegetational and climatic history at Kylen Lake, north-eastern Minnesota. Quaternary Res 16:322–355

Birks HJB (1986) Numerical zonation, comparison and correlation of Quaternary pollen-stratigraphical data. In: Berglund BE (ed) Handbook of Holocene palaeoecology and palaeohydrology. Wiley, Chichester, pp 743–774

Birks HJB (1987) Multivariate analysis of stratigraphic data in geology: a review. Chemometr Intell Lab Syst 2:109–126

Birks HJB (1992) Some reflections on the application of numerical methods in Quaternary palaeoecology. Publication of the Karelian Institute, University of Joensuu 102, pp 7–20

Birks HJB (1997) Environmental change in Britain – a long-term palaeoecological perspective. In: Mackay AW, Murlis J (eds) Britain's natural environment: a state of the nation review. ENSIS Publications, London, pp 23–28

Birks HJB (1998) Numerical tools in palaeolimnology – progress, potentialities, and problems. J Paleolimnol 20:307–332

Birks HJB (2007a) Numerical analysis methods. In: Elias SA (ed) Encyclopedia of Quaternary science. Volume 3. Elsevier, Oxford, pp 2515–2521

Birks HJB (2007b) Estimating the amount of compositional change in late-Quaternary pollen-stratigraphical data. Veg Hist Archaeobot 16:197–202

Birks HJB (2010) Numerical methods for the analysis of diatom assemblage data. In: Smol JP, Stoermer EF (eds) The diatoms: application for the environmental and earth sciences, 2nd edn. Cambridge University Press, Cambridge

Birks HJB (2012a) Chapter 2: Overview of numerical methods in palaeolimnology. In: Birks HJB, Lotter AF, Juggins S, Smol JP (eds) Tracking environmental change using lake sediments. Volume 5: Data handling and numerical techniques. Springer, Dordrecht

Birks HJB (2012b) Chapter 21: Conclusions and future challenges. In: Birks HJB, Lotter AF, Juggins S, Smol JP (eds) Tracking environmental change using lake sediments. Volume 5: Data handling and numerical techniques. Springer, Dordrecht

Birks HJB, Berglund BE (1979) Holocene pollen stratigraphy of southern Sweden: a reappraisal using numerical methods. Boreas 8:257–279

Birks HJB, Birks HH (2008) Biological responses to rapid climate change at the younger Dryas-Holocene transition at Kråkenes, western Norway. Holocene 18:19–30

Birks HJB, Gordon AD (1985) Numerical methods in Quaternary pollen analysis. Academic Press, London

Birks HJB, Jones VJ (2012) Chapter 3: Data-sets. In: Birks HJB, Lotter AF, Juggins S, Smol JP (eds) Tracking environmental change using lake sediments. Volume 5: Data handling and numerical techniques. Springer, Dordrecht

Birks HJB, Line JM (1994) Sequence splitting of pollen accumulation rates from the Holocene and Devensian late-glacial of Scotland. Diss Botanic 234:145–160

Bjerring R, Bradshaw EG, Amsinck SL, Johansson LS, Odgaard BV, Nielsen AB, Jeppesen E (2008) Inferring recent changes in the ecological state of 21 Danish candidate reference lakes (EU Water Framework Directive) using palaeolimnology. J Appl Ecol 45:1566–1575

Blaauw M, Heegaard E (2012) Chapter 12: Estimation of age-depth relationships. In: Birks HJB, Lotter AF, Juggins S, Smol JP (eds) Tracking environmental change using lake sediments. Volume 5: Data handling and numerical techniques. Springer, Dordrecht

Borcard D, Gillet F, Legendre P (2011) Numerical ecology with R. Springer, New York

Bradshaw EG, Rasmussen P, Odgaard BV (2005) Mid- to late-Holocene land-use change and lake development at Dallund Sø, Denmark: Synthesis of multiproxy data, linking land and lake. Holocene 15:1152–1162

Cumming BF, Smol JP, Kingston JC, Charles DF, Birks HJB, Camburn KE, Dixit SS, Uutala AJ, Selle AR (1992) How much acidification has occurred in Adirondack (New York, USA) lakes since pre-industrial times? Can J Fish Aquat Sci 49:128–141

Cushing EJ (1967) Late-Wisconsin pollen stratigraphy and the glacial sequence in Minnesota. In: Cushing EJ, Wright HE (eds) Quaternary paleoecology. Yale University Press, New Haven, pp 59–88

De'ath G (2002) Multivariate regression trees: a new technique for modeling species-environment relationships. Ecology 83:1105–1117

De'ath G, Fabricus KE (2000) Classification and regression trees: a powerful yet simple technique for ecological data analysis. Ecology 81:3178–3192

Dixit SS, Dixit AS, Smol JP (1992) Assessment of changes in lake water chemistry in Sudbury area lakes since pre-industrial times. Can J Fish Aquat Sci 49(Suppl 1):8–16

Dufrêne M, Legendre P (1997) Species assemblages and indicator species: the need for a flexible asymmetrical approach. Ecol Monogr 67:345–366

Dutilleul P, Cumming BF, Lontoc-Roy M (2012) Chapter 16: Autocorrelogram and periodogram analyses of palaeolimnological temporal series from lakes in central and western North America to assess shifts in drought conditions. In: Birks HJB, Lotter AF, Juggins S, Smol JP (eds) Tracking environmental change using lake sediments. Volume 5: Data handling and numerical techniques. Springer, Dordrecht

Enache MD, Paterson AM, Cumming BF (2011) Changes in diatom assemblages since pre-industrial times in 40 reference lakes from the Experimental Lakes Area (northwestern Ontario, Canada). J Paleolimnol 46:1–15
Gauch HG (1982) Noise reduction by eigenvector ordination. Ecology 63:1643–1649
Gordon AD (1982) Numerical methods in Quaternary palaeoecology. V. Simultaneous graphical representations of the levels and taxa in a pollen diagram. Rev Palaeobot Palynol 37:155–183
Gordon AD (1999) Classification, 2nd edn. Chapman & Hall/CRC, Boca Raton
Gordon AD, Birks HJB (1972) Numerical methods in Quaternary palaeoecology. I. Zonation of pollen diagrams. New Phytol 71:961–979
Gordon AD, Birks HJB (1974) Numerical methods in Quaternary palaeoecology. II. Comparison of pollen diagrams. New Phytol 73:221–249
Green DG (1982) Fire and stability in the postglacial forests of southwest Nova Scotia. J Biogeogr 9:29–40
Grimm EC (1987) CONISS: a FORTRAN 77 program for stratigraphically constrained cluster analysis by the method of incremental sum of squares. Comput Geosci 13:13–25
Grimm EC, Watts WA, Jacobson GL, Hansen BCS, Almquist HR, Dieffenbacher-Krall AC (2006) Evidence for warm wet Heinrich events in Florida. Quaternary Sci Rev 25:2197–2211
Grimm EC, Donovan JJ, Brown KJ (2011) A high-resolution record of climatic variability and landscape response from Kettle Lake, northern Great Plains, North America. Quaternary Sci Rev 30:2626–2650
Haberle SG, Bennett KD (2004) Postglacial formation and dynamics of north Patagonian rainforest in the Chonos Archipelago, southern Chile. Quaternary Sci Rev 23:2433–2452
Haberle SG, Tibby J, Dimitriadis S, Heijnis H (2006) The impact of European occupation on terrestrial and aquatic ecosystem dynamics in an Australian tropical rain forest. J Ecol 94:987–1002
Harkness DD, Miller BF, Tipping RM (1997) NERC radiocarbon measurements 1977–1988. Quaternary Sci Rev 16:925–927
Hedberg HD (ed) (1976) International stratigraphic guide. Wiley, New York
Hobbs WO, Telford RJ, Birks HJB, Saros JE, Hazewinkel RRO, Perren BB, Saulnier-Talbot É, Wolfe AP (2010) Quantifying recent ecological changes in remote lakes of North America and Greenland using sediment diatom assemblages. PLoS One 5:e10026. doi:10.1371/journal.pone.0010026
Jackson DA (1993) Stopping rules in principal components analysis: a comparison of heuristical and statistical approaches. Ecology 74:2204–2214
Jacobson GL, Grimm EC (1986) A numerical analysis of Holocene forest and prairie vegetation in central Minnesota. Ecology 67:958–966
Jacobson GL, Grimm EC (1988) Synchrony of rapid change in late-glacial vegetation south of the Laurentide ice sheet. Bull Buffalo Soc Nat Sci 33:31–38
Jacobson GL, Webb T, Grimm EC (1987) Patterns and rates of vegetation change during the deglaciation of eastern North America. In: Ruddiman WF, Wright HE (eds) North America and adjacent oceans during the last deglaciation. Volume K-3: The geology of North America. Geological Society of America, Boulder, pp 277–288
Jones VJ, Stevenson AC, Battarbee RW (1989) Acidification of lakes in Galloway, southwest Scotland: a diatom and pollen study of the post-glacial history of the Round Loch of Glenhead. J Ecol 77:1–23
Juggins S (2009) rioja: analysis of Quaternary science data. R package version 0.5-6. http://www.staff.ncl.ac.uk/staff/stephen.juggins/
Juggins S, Birks HJB (2012) Chapter 14: Quantitative environmental reconstructions from biological data. In: Birks HJB, Lotter AF, Juggins S, Smol JP (eds) Tracking environmental change using lake sediments. Volume 5: Data handling and numerical techniques. Springer, Dordrecht
Kovach WL (1989) Comparisons of multivariate analytical technique for use in pre-Quaternary plant paleocology. Rev Palaeobot Palynol 60:255–282

Laird KR, Fritz SC, Cumming BF, Grimm EC (1998) Early-Holocene limnological and climatic variability in the northern Great Plains. Holocene 8:275–285

Legendre P, Birks HJB (2012a) Chapter 7: Clustering and partitioning. In: Birks HJB, Lotter AF, Juggins S, Smol JP (eds) Tracking environmental change using lake sediments. Volume 5: Data handling and numerical techniques. Springer, Dordrecht

Legendre P, Birks HJB (2012b) Chapter 8: From classical to canonical ordination. In: Birks HJB, Lotter AF, Juggins S, Smol JP (eds) Tracking environmental change using lake sediments. Volume 5: Data handling and numerical techniques. Springer, Dordrecht

Legendre L, Legendre P (1983) Numerical ecology, 2nd English edn. Elsevier, Amsterdam

Legendre P, Dallot S, Legendre L (1985) Succession of species within a community: chronological clustering, with applications to marine and freshwater zooplankton. Am Nat 125:257–288

Lotter AF (1998) The recent eutrophication of Baldeggersee (Switzerland) as assessed by fossil diatom assemblages. Holocene 8:395–405

Lotter AF, Anderson NJ (2012) Chapter 18: Limnological responses to environmental changes at inter-annual to decadal time-scales. In: Birks HJB, Lotter AF, Juggins S, Smol JP (eds) Tracking environmental change using lake sediments. Volume 5: Data handling and numerical techniques. Springer, Dordrecht

Lotter AF, Birks HJB (2003) The Holocene palaeolimnology of Sägistalsee and its environmental history – a synthesis. J Paleolimnol 30:333–342

Lotter AF, Eicher U, Birks HJB, Sigenthaler U (1992a) Late-glacial climatic oscillations as recorded in Swiss lake sediments. J Quaternary Sci 7:187–204

Lotter AF, Ammann B, Sturm M (1992b) Rates of change and chronological problems during the late-glacial period. Clim Dyn 6:233–239

MacArthur RH (1957) On the relative abundance of bird species. Proc Natl Acad Sci USA 43:293–295

Massaferro J, Brooks SJ, Haberle SG (2005) The dynamics of chironomid assemblages and vegetation during the late-Quaternary at Laguna Facil, Chonos Archipelago, southern Chile. Quaternary Sci Rev 24:2510–2522

McAndrews JH (1966) Postglacial history of prairie, savanna, and forest in northwestern Minnesota. Mem Torrey Bot Club 22(2):1–72

Odgaard BV (1994) Holocene vegetation history of northern West Jutland, Denmark. Op Bot 123:1–171

Padmore J (1987) Program SHEFFPOLL – a program for the zonation of stratigraphical data sets. Technical report research report no. 304/87, Department of Probability & Statistics, Sheffield University, UK

Prentice IC (1980) Multidimensional scaling as a research tool in Quaternary palynology: a review of theory and methods. Rev Palaeobot Palynol 31:71–104

Prentice IC (1986) Multivariate methods for data analysis. In: Berglund BE (ed) Handbook of Holocene palaeoecology and palaeohydrology. Wiley, Chichester, pp 775–797

Simpson GL (2012) Chapter 15: Analogue methods in palaeolimnology. In: Birks HJB, Lotter AF, Juggins S, Smol JP (eds) Tracking environmental change using lake sediments. Volume 5: Data handling and numerical techniques. Springer, Dordrecht

Simpson GL, Birks HJB (2012) Chapter 9: Statistical learning in palaeolimnology. In: Birks HJB, Lotter AF, Juggins S, Smol JP (eds) Tracking environmental change using lake sediments. Volume 5: Data handling and numerical techniques. Springer, Dordrecht

Simpson GL, Shilland EM, Winterbottom JM, Keay J (2005) Defining reference conditions for acidified waters using a modern analogue approach. Environ Pollut 137:119–133

Smol JP (2008) Pollution of lakes and rivers: a paleoenvironmental perspective, 2nd edn. Blackwell, Oxford

Smol JP, Wolfe AP, Birks HJB et al. (2005) Climate-driven regime shifts in Arctic Lake ecosystems. Proc Natl Acad Sci USA 102:4397–4402

St. Jacques J-M, Cumming BF, Smol JP (2008a) A pre-European settlement pollen-climate calibration set for Minnesota, USA: developing tools for palaeoclimatic reconstructions. J Biogeogr 35:306–324

St. Jacques J-M, Cumming BF, Smol JP (2008b) A 900-year pollen-inferred temperature and effective moisture record from varved Lake Mina, west-central Minnesota, USA. Quaternary Sci Rev 27:781–796

ter Braak CJF, Prentice IC (1988) A theory of gradient analysis. Adv Ecol Res 18:271–317

ter Braak CJF, Šmilauer P (2002) CANOCO reference manual and CanoDraw for Windows user's guide: software for canonical community ordination (version 4.5). Microcomputer Power, Ithaca

Thompson R, Clark M, Boulton GS (2012) Chapter 13: Core correlation. In: Birks HJB, Lotter AF, Juggins S, Smol JP (eds) Tracking environmental change using lake sediments. Volume 5: Data handling and numerical techniques. Springer, Dordrecht

Walanus A, Nalepka D (1997) Palynological diagram drawing in Polish PolPal for windows. INQUA Commission for the Study of the Holocene: Sub-commission on data-handling methods. Newsletter 16:3

Walker D, Pittelkow Y (1981) Some applications of the independent treatment of taxa in pollen analysis. J Biogeogr 8:37–51

Walker D, Wilson SR (1978) A statistical alternative to the zoning of pollen diagrams. J Biogeogr 5:1–21

Wall AAJ, Magny M, Mitchell EAD, Vannière B, Gilbert D (2010) Response of testate amoeba assemblages to environmental and climatic changes during the Lateglacial-Holocene transition at Lake Lautrey (Jura Mountains, eastern France). J Quaternary Sci 25:945–956

Ward JH (1963) Hierarchical grouping to optimize an objective function. J Am Stat Assoc 58:236–244

Webb T (1973) A comparison of modern and pre-settlement pollen from southern Michigan, USA. Rev Palaeobot Palynol 16:137–156

Chapter 12
Estimation of Age-Depth Relationships

Maarten Blaauw and Einar Heegaard

Abstract An accurate and precise chronology is an essential pre-requisite for any palaeolimnological study. Chronologies give time-scales for events, and hence for rates for patterns and processes, and make it possible to compare and correlate events in different stratigraphical sequences. Palaeolimnology without chronology is history without dates.

As radiocarbon dating is so widely used in palaeolimnology, this chapter focuses on ^{14}C dating, and its associated errors and the calibration of ^{14}C ages to calibrated ^{14}C ages. Calibration is an essential step before constructing age-depth models. There are several approaches to establishing age-depth relationships – linear interpolation, polynomial regression, splines, mixed-effect models, and Bayesian age-depth modelling involving chronological ordering or wiggle-matching. The critical question of model selection is discussed and future developments are outlined, along with details of available software.

Keywords Age-depth modelling • Bayesian approaches • Calibration • Interpolation • Mixed-effects models • Monte Carlo simulations • Polynomials • Radiocarbon dates • Splines • Wiggle matching

M. Blaauw (✉)
School of Geography, Archaeology & Palaeoecology, Queen's University Belfast, Belfast, BT7 1NN, UK
e-mail: maarten.blaauw@qub.ac.uk

E. Heegaard
Department of Biology and Bjerknes Centre for Climate Research, University of Bergen, PO Box 7803, Bergen N-5020, Norway

Now at: Norwegian Forest and Landscape Institute, Fanaflaten 4, Fana N-5244, Norway
e-mail: einar.heegaard@skogoglandskap.no

H.J.B. Birks et al. (eds.), *Tracking Environmental Change Using Lake Sediments*, Developments in Paleoenvironmental Research 5, DOI 10.1007/978-94-007-2745-8_12,

Introduction

An accurate and precise chronology is an essential pre-requisite for any palaeolimnological or other palaeoecological investigation. Chronologies give time-scales for events, and hence rates for patterns and processes. They make it possible to compare the temporal course of events in different sequences. Palaeolimnology without chronology is history without dates.

Chronologies of deposits are generally obtained by drawing a curve through a number of dated depths. This curve is then used to estimate the ages of the dated as well as the undated depths (Fig. 12.1). However, producing a reliable chronology can prove challenging for several reasons. First, dates come with errors or uncertainties, and sometimes with a systematic yet imprecisely known offset. Further, constraints in research budget or dateable material (e.g., availability of suitable organic material for radiocarbon dating) can limit the number of dates to, typically, a handful for an entire Holocene sequence. Last but not least, age-depth curves need to be chosen so that they provide us with a likely reconstruction of the unknown true sedimentation history (e.g., gaps and abrupt or gradual changes in accumulation rate: Telford et al. 2004a).

Ages may be expressed in a number of forms, and it is important to be sure which is being used in any investigation. For most palaeolimnological and other palaeoecological applications, the units of time are years (yr, a), thousands of years (kyr, ka), etc. Different units are sometimes used to indicate absolute ages (e.g., 5 ka:

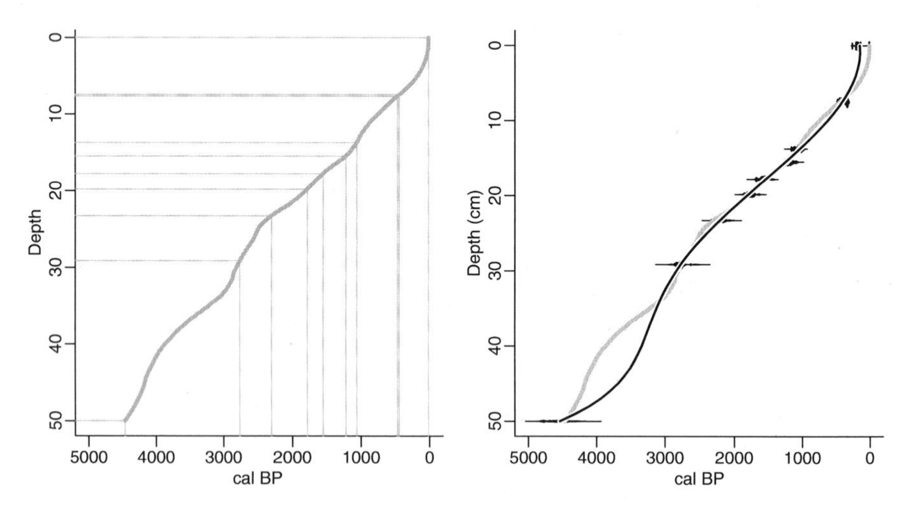

Fig. 12.1 Schematic representation of age-depth modelling. A simulated accumulation of sediment over four millennia (*left panel*) results in a 'true' age-depth curve (*grey line*). Ten depths of this sequence are ^{14}C dated (*thin grey lines*, with true calendar ages indicated) and calibrated (black 1 standard deviation ranges in *right panel*), after which a fourth-order polynomial is chosen as the age-depth model (*black curve*). Whereas this age-model generally resembles the true sedimentation history, at times it deviates by several centuries (especially in sections with few dates)

5000 years ago) or durations (e.g., 5 kyr: an interval lasting 5000 years). Care should be taken when comparing ages from different sources. Raw radiocarbon ages are usually expressed as '^{14}C BP' or 'years BP', where 'BP', 'before present', is defined as AD 1950 (chosen because it was a round number near the time when Willard Libby obtained the first ^{14}C dates). Calibrated ^{14}C ages are given in 'cal BP' or 'cal BC/AD' (sometimes one sees 'cal yr. BP', 'kcal BP', or variations thereof). Ice-core ages are usually given as years before AD 2000 (b2k), luminescence dates are often reported as years before sampling (e.g., Madsen et al. 2007), and ^{210}Pb dates are given as ages AD.

Radiocarbon Dating

The age of a particular depth can be estimated by submitting material from the defined layer to a laboratory for ^{14}C (Boaretto et al. 2002), ^{210}Pb (Appleby 2001), or other dating procedure. As radiocarbon dating is so widely used in palaeolimnology, we will focus here on this method. For in-depth reviews of radiocarbon dating, see Bowman (1990), Pilcher (1991), Walker (2005), and several chapters in the Encyclopedia of Quaternary Science (Elias 2007). A general overview of methods in preparing and obtaining ^{14}C dates, as well as several material-related problems such as contamination and lake reservoir effects, can be found in Björck and Wohlfarth (2001).

Carbon in atmospheric CO_2 is present as several isotopes, with ^{12}C being the most abundant (c. 98.9%), ^{13}C less abundant (c. 1.1%), and ^{14}C very rare (1 in every million million carbon atoms, comparable to a single grain in a cubic metre of sand). Atmospheric CO_2 is taken up by plants through photosynthesis, and then further distributed through the food chain in more or less unchanged isotope ratios.

While ^{12}C and ^{13}C are stable isotopes, radiocarbon is radioactive or unstable. ^{14}C atoms disintegrate gradually over time, thus resulting in ever-decreasing $^{14}C/^{12}C$ ratios within dead organic matter as time passes. Although ^{14}C is a very rare isotope, there are still about 50 million ^{14}C atoms among the c. 5×10^{19} atoms in 1 mg of recent carbon. As the internationally agreed half-life of ^{14}C is 5568 yr (Mook 1986), after 5568 yr this 1 mg of carbon will have about 25 million remaining ^{14}C atoms, while after $2 \times 5568 = 11{,}136$ yr it will have 12.5 million surviving ^{14}C atoms, and so forth until only several tens of thousands are left over after about 11 half-lives (60,000 yr, the current limit of radiocarbon dating). The age of a fossil can thus be inferred from its $^{14}C/^{12}C$ ratio. We express fossil $^{14}C/^{12}C$ ratios ('activity' $A = {}^{14}C/^{12}C$, after correction for the machine background) against the $^{14}C/^{12}C$ activity A_0 from a standard relative to AD 1950, and calculate the ^{14}C ages using the Libby half life: $-5568/ln(2)\ ln(A/A_0)$, which becomes $-8033\ ln(A/A_0)$.

We mentioned earlier that the isotope ratios remain largely unchanged while passing through the carbon cycle from atmospheric CO_2 to fossil organic material. In fact, most chemical processes cause some isotopic fractionation, favouring either the heavier or the lighter isotopes. From theoretical reasoning we know that the fractionation of $^{14}C/^{12}C$ should be twice that of $^{13}C/^{12}C$. Therefore most ^{14}C laboratories also measure the ratio of the stable carbon isotopes, $^{13}C/^{12}C$ (expressed as per mille deviations from zero, $\delta^{13}C$), followed by correcting the $^{14}C/^{12}C$ ratios to the oxalic acid standard of −26‰.

There are several ways to measure the $^{14}C/^{12}C$ content. The oldest one (conventional dating) consists of counting the radioactive decay events of the ^{14}C atoms. In order to obtain sufficient ^{14}C decay events, typically several tens to hundreds of grams of carbon are needed, as well as days to months of counting time. The most recent dating method is accelerator mass spectrometry (AMS) radiocarbon dating, which works by counting the number of ^{12}C, ^{13}C, and ^{14}C atoms directly in a large-scale, high-voltage mass spectrometer. As the method counts all atoms, the amount of material needed is much less (c. 0.1–10 mg carbon), as is the required counting time (minutes).

Errors

By the term 'error' we can mean 'uncertainty' as well as 'mistake', and we can distinguish between random and systematic errors (see Maher et al. 2012: Chap. 6). Random errors refer to the problem that, owing to random processes, machine measurements of, for example, a $^{14}C/^{12}C$ ratio will merely provide inexact approximations of the 'true' value (Fig. 12.2). Multiple measurements of the same sample will result in different estimates (precise measurements obtaining little scatter and thus small standard deviations). Even more, these measurements can be offset from the true value by a systematic amount (accuracy or bias: Fig. 12.2). Last but not least, measurements can differ from the true value owing to other mistakes, such as submission of the wrong material, contamination, or faults in the sample preparation.

There is no difference, from a data-handling point of view, in the random errors obtained by AMS and conventional methods. Standard deviations quoted with radiocarbon ages are obtained by methods that vary between dating laboratories. With decay counting (conventional dating), some laboratories assume that the decays are distributed as Poisson processes. In this case, the standard deviation $sd(N) = -8{,}033 \times (\sqrt{N})/N$ where N is the number of observed decays. Thus, the standard deviation is smaller for larger and younger samples, and longer counting times. Other laboratories count the sample in a series of short time periods (e.g., 100 min), and calculate a standard error of the mean from this series. Radiocarbon ages obtained by AMS depend upon counts of ^{14}C atoms arriving at a detector, which may be assumed to be a Poisson process, as for the detection of decaying ^{14}C atoms in the conventional dating procedure. For large numbers as is the case here, the Poisson distribution can be approximated by the normal distribution. However, the quoted errors on AMS ages depend also upon complex laboratory factors (including the expected errors attributable to chemical preparation of specific

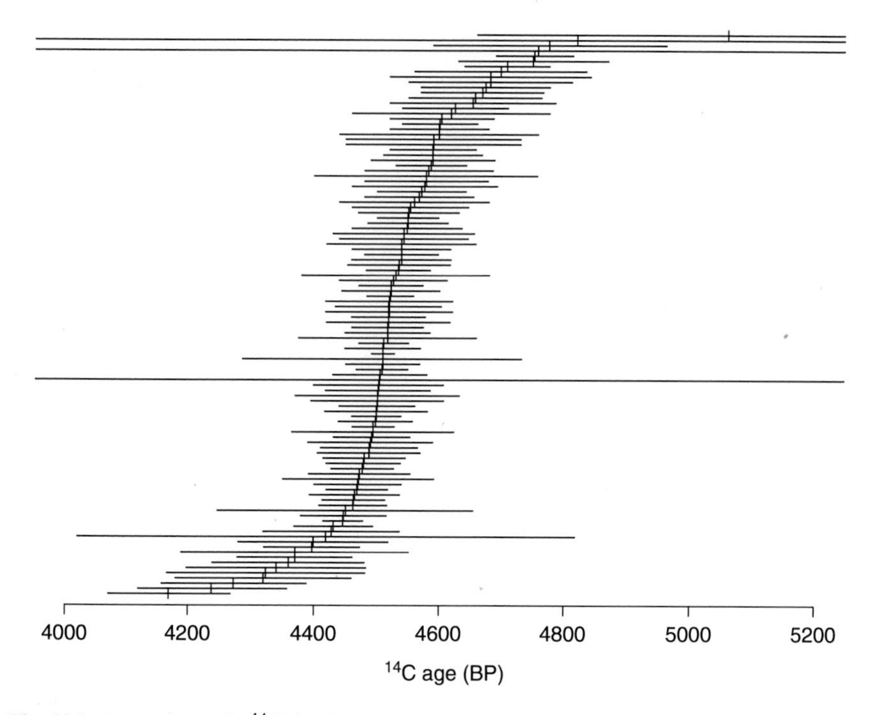

Fig. 12.2 A wood sample ^{14}C dated more than 100 times at a range of AMS and conventional laboratories. The distribution of ages appears Gaussian. Note the large range of reported 2 sd error sizes. Several error bars are too large to fit on the graph. Redrawn from Scott (2007)

types of organic material such as bulk sediment or macrofossils). Therefore, the use of a laboratory-specific error amplification factor is advocated (Scott 2007). In this chapter, we assume that the quoted standard deviation (abbreviated to sd or σ) for any radiocarbon age can be treated as a sample standard deviation, although it is probably often an over-optimistic estimate (Christen and Pérez 2009).

The Need for Radiocarbon Calibration

Radiocarbon dating of materials of known-age, notably tree-rings (dated independently to yearly precision through dendrochronology), has shown that past changes in the atmospheric $^{14}C/^{12}C$ ratio have resulted in radiocarbon years not being the same as calendar years (Fig. 12.3). Radiocarbon years generally are slightly longer than calendar years, but by a variable amount, causing periods with large fluctuations ('wiggles') or constant radiocarbon ages (radiocarbon 'plateaux'). This requires translation, or calibration, of ^{14}C ages into calendar years.

For the current ^{14}C calibration curve IntCal09 (Reimer et al. 2009), ages younger than 12.55 kcal BP are based on large numbers of ^{14}C dates from dendro-dated

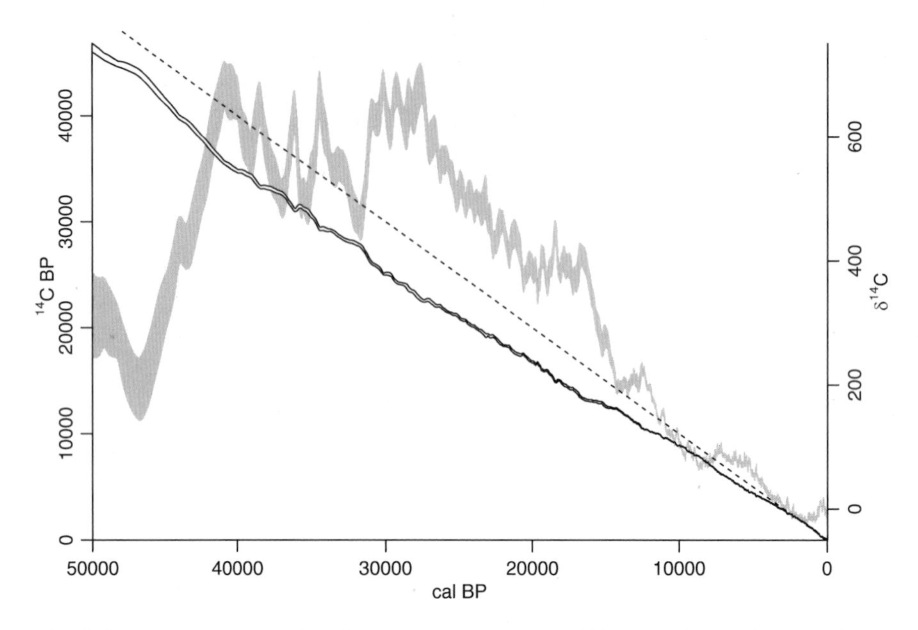

Fig. 12.3 Calendar versus radiocarbon ages over the past 50,000 years. The *black diagonal lines* indicate the 1 standard deviation (sd) envelope of the IntCal09 calibration curve for the Northern Hemisphere (Reimer et al. 2009). A 1:1 line (*'Libby line'*, *dashed*) is shown for comparison. From the slope of the calibration curve, past changes in atmospheric radiocarbon concentrations can be reconstructed (*grey shades* show the 1 sd envelope of $\delta^{14}C$; per mille values shown on the secondary y axis)

tree rings. The older glacial section, until 50 kcal BP, is based on fewer dates derived from marine archives dated with less precision/accuracy (e.g., owing to uncertainties in tuning and U/Th dating and an imprecisely known marine reservoir effect), and this is reflected in wider confidence intervals. The curve through all these ^{14}C dates with independent calendar-age estimates is assumed to be currently the best approximation of the 'true' radiocarbon age for each calendar year over the period considered. Separate curves are available for Northern Hemisphere (Reimer et al. 2009), Southern Hemisphere (McCormac et al. 2004), and marine ^{14}C dates (Reimer et al. 2009), and regional post-bomb curves have also been published (e.g., Hua and Barbetti 2004).

Atmospheric radiocarbon concentrations varied through time because of changes in both the production of ^{14}C as well as its utilisation in the carbon cycle (Burr 2007). Radiocarbon is a so-called cosmogenic isotope: high-energy cosmic rays entering the Earth's atmosphere cause neutrons to collide with atmospheric ^{14}N, transforming it into ^{14}C by displacing a proton out of its nucleus. The ^{14}C atoms are then oxidised to form $^{14}CO_2$.

The Earth and solar magnetic fields form a shield against cosmic rays entering the Earth's atmosphere. Decreases in solar activity or in the strength of Earth's magnetic field will weaken this shield, enabling more cosmic rays to enter and thus causing an increased production of cosmogenic isotopes (and *vice versa*). The Earth's magnetic field varies mostly on millennial time-scales, although some more abrupt changes have been identified (King and Peck 2001; Twyman 2007). Solar activity varies over time-scales from minutes to centuries, with periodicities of approximately 11, 22, 80–90, and 200 years being the most relevant for our discussion. Atomic bomb explosions in the 1950s and 1960s caused an enormous pulse in atmospheric ^{14}C concentration, followed by a gradual decline towards pre-bomb levels over the last few decades.

Besides the mechanisms acting as varying sources of ^{14}C, sinks of ^{14}C tend to vary as well. Atmospheric carbon forms only a very small part of the total carbon reservoir (600 Gt or 1% according to the latest IPCC report: see Denman et al. 2007) compared to that of plants, soils, sediments, and peat (2450 Gt, 5%), fossil fuels (3700 Gt, 8%), the surface ocean (900 Gt, 2%), and, especially, the deep ocean (37,100 Gt, 83%). Changes in fluxes of carbon between the atmosphere and these reservoirs will cause changes in atmospheric ^{14}C concentration. Uptake of atmospheric $^{14}CO_2$ by the ocean can increase or decrease with changes in ocean circulation/ventilation. Owing to radiocarbon decay, oceanic deep-water kept isolated from atmospheric CO_2 will contain ever declining ^{14}C ratios as time passes. Millennia-old ocean-water will thus be depleted in ^{14}C, and this water will cause dilution of atmospheric $^{14}CO_2$ levels when surfacing and exchanging carbon. Additionally, the burning of ^{14}C-free fossil fuel has caused a dilution of recent atmospheric $^{14}CO_2$ ratios (the so-called Suess effect).

Calibration Methods

Owing to past variations in atmospheric ^{14}C concentrations, ^{14}C dates need to be calibrated into calendar years in order to be interpreted. At first sight, the most obvious way to calibrate a ^{14}C date would be to find its intercept in the calibration curve, and look up its corresponding calendar age (similarly for its 1 or 2 sd range). However, this 'intercept' calibration method is prone to problems (Telford et al. 2004b). Wiggles and plateaux in the calibration often cause multiple calendar ages for a single ^{14}C age. For the date of 5380 ± 50 ^{14}C BP in Fig. 12.4, the result seems straightforward with only one corresponding calendar age around 6200 cal BP. However, if by chance the radiocarbon measurement would have resulted in a slightly different age (Fig. 12.2), we would probably have to deal with several intercepts on the calendar age. The same holds for the translation of the error bars from ^{14}C into calendar years. Moreover, using the intercept method it is hard to take into account the errors of the calibration curve. Therefore, the intercept method is not to be recommended for calibrating ^{14}C dates (Telford et al. 2004b), and is not now in use in modern ^{14}C calibration software.

Current calibration methods do not start by looking up the calendar age(s) of a ^{14}C age; instead the process is reversed. For a given calendar year (usually depicted with the theta symbol, θ), the corresponding radiocarbon age of the calibration

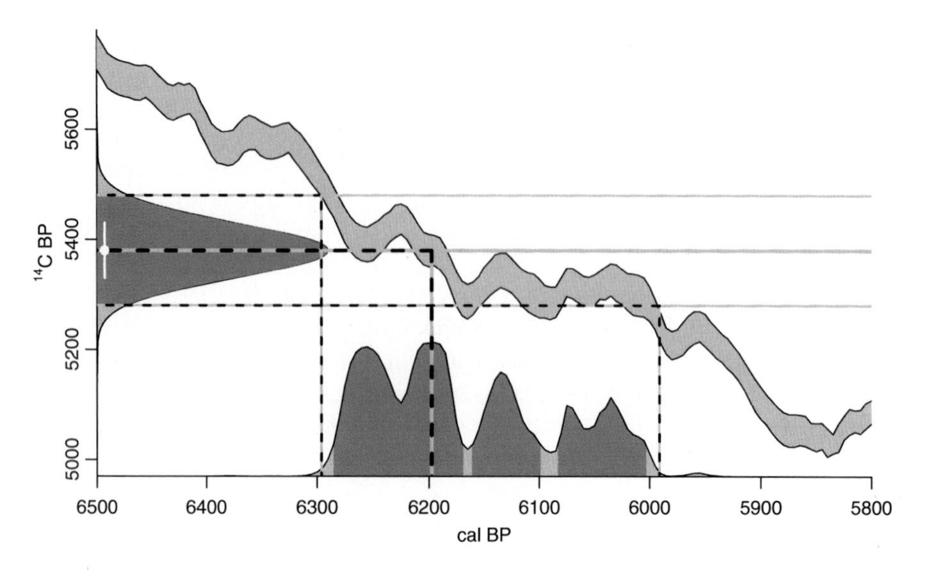

Fig. 12.4 The effect of calibrating the radiocarbon date 5380 ± 50 ^{14}C BP (*white circle* and *vertical bar*). *Dashed lines* illustrate the intercept method. With probabilistic calibration, for every calendar year the corresponding radiocarbon age of the ^{14}C calibration curve (*grey wiggly curve*, IntCal09, 2 standard deviation envelope) is looked up. This age is compared with that of the sample, and the corresponding probability is plotted on the vertical and horizontal axes (histograms). Then the highest posterior density ranges are calculated (*dark grey*, here 95%). The left-most calibrated range encompasses c. 55% of the probability range, the other two ranges cover 20% and 21%, respectively

curve μ is looked up, and compared with the measurement (here written as mean y_j with uncertainty σ_j for date *j*). The trick here is that whereas for every ^{14}C year there could be several calendar years in the calibration curve, there is always only one unique ^{14}C year μ (with uncertainty σ) for every calendar year θ. The calibration curve can thus be seen as a list of known calendar years θ, each with its corresponding radiocarbon year $\mu(\theta)$ and uncertainty $\sigma(\theta)$. A ^{14}C measurement y_j provides an estimate of the true $\mu(\theta)$ belonging to the calendar year when its organic material was deposited. The comparison of the measurement with the calibration curve is performed assuming a Gaussian/normal distribution of the measurement, $N(y_j, \sigma_j)$:

$$y_j \,|\theta \sim \mathrm{N}(\mu(\theta)), \sigma^2 \tag{12.1}$$

where σ^2 is a combination of the error in both the measurement and the calibration curve, $\sigma_j^2 + \sigma(\theta)^2$. The vertical bar between *y* and θ means the probability of the ^{14}C age *given* the calendar year. A calendar year θ with a ^{14}C year $\mu(\theta)$ close to the measurement y_j will be very likely. The further away from y_j, the less likely $\mu(\theta)$ and thus θ will become. In other words, the above formula means that from the distance between y_j and $\mu(\theta)$, the height of the normal probability distribution is calculated, and this height is transferred to the corresponding θ on the horizontal or cal BP axis. This process is done for each calendar year θ (within wide limits), which results in the often asymmetrical and multi-peaked calibrated distribution on the calendar axis (the histogram on the vertical axis in Fig. 12.4). The entire distribution is then normalised to 100%. Compared with the intercept method, probabilistic calibration is much less likely to behave problematically during periods with radiocarbon wiggles and plateaux. It also helps us choose likely point estimates for age-depth modelling, as we discuss later.

The calibrated distribution can be divided into one or more calibrated *ranges*; for example, at 1 or 2 standard deviation (sd) (Fig. 12.4). This is done by ranking the calendar ages θ according to their probabilities (heights of the calibrated distribution), starting with the most likely calendar ages. After normalising their cumulative sum of probabilities to 100%, those ages θ that fall within say the 68% or 95% range (1 or 2 sd, respectively) are retained. These calendar ages define the highest posterior density (hpd) ranges. Some of these ranges will encompass a higher proportion of the total range than others (see Fig. 12.4), but this does not necessarily allow us to neglect calibrated ranges with lower probabilities.

The above calibration method has been implemented in the most popular ^{14}C calibration software packages, e.g., CALIB (Stuiver and Reimer 1993: http://www.calib.org), BCal (Buck et al. 1999: http://bcal.shef.ac.uk), and OxCal (Bronk Ramsey 2007: https://c14.arch.ox.ac.uk/oxcal). Age-depth modelling with raw ^{14}C ages is not valid at all because of the non-linear relationship between radiocarbon years and calendar years. Modelling must therefore always be carried out with calibrated ages. If an age-offset is expected (e.g., owing to a hard-water effect: Björck and Wohlfarth 2001), the ^{14}C ages should be corrected before calibration (it is also advisable to increase the error σ to reflect uncertainties in the offset estimate) (see Blaauw et al. 2012 for a detailed case study).

Reduction to Single Point Estimates

Calibrated ^{14}C ranges often span several centuries, owing to both the non-linear relation between ^{14}C and calendar ages and to the uncertainties in the measurements. Calibration of even a high-precision ^{14}C measurement could result in a multi-peaked asymmetric distribution spanning several centuries (e.g., c. 330 calendar years for a date of 2450 ± 10 ^{14}C BP). Additionally, ^{14}C dates from lakes are often obtained from slices of sediment which must have accumulated over decades. Nevertheless, it is common practice to reduce ^{14}C ages to single calibrated year point-estimates, for example as input for age-depth models. There are several approaches to obtain such point estimates:

1. Take the mid-point between the pair of calendar ages that enclose the 68% (1 sd) or 95% (2 sd) range of the distribution for an interval as an estimate of the calendar age of the sample, and half the distance between this age and

either of the ages marking the confidence interval as the standard deviation of that estimate. This is probably the commonest approach, but the complexity of calibrated age distributions may mean that the chosen ‘date’ does not necessarily fall within a region of high probability.

2. Take the modal (maximum) value of the probability distribution of the calibrated age. This has the advantage of being the age that is most likely, but it can concentrate emphasis on one peak that happens to be fractionally higher than another in a distribution that is bimodal or even multi-modal. For a description of the spread/uncertainty we can use the tolerances, either as a fraction of the optimum (Pawitan 2001; Heegaard 2002) or as percentiles of the probability distribution (see point 5).
3. Use the weighted average of the probability distribution of the calibrated radiocarbon age, as recommended by Telford et al. (2004a, b). This provides a calibrated age estimate and an uncertainty measurement for the dated object. The uncertainty can be calculated as the weighted variance with a corresponding weighted standard deviation. The weights are obtained from the probability distribution.
4. Choose the calibrated range containing the highest proportion of the entire calibrated range (e.g., the leftmost range in Fig. 12.4), and choose the mid-point, the mode, or the weighted average of this range as the point-estimate. However, although less likely, the possibility that the ‘true’ calendar age of the ^{14}C date lies outside this range cannot be neglected.
5. Use the entire distribution to model errors by simulation. This shifts the emphasis away from a single-value calibration of a single radiocarbon date (which is often not possible) towards the reality of an irregular spread of probabilities. Values are generally simulated by sampling randomly from all calendar ages in a calibrated distribution, where the probability of a calendar age being sampled is proportional to its height in the probability distribution (see Eq. 12.1; also known as Monte Carlo sampling). Calendar ages closer to maxima will thus be more likely to be sampled, but even calendar ages outside the calibrated ranges can be sampled by chance.
6. Use additional information about the sequence such as tephra-layers, assumptions about the sedimentation history, or historical events recorded in proxies (e.g., the start of tree plantations reflected in pollen assemblages). Such approaches are discussed later in this chapter under Bayesian age-depth modelling.

Age-Depth Models

Critical numerical consideration of the problem of estimating age-depth relationships appears to have begun with Maher (1972). He rejected a simple linear interpolation approach because “small errors in the age of any or all the samples markedly affect the slopes of the intervening lines” and applied two-term polynomial, exponential, and power functions to his data. The power curve provided the best fit. He then obtained the upper and lower limits at the location of each radiocarbon date by multiplying the error on the radiocarbon-age determination by the error of the regression of the power function. Two new power functions were then obtained for the sets of upper and lower limits.

Throughout, we proceed with depth as the independent, predictor variable (x), and age as the dependent, response variable (y or θ) in the statistical model, because depth is 'controlled' and age is 'measured', and thus we obtain relationships that enable us to calculate ages as functions of depth. Although the statistical modelling has depth as the independent variable and age as the dependent variable, it is common to plot the resulting age-depth model with depth on the vertical axis (ordinate) and age on the horizontal axis (abscissa) so that the plot can be compared with palaeolimnological diagrams with depth on the vertical axis.

Analytical vs. Monte Carlo Age-Depth Models

In 'classical' statistics, exact solutions to problems can generally be found through applying analytical methods. For example, the slope of a straight line through a cloud of points can be found by least-squares linear regression (Birks 2012: Chap. 2), which consists of deriving the exact parameters by solving a set of closed equations. These analytical methods work very well when the data points follow normal distributions (e.g., uncalibrated ^{14}C dates), but they break down in more complex cases such as the multi-modal asymmetrical distributions of most calibrated ^{14}C dates. For such cases we should either assume that we can reduce the calibrated distributions to normal distributions, or work with approximations obtained by numerical simulations.

If we assume that calibrated ^{14}C dates can safely be approximated by a normal distribution (which in most cases is not true), then producing an age-depth model can be an easy task using dedicated software (e.g., PSIMPOLL Bennett 1994b or TILIA Grimm 1990) or spreadsheet programs such as EXCEL®. Even if the assumption of normally distributed calibrated ^{14}C ages is not met, this will probably cause errors on the order of decades to centuries. These errors could be considered acceptable for low-resolution chronological questions such as whether the sediment is of mid- or late-Holocene age. However, in many cases our research questions require a higher precision. Therefore we recommend working with calibrated distributions if possible.

Multi-modal calibrated distributions can be dealt with using numerical simulations (Fig. 12.5). For each radiocarbon date (and/or other dates with their quoted uncertainties), a point calendar age estimate is sampled from its calibrated distribution (with more likely calendar ages more likely to be sampled, see Eq. 12.1). These cal BP points and their known depths are then used to produce an age-depth model (see later), after which the sampling process and derivation of an age-depth model is repeated/iterated many times (e.g., 10,000). From all iterations together, inferences can be made about, for example, the most likely age-depth model or its uncertainty (e.g., through calculating a highest posterior density interval: Fig. 12.5c).

This is an application of the 'Monte Carlo' simulation method, also known as distribution sampling (Kleijnen 1974), and it enables us to obtain estimates of sample standard deviations for age estimates and accumulation rates. Such an approach has certain advantages and disadvantages compared to analytical calculations of confidence intervals. The solution is not exact, but bears a probabilistic relationship to the (unknown) exact solution in relation to the number of times the sampling was repeated. Random numbers are needed for drawing from the distribution of the input variables, so the solution will usually be slightly different with each run on the same data. The need for repeated sampling makes the technique slow relative to an analytical solution, but this is becoming less important with the widespread use of fast desk-top computers.

On the other hand, a solution will always be obtained with any data, through any age model. As Press et al. (1992) comment: "Offered the choice between mastery of a five-foot shelf of analytical statistics books and middling ability at performing statistical Monte Carlo simulations, we would surely choose to have the latter skill" (p. 691). Given the crudeness of the basic data (e.g., single estimates of the pollen concentration at a particular depth, single estimates of radiocarbon ages for a sample) it might be unwise to rely on statistically exact results.

Basic Age-Depth Models

We consider here three general approaches to age-depth modelling. The first approach, here called 'basic' age-depth modelling, considers only the errors on the individual age-determinations, and assumes that these are completely independent of each other and that there are no systematic errors affecting the whole sequence in a similar way. The second approach, 'mixed-effect models', incorporates a consideration of a wider range of possible errors, including the possibility of systematic errors. The third approach, Bayesian age-depth modelling, uses additional information from the sequences to derive an age-depth model with measures of its uncertainties.

We illustrate here the application of age-depth models to a radiocarbon-dated sequence of Holocene lake sediments from The Round Loch of Glenhead (Jones et al. 1989; Stevenson et al. 1990; Harkness et al. 1997; Birks and Jones 2012: Chap 3). The age-depth models used here are based on 20 radiocarbon dates (Table 12.1), and include a value of -35 ± 10 cal BP for the surface sediment (the core was sampled around AD 1985). They are developed here solely for simplicity and illustration, and should not be taken as a replacement of the earlier model (Jones et al. 1989), especially in the basal portion of the sequence (see also Birks 2012b: Chap 11).

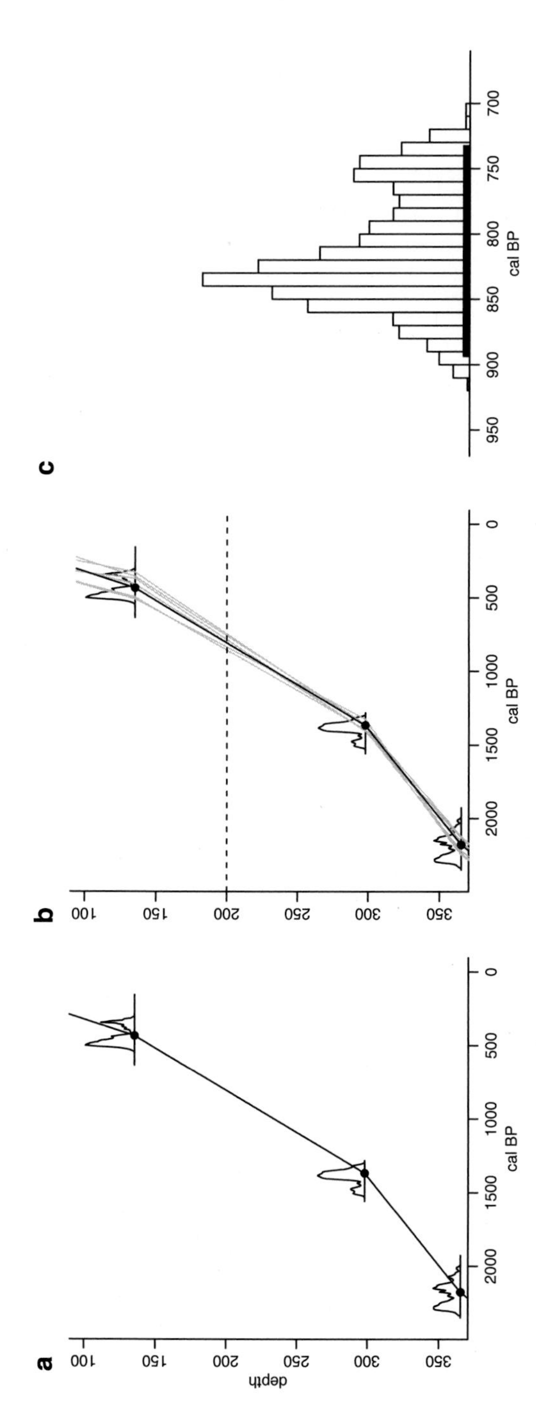

Fig. 12.5 Schematic representation of Monte Carlo-based age-depth modelling. From each calibrated distribution, a single calendar year point estimate is sampled, and an age-model is calculated (**a**). This sampling process is repeated many times, each iteration resulting in different point estimates for the calendar ages (several outcomes are shown in **b**). Each of these iterations will assign a calendar year to a depth, e.g., at 200 cm (*dashed horizontal line*). The age uncertainties can be estimated from a histogram of all simulated ages for this depth (**c**; *black bar* shows 95% of the highest posterior density error range)

Table 12.1 Radiocarbon dates from The Round Loch of Glenhead

Sample identified	Depth (cm)	^{14}C age	Error[a]
Surface	0	−35[b]	10
SRR-2810	39.5	2020	80
SRR-3258	44.5	1440	60
SRR-3259	50.5	1420	60
SRR-2811	55.5	1350	70
SRR-3260	63.5	730	60
SRR-3261	70.5	1690	60
SRR-2812	77.5	1910	70
SRR-3262	84.5	2010	70
SRR-3263	91.5	1570	60
SRR-2813	98.5	2550	70
SRR-3264	105.5	1810	60
SRR-3265	112.5	2720	60
SRR-2814	122.5	2250	70
SRR-2815	141.5	3970	70
SRR-2816	155.5	4660	70
SRR-2817	170.5	5180	80
SRR-2818	185.5	6390	80
SRR-2819	200.5	6890	70
SRR-2820	210.5	7250	70
SRR-2821	225.0	9280	80

Jones et al. (1989), Stevenson et al. (1990), Harkness et al. (1997), and Birks and Jones (2012): Chap. 3
[a]Error expressed as 1 standard deviation
[b]Surface age expressed in calendar years (cal BP)

Linear Interpolation

This is probably the most frequently used age-depth model, and the most obvious and basic way to start (Fig. 12.6a). Reported radiocarbon ages or, more sensibly, derived calibrated ages are plotted against depth with the neighbouring points connected by straight lines (often necessitating extrapolation to the base of the sequence). Estimates of accumulation rate are found from the gradients between adjacent pairs of points, and interpolated ages are read off (or calculated) for intermediate depths (see Bennett 1994a for an example).

This is a superficially crude approach, but does provide reasonable estimates for both ages and gradients. However, it takes no account of the errors on the radiocarbon or calibrated ages (although this can be dealt with using Monte Carlo simulation), and it turns out to be inadequate when confidence intervals on ages and slopes are obtained. Note also that the gradient will normally change at every radiocarbon age, which is far from necessarily being a reasonable reflection of what really happens as basins infill. However, it often turns out that the fit produced by linear interpolation is rarely badly wrong in comparison with other, more complex models (e.g., Bennett and Fuller 2002).

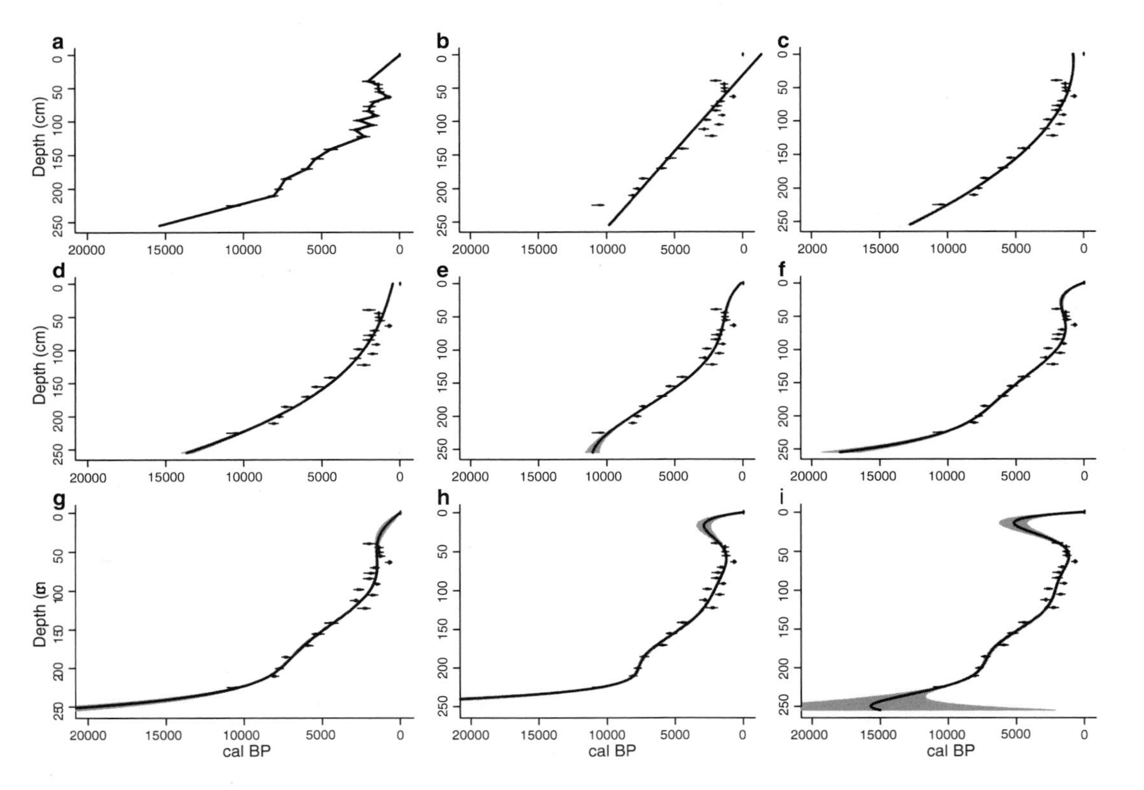

Fig. 12.6 Results of using linear interpolation (**a**) and polynomial (**b**, 2-terms; **c**, 3-terms; **d**, 4-terms; **e**, 5-terms; **f**, 6-terms; **g**, 7-terms; **h**, 8-terms; **i**, 9-terms) age-depth models with The Round Loch of Glenhead data. The ^{14}C dates were calibrated using the IntCal09 calibration curve (Reimer et al. 2009). Calendar ages are on the horizontal axis, depths on the vertical axis. *Thin black lines* show the 2 standard deviation error envelope estimated using Monte Carlo simulations from the calibrated distributions

Polynomials

Polynomials with the following general form are fitted to the data: $y = a + bx + cx^2 + dx^3$ etc., where x = depth (independent variable), y = age (dependent variable), and a, b, c, and d are regression coefficients (curve parameters) to be estimated. Polynomials may be considered by the number of terms they include:

$y = a + bx$ is a straight line (first-order or linear regression)
$y = a + bx + cx^2$ is a quadratic curve (second-order polynomial regression)
$y = a + bx + cx^2 + dx^3$ is a cubic curve (third-order polynomial regression)
etc.

The gradients of these curves for any depth x can be differentiated to obtain dy/dx, the rate of change of y at x (sediment accumulation rate (cm yr^{-1}) or deposition time (yrs cm^{-1})).

If $y = a + bx$, then $dy/dx = b$ (constant gradient for all x values)
If $y = a + bx + cx^2$, then $dy/dx = b + 2cx$
If $y = a + bx + cx^2 + dx^3$ then $dy/dx = b + 2cx + 3dx^2$
etc.

The idea of fitting a curve is to find a line that is a reasonable model of the actual data points. The curve does not necessarily have to pass through all the points because the points are only statistical estimates of the 'true' (unknown) age of the sample. For $y = a + bx$, we need to find values for a and b such that values of y calculated from the line at each x are as close as possible to the observed values of y. 'As close as possible' can be defined in many ways, of which the most usual is the 'least-squares' criterion (Birks 2012: Chap 2). This means minimising the sum of the squared distances for the dependent variable. The errors on the radiocarbon ages are incorporated as weightings on the dependent variable. It will normally be appropriate to include an age and error estimate for the top sample of a sequence (e.g., -35 ± 10 cal BP). The procedure for polynomials with more terms is conceptually identical, but the arithmetic for finding a, b, c, etc. is more complex.

The coefficients obtained enable a curve to be plotted and gradients to be calculated by differentiation. Curves become more 'flexible' with more terms. We want to use a polynomial that is as simple as possible (few terms), but is still a 'reasonable' fit, for example a model without age reversals, namely a minimal adequate model (Birks 2012: Chap 2) (e.g., Fig. 12.1).

Goodness-of-fit may be assessed from a t-test to determine if the individual terms are significant, an F-test of term improvement by model comparison, or by chi-square (χ^2) tests. The latter comprises the squared distances from the dependent variable to the fitted curve weighted by the squared errors on each age, and summed (Bennett 1994a). This approach assumes that the quoted errors on the radiocarbon ages are the population values. In practice, they are sample values from a single measurement exercise, and will tend to be slightly too small as estimators of the population value. The χ^2 obtained is zero for a perfect fit (i.e., the fitted curve passes through all the given data points), and this will always occur when the number of terms is equal to the number of data points. The χ^2 value may be assessed from standard tables or derived analytically, for its size is a function of the number of ages, the standard deviations of the ages, and the number of terms in the polynomial, to provide a measure of 'goodness-of-fit'. This measure is the probability that the observed difference between the fitted curve and the data points could have been obtained by chance if the fitted curve was the 'correct' solution. Thus, ideally, the goodness-of-fit should exceed 0.05, but values as low as 0.001 may, with caution, be acceptable. The 'goodness-of-fit' measure cannot make any judgement about the course of the fitted curve between or beyond the given points: assessment of this remains a matter for the palaeolimnologist to explore. The goodness-of-fit will be unacceptably low if one or more of the following conditions holds:

1. The model is wrong (the polynomial is a poor expression of the way that sediment has accumulated over time);
2. The errors on the radiocarbon ages are too small;
3. The errors on the radiocarbon ages are not normally distributed.

Condition (1) will often hold for polynomials where the number of terms is much less than the number of radiocarbon ages. Condition (2) could hold to at least some extent if sample values are used for the errors rather than population values, and it may be substantial depending on the extent to which laboratory errors have been exhaustively assessed in measuring radiocarbon ages. Where the goodness-of-fit is low for all numbers of terms, it may be worth increasing the quoted errors by a laboratory multiplier, as in the calibration of radiocarbon ages. An International Study Group (1982) found that the quoted errors on radiocarbon ages needed to be multiplied by a factor of between 2 and 3 if these errors are to cover the true variability of radiocarbon-age measurements when compared with material of known-age. However, most ^{14}C laboratories have since enhanced their methods for deriving age uncertainties (Blaauw et al. 2005; Scott 2007). Condition (3) is a major problem for using the approach described here with calibrated radiocarbon ages (although as explained above, Monte Carlo sampling can solve this problem).

Results using The Round Loch of Glenhead (RLGH) data are shown in Fig. 12.6, and the full table of goodness-of-fit values in Table 12.2. As the number of terms increases, the χ^2 value decreases, and the goodness-of-fit increases. However, there is little improvement after about 9-terms, and results start to become unpredictable with more terms (e.g., 17). The plotted results show similarly improving fit to the general trend of the dates as the number of terms increases, but the increasing flexibility of the curves introduces problems, especially with extrapolated portions. With this particular data-set, it is not possible to obtain lower values of goodness-of-fit because of the scatter of the data points in the upper part of the sequence. Given this, the 'best' model is probably one with a low number of terms, even if the goodness-of-fit is, statistically, rather poor.

Table 12.2 Goodness-of-fit of polynomial age-depth models with different number of terms for The Round Loch of Glenhead radiocarbon data

Terms	χ^2	Goodness-of-fit
2	2670.3	0.00
3	992.6	0.00
4	808.3	0.00
5	700.6	0.00
6	537.7	5.77×10^{-105}
7	534.1	5.36×10^{-105}
8	424.1	1.80×10^{-82}
9	391.0	3.01×10^{-76}
10	364.8	1.74×10^{-71}
11	350.1	3.73×10^{-69}
12	384.6	2.66×10^{-77}
13	341.1	7.29×10^{-69}
14	338.3	3.89×10^{-69}
15	337.5	7.56×10^{-70}
16	336.7	1.25×10^{-70}
17	336.0	1.05×10^{-71}
18	335.1	0.55×10^{-72}
19	334.0	3.02×10^{-73}
20	329.6	1.20×10^{-73}

The RLGH data are atypical in the sense of having a large number of radiocarbon dates by the general standards of Holocene palaeolimnological sequences. Results from Dallican Water, Shetland, with a more typical number of dates (6) are presented by Bennett (1994a). The sequence at Hockham Mere (eastern England) has a similar number of radiocarbon dates (23) to RLGH, but without any age-reversals (Bennett 1983). The goodness-of-fit results for polynomials with increasing numbers of terms are shown in Table 12.3. Here, there is a decline to low values, with a minimum reached at about 10-terms, and erratic values beyond that. Inspection of the plots for these polynomials (not shown) shows that 10-terms provide a good fit for all the sequence except for the interval between the uppermost pair of dates: the apparent sharp increase in accumulation rate at this site due to a change in the sediment type is hard to model with any number of terms.

Splines

A spline is a polynomial curve (see above) fitted between each pair of points, but whose coefficients are determined slightly non-locally: some information is used from other points than just the pair under immediate consideration (Birks 2012: Chap. 2). This non-locality is intended to make the fitted curve smooth overall, and not change gradient abruptly at each data point. The usual polynomial fitted between pairs of points is a cubic (four-term) polynomial, producing a cubic spline (Fig. 12.7a). This method also takes no account of the multi-modal errors

Table 12.3 Goodness-of-fit of polynomial age-depth models for the Hockham Mere radiocarbon data

Terms	χ^2	Goodness-of-fit
2	3770.7	0.00
3	2640.6	0.00
4	2491.2	0.00
5	393.4	2.20×10^{-72}
6	365.2	3.34×10^{-67}
7	41.7	4.42×10^{-04}
8	39.3	5.80×10^{-04}
9	37.8	5.50×10^{-04}
10	28.4	7.86×10^{-03}
11	26.7	8.45×10^{-03}
12	26.7	5.08×10^{-03}
13	48.2	5.73×10^{-07}
14	29.6	5.08×10^{-04}
15	32.6	7.40×10^{-05}
16	23.4	1.45×10^{-03}
17	24.8	3.74×10^{-04}
18	23.9	2.25×10^{-05}
19	26.1	2.97×10^{-05}
20	21.9	6.74×10^{-05}
21	22.6	1.22×10^{-05}
22	20.8	5.13×10^{-06}

on the radiocarbon ages (although this can be dealt with by using Monte Carlo simulations), and can produce 'ruffle-like' bends that include sections with negative deposition times. The RLGH data show this clearly in the upper part of the record, where there are many dates and age-reversals, but the curve in the lower part with more widely-spaced dates and no reversals is more acceptable. Note the 'wild' behaviour in the extrapolated portion at the base.

An alternative to regression splines are smooth splines (Fig. 12.7b, c) or locally weighted splines (Fig. 12.7d–f; often called LOESS or LOWESS splines (Birks 2012: Chap. 2)), where penalisation or smoothing terms are adjusted to constrain the degree of bending of the curves (Wood 2006). The data points can also be weighted according to their uncertainties. LOESS splines cannot be used to extrapolate beyond the dated range of a core (which is a dangerous practice anyway and should be avoided if at all possible).

Other Models

There is a wide variety of other age-depth models that can be used. Maher (1972) and Campbell (1996) advocated the use of a power function, which takes the general form $y = a + bx^c$. Campbell points out the potential utility of this function for loose top sediments where compaction increases with depth. The parameter a is the age at

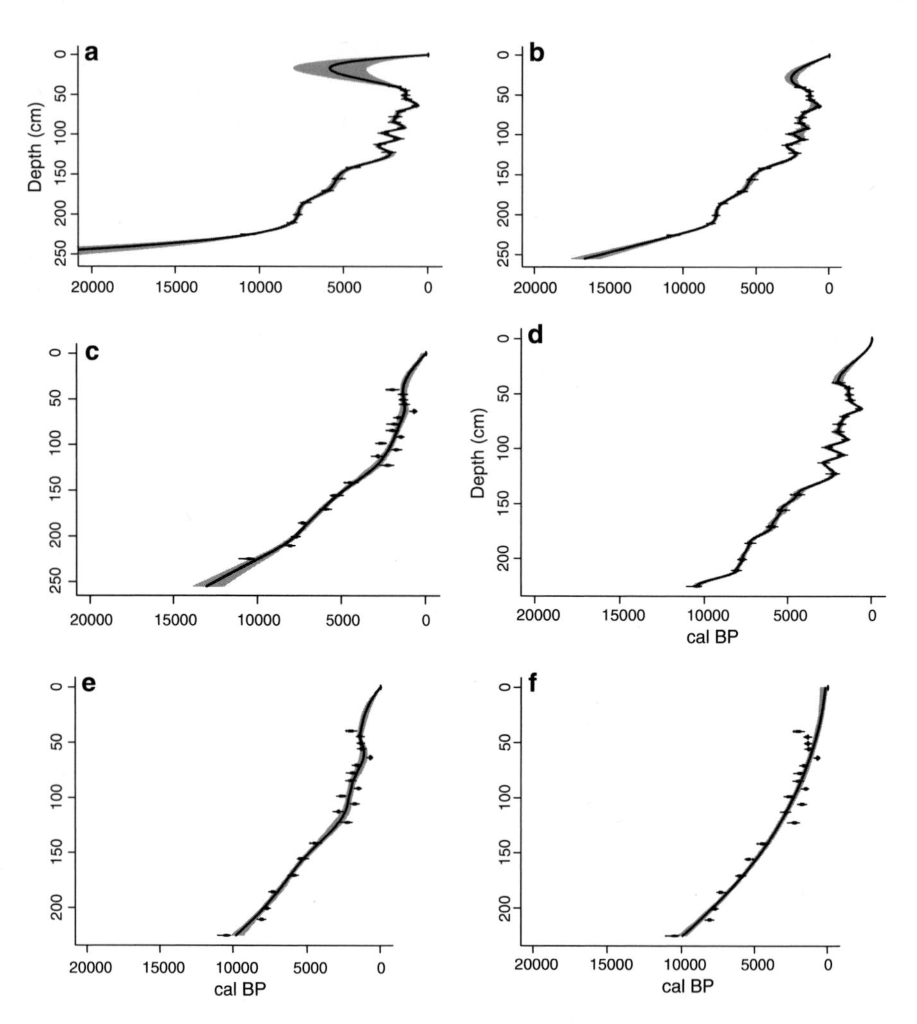

Fig. 12.7 Results of using cubic spline (**a**), smooth spline (smoothing parameter, (**b**) 0.1, (**c**) 0.5) and LOESS spline (smoothing parameter, (**d**) 0.1, (**e**) 0.5, (**f**) 1.5) age-depth models with The Round Loch of Glenhead data. The ^{14}C dates were calibrated using the IntCal09 calibration curve (Reimer et al. 2009). Calendar ages are on the horizontal axis, depths on the vertical axis. *Thin black lines* show the 2 standard deviation error envelope estimated using Monte Carlo simulations from the calibrated distributions. Extrapolation was not possible with the LOESS age-depth modelling (*d–f*)

the sediment-water interface (usually 0), b is the accumulation rate at the sediment-water interface, and c is the rate of compaction. Values for these parameters can be computed from statistical packages (see Campbell 1996).

The Bernshtein polynomial (also known as a Bézier curve) is a curve that passes smoothly through or near all the points, and is fitted by successive approximation. It is constrained so that it must pass through both endpoints (the highest and lowest ages). It will always fall within a polygon that bounds all the points. The curve uses all points for estimating an age for any given depth, so changing any point influences the entire curve. It tends to fit close to most points, but will effectively bypass an outlier. The curve also tends to be rather 'stiff', but this has the advantage that it does not behave wildly in extrapolated portions (cf. splines: see Bennett and Fuller 2002). Newman and Sproull (1981) discuss the properties of these curves in more detail.

Mixed-Effects Models

The approach here involves a consideration of the errors involved in the various components of the system (Birks 2012: Chap. 2). First, there is a variance in the age determination of an object from a particular depth. Second, the radiocarbon-age determination from a certain depth in a sequence is displaced from the true (radiocarbon) age by an amount that is unknown but which can be assumed to be random. Such displacement may occur because of contamination, incorporation of younger or older material, movement of material up or down the sequence, etc., and such processes can occur throughout the sequence. The calibrated age is then also displaced by a similar, unknown but randomly distributed, amount. It is assumed that this displacement is independent of the variance of the age determination itself.

It then follows that the variance of an age estimate at a particular depth is the sum of the variance of the age determination (within-object variance) and the variance of the age-depth relationship (between-object variance) (Heegaard et al. 2005). It should be noted that this procedure approximates the calibrated distributions to normal distributions.

The relationship between age and depth can be obtained by any regression procedure, such as generalised linear models or by various smoothing procedures, including splines. The variance of the resulting relationship is then the between-object variance. In very rare cases, there may be a relationship between age and the between-object variance, perhaps because of increasing displacements with increasing depth (Heegaard et al. 2005). We note that these procedures require the assumption that the calibrated distributions can be approximated by a normal or other statistical distribution.

Output from three alternative mixed-effects models (all excluding the data point of the surface being of recent age) is shown in Fig. 12.8. These results suggest that the age-depth relationship of RLGH is complex, including a reversal of sedimentation of the more recent deposits (Fig. 12.8a). However, there is a near-constant sedimentation rate below 110 cm, and the sedimentation rate increases slightly at about 60 cm. Assuming superimposed sedimentation and that contamination is a major source of error, a fit can be achieved without reversals by reducing the number of 'knots' (Fig. 12.8b). On the other hand, it is also possible to model the upper part of the sequence separately, for example by a two-part piece-wise regression (Fig. 12.8c). Although this piece-wise regression is actually a better fit statistically (Hastie et al. 2001), it results in negative accumulation rates for the upper part

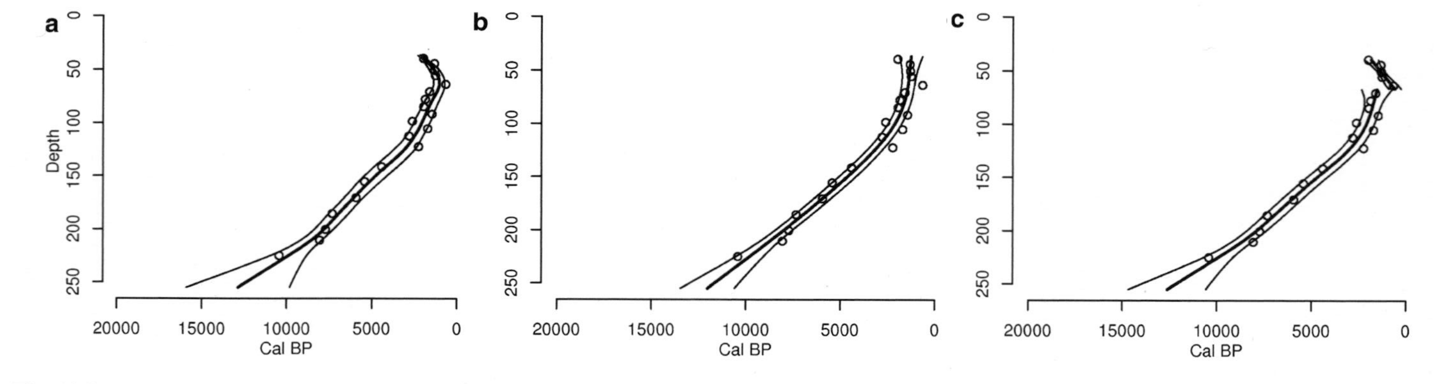

Fig. 12.8 Results of using mixed effect age-depth modelling on The Round Loch of Glenhead data, using the function Cagedepth.fun() (Heegaard et al. 2005). The younger and older ranges are 95% confidence intervals with: (**a**) the default setting, (**b**) increased smoothing by reducing the number of knots, and (**c**) a piece-wise estimation of the calibrated age to depth relationship

(probably a redeposited sediment or inwash of peat (Stevenson et al. 1990)). This solution suggests that the pattern of sediment accumulation in the upper part of the sequence was quite different from the main part. This would clearly require further investigation (see Stevenson et al. 1990).

Implementation

There are several implementations of the age-depth modelling techniques discussed above:

1. PSIMPOLL provides most of the methods described above. Confidence intervals are obtained by exact calculation (for linear interpolation and two-term polynomials). Although the program does not calibrate ^{14}C dates (it uses raw ^{14}C dates), it has the option to import calibrated dates from e.g., BCal (Buck et al. 1999). PSIMPOLL is available from http://chrono.qub.ac.uk/psimpoll/psimpoll.html.
2. TILIA provides linear interpolation, LOESS spline interpolation, and polynomials. Also here, ^{14}C dates are not calibrated. Errors on the dates are not taken into account. For availability of TILIA, see http://chrono.qub.ac.uk/datah/tilia.html, http://www.ncdc.noaa.gov/paleo/tiliafaq.html and http://museum.state.il.us/pub/grimm/tilia/.
3. Code for mixed-effect modelling is available at http://www.eecrg.uib.no/Homepages/agedepth_1.0.zip. The code is written in R (http://www.r-project.org/) which must be installed in order to run the procedure. It uses calibrated dates, but assumes that the calibrated distributions can be approximated using normal or other standard distributions. Earlier versions of the functions were named `Cagedepth.fun()` (Heegaard et al. 2005).
4. Clam is R code for 'classical age-modelling', which calibrates ^{14}C dates and provides a range of age-depth models including interpolation, linear/polynomial regression, and splines (Blaauw 2010). Also here, R needs to be installed for the code to run. The error estimates are based on Monte Carlo sampling from the calibrated distributions. Clam was used to produce Figs. 12.1, 12.3, 12.4, 12.5, 12.6, 12.7. For availability of Clam see http://www.chrono.qub.ac.uk/blaauw/wiggles/.

Bayesian Age-Depth Modelling

Over the past decade or so, Bayesian approaches have become an increasingly popular alternative to the 'classic' age-depth modelling methods described above. Bayesian statistics systematically combines data with other available information, during analysis and not afterwards. In other words, through combining *prior information* with the data, we arrive at inferences of the *posterior* distributions. Reviews of the techniques can be found in Christen (1994a), Buck et al. (1996), and Buck and Millard (2004). To take an example, when using classical statistical age-depth modelling one would try a range of models and settings (e.g., splines with different degrees of smoothness), and based on the outcomes of these trials a model is chosen that does not have any 'age-reversals'. Such decisions are often made subjectively, *ad hoc* and non-transparently. Instead, with a Bayesian approach, age-models would be constructed using the explicit prior constraint of positive accumulation rates.

Chronological Ordering

For her PhD research, Heather Ibbetson ^{14}C dated charcoal layers in order to find the ages of burning events in a soil in northern England (Ibbetson 2011). The dates were known to be ordered chronologically and stratigraphically, with date 1 the oldest/deepest, date 2 in the middle, and date 3 the youngest. However, at first sight the radiocarbon dates (1: 104 ± 15 ^{14}C BP, 2: 193 ± 15 ^{14}C BP, 3: 146 ± 15 ^{14}C BP; the errors have been decreased from 30 yr for reasons of clarity) seem confusing, inconclusive, and imprecise, and thus a waste of time, effort, and money (Fig. 12.9). Moreover, atmospheric radiocarbon concentrations fluctuated considerably during the last few centuries, resulting in wide ranges and multi-peaked distributions of the calibrated ^{14}C dates. However, when inspected closely using a Bayesian framework, the results make much more sense.

The key aspect here is to use additional information while interpreting the dates, in this case our stratigraphical knowledge of the site. As stated above, layering in the soil tells us that date 1 should be older than date 2, and date 2 older than date 3. Moreover, all dates are situated below, and should thus be older than, a well-defined layer which has been dated independently to the (AD) 1840s. Using this information together with the peculiar shape of the calibration curve, we can narrow down considerably the uncertainties of the calibrated ages. Each date falls into several distinct sections of the calibration curve (shown by overlaps of the calibration curve with the hatched bars in Fig. 12.9). As we know that all dates originate from before the 1840s, we can safely neglect the wiggles after 110 cal BP. Using the known chronological relations between the dates (e.g., date 1 should be older than the other dates, and should thus fall in an older section of the calibration curve than dates 2 and 3, the 'valley' around 220–260 cal BP being the most likely), we can manually deduce the most likely calendar age ranges for each date (horizontal position of dark boxes in Fig. 12.9).

We can express this more formally. We have $n = 3$ dates (each with a reported ^{14}C age with mean y_j, and error σ_j). From the soil layering, without looking at the ^{14}C ages we already know the *a priori* relationship of their calendar ages (θ) $\theta_1 > \theta_2 > \theta_3$, while all should be older than 110 cal BP. We thus need to calculate the posterior calendar age probabilities of the dates given the prior information and the ^{14}C ages. The process basically consists of sampling calendar ages $\theta_{1\text{–}3}$ from the calibrated distributions of each of the three ^{14}C dates, with more likely calendar ages more likely to be sampled (see Eq. 12.1), and only accepting those iterations where

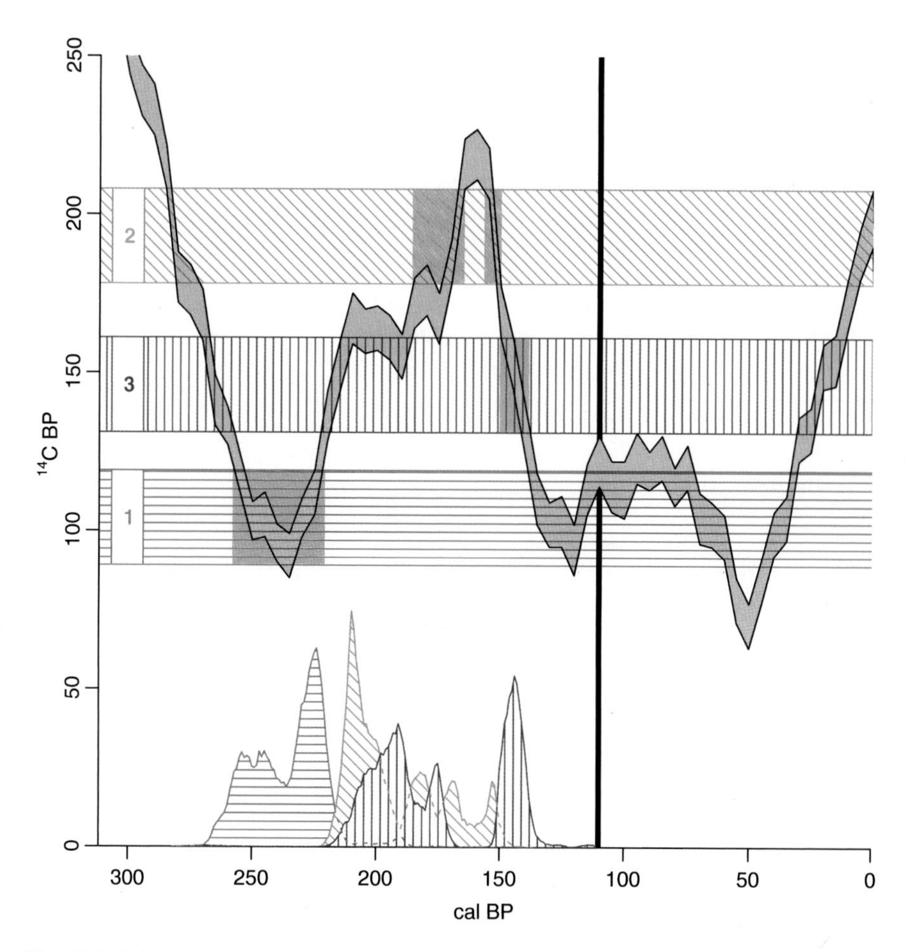

Fig. 12.9 Scheme of using stratigraphical information to constrain the calibrated ranges of three radiocarbon dates. Stratigraphical ordering indicates that date 1 should be older than date 2, and date 2 older than date 3, and all are older than a stratigraphical marker independently dated at AD 1840 (*black vertical bar*). However, the dates appear conflicting (*hatched bars* show apparent reversals, $2 > 3 > 1$) and imprecise owing to recurring ^{14}C ages in the calibration curve (*grey envelope*). If we use the chronological information (*see text*), we can use the wiggles in the calibration curve to reduce the likely calendar age ranges of the three dates (*grey blocks*). Errors on the ^{14}C dates are reduced for clarity. BCal posterior probabilities are shown as histograms (Adapted from a case study by Heather Ibbetson)

$\theta_1 > \theta_2 > \theta_3$ and all $\theta > 110$ cal BP. Any iterations with calendar-age reversals or with calendar ages younger than 110 cal BP, are rejected. The resulting *posterior* probability distribution thus takes into account the ^{14}C dates, the calibration curve, and the prior information. Calculations with the on-line Bayesian age-modelling software BCal (Buck et al. 1999) largely confirm our manual age assignments, although some additional local peaks have appeared (Fig. 12.9).

The hypothesis of superposition (dates further down in a core should be older than dates higher up) is easy to justify in palaeolimnology and indeed it provides the basis of palaeolimnology as a stratigraphical technique. This use of the chronological ordering of dates is therefore very worthwhile, leading to reduced uncertainties, especially in sediments dated at high resolution (such that neighbouring calibrated-age distributions are overlapping). Even so, the method does not provide calendar ages for non-dated levels (other than that the ages of non-dated levels should fall between the ages of their neighbouring levels). Often for age-depth modelling we need to make additional assumptions. For example, we can estimate ages for depths by interpolating linearly between the chronologically ordered posterior distributions of the dated depths (e.g., Wohlfarth et al. 2006).

Wiggle-Match Dating

Another popular Bayesian approach to age-depth modelling is known as ^{14}C wiggle-match dating (Kilian et al. 1995, 2000; Blaauw and Christen 2005). Here a sediment sequence is ^{14}C dated at close intervals (down to every cm), which should be close enough to reconstruct the decadal to centennial scale wiggles in the calibration curve (Fig. 12.10). Then, by assuming a constant sediment accumulation rate within sections of a sequence (piece-wise linear accumulation), the radiocarbon 'wiggles' of the sequence are matched to those of the calibration curve (this works best in periods with considerable wiggles such as the 'Little Ice Age'). With a faster accumulation rate, the sequence of dates will be compressed on the calendar scale, while a slower accumulation rate will expand the sequence. The best wiggle-match can then be found by trying different values of this 'accordion' parameter as well as by shifting the entire sequence on the calendar scale (or even on the ^{14}C scale if we want to estimate a reservoir effect; Kilian et al. 1995). Whereas the previous Bayesian method only uses the chronological *ordering* of dates in a deposit, here we also use the chronological *positioning*, i.e., the distances in depth and thus in time between the dated levels.

The assumption of a linear accumulation rate suggests a model of the kind $y = ax + b$. Here the accumulation rate a and horizontal placement b result in calendar ages θ for the depths d: $\theta_d = a \times d + b$. Prior information about the parameters a and b is available. Obviously, horizontal shifts should never result in 'future' calendar ages (nor in pre-Holocene ages for deposits known otherwise to be of Holocene age). Even more, it is often safe to assume that the accumulation rate a can never become negative. Similarly, very low and very high accumulation rates are very unlikely, so we can propose ranges of likely prior values for a (the ranges may differ between sites and between researchers with different opinions). Additional prior information can be incorporated, for example the depths of changes in accumulation rate, the size of any hiatuses between different sections of a core, etc. These priors can be translated into gamma distributions (Blaauw and

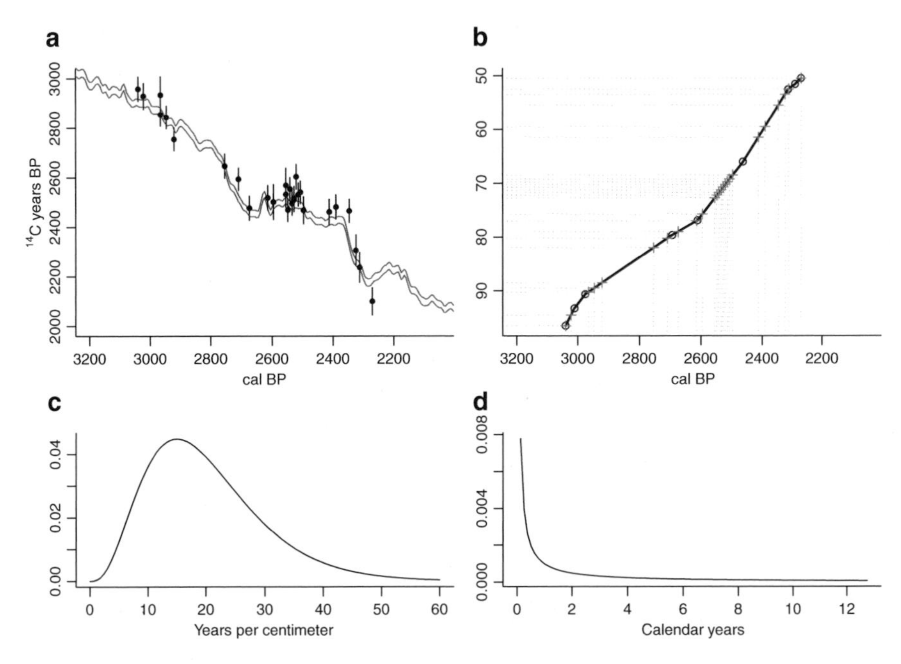

Fig. 12.10 Bayesian wiggle-match dating of a high-resolution ^{14}C dated sequence (part of peat core Eng-XV: Blaauw et al. 2004). The ^{14}C dates were fitted to the calibration curve (**a**) by calculating an age-depth model based on piece-wise linear accumulation of eight sections (**b**, *grey lines* show depths and assigned calendar ages of the dated levels). The prior information was set to favour accumulation rates between 10 and 25 year cm^{-1} (**c**), while any hiatus between the sections were assumed be of very short duration (**d**)

Christen 2005; Blaauw et al. 2007a). Gamma distributions are comparable to normal distributions, having a mean and a standard deviation, but they differ in that gamma values are always positive.

Bayesian 'wiggle-match' age-depth models are constructed by sampling repeatedly from the prior distributions, each time resulting in values for a, b, and additional parameters, and thus in an age-depth model M which translates depths d into calendar age estimates (Fig. 12.10). A calendar age θ_d will have been assigned to each dated depth d, and the likelihood of that calendar age again depends on the calibrated distribution of its date (see Eq. 12.1). The posterior probability of model M is proportional to the probability of the prior values times the likelihood of the data given the model (for details see Blaauw and Christen 2005; Blaauw et al. 2007a). Posterior distributions are found through Markov chain Monte Carlo simulations (MCMC, similar to general Monte Carlo iterations discussed earlier, but here individual iterations are sampled from preceding ones). The probability that a certain combination of prior values will be chosen depends on its combined posterior probability. Histograms of sufficiently many iterations (usually several million) will thus approximate the true posterior distributions (e.g., of accumulation rates, hiatus sizes, or calendar ages for any depth or proxy score) (Fig. 12.11).

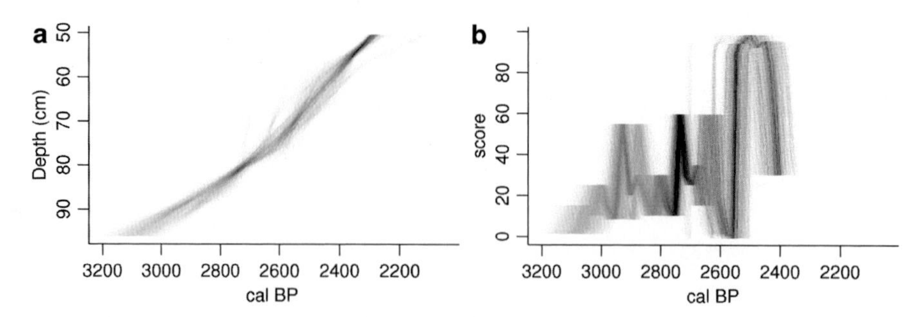

Fig. 12.11 Grey-scale graphs showing the chronological uncertainties of core Eng-XV (Fig. 12.10). Each of the millions of Markov Chain Monte Carlo iterations gave a slightly different calendar age to each depth, resulting in a posterior distribution of the calendar ages for each depth (**a**). *Darker colours* indicate more secure sections, while light shades indicate larger chronological uncertainties. Instead of depths, the proxy scores of these depths (here those macrofossils indicating moist conditions) can also be plotted against calendar age (**b**) (Blaauw et al. 2007b)

Other Models

Although wiggle-match dating can provide statistically plausible fits, it has been noted that the model of piece-wise linear accumulation might not always capture subtle changes in accumulation rate (Yeloff et al. 2006; Blockley et al. 2007; Haslett and Parnell 2008). Recently, updated models aiming to take into account such issues have become available. In OxCal the accumulation of sediment can be simulated by generating Poisson events akin to water dripping into a tube (Bronk Ramsey 2007). The user determines the stiffness of the age-depth model based on his/her knowledge of the sequence and the site. Another recent approach simulates sediment accumulation through sampling from Poisson and gamma distributions (Bchron: Haslett and Parnell 2008). In the latter approach, age uncertainties of depths tend to inflate considerably with increasing distances from dated levels.

Dealing with outlying ^{14}C dates can be problematic (Blaauw et al. 2005), and also here Bayesian statistics can help us. Usually decisions to label ^{14}C dates as outliers, and remove them, are based on *ad hoc* and subjective criteria. Bayesian outlier analysis forces the user to express a belief in the reliability of the individual ^{14}C dates (e.g., well-preserved identifiable leaves or corroded unrecognisable plant remains), and takes this prior information into account when producing the age-depth models (e.g., Christen 1994b; Blaauw and Christen 2005; Wohlfarth et al. 2008; Bronk Ramsey 2009). For every MCMC iteration, in case a date does not fit with the other dates and the applied age-depth model, it is simply neglected and labelled as outlier. The proportion of times a date has been labelled as outlying provides us with its posterior outlier probability, as well as with a measure of the age-model fit (Blaauw and Christen 2005). See Christen and Pérez (2009) for an alternative way to deal with outlying dates.

Software Packages

There are several programs to produce Bayesian age-depth models:

1. While BCal (Buck et al. 1999: http://bcal.shef.ac.uk/) is mainly aimed at archaeologists, it can be used for producing age-depth models based on chronological ordering of dates, and it provides the option for outlier analysis as well as the inclusion of relative or absolute dating information.
2. OxCal (Bronk Ramsey 2007: http://c14.arch.ox.ac.uk/) is a popular and versatile environment for Bayesian ^{14}C calibration and modelling. It includes a range of age-depth models, and complex ones can be created by combining modules and even sediment sequences. Outlier analysis is optional.
3. Bpeat (Blaauw and Christen 2005; Blaauw et al. 2007a) and Bacon (Blaauw and Christen 2011) use the model of piece-wise linear accumulation. The methods force the user to express his/her prior knowledge on accumulation rate and variability, possible hiatus size, and outlier probabilities. Although Bpeat was built for age-modelling of peat deposits, it has also been used for lake sediments (e.g., Wohlfarth et al. 2008). Bpeat and Bacon work through the R interface (http://www.r-project.org).
4. Bchron (Haslett and Parnell 2008) provides a model of monotonous sediment accumulation with outlier analysis. It does not use information on the accumulation rate, so age-models can include nearly vertical or horizontal sections. As with Bpeat, it uses R as an interface.

Discussion

Many different models are available for age-depth modelling, of which the most important are discussed above. Unfortunately, we do not know *a priori* how sediment accumulated over time, and hence we cannot know which model is the best one to use (Telford et al. 2004a). If the mode of sedimentation has changed, it may even be appropriate to use different models for different parts of the sequence, but, again, we do not know that in advance. So, how to proceed? We list below some recommendations:

1. Use multiple models, and observe how closely they approximate the data and your knowledge of the sedimentation conditions of the site (e.g., Blockley et al. 2007).
2. Use goodness-of-fit statistics to help determine which model, statistically, is closest to the data. However, this can be difficult as different models may use different measures of fit.
3. Beware of a model that passes too close (or even through) all the data points ('over-fitting'). This would imply that all the dates are exactly 'right', which is extremely unlikely given all the uncertainties involved (Birks 2012: Chap. 2).

4. The most parsimonious model in a statistical sense is frequently preferred (Birks 2012: Chap. 2), although more complex models might seem more plausible palaeoecologically (Yeloff et al. 2006; Blockley et al. 2007).
5. Always check if the dates are reasonable and check for outliers or systematic offsets.

Choice of Model

Age-depth modelling necessarily involves many sources of uncertainties, for example, multi-modal calibrated distributions, diverse dating sources (e.g., ^{14}C, tephra, ^{210}Pb, varves), outliers, qualitative stratigraphical information, and systematic offsets. Given this multidimensional problem, classical statistical methods often fail to provide reliable age-depth models. Fortunately, probabilistic/Bayesian statistics can take into account many of the intricacies of age-modelling.

Several of the age-depth modelling methods described here should be applicable where sediment stratigraphy can be assumed to have been approximately uniform, such as may be found in Holocene sediments from formerly glaciated rock basins in northern Europe, northern North America, or southern South America. It may be an inappropriate strategy where there is substantial lithostratigraphical variation along the sediment sequence. In such situations, it may be necessary to divide the sequence into more uniform sections, and carry out age-depth modelling on the individual sections.

The two most important aspects for providing reasonable age-depth relationships are the amount of information and the quality of the information (Telford et al. 2004a). With more dates available, the better the relationship and the more accurate and precise the model will be of the age-depth relationship. In this chapter we have presented three different approaches to the determination of the errors on age-depth models: by simulation (or exact calculation) from the dated objects themselves, by a mixed-effect model approach, or by a Bayesian approach. The two former procedures are reasonable in cases of minimal information (Telford et al. 2004a), whereas for data-sets with numerous dates, and combined information, a Bayesian approach is to be preferred.

It is important to appreciate that the different models presented here are alternative models of a site's unknown true sedimentation history, and it is not always possible to distinguish between them. It is therefore very important that a clear statement of the modelling procedure used should accompany all interpretations, and that the same procedure is used when sequences from different cores are being compared. Further, there may be slight differences in the interpretation of some models compared with what the estimated uncertainty is actually showing.

Conclusions and Future Developments

The greatest uncertainty about age-depth modelling remains the uncertainty associated with the choice of model. We do not have a way to determine which is the correct model. Perhaps even worse, we do not have a way of estimating the errors associated with these approaches. Adding the error associated with model selection (Johnson and Omland 2004) to other errors would be a major step forward.

General principles could exist for the way sedimentation develops over time within a site. Such models could guide us in choosing a particular type of age-depth model, for example linear or polynomial accumulation. Clymo (1984) developed a model for peat growth based on the combination of growth, compaction, and decay. Townsend et al. (2002) argued that many types of sedimentation should be regarded as basically linear over time, with the modification that compaction brings about apparent exponentially-decreasing sedimentation rates with age. Bayesian age-depth models such as Bpeat (Blaauw and Christen 2005), Bacon (Blaauw and Christen 2011), OxCal (Bronk Ramsey 2007), and Bchron (Haslett and Parnell 2008) provide such process-based models of sedimentation, but more developments (e.g., taking into account basic stratigraphical information) are needed.

As we wrote in the introduction, palaeolimnology without chronology is history without dates. Similarly, palaeolimnology with just one core is history without context. When multiple cores are compared for an analysis at site-wide, regional, or even broader scales, reliable age-models are needed for each of the cores. However, such comparisons often either neglect the age-model errors, or take advantage of them through tuning or synchronisation (stretching individual time-scales to fit proxy events with those in other time-scales). Compilations obtained through tuning often seem plausible and precise. However, tuning invokes the danger of creating a 'coherent myth', especially if inferences are made about leads or lags between synchronised events (Blaauw 2012). Bayesian methods have recently been developed to test objectively for the synchroneity of events, given the uncertainties of the proxy archives (Blaauw et al. 2007b, 2010; Parnell et al. 2008). Such tests warn us against tuning as well as against over-interpreting proxy records, in that their chronological uncertainties set the limit on the precision of research questions (e.g., proxy archives with centennial-scale uncertainties should not be used to investigate decadal time-scale questions).

The use of age-depth models with proper assessment of errors needs to be more widely practised. The overwhelming majority of age-depth modelling still uses uncritically simple linear interpolation between single data points, sometimes in ^{14}C years. Even though the original radiocarbon dates and their calibrated equivalents are typically quoted with their errors (and always have been), it is rare to see these errors used in modelling or in interpretations, leading to an over-reliance on the dates themselves. Even if we are not yet able to calculate the full extent of the error in age-depth models, making attempts at this has been a major step forward in itself, but there is still quite a long way to go.

Acknowledgements We are very grateful to Keith Bennett for generously providing the first draft of this chapter. We are also grateful to John Birks, Steve Juggins, Andy Lotter, and Richard Telford for comments on earlier drafts of this chapter. We thank Heather Ibbetson for kindly providing her case study. This is publication no. A336 from the Bjerknes Centre for Climate Research.

References

Appleby PG (2001) Chronostratigraphic techniques in recent sediments. In: Last WM, Smol JP (eds) Tracking environmental change using lake sediments. Volume 1: Basin analysis, coring, and chronological techniques. Kluwer Academic Publishers, Dordrecht, pp 171–203

Bennett KD (1983) Devensian late-glacial and Flandrian vegetational history at Hockham Mere, Norfolk, England. I. Pollen percentages and concentrations. New Phytol 95:457–487

Bennett KD (1994) Confidence intervals for age estimates and deposition times in late-Quaternary sediment sequences. Holocene 4:337–348

Bennett KD (1994b) PSIMPOLL version 2.23: a C program for analysing pollen data and plotting pollen diagrams. INQUA Commission for the Study of the Holocene: Working group on data-handling methods. Newsletter 11:4–6. http://chrono.qub.ac.uk/psimpoll/psimpoll.html

Bennett KD, Fuller JL (2002) Determining the age of the mid-Holocene *Tsuga canadensis* (hemlock) decline, eastern North America. Holocene 12:421–429

Birks HJB (2012a) Chapter 2: Overview of numerical methods in palaeolimnology. In: Birks HJB, Lotter AF, Juggins S, Smol JP (eds) Tracking environmental change using lake sediments. Volume 5: Data handling and numerical techniques. Springer, Dordrecht

Birks HJB (2012b) Chapter 11: Stratigraphical data analysis. In: Birks HJB, Lotter AF, Juggins S, Smol JP (eds) Tracking environmental change using lake sediments. Volume 5: Data handling and numerical techniques. Springer, Dordrecht

Birks HJB, Jones VJ (2012) Chapter 3: Data-sets. In: Birks HJB, Lotter AF, Juggins S, Smol JP (eds) Tracking environmental change using lake sediments. Volume 5: Data handling and numerical techniques. Springer, Dordrecht

Björck S, Wohlfarth B (2001) ^{14}C chronostratigraphic techniques in palaeolimnology. In: Last WM, Smol JP (eds) Tracking environmental change using lake sediments. Volume 1: Basin analysis, coring, and chronological techniques. Kluwer, Dordrecht, pp 205–245

Blaauw M (2010) Methods and code for 'classical' age-modelling of radiocarbon sequences. Quaternary Geochronol 5:512–518

Blaauw M (2012) Out of tune: the dangers of aligning proxy archives. Quaternary Sci Rev 36:38–49

Blaauw M, Christen JA (2005) Radiocarbon peat chronologies and environmental change. Appl Stat 54:805–816

Blaauw M, Christen JA (2011) Flexible paleoclimate age-depth models using an autoregressive gamma process. Bayes Anal 6:457-474

Blaauw M, van Geel B, Mauquoy D, van der Plicht J (2004) Radiocarbon wiggle-match dating of peat deposits: advantages and limitations. J Quaternary Sci 19:177–181

Blaauw M, Christen JA, Guilderson TP, Reimer PJ, Brown TA (2005) The problems of radiocarbon dating. Science 308:1551–1553

Blaauw M, Bakker R, Christen JA, Hall VA, van der Plicht J (2007a) A Bayesian framework for age-modelling of radiocarbon-dated peat deposits: case studies from the Netherlands. Radiocarbon 49:357–367

Blaauw M, Christen JA, Mauquoy D, van der Plicht J, Bennett KD (2007b) Testing the timing of radiocarbon-dated events between proxy archives. Holocene 17:283–288

Blaauw M, Wohlfarth B, Christen JA, Ampel L, Veres D, Hughen KA, Preusser F, Svensson A (2010) Were last glacial events simultaneous between Greenland and France? A quantitative comparison using non-tuned chronologies. J Quaternary Sci 25:387–394

Blaauw M, van Geel B, Kristen I, Plessen B, Lyaruu A, Engstrom DR, van der Plicht J, Verschuren D (2011) High-resolution ^{14}C dating of a 25,000-year lake-sediment record from equatorial East Africa. Quaternary Sci Rev 30:3043–3059

Blockley SPE, Blaauw M, Bronk Ramsey C, van der Plicht J (2007) Building and testing age models for radiocarbon dates in Lateglacial and early Holocene sediments. Quaternary Sci Rev 26:1915–1926

Boaretto E, Bryant C, Carmi I, Cook G, Gulliksen S, Harkness D, Heinemeier J, McClure J, McGee E, Naysmith P, Possnert G, Scott M, van der Plicht H, van Strydonck M (2002) Summary findings of the fourth international radiocarbon intercomparison (FIRI) (1998–2001). J Quaternary Sci 17:633–637

Bowman S (1990) Interpreting the past: radiocarbon dating. University of California Press, Berkeley

Bronk Ramsey C (2007) Deposition models for chronological records. Quaternary Sci Rev 27:42–60

Bronk Ramsey C (2009) Dealing with outliers and offsets in radiocarbon dating. Radiocarbon 51:1023–1045

Buck CE, Millard AR (eds) (2004) Tools for constructing chronologies: crossing disciplinary boundaries. Springer, London

Buck CE, Cavanagh WG, Litton CD (1996) Bayesian approach to interpreting archaeological data. Wiley, Chichester

Buck CE, Christen JA, James GN (1999) BCal: an on-line Bayesian radiocarbon calibration tool. Internet Archaeol 7. http://intarch.ac.uk/journal/issue7/buck_toc.html. Accessed 30 December 2011

Burr GS (2007) Radiocarbon dating: causes of temporal variations. Encyclopedia of Quaternary science. Elsevier, Oxford, pp 2931–2941

Campbell ID (1996) Power function for interpolating dates in recent sediment. J Paleolimnol 15:107–110

Christen JA (1994a) Bayesian interpretation of radiocarbon results. Unpublished PhD thesis, University of Nottingham

Christen JA (1994b) Summarizing a set of radiocarbon determinations: a robust approach. Appl Stat 43:489–503

Christen JA, Pérez S (2009) A new robust statistical model for radiocarbon data. Radiocarbon 51:1047–1059

Clymo RS (1984) The limits to peat growth. Philos Trans R Soc Lond B 303:605–654

Denman KL, Brasseur G, Chidthaisong A, Ciais P, Cox PM, Dickinson RE, Hauglustaine D, Heinze C, Holland E, Jacob D, Lohmann U, Ramachandran S, da Silva Dias PL, Wofsy SC, Zhang X (2007) Couplings between changes in the climate system and biogeochemistry. In: Climate change 2007: the physical science basis. Contribution of Working Group I to the Fourth Assessment Report of the Intergovernmental Panel on Climate Change. Cambridge University Press, Cambridge

Elias SA (ed) (2007) Encyclopedia of Quaternary science. Elsevier, Oxford

Grimm EC (1990) TILIA and TILIA GRAPH: PC spreadsheet and graphics software for pollen data. INQUA Commission for the Study of the Holocene: Working group on data-handling methods. Newsletter 4:5–7. http://museum.state.il.us/pub/grimm/tilia

Harkness DD, Miller BF, Tipping RM (1997) NERC radiocarbon measurements 1977–1988. Quaternary Sci Rev 16:925–927

Haslett J, Parnell A (2008) A simple monotone process with application to radiocarbon-dated depth chronologies. Appl Stat 57:399–418

Hastie T, Tibshirani R, Friedman J (2001) The elements of statistical learning. Springer, New York

Heegaard E (2002) The outer border and central border for species-environmental relationships estimated by non-parametric generalised additive models. Ecol Model 157:131–139

Heegaard E, Birks HJB, Telford RJ (2005) Relationships between calibrated ages and depth in stratigraphical sequences: an estimation procedure by mixed-effect regression. Holocene 15:612–618

Hua Q, Barbetti M (2004) Review of tropospheric bomb ^{14}C data for carbon cycle modeling and age calibration purposes. Radiocarbon 46:1273–1298

Ibbetson H (2011) The environmental history of a high South Pennine landscape: 1284 A.D. to present. PhD thesis, Queen's University Belfast

International Study Group (1982) An inter-laboratory comparison of radiocarbon measurements in tree rings. Nature 298:619–623

Johnson JB, Omland KS (2004) Model selection in ecology and evolution. Trends Ecol Evol 19:101–108

Jones VJ, Stevenson AC, Battarbee RW (1989) Acidification of lakes in Galloway, southwest Scotland: a diatom and pollen study of the post-glacial history of the Round Loch of Glenhead. J Ecol 77:1–23

Kilian MR, van der Plicht J, van Geel B (1995) Dating raised bogs: new aspects of AMS ^{14}C wiggle matching, a reservoir effect and climatic change. Quaternary Sci Rev 14:959–966

Kilian MR, van Geel B, van der Plicht J (2000) ^{14}C AMS wiggle matching of raised bog deposits and models of peat accumulation. Quaternary Sci Rev 19:1011–1033

King J, Peck J (2001) Use of paleomagnetism in studies of lake sediments. In: Last WM, Smol JP (eds) Tracking environmental change using lake sediments. Volume 1: Basin analysis, coring, and chronological techniques. Kluwer, Dordrecht, pp 371–389

Kleijnen JPC (1974) Statistical techniques in simulation. Part I. Dekker, New York

Madsen AT, Murray AS, Andersen TJ, Pejrup M (2007) Optical dating of young tidal sediments in the Danish Wadden Sea. Quaternary Geochronol 2:89–94

Maher LJ (1972) Absolute pollen diagram of Redrock Lake, Boulder County, Colorado. Quaternary Res 2:531–553

Maher LJ, Heiri O, Lotter AF (2012) Chapter 6: Assessment of uncertainties associated with palaeolimnological laboratory methods and microfossil analysis. In: Birks HJB, Lotter AF, Juggins S, Smol JP (eds) Tracking environmental change using lake sediments. Volume 5: Data handling and numerical techniques. Springer, Dordrecht

McCormac FG, Hogg AG, Blackwell PG, Buck CE, Higham TFG, Reimer PJ (2004) SHCal04 southern hemisphere calibration 0–11.0 cal kyr BP. Radiocarbon 46:1087–1092

Mook WG (1986) Recommendations/resolutions adopted by the 12th International Radiocarbon Conference. Radiocarbon 28:799

Newman WM, Sproull RF (1981) Principles of interactive computer graphics, 2nd edn. McGraw-Hill, Singapore

Parnell AC, Haslett J, Allen JRM, Buck CE, Huntley B (2008) A flexible approach to assessing synchroneity of past events using Bayesian reconstructions of sedimentation history. Quaternary Sci Rev 27:1872–1885

Pawitan Y (2001) In all likelihood: statistical modelling and inference using likelihood. Oxford Science, Oxford

Pilcher JR (1991) Radiocarbon dating for the Quaternary scientist. Quaternary Proc 1:27–33

Press WH, Teukolsky SA, Vetterling WT, Flannery BP (1992) Numerical recipes in C: the art of scientific computing, 2nd edn. Cambridge University Press, Cambridge

Reimer PJ, Baillie MGL, Bard E, Bayliss A, Beck JW, Blackwell PG, Bronk Ramsey C, Buck CE, Burr GS, Edwards RL, Friedrich M, Grootes PM, Guilderson TP, Hajdas I, Heaton TJ, Hogg AG, Hughen KA, Kaiser KF, Kromer B, McCormac FG, Manning SW, Reimer RW, Richards DA, Southon JR, Talamo S, Turney CSM, van der Plicht J, Weyhenmeyer CE (2009) IntCal09 and Marine09 radiocarbon age calibration curves, 0–50,000 years cal BP. Radiocarbon 51:1111–1150

Scott EM (2007) Radiocarbon dating: sources of error. In: Elias SA (ed) Encyclopedia of Quaternary science. Elsevier, Oxford, pp 2918–2923

Stevenson AC, Jones VJ, Battarbee RW (1990) The cause of peat erosion: a palaeolimnological approach. New Phytol 114:727–735

Stuiver M, Reimer PJ (1993) Extended ^{14}C data base and revised CALIB 3.0 ^{14}C age calibration program. Radiocarbon 35:215–230

Telford RJ, Heegaard E, Birks HJB (2004a) All age-depth models are wrong: but how badly? Quaternary Sci Rev 23:1–5
Telford RJ, Heegaard E, Birks HJB (2004b) The intercept is a poor estimate of a calibrated radiocarbon age. Holocene 14:296–298
Townsend PD, Parish R, Rowlands AP (2002) A new interpretation of depth-age profiles. Radiat Prot Dosim 101:315–319
Twyman RM (2007) Geomagnetic excursions and secular variations. In: Elias SA (ed) Encyclopedia of Quaternary science. Elsevier, Oxford, pp 717–720
Walker M (2005) Quaternary dating methods. Wiley, Chichester
Wohlfarth B, Blaauw M, Davies SM, Andersson M, Wastegård S, Hormes A, Possnert G (2006) Constraining the age of Lateglacial and early Holocene pollen zones and tephra horizons in southern Sweden with Bayesian probability methods. J Quaternary Sci 21:321–334
Wohlfarth B, Veres D, Ampel L, Lacourse T, Blaauw M, Preusser F, Andrieu-Ponel V, Kéravis D, Lallier-Vergès E, Björck S, Davies SM, de Beaulieu J-L, Risberg J, Hormes A, Kasper HU, Possnert G, Reille M, Thouveny N, Zander A (2008) Rapid ecosystem response to abrupt climate changes during the last glacial period in western Europe, 40–16 kyr BP. Geology 36:407–410
Wood SN (2006) Generalized additive models. Chapman & Hall, London
Yeloff D, Bennett KD, Blaauw M, Mauquoy D, Sillasoo Ü, van der Plicht J, van Geel B (2006) High precision C-14 dating of Holocene peat deposits: a comparison of Bayesian calibration and wiggle-matching approaches. Quaternary Geochronol 1:222–235

Chapter 13
Core Correlation

Roy Thompson, R. Malcolm Clark, and Geoffrey S. Boulton

Abstract The numerical procedure of sequence slotting aims to combine, in a mathematically optimal manner, two ordered sequences of stratigraphical data (e.g., loss-on-ignition, percentages of different biological taxa) into a single sequence, while preserving the ordering within each sequence and satisfying any other relevant external constraint such as volcanic tephra layers. The procedure provides a convenient means of core correlation in palaeolimnology and is illustrated by two examples. The first involves univariate pollen data from a lake-sediment core being matched with isotopic ice-core data and its associated chronology. The second involves core correlation of two to four cores with a dated master core from eight mountain and arctic lakes in Europe using dry weight and bulk organic matter (loss-on-ignition) data as the proxy variables to derive 3405 age estimates for the individual core samples.

Keywords Age estimation • Core correlation • GRIP ice-core • Lago Grande di Monticchio • Loss-on-ignition • MOLAR • Multiple cores • Sequence slotting • Splines

Introduction

The correlation of sediment sequences is a procedure used in many fields of study including the analyses of well-logs (e.g., Griffiths and Bakke 1990; Luthi and Bryant

R. Thompson (✉) • G.S. Boulton
School of GeoSciences, University of Edinburgh, Edinburgh, EH9 3JN, UK
e-mail: roy@ed.ac.uk; g.boulton@ed.ac.uk

R.M. Clark
School of Mathematical Sciences, Monash University, 3800, Victoria, Australia
e-mail: malcolm.clark@monash.edu

H.J.B. Birks et al. (eds.), *Tracking Environmental Change Using Lake Sediments*, Developments in Paleoenvironmental Research 5, DOI 10.1007/978-94-007-2745-8_13,

1997), marine sequences (e.g., Martinson et al. 1982; Kovach 1993; Lisiecki and Lisiecki 2002), ice cores (this chapter), loess deposits (e.g., Maher et al. 1994), peat bogs (e.g., Gardner 1999), and lake-sediments (e.g., Gordon and Birks 1974). Lake-sediment core correlation is often used to match multiple cores from different parts of a basin (e.g., littoral vs. profundal) because different cores give different palaeosignals (shallow-water vs. deep-water biology), or because different cores overlap in time or have different resolutions. Numerical correlation between basins can also be made on the basis of pollen and microfossil records (Pels et al. 1996; Gary et al. 2005). In addition, correlation methods also serve as a dating tool, for example aligning palaeomagnetic measurements with a regional master curve (Marwan et al. 2002; Barletta et al. 2010). Finally numerical methods can provide a quantitative measure of the degree of reproducibility of results from one core, or site, to another (Thompson 1991).

Numerical approaches to core correlation are becoming increasingly desirable with the more exacting demands of high-resolution chronologies required for studies of rapid environmental and climatic change; with the increasing need to combine duplicate or triplicate cores in order to provide a sufficient volume of material for multi-proxy studies (Birks and Birks 2006); with the ever greater quantities of palaeolimnological data to be handled; and especially in studies of sediment accumulation rate, rather than just sediment age, in investigations involving whole basin fluxes.

A wide range of numerical methodologies have been proposed (e.g., Gordon and Birks 1974: Gordon and Reyment 1979; Martinson et al. 1982; Ghose 1984; Waterman and Raymond 1987; Maher 1993, 1998; Marwan et al. 2002; Lisiecki and Lisiecki 2002; Gary et al. 2005; Hladil et al. 2010). However, core correlation is not always straightforward and as a result visual or graphical approaches (Shaw and Cubitt 1979; Shaw 1982; Edwards 1984) are still widely employed. Although these graphical methods are increasingly computer-aided, they remain worryingly subjective and often lack detail. Conversely many numerical approaches, while they have the advantage of being less subjective, are often unable to make use of the full range of stratigraphical information available, and so can generate inappropriate correlations.

We describe two case studies to illustrate the use of a practical numerical approach, named sequence slotting, to core correlation. Our first example matches palaeobotanical data from a lake sequence with a target chronology (the GRIP ice-core sequence). It was motivated by the EU projects PAGEPA (**Pa**laeohydrology and **ge**oforecasting for **p**erformance **a**ssessment in geosphere repositories for radioactive waste disposal) and BENCHPAR (**Bench**mark tests and guidance on coupled processes for **p**erformance **a**ssessment of nuclear waste **r**epositories), and by the desire to study the variation of temperature with latitude, within Europe, over the last 100,000 years. Our second example takes physical data from eight remote alpine and arctic lakes and generates within-lake correlations. The second case study was an integral part of the multidisciplinary EU project MOLAR (Measuring and modelling the dynamic response of remote mountain lake ecosystems to environmental change: A **mo**untain **la**ke **r**esearch programme). The aspirations of the MOLAR programme, and its study of pollution and environmental change as recorded in the sediments of mountain lakes, are encapsulated in the following sentences taken from the RTD Magazine for European research (2000):

> Far above or beyond the tree line, remote mountain lakes are strongholds of secluded nature, where there is virtually no direct contact with man, save perhaps for the occasional summer hiker. In their splendid isolation, their ecosystems are subject only to the wind that blows, the rain and snow that fall from the sky, and daily and seasonal temperature fluctuations. ... High altitude ... lakes are like sentinels at the furthest outpost of environmental and climatic change. Beneath their appearance of immutable solitude, they ... retain a vast 'memory' of onslaughts they have suffered across the centuries. Their sediments are archives of the effects of the industrial revolution, of acid rain, and ... the impact of climate change.

An overall goal of the work in MOLAR was to try to validate palaeoclimatic reconstructions, based on sediment-core measurements (Lami et al. 2000a; Battarbee et al. 2002a, b), by correlation with 200-year long instrumental records (Agustí-Panareda and Thompson 2002). Four organism groups that are preserved in lake sediments were chosen for palaeoclimatic reconstructions. Diatoms and chrysophytes were selected because they often dominate the primary production of arctic-alpine lakes and because variations in climate can control changes in abundance and species composition in these unpolluted waters. Similarly chironomid larvae are good potential indicators of temperature (Walker et al. 1991). Finally remains of cladocerans are common in arctic-alpine lake sediments and they too are potential indicators of climate. Multiple cores were needed to generate adequate quantities of sediment for the many analyses carried out at the eight MOLAR lakes. Sequence slotting was chosen as the preferred method of constructing within lake core-correlations and hence allowing the four palaeoclimatic proxies originating from different cores to be compared with one another and to be dated by correlation with a master chronology. Thus the specific objective of this case study was to produce quantitative within-lake core correlations to the master cores. In practical terms an age estimate was to be generated for each of the 3405 lake sediment horizons.

Theory and Method

The aim of sequence slotting (Gordon and Birks 1974; Gordon 1982; Birks and Gordon 1985; Thompson and Clark 1989) is to combine, in an optimal manner, two ordered sequences $A = \{A_1, A_2, \ldots, A_m\}$ and $B = \{B_1, B_2, \ldots, B_n\}$ of observations into a single sequence, while preserving the ordering within each sequence and satisfying any other relevant constraints. No assumption is made about the temporal variation of the measurements within either core. It is simply assumed that samples with similar measurements should be close together in the combined sequence, provided all order constraints are satisfied.

The procedure assumes that there is a well-defined measure d of ‘local discordance’, dissimilarity, or ‘distance’ between any two samples in either sequence that depends on the nature of the measurements. These measurements could be (i) multivariate (such as well logs), (ii) directional (palaeomagnetism), (iii) vector (magnetic direction and intensity), or (iv) compositional data (e.g., pollen or diatom counts). The ‘total discordance’ (Gordon and Reyment 1979) of any proposed combined sequence is then defined as the sum of the distances between consecutive samples in the pooled sequence, i.e., the combined path length (CPL). The optimal slotting of sequences A and B is that for which this CPL is minimised, subject to the stratigraphical order constraints within each sequence and any additional external constraints. Further details about CPL and the assessment of slottings are given in Thompson and Clark (1989) and Gordon et al. (1988).

This optimal slotting can be determined recursively by the following dynamic programming algorithm (Delcoigne and Hansen 1975).

Let $F(j, k; P)$ denote the minimum CPL corresponding to the optimal slotting of the sub-sequences $A = (A_1, A_2, \ldots, A_j)$ and $(B_1, B_2, \ldots, B_k)$, subject to the constraint that the last sample in this slotting belongs to the sequence P ($j = 1, 2, \ldots, m$; $k = 1, 2, \ldots, n$; $P = A, B$). Then:

$$F\,(1, 1; A) = d(A_1, B_1) = F\,(1, 1; B) \tag{13.1}$$

$$F(1, k; A) = d(A_1, B_k) + \sum_{i=1}^{k-1} d(B_i, B_{i+1}) \quad (k = 2, 3, \ldots, n) \tag{13.2}$$

$$F(j, 1; B) = d(A_j, B_1) + \sum_{i=1}^{j-1} d(A_i, A_{i+1}) \quad (j = 2, 3, \ldots, m) \tag{13.3}$$

$$\begin{aligned} F\,(j, k; A) = \min\{&F\,(j-1, k; A) + d\left(A_{j-1}, A_j\right), F\,(j-1, k; B) \\ &+ d\left(A_j, B_k\right)\}\,(j = 2, \ldots, m; k = 1, \ldots, n) \end{aligned} \tag{13.4}$$

$$\begin{aligned} F\,(j, k; B) = \min\{&F\,(j, k-1; A) + d\left(A_j, B_k\right), F\,(j, k-1; B) \\ &+ d\,(\mathrm{B}_{k-1}, B_k)\}\,(j = 1, \ldots, m; k = 2, \ldots, m) \end{aligned} \tag{13.5}$$

Given the initial values specified by steps (13.1), (13.2), and (13.3), the entire F-array can be evaluated by recursive use of steps (13.4) and (13.5). The minimum CPL for the entire sequences is given by

$$C^* = \min\left\{F\,(m, n; A)\,, F\,(m, n; B)\right\} \tag{13.6}$$

The optimal slotting can be traced by noting which of the two terms in braces on the right-hand sides of steps (13.4), (13.5), and (13.6) provide the minimum. Equality of these two terms at any stage indicates multiple slottings with the same optimal CPL.

Occasionally, there is additional stratigraphical evidence indicating how the two sequences must be slotted together. For example, there could be independent external evidence that sample A_s in sequence A must precede sample B_t in sequence B (but not necessarily immediately before it). This constraint can be achieved by defining

$$F\,(s,k;A) = M, k = t, t+1, \ldots, n$$

where M is a sufficiently large number ('machine infinity'), and then skipping these terms during the recursion. The constraint that B_t precedes A_s can be handled in a similar manner.

Such precedence constraints can be used, for example, when there is a distinctive marker layer (e.g., tephra or turbidites) in each sequence. Suppose, as an example, that A_9 and B_{15} are the only samples in either sequence from this marker bed. Then in any valid combined slotting, A_9 and B_{15} must be immediately adjacent. This condition can be achieved by imposing simultaneously the four logically equivalent precedence constraints: (i) A_8 precedes B_{15}, (ii) A_9 precedes B_{16}, (iii) B_{14} precedes A_9 and (iv) B_{15} precedes A_{10}.

When the sediment properties in part of either sequence do not vary much, the optimum slotting often contains long blocks of consecutive samples from the same sequence, so-called 'blocking'. This common practical problem can be avoided by imposing upper limits on these 'block lengths' (see Thompson and Clark 1989). The basic recursive algorithm can still be used, provided these block-size constraints are checked at every stage. For example, suppose that the maximum number of consecutive samples from sequence A is set at b, and that the optimal part-slotting corresponding to $F(j-1, k; A)$ ends in a block of b As. According to the minimisation step (13.4) of the usual recursive procedure, the optimal part-slotting corresponding to $F(j, k; A)$ could be obtained from that for $F(j-1, k; A)$ simply by appending A_j to the existing sequence. This, however, would produce a block of $(b+1)$ consecutive As. To avoid this, the next element in the optimal part-slotting must be B_{k+1}, the next available sample in sequence B. If $k = n$, there is no available B to control the number of consecutive A's. The resulting 'end-block' of consecutive A's is of no consequence, but could arise when sequence A is much longer than sequence B.

As an example, suppose that $j = 10$, $b = 3$, and the optimal part-slotting for $F(9, k; A)$ ends with $..B_k A_7 A_8 A_9$. Then in the next stage in the optimal part-slotting, we must have:

$$\ldots B_k A_7 A_8 A_9 B_{k+1} \text{ not } B_k A_7 A_8 A_9 A_{10}$$

This can be achieved by setting $F(10,k;A) = M$.

A natural measure of the degree of similarity of the two sequences is the standardised criterion

$$\delta = \frac{2 \times (\min CPL)}{L_1 + L_2} - 1 \tag{13.7}$$

where min CPL is the minimum CPL over all permissible slottings, and L_1 and L_2 are the path lengths through the individual sequences, given by

$$L_1 = \sum_{j=1}^{m-1} d(A_j, A_{j+1}) \quad \text{and} \quad L_2 = \sum_{k=1}^{n-1} d(B_k, B_{k+1})$$

Gordon (1982) shows that, provided the distance measure d satisfies the triangle inequality,

$$0 \leq \delta \leq 1 + 2 \times \min\left[d\left(A_m, B_1\right), d\left(A_1, B_n\right)\right] / \left(L_1 + L_2\right)$$

the lower bound is attained when either the two sequences are identical or, roughly speaking, the $m+n$ observations lie on a line in the appropriate metric space corresponding to the distance measure d.

A Windows®-based program (CPLSlot) to perform sequence slotting, correlation function fitting (spline or polynomials), and other methods of critical assessment of the resulting correlations based on the principles outlined here is available for free download from http://geography.lancs.ac.uk/cemp/resources/software/cplslot.htm. CPLSlot has the capacity for dealing with multivariate, directional, vector, and fossil count or proportional data.

Case Studies

Ice Chronology

The first case study involves the GRIP ice-core, a 3029 m long sequence drilled in central Greenland (Johnsen et al. 1997). It preserves a detailed record of past environmental changes, through the successive accumulation of annual increments of snow, along with entrapped atmospheric gases and dust impurities. The age and durations of past environmental events can be estimated by counting the annual ice-increments, consequently the GRIP ice-core can be used as a master sequence against which other records can be compared. Figure 13.1a plots the variations of δ^{18}O when averaged at 500-year intervals. The aim of our first case study is to match lake sediments from Lago di Grande Monticchio in southern Italy with the GRIP δ^{18}O data and hence to align the Monticchio pollen-stratigraphical sequence with the ice-core chronology. The percentage of pollen of woody taxa, as interpolated at 496 horizons from Monticchio, was provided as the basic lake-sediment data (Allen et al. 1999; Guiot et al. 1993) for the correlation work.

Figure 13.1 plots the Monticchio pollen data when sequence slotted with the GRIP ice-core chronology. It shows three correlations that are generated when using different stratigraphical constraints. Figure 13.1d plots the Monticchio data when

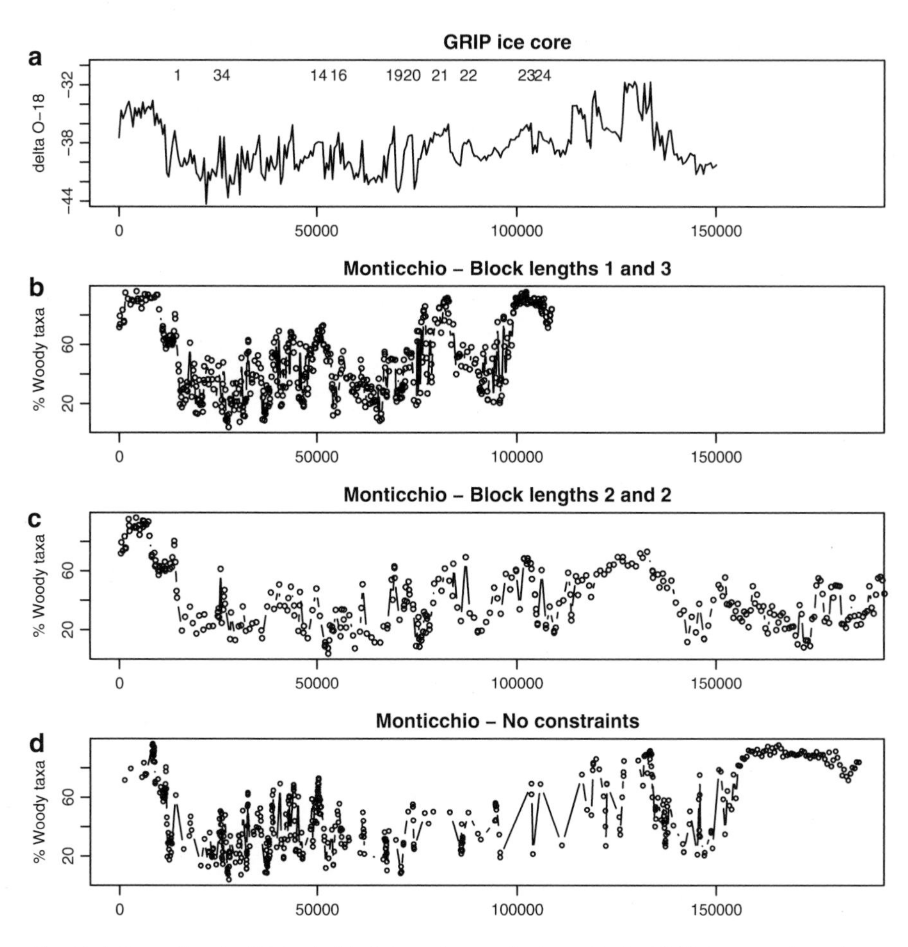

Fig. 13.1 Lago Grande di Monticchio lake-sediment slotting with the GRIP ice-core chronology. (**a**) $\delta^{18}O$ at 500-year intervals (301 data points) in the GRIP sequence. Main interstadial (or Dansgaard/Oeschger) events, marked by numbers between 1 and 24, are identified by light $\delta^{18}O$ layers. (**b–d**) Percentage woody pollen taxa (496 data horizons) in the Monticchio lake-sediment sequence. Sequence slotting constrained with (**b**) maximum block lengths of 1 and 3. i.e., a maximum hiatus duration of 500 years, and a maximum deposition rate of three Monticchio horizons per 500 year time interval; (**c**) maximum block lengths of 2 and 2. i.e., maximum hiatus of 1000 years, and a maximum deposition rate of two Monticchio horizons every 500 years. Here the deepest Monticchio sediments are dated as much older than the basal GRIP ice and so fall beyond the right-hand edge of the diagram; and (**d**) no maximum block lengths. Note how the alternative constraints lead to very different correlations. Slotting (**b**) is in excellent agreement with the Monticchio tephra chronology of Allen et al. (1999)

sequence slotted with the GRIP data of Fig. 13.1a using no external constraints. This slotting corresponds to the global minimum for the similarity between the two sequences as measured by expression 13.7. Although the match is visually attractive, and is the correlation that would be generated by the original sequence-slotting algorithm of Gordon and Reyment (1979), it is only achieved by allowing very variable local deposition rates and so it is unlikely to be correct. We can moderate the deposition rates through block-size constraints, i.e., constraints on the maximum number of consecutive samples in either sequence. A first attempt at a constrained match, Fig. 13.1c, has constraints that are too severe, and once again a poor match is generated. We see that the basal lake-sediment is forced to be far too old. Figure 13.1b shows the slotting for a more sensible combination of block-size constraints. Now the match is in good agreement with that of the Monticchio tephra chronology (Allen et al. 1999). We use the slotting of Fig. 13.1b as the final output from our sequence-slotting algorithm. It yields an age for each of the 496 Monticchio data horizons, two scaling parameters, and the degree of similarity (δ) between the GRIP ice-core and Monticchio pollen sequences.

Any worthwhile numerical core-correlation procedure must somehow address the critical issue of locating such local minima, rather than just the global minimum. That is, the core-correlation procedure must be capable of much more than minimisation alone. It must be able to handle additional stratigraphical information (e.g., deposition rate constraints, precedence constraints) and to incorporate formally this additional knowledge into the correlation process.

Mountain and Arctic Lakes

Between 1996 and 1998 at least three cores were collected at each of eight mountain and Arctic lakes in the MOLAR project. The cores were generally sliced at contiguous 0.2–0.25 cm intervals, providing a sampling resolution of between 1 and 5 years. Following a standard protocol, dry weight and bulk organic matter, as loss-on-ignition, were measured on all of the samples. Selected cores or samples were measured for organic carbon and nitrogen; sulphur; plant pigments (Lami et al. 2000b); and for the remains of diatoms, chrysophytes, cladocerans, and chironomids (Brancelj et al. 2000; Granados and Toro 2000; Kamenik et al. 2000; Lotter et al. 2000; Rautio et al. 2000; Ventura et al. 2000). A chronology was constructed for a master core at each lake (Table 13.1) using ^{210}Pb, ^{241}Am, and ^{137}Cs radiometric techniques (Appleby 2000). Dry weight and loss-on-ignition measurements at remote mountain lakes can reveal remarkably high-resolution stratigraphical fluctuations. These two parameters were measured on all 3405 samples (Table 13.1) thus providing a basis for within-lake core correlation.

The results of the sequence slotting for four of the eight lakes are illustrated in Figs. 13.2, 13.3, 13.4, and 13.5, while Fig. 13.6 shows the final time–depth diagrams for all eight lakes. Sequence slotting is carried out on ordered data; i.e., no depth or chronological information is used, only sample number. Thus the results of sequence slotting are best viewed in terms of sample order, i.e., sample position in a core. In Figs. 13.2, 13.3, 13.4, and 13.5 dry weight and loss-on-ignition are plotted with respect to position in the master core.

Table 13.1 Details of the eight MOLAR lakes included in this correlation study

Lake	Latitude (degrees)	Longitude (degrees)	Elevation (m a.s.l.)	Master core acronym	Core length (cm)	Number of cores	Number of samples
Gossenköllesee	47° 13′N	11° 01′E	2417	Gks97-3	32.2	5	510
Hagelseewli	46° 40′N	08° 02′E	2339	Hag96-2	34.0	5	370
Estany Redó	42° 39′N	00° 46′E	2240	Rcm97-1	42.6	3	404
Laguna Cimera	40° 15′N	05° 18′W	2140	Cim97-1	24.4	4	334
Jezerov Ledvicah	46° 20′N	13° 47′E	1830	Ledv96-4	29.8	3	444
Nižne Terianske Pleso	49° 10′N	20° 00′E	1775	Teri96-6	28.2	3	438
Øvre Neådalsvatn	62° 46′N	09° 00′E	728	Ovne97-5	28.6	4	636
Saanajärvi	69° 05′N	20° 52′E	679	Saan98-1	10.2	4	269

Latitude, longitude, elevation (metres above sea level), acronym for the master core and its length, the number of additional cores, and the number of samples are given

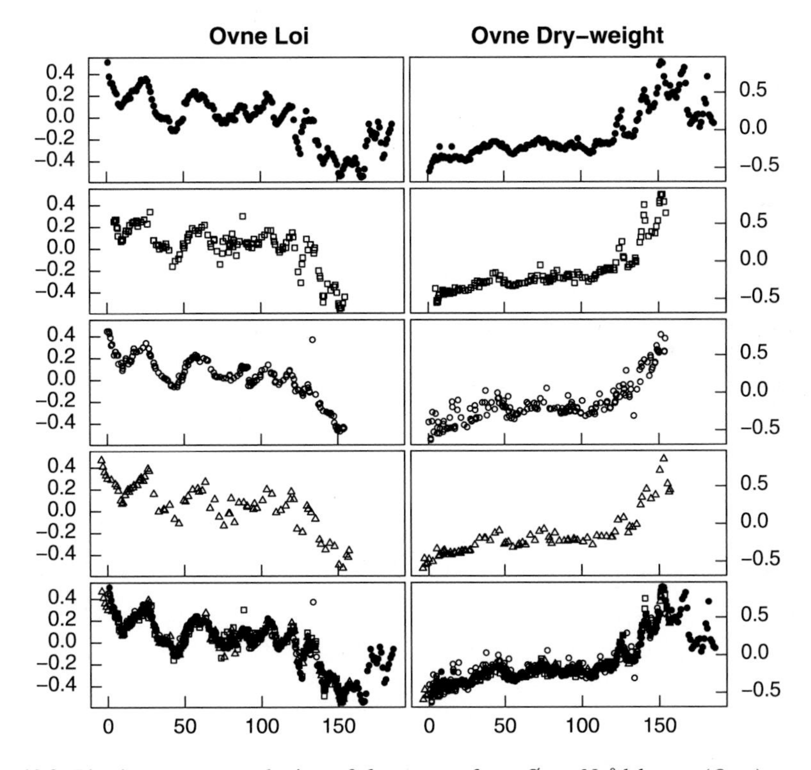

Fig. 13.2 Bivariate sequence slotting of four cores from Øvre Neådalsvatn (*Ovne*), central Norway. Measurements on the master core are plotted in the *uppermost panels* as solid symbols. Three daughter cores are plotted in the *three middle rows* using open symbols. The final stack of all four cores is shown in the *bottom panels*. The *left-hand panels* show loss-on-ignition while the *right-hand panels* show dry weight. Both variables have been standardised (see text). The *horizontal scale in all panels* refers to sample position in the master core with the core top at the *left-hand edge of the panel*. The sample positions in the master core are plotted using equal spacings

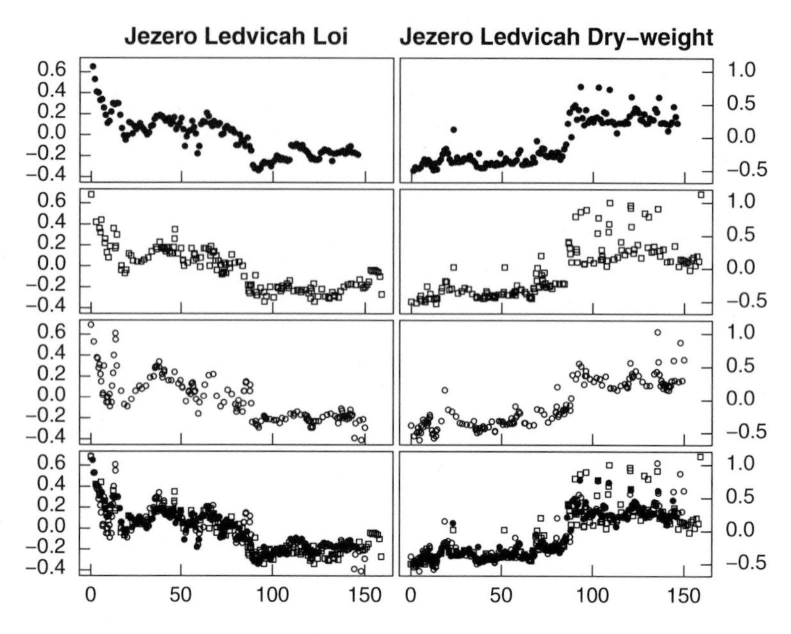

Fig. 13.3 Bivariate sequence slotting of three cores from Jezerov Ledvici, Slovenia. Measurements, symbols, and figure layout are as in Fig. 13.2

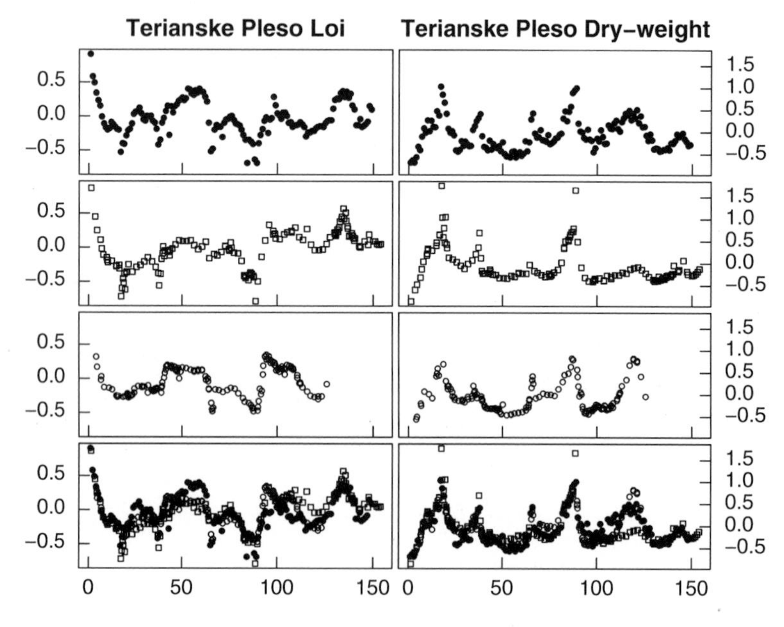

Fig. 13.4 Bivariate sequence slotting of three cores from Nižne Terianske Pleso, Slovakia. Measurements, symbols, and figure layout are as in Fig. 13.2

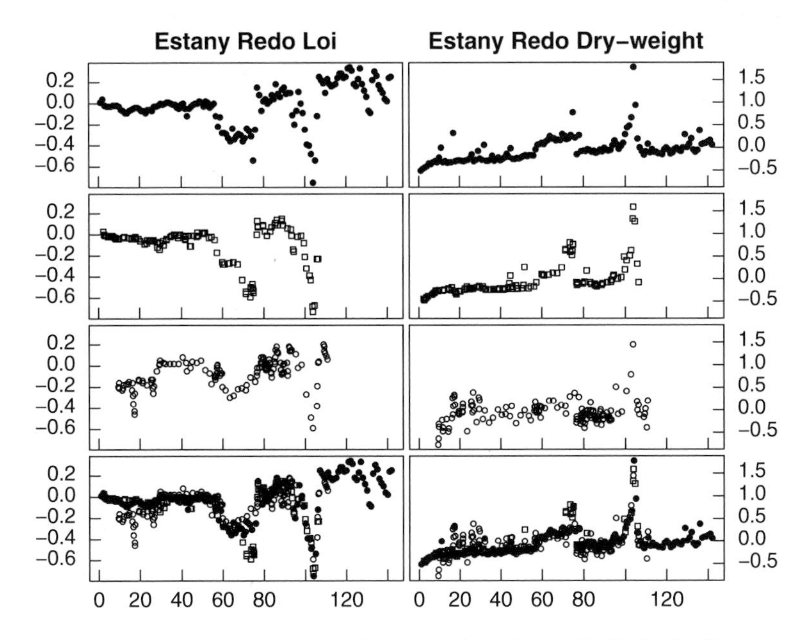

Fig. 13.5 Bivariate sequence slotting of three cores from Estany Redó, Spain. Measurements, symbols, and figure layout are as in Fig. 13.2

The slottings were achieved by minimising δ (in expression 13.7) subject to certain constraints. A maximum internal block length of two was imposed. This constrains the relative deposition rates in the two cores to be within a factor of 2 of each other. It also precludes hiatuses. The core tops were aligned using a combination of constraints of the type described above and explained in more detail in Clark (1985). One core from Laguna Cimera (Spain) needed some additional assistance in its alignment. This was easily achieved by using a combination of stratigraphical constraints to guide the slotting over a somewhat difficult section in the middle of the core. One core from Estany Redó (Spain) needed a more relaxed block-length constraint (of four). Otherwise the block-length constraint of two (in practical terms the most exacting possible) worked extremely well, generating visually appealing core correlations and confirming the absence of any major hiatuses.

Linear transformation (scaling) of the observations can slightly improve slotting, especially when cores are not of the same length. Scaling is also a simple method of weighting. Here it serves to give equal weight to the dry weight and to the loss-on-ignition data. For each slotting the master core was scaled using N(0,1) (i.e., transformed to z-scores (=standardised scores) with zero mean and unit standard deviation) and then the daughter core was scaled using N(μ, σ). The best combination of μ and σ was found for each core individually by using a simplex optimisation that minimised delta. [The Nelder-Mead simplex algorithm (Nelder and Mead 1965) provides a particularly simple way to minimise a fairly well-behaved function. It requires only function evaluations and so is often a good choice

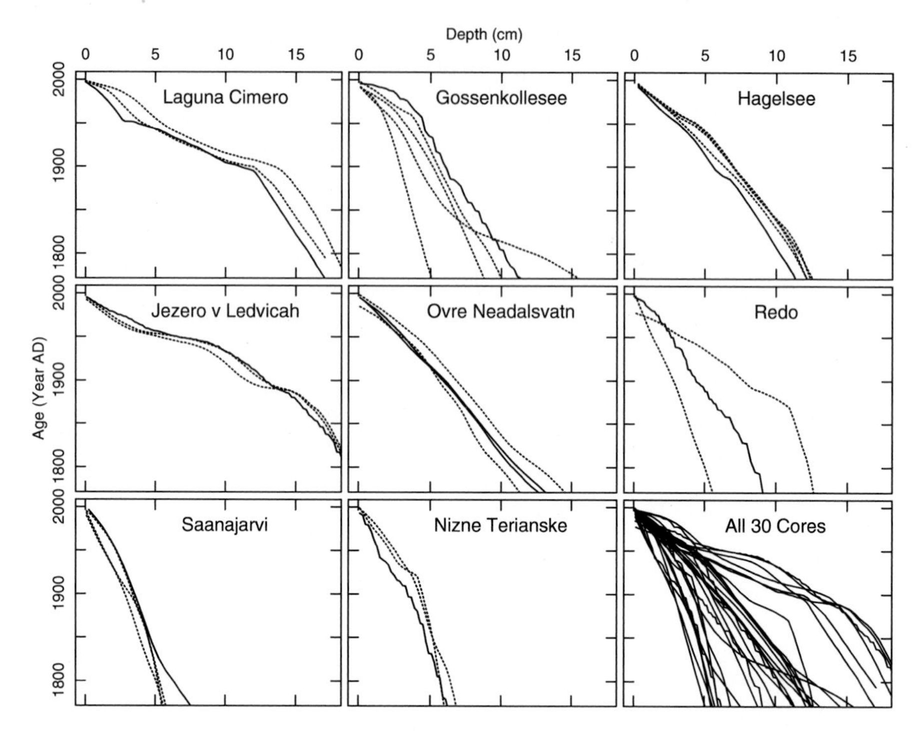

Fig. 13.6 Time–depth diagrams for the eight MOLAR lakes. Nine chronologies dated by ^{210}Pb are drawn as *solid lines*. The 21 chronologies, of the daughter cores, as derived by sequence slotting, are plotted as *dotted lines*. The *lower right-hand panel* plots a synthesis of the time–depth relationships of all 30 sediment sequences in the eight lakes

for simple minimisation problems. Because it does not use any gradient evaluations, it can take longer to find the minimum than other methods. But here, as there are only two parameters to optimise, the time-loss is not a problem (Wright 1996)]. The net result of all these sequence-slotting computations was that for every sample in the daughter cores an equivalent depth in the master core was obtained.

Finally a cubic-spline interpolation (Birks 2012: Chap. 2) was used to convert all the depths to ages. The ^{210}Pb measurements of Appleby (2000), on the master cores, provided the basis for the spline fitting. In addition the age of the topmost sediment was constrained by the coring date, and where necessary the age of the lowermost sediment, beyond the range of the radiometric ages, was found by extrapolation. The resulting time-depth profiles of all 31 cores are plotted in Fig. 13.6.

Other Palaeolimnological Applications

Sequence slotting has also been used in palaeolimnological studies to assess how similar different stratigraphical sequences are by computing δ, the measure of similarity for pairs of sequences. Sequences compared include oxygen-isotope curves from Swiss lakes and the Dye-3 Greenland ice-core (Lotter et al. 1992) and chironomid-inferred July air temperatures at five late-glacial sites in north-west England (Lang et al. 2010). In both cases, slottings were made without constraints and with some mild external stratigraphical constraints. Little difference was found in the slottings or in the values of the measure of the degree of similarity with and without any external constraints.

Conclusions

In the two case studies presented here, the mathematical method of sequence slotting has allowed a large number of measurements to be handled automatically and repeatably. The core correlations generated can be seen in Figs. 13.1, 13.2, 13.3, 13.4, 13.5, and 13.6 to be of high quality. Remarkably consistent time-depth profiles were found at all eight of the slowly depositing arctic or alpine lakes used in the MOLAR project. Three thousand four hundred and five age estimates were generated for climatic variation studies. Abiotic, multivariate, within-lake signals were correlated. In the BENCHPAR work biotic, univariate, lake-sediment data were matched to a master, isotopic, ice-core chronology. The BENCHPAR case-study explored a key issue for numerical core-correlation, namely that in practice the correct match is not necessarily that associated with the global optimum. Instead a local minimum must be identified which satisfies all the available stratigraphical constraints.

Acknowledgements MOLAR involved a large team of European researchers from over 20 institutes in the United Kingdom, Norway, Finland, Austria, Spain, France, Italy, Switzerland, Czech Republic, Slovak Republic, Poland, Slovenia, and Russia. The MOLAR project benefited from European Union support under the Environment and International Co-operation programmes and was funded by the European Commission Framework Programme IV: Environment and Climate with assistance from INCO (ENV4-CT95-0007/IC20-CT96-0021). We particularly thank all the members of the MOLAR project who collected the 31 cores or sampled and measured loss-on-ignition and dry weight on the 3405 samples. We similarly thank the BENCHPAR project for the gridded ice-core and Monticchio pollen data.

References

Agustí-Panareda A, Thompson R (2002) Reconstructing air temperature at eleven remote alpine and arctic lakes in Europe from 1781 to 1997 AD. J Paleolimnol 28:7–23

Allen JRM, Brandt U, Brauer A et al. (1999) Rapid environmental changes in southern Europe during the last glacial period. Nature 400:740–743

Appleby PG (2000) Radiometric dating of sediment records in European mountain lakes. J Limnol 59(Suppl 1):1–14

Barletta F, St-Ongea G, Stoner JS, Lajeunesse P, Locate J (2010) A high-resolution Holocene paleomagnetic secular variation and relative paleointensity stack from eastern Canada. Earth Planet Sci Lett 298:162–174

Battarbee RW, Grytnes J-A, Thompson R, Appleby PG, Catalan J, Korhola A, Birks HJB, Heegaard E, Lami A (2002a) Comparing palaeolimnological and instrumental evidence of climate change for remote mountain lakes over the last 200 years. J Paleolimnol 28:161–179

Battarbee RW, Thompson R, Catalan J, Grytnes J-A, Birks HJB (2002b) Climate variability and ecosystem dynamics of remote alpine and arctic lakes: the MOLAR project. J Paleolimnol 28:1–6

Birks HJB (2012) Chapter 2: Overview of numerical methods in palaeolimnology. In: Birks HJB, Lotter AF, Juggins S, Smol JP (eds) Tracking environmental change using lake sediments. Volume 5: Data handling and numerical techniques. Springer, Dordrecht

Birks HH, Birks HJB (2006) Multi-proxy studies in palaeolimnology. Veg Hist Archaeobot 15:235–251

Birks HJB, Gordon AD (1985) Numerical methods in Quaternary pollen analysis. Academic Press, London

Brancelj A, Šiško M, Lami A, Appleby P, Livingstone DM, Rejec-Brancelj I, Ogrin D (2000) Changes in the trophic level of an Alpine lake, Jezero v Ledvici (NW Slovenia), induced by earthquakes and climate change. J Limnol 59(Suppl 1):29–42

Clark RM (1985) A FORTRAN program for constrained sequence-slotting based on minimum combined path length. Comput Geosci 11:605–617

Delcoigne A, Hansen P (1975) Sequence comparison by dynamic programming. Biometrika 62:661–664

Edwards LE (1984) Insights on why graphic correlation (Shaw's method) works. J Geol 92: 583–597

Gardner AR (1999) The impact of Neolithic agriculture on the environments of south-east Europe. PhD thesis, University of Cambridge

Gary AC, Johnson GW, Ekart DD, Platon E, Wakefield MI (2005) A method for two-well correlation using multivariate biostratigraphical data. In: Powell AJ, Riding JB (eds) Recent developments in applied biostratigraphy. Special Publications Micropalaeontological Society, London, pp 205–217

Ghose BK (1984) STRETCH: a subroutine for stretching time series and its use in stratigraphic correlation. Comput Geosci 10:137–147

Gordon AD (1982) An investigation of two sequence-comparison statistics. Aust J Stat 24:332–342

Gordon AD, Birks HJB (1974) Numerical methods in Quaternary palaeoecology. II. Comparison of pollen diagrams. New Phytol 73:221–249

Gordon AD, Reyment RA (1979) Slotting of borehole sequences. Math Geol 11:309–327

Gordon AD, Clark RM, Thompson R (1988) The use of constraints in sequence slotting. In: Diday E (ed) Data analysis and informatics V. North-Holland, Amsterdam, pp 353–364

Granados I, Toro M (2000) Recent warming in a high mountain lake (Laguna Cimera, Central Spain) inferred by means of fossil chironomids. J Limnol 59(Suppl 1):109–119

Griffiths CM, Bakke S (1990) Interwell matching using a combination of petrophysically derived numerical lithologies and gene-typing techniques. In: Hurst A, Lovell MA, Morton AC (eds) Geological applications of wireline logs. Geological Society of London Special Publication No. 48. pp 33–151

Guiot J, de Beaulieu JL, Cheddadi R, David F, Ponel P, Reille M (1993) The climate in Western Europe during the last glacial/interglacial cycle derived from pollen and insect remains. Palaeogeogr Palaeoclim Palaeoecol 103:73–94

Hladil J, Vondra M, Cejchan P, Vich R, Koptikova L, Slavik L (2010) The dynamic time-warping approach to comparison of magnetic-susceptibility logs and application to Lower Devonian calciturbidites (Prague Synform, Bohemian Massif). Geologica Belgica 13(4):385–406

Johnsen SJ, Clausen HB, Dansgaard W et al. (1997) The delta ^{18}O record along the Greenland Ice Core Project deep ice core and the problem of possible Eemian climatic instability. J Geophys Res 102:26397–26410

Kamenik C, Koinig KA, Schmidt R, Appleby PG, Dearing JA, Lami A, Thompson R, Psenner R (2000) Eight hundred years of environmental changes in a high Alpine lake (Gossenköllesee, Tyrol) inferred from sediment records. J Limnol 59(Suppl 1):43–52

Kovach W (1993) Multivariate techniques for biostratigraphic correlation. J Geol Soc Lond 150:697–705

Lami A, Cameron N, Korhola A (2000a) Preface. J Limnol 59(Suppl 1):1–14

Lami A, Guilizzoni P, Marchetto A (2000b) High resolution analysis of fossil pigments, carbon, nitrogen and sulphur in the sediment of eight European Alpine lakes: the MOLAR project. J Limnol 59(Suppl 1):15–28

Lang B, Brooks SJ, Bedford A, Jones RT, Birks HJB, Marshall JD (2010) Regional consistency in Late-glacial chironomid-inferred temperature from five sites in north-west England. Quaternary Sci Rev 29:1528–1538

Lisiecki LL, Lisiecki PA (2002) Application of dynamic programming to the correlation of paleoclimate records. Paleoceanography 17. doi:10.1029/2001PA000733

Lotter AF, Eicher U, Birks HJB, Siegenthaler U (1992) Late-Glacial climatic oscillations as recorded in Swiss lake-sediments. J Quaternary Sci 7:187–204

Lotter AF, Hofmann W, Kamenik C et al (2000) Sedimentological and biostratigraphical analyses of short sediment cores from Hagelseewli (2339 m.a.s.l.) in the Swiss Alps. J Limnol 59(Suppl 1):53–64

Luthi SM, Bryant ID (1997) Well-log correlation using a back-propagation neural network. Math Geol 29:413–425

Maher LJ (1993) SLOTDEEP.EXE: manual correlation using the dissimilarity matrix. INQUA Commission for the Study of the Holocene: Working group on data-handling methods. Newsletter 9:21–26

Maher LJ (1998) SLOTDEEP v. 1.8 adds DC profiles to its DC map. INQUA Commission for the Study of the Holocene: Sub-commission on data-handling methods. Newsletter 18:4–7

Maher BA, Zhou LP, Thompson R (1994) Reconstruction of palaeoprecipitation values for the Chinese loess plateau from proxy magnetic data. In: Funnell BM, Kay RLF (eds) Palaeoclimate of the last glacial/interglacial cycle. NERC Earth Sci Direct Special Publication 94/2, pp 33–36

Martinson DG, Menke W, Stoffa P (1982) An inverse approach to signal correlation. J Geophys Res 87:4807–4818

Marwan N, Thiel M, Nowaczyk NR (2002) Cross recurrence plot based synchronization of time series. Nonlin Process Geophys 9:325–331

Nelder JA, Mead R (1965) A simplex method for function minimization. Comput J 7:308–313

Pels B, Keizer J, Young R (1996) Automated biostratigraphic correlation of palynological records on the basis of shapes of pollen curves and evaluation of next-best solutions. Palaeogeogr Palaeoclim Palaeoecol 124:17–37

Rautio M, Sorvari S, Korhola A (2000) Diatom and crustacean zooplankton communities, their seasonal variability and representation in the sediments of subarctic Lake Saanajärvi. J Limnol 59(Suppl 1):81–96

RTD Magazine for European research (2000) The message from the mountain lakes. RTD Info 28. http://ec.europa.eu/research/rtdinfo/en/28/environnement1.html. Accessed 7 June 2011

Shaw BR (1982) A short note on the correlation of geological sequences. In: Cubitt JM, Reyment RA (eds) Quantitative stratigraphic correlation. Wiley, Chichester, pp 7–11

Shaw BR, Cubitt JM (1979) Stratigraphic correlation of well logs: an automated approach. In: Gill D, Merriam DF (eds) Geomathematical and petrophysical studies in sedimentology. Pergamon Press, Oxford, pp 127–148

Thompson R (1991) Palaeomagnetic dating. In: Smart PD, Frances PD (eds) Quaternary dating methods – a user's guide. Technical guide 4. Quaternary Research Association, Cambridge, pp 177–194

Thompson R, Clark RM (1989) Sequence slotting for stratigraphic correlation between cores: theory and practice. J Paleolimnol 2:173–184

Ventura M, Camarero L, Buchaca T, Bartumeus F, Livingstone DM, Catalan J (2000) The main features of seasonal variability in the external forcing and dynamics of a deep mountain lake (Redó, Pyrenees). J Limnol 59(Suppl 1):97–108

Walker IR, Smol JP, Engstrom DR, Birks HJB (1991) An assessment of Chironomidae as quantitative indicators of past climatic change. Can J Fish Aquat Sci 48:975–987

Waterman MS, Raymond RJ (1987) The match game: new stratigraphic correlation algorithms. Math Geol 19:109–127

Wright MH (1996) Direct search methods: once scorned, now respectable. In: Griffiths DF, Watson GA (eds) Numerical Analysis 1995. Papers from the sixteenth Dundee biennial conference held at the University of Dundee, Dundee, June 27–30, 1995. Longman, Harlow, pp 191–208

Chapter 14
Quantitative Environmental Reconstructions from Biological Data

Steve Juggins and H. John B. Birks

Abstract Quantitative reconstructions of past environmental conditions (e.g., lake-water pH) are an important part of palaeolimnology. Such reconstructions involve three steps: (1) the development of a representative modern organism-environment training-set, (2) the development and application of appropriate numerical techniques to model the relationship between modern occurrences and abundances of the organisms in the training-set and their contemporary environment, and (3) the application of this model to stratigraphical palaeolimnological data to infer past environmental conditions, and model selection, testing, and evaluation and assessment of the final reconstruction. These three stages are discussed. Problems of spatial autocorrelation are outlined. The general approach is illustrated by a case-study. The assumptions and limitations of the calibration-function approach are presented, and violations of these assumptions are discussed in relation to different environmental reconstructions. Appropriate computer software is outlined, and future research areas are presented. The chapter challenges palaeolimnologists to be more critical of their environmental-inference models and to be alert to the problems and dangers of confounding variables, and of violating the main assumptions of the approach.

S. Juggins (✉)
School of Geography, Politics & Sociology, Newcastle University,
Newcastle-upon-Tyne NE1 7RU, UK
e-mail: stephen.juggins@ncl.ac.uk

H.J.B. Birks
Department of Biology and Bjerknes Centre for Climate Research, University of Bergen,
PO Box 7803, Bergen N-5020, Norway

Environmental Change Research Centre, University College London, Pearson Building,
Gower Street, London, WC1E 6BT, UK

School of Geography and the Environment, University of Oxford, Oxford, OX1 3QY, UK
e-mail: john.birks@bio.uib.no

H.J.B. Birks et al. (eds.), *Tracking Environmental Change Using Lake Sediments*,
Developments in Paleoenvironmental Research 5, DOI 10.1007/978-94-007-2745-8_14,

Keywords Artificial neural networks • Bayesian approaches • Calibration functions • Correspondence analysis regression • Cross-validation • Environmental reconstruction • Gaussian logit regression • Locally-weighted weighted-averaging • Maximum likelihood estimation • Model evaluation • Model selection • Modern analogue technique • Partial least squares • Principal components regression • Reconstruction evaluation • Reconstruction significance • Reconstruction validation • Spatial autocorrelation • Transfer functions • Weighted averaging • Weighted-averaging partial least squares

Introduction

Many palaeolimnological studies aim to reconstruct aspects of the past environment from the stratigraphical record preserved in lake sediments. These records may be lithostratigraphical (e.g., varve thickness), geochemical (e.g., chemical elements such as Ca, Mg, K, Na, or organic geochemical biomarkers), physical (e.g., sediment magnetic properties, optical properties, stable-isotope ratios), or biostratigraphical (e.g., diatoms, chironomids, cladocerans, chrysophytes). Environmental reconstructions in palaeolimnology and other branches of palaeoecology were, prior to the work of Imbrie and Kipp (1971), primarily qualitative and presented as 'acid', 'mildly basic', 'cool', 'temperate', 'moist', 'dry', etc. In 1971 Imbrie and Kipp revolutionised Quaternary palaeoecology by presenting, for the first time, a procedure for the quantitative reconstruction of past environmental variables from biostratigraphical fossil assemblages involving calibration or so-called transfer functions. Since this pioneering work on marine foraminifera in relation to ocean surface temperatures and salinity, the general approach of quantitative palaeoenvironmental reconstruction has been adopted in many areas of palaeoecology, palaeolimnology, and palaeoceanography with fossils as diverse as pollen, diatoms, chrysophytes, chironomids, cladocerans, ostracods, coleoptera, phytoliths, mosses, radiolaria, dinoflagellates, coccolithophores, testate amoebae, and foraminifera.

Much of the impetus for the development and refinement of methods for the quantitative environmental reconstruction in palaeolimnology came from the need to quantify changes in lake-water pH and nutrients resulting from recent freshwater acidification and eutrophication (Renberg and Hellberg 1982; Birks et al. 1990a; Bennion et al. 1996; Smol 2008; Battarbee et al. 2010; Hall and Smol 2010; Simpson and Hall 2012: Chap. 19). Quantitative palaeolimnological reconstructions now play an increasingly important role in studies of past climate change (e.g., Battarbee 2000; Brooks 2003; Verschuren 2003: Cumming et al. 2012: Chap. 20), in issues of lake management and reference conditions (e.g., Brenner et al. 1993; Battarbee et al. 2005a; Simpson et al. 2005; Reavie et al. 2006; Battarbee and Bennion 2011; Battarbee et al. 2011; Bennion and Simpson 2011; Bennion et al. 2011a, b; Simpson and Hall 2012: Chap. 19), ecosystem sensitivity and resilience (e.g., Paterson et al. 2002b), and lake ontogeny (e.g., Renberg et al. 1993; Engstrom et al. 2000), and in quantifying the impact of prehistoric societies on aquatic systems (e.g., Bindler et al. 2002; Bradshaw et al. 2005).

The quantitative interpretation of biostratigraphical records is based on the principle of uniformitarianism (Rymer 1978; Birks et al. 2010). That is, knowledge of an organism's present-day ecology and environmental preferences can be used to make inferences about past conditions. The use of quantitative methods allows palaeolimnologists to formalise this procedure by defining a series of equations or numerical procedures that relate a set of modern biological responses to one or more modern environmental parameters (Birks 1995, 1998, 2003; Birks et al. 2010):

$$\mathbf{Y}_m = f\,(\mathbf{X}_m) + \text{error} \tag{14.1}$$

where $\mathbf{Y}_m$ is a matrix of modern biological responses (i.e. taxon abundances), $\mathbf{X}_m$ is a matrix of contemporary physico-chemical environmental 'predictors', and $f\,(\,)$ is a set of ecological response functions.

If we understood and could quantify the various physical, chemical, and biological processes that determine biological distributions and abundances today we could derive the response functions $f\,(\,)$ directly. Such detailed autecological information is usually lacking so we have to solve Eq. 14.1 empirically by relating the modern distribution and abundance of taxa to contemporary environmental measurements. This usually involves a modern 'training' or 'calibration' data-set of biological census counts extracted from surface sediments together with environmental measurements from the same sites. The relationship between the modern biological and environmental data-sets is then used to solve $f\,(\,)$ and the resulting model applied to fossil assemblages to derive estimates of the environmental variable of interest for times in the past (Fig. 14.1).

Quantitative reconstruction thus involves three separate steps, (1) the development of the modern training-set, (2) the development of the numerical model to solve Eq. 14.1, and (3) the application of the model to the fossil biostratigraphical record and the evaluation of the resulting reconstruction. Each of these steps involves a number of decisions concerning training-set selection, data transformation and screening, choice of numerical method, model selection, testing, and evaluation, and assessment of the final reconstruction (Fig. 14.2). These steps are described in more detail below and then illustrated using a case study of recent lake acidification. We then examine the assumptions of the quantitative approach to environmental reconstruction and discuss ways in which they can be tested and explore the implications of their violation.

There are several recent reviews on quantitative environmental reconstructions from fossil assemblages (e.g., ter Braak 1995; Birks 1995, 1998, 2003, 2010; Birks and Seppä 2004; Kumke et al. 2004; Brewer et al. 2007; Guiot and de Vernal 2007; Birks et al. 2010). Our intention here is not to duplicate these reviews but to provide a critical discussion of the concepts, methodologies, and assumptions illustrated by recent studies using a range of modern and fossil palaeolimnological data-sets.

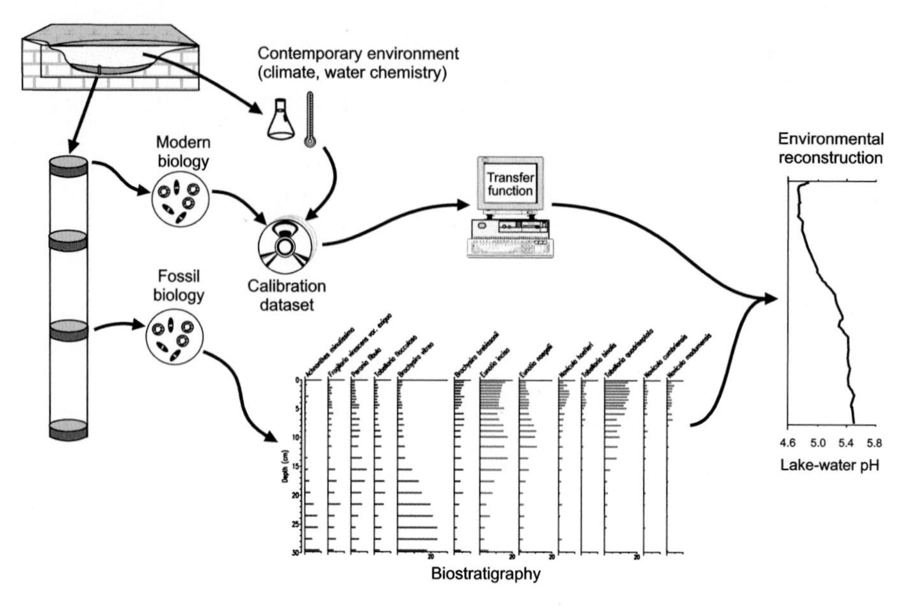

Fig. 14.1 Schematic diagram showing the steps involved in deriving a quantitative reconstruction from biostratigraphical palaeolimnological data using a modern training or calibration data-set

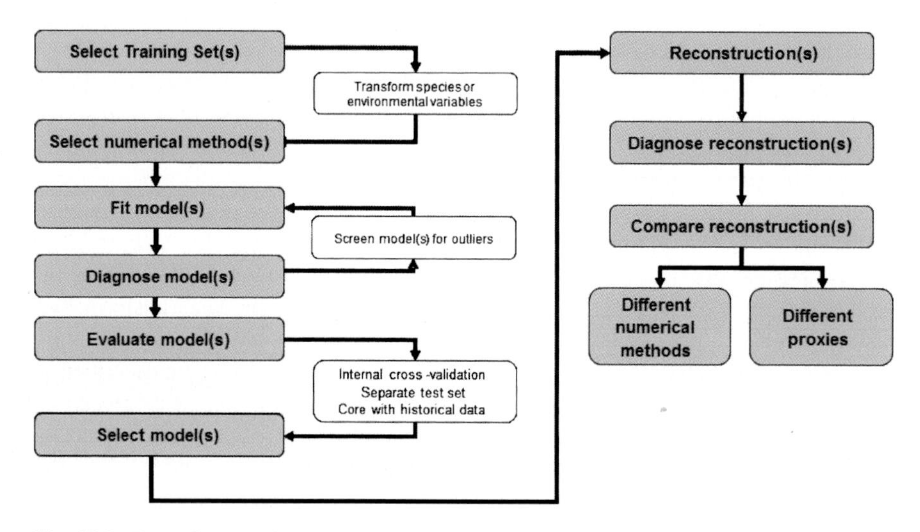

Fig. 14.2 Flow diagram showing the steps involved in deriving and evaluating quantitative reconstructions

Training-Set Development

The major requirement for developing quantitative reconstructions is the availability of an appropriate training-set of modern samples and associated environmental measurements. Indeed, the resulting accuracy and reliability of the reconstructions depends in part on the size, coverage, and quality of the training-set. Despite the high cost of sample collection and analysis, there has been little work on the optimal design of training-sets (Telford and Birks 2011a). The training-set will usually be derived from the same type of sedimentary environment as the fossil material to minimise the influence of taphonomic effects (Birks 1995) and should span the range of environmental values likely to be represented by the fossil material (Birks 1998). Where prior environmental data are available, stratified sampling can be used to ensure training-set lakes span the gradient(s) of interest (e.g., Bennion 1994). There are few guidelines as to the optimal size: 30 lakes is a useful rule-of-thumb for simple systems dominated by a single strong environmental gradient but larger data-sets will generally be needed to estimate accurately the ecological preferences of taxa across more complex gradients. Most calibration methods are sensitive to the distribution of the environmental variable in the training-set (e.g., Mohler 1983; ter Braak and Looman 1986; Telford and Birks 2011a), so samples should be evenly spaced along the gradient. However, without prior knowledge it is very difficult to control for $\mathbf{X}_m$ and $\mathbf{Y}_m$ during data collection and there will often be some bias in the distribution of environmental values.

Taxonomic consistency, both within the training-set, and between the modern training-set and the fossil material is critical. Where different analysts have been involved it is important to follow agreed taxonomic protocols and to harmonise differences between laboratories using agreed identification guides and nomenclature, slide exchanges, and a programme of analytical quality control (e.g., Munro et al. 1990; Kingston et al. 1992). Collection of adequate environmental data will often be the most expensive part of training-set construction because the biological data are usually calibrated against seasonal or annual means based on multiple measurements. Surface-sediment samples (0–1 cm) represent a time-averaged assemblage that typically span the previous 1–20 years. Where samples have been collected at existing monitoring sites they should be related to chemical or other limnological variables averaged over the same period (e.g., Clarke et al. 2003). Where new chemical or other environmental data are to be collected they should be as comprehensive as possible: at least seasonally for 1 year, but ideally monthly to obtain an accurate estimate of the mean for highly variable parameters (e.g., Stauffer 1988). In some cases it may be appropriate to use seasonal rather than annual averages, motivated by an understanding of the temporal dynamics of the species–environment relationship (e.g., Siver and Hamer 1992), or on purely empirical grounds of lower prediction error (Lotter et al. 1997; Schmidt et al. 2004). Environments with extremely high inter- and intra-annual variability pose special problems and in these cases it may be more appropriate to calibrate biological data from individual habitats to spot environmental measurements (Gasse et al. 1995, 1997).

The full range of taxa or environments cannot always be sampled in a single region and palaeolimnologists are increasingly collaborating to merge small local data-sets into larger regional or continental training-sets (e.g., Bennion et al. 1996; Wilson et al. 1996; Walker et al. 1997; Battarbee et al. 2000; Ginn et al. 2007). The benefits of merged data-sets include extended and more even sampling of environmental gradients and a better coverage of species distributions, both of which increase the likelihood of finding analogous modern samples. However, merging data-sets potentially increases noise as a result of increased genetic and ecological variability not captured by the taxonomy in use, and the influence of 'nuisance' or secondary environmental gradients (cf. Kucera et al. 2005). Thus an expanded training-set may be applicable to a wider range of fossil material but the error of any reconstructions based on it may be larger. Numerical methods based on 'dynamic' training-sets (Birks 1998) may be appropriate in this situation (see the section below on Numerical Methods).

Many environmental variables exhibit log-normal or other skewed distributions (Ott 1990; Limpert et al. 2001) and require either square-root or $\log_{10}$-transformation (see Juggins and Telford 2012: Chap. 5). Training-set biological data are usually expressed as percentages and may sometimes also be square-root or $\log_{10}(x + 1)$ transformed to improve their 'signal to noise' ratio (e.g., Prentice 1980; Lotter et al. 1997). Choice of transformation can be included as part of the model-selection process (e.g., Köster et al. 2004) although this approach has hidden problems discussed below. The choice of which taxa to include in the training-set has received little attention. Almost all published examples use the full taxon list, after deletion of rare taxa (typically those with two or fewer occurrences and a maximum value of less than 1%). A few experiments suggest that weighted-averaging models have lowest prediction errors when all but the very rare taxa are included (e.g., Cumming and Smol 1993; Birks 1994; Quinlan and Smol 2001). In some situations there may be ecological grounds for excluding certain taxa: for example Siver (1999) argues that inference models for epilimnetic water-chemistry should be based on planktonic taxa only, although comparisons suggest that models that also include benthic taxa have superior predictive power (Philibert and Prairie 2002b). Finally, Racca et al. (2003, 2004) argue that current criteria to screen taxa for inclusion are largely *ad hoc* and present a method for pruning taxa in an artificial neural network model (see below) on the basis of their predictive ability. Their procedure is analogous to backward elimination in multiple regression (Draper and Smith 1981, see Birks 2012: Chap. 2) and is used to build species-tailored models with enhanced performance. Although the root mean squared error of prediction (RMSEP, see below) stays constant even when 60–80% of the taxa are deleted, the ratio of RMSEP/RMSE (root mean squared error) declines, suggesting that the 'pruned' model is more robust than the model based on all taxa (Racca et al. 2003). More work is needed to see if their results hold for other numerical approaches. Wehrens (2011) discusses a range of potentially useful techniques for variable selection in the related field of chemometrics (see also Varmuza and Filzmoser 2009).

Most biological and environmental data contain outliers or atypical observations. In the context of calibration these are defined as observations that a have a poor relationship with the environmental variable of interest. Such outliers can have a strong influence on taxon coefficients and reduce the predictive ability of the final model. Birks et al. (1990a), Jones and Juggins (1995), and Lotter et al. (1997) describe criteria for identifying outliers, based on a comparison of observed and inferred environmental values, measures of the influence of each sample on the model coefficients (e.g., Cook's *D*: Cook and Weisberg 1982), and the fit to the species–environment relationship in ordination space (see below). As we discuss below, model performance is usually determined by comparing observed and inferred values for the training-set. Thus the resulting apparent performance of a model is strongly dependent on the extent and criteria of the screening procedure (e.g., Birks et al. 1990a). For this reason we prefer to take a conservative approach to sample deletion and initially remove only outliers that have a standardised residual (under internal cross-validation (CV)) that is greater in absolute value than 2 or 2.5. This corresponds to an expected distribution of about 5% and 1% of observations, respectively. Other observations should only be removed if there is additional justification, such as unusual values of secondary environmental variables. Finally, the pattern of outliers may be different for different numerical methods so training-set screening should be repeated for each method using exactly the same numerical criteria.

Numerical Methods

Introduction

Inferring one or more environmental parameters from biological data is a problem of multivariate calibration or inverse regression. This is a well-established area of statistics (e.g., Martens and Næs 1989; Næs et al. 2002; Varmuza and Filzmoser 2009) but palaeolimnological data possess a number of properties (Birks 2012: Chap. 2) that make calibration using traditional numerical methods problematical. First, the modern training data-sets usually contain many predictors (typically 50–300 taxa in the calibration), and the number of predictors often exceeds the number of samples. Second, the calibration predictors are often highly inter-correlated or collinear, and exhibit non-linear relationships with the environmental variables. Third, the predictor variables are usually expressed as percentages and have a constant sum constraint. Fourth, the matrix of predictors is often sparse and contains many zero values: typically between 50% and 75% of all entries are zero in a matrix of diatom abundance values. Fifth, the predictor variables are subject to both structured and unstructured noise: the former due to the influence of secondary environmental gradients and the latter the result of other unmeasured environmental factors, biotic interactions, taphonomic and stochastic effects, analytical and counting errors (see Maher et al. 2012: Chap. 6), etc. Finally, despite careful design, the training-set may exhibit an uneven distribution of samples along the environmental variable(s) of interest (Coudun and Gégout 2006; Telford and Birks 2011a), and these variables themselves may be subject to substantial measurement error.

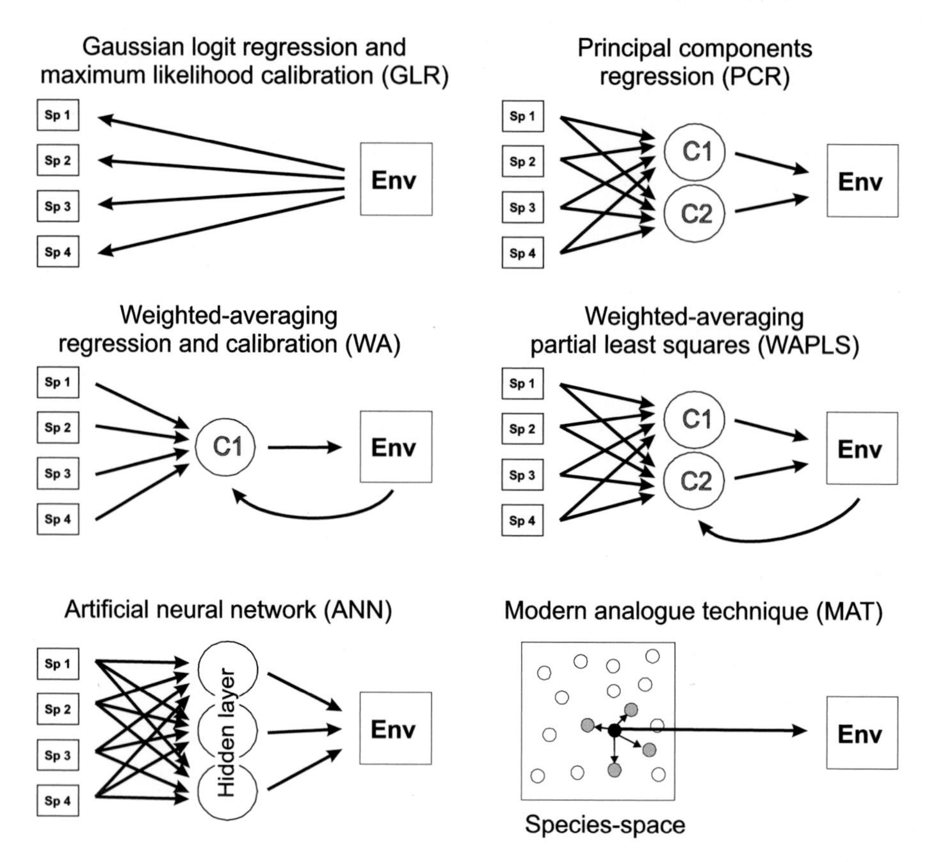

Fig. 14.3 Conceptual diagram illustrating the different approaches to multivariate calibration. Sp are biological taxa in the modern training-set and Env is the environmental variable of interest in the modern data. C are components

Several methods for quantitative reconstructions have been proposed that attempt to account for, either explicitly or implicitly, the particular numerical properties of palaeolimnological data. Some have a stronger ecological or statistical basis than others. As ter Braak (1995), Birks (1995, 2003), and Birks et al. (2010) discuss, there is a major conceptual distinction between inverse and classical approaches, and between methods that assume a linear response of species to an environmental gradient and methods that can account for the non-linear, unimodal relationships frequently observed in biological data. Figure 14.3 gives a schematic representation of the methods most commonly used in palaeolimnology.

Classical Methods

Given the causal relationship between biology and environment the seemingly most obvious way to solve Eq. 14.1 is to estimate the response functions f ( ) by a regression of $\mathbf{Y}_m$ on $\mathbf{X}_m$ (Birks 2012: Chap. 2). This is the so-called classical approach and can be solved using linear, non-linear, or multivariate regression; the choice of the particular regression model depending on the nature of the relationship between species and environment and the assumed error structure of the residuals.

With constant-sum biological data the appropriate regression model is the multinomial logit (ter Braak and van Dam 1989; ter Braak et al. 1993; Birks 1995; ter Braak 1995) in which the expected proportional abundance of each taxon is modelled as a non-linear function of the environment. This approach fits response curves to all species simultaneously and ensures that the sample totals sum to 1 but this has proved difficult in practice, especially with data-sets containing large numbers of taxa (ter Braak et al. 1993; ter Braak 1995). A simple compromise that provides an approximation to the multinomial model is to fit separate regressions for each taxon using logistic regression and include a quadratic term in the environment to fit Gaussian-like unimodal curves (ter Braak and van Dam 1989; Birks et al. 1990a), so-called Gaussian logit (or logistic) regression (GLR):

$$y(x) = \left[\exp\left(b_0 + b_1 x + b_2 x^2\right)\right] / \left[1 + \exp\left(b_0 + b_1 x + b_2 x^2\right)\right] \qquad (14.2)$$

where $y(x)$ is the expected proportional abundance of a taxon as a function of x. The significance of the quadratic (unimodal) vs. linear regression model can be tested using a quasi-likelihood F-test and the simplest curve used in the calibration (ter Braak and van Dam 1989; Birks 2012: Chap. 2).

The fitted response curves, f ( ), together with their error structure, form a statistical model of the modern biological data in relation to the environment. This model can then be 'inverted' to find the unknown environment $\hat{\mathbf{X}}_0$ from fossil samples $\mathbf{Y}_0$. Inversion in this case means finding the value of $\hat{\mathbf{X}}_0$ that is most likely given the observed biological data $\mathbf{Y}_0$, namely the maximum-likelihood estimate of $\hat{\mathbf{X}}_0$. This is obtained by maximising the log-likelihood function summed over all m taxa (Juggins 1992):

$$l = \sum_{k=1}^{m} \left(y_k \log(\mathrm{U}_k) + (1 - y_k)\log(1 - \mathrm{U}_k)\right) \qquad (14.3)$$

where U_k is the expected abundance of taxon y_k. In most cases the maximum of Eq. 14.3 lacks a direct analytical solution and $\hat{\mathbf{X}}_0$ is found using a grid search or optimisation procedure (Birks 1995, 2001).

Gaussian logit regression and maximum-likelihood calibration (GLR-ML) provides a statistically rigorous approach to quantitative reconstruction but it has only rarely been used in palaeolimnology. There are a number of reasons for this. First, the method is computationally demanding, although with the increase in the computing power of personal computers and the use of efficient cross-validation procedures (C2: Juggins 2007; rioja: Juggins 2009) this is no longer a critical constraint. Second, and most important, in early comparisons the computationally simpler inverse methods described below often performed as well or better than GLR-ML (Birks et al. 1990a; Cumming et al. 1992). However, recent comparisons by Köster et al. (2004) using C2 and by Birks (2001) using a corrected version of the WACALIB software suggest that GLR-ML can out-perform alternative methods. Because GLR-ML can model both unimodal and monotonic relationships it seems particularly useful for data-sets spanning relatively short gradients (Köster et al. 2004). More comparisons are needed to evaluate fully its merits but these new results indicate that GLR-ML can provide a useful, alternative reconstruction method (see also Telford and Birks 2005; Yuan 2007).

Inverse Methods

The inverse approach to calibration avoids the difficult inversion step required in classical methods by estimating the inverse of f ( ) directly from the training-set by an inverse regression of $\mathbf{X}_m$ on $\mathbf{Y}_m$ (Birks 2012: Chap. 2).

$$\mathbf{X}_m = g\,(\mathbf{Y}_m) + \text{error} \tag{14.4}$$

The estimate of the past environment, $\hat{\mathbf{X}}_0$, is then simply obtained by inserting the values of the fossil biology into Eq. 14.4: $\hat{\mathbf{X}}_0 = \mathrm{g}\,(\mathbf{Y}_0)$. A simple approach to inverse regression that performs a reconstruction from a single taxon is to use linear least-squares regression (e.g., Beerling et al. 1995; Finsinger and Wagner-Cremer 2009). In this model the environmental variable is the 'response' variable and the taxon abundance is the 'explanatory' variable (note that we do not imply that biology 'causes' environment but we simply 'invert' the roles of the variables for convenience of statistical modelling). With m taxa this approach extends to multiple least-squares linear regression (Birks 2012: Chap. 2):

$$x = b_0 + b_1 y_1 + b_2 y_2 \ldots b_m y_m \tag{14.5}$$

Modern training-sets usually have large numbers of taxa and the abundances of ecologically similar taxa are often highly correlated. This causes two problems for multiple linear regression: (1) when the number of taxa approaches or exceeds the number of samples the regression solution is indeterminate, and (2) multicollinearity among the explanatory variables leads to instability in the regression parameters (Montgomery and Peck 1982). In addition, Eq. 14.5 assumes a linear relationship between taxa and their environment that is rarely observed in nature. Practical implementations of the inverse approach attempt to address these problems in different ways and include multiple regression of species groups, principal components regression (PCR), partial least squares (PLS), and weighted-averaging (WA) based methods (Fig. 14.3). The first of these tackles the multicollinearity problem by combining the original taxa into a small number of species groups, or 'supertaxa', and uses these as predictors in a multiple regression. The groups may be derived from an *a priori* classification, such as F Hustedt's pH classification of diatoms (Flower 1986; ter Braak and van Dam 1989), or from a cluster analysis of taxa in the training-set (see Charles 1985; Legendre and Birks 2012a: Chap. 7). The second method, principal components regression, reduces the original biological variables to a small number of principal components and uses these as predictors in a multiple regression. PCR forms the basis of the Q-mode factor analysis method (IKFA) developed by Imbrie and Kipp (1971). Despite its widespread use in palaeoceanography (e.g., Gersonde et al. 2005), PCR suffers from a number of problems (Birks 1995). First, because the components used as predictors in PCR are linear combinations of the original variables, the method is only likely to perform well over short compositional gradients where there is a predominantly linear relationship between taxa and their environment. Second, the principal components are extracted to provide a summary of the main directions of variation in the biological data (see Legendre and Birks 2012b: Chap. 8). There is no guarantee that they will have any predictive power for the environmental variable of interest. Extensions to these approaches that can model non-linear species–environment relationships and extract components in a more efficient way so as to maximise the covariance with the environmental variable, lead to weighted averaging and partial least squares.

Weighted-Averaging (WA) Regression and Calibration

Weighted-averaging regression and calibration are motivated by the idea that species occupy different niches in environmental space, and that these niches can be characterised by parameters that describe the niche centre (u) and niche breadth (t). If species follow a unimodal distribution in relation to a particular environmental variable then the niche centres and breadths are the optima and tolerances of those distributions. Since species will tend to be most abundant at sites with an environmental value close to its optimum, an estimate of the optimum is obtained by a simple weighted average of the environmental values over the sites where the species is found, so-called WA regression:

$$\hat{u}_k = \sum_{i=1}^{n} y_{ik} x_i \Bigg/ \sum_{i=1}^{n} y_{ik} \qquad (14.6)$$

where $\hat{u}_k$ is the optimum of taxon k, y_{ik} is the abundance of taxon k in sample i, x_i is the environmental variable in sample i, and n is the number of sites. Similarly, a fossil sample will tend to be dominated by taxa whose optima are similar to the environmental conditions that prevailed when that sample was deposited. An intuitive estimate of the past environment $\hat{\mathbf{X}}_0$ is therefore given by a weighted average of the species optima in the sample, so-called WA calibration:

$$\hat{x}_0 = \sum_{k=1}^{m} y_{ik} \hat{u}_k \Bigg/ \sum_{k=1}^{m} y_{ik} \qquad (14.7)$$

where m is the number of taxa. Equations 14.6 and 14.7 present weighted averaging as a multivariate indicator-species approach to environmental reconstruction in which potentially all species have the same indicative value. Observations suggest, however, that some species may be better indicators than others: those with narrow tolerances should be more faithful indicators than those found over a wide range of environmental conditions. Equation 14.7 can be modified to weight taxa by the inverse of their squared tolerance (ter Braak and van Dam 1989), thus effectively down-weighting taxa with broad tolerances (low indicative values).

$$\hat{x}_0 = \sum_{k=1}^{m} \frac{y_{ik}\hat{u}_k}{t_k^2} \Big/ \sum_{k=1}^{m} \frac{y_{ik}}{t_k^2} \tag{14.8}$$

A simple weighted average estimate of the tolerance t for taxon k (= weighted standard deviation) is given by:

$$t_k = \left[\sum_{i=1}^{n} y_{ik}(x_i - \hat{u}_k)^2 \Big/ \sum_{i=1}^{n} y_{ik} \right]^{\frac{1}{2}} \tag{14.9}$$

Note that for an unbiased estimate of the tolerances, t_k in Eq. 14.9 should be divided by $(1-1/N2)^2$, where $N2$ is the effective number of occurrences of the taxon (Hill 1973; Birks 1995; ter Braak and Šmilauer 2002).

Tolerance down-weighting in weighted averaging (WAT) is intuitively reasonable but in practice it is seldom found to improve over simple WA (e.g., Birks et al. 1990a). This conclusion is due, in part, to a bug in early versions of WACALIB (3.5 or earlier). Recent comparisons by ourselves and others (e.g., Köster et al. 2004; Reid 2005) using C2 and a debugged version of WACALIB indicates that tolerance down-weighting can produce moderate improvements over simple WA with some data-sets.

In WA averages are calculated twice. This means that the range of the optima, $\hat{u}$, and consequently the range of the estimated environmental values, $\hat{\mathbf{X}}_i$, is shrunken with respect to the original gradient. To correct for this a simple linear regression is performed to 'deshrink' the original $\hat{\mathbf{X}}_i$ values using either classical regression in which the initial estimated values ($\hat{\mathbf{X}}_i$) from Eqs. 14.7 or 14.8 are regressed on the observed values (x_i):

$$\textbf{initial } \hat{x}_i = b_0 + b_1 x_1;\ \textbf{final } \hat{x}_i = (\textbf{initial } \hat{x}_i - b_0)/b_1 \tag{14.10}$$

or by inverse regression, where x_i are regressed on $\hat{\mathbf{X}}_i$:

$$x_i = b_0 + b_1 \textbf{ initial } \hat{x}_i;\ \textbf{final } \hat{x}_i = b_0 + b_1 \textbf{initial } \hat{x}_i \tag{14.11}$$

Birks et al. (1990a) discuss the relative merits of the two approaches. Inverse regression minimises the overall root mean squared error (RMSE) in the training-set (see below) but the mean squared error properties are not optimal for all x-values (Næs et al. 2002): it gives more accurate predictions for samples that lie close to the mean of the training-set but, because it pulls the inferred values closer to the mean than in classical regression, the inferred values are often over-estimated at low values and under-estimated at high values of x_i (see also Robertson et al. 1999). Classical regression is therefore preferable if accurate predictions are required for samples towards the ends of the training-set gradient (Birks et al. 1990a). The final estimates ($\hat{\mathbf{X}}_i$) from the two methods diverge with increasing distance from the mean. The rate of the divergence is given by $(1 - r^2)$, where r is the Pearson product–moment correlation coefficient between the observed and the inferred values (Draper and Smith 1981). For calibrations with a squared correlation of 0.7–0.8, commonly observed in palaeolimnological training-sets, the difference can be appreciable, especially for samples far from the mean.

WA has gained considerable popularity in palaeolimnology in recent years (Smol 2008). Birks (1995, 2003) and Birks et al. (2010) identify three reasons for this. First, the method is based on statistically sound theory (ter Braak and Barendregt 1986; ter Braak and Looman 1986): despite its apparent simplicity it has been shown to provide a good approximation to the maximum-likelihood calibration of Gaussian response curves described above (ter Braak and Barendregt 1986; ter Braak and Looman 1986), but is computationally much simpler. Second, the method is based on an underlying unimodal response model between species and environment that is predicted by niche theory and the species packing model, and that is frequently observed in real data. Finally, and most compelling, is that the method has good empirical predictive ability and, in comparisons using real and simulated data, performs as well or better than competing methods (e.g., ter Braak and van Dam 1989; Juggins 1992; Juggins et al. 1994; Birks 1995). Unfortunately WA also has three important weaknesses. First, estimates of species optima, and consequently the final inferred values, are sensitive to an uneven distribution of x_i values in the training-set (ter Braak and Looman 1986) although with large data-sets (>400 lakes), WA appears to be quite robust to this distributional sensitivity (Ginn et al. 2007; Telford and Birks 2011a). Second, WA regression and calibration suffers from 'edge effects' that lead to non-linear distortions at the gradient ends. This problem is particularly acute for training-sets with a long (>3 standard deviation units of compositional turnover; see Legendre and Birks 2012b: Chap. 8) and a single dominant gradient, and results in an over-estimation of optima at the low end of the gradient and an under-estimation at the high end. This in turn leads to biases in the predicted values (ter Braak and Juggins 1993). Finally, it is likely that there are additional variables that can influence species distribution in the training-set. The structure that results from these variables may be useful for predicting x_i, but is ignored by WA. These limitations are addressed by an extension of WA to weighted-averaging partial least squares regression (WAPLS).

Partial Least Squares (PLS) and Weighted-Averaging Partial Least Squares Regression and Calibration (WAPLS)

We saw above that one approach to the multicollinearity problem is to reduce, or compress, the original number of predictors into a small number of principal components and then use these in a multiple regression (PCR: Fig. 14.3). We have also seen that this is not an efficient solution as the components are chosen to maximise the variation within the biological data, irrespective of their predictive value for x_i. Thus the information in **Y** that is useful for predicting x_i is likely to be spread over several components, and these components will also likely be 'contaminated' by other sources of variation. This problem is addressed in PLS by using both the biological *and* environmental data to extract the components (Fig. 14.3). Whereas PCR extracts components to maximise *variance* in **Y**, PLS effectively combines the data-reduction and regression steps and extracts components that maximise the *covariance* between x and linear combinations of **Y**. Higher components are extracted using the same criterion but, as with PCA, are orthogonal to early components. PLS usually outperforms PCR, and because the components in PLS are directly related to the variability in x, PLS usually requires fewer components that PCR, resulting in a more parsimonious model (Næs et al. 2002).

A very important part of PLS model-building is choosing an appropriate number of components. It is possible to calculate as many PLS components as there are linearly independent rows or columns in the species data, but usually only a small number are used. This is because the biological and environmental data are never noise-free, and some of the higher components will describe this noise (Geladi and Kowalski 1986). If too many components are used, we start to model these intrinsic features of the training-set and the model becomes over-fitted. In such cases the model may provide an excellent fit to the training-set data but will generally have poor predictive power when applied to new observations (Næs et al. 2002). Conversely, with too few components the model is under-fitted and does not adequately account for the training-set data. If our interest is in model building for data exploration, the number of components can be estimated using an F-test (Geladi and Kowalski 1986), but, because of the danger of over-fitting, the quality of the fit to the training-data cannot be used to choose the number of components that gives the best *predictive* model. To do this the number of components must be chosen on the basis of how well the model predicts observations *not* included in the training-set. In practice this is achieved using some form of cross-validation (see below Model Selection and Evaluation).

By retaining just a few orthogonal components, PLS provides an elegant solution to the multicollinearity problem but, as in PCA, the resulting components are a linear combination of the original biological variables. Like PCR, the method is therefore only likely to perform well over short compositional gradients where we can expect (approximately) linear relationships between taxa and their environment (Birks 1995). In PCR this problem was addressed by Roux (1979) who exploited the ability of correspondence analysis (CA) to model unimodal relationships (ter Braak and

Prentice 1988) and replaced the PCA step in PCR with CA to give correspondence analysis regression (CAR). CAR can be seen as a non-linear, unimodal version of PCR, and, although it usually outperforms the latter (e.g., ter Braak et al. 1993; Birks 1995), it also suffers from the same limitation as PCR in that the axes are not guaranteed to contain information useful for predicting x_i. However, just as PLS extends PCR by selecting components that maximise covariance between x and linear combinations of **Y**, we can derive a technique from CAR that extracts components to maximise the covariance between x and the weighted averages of **Y**. Because the components are weighted averages of the taxon scores, this technique is called weighted-averaging partial least squares (WAPLS: ter Braak and Juggins 1993).

WAPLS combines the attractive features of WA (ability to model unimodal responses) with those of PLS (efficient extraction of components). As with PLS, in WAPLS the optimal number of components is chosen by cross-validation on the basis of the prediction error (see below "Model Selection and Evaluation"). If only one component is retained, WAPLS reduces to WA with an inverse deshrinking regression, with one small difference, namely in WAPLS the deshrinking regression is weighted by the sample total, although for percentage data with constant site totals this weighting is immaterial (ter Braak and Juggins 1993). WA with inverse deshrinking regression can thus also be seen as a component-based method closely related to WAPLS but with only one component (Fig. 14.3). The first WAPLS component is also equivalent to the first axis of a canonical correspondence analysis (CCA) with a single constraining environmental variable. The CCA-based reconstruction method of Stevenson et al. (1989) is therefore equivalent to WA with inverse deshrinking or a one-component WAPLS model.

The above derivation of WAPLS focuses on its similarity with PCR, CAR, and PLS. WAPLS also has a more heuristic description based on its similarity to WA: the first WAPLS component is equivalent to simple WA (with the above qualification). Subsequent components are equivalent to a WA of the *residuals* of x_i estimated from previous components. Each component has an accompanying set of species scores that are combined to update the estimates of the species 'optima' to improve their predictive ability. Note that although the modified coefficients may give a better prediction of x_i they should not be regarded as estimates of ecological optima as they no longer reflect the weighted centroids of species distributions. They are regression coefficients, not species optima.

Ter Braak and Juggins (1993) and ter Braak et al. (1993) compared the performance of WAPLS to WA and maximum-likelihood methods using a variety of simulated and real training-sets. Their results suggest that WAPLS offers substantial improvements over WA and linear PLS for data-sets with long compositional gradients and low noise. However, the advantage of WAPLS decreases with increasing structured and unstructured noise and modest reductions in prediction error of 10–20% are typical for most biological data-sets. With very noisy, species-rich data, WAPLS may fail to improve on WA. Interestingly, even over short gradients where we might expect linear methods to do well, WAPLS usually outperforms PLS, or can equal it with fewer components, and so provide a more parsimonious model (e.g., Seppä et al. 2004). Presumably even short gradients contain some non-linear responses and additional PLS components are required to model these adequately.

Where WAPLS provides a reduction in prediction error over WA, it usually does so for one of three reasons. First, by adjusting the coefficients for species with the 'optima' at the ends of the gradient, WAPLS can help reduce or eliminate the 'edge effects' described above that plague all WA-based models. Second, it is often the case that additional environmental variables influence species distributions, and this influence may result in a structured pattern in the residuals. WAPLS may be able to exploit this structure to improve the estimates of the species parameters in the final calibration function. Third, WAPLS may improve the fit of outliers, or gross errors. These three situations are illustrated in Fig. 14.4 using different data-sets. Characteristics of the data-sets are listed in Table 14.1.

The first example uses an unpublished diatom pH training-set of 96 lakes (Bergen data-set: Birks et al. unpublished). Detrended canonical correspondence analysis (DCCA) (Birks 2012: Chap. 2; Lotter and Anderson 2012: Chap. 18; Simpson and Hall 2012: Chap. 19) indicates that species distributions and abundances are influenced by a single strong compositional or turnover gradient related to lake-water acidity (gradient length = 3.8 standard deviations (SD)) and that secondary gradients have minimal effect (lambda 1/lambda 2 = 1.76, see below). Figure 14.4 shows the relationship between observed and predicted pH for the first WAPLS component. The overall relationship between observed and predicted pH is strong (RMSEP = 0.44, see below "Model Selection and Evaluation") but edge effects are clearly apparent with an over-estimation of lake-water pH in more acid lakes. The second WAPLS component has a cross-validation RMSEP of 0.37 pH units (Table 14.1). This represents a reduction in prediction error of 16% and is achieved primarily by 'straightening out' the trend in the residuals, although this is not perfect and there is still a slight tendency to over-estimate values at the low end of the gradient (Fig. 14.4).

The second example is a diatom training-set developed by Clarke (2001) and used to reconstruct total nitrogen (TN) in Danish coastal waters (Danish data-set: Clarke et al. 2003). DCCA of these data indicate modest diatom turnover along the TN gradient (gradient length = 1.8 SD), and that TN is not the primary environmental variable controlling diatom composition and abundance (lambda 1/lambda 2 = 0.5). Figure 14.4 shows the relationship between observed and predicted TN for WAPLS components 1 and 2 and reveals a clear bias in the residuals reflecting over-estimation at low values and under-estimation at high values. In this example the second component is not able to correct for the bias although it does provide a modest reduction in prediction error of c. 13% from 0.15 to 0.13 $\log_{10}$ μgL^{-1} TN. A clue to how this improvement is achieved can be found by examining the relationship between the WAPLS sample scores and other environmental variables: WAPLS component 1 scores are negatively correlated with water-depth ($r = -0.55$), reflecting the negative correlation between TN and depth in the observed data ($r = -0.43$), but WAPLS component 2 scores are positively correlated with depth ($r = 0.41$). It appears that the second WAPLS component has exploited structure in the residuals related to water-depth and used this to improve

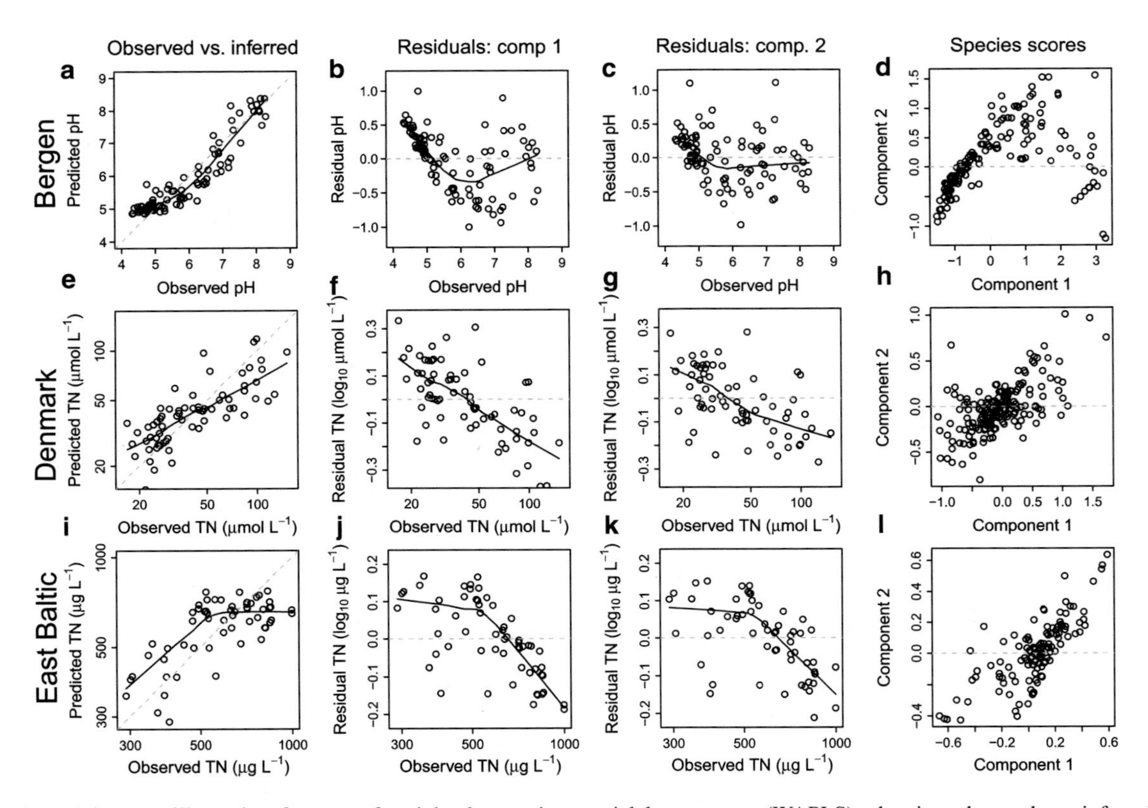

Fig. 14.4 Example training-sets illustrating features of weighted-averaging partial least square (WAPLS), showing observed vs. inferred environmental variables (**a**, **e**, **i**) and residuals for WAPLS components 1 (**b**, **f**, **j**) and 2 (**c**, **g**, **k**) vs. observed values, and species scores plotted on WAPLS components 1 and 2 (**d**, **h**, **l**). In plots a–k, a LOWESS smoother (span = 0.67) has been fitted to highlight trends. The *thin dashed lines* are 1:1 or y-axis origin lines

Table 14.1 Summary statistics of three diatom sets used to illustrate features of weighted-averaging partial least squares (WAPLS)

	Bergen	Denmark	E. Baltic
No. sites	96	67	58
No. taxa	129	180	122
Gradient length	3.8	1.8	1.8
Lambda 1	0.6	0.2	0.2
Lambda 1/lambda 2	1.76	0.5	0.6
RMSEP (WAPLS comp. 1)	0.44	0.15	0.010
RMSEP (WAPLS comp. 2)	0.37	0.13	0.093

Lambda 1 is the eigenvalue of the first and only constrained canonical axis and Lambda 2 is the eigenvalue of the first unconstrained ordination axis in a detrended canonical correspondence analysis (*DCCA*). *RMSEP* root mean squared error of prediction, *comp.* component

(slightly) the TN predictions. The third example is an unpublished diatom-TN data-set also from the Baltic (E. Baltic data-set), created by merging sites in Weckström et al. (2004) from Finland with those in Eastern Sweden analysed by Elinor Andrén. The features of this data-set are similar to the previous set except here the second component only represents a small improvement in RMSEP (7%) which appears to be related to the improved fitting of a few isolated samples rather than any overall systematic improvement in the model (Fig. 14.4). This last example is problematic and probably represents over-fitting, in this case, of a few outliers. Cross-validation should guard against this but it is not always successful and it can sometimes be difficult to differentiate over-fitting from real improvement. We return to this problem below (Model Selection and Evaluation).

The right-hand column of Fig. 14.4 also shows plots of the species scores for WAPLS components 1 and 2. These give a valuable insight into the modifications of the species coefficients that yield the improved fit for component 2. For the Bergen data-set (Fig. 14.4) there is a clear pattern in the component 2 scores that compensate for the over-prediction at low pH with component 1. For the Denmark and East Baltic data-sets there is also an overall trend that attempts to compensate for the bias in the component 1 model but there is much scatter and a number of outlying points: in these situations careful examination is needed to ensure that the scores and the resulting taxon modifications are ecologically sensible.

WAPLS is attractive because it often offers improvements over WA in terms of lower prediction error but, as the above examples illustrate, it does have two disadvantages. The first is that it needs careful model *selection* to avoid over-fitting. The second is that it needs careful model *diagnosis* to understand exactly why higher components improve the fit to the training data. Where higher components use residual structure due to additional environmental variables, it is important that the joint distribution of these variables in the past should be the same as in the modern data-set (ter Braak and Juggins 1993). This aspect of WAPLS model building has so far received little attention and we return to its consequences below in the section on Assumptions and Limitations.

Artificial Neural Networks (ANN)

ANNs are a family of numerical models that learn and predict from a set of data by mimicking the way the human neural network processes information (Birks 2012: Chap. 2; Simpson and Birks 2012: Chap. 9). There are a number of different types of ANNs: for calibration a feed-forward network and back-propagation learning algorithm is usually used (Næs et al. 1993, 2002). A feed-forward network consists of a set of interconnected processing units, or neurons, arranged in three layers. The input layer takes data from the input variables and feeds it to one or more intermediate or hidden layers where it is combined into the output layer to give the output variable (Fig. 14.3). The variables represented by the hidden and output layers are non-linear functions of their inputs, and thus the network is able to model the output as an arbitrary non-linear function of the inputs. Basheer and Hajmeer (2000) describe the design and training of ANNs in more detail and Næs et al. (2002) discuss their use in calibration. ANNs have been frequently used for reconstruction in palaeoceanography (e.g., Malmgren et al. 2001; Kucera et al. 2005) and, more rarely, in palaeolimnology (e.g., Racca et al. 2001, 2003, 2004).

ANNs have one main advantage over existing methods: namely that no prior assumptions about the relationship between species and environment are needed. ANNs can model any arbitrary mix of linear and non-linear responses. However, they also have several disadvantages. First, it is difficult to interpret network coefficients in any ecologically meaningful way – the ANN is essentially a 'black-box' predictor (Olden and Jackson 2002; Olden et al. 2004). Second, and most important, ANNs are very flexible functions and with large numbers of taxa they are very prone to over-fitting (Simpson and Birks 2012: Chap. 9). Careful cross-validation is critical: Næs et al. (2002), Telford et al. (2004), and Telford and Birks (2005) recommend the use of a separate optimisation data-set and an independent test-set, although this is seldom available in palaeolimnology (see below "Model Selection and Evaluation"). Despite these limitations ANNs show some promise and may be useful in data-sets showing a variety of linear and unimodal species responses (Köster et al. 2004).

Modern Analogue Technique (MAT)

All methods of quantitative reconstruction described in this chapter invoke the principle of uniformitarianism, namely the use of modern organism–environment relationships as a model for interpreting fossil assemblages (Birks et al. 2010). Under this principle the simplest and most intuitive approach to environmental reconstruction is a direct space-for-time substitution: if we assume that similar biological assemblages are deposited under similar environmental conditions, it follows that the environment of samples in the modern training-set that have similar species composition to a fossil sample can be used as a direct estimate of the environment of that fossil sample (Jackson and Williams 2004; Simpson 2012: Chap. 15). Where there is more than one close match in the training-set we can take an average, or consensus, of the k most similar samples (Fig. 14.3). In the pattern-recognition literature this technique is most often used to solve discrimination problems and goes under the name of k-nearest neighbours (k-NN: ter Braak 1995; Webb 2002). In palaeoecology it is known as the modern analogue technique (MAT: Prell 1985; Simpson 2007, 2012: Chap. 15).

The starting point for MAT is to calculate a measure of dissimilarity between each fossil sample and each sample in the modern training-set. For relative abundance biological data, squared chord distance (= squared Hellinger distance in Legendre and Birks 2012b: Chap. 8) is often used and although it has been criticised as a measure of compositional dissimilarity (Faith et al. 1987), it possesses good signal-to-noise properties (Prentice 1980; Overpeck et al. 1985) and generally provides a good approximation of ecological similarity at the small distances important in MAT (Gavin et al. 2003). The squared chord distance between samples i and j is given by:

$$d_{ij} = \sum_{k=1}^{m} \left(p_{ik}^{1/2} - p_{jk}^{1/2} \right)^2 \tag{14.12}$$

where p_{ik} is the proportion of taxon k in sample i. Birks et al. (1990a) use the related squared χ^2 distance, which Bennett and Humphry (1995) argue performs better with certain data-sets, although there is usually very little difference in practice.

The matrix of dissimilarities is then searched to find the modern sample(s) with the smallest dissimilarities for each fossil sample. The environmental reconstruction can be based on the single most similar modern sample but it is usual to take the mean or weighted mean of the k closest matches. The use of a weighted mean follows from the assumption that modern samples that are more similar in biological composition to a fossil sample are also likely to be more similar in environment. Thus the final estimate of $\hat{\mathbf{X}}_0$ is given by a weighted mean of the k nearest analogues, using the inverse of d as weights:

$$\hat{x}_0 = \sum_{i=1}^{k} \frac{x_i}{d_{0i}} \Bigg/ \sum_{i=1}^{k} \frac{1}{d_{0i}} \tag{14.13}$$

MAT is widely used in palaeoceanography (e.g., Kucera et al. 2005) and palynology (e.g., Davis et al. 2003) but is rarely employed by palaeolimnologists (Birks 1998; Simpson 2007). The main reason for this is that MAT generally requires a large training-set. The other methods considered above fit a single species–environment model to the whole training-set and are apparently able to generalise relationships, often with only a modest number of samples (e.g., Bennion 1994). MAT is a form of inverse non-parametric regression via smoothing (ter Braak 1995) and rather than fitting a single *global* model it proceeds by fitting a *local* model to small subsets of the training data for each fossil sample. For MAT to be effective we thus require reasonable coverage of samples in local space, which implies a well-populated network of samples across the environmental gradient(s) of interest (Gonzales et al. 2009).

The choice of k, the number of samples in the local model, is somewhat arbitrary: values of 5–10 are usual, depending on the size and diversity of the training-set. The choice can be important as a small number of analogues will tend to produce noisy or 'spiky' reconstructions whereas a larger number will damp out fine-scale variation and produce a 'flatter' profile. A plot of prediction error (under cross-validation) for different values of k can provide a more objective guide for a particular training-set (Simpson 2012: Chap. 15).

Because MAT depends on finding analogous modern samples, it can be expected to yield unreliable reconstructions when good modern analogues do not exist. This of course begs the question of what constitutes a 'good analogue' (Simpson 2012: Chap. 15). The definition is complex and involves total taxonomic composition, the relative abundance of dominants, the identification of specialists and generalists, and life-form and functional types. These attributes are not easy to measure and even more difficult to encapsulate in a single dissimilarity measure. Extreme non-analogue situations with few taxa in common between modern and fossil samples are easy to identify from a comparison of species lists. The identification of good analogues is more difficult and we return to this question below when we discuss evaluating and validating the reconstructions. MAT may also yield unreliable reconstructions in the presence of multiple analogues, that is, similar modern biological assemblages with different environmental values (Birks et al. 2010). This situation is easily identified by inspecting the range or weighted standard deviation of x_i (weighted by $1/d$) among the selected analogues. Indeed, high variability among the analogues is a useful diagnostic in highlighting problematic cases (ter Braak 1995) or in suggesting that the biology is not actually responsive to the environmental variable, at least in the region of the gradient represented by the fossil sample.

Despite these limitations, comparisons (e.g., Juggins et al. 1994; Paterson et al. 2002a) suggest that when large training-sets of several hundred samples are available there is often little difference in prediction error between MAT and WA-based methods. Even with moderately-sized training-sets MAT can provide a useful alternative reconstruction for comparison with those from other methods (see below). Simpson (2007, 2012: Chap. 15) discusses MAT in further detail and explains its use in the interpretation of biostratigraphical data and in lake-restoration applications.

Locally-Weighted Weighted-Averaging (LWWA) Regression and Calibration

Most of the techniques described above generalise species–environment relationships well from moderately sized data-sets but perform less well with large heterogeneous training-sets, because of the effects of secondary gradients and other sources of noise. MAT, on the other hand, works well with larger data-sets because it can model local relationships but can produce 'noisy' reconstructions because it models too much local structure in the training-set. Locally-weighted weighted-averaging seeks to exploit the best features of both methods and selects a local training-set of size *k* for a fossil sample using the distance criteria of MAT. WA is then used to develop a reconstruction based on this local training-set and the process repeated for each fossil sample. The size of *k*, as with MAT, can be determined by cross-validation. Our unpublished experiments suggest a value of 30–50 is appropriate. The technique is 'locally-weighted' because in the WA part, training-set samples have weight 1 or 0: other non-zero weights could be applied to differentially down-weight more distance samples (e.g., Næs and Isaksson 1992; Næs et al. 2002). Similarly, the WA part could be replaced with other methods such as WAPLS or GLR-ML (Hübener et al. 2008).

LWWA creates a dynamic training-set that is tailored to each fossil sample (Birks 1998). Our unpublished and published comparisons (e.g., Battarbee et al. 2005b; Huber et al. 2010) with large merged data-sets (e.g., European Diatom Database Initiative (EDDI): Battarbee et al. 2000) suggest that it can perform as well as traditional methods applied to smaller regional data-sets. That is, it can exploit the advantage of increased environmental and biological coverage given by very large training-sets without suffering the disadvantage of increased prediction error. More comparisons are needed to evaluate fully the method but initial results suggest that it provides a useful way to exploit large, environmentally diverse training-sets.

Bayesian Methods

The methods described above have different statistical and ecological motivations but they have one thing in common: they are all so-called frequentist methods and make the assumption that the model parameters (e.g., WA optima, WAPLS coefficients, etc.) are fixed and can be estimated from observations (the measured data) distributed randomly about the fitted values (Holden et al. 2008). Conversely, a Bayesian approach does not rely on an explicit model of the relationships between species and environment but assumes that the model is unknown and to be estimated from the measured data which are fixed. Specifically, the Bayesian approach uses measured information to modify some prior belief about the environmental values (Robertson et al. 1999). This additional information is derived from a training-set and expressed as a conditional probability density function, which is combined with the prior probability density function to give a posterior density function using Bayes theorem.

Bayesian approaches have been applied to several palaeoecological problems: Toivonen et al. (2001) describe a Bayesian model with a conditional probability density function based on a unimodal model. Vasko et al. (2000) and Korhola et al. (2002) extend this to include a more realistic multinomial Gaussian response model and apply it to chironomid-based temperature reconstructions. Haslett et al.

(2006) further develop these ideas to more generalised modelling of pollen-climate response surfaces, and Holden et al. (2008) describe a computationally efficient approach based on probability weighting of species response curves. Li et al. (2010) develop a Bayesian hierarchical model to reconstruct past temperatures that integrates information from different sources such as climate proxies with different temporal resolution.

Although the prediction errors for Bayesian methods are of similar order to conventional approaches they have a major advantage in their coherent and explicit handling of uncertainty (Birks et al. 2010; Li et al. 2010). However, the lack of available software and the huge computational burden of most existing models (days to weeks for a single reconstruction) currently prevent more widespread use.

Model Selection and Evaluation

Table 14.2 summarises the advantages and disadvantages of the numerical methods described above. One key conclusion is that there is no single 'best' or 'optimal' method that can be recommended: differences in training-set size, taxonomic diversity, and form and complexity of the species–environment relationship make some numerical techniques more useful than others for particular data-sets. Model selection thus becomes a crucial step in any quantitative reconstruction (Xu et al. 2010).

The basic requirement for a reconstruction model is that it is statistically significant, accurate and precise, reliable, and makes ecological and palaeoecological sense. Model accuracy and precision are measured by the prediction error but because we usually do not know the value of x in the past, we cannot directly calculate the true error of the predictions for a particular fossil sequence. Instead we estimate the prediction error for the modern training-set and assume that this reflects the true prediction error. Prediction error is usually calculated as the root mean squared error (RMSE), defined as the square root of the mean squared differences between the observed and inferred environmental values (Wallach and Goffinet 1989; Power 1993):

$$\text{RMSE} = \left[\frac{1}{n} \sum_{i=1}^{n} (x_i - \hat{x}_i)^2 \right]^{\frac{1}{2}} \tag{14.14}$$

The RMSE provides a useful overall summary of the model's predictive ability and has the advantage that it is given in the same units as the original environmental values (Næs et al. 2002). The correlation (r) and/or coefficient of determination (r^2) between x_i and $\hat{x}_i$ are also often calculated. They can be useful in comparing models for different environmental variables but because they are dependent on both the magnitude of the model error *and* the variation in x_i and $\hat{x}_i$ they should be interpreted with care: models with the same prediction errors can have very different

Table 14.2 Advantages and disadvantages of numerical methods described in the text

Method	Advantages	Disadvantages
Simple two-way weighted averaging (WA)	Computationally simple, good underlying statistical theory and ecologically plausible; consistently performs well in comparisons and tests and is generally well-behaved in non-analogue situations, zero values ignored	Suffers from edge-effects (bias at gradient ends); the somewhat arbitrary choice of classical or inverse deshrinking can have a large effect on the final predictions; disregards residual correlations in the biological data (see text)
Tolerance down-weighted WA (WAT)	Giving taxa with narrow tolerances greater weight is ecologically realistic and can improve predictions in some cases	Can extrapolate if fossil assemblages contain taxa that are rare and have narrow tolerances in the training-set
Gaussian logit regression and maximum likelihood calibration (GLR-ML)	Statistically rigorous and ecologically realistic based on explicit species response model; good empirical predictive power	Curve fitting sometimes fails with low numbers of observations; susceptible to outliers in species data; zero values influential
Principal components regression (PCR), including Imbrie and Kipp Factor Analysis (IKFA)	Based on a linear species response model which may be appropriate for short environmental gradients or non-biological data	Extraction of components is inefficient for prediction: PLS or WAPLS will always provide a better fit with fewer components
Partial least squares regression (PLS)	Based on a linear species response model which may be appropriate for short environmental gradients or non-biological data; efficient extraction of components for prediction	Cross-validation and careful diagnosis are needed to select appropriate model complexity (number of components); prone to extrapolation under non-analogue conditions, and WAPLS will usually provide a better fit with fewer components
Weighted-averaging partial least squares regression (WAPLS)	Underlying unimodal species response model, with additional components extracted to maximise covariance with environmental variable; good empirical predictive power and improves over WA with many data-sets	Cross-validation and careful diagnosis are needed to select appropriate model complexity and check that model is ecologically plausible; can extrapolate under non-analogue conditions

(continued)

Table 14.2 (continued)

Method	Advantages	Disadvantages
Modern analogue technique (MAT)	Ecologically plausible; able to model local species–environment relationships in large training-sets	Needs large training-sets and careful choice of number of analogues. Tends to produce spiky reconstructions with small number of analogues and flat profiles with larger number of analogues
Locally-weighted weighted averaging (LWWA)	Ecologically plausible; for large heterogeneous data-sets it can provide a good compromise between local modelling of MAT and global modelling of WA; good empirical predictive power with large data-sets	Needs large training-sets and careful choice of number of samples in local training-set; resulting model more difficult to interpret

values of *r*, depending on the length of the environmental gradient spanned by the training-set (Birks 1995).

Measures of model performance derived from the training-set are almost certainly under-estimates because they are essentially an estimate of the model error and *not* the prediction error (Næs et al. 2002). A more realistic estimate of the prediction error when the model is applied to new data is provided by some form of cross-validation. Ideally this would involve testing the model using an independent test data-set in a so-called *external* cross-validation (Birks 2012: Chap. 2). However, we rarely have the luxury of an additional test data-set – these samples could be used more profitably in the training-set (though see Lotter et al. 1999 for an interesting approach). Instead we use *internal* cross-validation of the training-set to simulate the likely errors when the model is applied to new data. There are three main types of internal cross-validation available; namely *k*-fold leave-out, leave-one-out, and bootstrapping. In *k*-fold leave-out the order of the samples is randomised and a fixed proportion (e.g., 20%) is left out of the training-set in turn (giving, in this case, five-fold leave-out). The calibration function based on the remaining samples is then applied to the left-out samples, which act as the test-set, and the squared errors accumulated to form the RMSE of prediction (RMSEP). Note that the RMSEP is distinguished from the so-called apparent RMSE calculated solely from the training-set. Leave-one-out is a special case of *k*-fold leave-out where each sample in turn is left out to form a single sample test-set (Manly 1997). Larger training-sets usually include some inherent replication, so as training-set size increases, leave-one-out RMSEP becomes a less reliable estimate of true prediction error (Næs et al. 2002). *K*-fold leave-out provides a more rigorous test but can be overly pessimistic because the training-sets in each step are of a correspondingly smaller size (Webb 2002). This problem is addressed in bootstrap cross-validation by selecting samples from the original training-set (of size *n*) with replacement to give a new bootstrap training-set also of size *n* (Stine 1990; Mooney and Duval 1993; Manly 1997).

Since samples are selected with replacement, some will be included more than once, and, on average, approximately one third will remain unselected. These become the bootstrap test data-set. This procedure is repeated a large number of times (typically 1000) and the squared errors for the test samples accumulated across all bootstrap cycles into the bootstrap RMSEP (Birks et al. 1990a; Simpson and Birks 2012: Chap. 9).

As a measure of model performance the RMSEP incorporates both random and systematic components of the error, represented by the standard error of the predicted residuals $(x_i - \hat{x}_i)$ (SEP) and the mean bias (MB) or mean of $(x_i - \hat{x}_i)$, respectively (Birks et al. 1990a). SEP and MB are related to the RMSEP by:

$$\mathrm{RMSEP}^2 = \mathrm{SEP}^2 + \mathrm{MB}^2 \tag{14.15}$$

In practice the overall mean bias calculated for the training-set is usually close to zero but models often show a tendency to over- or under-estimate along particular parts of the gradient. This form of systematic error is quantified by the maximum bias, calculated by subdividing the gradient into a number (usually 10) of equal-spaced segments and calculating the mean bias for each segment (ter Braak and Juggins 1993). The maximum bias is the largest absolute value of the mean bias for the ten segments. These types of systematic error are also easily detected using graphical methods and inspection of the plot of either $\hat{x}_i$ or residuals $(x_i - \hat{x}_i)$ versus x_i is an important tool in model validation. Observations should fall close to the 1:1 line and show no trend in the residual plot (e.g., Fig. 14.4). Racca and Prairie (2004) (see also Piñeiro et al. 2008) have argued that instead of plotting residuals against x_i they should be plotted against $\hat{x}_i$ so that we evaluate any bias in the residuals relative to the model's predictions, not the original observations. When this is done the bias observed in many inverse regression models disappears, leading to the situation in which a model that is clearly biased when evaluated against observed values apparently produces unbiased predictions (e.g., Cameron et al. 1999; Simpson 2012: Chap. 15). This paradox is explained by the fact that a plot of residuals against observed values reveals overall, or external, model bias when the model is asked to predict reality, whereas the plot of residuals against predictions only tells us about potential biases within the model itself, assuming that the model is correct. Because the analysis of prediction bias under this assumption fails to consider overall model bias, we believe it is not a useful tool for model evaluation. Instead we prefer to plot residuals against observed values.

Low prediction error and maximum bias are useful criteria to help discriminate between different candidate models. In some cases there may be an obvious winner but often the difference between models is small. In this situation the principle of parsimony in statistical modelling (Birks 2012: Chap. 2) dictates that we choose the 'minimal adequate model', that is, the one with the fewest parameters (Crawley 2005). This is especially important in identifying appropriate model complexity (i.e., number of components) in PLS or WAPLS to prevent over-fitting (see above). In WAPLS a plot of the prediction error against component number usually shows a reduction in error after the first few components, then stabilises around the optimal number before it increases as we start to over-fit. The number of components that produces the minimum prediction error is not always well defined, and where a model with fewer components gives a similar prediction error, it is preferable to accept this as the most parsimonious solution. Birks (1998) suggests a reduction in RMSEP of at least 5% is needed for a component to be considered useful.

Figure 14.5 shows the change in RMSEP and maximum bias with increasing WAPLS components for the three data-sets described in the section Numerical Methods. The Bergen data-set shows a clear reduction in cross-validation RMSEP of 16% between WAPLS components 1 and 2 and a subsequent gradual rise for higher components. Maximum bias shows a similar large reduction for component 2 and a further modest drop for component 3. The residual plot also shows an overall systematic improvement between components 1 and 2 (Fig. 14.5). Taken together these results indicate that a 2-component WAPLS model is appropriate. The Danish data-set shows a similar reduction of RMSEP of 13% for component 2 and a further modest reduction of 3% for component 3. However, the improvement for component 3 is less than the rule-of-thumb threshold of 5% so we select the 2-component WAPLS model. In the East Baltic example there is a reduction of RMSEP of 7% for component 2, which although modest is above the rule-of-thumb threshold. There is also a large reduction in maximum bias between components 1 and 2 so it is very tempting to select the 2-component WAPLS model. However, as we saw above, the reduction in prediction error appears to be caused by the improved fitting of a few individual samples rather than to an overall improvement in model performance. In this case, model selection is difficult, and the arbitrary threshold of 5% does not guard against selecting an over-fitted model. A simple solution, and one that has rarely been applied to the problem of model selection in palaeolimnology, is to use a randomisation t-test to test the equality of predictions from two models (van der Voet 1994). Using the test to compare WAPLS components 1 and 2 yields p-values of <0.001, 0.009, and 0.091 for the Bergen, Denmark, and East Baltic data-sets, respectively, indicating that for Bergen and Denmark a 2-component WAPLS model has significantly different (lower) RMSEP than the 1-component WAPLS model (999 randomisations). The 3-component Danish model is not significant ($p = 0.241$). The test also indicates that the second WAPLS component for the East Baltic data-set is not significant despite its lower RMSEP. The latter example in particular illustrates that a simple comparison of RMSEP may lead to the selection of an inappropriate model and lead to a correspondingly over-optimistic impression of the prediction error. In this case the randomisation t-test provides valuable additional information that can help to discriminate 'hidden' over-fitting from real systematic model improvement. Other palaeolimnological uses of this test include Racca et al. (2001), Adler et al. (2010), and Velle et al. (2011a).

Current practice in palaeolimnology is to use the cross-validation RMSEP in both model optimisation, that is to determine the number of WAPLS components, MAT analogues, or ANN hidden neurons, etc., and as a measure of model performance. However, as Telford et al. (2004) and Telford and Birks (2005, 2009) point out, in this situation the RMSEP is not independent of model choice. For an unbiased estimate of RMSEP we should ideally use separate optimisation and test data-sets

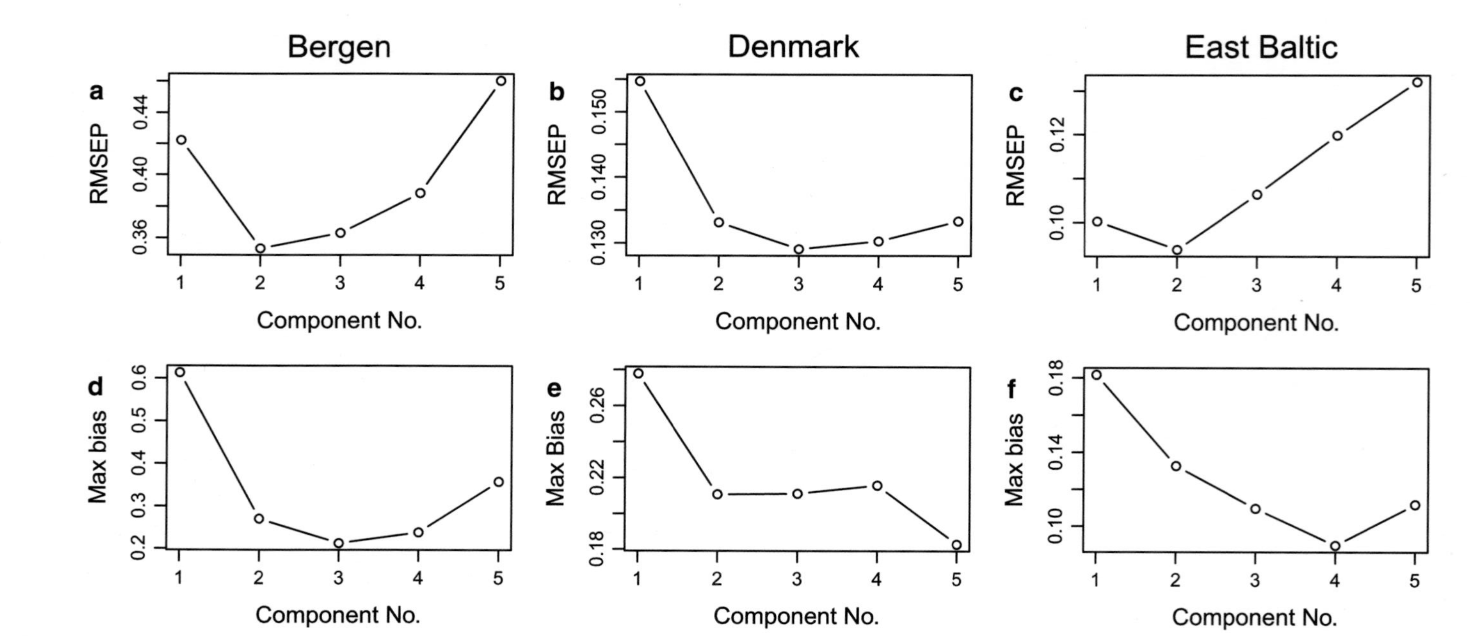

Fig. 14.5 Root mean squared error of prediction (RMSEP) (**a–c**) and maximum bias (**d–f**) vs. weighted-averaging partial least squares (WAPLS) component number for Bergen, Denmark, and East Baltic modern data-sets. See text for details

(Webb 2002). Where these are lacking, double cross-validation (CV) provides a more robust way of calculating prediction error. In double CV the data are split into test- and calibration-sets in an outer CV loop, and each calibration-set is then split into a training- and test-set in an inner CV loop. The inner loop is used to optimise the model (e.g., select number of components) and the outer loop used to estimate the RMSEP. This strategy is particularly useful for optimising component-based models (Varmuza and Filzmoser 2009) but has so far not been used in palaeolimnology.

In addition to good empirical performance measured by RMSEP we also require that our reconstruction model be reliable and robust, that is it will show a similar performance when applied to fossil data. Where historical measurements are available, model predictions for well-dated recent samples can be compared against time-series of environmental measurements (Cumming et al. 2012: Chap. 20). Such comparisons are usually qualitative, and can be used to validate a model (e.g., Fritz 1990; Bradshaw and Anderson 2001; Larocque and Hall 2003; Kamenik and Schmidt 2005), to reveal problems (e.g., Fritz et al. 1993; Sayer 2001; Battarbee et al. 2002; Bigler and Hall 2003), to guide model selection (e.g., Köster et al. 2004), or even to develop a within-site calibration-in-time model (e.g., Larocque-Tobler et al. 2011). The problems of a within-lake calibration-in-time model are that because of temporal autocorrelation, it is unclear how many independent n observations there are in estimating RMSEP or r^2 and how applicable the single within-lake model is to other lakes.

Parsimony can guide choice among nested models (e.g., WAPLS models with different numbers of components) but does not help in discriminating among different families of models (e.g., WAPLS, MAT, GLR-ML). In this case, where a number of models all perform well, we suggest applying them all, unless there are good theoretical grounds for selecting only one or two (see below). Table 14.3 lists some errors that can arise (and have arisen) in developing modern calibration data-sets in palaeolimnology and suggests possible solutions (Birks et al. 2010). Table 14.4 summarises some problems that can arise in palaeoenvironmental reconstruction from fossil assemblages, how the problems can be detected, and how they can be solved (Birks et al. 2010).

Spatial Autocorrelation and Environmental Reconstruction

The estimation of the predictive power and performance of a training-set in terms of RMSEP, r^2, mean bias, maximum bias, etc. (Birks 1995) by any form of cross-validation assumes that the test-set (one or many samples) is statistically independent of the training-set whose performance is being evaluated. Cross-validation in the presence of spatial autocorrelation seriously violates this assumption (Telford and Birks 2005, 2009). Positive spatial autocorrelation is the tendency of samples close to each other geographically to resemble one another more than randomly selected samples (Shurin et al. 2009). Using a large foraminiferal training-set from

Table 14.3 Some errors that can arise (and have arisen) in developing modern organism–environment calibration data-sets and possible solutions (from Birks et al. 2010)

Errors	Possible solutions
Modern data from different sedimentary environments with contrasting taphonomies	Data screening
Not all samples are modern samples	Data screening and examination of meta-data
In very large data-sets, same samples may be duplicated by accident	Data screening
Inconsistent taxonomy or low taxonomic resolution	Analytical quality control and recounting
Inconsistent and/or poor environmental data	Data screening, new data
Model over-fitting solely to minimise RMSEP	van der Voet (1994) randomisation test, more rigorous cross-validation, graphical plots of RMSEP, etc.
Ignoring spatial autocorrelation in assessing model performance	Telford and Birks (2009) deletion and *h*-block methods
No estimates of sample-specific errors of reconstruction	Bootstrapping
Model performance not based on cross-validation. Apparent statistics only	Use cross-validation (leave-one-out, split sampling, *n*-fold cross-validation)
No numerical evaluation of reconstruction	Apply reconstruction diagnostic measures and statistical tests of reconstruction significance
No validation of reconstruction using independent records	Not always possible but literature or internet searches can be useful
Reconstructing two or more variables that may, in reality, not be reconstructable	Careful use of partial constrained ordinations, consider not only marginal effects but also conditional effects, simulations, randomisation, and significance tests

RMSEP root mean squared error of prediction

the North Atlantic and a test-set 10% of the size of the full data-set for cross-validation, Telford and Birks (2005) compared the RMSEP in GLR-ML, WA, WAPLS, MAT, and ANN models. They found that the lowest RMSEPs for sea-surface temperature (SST) were in the MAT, ANN, and WALPS models. If the South Atlantic data-set was used as a test-set, where there can be no spatial autocorrelation with the North Atlantic training-set, the lowest RMSEPs are produced for GLR-ML and WA. These results are interpreted as the result of spatial autocorrelation resulting in the apparently superior performance of ANN and MAT in the North Atlantic, both of which involve local non-parametric estimation, when applied to a test-set *within* the same geographical areas as the training-set. In contrast, WA and GLR-ML can only model the variation in the foraminiferal assemblages that is correlated with SST. They involve global parametric estimation and are therefore much more robust to the spatial structure in the data, and therefore perform best with a spatially independent test-set, namely the South Atlantic. Is spatial autocorrelation an important problem in palaeolimnology?

Table 14.4 Some problems that can arise in environmental reconstructions from biological assemblage data, the detection of these problems, and potential solutions (from Birks et al. 2010)

Problem	Detection	Potential solution
High amount of noise in reconstruction	Visual examination of time series; calculation of measures of dispersion for time series	Smoothing using locally weighted regression (Birks 2012: Chap. 2); use simpler model to reduce over-fitting
Systematic bias	Comparison with other quantitative reconstructions	Express reconstructed values as differences from mean or modern reconstructed values; use WA with classical deshrinking
Non-analogue situations	Reconstruction diagnostic statistics and evaluation methods	None; try WA or WAPLS
Variable other than the one of interest driving the biotic changes	Comparison with independent physical, chemical, or biological proxies	None
Poor model performance statistics	High RMSEP, bias, etc.	Careful numerical and ecological examination of data
Large sample-specific errors of reconstruction	Large error bars	Examine reconstruction diagnostics and test for non-analogue assemblages

WA weighted averaging, *WAPLS* weighted-averaging partial least squares, *RMSEP* root mean squared error of prediction

Telford and Birks (2009) examined several training-sets including a large diatom-pH data-set from the north-eastern US (Dixit et al. 1999). They developed a simple test to detect spatial autocorrelation by (1) deleting samples at random and deriving a new calibration-function and estimating its performance statistics and (2) deleting sites geographically close to the test sample and deriving a new calibration-function and its performance statistics based on the remaining samples. If strong spatial autocorrelation is present, deleting geographically close sites will preferentially delete the environmentally closest sites. With autocorrelated data these will bias the apparent 'good' performance statistics of the calibration-function, and their deletion should drastically decrease the performance of the calibration-function. In contrast, random deletion should have much less effect on the performance of the calibration-function. This test showed no difference in calibration-function performance between the randomly-deleted and neighbour-deleted diatom data-sets, suggesting that spatial autocorrelation is not a problem in this diatom-pH data-set. There is similarly little evidence for significant autocorrelation in a modern Norwegian-Svalbard chironomid-environmental calibration-set from over 150 lakes (RJ Telford, SJ Brooks, and HJB Birks, unpublished results) or in several small intra-lake chironomid calibration-sets (Velle et al. 2011b).

Telford and Birks (2009) have also developed a method for cross-validating a calibration function in the presence of spatial autocorrelation. The method is based on *h*-block cross-validation where a test sample is deleted from the training-set along with *h* observations within a certain radius of the test sample. Not surprisingly, there is no difference in the RMSEP between conventional leave-one-out cross-validation and *h*-block cross-validation with the diatom-pH data-set as there is no spatial autocorrelation in these data. On the other hand, there is a large increase in RMSEP in *h*-block cross-validation for foraminifera-SST data-sets and for pollen-climate data-sets where there is high spatial autocorrelation in the environmental data. Palaeolimnologists should be aware of the problems of spatial autocorrelation, especially in studies of within-lake variation (e.g., Heiri et al. 2003) and in the derivation of within-lake calibration-functions for variables such as water-depth and distance to littoral vegetation in medium or large lakes (e.g., Luoto 2010). They should test for spatial autocorrelation and, if present, they should use appropriate techniques like *h*-block cross-validation to establish robust calibration-function performance statistics in the presence of autocorrelation.

Reconstruction Testing, Evaluation, and Validation

All numerical procedures will produce a quantitative reconstruction when given a training-set and a fossil biostratigraphy that have at least some taxa in common. It is therefore crucial to be able to evaluate the reliability of the reconstructed values. To paraphrase the statistician GPE Box, "all reconstructions are wrong, but some reconstructions may be useful". The challenge is to identify the useful ones! This is a difficult task and, as Birks (1995, 1998, 2003) points out, one that has so far received sparse attention until recently.

Usually we require a reconstruction to be statistically significant, accurate, and precise. That is, it should explain more of the variance in the fossil data-set than most reconstructions derived from calibration-functions trained on random environmental data (Telford and Birks 2011b; Birks et al. 2012), it should be an accurate reflection of temporal trends, and it should provide an accurate and precise estimate of the absolute values of the reconstructed variable. In some cases the identification of major trends and change points may provide valuable information, for example in studies of lake ontogeny (e.g., Fritz et al. 2004). In other cases the absolute values of x may be essential, for example in data-model comparisons or identification of pre-disturbance conditions. Thus the definition of 'reliability' may vary with context.

It is also useful to distinguish validation of the reconstruction from evaluation of the reconstruction. Validation requires comparison of the reconstructed values against historical measurements (cf. Meyer and Butler 1993). Unfortunately the latter are rarely available – if they were, we would not need the reconstructed values in the first place. Comparison with historical data does, of course, provide a powerful tool for *model* validation and selection (see above and Cumming et al. 2012: Chap. 20). But while good agreement between reconstructions and historical data may increase our confidence in the model it does not validate other reconstructions produced by the same model.

Given that true validation of a reconstruction is rarely achievable and probably impossible for periods beyond the recent past, we must resort to methods that provide an indirect and more qualitative assessment of reliability. There are four approaches currently available: (1) RMSEP and sample-specific errors for each fossil sample, (2) numerical 'goodness-of-fit' or 'analogue' measures, (3) comparison of reconstructions produced using different numerical procedures, and (4) comparison of reconstructions derived from different proxies. We evaluate each of these below. However, we will consider first how to assess the overall statistical significance of a quantitative reconstruction.

Assessing the Statistical Significance of a Quantitative Reconstruction

As it is always possible to obtain quantitative results for a reconstruction of any environmental variable, regardless of its ecological relevance or significance, some global test of reconstruction utility is needed. Telford and Birks (2011b) and Birks et al. (2011) propose that a single down-core reconstruction as a whole should explain more of the variation in the fossil data in a constrained or canonical ordination (see Legendre and Birks 2012b: Chap. 8) than a calibration-function trained on random environmental data applied to the same fossil data (Telford and Birks 2011b) or a reconstruction performed using randomised species coefficients (Birks et al. 2012). Multiple independent reconstructions should *each* explain more of the variation in the fossil data than a random variable after the other reconstructions have been partialled out statistically as covariables (see Birks 2012: Chap. 2) (Telford and Birks 2011b).

Telford and Birks' (2011b) procedure, available in the R package palaeoSig (Telford 2011), involves estimating the proportion of the variance in the fossil data explained by the reconstruction derived from any numerical calibration procedure discussed in this chapter using redundancy analysis (RDA) (see Borcard et al. 2011; Legendre and Birks 2012b: Chap. 8) as compositional turnover is low in the majority of fossil palaeolimnological data-sets (<1.5–2.0 standard deviations). Then, using the biological part of the modern training-set used for the reconstruction, environmental reconstructions (usually 999) are generated from calibration-functions trained on random environmental variables drawn from a uniform distribution. The proportion of the variance explained by these random-based distributions is estimated by RDA. If the reconstructed environmental variable based on the training-set with the observed environmental variable explains more of the variance than 95% of the random reconstructions, the reconstruction is considered to be statistically significant ($\alpha = 0.05$). The proportion of the variance explained by the first axis of a principal component analysis (see Borcard et al.

2011; Legendre and Birks 2012b: Chap. 8) is also estimated, as this represents the maximum proportion of the variance in the fossil data that any single reconstruction or underlying latent variable could possibly explain (Telford and Birks 2011b).

If there are multiple reconstructions from the same fossil data, a forward selection procedure (see Birks 2012: Chap. 2) is adopted. First, the reconstruction that explains the most variance is accepted and entered. Then the other reconstructions are tested to determine if they explain significantly more variance than the random reconstructions when the first accepted reconstruction is partialled out (Telford and Birks 2011b). This selection procedure is repeated until no significant reconstructions remain.

Telford and Birks (2011b) used 999 random environmental variables to produce the null distribution for comparison with the observed reconstruction. Statistical significance values are estimated as the fraction of random variables that explain as much as or more of the variance in the fossil data than the observed environmental variable.

This procedure indicates that diatom-based pH reconstructions and chironomid-based late-glacial temperature reconstructions are statistically significant, but that many chironomid-based reconstructions of within-lake variables such as water-depth are not statistically significant (see also Velle et al. 2011b). Such reconstructions may, however, be palaeolimnologically significant and we now consider ways of evaluating and validating reconstructions, whether they be statistically significant or not, as they may still be useful palaeolimnologically.

Birks et al. (2012) propose a similar procedure to test the significance of diatom-inferred climate and lake-water chemistry reconstructions in a late-glacial sequence from northern Norway but instead of using multiple random environmental data-sets to generate the null distribution they permute the species coefficients from the original calibration-function and perform the reconstruction using original but randomised coefficients. In this case, results indicate that the diatom-pH reconstruction is statistically significant but that reconstructions of ice-free period are not.

RMSEP and Sample-Specific Error Estimates

Estimating the uncertainty of individual reconstructed values is a useful first step: reconstructed values with low uncertainty are usually considered to be more reliable. An estimate of the error associated with each fossil sample is given by the RMSEP of the training-set, derived using a separate test-set, or by internal cross-validation (see above). Approximate 95% confidence intervals (CI) for $\hat{x}_0$ are given by multiplying the RMSEP by a value from Student's *t*-distribution with appropriate degrees of freedom. This value is close to 2 for moderately large samples yielding 95% CIs of $\hat{x}_0 \pm 2 \times \text{RMSEP}$ (Næs et al. 2002). When the modelling has been carried out on transformed environmental data, the RMSEP (or CIs) around $\hat{x}_0$ must be appropriately back-transformed. For $\log_{10}$-transformed data the upper and lower RMSEP are given by antilog ($\hat{x}_0 \pm$ RMSEP) (Sokal and Rohlf 1981).

When we use the training-set RMSEP as a measure of the reconstruction error we assume that there are no sources of additional 'hidden error' as a result of non-analogue problems and that the environmental responses of fossil taxa are well-described by the modern training-set. Sample-specific error estimation provides a relatively simple way to simulate the additional likely reconstruction error if this is not the case. The method is based on the decomposition of the RMSEP for fossil samples into two components. The first, s_1, represents errors in the reconstructed values due to uncertainty in the estimates of model parameters, and the second, s_2, is due to errors in the observed x_i and model mis-specification (Birks et al. 1990a). Both of these parameters are estimated by bootstrap resampling: s_1 is the standard deviation of the reconstructed values obtained at each bootstrap cycle. It can vary from sample to sample and will be larger for fossil samples consisting of taxa that have few occurrences in the training-set. The second component represents the difference between the model predictions and observed values and is calculated as the RMSEP for the training-set samples. In practice the sample-specific error is usually dominated by s_2 with s_1 often comprising only 10–20% of the overall error. In extreme non-analogue situations where abundant fossil taxa are very infrequent and have low abundance in the training-set s_1 may reach the same magnitude as s_2. A separate stratigraphic plot of s_1 is a useful diagnostic in highlighting such problematic samples.

Sample-specific errors are currently the best quantitative estimate we have of the uncertainty in the reconstructed values for fossil samples. They may, however, be overly optimistic or overly pessimistic for the following reasons. First, fossil samples almost always reflect climatic, hydrochemical, and limnological conditions that are in some way different from those sampled by the modern training-set. The sample-specific RMSEP is therefore likely to be an underestimate of the true RMSEP because it does not quantify all the sources of error that result from a lack of environmental and biological analogy between training and fossil samples. Second, s_2, which usually dominates the sample-specific error, is comprised of components related to error due to model mis-specification, within-lake, and other biological sampling variability, and variability in the modern environmental data. Reconstructions based on replicate cores suggest that biological variability may represent a maximum of 15% of total RMSEP (e.g., Heiri et al. 2003; Battarbee et al. 2005a). There is, however, often substantial error in modern environmental data and this can account for 30–40%, and sometimes up to 70%, of the total RMSEP (e.g., Nilsson et al. 1996; Birks 1998; Dabakk et al. 1999; Brooks and Birks 2001). This finding, and our unpublished simulations, suggests that with noisy environmental data the true reconstruction error may be substantially smaller than the training-set RMSEP would suggest (cf. McCune 1997).

Sample-specific errors are relatively easy to estimate for most quantitative reconstructions. Interpretation of these errors is far more difficult. For many training-sets the RMSEP represents 10–20% of the sampled gradient and is often of the same magnitude as the down-core changes being reconstructed. If the errors are expressed as 95% confidence intervals, they will almost certainly encompass all but the most profound environmental changes or impacts. The consequence of this is that reconstructions that are statistically significant and show consistent temporal trends, and which have a clear palaeoenvironmental and biostratigraphical interpretation, may often have continuously overlapping error bars. This paradox arises because the RMSEP or sample-specific error is an estimate of the likely error when the model is applied to new, *independent*, samples. Interpretation of the down-core errors is problematic because fossil samples in a time-series are temporally autocorrelated and consequently are not statistically independent. The use of sample-specific errors as a criterion for evaluating either the trends or absolute values of a reconstruction is therefore surprisingly difficult and full of problems for the unwary.

Goodness-of-Fit and Analogue Measures

A reconstructed environmental value is likely to be more reliable if the fossil sample has a close modern analogue in the modern training-set (ter Braak 1995). The similarity or fit between fossil and modern samples should therefore provide a simple heuristic for evaluating the reconstruction. Birks (1998) discusses three numerical criteria. The first is the 'goodness-of-fit' statistic assessed by fitting the fossil samples passively onto an ordination axis constrained by the environmental variable being reconstructed (see Simpson and Hall 2012: Chap. 19). The goodness-of-fit is then quantified as the squared residual distance of the fossil sample to the constrained axis. The second criterion is a simple analogue statistic that measures the dissimilarity between each fossil sample and its closest analogue in the modern training-set (see above "Modern Analogue Technique (MAT)" and Simpson 2012: Chap. 15). Although both these measures make palaeoecological sense, there is a major problem in applying them in practice, namely the definition of a cut-off value to define a 'good fit' or 'good' analogue, beyond which we deem the reconstruction unreliable (Simpson 2012: Chap. 15). For the squared residual distance goodness-of-fit measure, Birks et al. (1990a) regarded fossil samples with a residual distance greater or equal to the residual distance of the extreme 5% of the training-set to have a 'very poor' fit, and those greater or larger than the extreme 10% to have a 'poor' fit. Similarly, Jones and Juggins (1995) used the 5th percentile of the distribution of dissimilarities between all modern training samples to define the cut-off for a good analogue (Bartlein and Whitlock 1993; Simpson et al. 2005). Unfortunately, in both cases the rules used to define the cut-off are rather arbitrary, primarily because strict statistical criteria are lacking (although see Gavin et al. 2003 for a more robust approach in palynology and Simpson 2012: Chap. 15). Furthermore, there is actually little empirical evidence of a relationship between 'fit' or 'distance' between fossil and training-set samples and increased error or unreliability, provided fossil taxa are well-represented in the training-set. Indeed, experiments with simulated data suggest that many methods, with the exception of MAT, perform surprisingly well under mild non-analogue situations (ter Braak et al. 1993; ter Braak 1995).

Given the difficultly of interpreting the above statistics we recommend using the following three simple measures to represent aspects of the similarity between modern and fossil samples that directly affect reconstruction error (Birks 1998). These are the percentages of the total fossil assemblage that consists of taxa (1) absent from the modern training-set, (2) that are poorly represented in the modern training-set, for example, by less than ten occurrences, and (3) that have fossil abundances greater than their maximum value in the training-set. The first two measures will highlight samples containing taxa that are poorly represented in the training-set. The last will highlight non-analogue situations where WAPLS, in particular, may be prone to extrapolation.

Comparison of Reconstructions Using Different Numerical Methods

The numerical methods described above are different mathematically and model (either implicitly or explicitly) species responses in different ways. They should been seen as complementary, not competitive (Racca et al. 2001), especially when there is no clear 'winner' in terms of lower RMSEP or statistical significance. In these cases it is useful to compare reconstructions from a range of techniques. If down-core biological changes are primarily driven by changes in the reconstructed environmental variable, we would expect reconstructions from different models to follow similar trajectories, even if they differ in their absolute values. If different methods produce widely divergent reconstructions, it would suggest that fluctuations in the dominant taxa are not primarily related to changes in the reconstructed variable, the statistical significance of the reconstructions must be tested, and the reconstructions treated with caution. Similarity in reconstructions based on different methods does not imply validation but it does tell us that they are free from technique-specific bias. Walker et al. (1997), Birks (2003), Köster et al. (2004), and Xu et al. (2010) provide examples of such comparisons. Where different models produce consistent, rather than conflicting reconstructions, these may be combined into a single, consensus reconstruction (e.g., Racca et al. 2001; Birks 2003; Barrows and Juggins 2005).

Kucera et al. (2005) outline a useful conceptual model for assessing the reliability of palaeoenvironmental reconstructions based on a scatter-plot of dissimilarity d, against method divergence Δ, represented by the standard deviation of the different estimates for each fossil sample. Samples with low d and Δ have convergent estimates and good analogues and are deemed reliable. Samples with low d but high Δ also have good analogues but suffer from technique-specific bias. Interpretation of samples in other parts of the diagrams follows similarly. Although one is still faced with the problem of defining critical values of d and Δ, this simple method provides a tool that focuses attention on the interpretation of the differences between reconstructions from different techniques. A method for examining the temporal change in the correlation between two reconstructions is described by Aykroyd et al. (2001) with a palaeolimnological example in Korhola et al. (2002).

Comparison of Reconstructions Using Different Proxies

In the absence of historical measurements the most powerful method of evaluation is to compare reconstructions derived independently from different proxies. Although they can be extremely time-consuming, such comparisons are becoming increasingly important (e.g., Lotter 2003; Birks and Birks 2006). Examples of comparison among reconstructions using a range of proxies including diatoms, chrysophytes, chironomids, cladocerans, pollen, plant macrofossils, stable isotopes, and sediment optical properties are given by Battarbee (2000), Birks and Ammann (2000), Birks et al. (2000), Rosén et al. (2001, 2003), Bigler et al. (2002), Dixit et al. (2002), Larocque and Bigler (2004), Verschuren et al. (2004), Heiri and Lotter (2005), Peyron et al. (2005), Smol (2008), and Hausmann et al. (2011).

One observation of many published multi-proxy studies is that, while the different reconstructions usually agree in terms of highlighting major trends and change-points, there can be substantial disagreement in the absolute values of the reconstructions between proxies (e.g., Birks and Ammann 2000; Birks et al. 2000). This is in some ways inevitable, especially in climate reconstructions, as different proxies have different sensitivity to the environmental variable of interest and to any confounding variables. In this way a partial validation of the records follows from an attempt to understand and explain the biases inherent in each proxy (Birks and Birks 2006).

When multi-proxy reconstructions are lacking we are forced to resort to the above numerical criteria for evaluating the reconstruction. However, while these criteria can help in identifying non-analogue situations, their interpretation is often problematic. Numerical evaluation should therefore be seen not as an end in itself but as a source of information to supplement an interpretation based on an understanding of the ecological mechanisms underlying the observed changes. Korhola et al. (2002) provide an example of the use of both numerical and ecological criteria to evaluate a chironomid-based climate reconstruction for northern Fennoscandia.

Case Study

In this section we illustrate the issues of model selection and reconstruction evaluation by reconstructing the recent acidification history of The Round Loch of Glenhead (RLGH), a small soft-water lake in Galloway, south-west Scotland (see Birks and Jones 2012: Chap. 3). A number of diatom-based reconstructions have been published for this site (Flower and Battarbee 1983; Jones et al. 1986; Birks et al. 1990a, b; Allott et al. 1992). The summary diatom biostratigraphy for the most recent, core KO5 collected in 1989, is illustrated in Fig. 14.6 and shows clearly

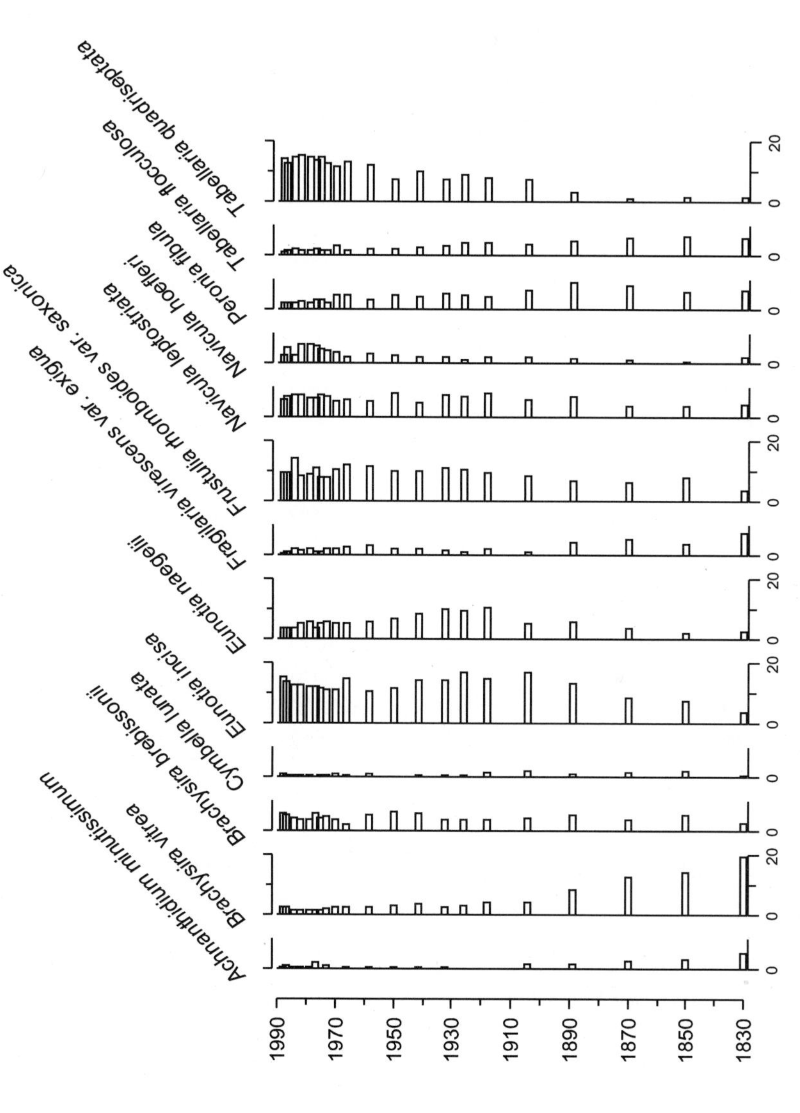

Fig. 14.6 Summary diatom stratigraphy of The Round Loch of Glenhead core KO5. Data are expressed as percentages of the total diatoms counted. The vertical axis is age in years AD

the replacement of a flora dominated by *Brachysira vitrea* by one dominated by *Tabellaria quadriseptata* as a result of acidification from the late nineteenth century onwards. Previous pH reconstructions for this site differ slightly in detail but all show a gradual decline from around pH 5.6 in the mid nineteenth century to the 1989 value of c. pH 4.7. Much of the earlier work at this and other acidified lakes focused on identifying the magnitude and timing of chemical changes in an effort to identify causal mechanisms and the spatial extent of freshwater acidification (e.g., Battarbee et al. 2010). More recently the emphasis has shifted to quantifying baseline, or pre-impact, chemistry for use in setting targets for lake recovery and restoration (Battarbee et al. 2010; Simpson and Hall 2012: Chap. 19). In the case of RLGH, the diatom-based estimate of pre-acidification pH is substantially lower than that derived from dynamic hydrochemical modelling using the modelling of acidification of groundwater in catchments (MAGIC) model (Jenkins et al. 1990). This discrepancy, and the renewed need for accurate hindcasts of baseline conditions, prompted Battarbee et al. (2005b) to re-analyse the RLGH data using a range of different methods and to re-evaluate the reconstructions for the pre-acidification levels (see also Battarbee et al. 2008).

In addition to the core data, Battarbee et al. (2005b) used three separate training-sets: Surface Waters Acidification Programme (SWAP), consisting of 178 lakes from the UK, Sweden, and Norway (Stevenson et al. 1991), an unpublished set of 163 UK lakes, and a combined European data-set of 693 lakes from the European Diatom Database Initiative (EDDI: Battarbee et al. 2000). WA, WAPLS, MAT, and GLR-ML were applied to each data-set-method combination and evaluated using the following criteria: (1) bootstrap cross-validation RMSEP and maximum bias (1000 bootstrap cycles), (2) RMSEP and maximum bias calculated from an independent test-set of 20 UK lakes, and (3) comparison of monitored pH with that inferred from sediment-trap diatom assemblages for the period 1991–2002. LWWA was also applied to the larger and more heterogeneous EDDI modern data-set. Each training-set was screened for outliers and samples deleted if they had a standardised cross-validation residual from a WA model (with inverse deshrinking) greater than 2.5 in absolute value (see above). This is a fairly conservative screening and only removes gross outliers: for the WA model it resulted in the deletion of 4, 3, and 13 samples from the SWAP, UK, and EDDI training-sets, respectively. The data-sets were also screened to remove rare taxa, defined as those with less than two occurrences and with a maximum relative abundance of less than 1%.

The distribution of training-set samples along the pH gradient is illustrated in Fig. 14.7. They have a good coverage from pH 4.5 to 7.0 although very acid waters are poorly represented in all three, as are waters above pH 7 in the SWAP and UK data-sets, and above pH 7.5 in the EDDI data-set. The test-set spans pH 4.5–7.0 with some bias towards waters below pH 5.0. Figure 14.7 also shows performance measures for WAPLS plotted against component number, and for MAT and LWWA plotted against number of analogues used in the 'local' model (see above). The second WAPLS component yields the lowest RMSEP for all three data-sets although the reduction is small for SWAP and UK. For the

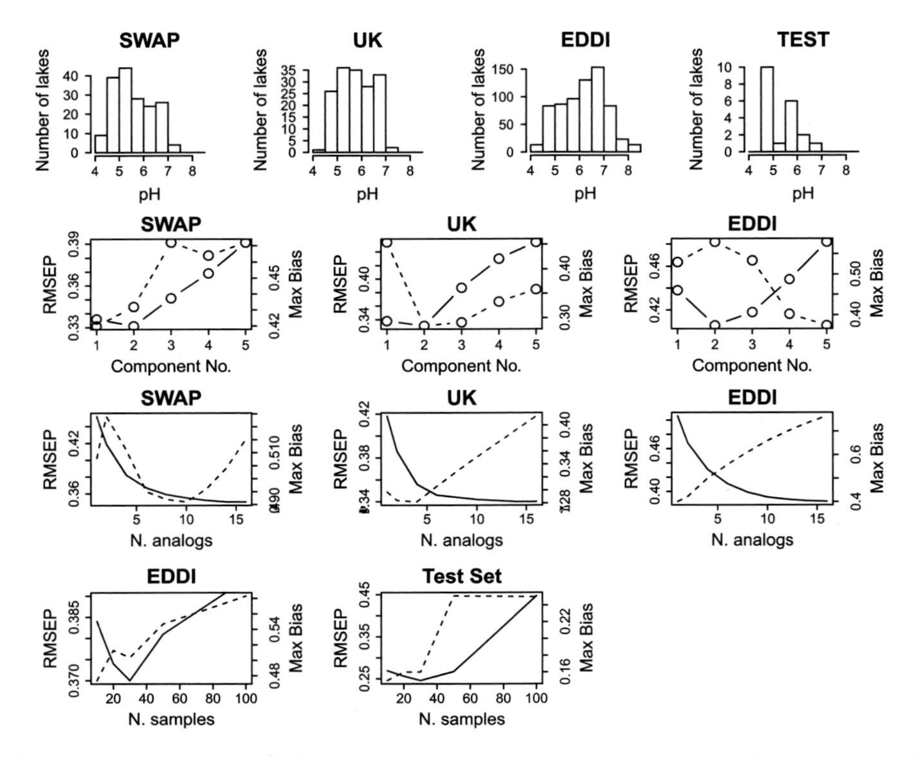

Fig. 14.7 Summary statistics of the three training-sets and the independent test-set used in the Case Study in this chapter. The *first row* shows histograms of the distribution of lakes along the lake-water pH gradient in the four data-sets. The *second row* shows root mean squared error of prediction (RMSEP) and maximum bias (Max Bias) plotted against weighted-averaging partial least squares (WAPLS) component number for the three training-sets. The *third row* shows RMSEP and maximum bias for the modern analogue technique (MAT) for the three training-sets. The *last row* shows RMSEP and maximum bias for locally-weighted weighted-averaging (LWWA) and for the 20-lake independent test-set plotted against the number of samples used in the LWWA model. No. and *N* number

Table 14.5 Root mean squared error of prediction (RMSEP) and maximum bias performance statistics for the three diatom-pH training-sets used in the Case Study (SWAP, UK, EDDI)

	SWAP	UK	EDDI
No. sites	174	161	680
No. taxa	272	237	477
Internal cross-validation			
WA (inverse DS)	0.335 (0.388)	0.338 (0.350)	0.438 (0.483)
WA (classical DS)	0.345 (0.206)	0.348 (0.304)	0.473 (0.409)
WAPLS (2 component)	Not significant	0.331 (0.284)	0.406 (0.530)
GLR-ML	0.358 (0.381)	0.353 (0.294)	0.427 (0.299)
MAT (5)	0.374 (0.506)	0.349 (0.293)	0.420 (0.531)
MAT (10)	0.355 (0.491)	0.342 (0.351)	0.392 (0.660)
LWWA (30)	N/A	N/A	0.370 (0.504)
Independent test-set (external cross-validation)			
WA (inverse DS)	0.294 (0.366)	0.234 (0.257)	0.390 (0.320)
WA (classical DS)	0.373 (0.369)	0.294 (0.279)	0.520 (0.476)
WAPLS (2 component)	N/A	0.287 (0.252)	0.439 (0.287)
GLR-ML	0.347 (0.284)	0.279 (0.250)	0.427 (0.273)
MAT (5)	0.242 (0.401)	0.171 (0.106)	0.204 (0.205)
MAT (10)	0.239 (0.406)	0.171 (0.224)	0.186 (0.083)
LWWA (30)	N/A	N/A	0.246 (0.160)

Maximum bias shown in parentheses. *DS* deshrinking, *N/A* not available. *WA* weighted-averaging regression and calibration, *WAPLS* weighted-averaging partial least squares regression and calibration, *GLR-ML* Guassian logit regression and maximum likelihood calibration, *MAT* modern analogue technique, *LWWA* locally-weighted weighted-averaging regression and calibration, *SWAP* Surface Waters Acidification Programme, *EDDI* European Diatom Database Initiative

SWAP data, the randomisation *t*-test suggests this is not statistically significant ($p = 0.112$, 999 permutations). For the UK data, the second WAPLS component is also associated with a substantial reduction in maximum bias and is marginally significant ($p = 0.023$). The second WAPLS component is highly significant for EDDI ($p < 0.001$). Plots of performance measures against the number of analogues for MAT show a gradual decrease in RMSEP with increasing number although there is little change after five analogues. Maximum bias increases with increasing number of analogues, primarily as a result of poor prediction at the gradient ends. Finally, for LWWA a compromise between low RMSEP and low maximum bias is given by a local model size of 30 samples for both the EDDI and the test data-set. These results therefore indicate the following models are appropriate: a 2-component WAPLS model for UK and EDDI, and a 30-sample LWWA model. The picture is less clear for MAT so we apply both 5 and 10-analogue models (MAT5 and MAT10, respectively).

Performance statistics of the various models are listed in Table 14.5. There are three important features. First, for the SWAP and UK data-sets, the internal cross-validation RMSEPs fall between 0.33 and 0.36 for all methods except MAT5. Maximum bias is also surprisingly similar for most methods, and apart from MAT using SWAP, there is very little to choose between methods. Second, RMSEPs for all methods are higher for the EDDI data-set. This is not surprising given the more heterogeneous nature of this training-set. In this case, LWWA outperforms other methods, although MAT10 and WAPLS are reasonable. Third, RMSEPs are generally lower for the test-set, and show far more variation between methods. In this case, WA and MAT for all three training-sets perform well, as does EDDI-LWWA. WAPLS and GLR are relatively poor performers although their RMSEP is no higher than that estimated from the training-set.

Comparisons of the measured pH and diatom-inferred pH for the sediment-trap data are shown in Fig. 14.8. Measured pH exhibits seasonal variability of 0.2–0.5 pH units and there is a statistically significant trend towards increased mean annual pH from the mid 1990s (Davies et al. 2005). The various reconstructions can be grouped into one of three types: (1) tracking trend well but under-estimation of absolute values by c. 0.2 pH units (SWAP, some EDDI and UK), (2) grossly under-estimating measured pH and failing to track trend or inter-annual variability (EDDI-GLR-ML and EDDI-WAPLS, UK-GLR-ML), and (3) tracking trend and matching absolute values reasonably well (UK-WA, UK-MAT5, EDDI-LWWA). The poor performance of EDDI-WAPLS and EDDI-GLR-ML is expected for the reasons mentioned above. Similarly, the comparisons indicate that some methods are at least able to reconstruct trends and absolute values in acid waters reasonably well. What is less clear is which methods will yield the most accurate reconstructions for the pre-acidification levels of The Round Loch, when the likely pH was 5.5 or higher. Combining the results from the cross-validation, test data-set, and sediment traps we conclude that EDDI-LWWA, UK-WA, and UK-MAT5 would appear to be the most appropriate methods, and that SWAP-WA, UK-WAPLS, and UK-GLR-ML will provide useful alternative reconstructions for comparison.

We now apply the selected methods to The Round Loch of Glenhead core. Less than 5% of the individuals in each level come from taxa with less than ten effective numbers of occurrences (N2: Hill 1973) in the training-sets and only four levels (1918–1941) have taxa that are marginally more abundant in the core. Squared chi-square distances between fossil samples and their closest analogue in the training-sets are all below 0.46. This value is less than the first percentile of the matrix of dissimilarities between all modern samples (0.54). Taken together these measures indicate that The Round Loch samples consist of taxa that are very well represented in the modern training-sets. Results of the six reconstructions are shown in Fig. 14.9. Sample-specific RMSEPs for each method vary only very slightly down-core and range from 0.310 for UK-WAPLS to 0.360 for EDDI-LWWA. We omit them from Fig. 14.9 for clarity. Not surprisingly, given the strong down-core species shift, all reconstructions follow the same trajectory although there are consistent differences of up to 0.3 pH units between methods. Estimates for the 1830-level vary from 5.44 (EDDI-LWWA) to 5.69 (UK-WAPLS) with a mean of pH 5.57. SWAP-WA, used in previous publications (e.g., Birks et al. 1990a, b), is one of the lowest estimates (5.48). Figure 14.9 also shows the hindcast-pH derived

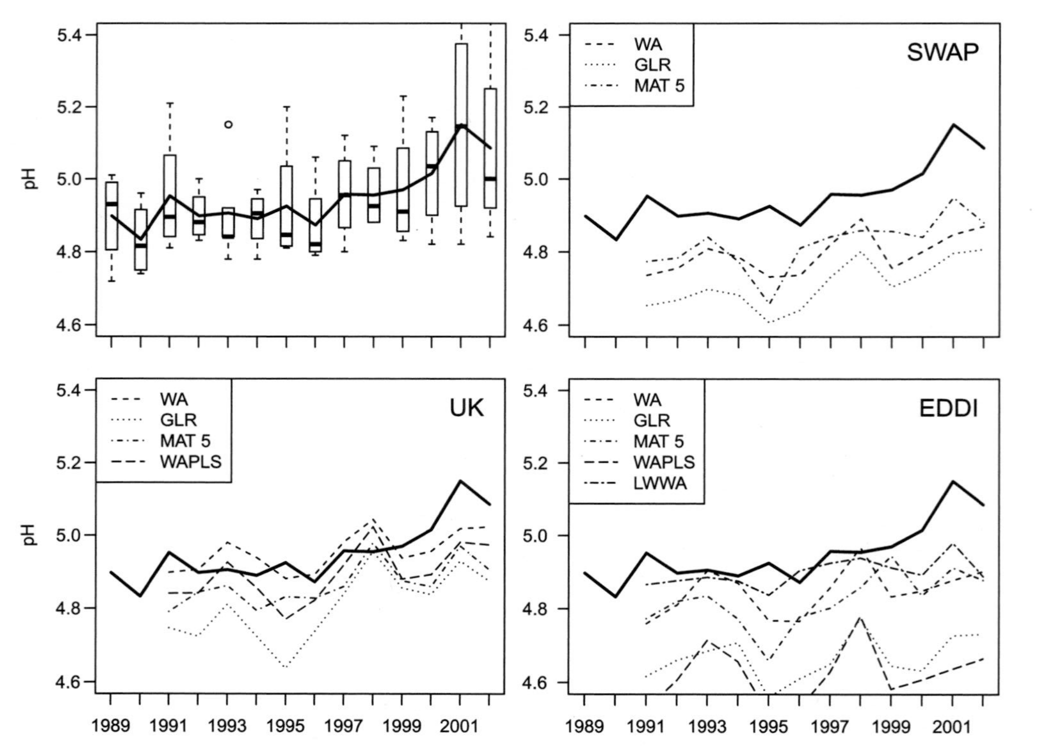

Fig. 14.8 Ranges of lake-water pH at The Round Loch of Glenhead from 1989 to 2002 shown as *box plots* (*top left*). The remaining *three panels* show diatom-inferred pH values based on weighted averaging (WA), Gaussian logit regression and maximum-likelihood calibration (GLR-ML), and modern analogue technique (MAT) using five analogues, weighted-averaging partial least squares (WAPLS) and locally-weighted weighted-averaging (LWWA) using the three calibration-sets detailed in Table 14.3. The mean of the measured pH values is plotted as a *thick solid line*

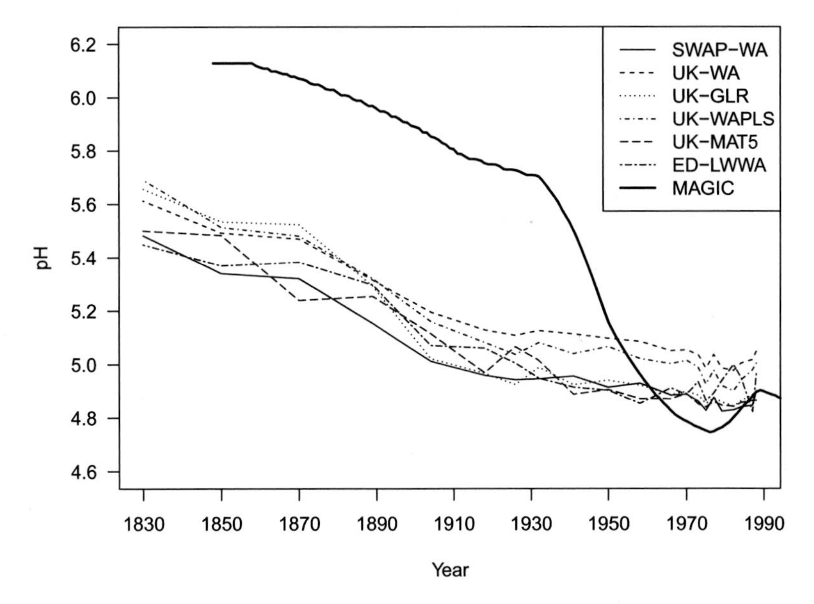

Fig. 14.9 Diatom-inferred pH based on The Round Loch of Glenhead core KO5 (see Fig. 14.6) plotted against age (Year AD). The various reconstructions are based on different training-sets (Surface Waters Acidification Programme (SWAP), UK, European Diatom Database Initiative (EDDI) (= ED)) or method (weighted averaging (WA), Gaussian logit regression and maximum-likelihood calibration (GLR-ML), weighted-averaging partial least squares (WAPLS), modern analogue technique (MAT) with five analogues) or locally-weighted weighted-averaging (LWWA). The *solid line* shows the hindcast pH derived from the modelling of acidification of groundwater in catchments (MAGIC) hydrochemical model (Jenkins et al. 1990)

from the MAGIC hydrochemical model and highlights the discrepancy between diatom-inferred and modelled lake-water chemistry. The new reconstructions raise the previous estimates slightly but they are still substantially lower than the MAGIC hindcasts which suggests the pre-acidification pH of the lake was over 6.0. The consistency of the diatom-based estimates is encouraging and they pass all the methods of evaluation described above, but do we have enough faith in them to question the hydrochemical model? Figure 14.10 shows the relationship between pH and squared chi-square distance for the ten closest analogues to the AD 1830 sample. The pH of the analogues varies from c. 5.1 to c. 6.1 although all the very close analogues (distance <0.55) come from lakes with a pH of less than 5.8. This additional analysis does not validate the reconstructed values but it does increase our confidence in them, and leads us to question the hydrochemical hindcasts (Battarbee et al. 2005b).

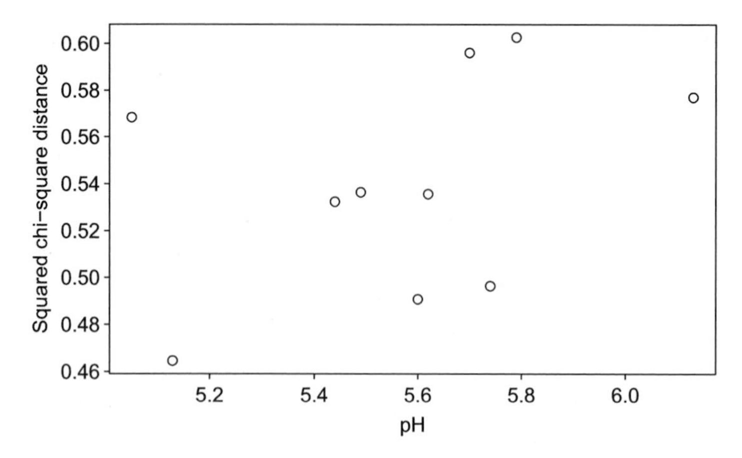

Fig. 14.10 Scatter-plot showing the relationship between measured pH and squared chi-square distance for the ten closest modern analogues of The Round Loch of Glenhead AD 1830 core sample. All but one of the analogues have a measured pH less than 5.8

Assumptions and Limitations

Quantitative palaeoenvironmental reconstructions involve a number of statistical, ecological, and palaeoecological assumptions (Imbrie and Kipp 1971; Birks et al. 1990a, 2010). These are rarely completely fulfilled but the effect of violation on the accuracy and precision of the reconstructed values is poorly understood. In this section we critically analyse the assumptions and highlight situations where they may limit the use of quantitative reconstructions in palaeolimnology.

The main assumptions are (Imbrie and Kipp 1971; Imbrie and Webb; 1981; Birks et al. 1990a, 2010; Birks 1995):

1. The taxa in the modern training-set are systematically related to the environment in which they live.
2. The environmental variable(s) to be reconstructed is, or is linearly, or at least monotonically, related to, an ecologically important determinant in the system of interest.
3. The taxa in the training-set are the same biological entities as in the fossil data and their ecological responses to the environmental variable(s) of interest have not changed over the time represented by the fossil assemblage.
4. The mathematical methods adequately model the biological responses to the environmental variable(s) of interest and yield numerical models that allow accurate and unbiased reconstructions.
5. Environmental variables other than the one of interest have negligible influence, or their joint distribution with the environmental variable does not change with time.

6. In model validation and in the estimation of prediction errors by cross-validation, the test-data are statistically independent of the training-set (Telford and Birks 2005).

Assumption 1 actually follows from an analysis of assumption 2. Assumption 3 is based on the principal of uniformitarianism and is usually assumed to be met for most studies covering the Holocene, although it may be questionable for earlier periods. Criteria for assessing assumption 4 have been discussed above in the sections on "Model Selection and Evaluation" and on "Reconstruction Testing, Evaluation, and Validation". Criteria for assessing assumption 6 have been outlined above in the section on "Spatial Autocorrelation and Environmental Reconstruction".

Assumptions 2 and 5 are critical but are not often challenged. Assumption 2 begs the fundamental question of what variables can (or cannot) be reconstructed using a particular training-set? Diatom-based models, for example, have been developed to reconstruct a wide range of hydrochemical, climatic, and limnological variables including lake-water pH (Birks et al. 1990a), alkalinity (Köster et al. 2004), salinity (Fritz et al. 1991), specific conductivity (Gasse et al. 1995), cation concentrations (Dixit et al. 2002), cation and anion ratios (Gasse et al. 1995), total nitrogen (Werner and Smol 2005), total phosphorus (Bennion et al. 1996), chlorophyll-*a* (Jones and Juggins 1995), nitrate (Curtis et al. 2009), dissolved organic carbon (Birks et al. 1990b; Curtis et al. 2009; Hausmann et al. 2011), total organic carbon (Rosén 2005), dissolved inorganic carbon (Rühland and Smol 2002), dissolved CO_2 (Philibert and Prairie 2002a), aluminium (Birks et al. 1990b), nickel (Dixit et al. 2002), water temperature (Pienitz et al. 1995; Huber et al. 2010), air temperature (Lotter et al. 1997; Bigler and Hall 2003), date of autumn mixing (Schmidt et al. 2004), water depth (Yang et al. 2003), duration of ice cover (Thompson et al. 2005; Curtis et al. 2009), wind activity (Hausmann et al. 2011), and duration of snow cover (Mackay et al. 2005). Chironomid assemblages have been used to reconstruct not only summer water or air temperature, but also hypolimnetic anoxia, chlorophyll-*a*, total phosphorus, salinity, water depth (Velle et al. 2010, 2011b), distance from littoral vegetation (Luoto 2010), and even the 1/0 binary lotic index of stream flow (Luoto 2010)! What is the basis for these models?

The choice of environmental variable is usually made in one of two ways. First, it may be defined *a priori* by wider project aims and the training-set designed specifically to sample the gradient of interest. Second, the environmental variable may be selected *post hoc* after it is found to be an important biological determinant during an analysis of the training-set. In both cases assumption 2, namely that the environmental variable explains, in a statistical sense, a significant portion of the variation in the biological data, can be tested using constrained ordination and associated Monte Carlo permutation tests (Borcard et al. 2011; Telford and Birks 2011b; Legendre and Birks 2012b: Chap. 8), with the environmental variable of interest as the single constraining variable. Similarly, the relative strength or importance of the environmental variable can be estimated by comparing the eigenvalue of the first axis of a detrended canonical correspondence analysis

(DCCA) using a single constraining variable with that of the first unconstrained DCA axis (e.g., Lotter et al. 1997). Environmental factors are inherently interrelated and assumption 2 is often qualified by the additional requirement that the variable of interest explains a significant, *and independent*, portion of the variation in the biological data. This assumption is tested implicitly if variables are stepwise forward-selected in a constrained ordination or explicitly using partial ordination (Borcard et al. 2011; Legendre and Birks 2012b: Chap. 8; see Gasse et al. 1995 for an example). In this, both marginal and unique conditional effects of the environmental variables of interest should be assessed and tested (ter Braak and Verdonschot 1995) (see Curtis et al. 2009 for an example of environmental variable selection and assessment).

There are two important consequences that lead from the above discussion. The first is that assumption 2 only requires a correlation between environment and biology, not a causal relationship. For some species–environment groups there may be experimental evidence that suggests causal effects or demonstrates the physiological basis for the underlying response. For others there may be a long history of observational data that strongly suggests a direct environmental effect. Where this is not the case it should be recognised that the model is based on correlation only and lacks a sound ecological basis. This is not necessarily a problem – the model may still have good predictive ability – but it does require an additional assumption that the relationship between the measured environmental variable and the underlying *causal* ecological gradient has not changed through time. This requirement is further explored below. The second consequence is that models developed for variables selected *post hoc* have an inherently weaker basis than those for variables selected *a priori* as part of a hypothesis-testing approach (cf. Hallgren et al. 1999). This is because of the circularity in cross-validating a model that has already been found to be statistically significant. Models developed for variables selected using an automated step-wise procedure are especially problematic (Crawley 2005). Such an approach borders on 'data-dredging' and models generated in this way need very careful validation using an additional independent data-set (cf. Johnson and Omland 2004).

We illustrate the above concerns with a second example from The Round Loch of Glenhead (Birks et al. 1990b). Analysis of the SWAP training-set (Stevenson et al. 1991) using canonical correspondence analysis (CCA) with forward selection and associated Monte Carlo permutation tests indicated that ten hydrochemical variables made a significant and independent contribution to explaining the variation in the modern diatom composition and relative abundances. The first two CCA axes reflected pH and dissolved organic carbon (DOC) gradients, respectively. Total aluminium (Al) was significantly correlated to both axes. On this basis Birks et al. (1990b) concluded that pH, DOC, and Al were potentially reconstructable and used WA-based models to hindcast changes in these variables for the last 10,000 years. Here we re-evaluate the validity of these reconstructions.

First, we quantify the relative explanatory power of pH, DOC, and Al using CCA and partial CCA to partition the variation in the training-set diatom data into components related to (1) the marginal effect, or total variation explained by each variable and (2) the conditional effect, or unique and independent contribution of each variable (cf. Borcard et al. 1992; ter Braak and Verdonschot 1995). Results indicate that pH, DOC, and Al explain 6.9%, 3.7%, and 1.5% of the variation in the modern diatom data individually, but only make unique contributions of 5.2%, 2.9%, and 1.0%. The unique contributions are highly significant for pH and DOC ($p < 0.001$) but only marginally so for Al ($p = 0.041$). The explained variations for pH and DOC are small but such values are typical for large, species-rich data-sets. The unique effect of Al is very small and there is only weak evidence that it has a significant and independent effect. The independence or otherwise of the Al reconstruction is further explored by comparing the correlations between the three variables in the water-chemical data of the training-set and in the core predictions. For the training-set, the correlations between observed pH and DOC, and pH and A1, are 0.26 and −0.44, respectively. Correlations for the corresponding diatom-inferred values are 0.60 and −0.83, and 0.35 and −0.98 for the core reconstructions. Thus while there are only weak correlations among these variables in the training-set, the correlation is stronger for the diatom-inferred values, especially in the case of Al. Indeed, 96% of the down-core variation in reconstructed Al concentration is explained by reconstructed pH. We conclude therefore that there is a relatively small confounding effect between pH and DOC but that the Al reconstructions cannot be considered independent of pH. In this context it is interesting to note that in the Bergen diatom training-set described above, which was specifically designed to span gradients of pH and labile aluminium species, Al is not independently significant statistically.

The problem of the confounding effects of correlated environmental variables pervades almost all quantitative reconstructions. In almost all lake systems, the composition of sedimentary assemblages is a complex function of multiple chemical and limnological variables that are intimately inter-linked through the interaction of climate, catchment, and lake processes. The first part of assumption 5, that environmental variables other than the one of interest have negligible influence, is therefore almost never met. The second part, that the joint distribution of additional variables with the one of interest does not change with time, is also violated in many cases. Input of strong acid anions from acid precipitation and nutrients from agricultural runoff have had profound effects on the relationships between climate, alkalinity, nutrients, and DOC for training-set lakes in impacted areas for the recent past. Changes in catchment vegetation and soils, and climate-driven physical limnology, will have also modified the joint distribution of these variables on longer timescales. The likely importance of these effects can be estimated using variation partitioning (see Legendre and Birks 2012b: Chap. 8). If the reconstructed variable represents a single dominant signal in the biology-environment relationship, we can consider assumption 5 to have been met. However, in many cases the variable of interest is not a dominant signal, and may not even be the primary gradient. For example, in the diatom training-set described by Lotter et al. (1997) for the Swiss Alps, summer temperature, total phosphorus, and alkalinity individually explained 9.1%, 7.5%, and 7.8%, respectively, of the variation in the diatom data, but had conditional effects of only 2.9%, 2.6%, and 1.6%, respectively. In this case, approximately two thirds of the signal for each variable was confounded with other factors. A Monte Carlo permutation test confirms that these unique components are statistically significant ($p < 0.010$) demonstrating statistical independence, but as Anderson (2000) points out, the final model used to reconstruct such variables is not a partial model. As a result, when assumption 5 is not met, down-core changes in biota that are related to a confounding factor, but independent of the reconstructed variable, may nevertheless cause corresponding spurious changes in the resulting reconstruction.

The consequences of violating assumption 5 are illustrated in the following example. Clarke (2001) and Clarke et al. (2003) developed a diatom-based calibration-function to infer total nitrogen (TN) in Danish coastal waters and applied the model to an 80 cm ^{210}Pb-dated core from Roskilde Fjord to reconstruct changes in TN for the last ~150 years. Figure 14.11 shows the summary diatom stratigraphy and the TN reconstruction. The record shows a recent increase in planktonic taxa starting in the mid-1900s (35 cm depth) and a corresponding increase in reconstructed TN at the top of the core. The core-top value of 90 μmol L^{-1} agrees well with the contemporary measured TN concentration of 84 μmol L^{-1} and the historical values of c. 50 μmol L^{-1} are consistent with hindcasts from mass-balance models, giving us some confidence in the reconstructions.

Constrained ordination and variation partitioning in this example indicate that the reconstructed variable, TN, is a secondary gradient and diatom distribution and abundance are primarily related to water salinity and water depth. Individually, depth and salinity account for 9.0% and 6.5% of the variation in the diatom data, and TN only 4.7%. The unique effect of TN is highly significant ($p < 0.001$) but accounts for only 2.5% of the total variation in the diatom data, indicating that approximately half the total TN signal is confounded with depth and salinity. However, as water depth and salinity have not changed significantly at the core site over the period of interest, we can assume that floristic changes are primarily a response to changing nutrient concentrations. The unique effects of salinity and depth are 6.6% and 5.3%, respectively, suggesting that reconstructions for these variables should not be overly influenced by confounding effects. To test this we developed predictive models for these variables using WA. Both models show strong correlations between measured and inferred values for the training-set (cross-validation $r^2 = 0.75$ and 0.67 for salinity and depth, respectively, compared to 0.70 for TN). Down-core reconstructions for these variables are also included in Fig. 14.10 and indicate a gradual increase in depth at the site of c. 4 m and a decrease in salinity of c. 3 g L^{-1}. These reconstructed changes are artefacts because recent nutrient increases have almost certainly modified the joint distribution of TN, salinity, and depth at this site. As a result, changes in benthic and planktonic taxa that are driven by nutrient increases reconstruct as spurious changes in depth and salinity.

In this example salinity, depth, and TN meet all the requirements for a variable to be reconstructed and all yield models with good predictive ability. The example provides, however, an appropriate cautionary tale: in this case we have sufficient prior knowledge to dismiss the depth and salinity reconstructions but usually such information is lacking.

Since the pioneering work on quantitatively reconstructing lake-water pH from diatom assemblages by ter Braak and van Dam (1989) and Birks et al. (1990a), there have been many advances in the development of techniques of deriving mathematically calibration functions (e.g., WAPLS, ANN, LWWA), and in the

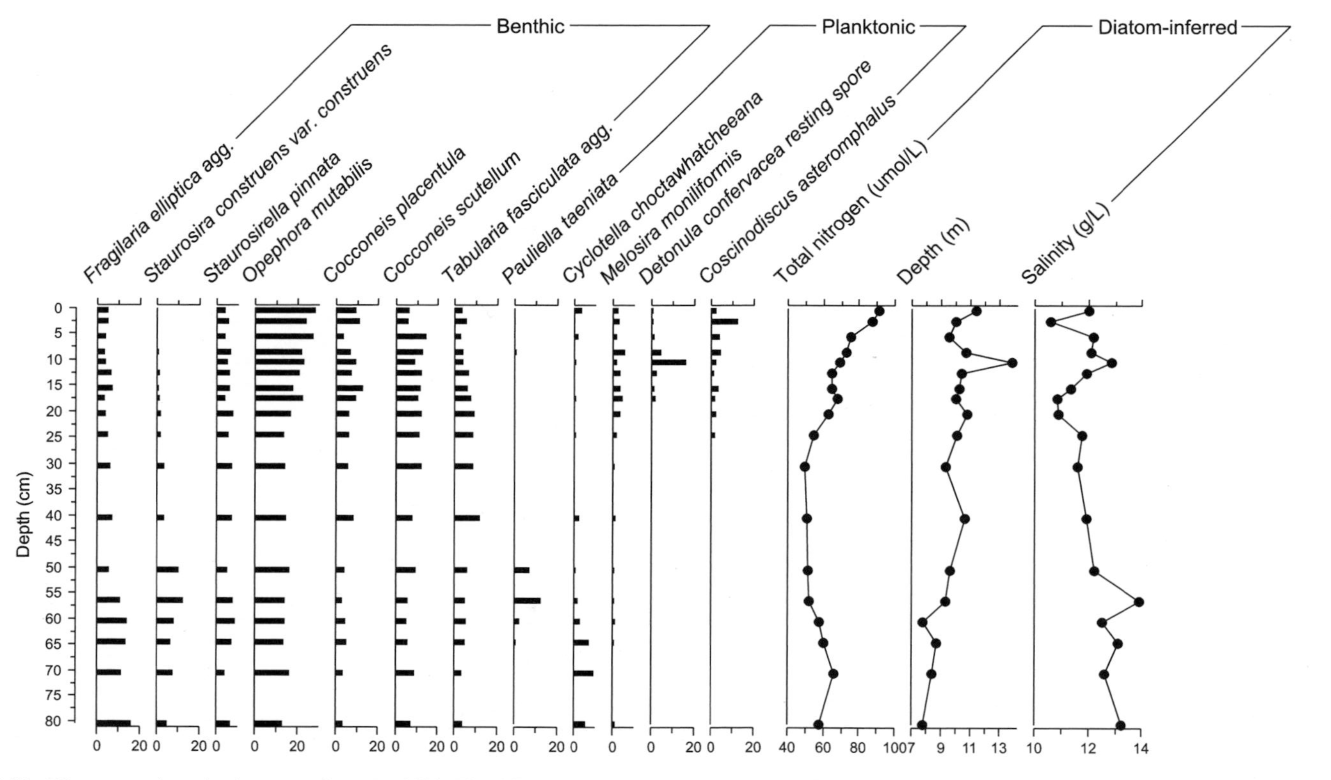

Fig. 14.11 Diatom stratigraphy in a core from Roskilde Fjord in Denmark, along with the diatom-inferred reconstructions of total nitrogen, water depth, and salinity for the last 150 years (based in part on Clarke et al. 2003). Data are expressed as percentages of the total diatoms counted

creation and application of training-sets from many parts of the Earth (e.g., Smol 2008). Despite all these activities, surprisingly little work has been done on the underlying assumptions and limitations of transfer functions (but see Telford and Birks 2005, 2009, 2011a, b). The examples discussed above highlight the need to 'return to basics' and to understand better the limitations of quantitative reconstructions when these assumptions are violated.

Software

There are number of specialised and more general-purpose software packages useful for the analysis of palaeoecological training-sets. Software for analysis of species–environment relationships using regression, ordination, and cluster analyses is mentioned by Legendre and Birks (2012a, b: Chaps. 7 and 8). Specialised software for developing calibration-functions and performing reconstructions includes WACALIB (WA, WAT, bootstrap sample-specific errors: Line et al. 1994), C2 (WA, WAT, PCAR, PLS, WAPLS, LWWA, GLR, MAT, bootstrap sample-specific errors: Juggins 2007), paleoNet (ANN: Racca et al. 2007) and PaleoToolBox (PCAR (IKFA), MAT: Sieger et al. 1999). Although these packages offer a convenient way to fit and evaluate a range of models, we increasingly use the R language for statistical computing (R Development Core Team 2010) as it provides a single, powerful environment for all stages of analysis, including data-set compilation and screening, exploratory data analysis, model fitting and evaluation, and graphical output. The following R packages are particularly useful for developing quantitative reconstructions: vegan (ordination and cluster analysis, dissimilarity measures: Oksanen et al. 2011); analogue (WA, MAT: Simpson and Oksanen 2009); rioja (WA, WAT, PLS, WAPLS, MAT, GLR: Juggins 2009); paltran (WA, WAPLS, LWWA: Adler 2010); paleoMAS (MAT: Correa-Metrio et al. 2011), and palaeoSig (significance testing of reconstructions: Telford 2011).

Conclusions and Future Work

The last two decades has witnessed considerable progress in the development of numerical methods and high-quality training-sets for organism-based palaeolimnological reconstructions. The development of user-friendly software has given all palaeolimnologists access to a range of numerical procedures and we observe a healthy shift from a single model to a multi-model approach that recognises in many cases there is no single best numerical method. This shift also includes an important move away from an obsession with the lowest RMSEP to a more rigorous consideration of uncertainty including bias, and by using multi-model reconstructions, the increasing recognition that model selection contributes to the uncertainty. The recent work described above, embedding reconstruction procedures within a Bayesian methodology, offers an elegant solution for modelling multiple sources of evidence and their associated uncertainty within a single, coherent framework and is a priority for future work.

We also observe a trend over the last two decades to reconstruct a wide range of water quality and other environmental parameters. Given the complex set of interacting influences on taxon distribution and abundance, method robustness in the face of confounding variables becomes crucial. Many of the early applications that provided the impetus for method development were based on modelled responses to very strong anthropogenic gradients over the recent past (e.g., pH decline due to lake acidification). Applications reconstructing weaker, secondary gradients over longer time periods are sometimes problematic: different methods may give conflicting reconstructions, and where we do have historical monitoring data, models are sometimes unable to reconstruct accurately absolute values, and in some cases even trends. The problem of confounding variables has barely been addressed but almost certainly means that reconstructions for some variables and proxies will be problematic at best. We should not 'throw out the baby with the bathwater' but our examples challenge palaeolimnologists to be more critical of their models and associated reconstructions and to be alert to the effects of confounding variables.

Acknowledgements We thank Richard J. Telford, Cajo ter Braak, John Smol, and Hilary Birks for helpful discussions and/or reading an early draft of the Chapter and Cathy Jenks for invaluable editorial help. This is publication A360 from the Bjerknes Centre for Climate Research.

References

Adler S (2010) paltran: WA, WA-PLS, MW for paleolimnology. R package version 1.3-0. http://CRAN.R-project.org/package=paltran

Adler S, Hübener T, Dressler M, Lotter AF, Anderson NJ (2010) A comparison of relative abundance versus class data in diatom-based quantitative reconstructions. J Environ Manage 91:1380–1388

Allott TEH, Harriman R, Battarbee RW (1992) Reversibility of lake acidification at the Round Loch of Glenhead, Galloway, Scotland. Environ Pollut 77:219–225

Anderson NJ (2000) Diatoms, temperature and climate change. Eur J Phycol 35:307–314

Aykroyd RG, Lucy D, Pollard AM, Carter AHC, Robertson I (2001) Temporal variability in the strength of proxy-climate correlations. Geophys Res Lett 28:1559–1562

Barrows TT, Juggins S (2005) Sea-surface temperatures around the Australian margin and Indian Ocean during the Last Glacial Maximum. Quaternary Sci Rev 24:1017–1047

Bartlein PJ, Whitlock C (1993) Paleoclimatic interpretation of the Elk Lake pollen record. In: Bradbury JP, Dean WE (eds) Elk Lake, Minnesota: evidence for rapid climatic change in the north-central United States. Geological Society of America, Boulder, pp 275–293

Basheer JA, Hajmeer M (2000) Artificial neural networks: fundamentals, computing, design, and application. J Microbiol Methods 43:3–31

Battarbee RW (2000) Palaeolimnological approaches to climate change, with special regard to the biological record. Quaternary Sci Rev 19:107–124

Battarbee RW, Bennion H (2011) Palaeolimnology and its developing role in assessing the history and extent of human impact on lake ecosystems. J Paleolimnol 45:399–404

Battarbee RW, Juggins S, Gasse F, Anderson NJ, Bennion H, Cameron N (2000) European Diatom Database Initiative (EDDI): an information system for palaeoenvironmental reconstruction. In: European climate science conference, Vienna, 19–23 Oct 1998

Battarbee RW, Grytnes J-A, Thompson R, Appleby PG, Catalan J, Korhola A, Birks HJB, Heegaard E, Lami A (2002) Comparing palaeolimnological and instrumental evidence of climate change for remote mountain lakes over the last 2000 years. J Paleolimnol 28:161–179

Battarbee RW, Anderson NJ, Jeppesen E, Leavitt PR (2005a) Combining palaeolimnological and limnological approaches in assessing lake ecosystem response to nutrient reduction. Freshw Biol 50:1772–1780

Battarbee RW, Monteith DT, Juggins S, Evans CD, Jenkins A, Simpson GL (2005b) Reconstructing pre-acidification pH for an acidified Scottish loch: a comparison of palaeolimnological and modelling approaches. Environ Pollut 137:135–149

Battarbee RW, Monteith DT, Juggins S, Simpson GL, Shilland EW, Flower RJ, Kreiser AM (2008) Assessing the accuracy of diatom-based transfer functions in defining reference conditions for acidified lakes in the United Kingdom. Holocene 18:57–67

Battarbee RW, Charles D, Bigler C, Cumming BF, Renberg I (2010) Diatoms as indicators of surface water acidity. In: Smol JP, Stoermer EF (eds) The diatoms: applications for the environmental and earth sciences, 2nd edn. Cambridge University Press, Cambridge, pp 98–121

Battarbee RW, Simpson GL, Bennin H, Curtis C (2011) A reference typology of low alkalinity lakes in the UK based on pre-acidification diatom assemblages from lake sediment cores. J Paleolimnol 45:489–505

Beerling DJ, Birks HH, Woodward FI (1995) Rapid late-glacial atmospheric CO_2 changes reconstructed from the stomatal density record of fossil leaves. J Quaternary Sci 10:379–384

Bennett KD, Humphry RW (1995) Analysis of late-glacial and Holocene rates of vegetational change at two sites in the British Isles. Rev Palaeobot Palynol 85:263–287

Bennion H (1994) A diatom-phosphorus transfer function for shallow, eutrophic ponds in southeast England. Hydrobiologia 275(276):391–410

Bennion H, Simpson GL (2011) The use of diatom records to establish reference conditions for UK lakes subject to eutrophication. J Paleolimnol 45:469–488

Bennion H, Juggins S, Anderson NJ (1996) Predicting epilimnetic phosphorus concentrations using an improved diatom-based transfer function and its application to lake eutrophication management. Environ Sci Technol 30:2004–2007

Bennion H, Battarbee RW, Sayer CD, Simpson GL, Davidson TA (2011a) Defining reference conditions and restoration targets for lake ecosystems using palaeolimnology: a synthesis. J Paleolimnol 45:533–544

Bennion H, Simpson GL, Anderson NJ, Clarke G, Dong X, Hobæk A, Guilizzoni P, Marchetto A, Sayer CD, Thies H, Tolotti M (2011b) Defining ecological and chemical reference conditions and restoration targets for nine European lakes. J Paleolimnol 45:415–431

Bigler C, Hall RI (2003) Diatoms as quantitative indicators of July temperature: a validation attempt at century-scale with meteorological data from northern Sweden. Palaeogeogr Palaeoclim Palaeoecol 189:147–160

Bigler C, Larocque I, Peglar SM, Birks HJB, Hall RI (2002) Quantitative multiproxy assessment of long-term patterns of Holocene environmental change from a small lake near Abisko, northern Sweden. Holocene 12:481–496

Bindler R, Korsman T, Renberg I, Hogberg P (2002) Pre-industrial atmospheric pollution: was it important for the pH of acid-sensitive Swedish lakes? Ambio 31:460–465

Birks HH, Ammann B (2000) Two terrestrial records of rapid climatic change during the glacial-Holocene transition (14,000-9,000 calendar years BP) from Europe. Proc Natl Acad Sci USA 97:139–1394

Birks HH, Birks HJB (2006) Multi-proxy studies in palaeolimnology. Veg Hist Archaeobot 15:235–251

Birks HH, Battarbee RW, Birks HJB (2000) The development of the aquatic ecosystem at Krakenes Lake, western Norway, during the late glacial and early Holocene - a synthesis. J Paleolimnol 23:91–114

Birks HH, Jones VJ, Brooks SJ, Peglar SM, Telford RJ, Juggins S, Birks HJB (2012) From cold to cool in northernmost Norway: late-glacial multiproxy environmental reconstructions from Jansvatnet, Hammerfest. Quaternary Sci Rev 33:100–120

Birks HJB (1994) The importance of pollen and diatom taxonomic precision in quantitative palaeoenvironmental reconstructions. Rev Palaeobot Palynol 83:107–117

Birks HJB (1995) Quantitative palaeoenvironmental reconstructions. In: Maddy D, Brew J (eds) Statistical modelling of Quaternary science data. Technical guide 5. Quaternary Research Association, Cambridge, pp 161–254

Birks HJB (1998) Numerical tools in palaeolimnology - progress, potentialities, and problems. J Paleolimnol 20:307–322

Birks HJB (2001) Maximum likelihood environmental calibration and the computer program WACALIB - a correction. J Paleolimnol 25:111–115

Birks HJB (2003) Quantitative palaeoenvironmental reconstructions from Holocene biological data. In: Mackay A, Battarbee RW, Birks HJB, Oldfield F (eds) Global change in the Holocene. Arnold, London, pp 107–123

Birks HJB (2010) Numerical methods for the analysis of diatom assemblage data. In: Smol JP, Stoermer EF (eds) The diatoms – applications for the environmental and earth sciences. Cambridge University Press, Cambridge, pp 23–54

Birks HJB (2012) Chapter 2: Overview of numerical methods in palaeolimnology. In: Birks HJB, Lotter AF, Juggins S, Smol JP (eds) Tracking environmental change using lake sediments. Volume 5: Data handling and numerical techniques. Springer, Dordrecht

Birks HJB, Jones VJ (2012) Chapter 3: Data-sets. In: Birks HJB, Lotter AF, Juggins S, Smol JP (eds) Tracking environmental change using lake sediments. Volume 5: Data handling and numerical techniques. Springer, Dordrecht

Birks HJB, Seppä H (2004) Pollen-based reconstructions of late-Quaternary climate in Europe – progress, problems, and pitfalls. Acta Palaeobot 44:317–334

Birks HJB, Line JM, Juggins S, Stevenson AC, ter Braak CJF (1990a) Diatoms and pH reconstruction. Philos Trans R Soc Lond B 327:263–278

Birks HJB, Juggins S, Line JM (1990b) Lake surface-water chemistry reconstructions from palaeolimnological data. In: Mason BJ (ed) The Surface Waters Acidification Programme. Cambridge University Press, Cambridge, pp 301–313

Birks HJB, Heiri O, Seppä H, Bjune AE (2010) Strengths and weaknesses of quantitative climate reconstructions based on late-Quaternary biological proxies. Open Ecol J 3:68–110

Borcard D, Legendre P, Drapeau P (1992) Partialling out the spatial component of ecological variation. Ecology 73:1045–1055

Borcard D, Gillet F, Legendre P (2011) Numerical ecology with R. Springer, New York

Braak CJF, Šmilauer P (2002) CANOCO reference manual and CanoDraw user's guide – software for canonical community ordination (version 4.5). Microcomputer Power, Ithaca

Bradshaw EG, Anderson NJ (2001) Validation of a diatom-phosphorus calibration set for Sweden. Freshw Biol 46:1035–1048

Bradshaw EG, Rasmussen P, Nielsen H, Anderson NJ (2005) Mid- to late-Holocene land-use change and lake development at Dallund Sø, Denmark: trends in lake primary production as reflected by algal and macrophyte remains. Holocene 15:1130–1142

Brenner M, Whitmore TJ, Flannery MS, Binford MW (1993) Paleolimnological methods for defining target conditions in lake restoration: Florida case studies. Lake Reserv Manage 7: 209–217

Brewer S, Guiot J, Barboni D (2007) Use of pollen as climate proxies. In: Elias SA (ed) Encyclopedia of Quaternary science. Elsevier, Oxford, pp 2497–2508

Brooks SJ (2003) Chironomid analysis to interpret and quantify Holocene climate change. In: Mackay A, Battarbee RW, Birks HJB, Oldfield F (eds) Global change in the Holocene. Arnold, London, pp 328–341

Brooks SJ, Birks HJB (2001) Chironomid-inferred air temperatures from Lateglacial and Holocene sites in north-west Europe: progress and problems. Quaternary Sci Rev 20:1723–1741

Cameron NG, Birks HJB, Jones VJ, Berge F, Catalan J, Flower RJ, Garcia J, Kawecka B, Koinig KA, Marchetto A, Sanchez-Castillo P, Schmidt R, Sisko M, Solovieva N, Stefkova E, Toro M (1999) Surface-sediment and epilithic diatom pH calibration sets for remote European mountain lakes (AL:PE Project) and their comparison with the Surface Waters Acidification Programme (SWAP) calibration set. J Paleolimnol 22:291–317

Charles DF (1985) Relationships between surface sediment diatom assemblages and lakewater characteristics in Adirondack lakes. Ecology 66:994–1011

Clarke A (2001) A diatom-based transfer function to infer historical changes in total nitrogen from coastal sediments: a case study from Roskilde Fjord, Denmark. Unpublished PhD thesis, University of Newcastle, Newcastle-upon-Tyne, UK

Clarke A, Juggins S, Conley D (2003) A 150-year reconstruction of the history of coastal eutrophication in Roskilde Fjord, Denmark. Mar Pollut Bull 46:1615–1629

Cook DR, Weisberg S (1982) Residuals and influence in regression. Chapman & Hall, London

Correa-Metrio A, Urrego DH, Cabrera KR, Bush MB (2011) paleoMAS: paleoecological analysis. R package version 2.0. http://CRAN.R-project.org/package=paleoMAS

Coudun C, Gégout J-C (2006) The derivation of species response curves with Gaussian logistic regression is sensitive to sampling intensity and curve characteristics. Ecol Model 199: 164–175

Crawley M (2005) Statistics. An introduction using R. Wiley, Chichester

Cumming BF, Smol JP (1993) Development of diatom-based salinity models for paleoclimatic research from lakes in British Columbia (Canada). Hydrobiologia 269(270):179–196

Cumming BF, Smol JP, Birks HJB (1992) Scaled chrysophytes (Chrysophyceae and Synurophyceae) from Adirondack Drainage Lakes and their relationship to environmental variables. J Phycol 28:162–178

Cumming BF, Laird KR, Fritz SC, Verschuren D (2012) Chapter 20: Tracking Holocene climatic change with aquatic biota from lake sediments: case studies of commonly used numerical techniques. In: Birks HJB, Lotter AF, Juggins S, Smol JP (eds) Tracking environmental change using lake sediments. Volume 5: Data handling and numerical techniques. Springer, Dordrecht

Curtis CJ, Juggins S, Clarke G, Battarbee RW, Kernan M, Catalan J, Thompson R, Posch M (2009) Regional influence of acid deposition and climate change in European lakes assessed using diatom transfer functions. Freshw Biol 54:2555–2572

Dabakk E, Nilsson M, Geladi P, Wold S, Renberg I (1999) Sampling reproducibility and error estimation in near infrared calibration of lake sediments for water quality monitoring. J NIR Spectrosc 7:241–250

Davies JJL, Jenkins A, Monteith DT, Evans CD, Cooper DM (2005) Trends in surface water chemistry of acidified UK freshwaters, 1988–2002. Environ Pollut 137:27–39

Davis B, Brewer S, Stevenson A, Guiot J (2003) The temperature of Europe during the Holocene reconstructed from pollen data. Quaternary Sci Rev 22:1701–1716

Dixit SS, Smol JP, Charles DF, Hughes RM, Paulsen SG, Collins GB (1999) Assessing water quality changes in the lakes of the northeastern United States using sediment diatoms. Can J Fish Aquat Sci 56:131–152

Dixit SS, Dixit AS, Smol JP (2002) Diatom and chrysophyte functions and inferences of post-industrial acidification and recent recovery trends in Killarney lakes (Ontario, Canada). J Paleolimnol 27:79–96

Draper NR, Smith H (1981) Applied regression analysis. Wiley, New York

Engstrom DR, Fritz SC, Almendinger JE, Juggins S (2000) Chemical and biological trends during lake evolution in recently deglaciated terrain. Nature 408:161–166

Faith DP, Minchin PR, Belbin L (1987) Compositional dissimilarity as a robust measure of ecological distance. Vegetatio 69:57–68

Finsinger W, Wagner-Cremer F (2009) Stomatal-based inference models for reconstruction of atmospheric CO_2 concentration: a method assessment using a calibration and validation approach. Holocene 19:757–764

Flower RJ (1986) The relationship between surface sediment diatom assemblages and pH in 33 Galloway lakes: some regression models for reconstructing pH and their application to sediment cores. Hydrobiologia 143:93–103

Flower R, Battarbee R (1983) Diatom evidence for recent acidification of two Scottish Lochs. Nature 305:130–133

Fritz SC (1990) Twentieth-century salinity and water-level fluctuations in Devils Lake, North Dakota: test of a diatom-based transfer function. Limnol Oceanogr 35:1171–1781

Fritz SC, Juggins S, Battarbee RW, Engstrom DR (1991) Reconstruction of past changes in salinity and climate using a diatom-based transfer function. Nature 352:706–708

Fritz SC, Kingston JC, Engstrom DR (1993) Quantitative trophic reconstruction from sedimentary diatom assemblages: a cautionary tale. Freshw Biol 30:1–23

Fritz SC, Engstrom DR, Juggins S (2004) Patterns of early lake evolution in boreal landscapes: a comparison of stratigraphic inferences with a modern chronosequence in Glacier Bay, Alaska. Holocene 14:828–840

Gasse F, Juggins S, Ben Khelifa L (1995) Diatom-based transfer functions for inferring past hydrochemical characteristics of African lakes. Palaeogeogr Palaeoclim Palaeoecol 117:3–54

Gasse F, Barker P, Gell PA, Fritz SC, Chalie F (1997) Diatom-inferred salinity in palaeolakes: an indirect tracer of climate change. Quaternary Sci Rev 16:547–563

Gavin DG, Oswald WW, Wahl ER, Williams JW (2003) A statistical approach to evaluating distance metrics and analog assignments for pollen records. Quaternary Res 60:356–367

Geladi P, Kowalski BR (1986) Partial least squares regression: a tutorial. Anal Chim Acta 185:1–17

Gersonde R, Crosta X, Abelmann A, Armand L (2005) Sea-surface temperature and sea ice distribution of the Southern Ocean at the EPILOG Last Glacial Maximum - a circum-Antarctic view based on siliceous microfossil records. Quaternary Sci Rev 24:869–896

Ginn BK, Cumming BF, Smol JP (2007) Diatom-based environmental inferences and model comparisons from 494 northwestern North American lakes. J Phycol 43:647–661

Gonzales LM, Williams JW, Grimm EC (2009) Expanded response-surfaces: a new method to reconstruct paleoclimates from fossil pollen assemblages that lack modern analogues. Quaternary Sci Rev 28:3315–3332

Guiot J, de Vernal A (2007) Transfer functions: methods for quantitative paleoceanography based on microfossils. In: Hillaire-Marcel C, de Vernal A (eds) Proxies in late Cenozoic paleoceanography. Elsevier, Amsterdam, pp 523–563

Hall RI, Smol JP (2010) Diatoms as indicators of lake eutrophication. In: Smol JP, Stoermer EF (eds) The diatoms – applications for the environmental and earth sciences. Cambridge University Press, Cambridge, pp 122–151

Hallgren E, Palmer M, Milberg P (1999) Data diving with cross-validation: an investigation of broad-scale gradients in Swedish weed communities. Ecology 87:1037–1051

Haslett J, Whiley M, Bhattacharya S, Salter-Townsend M, Wilson S, Allen J, Huntley B, Mitchell F (2006) Bayesian palaeoclimate reconstruction. J R Stat Soc A 169:395–438

Hausmann S, Larocque-Tobler I, Richard PJH, Pienitz R, St-Onge G, Fye F (2011) Diatom-inferred wind activity at Lac du Sommet, southern Québec, Canada: a multiproxy paleoclimate reconstruction based on diatoms, chironomids and pollen for the past 9500 years. Holocene 21:925–938

Heiri O, Lotter AF (2005) Holocene and Lateglacial summer temperature reconstruction in the Swiss Alps based on fossil assemblages of aquatic organisms: a review. Boreas 34:506–516

Heiri O, Birks HJB, Brooks SJ, Velle G, Willassen E (2003) Effects of within-lake variability of fossil assemblages on quantitative chironomid-inferred temperature reconstruction. Palaeogeogr Palaeoclim Palaeoecol 199:95–106

Hill MO (1973) Diversity and evenness: a unifying notation and its consequences. Ecology 54: 427–432

Holden P, Mackay A, Simpson GL (2008) A Bayesian palaeoenvironmental transfer function model for acidified lakes. J Paleolimnol 39:551–566

Hübener T, Dressler M, Schwarz A, Langner K, Adler S (2008) Dynamic adjustment of training-sets ('moving-window' reconstruction) by using transfer functions in paleolimnology – a new approach. J Paleolimnol 40:79–95

Huber K, Weckström K, Drescher-Schneider R, Knoll J, Schmidt J, Schmidt R (2010) Climate changes during the last glacial termination inferred from diatom-based temperatures and pollen in a sediment core from Längsee (Austria). J Paleolimnol 43:131–147

Imbrie J, Kipp NG (1971) A new micropaleontological method for quantitative paleoclimatology: application to a Late Pleistocene Caribbean core. In: Turekian KK (ed) The Late Cenozoic glacial ages. Yale University Press, New Haven, pp 77–181

Imbrie J, Webb T (1981) Transfer functions: calibrating micropaleontological data in climatic terms. In: Berger A (ed) Climatic variations and variability: facts and theories. Reidel, Dordrecht, pp 125–134

Jackson ST, Williams JW (2004) Modern analogs in Quaternary paleoecology: here today, gone yesterday, gone tomorrow? Annu Rev Earth Planet Sci 32:495–537

Jenkins A, Whitehead PG, Cosby BJ, Birks HJB (1990) Modelling long-term acidification: a comparison with diatom reconstructions and the implications for reversibility. Philos Trans R Soc Lond B 327:435–440

Johnson JB, Omland KS (2004) Model selection in ecology and evolution. Trends Ecol Evol 19:101–108

Jones VJ, Juggins S (1995) The construction of a diatom-based chlorophyll a transfer function and its application at three lakes on Signy Island (maritime Antarctic) subject to differing degrees of nutrient enrichment. Freshw Biol 34:433–445

Jones VJ, Stevenson AC, Battarbee RW (1986) Lake acidification and the land use hypothesis: a mid-postglacial analogue. Nature 322:157–158

Juggins S (1992) Diatoms in the Thames Estuary, England: ecology, palaeoecology, and salinity transfer function. Bibliotheca Diatomologica 25:1–216

Juggins S (2007) C2 Version 1.5 User Guide. Software for ecological and palaeoecological data analysis and visualisation. University of Newcastle, Newcastle-upon-Tyne

Juggins S (2009) rioja: analysis of Quaternary science data, R package version 0.5-6. http://cran.r-project.org/package=rioja

Juggins S, Telford RJ (2012) Chapter 5: Exploratory data analysis and data display. In: Birks HJB, Lotter AF, Juggins S, Smol JP (eds) Tracking environmental change using lake sediments. Volume 5: Data handling and numerical techniques. Springer, Dordrecht

Juggins S, Battarbee RW, Fritz SC (1994) Diatom/salinity transfer functions and climate change: an assessment of methods and application to two Holocene sequences from the northern Great Plains. In: Funnell BM, Kay RLF (eds) Palaeoclimate of the last glacial/interglacial cycle. NERC Earth Sciences Directorate, Swindon, pp 37–41

Kamenik C, Schmidt R (2005) Chrysophyte resting stages: a tool for reconstructing winter/spring climate from Alpine lake sediments. Boreas 34:477–489

Kingston JC, Cumming BF, Uutala AJ, Smol JP, Camburn KE, Charles DF, Dixit SS, Kreis RG Jr (1992) Biological quality control and quality assurance: a case study in paleolimnologal biomonitoring. In: McKenzie DH, Hyatt DE, McDonald VJ (eds) Ecological indicators. Volume 1. Elsevier Science Publishers, London, pp 1542–1543

Korhola A, Vasko K, Toivonen HTT, Olander H (2002) Holocene temperature changes in northern Fennoscandia reconstructed from chironomids using Bayesian modelling. Quaternary Sci Rev 21:1841–1860

Köster D, Racca JMJ, Pienitz R (2004) Diatom-based inference models and reconstructions revisited: methods and transformations. J Paleolimnol 32:233–245

Kucera M, Weinelt M, Kiefer T, Pflaumann U, Hayes A, Chen MT, Mix AC, Barrows TT, Cortijo E, Duprat J, Juggins S, Waelbroeck C (2005) Reconstruction of sea-surface temperatures from assemblages of planktonic foraminifera: multi-technique approach based on geographically constrained calibration data sets and its application to glacial Atlantic and Pacific Oceans. Quaternary Sci Rev 24:951–998

Kumke T, Schölzel C, Hense A (2004) Transfer functions for paleoclimate reconstructions - theory and methods. In: Fischer H, Kumke T, Lohmann G, Flöser G, Miller H, von Storch H, Negendank JFW (eds) The climate in historical times: towards a synthesis of Holocene proxy data and climate models. Springer, Berlin, pp 229–244

Larocque I, Bigler C (2004) Similarities and discrepancies between chironomid- and diatom-inferred temperature reconstructions through the Holocene at Lake 850, northern Sweden. Quaternary Int 122:109–121

Larocque I, Hall RI (2003) Chironomids as quantitative indicators of mean July air temperature: validation by comparison with century-long meteorological records from northern Sweden. J Paleolimnol 29:475–493

Larocque-Tobler I, Grosjean M, Kamenik C (2011) Calibration-in-time versus calibration-in-space (transfer function) to quantitatively infer July air temperature using biological indicators (chironomids) preserved in lake sediments. Palaeogeogr Palaeoclim Palaeoecol 299:281–288

Legendre P, Birks HJB (2012a) Chapter 7: Clustering and partitioning. In: Birks HJB, Lotter AF, Juggins S, Smol JP (eds) Tracking environmental change using lake sediments. Volume 5: Data handling and numerical techniques. Springer, Dordrecht

Legendre P, Birks HJB (2012b) Chapter 8: From classical to canonical ordination. In: Birks HJB, Lotter AF, Juggins S, Smol JP (eds) Tracking environmental change using lake sediments. Volume 5: Data handling and numerical techniques. Springer, Dordrecht

Li B, Nychka DW, Amman CM (2010) The value of multiproxy reconstruction of past climate. J Am Stat Assoc 105:883–895

Limpert E, Stahel WA, Abbt M (2001) Log-normal distributions across the sciences: keys and clues. Bioscience 51:341–352

Line JM, ter Braak CJF, Birks HJB (1994) WACALIB version 3.3 - a computer program to reconstruct environmental variables from fossil assemblages by weighted averaging and to derive sample-specific errors of prediction. J Paleolimnol 10:147–152

Lotter AF (2003) Multi-proxy climate reconstructions. In: Mackay A, Battarbee RW, Birks HJB, Oldfield F (eds) Global change in the Holocene. Arnold, London, pp 373–383

Lotter AF, Anderson NJ (2012) Chapter 18: Limnological responses to environmental changes at inter-annual to decadal time-sclaes. In: Birks HJB, Lotter AF, Juggins S, Smol JP (eds) Tracking environmental change using lake sediments. Volume 5: Data handling and numerical techniques. Springer, Dordrecht

Lotter AF, Birks HJB, Hofmann W, Marchetto A (1997) Modern diatom, cladocera, chironomid, and chrysophytes cyst assemblages as quantitative indicators for the reconstruction of past environmental conditions in the Alps. I. Climate. J Paleolimnol 18:395–420

Lotter AF, Walker IR, Brooks SJ, Hofmann W (1999) An intercontinental comparison of chironomid palaeotemperature inference models: Europe vs North America. Quaternary Sci Rev 18:717–735

Luoto T (2010) Hydrological change in lakes inferred from midge assemblages through the use of an intralake calibration set. Ecol Monogr 80:303–329

Mackay AW, Ryves DB, Battarbee RW, Flower RJ, Jewson D, Rioual P, Sturm M (2005) 1000 years of climate variability in central Asia: assessing the evidence using Lake Baikal (Russia) diatom assemblages and the application of a diatom-inferred model of snow cover on the lake. Global Planet Change 46:281–297

Maher LJ, Heiri O, Lotter AF (2012) Chapter 6: Assessment of uncertainties associated with palaeolimnological laboratory methods and microfossil analysis. In: Birks HJB, Lotter AF, Juggins S, Smol JP (eds) Tracking environmental change using lake sediments. Volume 5: Data handling and numerical techniques. Springer, Dordrecht

Malmgren BA, Kucera M, Nyberg J, Waelbroeck C (2001) Comparison of statistical and neural network techniques for estimating past sea-surface temperatures from planktonic foraminifer census data. Paleoceanography 16:520–530

Manly BFJ (1997) Randomization, bootstrap and Monte Carlo methods in biology, 2nd edn. Chapman & Hall, London

Martens H, Næs T (1989) Multivariate calibration. Wiley, Chichester

McCune B (1997) Influence of noisy environmental data on canonical correspondence analysis. Ecology 78:2617–2623

Meyer DG, Butler DG (1993) Statistical validation. Ecol Model 68:21–32

Mohler CL (1983) Effect of sampling pattern on estimation of species distributions along gradients. Vegetatio 54:97–102

Montgomery DC, Peck EA (1982) Introduction to linear regression analysis. Wiley, New York

Mooney CZ, Duval RD (1993) Bootstrapping: a non-parametric approach to statistical inference. Sage, Newbury Park

Munro MAR, Kreiser AM, Battarbee RW, Juggins S, Stevenson AC, Anderson DS, Anderson NJ, Berge F, Birks HJB, Davis RB, Flower RJ, Fritz SC, Haworth EY, Jones VJ, Kingston JC, Renberg I (1990) Diatom quality control and data handling. Philos Trans R Soc Lond B 327: 257–261

Næs T, Isaksson T (1992) Locally weighted regression in diffuse near-infrared transmittance spectroscopy. Appl Spectrosc 46:34–43

Næs T, Kvaal K, Isaksson T, Miller C (1993) Artificial neural networks in multivariate calibration. J Near IR Spectrosc 1:1–11

Næs T, Isaksson T, Fearn T, Davies T (2002) A user-friendly guide to multivariate calibration and classification. NIR Publications, Chichester

Nilsson MB, Dabakk E, Korsman T, Renberg I (1996) Quantifying relationships between near-infrared reflectance spectra of lake sediments and water chemistry. Environ Sci Technol 30:2586–2590

Oksanen J, Blanchet FG, Kindt R, Legendre P, O'Hara RB, Simpson GL, Solymos P, Stevens MHM, Wagner H (2011) vegan: community ecology package. R package version 1.17-8. http://CRAN.R-project.org/package=vegan

Olden JD, Jackson DA (2002) Illuminating the "black box": a randomization approach for understanding variable contributions in artificial neural networks. Ecol Model 154:135–150

Olden JD, Joy MK, Death RG (2004) An accurate comparison of methods for quantifying variable importance in artificial neural networks using simulated data. Ecol Model 178:389–397

Ott WR (1990) A physical explanation of the lognormality of pollutant concentrations. J Air Waste Manage Assoc 40:1378–1383

Overpeck JT, Webb T, Prentice IC (1985) Quantitative interpretation of fossil pollen spectra: dissimilarity coefficients and the method of modern analogs. Quaternary Res 23:87–108

Paterson AM, Cumming BF, Dixit SS, Smol JP (2002a) The importance of model choice on pH inferences from scaled chrysophyte assemblages in North America. J Paleolimnol 27:379–391

Paterson AM, Morimoto DS, Cumming BF, Smol JP, Szeicz JM (2002b) A paleolimnological investigation of the effects of forest fire on lake water quality in northwestern Ontario over the past ca. 150 years. Can J Bot 80:1329–1336

Peyron O, Begeot C, Brewer S, Heiri O, Magny M, Millet L, Ruffaldi P, van Campo E, Yu G (2005) Late-Glacial climatic changes in eastern France (Lake Lautrey) from pollen, lake-levels, and chironomids. Quaternary Res 64:197–211

Philibert A, Prairie YT (2002a) Diatom-based transfer functions for western Quebec lakes (Abitibi and Haute Mauricie): the possible role of epilimnetic CO_2 concentration in influencing diatom assemblages. J Paleolimnol 27:465–480

Philibert A, Prairie YT (2002b) Is the introduction of benthic species necessary for open-water chemical reconstruction in diatom-based transfer functions? Can J Fish Aquat Sci 59:938–951

Pienitz R, Smol JP, Birks HJB (1995) Assessment of fresh-water diatoms as quantitative indicators of past climatic change in the Yukon and Northwest-Territories, Canada. J Paleolimnol 13: 21–49

Piñeiro G, Perelman S, Guerschman JP, Paruelo JM (2008) How to evaluate models: observed vs. predicted or predicted vs. observed? Ecol Model 216:316–332

Power M (1993) The predictive validation of ecological and environmental models. Ecol Model 68:33–50

Prell WL (1985) The stability of low-latitude sea-surface temperatures: an evaluation of the CLIMAP reconstruction with emphasis on the positive SST anomalies. Special Publication TRO 25, US. Department of Energy, Washington, DC
Prentice IC (1980) Multidimensional scaling as a research tool in Quaternary palynology - a review of theory and methods. Rev Palaeobot Palynol 31:71–104
Quinlan R, Smol JP (2001) Setting minimum head capsule abundance and taxa deletion criteria in chironomid-based inference models. J Paleolimnol 26:327–342
R Development Core Team (2010) R: a language and environment for statistical computing. R Foundation for Statistical Computing, Vienna, Austria. http://www.R-project.org/
Racca JMJ, Prairie Y (2004) Apparent and real bias in numerical transfer functions in palaeolimnology. J Paleolimnol 31:117–124
Racca JMJ, Philibert A, Racca R, Prairie YT (2001) A comparison between diatom-based pH inference models using artificial neural networks (ANN), weighted averaging (WA) and weighted averaging partial least squares (WA-PLS) regressions. J Paleolimnol 26:411–422
Racca JMJ, Wild M, Birks HJB, Prairie YT (2003) Separating wheat from chaff: diatom taxon selection using an artificial neural network pruning algorithm. J Paleolimnol 29:123–133
Racca JMJ, Gregory-Eaves I, Pienitz R, Prairie YT (2004) Tailoring palaeolimnological diatom-based transfer functions. Can J Fish Aquat Sci 61:2440–2454
Racca JMJ, Racca R, Pienitz R, Prairie YT (2007) paleoNet: new software for building, evaluating and applying neural network based transfer functions in paleoecology. J Paleolimnol 38: 467–472
Reavie ED, Neill KE, Little JL, Smol JP (2006) Cultural eutrophication trends in three southeastern Ontario lakes: a paleolimnological perspective. Lake Reserv Manage 22:44–58
Reid M (2005) Diatom-based models for reconstructing past water quality and productivity in New Zealand lakes. J Paleolimnol 33:13–38
Renberg I, Hellberg T (1982) The pH history of lakes in southwestern Sweden, as calculated from the subfossil flora of the sediments. Ambio 11:30–33
Renberg I, Korsman T, Anderson NJ (1993) A temporal perspective of lake acidification in Sweden. Ambio 22:264–271
Robertson I, Lucy D, Baxter L, Pollard AM, Aykroyd RG, Barker AC, Carter AHC, Switsur VR, Waterhouse JS (1999) A kernel-based Bayesian approach to climatic reconstruction. Holocene 9:495–500
Rosén P (2005) Total organic carbon (TOC) of lake water during the Holocene inferred from lake sediments and near-infrared spectroscopy (NIRS) in eight lakes from northern Sweden. Biogeochemistry 76:503–516
Rosén P, Segerstrom U, Eriksson L, Renberg I, Birks HJB (2001) Holocene climatic change reconstructed from diatoms, chironomids, pollen and near-infrared spectroscopy at an alpine lake (Sjuodjljaure) in northern Sweden. Holocene 11:551–562
Rosén P, Segerstrom U, Eriksson L, Renberg I (2003) Do diatom, chironomid, and pollen records consistently infer Holocene July air temperature? A comparison using sediment cores from four alpine lakes in northern Sweden. Arct Antarct Alp Res 35:279–290
Roux M (1979) Estimation des paleoclimats d'apres l'ecologie des foraminiferes. Les Cahiers de l'Analyse des donnees IV:61–79
Rühland KM, Smol JP (2002) Freshwater diatoms from the Canadian arctic treeline and development of paleolimnological inference models. J Phycol 38:249–264
Rymer N (1978) The use of uniformitarianism and analog in palaeoecology, particularly pollen analysis. In: Walker D, Guppy JC (eds) Biology and Quaternary environments. Australian Academy of Sciences, Canberra, pp 245–257
Sayer CD (2001) Problems with the application of diatom-total phosphorus transfer functions: examples from a shallow English lake. Freshw Biol 46:743–757
Schmidt R, Kamenik C, Kaiblinger C, Hetzel M (2004) Tracking Holocene environmental changes in an alpine lake sediment core: application of regional diatom calibration, geochemistry, and pollen. J Paleolimnol 32:177–196

Seppä H, Birks HJB, Odland A, Poska A, Veski S (2004) A modern pollen-climate calibration set from northern Europe: developing and testing a tool for palaeoclimatological reconstructions. J Biogeogr 31:251–267
Shurin JB, Cottenie K, Hillebrand H (2009) Spatial autocorrelation and dispersal limitation in freshwater organisms. Oecologia 159:151–159
Sieger R, Gersonde R, Zielinski U (1999) A new extended software package for quantitative paleoenvironmental reconstructions. EOS, Transactions, American Geophysical Union Electronic Supplement, 11 May 1999
Simpson GL (2007) Analogue methods in palaeoecology: using the analogue package. J Stat Softw 22:1–29
Simpson GL (2012) Chapter 15: Modern analogue techniques. In: Birks HJB, Lotter AF, Juggins S, Smol JP (eds) Tracking environmental change using lake sediments. Volume 5: Data handling and numerical techniques. Springer, Dordrecht
Simpson GL, Birks HJB (2012) Chapter 9: Statistical learning in palaeolimnology. In: Birks HJB, Lotter AF, Juggins S, Smol JP (eds) Tracking environmental change using lake sediments. Volume 5: Data handling and numerical techniques. Springer, Dordrecht
Simpson GL, Hall RI (2012) Chapter 19: Human impacts – applications of numerical methods to evaluate surface-water acidification and eutrophication. In: Birks HJB, Lotter AF, Juggins S, Smol JP (eds) Tracking environmental change using lake sediments. Volume 5: Data handling and numerical techniques. Springer, Dordrecht
Simpson GL, Oksanen J (2009) analogue: analogue and weighted averaging methods for palaeoecology. R package version 0.6-8. http://cran.r-project.org/package=analogue
Simpson GL, Shilland EM, Winterbottom JM, Keay J (2005) Defining reference conditions for acidified waters using a modern analogue approach. Environ Pollut 137:119–133
Siver PA (1999) Development of paleolimnological inference models for pH, total nitrogen and specific conductivity based on planktonic diatoms. J Paleolimnol 21:45–59
Siver PA, Hamer JS (1992) Seasonal periodicity of Chrysophyceae and Synurophyceae in a small New-England lake - implications for paleolimnological research. J Phycol 28:186–198
Smol JP (2008) Pollution of lakes and rivers: a paleoenvironmental perspective, 2nd edn. Blackwell, Oxford
Sokal RR, Rohlf FJ (1981) Biometry, 2nd edn. Freeman, San Francisco
Stauffer RE (1988) Sampling strategies and associated errors in estimating epilimnetic chlorophyll in eutrophic lakes. Water Resource Res 24:1459–1469
Stevenson AC, Birks HJB, Flower RJ, Battarbee RW (1989) Diatom-based pH reconstruction of lake acidification using canonical correspondence analysis. Ambio 18:228–233
Stevenson AC, Juggins S, Birks HJB, Anderson DS, Anderson NJ, Battarbee RW, Berge F, Davis RB, Flower RJ, Haworth EY, Jones VJ, Kingston JC, Kreiser AM, Line JM, Munro MAR, Renberg I (1991) The Surface Waters Acidification Project palaeolimnology programme: modern diatom/lake-water chemistry data-set. ENSIS, London
Stine R (1990) An introduction to bootstrap methods: examples and ideas. In: Fox J, Long JS (eds) Modern methods of data analysis. Sage, Newbury Park, pp 325–373
Telford RJ (2011) palaeoSig: significance tests for palaeoenvironmental reconstructions. R package version 1.0. http://cran.r-project.org/package/palaeoSig/index.html
Telford RJ, Birks HJB (2005) The secret assumption of transfer functions: problems with spatial autocorrelation in evaluating model performance. Quaternary Sci Rev 24:2173–2179
Telford RJ, Birks HJB (2009) Design and evaluation of transfer functions in spatially structured environments. Quaternary Sci Rev 28:1309–1316
Telford RJ, Birks HJB (2011a) Effect of unequal sampling along the environmental gradient on transfer functions. J Paleolimnol 46:99–106
Telford RJ, Birks HJB (2011b) A novel method for assessing the statistical significance of quantitative reconstructions inferred from biotic assemblages. Quaternary Sci Rev 30:1271–1278
Telford RJ, Andersson C, Birks HJB, Juggins S (2004) Biases in the estimation of transfer function prediction errors. Paleoceanography 19:PA4014

ter Braak CJF (1995) Non-linear methods for multivariate statistical calibration and their use in palaeoecology: a comparison of inverse (*k*-nearest neighbours, partial least squares and weighted averaging partial least squares) and classical approaches. Chemometr Intell Lab Syst 28:165–180

ter Braak CJF, Barendregt LG (1986) Weighted averaging of species indicator values: its efficiency in environmental calibration. Math Biosci 78:57–72

ter Braak CJF, Juggins S (1993) Weighted averaging partial least squares regression (WA-PLS): an improved method for reconstructing environmental variables from species assemblages. Hydrobiologia 269:485–502

ter Braak CJF, Looman CWN (1986) Weighted averaging, logistic regression and the Gaussian response model. Vegetatio 65:3–11

ter Braak CJF, Prentice IC (1988) A theory of gradient analysis. Adv Ecol Res 18:271–317

ter Braak CJF, van Dam H (1989) Inferring pH from diatoms: a comparison of old and new calibration methods. Hydrobiologia 178:209–223

ter Braak CJF, Verdonschot PFM (1995) Canonical correspondence analysis and related multivariate methods in aquatic ecology. Aquat Sci 57:255–289

ter Braak CJF, Juggins S, Birks HJB, van der Voet H (1993) Weighted averaging partial least squares regression (WA-PLS): definition and comparison with other methods for species-environment calibration. In: Patil GP, Rao CR (eds) Multivariate environmental statistics. Elsevier Science Publishers, Amsterdam, pp 525–560

Thompson R, Price D, Cameron N, Jones VJ, Bigler C, Rosen P, Hall RI, Catalan J, Garcia J, Weckstrom J, Korhola A (2005) Quantitative calibration of remote mountain-lake sediments as climatic recorders of air temperature and ice-cover duration. Arct Antarct Alp Res 37:626–635

Toivonen H, Manilla H, Korhola A, Olander H (2001) Applying Bayesian statistics to organism-based environmental reconstruction. Ecol App 11:618–630

van der Voet H (1994) Comparing the predictive accuracy of models using a simple randomization test. Chemometr Intell Lab Syst 25:313–323

Varmuza K, Filzmoser P (2009) Introduction to multivariate statistical analysis in chemometrics. CRC Press, Boca Raton

Vasko K, Toivonen H, Korhola A (2000) A Bayesian multinomial Gaussian response model for organism-based environmental reconstruction. J Paleolimnol 24:243–250

Velle G, Brodersen KP, Birks HJB, Willassen E (2010) Midges as quantitative temperature indicator species: lessons for palaeoecology. Holocene 20:989–1002

Velle G, Kongshavn K, Birks HJB (2011a) Minimizing the edge-effect in environmental reconstructions by trimming the calibration-set: chironomid-inferred temperatures from Spitsbergen. Holocene 21:417–430

Velle G, Telford RJ, Birks HJB (2011b) Validity of intra-site transfer functions. J Paleolimnol (submitted)

Verschuren D (2003) Lake-based climate reconstruction in Africa: progress and challenges. Hydrobiologia 500:315–330

Verschuren D, Cumming BF, Laird KR (2004) Quantitative reconstruction of past salinity variations in African lakes: assessment of chironomid-based inference models (Insecta: Diptera) in space and time. Can J Fish Aquat Sci 61:986–998

Walker IR, Levesque AJ, Cwynar LC, Lotter AF (1997) An expanded surface-water palaeotemperature inference model for use with fossil midges from eastern Canada. J Paleolimnol 18:165–178

Wallach D, Goffinet B (1989) Mean squared error of prediction as a criterion for evaluating and comparing system models. Ecol Model 44:299–306

Webb A (2002) Statistical pattern recognition, 2nd edn. Arnold, London

Weckström K, Juggins S, Korhola A (2004) Quantifying background nutrient concentrations in coastal waters: a case study from an urban embayment of the Baltic Sea. Ambio 33:324–327

Wehrens R (2011) Chemometrics with R. Springer, New York

Werner P, Smol JP (2005) Diatom-environmental relationships and nutrient transfer functions from contrasting shallow and deep limestone lakes in Ontario, Canada. Hydrobiologia 533:145–173

Wilson SE, Cumming BF, Smol JP (1996) Assessing the reliability of salinity inference models from diatom assemblages: an examination of a 219-lake data set from western North America. Can J Fish Aquat Sci 53:1580–1594

Xu Q, Li Y, Bunting MJ, Tian F, Liu J (2010) The effects of training set selection on the relationship between pollen assemblages and climate parameters: implications for reconstructing past climate. Palaeogeogr Palaeoclim Palaeoecol 1289:123–133

Yang XD, Kamenik C, Schmidt R, Wang SM (2003) Diatom-based conductivity and water-level inference models from eastern Tibetan (Qinghai-Xizang) Plateau lakes. J Paleolimnol 30:1–19

Yuan LL (2007) Maximum likelihood method for predicting environmental conditions from assemblage composition: the R package bio.infer. J Stat Softw 22:1–20

Chapter 15
Analogue Methods in Palaeolimnology

Gavin L. Simpson

Abstract Analogue methods in palaeolimnology consist of the modern analogue technique (MAT) as a means of reconstructing quantitatively past environments from proxy stratigraphical biological data and analogue matching (AM) as a means of comparing fossil assemblages with modern assemblages to inform environmental conservation and restoration of degraded lakes. The mathematics of MAT are presented and problems of spatial autocorrelation on MAT's performance statistics are reviewed.

Analogue matching using one or more proxies (e.g., diatoms, cladocerans) and the choice of appropriate dissimilarity measures are discussed. Various approaches to answering the question how many analogues (k) should be used for environmental reconstructions or to set restoration targets are discussed. These include choosing k to optimise some error function such as root mean squared error of prediction, finding 'jumps' in the dissimilarity values, examining the reference distributions of all the modern dissimilarities, Monte Carlo resampling, constructing receiver operating characteristic (ROC) curves, and applying logistic regression analyses. The use of analogue matching as a tool to help evaluate palaeoenvironmental reconstructions is outlined. Suitable software and directions for future work are discussed.

Keywords Analogue matching • Analogue selection criteria • Dissimilarity coefficients • Environmental reconstruction • Logistic regression • Modern analogue technique • Monte Carlo resampling • Receiver operating characteristic (ROC) curves • Reconstruction evaluation • Software • Spatial autocorrelation

---

G.L. Simpson (✉)
Environmental Change Research Centre, University College London, Pearson Building, Gower Street, London WC1E 6BT, UK
e-mail: gavin.simpson@ucl.ac.uk

H.J.B. Birks et al. (eds.), *Tracking Environmental Change Using Lake Sediments*, Developments in Paleoenvironmental Research 5, DOI 10.1007/978-94-007-2745-8_15,

Introduction

The result, therefore, of our present enquiry is, that we find no vestige of a beginning, — no prospect of an end. James Hutton, 1795, Theory of the Earth

A long-standing assumption in geology is that in nature there are fundamental physical laws or rules of behaviour and that even though the rates of change of environmental processes may vary over time, these underlying rules remain constant. This is Huttonian Uniformitarianism and gives rise to the notion that *the present is the key to the past* – that we can understand the past history of the Earth through analogy to processes occurring today.

Reasoning by analogy is also a key assumption in the science of palaeoecology (Rymer 1978; Birks et al. 2010). Palaeoecologists observe the past through analysis of proxies recovered from historical archives such as sediment cores from lakes. The interpretation of changes in these proxies over time is largely driven by our observations of those same proxies in the present across spatial environmental gradients of interest, such as lake-water pH or climate. In this sense, we substitute spatial information, gleaned from observations of the modern world, for temporal information obtained from our historical archives; a process called *space-for-time substitution*.

This method of working pervades our studies in palaeoecology, including palaeolimnology, but none more so than the techniques of analogue matching (AM) and the modern analogue technique (MAT), which I collectively refer to here as *analogue methods*. Here, I outline the theory and practice of analogue methods in palaeolimnology and provide examples of how these methods may be applied to palaeoenvironmental data. First, I discuss the modern analogue technique as a method for palaeoenvironmental reconstruction. Second, analogue matching is described with examples of how the approach is being used to inform environmental conservation and restoration of degraded lakes. Third, I discuss several technical issues relating to computations involved in analogue methods. Finally, I discuss how analogue methods may be used for evaluating palaeoenvironmental reconstructions.

The Modern Analogue Technique (MAT)

A simple method of inferring past environmental conditions from fossil species assemblages is the modern analogue technique, or MAT (Overpeck et al. 1985; Prell 1985). In the MAT, an inference about the prevailing environmental conditions at the time of the observed fossil assemblage is that those same environmental conditions occur today where we find similar modern species assemblages. Sites where modern species assemblages are observed that are similar to past assemblages are known as *modern analogues*. Where more than one modern analogue can be identified we take, as the inference of past conditions, a consensus of the present-day environmental conditions of all the modern analogues, such as the average.

The MAT has been used widely in palaeoceanographic (e.g., Pflaumann et al. 2003; Kucera et al. 2005) and palynological studies (e.g., Davis et al. 2003; Gavin et al. 2003; Sawada et al. 2004; Williams and Shuman 2008), though it has, so far, been little used in the field of palaeolimnology (Birks 1998).

Here, for completeness, I recap the basic theory and notation of calibration or 'transfer' functions, with specific reference to the MAT. For a more detailed discussion of calibration, see Birks et al. (2010) and Juggins and Birks (2012: Chap. 14).

MAT is an inverse multivariate calibration approach, arising from the regression of **X** on **Y**

$$\mathbf{X} = g\left(\mathbf{Y}\right) + \varepsilon \tag{15.1}$$

where **Y** is an $n \times m$ matrix of counts on m species and **X** is an $n \times p$ matrix of p environmental variables on n samples or sites. The calibration function $g(\cdot)$ is determined by some function of the p environmental variables for the k closest samples identified from the training-set. In the statistical literature, the MAT is more commonly used for problems of classification or discrimination, where it is known as k-nearest neighbours (k-NN) (Webb 2002; Fielding 2007). k-NN is also one of several approaches to regression via smoothing (ter Braak 1995).

The dissimilarity between each fossil sample and the training-set samples is calculated using one of many dissimilarity coefficients. Reconstructed environmental values for the fossil samples can be obtained by taking an average or weighted average of the k closest modern analogues. In the weighted case, it is logical to use as the weights the dissimilarity between the fossil sample and each of the k closest analogues, so that modern analogues that are more similar to the fossil sample have a greater influence on the prediction for that fossil sample than modern analogues that are less similar to the fossil sample.

As dissimilarities are used, we need to invert the measured dissimilarities ($1/d_{jk}$) to give larger weights to samples with lower dissimilarity. Williams and Shuman (2008) investigated the effect of strongly weighting predictions by dissimilarity ($1/d_{jk}^2$) for a large North American pollen data-set, but found that, for this data-set, strong weighting by the squared dissimilarity reduced the predictive ability of the MAT model, whilst little difference in predictions was observed whether an unweighted average or an average weighted by $1/d_{jk}$ was used.

MAT model performance statistics, such as root mean squared error of prediction (RMSEP), and sample-specific error estimates can be derived using cross-validation in the same manner as other calibration functions (see Juggins and Birks 2012: Chap. 14). It is worth noting that, unlike other calibration-function methods, MAT does not have so-called apparent error statistics (*sensu* Birks 1995). Instead, the basic model performance measures from the MAT are leave-one-out statistics, arising from the fact that the prediction for a selected training-set sample is determined by the training-set samples *excluding* the selected sample.

An alternative measure of reconstruction uncertainty, suggested by ter Braak (1995), is to use the standard error of the estimates calculated as the weighted variance of the values of the environmental variable (x_i) over the k analogues

$$s^2 = \frac{\sum_{i=1}^{k} w_i}{\left(\sum_{i=1}^{k} w_i\right)^2 - \sum_{i=1}^{k} w_i^2} \sum_{i=1}^{k} w_i (x_i - \mu)^2 \tag{15.2}$$

where w_i are weights, the inverse of the dissimilarity between training-set sample i and the fossil sample $(1 / d_{jk})$ and μ is the weighted mean of x_i. This measure has several appealing features. It is independent of the magnitude of dissimilarity to the k closest analogues and, if the magnitude of s^2 is large relative to those computed for training-set samples, it is an indication that no close modern analogues exist. Furthermore, the fossil sample may be similar to several training-set samples that vary strongly in terms of their environments, leading to large values of s^2 (ter Braak 1995).

A MAT reconstruction of pH at The Round Loch of Glenhead (RLGH) using the Surface Waters Acidification Programme (SWAP) diatom-pH training-set is illustrated in Fig. 15.1. This reconstruction uses the $k = 11$ closest analogues for each fossil sample and the fitted values are based on the average of the pH values for the 11 analogues selected at each time point. The value of k was chosen as the model with the lowest RMSEP assessed via bootstrapping the training-set samples. The error bars are bootstrap-derived sample-specific errors for each fossil sample computed using 1000 bootstrap samples (Juggins and Birks 2012: Chap. 14). The reconstructed values indicate that lake-water pH has varied considerably over the Holocene, but that the recent anthropogenic acidification due to acid deposition has reduced pH to values never before encountered in the history of the loch (Jones et al. 1986, 1989).

Several alternatives to the basic MAT have been developed, most notably SIMMAX (Pflaumann et al. 1996) and the revised analogue method (RAM: Waelbroeck et al. 1998). The SIMMAX approach implicitly uses the scalar product of the normalised species data as a dissimilarity coefficient and weights the modern analogue samples based on their geographical distance (Pflaumann et al. 1996). The RAM combines the response-surface approach to environmental reconstruction (Bartlein et al. 1986) with the MAT in which analogues are selected not from the original data but from an interpolated response surface. The number of analogues used in RAM reconstructions is determined by identifying dissimilarity 'jumps' (Waelbroeck et al. 1998).

SIMMAX and RAM appear to produce calibration functions that have lower RMSEP than the MAT and other techniques such as weighted averaging (WA) and weighted-averaging partial least squares (WAPLS) (see Juggins and Birks 2012: Chap. 14). Recently, this improvement was demonstrated to be the result of biases induced in SIMMAX and RAM through failure to ensure statistical independence of samples during cross-validation, geographically weighting the analogues, and

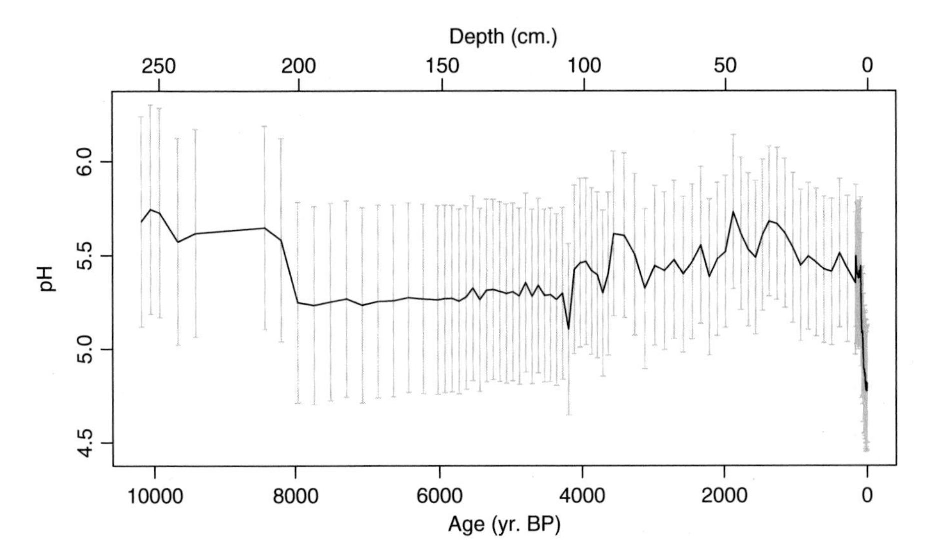

Fig. 15.1 MAT pH reconstruction for The Round Loch of Glenhead using the Surface Waters Acidification Programme (SWAP) diatom-pH training-set and $k = 11$ closest analogues. The error bars are drawn at $\pm$ the sample-specific bootstrap root mean squared error of prediction (RMSEP)

using the test-set to determine the number of analogues to retain in reconstructions (Telford et al. 2004).

Spatial autocorrelation in training-set data and its influence on calibration function performance has, until recently (Telford and Birks 2005, 2009; Telford 2006), been overlooked. Cross-validation assumes independence of samples; an assumption that is violated in the presence of spatial autocorrelation leading to an over-optimistic RMSEP and coefficient of determination (r^2). Telford and Birks (2005), using an Atlantic foraminifera training-set, showed that MAT utilises spatial structure in the species data that is uncorrelated with the environmental variable being modelled to improve model predictions. This residual spatial structure is likely to be related to spatial structure in environmental variables other than the one being reconstructed. When a spatially independent test-set was used to assess MAT model performance, MAT performed no better than WA or WAPLS, suggesting previously reported model performance statistics for MAT in the palaeoceanographic literature are over-optimistic (Telford and Birks 2005). These findings suggest that autocorrelation is a problem for calibration functions, and MAT in particular, where the environmental variable being modelled is spatially smooth.

The degree to which this problem affects other environments and proxies has recently been addressed by Telford and Birks (2009). In training-sets with strong spatial autocorrelation, MAT generally outperforms other reconstruction techniques, whilst it underperforms where weak autocorrelation is present. The most likely explanation for this result is that MAT finds a fit between the species and the environment that is local rather than global and as a result the technique is not robust to the presence of autocorrelation (Telford and Birks 2009). As such, Telford and Birks (2009) recommend that MAT not be used with training-sets where strong autocorrelation is present, and propose a simple, graphical means for evaluating the influence of autocorrelation on calibration functions and on MAT results.

Analogue Matching

In contrast to the MAT, where the ultimate aim is to produce a robust environmental reconstruction, analogue matching (AM) has no such purpose. Instead, AM is concerned solely with identifying sites in the present that are the closest match to the species assemblage identified in the past for one or more target sites. At their heart, MAT and AM are exactly the same approach, yet they emphasise fundamentally different roles for the modern analogues.

In MAT we optimise what we consider 'similar' in a manner that provides a calibration function with good predictive ability and low RMSEP. In AM the prime concern is to discriminate accurately between similar and dissimilar sites. In many cases the two approaches may result in the same set of close modern analogues being retained, but this need not be the case. What makes a set of analogues good for reconstructing pH, for example, may not be the best set of analogues when one considers the coherency of the species assemblage.

Where AM has begun to develop a role within the palaeolimnologist's toolbox is in the area of ecosystem restoration (e.g., Simpson et al. 2005). AM can be used to set ecological and biological restoration targets that complement the chemical targets derived from the application of calibration functions, for example. For some organism groups that preserve well in lake sediments, such as diatoms, chironomids, and littoral cladocerans, a straight palaeolimnological evaluation of reference-period sediments can provide a suitable biological restoration target (e.g. Ayres et al. 2008; Bennion and Simpson 2011; Bennion et al. 2011a, b).

However, aquatic-ecosystem managers are often interested in organism groups that do not preserve readily in lake sediments or where only incomplete assemblages are preserved. The rationale behind AM is that those lakes from a modern training-set that are most similar to a fossil assemblage are also likely to be similar in terms of these other organism groups. If close modern analogues for the pre-disturbance period of impacted lakes can be identified, aquatic-ecosystem managers can study the modern analogues to define ecological restoration targets that encompass organism groups that do not preserve in lake sediments (Simpson et al. 2005).

Using a modern training-set of 194 lakes from the UK, Norway, and Sweden where diatom counts of the surface sediments were available, Flower et al. (1997) identified close modern analogues that were floristically most similar to the diatom assemblages from discrete periods in the history of two acidified lakes from the acid-sensitive region of Galloway, south-west Scotland; The Round Loch of Glenhead (RLGH) and Loch Dee (LDEE). Three and seven analogues, respectively, were identified from the modern training-set for RLGH and LDEE, using the squared χ^2 distance coefficient. Of these, one and six analogues, respectively, were considered 'very good' analogues on the basis of the dissimilarity to the reference diatom assemblage being less than 0.57. The remaining analogues were considered 'good' and had a dissimilarity less than 0.65. These critical values of the χ^2 distance were determined from the distribution of all pair-wise dissimilarities for training-set samples and represent the first and second percentiles, respectively, of that distribution.

Despite being very good close modern analogues for RLGH and LDEE in terms of diatom assemblages, the closest matches to both lakes were poor matches for current lake-water calcium (Ca^{2+}) concentrations in RLGH and LDEE. The closest analogues had elevated calcium concentrations compared to present-day values in the two lochs (Flower et al. 1997). As calcium concentration is an important factor for other aquatic organisms such as fish and macroinvertebrates (Jeziorski et al. 2008) this discrepancy is difficult to ignore. This issue arises because diatoms are poor indicators of lake-water Ca^{2+} and as such there is little discrimination in diatom species composition between softwater lakes with differing Ca^{2+} concentrations. Flower et al. (1997) suggested that matching on additional criteria could be one way to solve these problems.

One suggested improvement was to base the matching on additional biological groups. One such group is the Cladocera, a group of crustacean zooplankton, the littoral members of which are very well preserved in lake sediments (Korhola and Rautio 2001). Littoral cladocerans are good indicators of a variety of limnological conditions, such as Ca^{2+} concentrations, habitat diversity and structure, such as aquatic macrophyte assemblages, and substrate availability (Duigan and Kovach 1991; Korhola and Rautio 2001). Simpson et al. (2005) adopted this approach and, using a training-set of diatom and cladoceran counts from 83 acid-sensitive lakes in Scotland and Wales, were able to identify close modern analogues for eight out of ten lakes from the UK Acid Waters Monitoring Network (AWMN) (Simpson et al. 2005).

Subsequent investigation of the close modern analogues for individual AWMN lakes showed a high degree of similarity in terms of hydrochemistry and aquatic macrophyte and macroinvertebrate assemblages. The selected modern analogues were much better matches for the AWMN in terms of Ca^{2+} concentrations than those identified using diatoms alone. Monitoring data from the RLGH also showed that species absent from the lake at the start of monitoring in 1988, but predicted to be present prior to acidification on the basis of their presence in the majority of the modern analogue lakes, were starting to return to the RLGH as the lake is now recovering from the affects of acidic deposition (Simpson et al. 2005; Battarbee 2010; Battarbee et al. 2011). Taken together, these results illustrate the power of the AM approach to setting ecological restoration targets in acidified lakes.

One issue that Simpson et al. (2005) did not address adequately at the time was how to calculate a dissimilarity that combined information from both diatom and cladoceran species assemblages. A naïve solution is to simply treat the diatom and cladoceran taxa as equal and join the two data-sets together prior to calculating the dissimilarities (Simpson et al. 2005). This is not ideal because this treats a diatom taxon and a cladoceran taxon as carrying the same information, and, as the diatom assemblage is much more species rich than the cladoceran assemblage, the diatom data will dominate the dissimilarity.

Two solutions to this problem suggest themselves. The first is to use a coefficient that can deal with different species data, such as Gower's (1971) general dissimilarity coefficient described below. Using weights, one could attempt to downweight some of the rarer diatom taxa such that the sum of weights applied to diatom taxa and those applied to cladoceran taxa were equal, thus giving both groups equal weight in the resulting dissimilarity measure.

The second solution is to adapt the approach of Melssen et al. (2006) to combine dissimilarities in a flexible manner. The dissimilarities for the two sets of variables (e.g., diatoms and cladocerans) are computed separately using appropriate dissimilarity coefficients. The resulting dissimilarities are scaled separately so that the maximum observed dissimilarity for each is equal to 1. The overall dissimilarity is a weighted sum of the scaled dissimilarities for the two sets

$$d_{fused_{jk}} = wd_{x_{jk}} + (1-w)\,d_{y_{jk}} \tag{15.3}$$

where $d_{fused_{jk}}$ is the combined dissimilarity, and $d_{x_{jk}}$ and $d_{y_{jk}}$ the dissimilarities for descriptor sets x and y, between samples j and k, with $0 \leq w \leq 1$, the relative weighting of the two dissimilarities. Selecting $w = 0.5$ ensures equal weights for the two sets of dissimilarities. Scaling the individual dissimilarities so that the maximum dissimilarity is equal to 1 accommodates different units of measurement for the sets of variables. This also allows one to use a different dissimilarity coefficient for the two sets. This procedure readily generalises to N sets of dissimilarities

$$d_{fused_{jk}} = \sum_{i=1}^{N} w_i\, d_{i_{jk}} \tag{15.4}$$

under the constraint that $\sum_{i=1}^{N} w_i = 1$.

Dissimilarity and Dimensionality

A critical consideration in analogue methods is to determine how close are a given pair of species assemblages to one another. This is achieved quantitatively using dissimilarity or distance coefficients, which measure the floristic or faunistic (dis)similarity between two sets of variables. A vast array of dissimilarity coefficients have been devised for various problems in ecology (Legendre and Legendre 1998), but four commonly chosen coefficients for quantitative data are the chord distance (Prentice 1980; Overpeck et al. 1985)

$$d_{jk} = \sqrt{\sum_{i=1}^{m} \left(x_{ij}^{0.5} - x_{ik}^{0.5}\right)^2} \tag{15.5}$$

the χ^2 distance (Birks et al. 1990)

$$d_{jk} = \sqrt{\sum_{i=1}^{m} \frac{\left(x_{ij} - x_{ik}\right)^2}{x_{ij} + x_{jk}}} \tag{15.6}$$

the information statistic (Overpeck et al. 1985)

$$d_{jk} = \sum_{i=1}^{m} \left(x_{ij} \log\left(\frac{2x_{ij}}{x_{ij} + x_{ik}}\right) + x_{ik} \log\left(\frac{2x_{ik}}{x_{ij} + x_{ik}}\right)\right) \tag{15.7}$$

and the Bray-Curtis distance (Faith et al. 1987)

$$d_{jk} = \frac{\sum_{i=1}^{m} \left|x_{ij} - x_{ik}\right|}{\sum_{i=1}^{m} \left(x_{ij} + x_{ik}\right)} \tag{15.8}$$

where x_{ij} is the proportion of taxon i in sample j, and d_{jk} the resulting dissimilarity between samples j and k. When proportional data are used, the chord distance is equivalent to the Hellinger distance (Legendre and Gallagher 2001; Legendre and Birks 2012: Chap. 8).

Prentice (1980), Overpeck et al. (1985), and Gavin et al. (2003) found the chord distance (or its squared form) to be particularly useful for closed compositional data of the type commonly found in palaeoecological studies as it has good signal to noise properties, flexibly weighting individual taxa by downweighting the influence of rare species, the noise, to emphasise the major patterns in the data, the signal (Overpeck et al. 1985). Another useful property is that the chord distance reaches an upper bound when the two samples have no species in common, and takes values of 0 to $\sqrt{2}$, when **Y** are proportional compositional data.

Two further signal-to-noise measures identified by Overpeck et al. (1985), the χ^2 distance and the information statistic, have similar properties to the chord distance and for most data-sets can be used interchangeably (though see Bennett and Humphry (1995) for a situation where this is not the case).

Faith et al. (1987) criticised the chord distance as a weak measure of compositional dissimilarity, recommending instead measures such as the Bray-Curtis distance, which use Manhattan-like absolute differences of the ith variable in j and k

rather than the squared differences of Euclidean-like measures, such as the chord distance, which magnify between-sample differences.

It is critical to choose a dissimilarity coefficient that is suited to the type of data to hand. The coefficients described above are all for use with quantitative community data, whilst others, such as the Jaccard coefficient, are suited to presence–absence data. The Jaccard coefficient, here expressed as a dissimilarity, is defined as

$$d_{jk} = \frac{b + c}{a + b + c} \tag{15.9}$$

where a is the number of species present in both j and k, b is the number of species present in j but not k, and c the number of species present in k but not j. Notice that the Jaccard coefficient ignores information about species that are absent from both j and k (commonly denoted d). The simple matching coefficient does not ignore these double absences

$$d_{jk} = \frac{b + c}{a + b + c + d} \tag{15.10}$$

but may be less well suited to assemblage data.

Where greater flexibility in weighting different taxa or where matrix **Y** is not species data but environmental data with variables measured in different units, a coefficient that treats each variable separately and allows weighting of individual variables, such as Gower's (1971) coefficient for mixed data, may be most suitable. Gower's coefficient, in its expanded form, is

$$d_{jk} = \frac{\sum_{i=1}^{m} w_i d_{ijk}}{\sum_{i=1}^{m} w_i} \tag{15.11}$$

where d_{ijk} is the dissimilarity between samples j and k for variable i and w_i is the weight associated with the ith variable, also known as Kronecker's Delta (Legendre and Legendre 1998). For presence–absence data or semi-quantitative data such as ordinal or nominal classes, d_{ijk} is 1 if j and k match on the ith variable (i.e., species i is present in both samples, or both samples are in the same class) and 0 otherwise. For quantitative data, d_{ijk} is calculated as follows

$$d_{jk} = 1 - \frac{|x_{ij} - x_{ik}|}{R_i} \tag{15.12}$$

where R_i is the value to standardise the ith variable by, usually taken to be the observed range of the ith variable. Podani (1999) suggests an extension to Gower's coefficient for ordinal variables that retains more of the encoded semi-quantitative information of the original data.

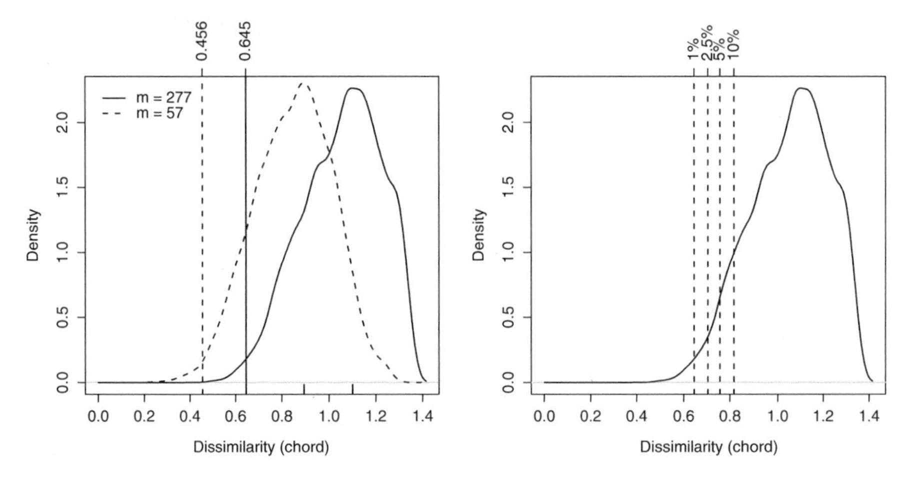

Fig. 15.2 Kernel density estimates of the distribution of pair-wise dissimilarities for the Surface Waters Acidification Programme (SWAP) diatom-pH training-set, showing the effect of species richness on dissimilarity values (*left*), and the choice of percentiles for selecting d_{crit} (*right*)

Regardless of which measure is used, however, the general form of these coefficients is to calculate some summation of the differences between samples j and k for a set of variables or taxa $i = 1, \ldots, m$. As such, it must be remembered that the number of taxa over which comparisons are made, m, has a bearing on the resulting dissimilarity, with species-rich samples invariably yielding higher values of d_{jk} than species-poor samples. This effect can be seen in Fig. 15.2, which shows the distribution of pair-wise dissimilarities for the SWAP training-set using all 277 taxa and a subset of 57 taxa that are present in $\geq$40 samples and have a maximum abundance $\geq$5%. In the smaller data-set the distribution of pair-wise dissimilarities is shifted to the left, to lower values than those observed using the full SWAP data-set. Therefore, one must be careful to not compare directly dissimilarities between sites calculated using two or more sets of variables, such as diatoms and pollen.

The Curse of Dimensionality

The curse of dimensionality (Bellman 1961) describes the problem of defining localness in high dimensions; neighbourhoods with a fixed number of samples become less local as the number of dimensions increases (Hastie and Tibshirani 1990). It is common for the dimensionality of palaeoecological data-sets to be high, especially with diverse proxies such as diatoms. In the SWAP and RLGH example presented here, there are 277 dimensions (species) and only 167 sites in the modern training-set. However, MAT and AM have been applied routinely in palaeoecology without any prior dimension-reduction.

Despite this, MAT and AM appear to defy the curse of dimensionality. This may be, as Härdle (1990) shows, because the relevant dimensionality is not m, the number of species, but p, the number of environmental variables (ter Braak 1995). Ter Braak (1995) also suggests that this defiance of the curse is due to dissimilarity coefficients simply summing over dimensions.

How Similar Is Similar Enough?

A logical question that arises, therefore, is how many analogues should one retain for inferences or to set restoration targets? For the MAT, various methods to derive this have been determined, though several are *ad hoc* owing to a lack of available formal statistical inference to guide the choice of k.

Choosing k should be viewed with as much importance as model selection in regression or component-based methods of calibration, such as PLS and WAPLS (Birks 2012: Chap. 2; Juggins and Birks 2012: Chap. 14). As k increases, so does the complexity of the MAT model. At low values of k substantial error or bias is likely to occur, whereas at larger values of k, MAT models will incorporate information from samples that are potentially not good matches and consequently overall performance will be low (Fig. 15.3). This represents a trade-off between having too simple a model and avoiding including information from non-analogous samples.

Choosing k to Optimise RMSEP

Plots of root mean squared error of prediction (RMSEP) against k can be used as a guide to the choice of a suitable value for k. Figure 15.3 shows RMSEP for values of k for the SWAP diatom-pH training-set under both leave-one-out (LOO) and bootstrap cross-validation (see Juggins and Birks 2012: Chap. 14 for details of these techniques). RMSEP is high when k is small and declines rapidly for values of k up to 6. The optimal choice of k is 10 when assessed using LOO and 11 using bootstrapping, though there is little to choose between the performance of the models in this case using 6 and 11 analogues.

Telford et al. (2004) demonstrated that choosing k *post hoc* by selecting the k with lowest RMSEP for the training-set can be biased, and that in some cases this bias can be quite large. The solution to this problem is to use an optimisation set alongside the usual training- and test-sets (Telford et al. 2004). The model is built on a subset of the training data retaining a small optimisation set of samples from the training-set. Cross-validated (leave-one-out or bootstrap) predictions for each of the samples in the optimisation set are computed, and the value of k that produces the lowest RMSEP for the optimisation set is then used for subsequent model predictions.

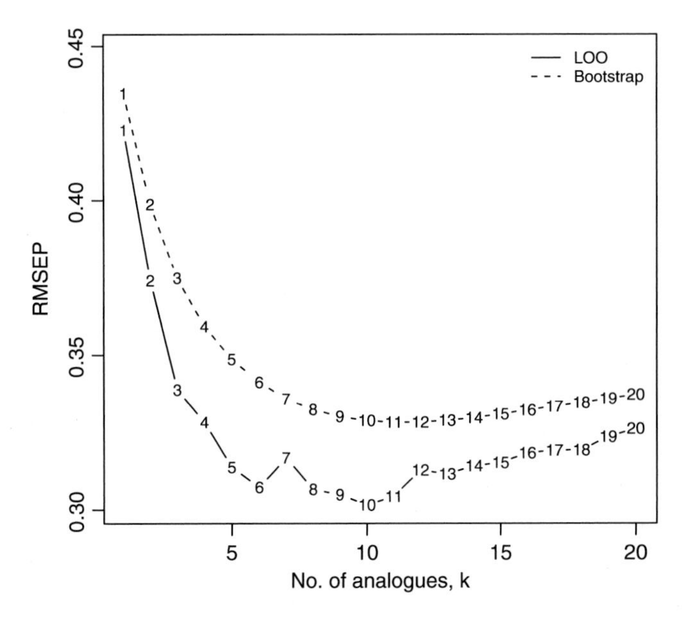

Fig. 15.3 Scree-plot of modern analogue technique (MAT) model leave-one-out (LOO) and bootstrap root mean squared error of prediction (RMSEP) as a function of k, the number of analogues retained for prediction, for the Surface Waters Acidification Programme (SWAP) diatom-pH training-set

Choosing k via Dissimilarity Jumps

An alternative approach to the choice of k is the 'jump' approach of Waelbroeck et al. (1998). In the 'jump' approach, a small number of analogues is retained, say the 10 or 20 closest matches, for each test sample. These analogues are ordered in terms of increasing dissimilarity to the test sample and the dissimilarity values for these analogues are differenced to yield the change in the dissimilarity of the current analogue to the test sample and that of the previous analogue. Where the proportional increase in dissimilarity is greater than a threshold, α, a 'jump' in dissimilarity is said to occur and the preceding analogues only are retained to provide the prediction for the test sample, retaining at a minimum the two closest analogues (Waelbroeck et al. 1998). Values for α are varied (Telford et al. (2004) used values of α ranging from 0.02 to 1, for example) and the value of α that minimises the RMSEP of the model under cross-validation is used for further analysis. Waelbroeck et al. (1998) found that values of α of between 0.1 and 0.3 are usually retained.

The two contrasting approaches to model selection above (RMSEP or dissimilarity jumps) boil down to selecting either a set number of k analogues and consequently a variable threshold α or a fixed threshold and variable number of analogues for each test sample. One deficiency of both approaches is that they are a step removed from the compositional dissimilarity between analogues and the test sample.

It would be more intuitive to retain only those samples that were sufficiently similar to the test sample for purposes of prediction. Despite the appeal of this, however, a number of difficulties exist, most notably how to define *sufficiently similar*? One could approach this problem by replacing α in the 'jump' method above with a dissimilarity critical threshold, d_{crit}, and vary d_{crit} between suitably chosen values to determine the value of d_{crit} that minimises RMSEP, again under cross-validation.

A further issue is that a prediction cannot be made for a test sample that has no sufficiently similar analogues using this method. This may be advantageous, with the user avoiding producing misleading reconstructions, but in practice producing a prediction is desirable as long as the uncertainty or utility of that prediction can be determined and presented alongside the predictions themselves. I return to this point when I discuss using analogue methods for the evaluation of reconstructions.

Reference Distributions of Dissimilarities

In AM there is no objective function to minimise (e.g., RMSEP) and as such the methods discussed above are not appropriate for the selection of the number of analogues to retain or a suitable dissimilarity threshold. The approach that has traditionally been used in AM, and also in the MAT for reconstruction evaluation, is to take as the critical dissimilarity a low quantile (percentile) of the distribution of pair-wise dissimilarities observed from the training-set samples (Overpeck et al. 1985; Flower et al. 1997).

The thinking behind this approach is that no two assemblages will ever be perfectly similar or dissimilar and that for a training-set covering a reasonable environmental space most sites will be, on average, moderately dissimilar to one another. Samples that are very similar or very dissimilar are unlikely to be observed in such a training-set. The observed distribution of pair-wise dissimilarities for the training-set is used to represent this 'likelihood' of two samples being 'good' analogues for one another, and the low percentiles of this distribution reflect the low 'likelihood' of observing two very similar samples. Note that 'likelihood' carries no statistical connotation here; at best, these percentiles are guides to selecting a suitable value for d_{crit}. One or more of the 1st, 2nd, 2.5th, 5th, and 10th percentiles have been used, depending on the absolute values of the dissimilarity (Birks et al. 1990). Often several percentiles have been used where varying grades of analogue closeness are desired so that the 1st percentile could represent close analogues, the 2.5th very good analogues, and the 5th percentile good analogues for example (e.g., Jones and Juggins 1995; Simpson et al. 2005).

The shape of the above distribution also needs to be taken into account when selecting a suitable d_{crit}. For the SWAP training-set, the selection of percentiles is illustrated in the right panel of Fig. 15.2. The distribution of pair-wise dissimilarities is strongly left skewed, with many samples being quite dissimilar to one another. This results in lower percentiles being observed at relatively high dissimilarities and therefore moderate differences in species assemblages. To guard against selecting an inappropriate percentile with too high a dissimilarity, a lower percentile than that usually chosen should be selected. For the full SWAP data-set, for example, the 1st percentile would be an appropriate upper limit on the choice for d_{crit}. Values for d_{crit} chosen using empirical quantiles are data-set-dependent and should be estimated for each study and data-set to hand.

Monte Carlo Resampling

The dissimilarity critical threshold, d_{crit}, could also be derived from an empirical distribution of dissimilarities derived via resampling of the observed pair-wise dissimilarities for the training-set (Sawada et al. 2004; Simpson 2007). The rationale behind this approach is that a 'good' analogue is one that is unlikely to occur by chance, i.e., the probability of observing a pair of samples that are very similar to one another is low. Monte Carlo methods can be used to produce a simulated distribution of pair-wise dissimilarities that one might expect to observe for the total population of possible samples by sampling with replacement from the observed training-set (Manly 1997). At random, a pair of samples is selected with replacement from the modern training-set and the dissimilarity between this pair of samples recorded. This is repeated a large number of times to produce the Monte Carlo distribution of dissimilarities. A low percentile of this distribution may then be selected for d_{crit}. Furthermore, the number of resampled pair-wise dissimilarities less than a given level of dissimilarity is the Monte Carlo p-value for that level.

Figure 15.4 shows the results of applying Monte Carlo resampling to the SWAP training-set, drawing 10,000 sample-pairs with replacement. The left panel shows both the Monte Carlo and observed distributions, with the Monte Carlo distribution being multi-modal and also having lower values of d_{jk} for the same percentile of the observed distribution (Table 15.1). The right panel of Fig. 15.4 shows the cumulative probability distribution of the Monte Carlo sample of dissimilarities. The shaded region of this curve indicates the dissimilarity values expected to occur by chance five times in 100, equivalent to a Monte Carlo p-value of ≤ 0.05. From Fig. 15.4, the chord distance at $p \leq 0.05$ is ≤ 0.692.

Receiver Operating Characteristic (ROC) Curves

A recent development in analogue methods is the use of receiver operating characteristic (ROC) curves to determine an optimal value for d_{crit} (Gavin et al. 2003; Wahl 2004). This value is optimal in the sense that it best discriminates between analogue and non-analogue samples. However, the use of ROC curves is

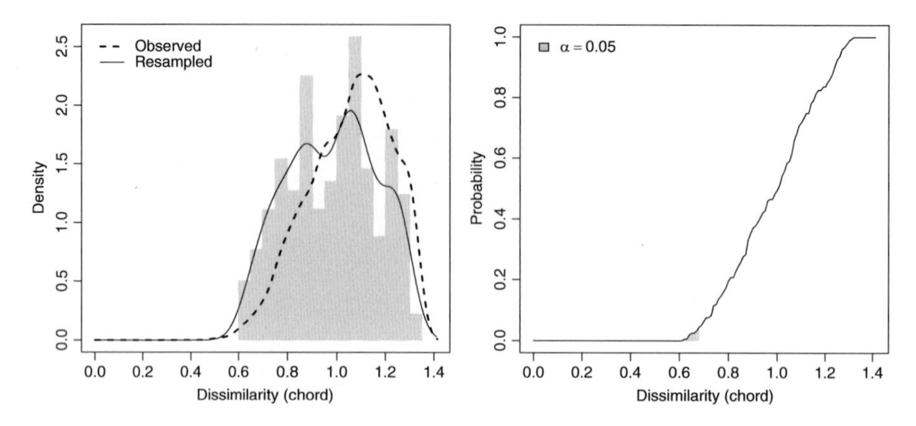

Fig. 15.4 Monte Carlo resampling applied to the Surface Waters Acidification Programme (SWAP) training-set showing (i) a comparison of the observed (*dashed line*) and resampled (*solid line* and *grey bars*) distribution of pair-wise dissimilarities (*left*), and (ii) the empirical cumulative distribution function of resampled pair-wise dissimilarities as a function of dissimilarity (*right*). The *grey shaded* area on the right plot is the area corresponding to a Monte Carlo *p*-value of ≤ 0.05

Table 15.1 Percentiles of the distribution of observed pair-wise dissimilarities for the Surface Waters Acidification Programme (SWAP) training-set and those derived from randomly sampling 10,000 sample pairs, with replacement, from the training-set

	1%	2.5%	5%	10%
Observed	0.645	0.690	0.758	0.817
Resampled	0.633	0.663	0.690	0.734

only applicable to situations where training-set samples can be assigned *a priori* to groups or types, such as vegetation types, for example, in the case of a pollen-based training-set. A pair of samples is considered to be analogues only if they belong to the same group.

Figure 15.5 shows some results of applying the ROC curve methodology to the North American Modern Pollen Database (NAMPD Version 1.7; Whitmore et al. 2005), which comprises 4525 samples where the potential vegetation (in the absence of significant anthropogenic land-use; hereafter termed *biome*) was determined from global land-cover data (Whitmore et al. 2005). Here, I only consider the 2157 pollen samples collected from lacustrine environments. Two samples are considered analogues if they are located in the same biome and non-analogues if they derive from different biomes. For each site, both the dissimilarity to the closest sample that occurs in the same biome and the closest sample not located within the same biome are determined.

If dissimilarity is a good discriminator between samples in like and unlike biomes we would expect to find more analogue pairings at low dissimilarities and a greater number of non-analogue pairings at larger dissimilarities. The upper-right panel of Fig. 15.5 illustrates this for a combined analysis of the NAMPD samples over all

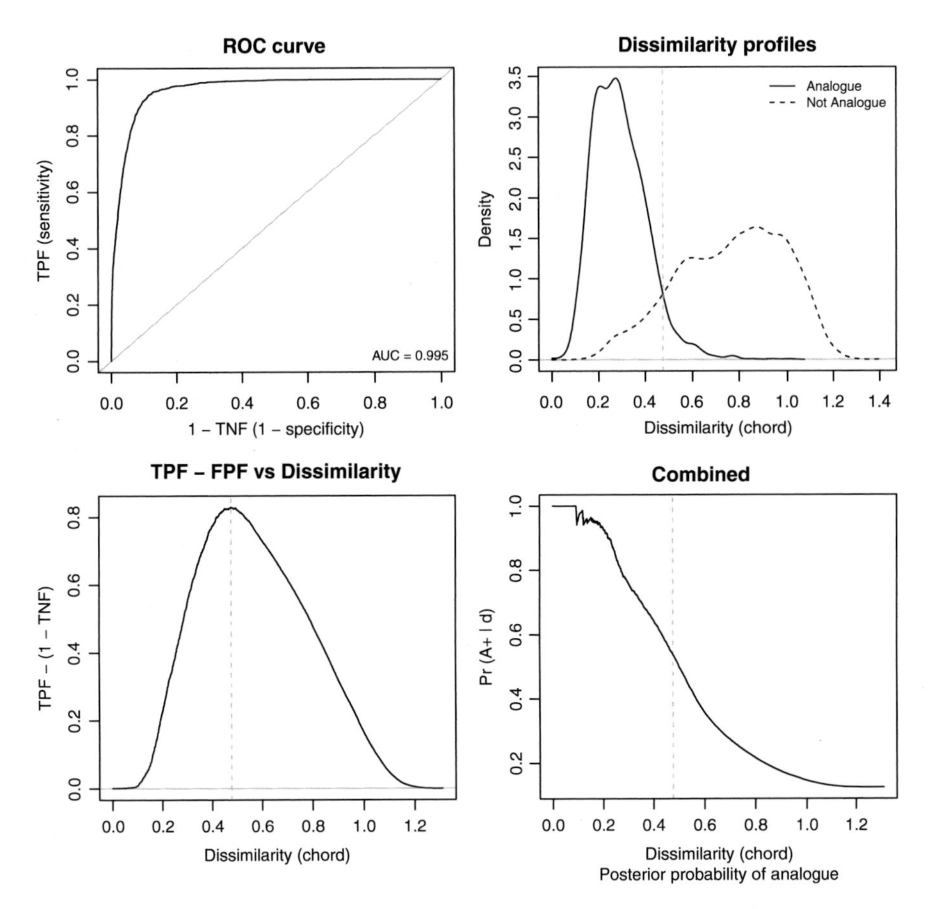

Fig. 15.5 Receiver operating characteristic (ROC) curve analysis applied to the North American Modern Pollen Database (NAMPD) using the chord distance. The results shown are for an analysis over all biomes, combined. Clockwise from *top left*: (i) the ROC curve and area under curve (AUC) statistic, (ii) kernel density estimates of the distribution of pair-wise dissimilarities for analogue and non-analogue samples, (iii) posterior probability that two samples are analogues as a function of dissimilarity, and (iv) plot of True Positive Function (TPF) – False Positive Function (FPF) as a function of dissimilarity. The *dotted* vertical in the plots is the optimal dissimilarity threshold indicated by the ROC curve (d_{crit}), which for the combined analysis shown is $d_{crit} = 0.457$

biomes. Far greater numbers of true analogues are found at lower dissimilarities than at high values.

ROC curves compare two different types of error that arise when evaluating analogue/non-analogue samples for a given value of d_{crit}: (i) *false positive error* and (ii) *false negative error*. False positive errors occur when two samples that are not analogues (i.e., come from different groups) are said to be analogues on the basis of d_{crit}. False negative error represents the converse, when two samples that are in the same group, and are therefore analogues, are said to be non-analogues given

d_{crit}. The optimal value for d_{crit} is the value that jointly minimises these two types of error.

A ROC curve is drawn using two measures of performance that are related to the two types of error described above; (i) sensitivity, the proportion of true analogues out of all analogues for a given value of d_{crit}; and (ii) specificity, the proportion of true non-analogues out of all non-analogues for the same value of d_{crit}. Sensitivity is also known as the True Positive Fraction (TPF) and specificity as the True Negative Fraction (TNF), with the False Positive Fraction (FPF) being defined as $1 - specificity$ (or $1 - TNF$). Sensitivity is drawn on the y-axis of the ROC curve plot and $1 - specificity$ on the x-axis.

The point on the ROC curve closest to the upper left corner of the plot corresponds to the optimal dissimilarity threshold d_{crit}. At this point the slope of the ROC curve is maximal and the difference between sensitivity and $1 - specificity$ is greatest (Fig. 15.5, lower-left panel). This point is also the point where the distributions of true analogues and true non-analogues cross (Fig. 15.5, upper-right panel).

The upper left panel of Fig. 15.5 shows the ROC curve for the NAMPD data-set for all biomes combined. The 1:1 line represents a naïve predictor of analogue status, namely the result of predicting that all pairings are analogues regardless of d_{crit}. The ROC curve in Fig. 15.5 lies close to the upper left corner of the plot, indicating strong discrimination of analogues/non-analogues by the chord distance.

The area under the ROC curve (AUC) is a measure of the degree to which the differences in species composition discriminates between analogue and non-analogue samples. The AUC is equivalent to the Mann–Whitney U statistic (Henderson 1993). An AUC value of 0.5 indicates that determining whether two samples are analogues on the basis of the dissimilarity between the samples is no better than random guessing. In the NAMPD example, the AUC is 0.996, indicating strong discrimination between true-analogues and true non-analogues using the dissimilarity between samples.

Table 15.2 shows the results of applying ROC curve analysis individually to the nine biomes in the pollen data as well as to all biomes combined. Overall, the optimal chord distance threshold is 0.476 with thresholds for the individual biomes ranging from 0.333 (Forest-Tundra) to 0.632 (Coastal). In all cases the AUC statistics are very high with low standard errors resulting in highly significant differences between the distributions of dissimilarity values for analogue and non-analogue pairings, although this is not unexpected given the large sample sizes.

The posterior probability that any two samples are analogues can be calculated from the ratio of TPF to FPF, $LR(+) = TPF/FPF$ (Henderson 1993). LR(+) is a likelihood ratio and can be converted to the posterior odds that two samples are analogues ($O^{+}_{post.}$) via

$$O^{+}_{post.} = LR(+) \times O^{+}_{pri.} \qquad (15.13)$$

Table 15.2 Summary results from the receiver operating characteristic (ROC) curve analysis of the North American Modern Pollen Database (NAMPD) samples

	Opt. Dis.	AUC	SE	In	Out	*p*-Value
Arctic	0.438	0.998	0.002	217	1940	$<2.22e-16$
Boreal forest	0.338	0.981	0.004	593	1564	$<2.22e-16$
Coastal	0.632	0.999	0.002	89	2068	$<2.22e-16$
Conifer/Hardwood	0.493	0.997	0.003	207	1950	$<2.22e-16$
Deciduous forest	0.451	0.986	0.004	442	1715	$<2.22e-16$
Desert	0.541	1.000	0.002	47	2110	$<2.22e-16$
Forest-Tundra	0.333	0.994	0.003	273	1884	$<2.22e-16$
Mountain vegetation	0.444	0.999	0.002	100	2057	$<2.22e-16$
Praries	0.475	0.995	0.004	189	1968	$<2.22e-16$
Combined	0.476	0.995	0.001	2157	17256	$<2.22e-16$

Opt. Dis. is the optimal dissimilarity, *SE* is the standard error of the Area Under the ROC Curve (*AUC*), *In* and *Out* are the number of analogue and non-analogue pairings, respectively

where $O^{+}_{\text{pri.}}$, the prior odds, is

$$O^{+}_{\text{pri.}} = \frac{\text{Pr}^{+}_{\text{pri.}}}{1 - \text{Pr}^{+}_{\text{pri.}}} \tag{15.14}$$

and $\text{Pr}^{+}_{\text{pri.}}$ is the prior probability of any two samples being analogous (Brown and Davis 2006). $\text{Pr}^{+}_{\text{pri.}}$ may be set at 0.5 (i.e., a 50% probability of two samples being analogues) or may be determined from the observed probability that two samples are analogues (i.e., in the same group) in the modern training-set.

The posterior odds of two samples being analogues conditional upon dissimilarity is converted to a posterior probability that the two samples are analogues via

$$\text{Pr}^{+}_{\text{post.}} = \frac{O^{+}_{\text{post.}}}{1 - O^{+}_{\text{post.}}} \tag{15.15}$$

In analyses where the number of true analogue sample pairs is very low compared with the number of true non-analogue pairs, such as the NAMPD example, it is essential that the prior probability, $\text{Pr}^{+}_{\text{pri.}}$, observed from the training-set is used to compute the posterior odds and posterior probabilities. If an equal probability of 0.5 is used, it will greatly over-estimate the posterior odds and probabilities and hence confidence in the discrimination between analogue and non-analogue samples.

Table 15.3 shows the optimal dissimilarity and the posterior probability that two samples are analogues for that distance for each of the biomes in the NAMPD example plus an overall assessment across all biomes. Probabilities range from 0.303 to 0.789, with only slightly greater than 50% probability that two samples are analogues when the optimal cut-off of 0.457 is used in the overall analysis.

Table 15.3 Optimal dissimilarities and associated posterior probability that two samples are analogues at this dissimilarity for receiver operating characteristic (ROC) curve (π_{ROC}) and logistic regression model (π_{logit}) analysis of the North American Modern Pollen Database (NAMPD) samples

	Optimal ROC d_{jk}	π_{ROC}	π_{logit}	d_{jk} ($\eta = 0.9$)
Arctic	0.438	0.779	0.234	0.317
Boreal forest	0.338	0.789	0.367	0.142
Coastal	0.632	0.420	0.040	0.238
Conifer/Hardwood	0.493	0.469	0.047	0.279
Deciduous forest	0.451	0.714	0.311	0.276
Desert	0.541	0.303	0.017	0.278
Forest-Tundra	0.333	0.512	0.129	0.123
Mountain vegetation	0.444	0.611	0.141	0.200
Praries	0.475	0.394	0.077	0.145
Combined	0.476	0.539	0.110	0.160

d_{jk} ($\eta = 0.9$) is the dissimilarity at which the posterior probability that two samples are analogues is equal to 0.9

Logistic Regression Modelling

Whilst the ROC curve is a simple, graphical technique for assessing the ability of dissimilarity to discriminate between analogue and non-analogue comparisons, several additional computational steps must be performed to extract the meaningful measures of interest; namely the optimal d_{crit} threshold and the posterior probability that two samples are analogues at the selected threshold. Furthermore, the way ROC curves have been presented in the palaeoecological literature does not account for uncertainty in analysis, and hence uncertainty in future assignments in the MAT or AM where d_{crit} is used.

The data used to create the ROC curve are effectively a vector of 1's and 0's and a vector of associated dissimilarities between the pair of samples. A 1 indicates that the pair of samples come from the same group and are hence true analogues, whereas a 0 indicates that the samples are not drawn from the same group and represent true non-analogues. The upper right panel of Fig. 15.5 shows the probability densities of the chord distances for true analogues and true non-analogues.

An alternative way to think of these data is as binomial observations y taking values 0 and 1 with associated dissimilarities x. Logistic regression (see Birks 2012: Chap. 2) can be used to model the relationship between these binomial observations and the associated pair-wise dissimilarities. In logistic regression, we attempt to model the probability π that $y = 1$ conditional upon the covariate, x.

Logistic regression is a special case of the Generalized Linear Model (GLM: McCullagh and Nelder 1989) where the response, y, has a binomial distribution and the linear predictor is related to the response through the *logit* link function (Birks 2012: Chap. 2). The logit link function implies the concept of the *odds* of an event being observed. The odds that two samples are true analogues is the ratio of the probability that two samples are true analogues (π) to its converse

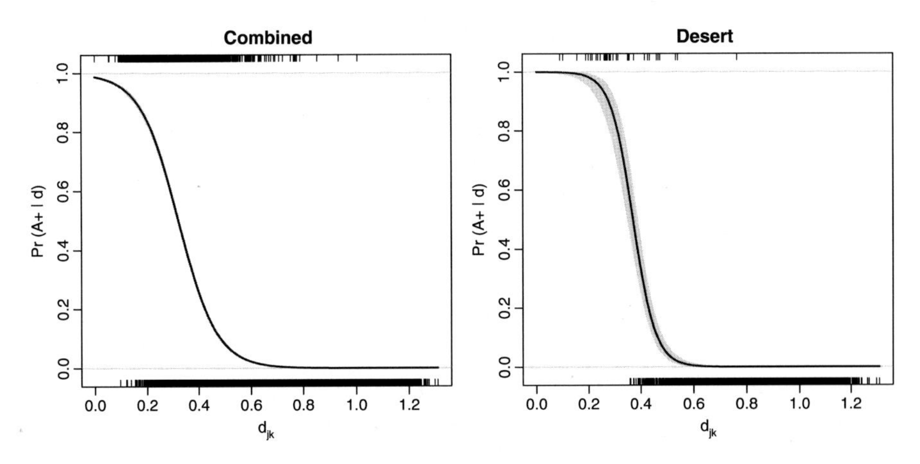

Fig. 15.6 Fitted logistic regression model curves for the all biomes (*left*) and the Desert biome (*right*) describing the relationship between the posterior probability that two samples are analogues and chord distance. The *shaded grey bands* on the plots are 95% point-wise confidence intervals on the fitted curve. In the combined analysis this confidence interval is so small as to not be visible. Rug-plots indicate the distribution of analogue and non-analogue samples as a function of dissimilarity

$(1-\pi)$. To complete the logit transformation, the natural logarithm of the odds is taken

$$\text{logit}(\pi) = \log\left[\frac{\pi}{1-\pi}\right] = \beta_0 + \beta_1 d_{\text{crit}} \tag{15.16}$$

where $\log\left[\frac{\pi}{1-\pi}\right]$ is the *log odds*.

Returning briefly to the NAMPD example, Fig. 15.6 shows a logistic regression model fit to the combined analysis (over all biome types) and for the Desert biome in particular. Rug-plots (Juggins and Telford 2012: Chap. 5) illustrate the dissimilarity at which true analogue ($y = 1$) and true non-analogue ($y = 0$) pairings were observed. The solid line represents the fitted function and is contained within a 95% confidence region; on the combined plot, this confidence interval is so small as to not be visible.

One feature that stands out immediately from the two plots is that the probability that two samples are true analogues only takes high values at very low dissimilarities. The probabilities from the logistic regression model fit are low when computed for the optimal dissimilarity suggested by the ROC curve analysis (π_{logit} in Table 15.3). These results suggest that the posterior probability that two samples are analogues at the optimal dissimilarity determined via the ROC curve method is overly optimistic when compared to the more direct estimation via logistic regression.

Using the fitted logistic regression model it is trivial to compute the dissimilarity at which a user-specified probability that a sample pair are analogues is attained. If we define η as the specified probability, the dissimilarity ($\hat{d}_{jk}$) at which the fitted probability is equal to η is computed using the estimates of the model coefficients of Eq. 15.16

$$\hat{d}_{jk} = \frac{\eta - \hat{\beta}_0}{\hat{\beta}_1} \qquad (15.17)$$

The dissimilarities at which $\eta = 0.9$ for each biome in the NAMPD example are shown in Table 15.3. These dissimilarities are considerably lower than the optimal dissimilarities suggested by the ROC curve analysis, on the order of 50% or more lower.

The logistic regression approach to determining a dissimilarity threshold will be more familiar to many palaeoecologists and the values of interest (dissimilarity threshold and π) are more easily obtained than via the ROC curve method. However, fitting logistic regression models is not without its difficulties.

Of greatest concern for the palaeoecologist using logistic regression is the problem of *separability*, a situation that arises when there is no overlap in the dissimilarity values for analogues (1) and non-analogues (0). Paradoxically, this problem arises when the model does too good a job and perfect predictions can be made. In such circumstances the maximum likelihood estimates of the regression coefficients may not exist or be subject to large uncertainties (Firth 1993). A simple way to diagnose such problems (if software used to fit the models does not issue appropriate warnings) is to examine the regression coefficients and their standard errors for excessively large values. Also, it is not uncommon for the fitted probabilities to become numerically 1 or 0, leading to infinite odds. This is, however, often not a problem if separability is not an issue.

Evaluation of Environmental Reconstructions

Techniques used in the construction of MAT calibration functions can also play a role in evaluating reconstructions produced using other calibration-function methodologies, such as weighted averaging or maximum likelihood-based methods (Juggins and Birks 2012: Chap. 14).

Intuitively, more faith can be placed in the reconstructed values for those samples that have one or more close modern analogues in the modern training-set. Where samples have no close modern analogues, the calibration function is potentially extrapolating beyond the bounds of the species-environment relationships found within the training-set.

There are several ways in which such information can be conveyed. The simplest is to indicate in some manner those fossil samples that do have a close modern analogue in the training-set. Additional levels of granularity (e.g., distinguishing between close, very good, and good modern analogues) can be achieved through the use of different markers to indicate the level of match and associated dissimilarity thresholds. Commonly, low percentiles of the distribution of pair-wise dissimilarities computed for the training-set are used.

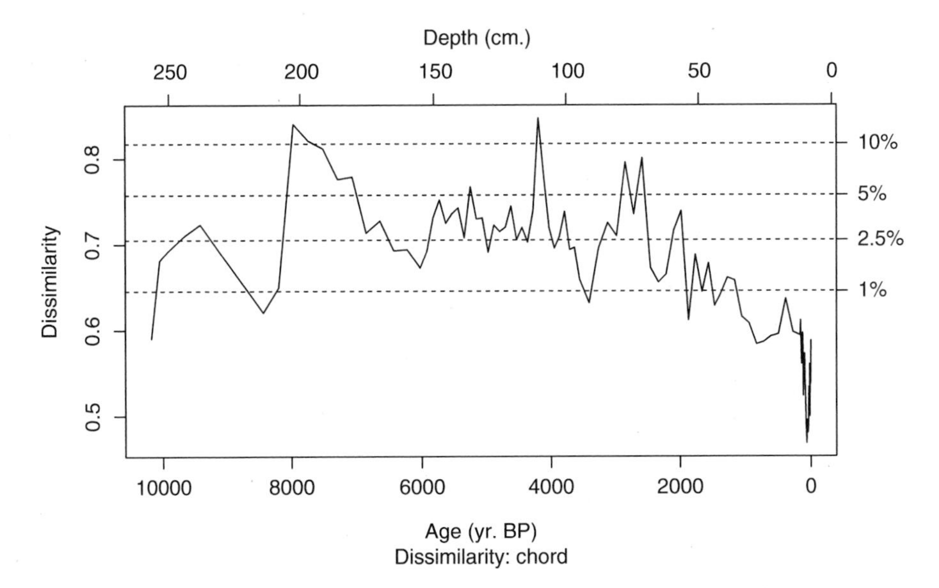

Fig. 15.7 Minimum dissimilarity (chord distance) between each fossil sample and samples in the Surface Waters Acidification Programme (SWAP) training-set for The Round Loch of Glenhead core. The *dashed horizontal lines* are percentiles of the pair-wise distribution of dissimilarities for the training-set. The percentiles are indicated on the right-hand axis of the plot

Alternatively, a plot of the minimum dissimilarity between each fossil sample and the training-set samples as a function of depth/age may be produced. Again, markers representing dissimilarity thresholds can be drawn on the plot to aid interpretation, and low percentiles can be used for this purpose in a similar manner to the simple marker method above. Figure 15.7 shows the minimum dissimilarity between each fossil sample in the RLGH core and a training-set sample. Here, the 1st, 2.5th, 5th, and 10th percentiles derived from the SWAP training-set have been superimposed on the plot. It is clear that, with few exceptions, samples from the RLGH core older than ~2000 yr BP have no close modern analogues in the SWAP training-set and thus one must treat the reconstructed values from any calibration-function methodology with a greater degree of uncertainty than reconstructed values for those samples that do have close modern analogues in the training-set.

Software

A wide range of software has been written to analyse data using analogue methods, in particular for MAT. Sawada (2006) lists several applications, including ANALOG (Schweitzer 1995, 1999), MODPOL (Maher 2000), SIMMAX (Pflaumann et al. 1996), RAM98 (Waelbroeck et al. 1998), PaleoToolBox (Sieger et al. 1999), C2 (Juggins 2007), and MATTOOLS (Sawada 2006). All of these applications, except

C2, are available free of charge from the internet. C2 is freely downloadable for unrestricted use on data-sets with 75 samples or fewer; a licence fee is payable for use on larger data-sets. Of the applications listed, only MATTOOLS is open source, though technically only via a Windows binary for the R statistical language (R Development Core Team 2009), limiting deployment on other operating systems. For ROC curve estimation, MATTOOLS and ROCKIT (Metz et al. 1998) are available.

An alternative to these applications is analogue (Simpson 2007; Simpson and Oksanen 2010), a freely available, open-source package for R distributed under the GNU General Public Licence version 2. All of the figures and example analyses presented here were generated using analogue version 0.6-23. This or later versions of the package can be downloaded from the Comprehensive R Archive Network (CRAN; http://www.cran.r-project.org/package=analogue). Scripts to reproduce the examples can be found at http://analogue.r-forge.r-project.org/dperbook/.

Conclusions and Future Work

Analogue methods, whilst remaining largely unchanged since their introduction to the literature (Overpeck et al. 1985; Prell 1985), have seen a recent burst of activity prompted by the novel use of ROC curves to define critical dissimilarity thresholds (Gavin et al. 2003; Wahl 2004). MAT calibration functions remain a valuable and powerful technique for palaeoenvironmental reconstruction, despite recent work suggesting problems with spatially auto-correlated environmental data (Telford and Birks 2005), and the underlying theory is increasingly being used in applied palaeoecology to define reference conditions for use in setting recovery and restoration targets (Bennion and Simpson 2011; Bennion et al. 2011a, b).

The one area of analogue methods that remains largely unsolved and deserving of attention is a general approach to determining dissimilarity thresholds for selecting analogues. As we have seen, ROC curves and logistic regression can provide statistically-based thresholds when data can be grouped *a priori*, but no satisfactory method exists for data not belonging to appropriate groups. Simulating assemblages that have similar properties to the observed training-set (species optima and tolerance ranges) and computing their dissimilarity is perhaps one avenue by which realistic simulation distributions can be derived from which thresholds may be determined.

Another avenue may be found in parametrising an appropriate multivariate probability distribution, such as the multinomial or Dirichlet, again with suitable optima, tolerances, and covariances/correlations between species, followed by computing the dissimilarity between randomly sampled pairs of observations from that distribution. Indeed, Overpeck et al. (1985), in their appendix entitled "Some theoretical properties of dissimilarity coefficients", point to the relationship between certain coefficients and the multinomial distribution.

The issue of fossil assemblages that lack close modern analogues is an area that is now starting to receive attention by ecologists and palaeoecologists (Jackson and Williams 2004). Understanding these non-analogue assemblages and how the underlying climatic and other environmental factors led to their formation are important questions in the fields of ecology and evolutionary biology. This is all the more important given the unprecedented human impact upon ecosystems through pollution and global warming observed in the past few decades (Fox 2007; Williams and Jackson 2007; Williams et al. 2007). The Earth is rapidly heading towards dramatic ecological changes arising from the unique environmental conditions engineered by anthropogenic disturbance of our fragile ecosystems.

Acknowledgements I am indebted to Viv Jones for providing the Round Loch of Glenhead sediment core data and age-depth estimates, and to Jack Williams and Konrad Gajewski for permission to use the North American Modern Pollen Database data in the analogue software package for R that was used for all the analyses illustrated in the text. Roger Flower, Steve Juggins, John Birks, Helen Bennion, and Rick Battarbee have, over the years, provided useful discussions and technical help that have improved my understanding of analogue methods and how best to explain the concepts. The writing of this chapter and the accompanying analogue software was conducted whilst participating in the European Union 6th Framework Integrated Project *Euro-limpacs* (GOCE-CT-2003-505540).

References

Ayres KR, Sayer CD, Skeate ER, Perrow Martin R (2008) Palaeolimnology as a tool to inform shallow lake management: an example from Upton Great Broad, Norfolk, UK. Biodivers Conserv 17:2153–2168

Bartlein PJ, Prentice IC, Webb T III (1986) Climatic response surfaces from pollen data for some eastern North American taxa. J Biogeogr 13:13–57

Battarbee RW (2010) Are our acidified upland waters recovering? Freshw Biol Assoc News 52:5–6

Battarbee RW, Curtis CJ, Shilland EW (2011) The Round Loch of Glenhead: recovering from acidification, climate change monitoring and future threats. Scottish Natural Heritage Commissioned Report No. 469

Bellman RE (1961) Adaptive control processes. Princeton University Press, Princeton

Bennett KD, Humphry RW (1995) Analysis of late-glacial and Holocene rates of vegetational change at two sites in the British Isles. Rev Palaeobot Palynol 85:263–287

Bennion H, Simpson GL (2011) The use of diatom records to establish reference conditions for UK lakes subject to eutrophication. J Paleolimnol 45:469–488

Bennion H, Battarbee RW, Sayer CD, Simpson GL, Davidson TA (2011a) Defining reference conditions and restoration targets for lake ecosystems using palaeolimnology: a synthesis. J Paleolimnol 45:533–544

Bennion H, Simpson GL, Anderson NJ, Clarke G, Dong X, Hobæk A, Guilizzoni P, Marchetto A, Sayer CD, Thies H, Tolotti M (2011b) Defining ecological and chemical reference conditions and restoration targets for nine European lakes. J Paleolimnol 45:415–431

Birks HJB (1995) Quantitative palaeoenvironmental reconstructions. In: Maddy D, Brew S (eds) Statistical modelling of Quaternary science data. Quaternary Research Association, Cambridge, pp 161–254

Birks HJB (1998) Numerical tools in palaeolimnology – progress, potentialities, and problems. J Paleolimnol 20:307–332

Birks HJB (2012) Chapter 2: Overview of numerical methods in palaeolimnology. In: Birks HJB, Lotter AF, Juggins S, Smol JP (eds) Tracking environmental change using lake sediments. Volume 5: Data handling and numerical techniques. Springer, Dordrecht

Birks HJB, Line JM, Juggins S, Stevenson AC, ter Braak CJF (1990) Diatoms and pH reconstruction. Philos Trans R Soc Lond B 327:263–278

Birks HJB, Heiri O, Seppä H, Bjune AE (2010) Strengths and weaknesses of quantitative climate reconstructions based on late-Quaternary biological proxies. Open Ecol J 3:68–110

Brown C, Davis H (2006) Receiver operating characteristics curves and related decision measures: a tutorial. Chemometr Intell Lab Syst 80:24–38

Davis B, Brewer S, Stevenson A, Guiot J, Contributors D (2003) The temperature of Europe during the Holocene reconstructed from pollen data. Quaternary Sci Rev 22:1701–1716

Duigan CA, Kovach WL (1991) A study of the distribution and ecology of littoral freshwater chydorid (Crustacea, Cladocera) communities in Ireland using multivariate analyses. J Biogeogr 18:267–280

Faith DP, Minchin PR, Belbin L (1987) Compositional dissimilarity as a robust measure of ecological distance. Vegetatio 69:57–68

Fielding AH (2007) Cluster and classification techniques for the biosciences. Cambridge University Press, Cambridge

Firth D (1993) Bias reduction of maximum likelihood estimates. Biometrika 80:27–38

Flower RJ, Juggins S, Battarbee RW (1997) Matching diatom assemblages in lake sediment cores and modern surface sediment samples: the implications for lake conservation and restoration with special reference to acidified systems. Hydrobiologia 344:27–40

Fox D (2007) Back to the no-analog future? Science 316:823–825

Gavin DG, Oswald WW, Wahl ER, Williams JW (2003) A statistical approach to evaluating distance metrics and analog assignments for pollen records. Quaternary Res 60:356–367

Gower J (1971) A general coefficient of similarity and some of its properties. Biometrics 27: 857–871

Härdle W (1990) Applied nonparametric regression. Cambridge University Press, Cambridge

Hastie T, Tibshirani R (1990) Generalized additive models. Chapman & Hall, London

Henderson RA (1993) Assessing test accuracy and its clinical consequences: a primer for receiver operating characteristic curve analysis. Ann Clin Biochem 30:521–539

Jackson ST, Williams JW (2004) Modern analogues in Quaternary paleoecology: here today, gone yesterday, gone tomorrow? Annu Rev Earth Planet Sci 32:495–537

Jeziorski A, Yan ND, Paterson AM, DeSellas AM, Turner MA, Jeffries DS, Keller W, Weeber RC, McNicol RC, Palmer ME, McIver K, Arseneau K, Ginn BK, Cumming BF, Smol JP (2008) The widespread threat of calcium decline in fresh waters. Science 322:1374–1377

Jones VJ, Juggins S (1995) The construction of a diatom-based chlorophyll *a* transfer function and its application at three lakes on Signy Island (maritime Antarctic) subject to differing degrees of nutrient enrichment. Freshw Biol 34:433–445

Jones VJ, Stevenson AC, Battarbee RW (1986) Lake acidification and the land-use hypothesis – a mid post-glacial analog. Nature 322:157–158

Jones VJ, Stevenson AC, Battarbee RW (1989) Acidification of lakes in Galloway, south west Scotland: a diatom and pollen study of the post-glacial history of the Round Loch of Glenhead. J Ecol 77:1–23

Juggins S (2007) C2 Version 1.5 user guide. Software for ecological and palaeoecological data analysis and visualisation. Newcastle University, Newcastle-upon-Tyne

Juggins S, Birks HJB (2012) Chapter 14: Quantitative environmental reconstructions from biological data. In: Birks HJB, Lotter AF, Juggins S, Smol JP (eds) Tracking environmental change using lake sediments. Volume 5: Data handling and numerical techniques. Springer, Dordrecht

Juggins S, Telford RJ (2012) Chapter 5: Exploratory data analysis and data display. In: Birks HJB, Lotter AF, Juggins S, Smol JP (eds) Tracking environmental change using lake sediments. Volume 5: Data handling and numerical techniques. Springer, Dordrecht

Korhola A, Rautio M (2001) Cladocera and other branchiopod crustaceans. In: Smol JP, Birks HJB, Last WM (eds) Tracking environmental change using lake sediments. Volume 4: Zoological indicators. Kluwer Academic Publishers, Dordrecht, pp 5–41

Kucera M, Weinelt M, Kiefer T, Pflaumann U, Hayes A, Weinelt M, Chen M, Mix A, Barrows T, Cortijo E, Duprat J, Juggins S, Waelbroeck C (2005) Reconstruction of sea-surface temperatures from assemblages of planktonic foraminifera: multi-technique approach based on geographically constrained calibration data-sets and its application to glacial Atlantic and Pacific Oceans. Quaternary Sci Rev 24:951–998

Legendre P, Birks HJB (2012) Chapter 8: From classical to canonical ordination. In: Birks HJB, Lotter AF, Juggins S, Smol JP (eds) Tracking environmental change using lake sediments. Volume 5: Data handling and numerical techniques. Springer, Dordrecht

Legendre P, Gallagher ED (2001) Ecologically meaningful transformations for ordination of species data. Oecologia 129:271–280

Legendre P, Legendre L (1998) Numerical ecology, 2nd English edn. Elsevier Science, Amsterdam

Maher LJ (2000) MODPOL.EXE: a tool for searching for modern analogs of Pleistocene pollen data. INQUA Commission for the Study of the Holocene: Sub-commission on data-handling methods. Newsletter 20. http://www.chrono.qub.ac.uk/inqua/news20/n20-ljm.htm. Accessed 4 Jan 2012

Manly BFJ (1997) Randomization, bootstrap and Monte Carlo methods in biology, 2nd edn. Chapman & Hall/CRC, London

McCullagh P, Nelder J (1989) Generalized linear models. Chapman and Hall/CRC, London

Melssen W, Wehrens R, Buydens L (2006) Supervised Kohonen networks for classification problems. Chemometr Intell Lab Syst 83:99–113

Metz C, Herman B, Roe C (1998) Statistical comparison of two ROC-curve estimates obtained from partially paired data-sets. Med Decis Mak 18:112–121

Overpeck JT, Webb T, Prentice IC (1985) Quantitative interpretation of fossil pollen spectra – dissimilarity coefficients and the method of modern analogs. Quaternary Res 23:87–108

Pflaumann U, Dupart J, Pujol C, Labeyrie LD (1996) SIMMAX: a modern analog technique to deduce Atlantic sea surface temperatures from planktonic foraminfera in deep-sea sediments. Palaeoceanography 11:15–35

Pflaumann U, Sarnthein M, Chapman M, d' Abreu L, Funnell B, Huels M, Kiefer T, Maslin M, Schulz H, Swallow J, van Kreveld S, Vautravers M, Vogelsang E, Weinelt M (2003) Glacial North Atlantic: sea-surface conditions reconstructed by GLAMAP 2000. Paleoceanography 18:1065

Podani J (1999) Extending Gower's general coefficient of similarity to ordinal characters. Taxon 48:331–340

Prell W (1985) The stability of low-latitude sea-surface temperatures: an evaluation of CLIMAP reconstructions with emphasis on positive SST anomalies. Tech. Rep. TR 025. U.S. Department of Energy, Washington, DC

Prentice IC (1980) Multidimensional scaling as a research tool in Quaternary palynology – a review of theory and methods. Rev Palaeobot Palynol 31:71–104

R Development Core Team (2009) R: a language and environment for statistical computing. R Foundation for Statistical Computing, Vienna. ISBN 3-900051-00-3. http://www.R-project.org

Rymer L (1978) The use of uniformitarianism and analogy in palaeoecology, particularly pollen analysis. In: Walker D, Guppy J (eds) Biology and Quaternary environments. Australian Academy of Science, Canberra, pp 245–257

Sawada M (2006) An open source implementation of the modern analog technique (MAT) within the R computing environment. Comput Geosci 32:818–833

Sawada M, Viau AE, Vettoretti G, Peltier WR, Gajewski K (2004) Comparison of North-American pollen-based temperature and global lake-status with CCCma AGCM2 output at 6 ka. Quaternary Sci Rev 23:225–244

Schweitzer P (1995) ANALOG: a program for estimating paleoclimate parameters using the method of modern analogs. INQUA Commission for the Study of the Holocene: Working group on data-handling methods. Newsletter 13. http://www.chrono.qub.ac.uk/inqua/news13/n13-pns.htm. Accessed 4 Jan 2012.

Schweitzer P (1999) ANALOG: a program from estimating palaeoclimate parameters using the method of modern analogs. U.S. Geological Survey Open-File Report 94–645, United States Geological Survey. http://pubs.usgs.gov/of/1994/of94-645/. Accessed 4 Jan 2012.

Sieger R, Gersonde R, Zielinksi U (1999) A new extended software package for quantitative paleoenvironmental reconstructions. EOS, Transactions of the Geophysical Union 80(19):223–223, doi:10.1029/99EO00171

Simpson G (2007) Analogue methods in palaeoecology: using the analogue package. J Stat Softw 22:1–29

Simpson GL, Oksanen J (2010) analogue: analogue and weighted averaging methods for palaeoecology. R package version 0.6-23. http://CRAN.R-project.org/package=analogue

Simpson GL, Shilland EM, Winterbottom JM, Keay J (2005) Defining reference conditions for acidified waters using a modern analogue approach. Environ Pollut 137:119–133

Telford RJ (2006) Limitations of dinoflagellate cyst transfer function. Quaternary Sci Rev 25:1375–1382

Telford RJ, Birks HJB (2005) The secret assumption of transfer functions: problems with spatial autocorrelation in evaluating model performance. Quaternary Sci Rev 24:2173–2179

Telford RJ, Birks HJB (2009) Evaluation of transfer functions in spatially structured environments. Quaternary Sci Rev 28:1309–1316

Telford RJ, Andersson C, Birks HJB, Juggins S (2004) Biases in the estimation of transfer function prediction errors. Paleoceanography 19:PA4014

ter Braak CJF (1995) Non-linear methods for multivariate statistical calibration and their use in palaeoecology: a comparison of inverse (*k*-nearest neighbours, partial least squares and weighted averaging partial least squares) and classical approaches. Chemometr Intell Lab Syst 28:165–180

Waelbroeck C, Labeyrie L, Duplessy J-C, Guiot J, Labracherie M, Leclaire H, Duprat J (1998) Improving past sea surface temperature estimates based on planktonic fossil faunas. Paleoceanography 13:272–283

Wahl ER (2004) A general framework for determining cutoff values to select pollen analogs with dissimilarity metrics in the modern analog technique. Rev Palaeobot Palynol 128:263–280

Webb A (2002) Statistical pattern recognition, 2nd edn. Arnold, London

Whitmore J, Gajewski K, Sawada M, Williams J, Shuman B, Bartlein P, Minckley T, Viau A, Webb T, Shafer S, Anderson P, Brubaker L (2005) Modern pollen data from North American and Greenland for multi-scale paleoenvironmental applications. Quaternary Sci Rev 24:1828–1848

Williams JW, Jackson ST (2007) Novel climates, no-analog communities, and ecological surprises. Front Ecol Environ 5:475–482

Williams JW, Shuman B (2008) Obtaining accurate and precise environmental reconstructions from the modern analog technique and North American surface pollen dataset. Quaternary Sci Rev 27:669–687

Williams JW, Jackson ST, Kurtzbach JE (2007) Projected distributions of novel and disappearing climates by 2100 AD. Proc Natl Acad Sci USA 104:5738–5742

Chapter 16
Autocorrelogram and Periodogram Analyses of Palaeolimnological Temporal-Series from Lakes in Central and Western North America to Assess Shifts in Drought Conditions

Pierre Dutilleul, Brian F. Cumming, and Melinda Lontoc-Roy

Abstract The use of appropriate statistical techniques of temporal-series analysis is becoming increasingly important for palaeolimnological studies. After a preamble outlining the two main approaches to statistical analysis of time-series, we focus on a classical time-domain technique using distance classes, autocorrelogram analysis, and a novel frequency-domain technique, multi-frequential periodogram analysis (MFPA). Both are appropriate for unequally spaced observations such as many palaeolimnological temporal data. In previous studies, broad-scale shifts in the mean climatic conditions over millennia have been inferred from changes in diatom assemblages in lake sedimentary records. Here we examine if the temporal characteristics of diatom-inferred changes in climatic conditions: (1) vary in three Dakota or Minnesota lakes before and after AD 1000–1300; and (2) vary among five dominant zones over the past 5500 years in a lake from southern British Columbia. In addition, we assess if the periodic components of temporal-series differ among lakes. Consistent with the major changes in the inferred climatic conditions, autocorrelogram analyses of the total series from all the lakes essentially exhibit heterogeneity in the mean due to broad-scale trends or differences in the mean level between zones. Furthermore, the autocorrelograms of the different portions suggest that the temporal structure varies between zones. The MFPA refines this interpretation by detecting periodic components in all of the portions of the palaeolimnological temporal-series. Specifically, the MFPA results support

P. Dutilleul (✉) • M. Lontoc-Roy
Laboratory of Applied Statistics, Department of Plant Science, Faculty of Agricultural and Environmental Sciences, McGill University, Macdonald Campus, Ste-Anne-de-Bellevue, Québec H9X 3V9, Canada
e-mail: pierre.dutilleul@mcgill.ca

B.F. Cumming
Paleoecological Environmental Assessment and Research Laboratory (PEARL), Department of Biology, Queen's University, Kingston, ON K7L 3N6, Canada
e-mail: brian.cumming@queensu.ca

H.J.B. Birks et al. (eds.), *Tracking Environmental Change Using Lake Sediments*, Developments in Paleoenvironmental Research 5, DOI 10.1007/978-94-007-2745-8_16,

the assertion that the periodic components of the temporal-series from all lakes tend to vary between the dominant zones of inferred limnological and climatic regimes. Many of the periodicities identified by MFPA are suggestive of connections to changes in solar activity, whereas some others may be lunar-related.

Keywords Autocorrelogram • British Columbia • Diatoms • Distance classes • Multi-frequential periodogram • Northern Great Plains • Palaeolimnology • Periodicities • Temporal-series analysis

Introduction

The sedimentary record of lakes offers many opportunities to study variability in climatic conditions over millennia. The relatively long time-frame covered by sedimentary records, in conjunction with a good spatial coverage of lakes, make palaeolimnological techniques an attractive approach to study changes in climatic conditions. However, there are plenty of challenges associated with using information in the sedimentary records of lakes to discern past climatic conditions and variability. Foremost among these challenges is demonstrating that the proxy records preserved in the sedimentary record of a lake can adequately track modern and past climatic conditions (see Cumming et al. 2012: Chap. 20). A number of palaeolimnological approaches are available for tracking climate-related variables from lake sediments (e.g., Fritz 1996, 2008; Battarbee 2000; Smol and Cumming 2000; Battarbee et al. 2002; Douglas and Smol 2010; Fritz et al. 2010; Lotter et al. 2010). Additionally, a number of physical and geochemical techniques can be used to extract past variations in limnologic and climatic conditions from the sedimentary record of lakes (e.g., Dean 2002; Dean et al. 2002; Moy et al. 2002; Pienitz et al. 2009).

Once appropriate proxies of past climatic conditions have been analysed from lake sediments, significant challenges still exist when it comes to the analysis of the temporal structure of these data. Although most investigators strive to obtain equally spaced sampling intervals and a similar temporal integration in each sample (i.e., the number of years in each interval), it often turns out that both features vary within cores and between sites. Such potential complications will be less problematic if the sediments from the lake under study are varved (i.e., annually laminated) (e.g., Bradbury et al. 2002; Dean 2002; Dean et al. 2002). Unfortunately, climatically-sensitive lakes that happen to be varved are extremely rare. Consequently, palaeolimnological data from most sites are best described as temporal-series, since the term 'time-series' is normally reserved for partial realisations of a discrete-time stochastic process (Priestley 1981), that is, observations that are made repeatedly on the same random variable at equal spacings in time (Ord 1988). In more formal terms, in a time-series of length n, the time index t is equal to $1, 2, \ldots, n$, whereas in a temporal-series, the n values of the time index are $t_1, t_2, \ldots, t_n$ without the constraint that $t_2 - t_1 = t_3 - t_2$ etc.

Analyses of temporal-series of lacustrine sediments from all over the world have provided insights into the dynamic nature of climate. Lake records from North America (Anderson 1993; Laird et al. 1996; Bradbury et al. 2002; Cumming et al. 2002; Dean 2002; Dean et al. 2002), Central America (Hodell et al. 2001), South America (Moy et al. 2002), and Africa (Halfman et al. 1994; Stager et al. 1997; Verschuren et al. 2000) have all identified periodicities that may be related to known variations in solar activity, as reflected in records of cosmogenically produced isotopes such as ^{14}C or ^{10}Be. Furthermore, many of these studies showed that the periodic components of the series have changed over time (e.g., Bradbury et al. 2002; Moy et al. 2002). Lastly, evidence is accumulating that millennial-scale 'state shifts' in limnological conditions may be an important mode of variability in many of the records (Laird et al. 1996; Verschuren et al. 2000; Cumming et al. 2002; Moy et al. 2002; Stager et al. 2011). These changes are similar to the oscillations that have been described in the North Atlantic (Bond et al. 2001), and also correspond to periods of major glacial advances in the Northern Hemisphere (Cumming et al. 2002).

Our main objective here is to present two statistical methods that should prove to be useful in palaeolimnological research: autocorrelogram analysis and a special form of periodogram analysis described as 'multi-frequential', which are both specifically designed to deal with unequally spaced observations in time. We will illustrate these methods using previously published palaeolimnological records from central (Laird et al. 1996; Fritz et al. 2000; Bradbury et al. 2002) and western (Cumming et al. 2002) North America. The presentation and the illustration of the autocorrelogram and periodogram analyses are preceded by a general preamble.

Statistical Background

Time-Domain Approach

A natural approach to the statistical analysis of time-series data is to study them in the domain in which they were collected, that is, time. Note that in palaeolimnology, the value of the time index corresponds to the vertical spatial location on a sediment core that is dated by varves or ^{14}C dating (Blaauw and Heegaard 2012: Chap. 12). The time-domain approach is based on an extension of the linear additive model

$$X_i = \mu + \varepsilon_i \text{ for } i = 1, 2, \ldots, n, \tag{16.1}$$

where the random variable X for individual i, X_i, is decomposed into the population mean μ (i.e., a constant that is generally unknown) and a random 'error', ε_i, which has the same probability distribution as X_i except that ε_i has a zero population mean. In particular, the population variance of the error is equal to the population variance of X, Var[X], which is classically denoted as σ^2. The equation above extends to discrete-time stochastic processes (of which time-series are partial realisations) as follows:

$$X(t) = \mu(t) + \varepsilon(t) \text{ for } t = 1, 2, \ldots, \tag{16.2}$$

where $X(t)$ denotes the random variable X at time t, and the population mean $\mu(t)$ or the population variance $\sigma^2(t)$, or both, can change with the time index t if no assumption is made.

Thus, the structure of a time-series, or of a temporal-series if the observations are unequally spaced, can be in the mean function or in the heterogeneity of the variance (also called 'heteroscedasticity'). It can also be in the autocorrelation of the random errors $\varepsilon(t)$. To introduce the concept of autocorrelation, we start with the correlation Corr[X, Y] between two random variables X and Y. By definition, it is the standardised covariance between X and Y

$$\frac{\text{Cov}\,[X, Y]}{\sqrt{\text{Var}\,[X]\,\text{Var}\,[Y]}} \tag{16.3}$$

By replacing in Eq. 16.3 above the two random variables X and Y by the same random variable X at times t and $t+k$, $X(t)$ and $X(t+k)$ (with k, a positive integer), one obtains the kth-order autocorrelation coefficient Corr[$X(t), X(t+k)$]. To estimate it correctly, a number of assumptions are required, as shown below.

Classically, Pearson's simple linear correlation statistic r is evaluated from n pairs of observations (X_i, Y_i) ($i = 1, \ldots, n$) made on two random variables X and Y, as follows:

$$\frac{\frac{1}{n-1}\sum_{i=1}^{n}(X_i - \bar{X})(Y_i - \bar{Y})}{\sqrt{S_X^2 S_Y^2}} \tag{16.4}$$

where $\bar{X}$ and $\bar{Y}$ denote the sample means for X and Y, and S_X^2 and S_Y^2 are the corresponding sample variances. If the n pairs of observations above are replaced by the $n-1$ pairs of observations $(X(t), X(t+1))$ ($t = 1, \ldots, n-1$), that is, the first $n-1$ observations of the time-series collected for X are paired with the $n-1$ last observations of the same time-series, the simple linear correlation statistic becomes

$$\frac{\frac{1}{n-1}\sum_{t=1}^{n-1}(X(t) - \bar{X})(X(t+1) - \bar{X})}{S_X^2} \tag{16.5}$$

It is obvious that to be correct, the use of such an autocorrelation coefficient estimate requires that the sample mean $\bar{X}$ be a good estimator of all population means $\mu(t)$ and that the sample variance S_X^2 be a good estimator of all population

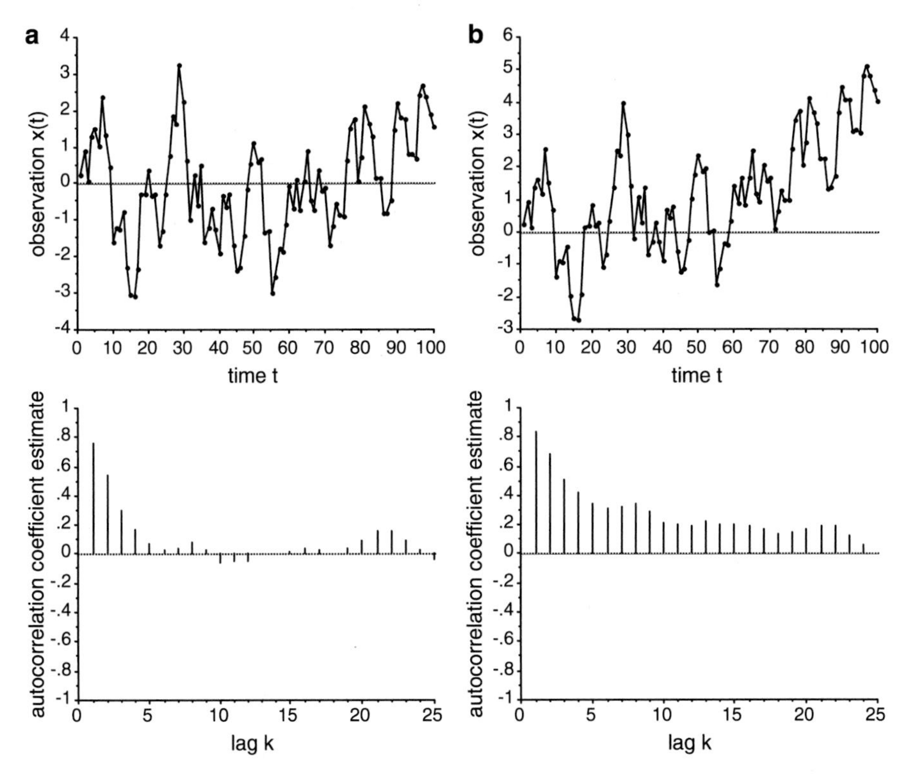

Fig. 16.1 (**a**) Partial realisation (n = 100) of a time-discrete first-order autoregressive process with parameter value 0.75 and the corresponding autocorrelogram (*below*); for such a process, the theoretical autocorrelation function is given by $(0.75)^k$, and in this case, the autocorrelation coefficient estimates reflect such an exponential decrease because $\mu(t) = 0$ for all t. (**b**) Partial realisation ($n = 100$) of the same time-discrete stochastic process, on which a linear trend defined by $\mu(t) = 0.025\,t$ is now superimposed, and the corresponding autocorrelogram (*below*); the autocorrelation coefficient estimates are biased upwards because of the lack of stationarity resulting from the trend

variances $\sigma^2(t)$. This is true in particular when the time-discrete stochastic process is weakly stationary, that is, when $\mu(t) = \mu$ for all t, $\sigma^2(t) = \sigma^2$ for all t, and the value of $\mathrm{Corr}[X(t), X(t+k)]$ depends only on k. Figure 16.1 illustrates this with one partial realisation of a weakly stationary first-order autoregressive process and the corresponding autocorrelogram in which the autocorrelation coefficient estimates are plotted against the lag k (panel a) and one partial realisation of a time-discrete stochastic process with the same theoretical autocorrelation function but for which $\mu(t) = 0.025t$, with the apparent effect of the lack of stationarity (i.e., trend) on the autocorrelation coefficient estimates (panel b). Recommended readings on these topics include the books by Priestley (1981), Diggle (1990), and Dutilleul (2011), and the papers by Legendre and Dutilleul (1992), Dutilleul (1995), and Alpargu and Dutilleul (2001, 2003a, b).

Frequency-Domain Approach

The alternative to the time-domain approach in the statistical analysis of time-series follows from the modelling of the mean function $\mu(t)$ as a sum of K cosine and sine waves, each of them being a function of a frequency ω_k ($k = 1, \ldots, K$) and time t:

$$\mu(t) = \sum_{k=1}^{K} \{a_k \cos(\varpi_k.t) + b_k \sin(\varpi_k.t)\} \tag{16.6}$$

At the basis of this frequency-domain approach, the Fourier series development of the observed time-series $x(1), \ldots, x(n)$ is performed at Fourier frequencies $2\pi p/n$ ($p = 0, 1, 2, \ldots, [n/2]$, with [.], the integer part operator), providing the following coefficients:

$$a_0^{'} = \frac{1}{n} \sum_{t=1}^{n} x(t) \text{ for } p = 0: \tag{16.7.1}$$

$$a_p^{'} = \frac{2}{n} \sum_{t=1}^{n} x(t) \cos\left(\frac{2\pi p}{n} \cdot t\right) \text{ for } p = 1, 2, \ldots, \left[\frac{n}{2}\right] \text{ if } n \text{ is odd;} \tag{16.7.2}$$

$$b_p^{'} = \frac{2}{n} \sum_{t=1}^{n} x(t) \sin\left(\frac{2\pi p}{n} \cdot t\right) \text{ for } p = 1, 2, \ldots, \left[\frac{n}{2}\right] \text{ if } n \text{ is odd.} \tag{16.7.3}$$

(*Note*: If n is even instead of odd, the calculation of the coefficient $a'_{[n/2]}$ is slightly different and the coefficient $b'_{[n/2]}$ is zero; Priestley 1981.) Some of the coefficients above are ordinary least-squares estimates of the coefficients a_k and b_k involved in the modelling of the mean function $\mu(t)$ only if $\omega_k = 2\pi p/n$ for some $p = 0, 1, 2, \ldots, [n/2]$ and the observations are equally spaced in time (Diggle 1990).

The plot of quantities ${a'_p}^2 + {b'_p}^2$ (multiplied by a constant in some packages and textbooks) against p provides the classical uni-frequential periodogram of Schuster (1898). Periodograms are primarily designed for the analysis of discrete spectra, that is, when the periodic components of a time-series are modelled by cosine and sine waves in the absence of autocorrelation. Periodograms are also used to detect periodic components in the analysis of mixed spectra, that is, when there is autocorrelation in addition to periodic components (Dutilleul 2001, 2011). In the case of purely continuous spectra (i.e., when there is only autocorrelation and no periodic component), a spectral density function can be defined from the autocorrelation function under the assumptions of weak stationarity. As spectral estimators, smoothed periodograms are then more appropriate. Figure 16.2 illustrates the use of Schuster's periodogram compared to a smoothed periodogram with a triangular spectral window, for a discrete spectrum (panel a) and a mixed spectrum (panel b).

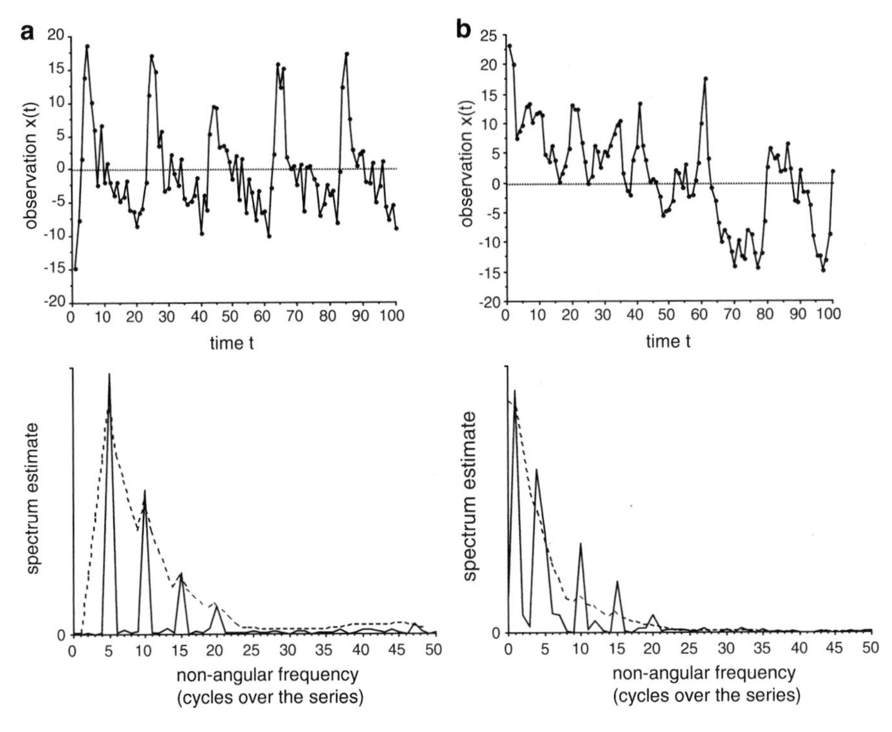

Fig. 16.2 Two time-series of length 100 composed of a periodic signal and random errors, with (*below*) the corresponding Schuster's periodogram (*solid line*) and smoothed periodogram with a triangular spectral window of width 7 (*dashed line*). In both cases, the periodic signal is made of 4 cosine and sine waves with periods 20, 10, 20/3, and 5, and coefficients 4, 3, 2, and 1. (**a**) Random errors are not autocorrelated. (**b**) Random errors follow a time-discrete first-order autoregressive process with parameter value 0.9. In both cases, the Schuster periodograms show the four periodic components relatively well because they correspond to Fourier frequencies (i.e., 5, 10, 15, and 20 cycles over the series). The smoothed periodogram (*dashed line*) captures the continuous part of the spectrum (*left*) better than Schuster's periodogram (*solid line*), at the expense of the discrete part as the peaks corresponding to periodic components with smaller coefficients almost vanish because of the smoothing. The latter feature is observed in both cases

Autocorrelograms and Periodograms

In the following sections, we present two statistical methods that we think will prove very useful in palaeolimnological studies for a number of reasons outlined below. We apply them to a number of data-sets and discuss the results.

The first statistical method is the autocorrelogram analysis based on Moran's *I* statistic (Moran 1950), which was originally used in a spatial context and is evaluated at a number of 'distance classes'. In other words, this further extension of Pearson's simple linear correlation coefficient is evaluated using all pairs of observations collected at sampling points separated by a distance not greater than a certain quantity (i.e., the upper bound of the class) and not smaller than another (i.e., the lower bound of the class). Its use here means that the pairs $(x(t_i), x(t_j))$ and $(x(t_i'), x(t_j'))$ (instead of $(x(t), x(t+1))$ and $(x(t+1), x(t+2))$) will be used for the evaluation of Moran's I statistic at a given distance class if the distances $|t_i - t_j|$ and $|t_i' - t_j'|$ (in years for palaeolimnological temporal-series) fall in the same class (instead of lag 1). Note that smoothed periodograms (whose use is classically restricted to equally spaced observations) can be calculated from autocorrelation coefficients thus estimated because of the links existing between autocorrelogram and periodogram statistics (Priestley 1981). The originality of this statistical method lies in the use of distance classes for analysing temporal-series in the time-domain approach.

The second statistical method involves the use of multi-frequential periodograms for spectral analysis in the frequency domain. Multi-frequential periodograms are general enough to allow the analysis of unequally spaced observations in time. Their analysis is not restricted to Fourier frequencies, and they are particularly powerful in detecting periodic components that are close together (Dutilleul 2001). On the other hand, periodograms are said to be 'inconsistent estimators' of the spectral density function. This is true, but what does it mean? By definition, a consistent estimator in statistics is unbiased (i.e., its expected value is equal to the parameter value), and has a variance that decreases to zero asymptotically (i.e., when the sample size tends to infinity). Schuster's periodogram and the multi-frequential periodogram are, in fact, unbiased asymptotically (Priestley 1981; Brillinger 1983; Dutilleul 2001). Their variance does not decrease to zero asymptotically, however, in particular at the frequencies corresponding to periodic components. But is this really a problem? As far as the detection of deterministic periodic components is concerned, the answer is 'no' because the spectral density function does not exist (i.e., it takes an infinite value) at the corresponding frequencies (Priestley 1981). Accordingly, the use of multi-frequential periodograms is appropriate in the context of discrete spectra for which they were actually designed as well as for the detection of periodic components in mixed spectra (Dutilleul 2001, 2011).

Because of its use in palaeoceanography and palaeoecology, we should mention the program SPECTRUM of Schulz and Stattegger (1997), whose approach in the frequency domain is different from ours in several respects. This program has been helpful to palaeolimnologists in the recent past because of the procedure WOSA (i.e., Welch-Overlapping-Segment-Averaging), which is available for series of unequally spaced observations in time. However, the spectral tools in SPECTRUM are uni-frequential and their use is restricted to Fourier frequencies and seem to require longer series than those available in our study, once diatom assemblage zones were defined. SPECTRUM proposes a coherency analysis (i.e., a correlation analysis in the frequency domain) that can be performed between two series with arbitrary spacing of the samples, which may be practically possible but does not meet the basic prerequisite of a correlation analysis. No interpolation is performed in SPECTRUM, so the spectrum is not under-estimated at high frequencies. Instead, the overlapping and averaging performed in WOSA introduces a strong positive correlation between successive ordinates of the estimated spectrum. The consequences are a global flattening and a broadening of the peaks corresponding to periodic components, with eventually a lack of precision in their estimation and the impossibility of separating periodic components close together. This is very clear in the examples given by Schulz and Stattegger (1997). By providing consistent estimators of the spectral density function, the procedure WOSA in SPECTRUM thus appears more appropriate for the analysis of continuous spectra and can be viewed as a complement to multi-frequential periodogram analysis. Note that the spectral density function is simply another way to look at the autocorrelation function (Priestley 1981).

Data and Methods

Study Sites: Historical Background

All of the temporal-series analysed in our study are based on inferences of past limnological variables from diatom assemblages in well-dated sediment cores. Diatoms are a common component of the algal flora and have been extensively used to infer changes in biologically important variables, such as salinity (Wilson et al. 1996; Fritz et al. 2010), lake depth, and nutrients (Hall and Smol 2010). In all of these records, there are major periods of dominant diatom floras (and inferred environmental variables), suggesting that millennial-scale shifts in moisture regimes are a common phenomenon. For example, based on floristic changes and estimates of salinity from Moon Lake, North Dakota, extreme droughts of greater intensity and duration than those of the 1930s were much more frequent between 2400 and 800 calendar years BP (Before Present) in comparison to after 800 years BP (Laird et al. 1996). These changes are related to declines in the lake water-level and concurrent increases in salinity. Similarly, abrupt changes in the diatom species composition were found to occur in Coldwater Lake, North Dakota (Fritz et al. 2000), and Elk Lake, Minnesota (Bradbury et al. 2002), at approximately the same time (Laird et al. 2003). However, in Elk Lake, changes in the diatom assemblages are likely to be related to changes in nutrients that are driven by changes in seasonality (i.e., high nutrient levels during times of vigorous mixing and storminess, lower nutrient levels during times of lake stratification) (Bradbury et al. 2002). Thus, estimates of TP (total phosphorus content) from the diatom assemblages (see Juggins and Birks 2012: Chap. 14) can provide a simplification of the more complex changes in the multivariate diatom assemblages, which themselves are an integration of changes in climatic conditions.

Changes in lake water-level and salinity were inferred from a diatom-based palaeoclimatic record for Big Lake, British Columbia (Cumming et al. 2002). Similar to the prairie-lake records, this record shows strong state shifts in diatom assemblages, but over a longer duration. Consequently, this allowed the assessment of persistent millennial-scale dynamics over the past 5500 years (Cumming et al.

2002). Over this timeframe, five distinct diatom assemblage zones were delineated. The diatom floras were distinct and relatively stable in these zones, with the length of a zone ranging from 1140 to 1400 years and a mean length of 1220 years. Between zones, mean salinity and lake depth oscillated back and forth between stable periods of higher salinity and lower water-levels and of lower salinity and higher water-levels. It is important to note that at 3770 years BP a major increase in diatom-inferred salinity occurred, which was driven by the disappearance of the most recent taxa in this core. However, variations in salinity (as well as lake depth) also occurred after this point, and are represented by strong changes between benthic (lower water-level) and planktonic taxa (higher water-level). In most cases, transitions were abrupt, often occurring in less than a few decades. However, within each of the zones distinct decadal- to multi-centennial fluctuations also occurred, with some portions exhibiting greater decadal to multi-decadal scale variability.

Data Handling

The diatom-inferred salinity series for Moon Lake extends from ~354 BC to AD 1980 (~2334 years, average resolution of 5.5 years, 424 samples: Laird et al. 1996, 1998). The diatom-inferred salinity series for Coldwater Lake extends from ~39 BC to AD 1940 (~1979 years, average resolution of 10.2 years, 194 samples: Fritz et al. 2000). The diatom-inferred phosphorus series from the varved lake sediments of Elk Lake extends from ~AD 450 to AD 1982 (~1532 years, average resolution of 5.5 years, 281 samples: Bradbury et al. 2002). The diatom-inferred salinity series from Big Lake extends from 3583 BC to AD 1991 (~5574 years, average resolution of 17 years, 325 samples: Cumming et al. 2002). In all the temporal-series, log-transformed data (i.e., diatom-inferred log salinity or log TP) were used in the autocorrelogram and periodogram analyses, since all the diatom-based inference models were log-based.

As mentioned above, major changes in effective moisture were identified in all of the lake sedimentary records from North Dakota (Laird et al. 2003) and British Columbia (Cumming et al. 2002). The times of occurrence of these changes were identified by constrained incremental sum-of-squares cluster analysis (CONISS) performed on the diatom assemblages with chord distance (=Hellinger distance) as the measure of dissimilarity (Laird et al. 1998; Bradbury et al. 2002; Cumming et al. 2002; Legendre and Birks 2012: Chap. 8; Birks 2012: Chap. 11), using the program TILIA v. 1.16 (Grimm 1987). The resulting diatom assemblage zones (i.e., portions of temporal-series) have been interpreted as relatively stable periods of similar climatic and limnologic conditions (Laird et al. 1996, 1998; Bradbury et al. 2002; Cumming et al. 2002). The significance and robustness of the zones identified by constrained cluster analysis were further examined by comparison to both binary and optimal splitting zonation methods (Bennett 1996), using the program PSIMPOLL v. 4.10 (see Bennett 2002; Birks 2012: Chap. 11). The primary splits detected by these techniques identified the same major changes, as did the constrained cluster analysis (Laird et al. 2003). Major changes in the Dakota and

Minnesota lakes (Moon, Coldwater, and Elk lakes) occurred between ~AD 1000 and ~AD 1300 (~950–650 BP). For the longer Big Lake record, transitions between zones occurred at 4960, 3770, 2300 and 1140 years BP.

Hypotheses

We primarily hypothesised that the periodic components in the temporal-series of salinity (Big, Coldwater, and Moon lakes) and total phosphorus (Elk Lake) varied between zones. We also hypothesised that the periodic components might differ among lakes. To assess these hypotheses, we performed autocorrelogram and periodogram analyses on diatom-based inferences of salinity (Big, Coldwater, and Moon lakes) and phosphorus (Elk Lake), since these variables effectively summarised the limnological changes in these lakes. The four palaeolimnological temporal-series and their respective portions are displayed in Fig. 16.3.

Statistical Techniques

In a preliminary step, we used autocorrelograms to assess the structure of the temporal-series of diatom-inferred salinity (Big, Coldwater, and Moon lakes) and phosphorus (Elk Lake). In fact, this type of analysis can be carried out with unequally spaced observations (in space, Legendre and Fortin 1989; in time, Dutilleul 1995). Autocorrelogram analysis was performed on the total series for each lake and for portions of the series divided according to the diatom-assemblage zones interpreted as representing different climatic conditions (two for each of the Dakota and Minnesota sites and five for the longer record from British Columbia; see above). Moran's I autocorrelation statistic (Moran 1950) was evaluated using The R Package (Legendre and Vaudor 1991). In an autocorrelogram, sample autocorrelation coefficients are plotted on the ordinate against increasing distance classes on the abscissa. Values of Moran's I generally range from -1 to $+1$. If greater than $+1$ or smaller than -1, the value was truncated. A value of zero indicates the absence of autocorrelation at the corresponding distance class. Positive or negative values indicate that two observations collected at the distance considered are, respectively, more similar or more dissimilar than expected for uncorrelated sample data. In this study, we used equal-width distance classes for comparison purposes (between zones and among lakes), although equal-frequency distance classes are known to provide a more even precision in estimation and even greater power in the testing of significance throughout the autocorrelogram (Dutilleul and Legendre 1993). (See Cliff and Ord (1981) and Upton and Fingleton (1985) for a formal presentation of autocorrelogram analysis.)

Following completion of the autocorrelogram analyses, the diatom-inferred salinity or phosphorus temporal-series were analysed in the frequency domain in

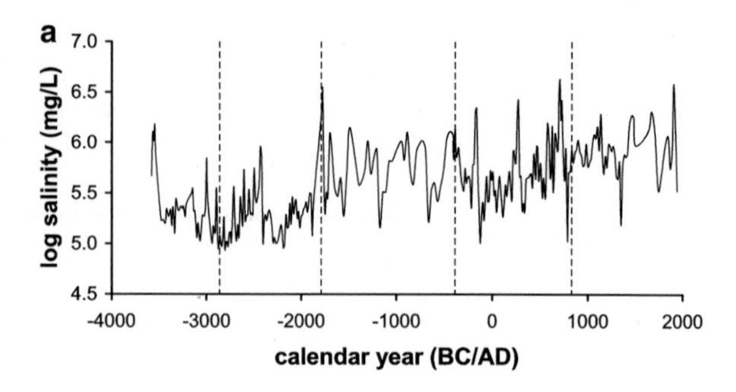

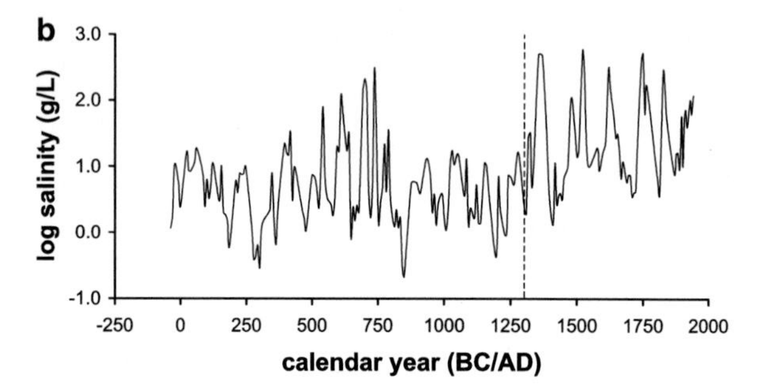

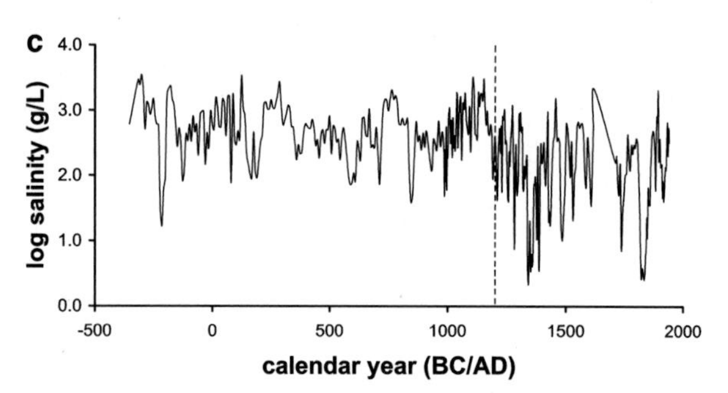

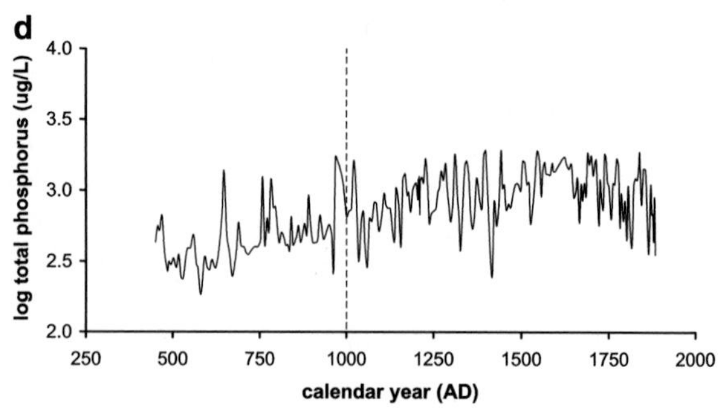

Fig. 16.3 Data plots over time of the log-transformed diatom-inferred salinity concentrations for Big Lake (**a**), Coldwater Lake (**b**), and Moon Lake (**c**), and of the log-transformed diatom-inferred total phosphorus concentrations for Elk Lake (**d**). A *dashed line* indicates where a shift in drought conditions has occurred

Table 16.1 Basic descriptive statistics of the palaeolimnological temporal-series portions

Lake	Portion	Number of observations	Length (years)	Time interval between observations: Shortest (years)	Mean width (years)	SD (years)
Big	1	49	709	11.24	14.68	10.31
	2	73	997	11.24	14.75	8.43
	3	46	1395	7.5	29.70	17.11
	4	95	1205	7.5	12.93	3.45
	5	52	1109	14.84	21.54	17.17
Coldwater	1	129	1319	7.53	10.39	5.28
	2	65	639	5.04	10.02	6.28
Moon	1	234	1553	1.92	6.51	2.04
	2	175	741	2.57	4.23	6.79
Elk	1	82	547	4	6.71	1.67
	2	166	887	0.5	5.34	2.51

order to detect their main periodic components. We used a statistical method that was designed for this purpose: multi-frequential periodogram analysis (MFPA, Dutilleul 1990). Details of this spectral technique can be found in Dutilleul (1998, 2001). Extensions of MFPA include the detection of periodic components in multivariate series, with or without replication. Here, we performed periodogram analysis separately on the different portions of the series in accordance with our primary hypothesis and because stationarity at first order (i.e., homogeneity in the mean) of the underlying temporal stochastic process is a prerequisite of spectral techniques as well as autocorrelogram analysis. The MFPA for equally spaced observations is available in the Periodmod v. 1.2 software (available on request with technical notes from the first author; see also Dutilleul 2011 and software on the companion CD-ROM). Because observations were unequally spaced in the palaeolimnological temporal-series (Table 16.1), we wrote a computer program in SAS/IML language (SAS Institute Inc. 1999) to perform the periodogram analyses. The main features of the MFPA are highlighted below, together with specific aspects of its application in our study.

MFPA belongs to the frequency-domain approach to the analysis of discrete-time stochastic processes. In MFPA, the periodic components (i.e., the discrete part of the spectrum) of a partial realisation (i.e., a time-series or a temporal-series) are detected through the estimation of the corresponding frequencies in a step-wise procedure. An important feature of the statistical method is the re-estimation of all frequencies at each step, by iterative maximisation of the multi-frequential periodogram statistic using an appropriate initial solution. A test of significance is performed to assess the increased explanation of the series' total variation at each step. By definition, the value of the multi-frequential periodogram statistic for a given vector of frequencies is exactly equal to the sum-of-squares of the corresponding trigonometric model fitted to the time-series or the temporal-series by least squares, whether the frequencies in question are Fourier frequencies or not. The multi-frequential periodogram statistic is more robust against autocorrelation (i.e., the continuous part of the spectrum) than the classical uni-frequential periodogram in the analysis of mixed spectra. MFPA is superior at detecting hidden periodicities when frequencies are close together, is applicable to unequally spaced observations, and can be used to determine whether a periodicity is present or absent in a portion of the series (e.g., Tardif et al. 1998). Consequently, MFPA will likely prove to be useful in palaeolimnology.

Results

Autocorrelogram Analyses

As it might be expected, all of the total temporal-series exhibit significant temporal autocorrelation (Fig. 16.4), but the meaning of this autocorrelation is questionable. One of the assumptions of autocorrelogram analysis is stationarity at the first order (Priestley 1981: p. 106), whereas the sample autocorrelation coefficients in Fig. 16.4 were calculated under heterogeneity in the mean of the underlying temporal stochastic process (see the different mean levels for different portions of the series in Fig. 16.3). This heterogeneity of the mean also affects the estimation of the variance of sample autocorrelation coefficients in the tests of significance (Dutilleul and Legendre 1993). In other words, the autocorrelograms in Fig. 16.4 reflect essentially the lack of stationarity of the total series. Specifically, the autocorrelograms for Big Lake and Elk Lake (Fig. 16.4a, d) reflect the curvilinear trends in the total series for these two lakes. For Coldwater Lake and Moon Lake (Fig. 16.4b, c), the autocorrelograms are consistent with the differences in the mean level between portions of the series (i.e., lower mean level for portion 2 in Coldwater Lake, higher mean level for portion 2 in Moon Lake). These large inferred shifts in drought conditions result in positive autocorrelation at short distances, negative autocorrelation at distances greater than the length of the first portion, and almost-zero autocorrelation in-between. Globally, the autocorrelograms in Fig. 16.4 show evidence of major changes in the inferred climatic conditions for all four lakes.

The analysis of separate autocorrelograms for different zones allows the pattern of variability in some portions of the palaeolimnological temporal-series to become more apparent, since the 'autocorrelation' and lack of stationarity of the total series no longer dwarf such autocorrelograms. In view of the autocorrelograms in Figs. 16.5 and 16.6, the structure of the palaeolimnological temporal-series appears to vary between portions, especially for Big Lake and Coldwater Lake. For Big Lake, (i) peaks of positive autocorrelation and troughs of negative autocorrelation are regularly distributed (c. every 200 years) in the autocorrelogram of the earliest portion (Fig. 16.5a); (ii) the autocorrelogram of the latest but one portion is characterised by only two peaks (and two troughs) separated by about 600 years – this may be due to persistent lack of stationarity in that particular portion (Fig. 16.5d); and (iii) the autocorrelograms of other portions show no clear pattern,

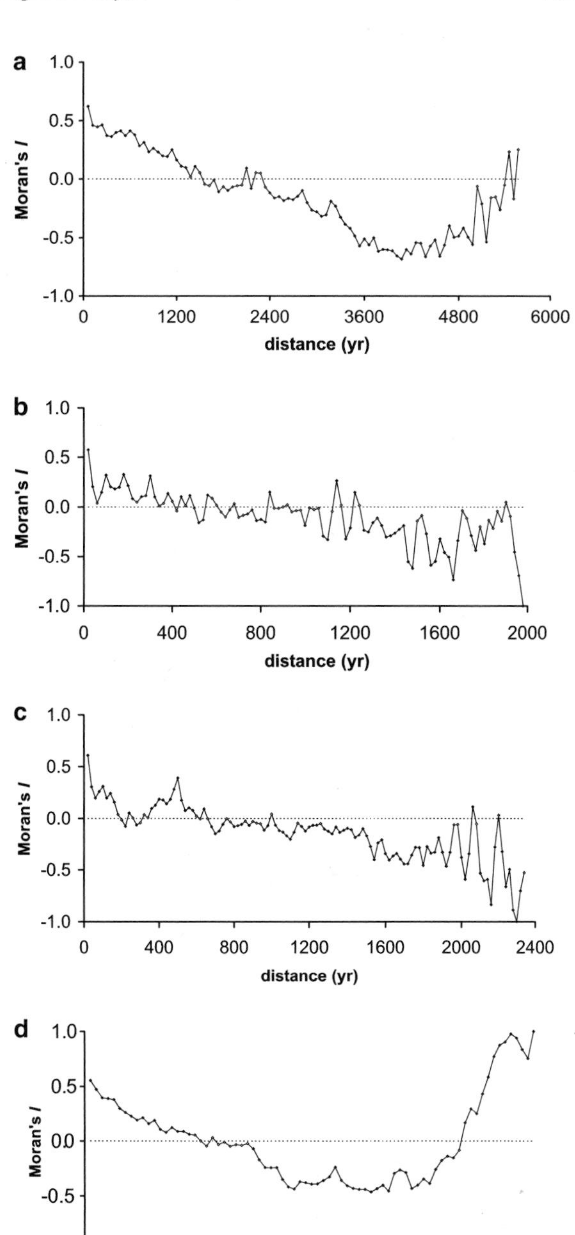

Fig. 16.4 Autocorrelograms using Moran's I statistic and equal-width distance classes for the whole palaeolimnological temporal-series from Big Lake (**a**), Coldwater Lake (**b**), Moon Lake (**c**), and Elk Lake (**d**). Significant values of Moran's I statistic are indicated by *a filled symbol*, and non-significant values by *an empty symbol* (significance level: 0.05, without Bonferroni correction)

with significant autocorrelation over various ranges of distances (Fig. 16.5b, c, e). For Coldwater Lake, (iv) the autocorrelogram of the first portion shows no clear pattern and (v) the autocorrelogram of the second portion resembles that of the earliest portion of Big Lake, with a smaller distance (c. 150 years) between

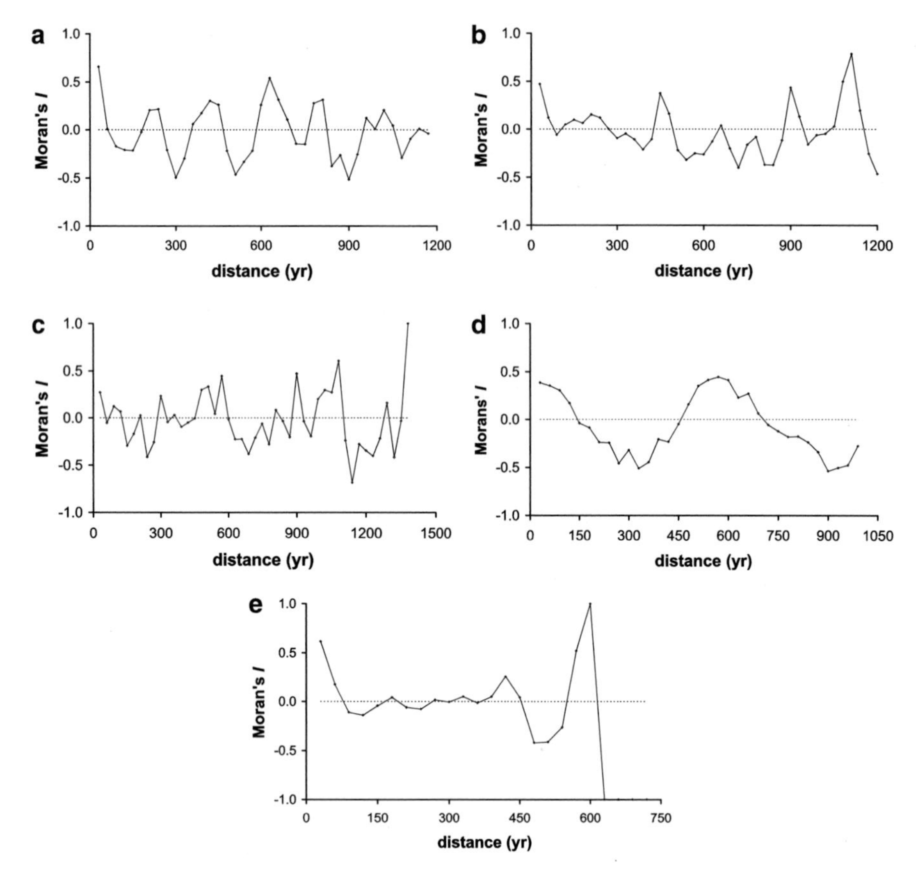

Fig. 16.5 Autocorrelograms using Moran's *I* statistic and equal-width distance classes for the five portions of the palaeolimnological temporal-series from Big Lake. Panels (**a**) to (**e**) correspond to the five portions, with (**a**) representing the earliest discernible zone of drought conditions and (**e**) the latest one. Significant values of Moran's *I* statistic are indicated by *a filled symbol*, and non-significant values by *an empty symbol* (significance level: 0.05, without Bonferroni correction)

peaks (and troughs) (Fig. 16.6a.1, a.2). Peaks and troughs in autocorrelograms may correspond to periodicities, but unlike periodogram analysis (see below), autocorrelogram analysis is not a statistical method designed for periodicity analysis when multiple periodicities are superimposed (Dutilleul 1995). To be complete, large distance classes were under-represented compared to intermediate distance classes in terms of pairs of observations used to evaluate Moran's *I* statistic, resulting in a larger variability of the statistic values as the distances get closer to the series length (i.e., fewer and fewer pairs of observations are used in the evaluation), and distance class widths other than those in Figs. 16.4, 16.5 and 16.6 were used for autocorrelogram analysis, with similar results (not reported here) in terms of pattern and structure.

Fig. 16.6 Autocorrelograms using Moran's *I* statistic and equal-width distance classes for the two portions of the palaeolimnological temporal-series from Coldwater, Moon, and Elk lakes. Panels (**a.1**)–(**a.2**), (**b.1**)–(**b.2**), and (**c.1**)–(**c.2**) correspond to the pre-shift and post-shift zones of drought conditions for Coldwater Lake, Moon Lake, and Elk Lake, respectively. Significant values of Moran's *I* statistic are indicated by *a filled symbol*, and non-significant values by *an empty symbol* (significance level: 0.05, without Bonferroni correction)

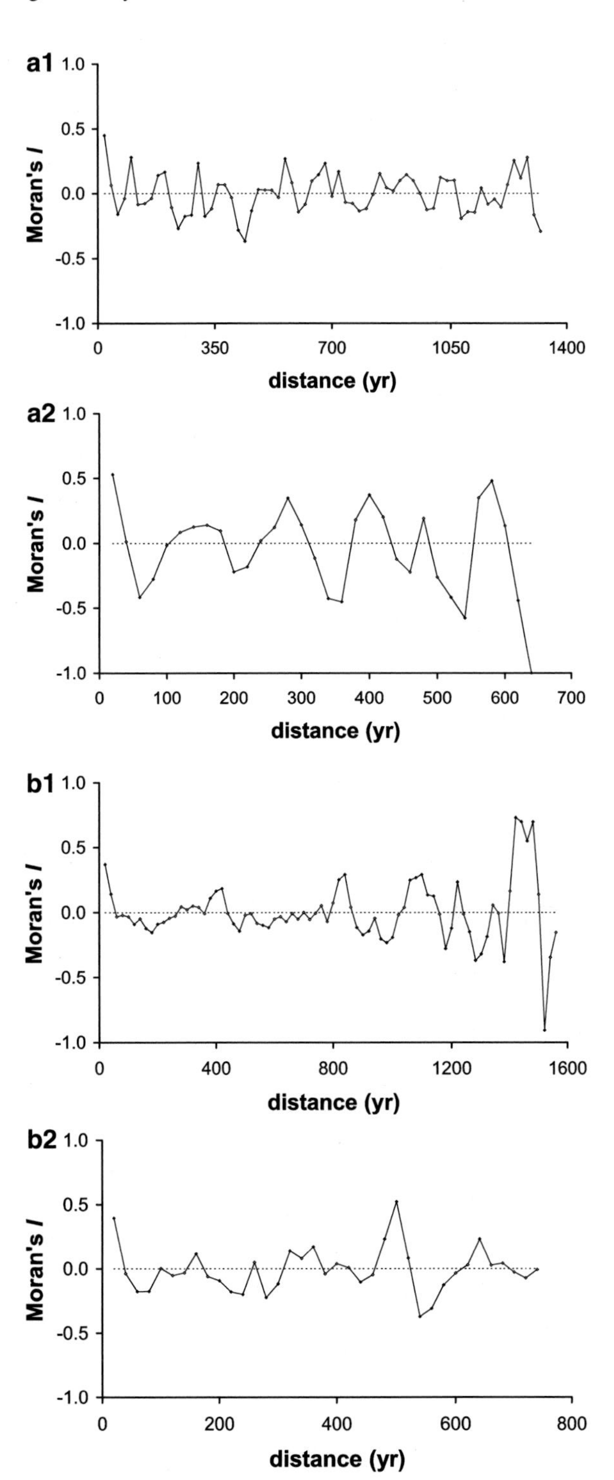

Fig. 16.6 (continued)

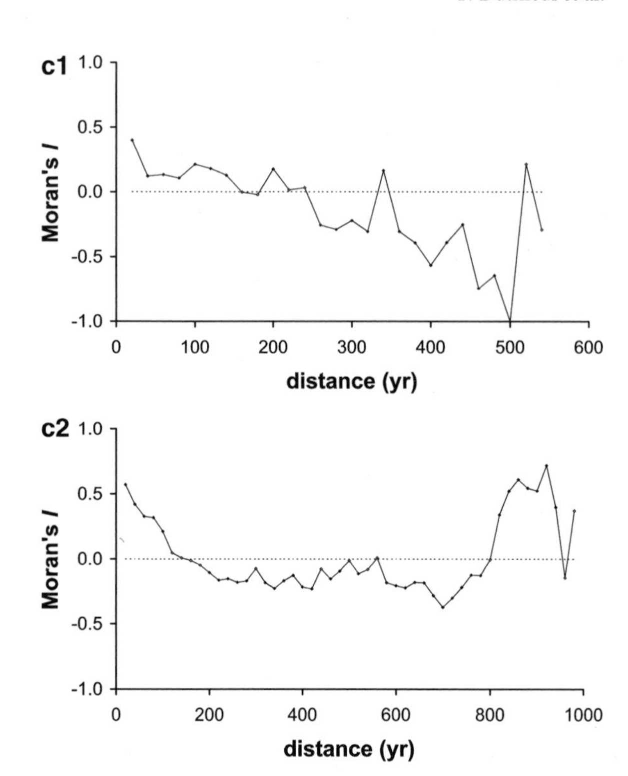

Periodogram Analyses

The MFPA detected periodic components in all of the portions of the palaeolimnological temporal-series analysed (Table 16.2). These MFPA results confirm those of autocorrelogram analyses to some extent, but definitely also refine them. In particular, the long periodicities foreseen in the autocorrelograms of Big 1, Big 4 (Fig. 16.5) and Coldwater 2 (Fig. 16.6) are the result of the superimposition of periodicities of 133.8 and 181.8 years, 109.5, 150.6, 219.1, and 482.0 years, and 118.3, 136.0, and 220.3 years, respectively. Moreover, as we shall see below, the autocorrelograms with no clear pattern follow from the presence of a broad range of periodicities (long, intermediate, and short). Frequency estimates <2 cycles over a portion of series correspond to the continuous part of the spectrum (i.e., autocorrelation), and will not be discussed further.

In addition to long periodicities, a good number of intermediate periodicities were estimated by MFPA in the range of about 30–90 years (Table 16.2). Their number varies with the series portion. For example, only two such periodicities were found for Big 4 against four or more for Big 1, Big 2, Big 3, and Big 5. The number of intermediate periodicities also differs among lakes, with only one such periodicity in Moon 1 (characterised by several long periodicities) and Elk 2 (characterised by many short periodicities) while three or more were pointed out

Table 16.2 Results of the multi-frequential periodogram analysis (MFPA) for unequally spaced observations, using the actual shortest interval of time between successive observations (see Table 16.1) to define the upper limit for frequency estimates

Lake	Portion	Step 1	Step 2	Step 3	Step 4	Step 5	Step 6	Step 7	Step 8	Step 9	Step 10	*P*
Big	1	0.7	0.6	5.3	0.9	3.9	20.1	29.5	15.2	8.4	7.8	0.018
		N/A	N/A	133.8	N/A	181.8	**35.3**	<u>24.0</u>	46.6	84.4	90.9	
		N/A	N/A	102.9	N/A	N/A	**34.8**	<u>23.8</u>	45.1	55.6	54.9	
		N/A	N/A	191.0	N/A	N/A	**35.8**	<u>24.3</u>	48.3	174.8	263.4	
Big	2	1.8	3.4	27.2	18.4	10.2	17.3	13.7	15.3	7.9	37.7	0.014
		N/A	293.2	**36.6**	54.2	97.7	57.6	72.8	65.2	126.2	<u>26.4</u>	
		N/A	274.1	**36.3**	53.2	95.2	56.2	71.4	63.7	120.6	<u>26.2</u>	
		N/A	315.2	**37.0**	55.2	100.4	59.1	74.2	66.6	132.4	<u>26.7</u>	
Big	3	84.9	38.6	7.9	15.6	34.3	87.8	62.8	10.6	17.0	14.6	0.016
		<u>16.43</u>	**36.1**	176.6	89.4	40.7	<u>15.9</u>	**22.2**	131.6	82.1	95.5	
		<u>16.39</u>	**36.0**	159.7	82.4	39.9	<u>15.8</u>	**21.9**	121.5	77.7	87.9	
		<u>16.69</u>	**36.8**	200.4	99.1	42.0	<u>16.1</u>	**22.8**	145.6	88.2	106.2	
Big	4	0.8	5.5	8.0	11.0	2.5	77.5	76.4	33.3	19.2	45.5	0.003
		N/A	219.1	150.6	109.5	482.0	<u>15.55</u>	<u>15.77</u>	**36.2**	62.8	26.5	
		N/A	213.2	147.8	107.9	402.9	<u>15.49</u>	<u>15.71</u>	**35.9**	61.9	26.4	
		N/A	225.3	153.6	111.3	599.7	<u>15.61</u>	<u>15.84</u>	**36.5**	63.7	26.6	
Big	5	5.7	2.5	4.5	22.9	14.4	15.2	15.3	29.2	8.1	18.3	0.023
		194.6	443.6	246.4	48.4	77.0	73.0	72.5	<u>**38.0**</u>	136.9	60.6	
		189.9	411.7	233.4	47.9	59.0	57.0	61.2	<u>**37.5**</u>	129.1	58.1	
		199.4	480.9	261.1	49.0	110.9	101.3	88.8	<u>**38.5**</u>	144.6	63.3	
Coldwater	1	14.2	1.7	7.9	40.8	27.7	53.7	31.7	4.3	16.6	76.6	0.007
		92.9	N/A	167.0	32.3	47.6	24.6	41.6	306.7	79.5	<u>17.2</u>	
		91.7	N/A	162.1	32.1	47.2	24.4	41.3	287.5	78.0	<u>17.1</u>	
		94.1	N/A	172.1	32.5	48.1	24.7	41.9	328.8	81.0	<u>17.3</u>	

(continued)

Table 16.2 (continued)

Lake	Portion	Step 1	Step 2	Step 3	Step 4	Step 5	Step 6	Step 7	Step 8	Step 9	Step 10	*P*
Coldwater	2	4.7	6.7	12.4	14.9	2.9	8	0.5	39	5.4	49.4	0.006
		136.0	95.4	51.5	42.9	220.3	79.9	N/A	16.4	118.3	**12.9**	
		114.5	84.5	50.9	42.4	174.0	76.4	N/A	16.2	105.1	**12.8**	
		167.4	109.5	52.2	43.3	300.4	83.6	N/A	16.5	135.4	**13.0**	
Moon	1	4.9	7.2	11.7	5.1	226.3	234.4	198	33.7	190.1	207.9	0.001
		316.9	215.7	132.7	304.5	**6.86**	**6.63**	7.84	46.1	8.17	**7.47**	
		295.2	210.8	130.5	282.5	**6.86**	**6.62**	7.83	45.7	8.16	**7.46**	
		342.1	220.9	135.0	330.3	**6.87**	**6.63**	7.86	46.4	8.18	**7.48**	
Moon	2	4.6	9.2	1.9	15.2	8.3	22.6	115.7	12	29.8	111.9	0.001
		161.1	80.5	N/A	48.7	89.3	32.8	**6.40**	61.7	24.9	**6.62**	
		157.0	79.4	N/A	48.3	85.5	32.5	**6.39**	60.2	24.7	**6.61**	
		165.4	81.7	N/A	49.2	93.4	33.1	**6.42**	63.3	25.1	**6.64**	
Elk	1	0.4	11.6	5	3.3	54.7	66	63.1	15.1	7.6	49.2	0.014
		N/A	47.2	109.4	165.8	**10.0**	8.29	8.67	**36.2**	72.0	**11.1**	
		N/A	46.4	104.5	152.4	**9.9**	8.24	8.63	**35.4**	68.1	**11.0**	
		N/A	47.9	114.7	181.7	**10.1**	8.34	8.71	**37.1**	76.2	**11.2**	
Elk	2	0.1	129.4	193.6	13.0	572.2	857.8	127.9	126.6	384.9	790.1	0.001
		N/A	**6.85**	4.58	68.2	1.55	1.03	**6.94**	**7.01**	2.30	1.12	
		N/A	**6.84**	4.58	67.3	1.55	1.03	**6.92**	**6.99**	2.30	1.12	
		N/A	**6.87**	4.59	69.1	1.55	1.03	**6.95**	**7.02**	2.31	1.12	

For a given portion of paleolimnological temporal-series, the four lines of results are for the frequency estimates (expressed in number of cycles; line 1), the corresponding period estimates (years; line 2), and the lower bound and upper bound of an approximate 95% confidence interval for the true value of the period, when applicable (years; lines 3 and 4). N/A = not applicable, that is, the frequency estimate is smaller than two cycles over the portion of paleolimnological temporal-series analysed and/or the number of observations is too small to provide reliable boundaries. *P* is the probability of significance at the end of the step-wise procedure. Underlined period values are equal to two to three times the shortest interval of time between successive observations (see Table 16.1). Bold-faced period values are discussed in the text in Coldwater 1, Coldwater 2, Moon 2, and Elk 1. There is some evidence for the presence of periodicities with a length comparable to the sunspot cycle (i.e., 11 years on average, within a range of 9–13 years), but primarily for the prairie lakes (i.e., Coldwater 2 and Elk 1) at this stage of our analyses. For Big Lake, only the third portion of the series shows a periodicity of c. 22 years, which is also solar-related (Dutilleul and Till 1992). Together with Elk 2, Moon 1 and Moon 2 showed several periodicities of about 7 years, which might be of climatic origin (see Discussion). Recall that the length of the shortest time interval between observations is greater than 7 years for Big 3 and Big 4, and greater than 11 years for Big 1, Big 2, and Big 5 (Table 16.1). In all series portions except Moon 1, MFPA estimated higher frequencies corresponding to shorter periodicities of <3 times the actual shortest interval of time between successive observations (e.g., 24.0 years for Big 1 for which the shortest time interval between observations is 11.24 years: Table 16.1).

Discussion

Our discussion is in three parts. First, we address the question of the aliasing effect in the spectral analysis of palaeolimnological temporal-series. Second, we make concluding remarks about the biological hypotheses stated at the beginning of the study. Third, we draw general conclusions on the appropriateness of autocorrelogram and periodogram analyses and other statistical methods for the analysis of palaeolimnological temporal-series.

The aliasing effect occurs in spectral analysis when a continuous-time stochastic process is sampled at too coarse a scale, so that a frequency (periodic) component appears lower (longer) than it should. Figure 4.10 in Priestley (1981: p. 224) illustrates this effect very well. Priestley (1981: p. 507) recommends that time-series analysts choose a sufficiently small sampling interval to reduce the aliasing effect to negligible proportions. This recommendation inspired us to do a second MFPA, in which we used a hypothetical 1-year interval as the shortest time interval between observations. This drastically changed the results, with the finding of numerous periodicities of about 11, 7, 5.5, and 3.5 years. The detection of periodicities shorter than the actual shortest time interval between observations is questionable. To address this question, we see two options: if possible, re-sample the sediment cores using a finer resolution and analyse the new data thus collected (i.e., this was not possible at the time of our study), or apply Priestley's recommendation directly in future studies.

We assessed if the periodic components of temporal-series from our three Dakota or Minnesota lakes and one lake in British Columbia varied between major shifts in limnological and climatic regimes. Our results strongly support the assertion that the temporal pattern and periodic components are different among zones. In particular, the periodic components are quite different in zones of differing limnological conditions. This is likely to be the result of the interplay of different climatic forcing mechanisms in conjunction with local characteristics of the various lakes. It must be noted that the record for Elk Lake was dated by varves, whereas those for the Moon, Coldwater, and Big lakes were dated by ^{14}C. The dates for Big Lake have been verified by comparison with known tephra layers. Unfortunately, there are no such independent horizons for the other records. We doubt that the dates would change the longer periodicities detected in our study, but they could change some of the shorter periodicities. A sensitivity analysis would be the only way to answer that question.

Several of the periodicities that were found in the records (i.e., 5.5, 11, and 22 years) are suggestive of a connection to changes in solar activity. Among them, the c. 11-year sunspot cycle is perhaps the best known. The link between periodicities and solar variability is not new, and is still much debated (e.g., Haigh 2001; Rind 2002). Furthermore, centennial-scale changes in proxy drought records have also been related to solar variability in many regions, including the Northern Great Plains, USA (Laird et al. 1996; Dean 1997; Yu and Ito 1999), and the Yucatan Peninsula, Mexico (Hodell et al. 2001) as well as equatorial East Africa (Verschuren et al. 2000). The finding of a c. 36-year periodicity in all five portions of the British Columbia lake series and in one portion of a prairie-lake series supports the 'lunar hypothesis' of the Swedish oceanographer Otto Petterson at the beginning of the twentieth century (Carson 1963). However, the total absence of the fundamental 18.6-year lunar periodicity and the disappearance of the 36-year periodicity in our second MFPA (results not reported here) prevent us from drawing too definitive conclusions about a lunar-related periodicity. In north-central United States, proxy variables from lake sediments suggest that periods of aridity are cyclic, and are a dominant feature of late-Holocene climate (Dean 1997, 2002; Yu and Ito 1999; Dean and Schwalb 2000). The 7-year periodicity found here is similar to that found for the precipitation series in a dendroclimatological study conducted in North Africa (Dutilleul and Till 1992). Highly significant spectral peaks observed for the New England varve chronology record between 17,500 and 13,500 calendar years BP were interpreted as being within the modern El Niño Southern Oscillation (ENSO) bandwidth of 3–7 years (Rittenour et al. 2000).

Conclusions

The use of appropriate statistical techniques of temporal-series analysis is becoming increasingly important for palaeolimnological studies. After a preamble outlining the two main approaches to statistical analysis of time-series, we focused on a classical time-domain technique using distance classes, autocorrelogram analysis, and a novel frequency-domain technique, multi-frequential periodogram analysis (MFPA). Both are appropriate for unequally spaced observations such as many palaeolimnological temporal data. In previous studies, broad-scale shifts in the mean climatic conditions over millennia have been inferred from changes in diatom assemblages in lake sedimentary records. Here we examined if the temporal characteristics of diatom-inferred changes in climatic conditions: (1) vary in three

Dakota or Minnesota lakes before and after ~AD 1000–1300; and (2) vary among five dominant zones over the past 5500 years in a lake from southern British Columbia. In addition, we assessed if the periodic components of temporal-series differ among lakes. Consistent with the major changes in the inferred climatic conditions, autocorrelogram analyses of the total series from all the lakes essentially exhibit heterogeneity in the mean due to broad-scale trends or differences in the mean level between zones. Furthermore, the autocorrelograms of the different portions suggest that the temporal structure varies between zones. The MFPA refines this interpretation by detecting periodic components in all of the portions of the palaeolimnological temporal-series. Specifically, the MFPA results support the assertion that the periodic components of the temporal-series from all lakes tend to vary between the dominant zones of inferred limnological and climatic regimes. Many of the periodicities identified by MFPA are suggestive of connections to changes in solar activity, whereas some others may be lunar-related.

The detection of periodicities in lake sedimentary records is in its infancy and further investigation, preferably in the frequency domain, is needed. In fact, our study was undertaken to provide an introduction to statistical methods that are not commonly used by palaeolimnologists. We focused on multi-frequential periodogram analysis and, to a lesser degree, on autocorrelogram analysis. The value of spectral techniques is that they are multi-scale, whereas autocorrelograms are evaluated at lags or distance classes. The frequency-domain approach followed in MFPA is definitely superior to the time-domain approach for the detection of single or multiple hidden periodicities (Dutilleul 1998). Other spectral techniques are designed for the analysis of continuous spectra (e.g., maximum entropy: Priestley 1981; WOSA in the program SPECTRUM: Schulz and Stattegger 1997), or are generally restricted to equally spaced observations (e.g., wavelet analysis: Abramovich et al. 2000). For its part, MFPA is available for unequally spaced observations, which is an advantage in the field of palaeolimnology, and its multivariate version (Dutilleul 2001, 2011) allows for spatio-temporal analysis.

References

Abramovich F, Bailey TC, Sapatinas T (2000) Wavelet analysis and its statistical applications. Statistician 49:1–29

Alpargu G, Dutilleul P (2001) Efficiency analysis of ten estimation procedures for quantitative linear models with autocorrelated errors. J Stat Comput Simul 69:257–275

Alpargu G, Dutilleul P (2003a) To be or not to be valid in testing the significance of the slope in simple quantitative linear models with autocorrelated errors. J Stat Comput Simul 73:165–180

Alpargu G, Dutilleul P (2003b) Efficiency and validity analyses of two-stage estimation procedures and derived testing procedures in quantitative linear models with AR(1) errors. Commun Stat B 32:799–833

Anderson RY (1993) The varve chronometer in Elk Lake: record of climatic variability and evidence for solar-geomagnetic-^{14}C-climate connection. In: Bradbury JP, Dean WE (eds) Elk Lake, Minnesota: evidence of rapid climate change in the North-Central United States. Geological Society of America, Boulder, Special Paper 276, pp 45–67

Battarbee RW (2000) Palaeolimnological approaches to climate change, with special regard to the biological record. Quaternary Sci Rev 19:107–124

Battarbee RW, Grytnes J-A, Thompson R, Appleby PG, Catalan J, Korhola A, Birks HJB, Heegaard E, Lami A (2002) Comparing palaeolimnological and instrumental evidence of climate change for remote mountain lakes over the last 200 years. J Paleolimnol 28:161–179

Bennett KD (1996) Determination of the number of zones in a biostratigraphical sequence. New Phytol 132:155–170

Bennett KD (2002) Documentation for PSIMPOLL v. 4.10 and PSCOMB v. 1.03. C programs for plotting pollen diagrams and analysing pollen data. http://www.chrono.qub.ac.uk/psimpoll/psimpoll.html/

Birks HJB (2012) Chapter 11: Analysis of stratigraphical data. In: Birks HJB, Lotter AF, Juggins S, Smol JP (eds) Tracking environmental change using lake sediments. Volume 5: Data handling and numerical techniques. Springer, Dordrecht

Blaauw M, Heegaard E (2012) Chapter 12: Estimation of age-depth relationships. In: Birks HJB, Lotter AF, Juggins S, Smol JP (eds) Tracking environmental change using lake sediments. Volume 5: Data handling and numerical techniques. Springer, Dordrecht

Bond G, Kromer B, Beer J, Muscheler R, Evans MN, Showers W, Hoffmann S, Lotti-Bond R, Hajdas I, Bonani G (2001) Persistent solar influence on North Atlantic climate during the Holocene. Science 294:2130–2136

Bradbury JP, Cumming BF, Laird KR (2002) A 1500-year record of climatic and environmental change in Elk Lake, Clearwater County, Minnesota III: measures of past primary productivity. J Paleolimnol 27:321–340

Brillinger DR (1983) The finite Fourier transform of a stationary process. In: Brillinger DR, Krishnaiah PR (eds) Handbook of statistics. Volume 3. North-Holland, Amsterdam, pp 21–37

Carson R (1963) The sea around us. Signet Science Library Books, New York

Cliff AD, Ord JK (1981) Spatial processes: models and applications. Pion, London

Cumming BF, Laird KR, Bennett JR, Smol JP, Salomon AK (2002) Persistent millennial-scale shifts in moisture regimes in western Canada during the past six millennia. Proc Natl Acad Sci USA 99:16117–16121

Cumming BF, Laird KR, Fritz SC, Verschuren D (2012) Chapter 20: Tracking Holocene climate change with aquatic biota from lake sediments: case studies of commonly used numerical techniques. In: Birks HJB, Lotter AF, Juggins S, Smol JP (eds) Tracking environmental change using lake sediments. Volume 5: Data handling and numerical techniques. Springer, Dordrecht

Dean WE (1997) Rates, timing, and cyclicity of Holocene eolian activity in north-central United States: evidence from varved lake sediments. Geology 25:331–334

Dean WE (2002) A 1500-year record of climatic and environmental change in Elk Lake, Minnesota II: geochemistry, mineralogy, and stable isotopes. J Paleolimnol 27:301–319

Dean WE, Schwalb A (2000) Holocene environmental climatic change in the northern Great Plains as recorded in the geochemistry of sediments in Pickerel Lake, South Dakota. Quaternary Int 67:5–30

Dean WE, Anderson RY, Bradbury JP, Anderson DM (2002) A 1500-year record of climatic and environmental change in Elk Lake, Minnesota I: varve thickness and gray-scale density. J Paleolimnol 27:287–299

Diggle PJ (1990) Time-series: a biostatistical introduction. Clarendon, Oxford

Douglas MSV, Smol JP (2010) Freshwater diatoms as indicators of environmental change in the High Arctic. In: Smol JP, Stoermer EF (eds) The diatoms: applications for the environmental and earth sciences, 2nd edn. Cambridge University Press, Cambridge, pp 249–266

Dutilleul P (1990) Apport en analyse spectrale d'un périodogramme modifié et modélisation des séries chronologiques avec répétitions en vue de leur comparaison en fréquence. DSc dissertation, Département de mathématique, Université catholique de Louvain, Belgium

Dutilleul P (1995) Rhythms and autocorrelation analysis. Biol Rhythm Res 26:173–193

Dutilleul P (1998) Incorporating scale in ecological experiments: data analysis. In: Peterson DL, Parker VT (eds) Ecological scale: theory and applications. Columbia University Press, New York, pp 387–425

Dutilleul P (2001) Multi-frequential periodogram analysis and the detection of periodic components in time-series. Commum Stat A 30:1063–1098
Dutilleul P (2011) Spatio-temporal heterogeneity: Concepts and analyses. Cambridge University Press, Cambridge
Dutilleul P, Legendre P (1993) Spatial heterogeneity against heteroscedasticity: an ecological paradigm versus a statistical concept. Oikos 66:152–171
Dutilleul P, Till C (1992) Evidence of periodicities related to climate and planetary behaviors in ring-width chronologies of Atlas cedar (*Cedrus atlantica*) in Morocco. Can J For Res 22: 1469–1482
Fritz SC (1996) Palaeolimnological records of climate change in North America. Limnol Oceanogr 41:882–889
Fritz SC (2008) Deciphering climatic history from lake sediments. J Paleolimnol 39:5–16
Fritz SC, Ito E, Yu Z, Laird KR, Engstrom DR (2000) Hydrologic variation in the northern Great Plains during the last two millennia. Quaternary Res 53:175–184
Fritz SC, Cumming BF, Gasse F, Laird KR (2010) Diatoms as indicators of hydrologic and climatic change in saline lakes. In: Smol JP, Stoermer EF (eds) The diatoms: applications for the environmental and earth sciences, 2nd edn. Cambridge University Press, Cambridge, pp 186–208
Grimm EC (1987) CONISS: a FORTRAN 77 program for stratigraphically constrained cluster analysis by the method of incremental sum of squares. Comput Geosci 13:13–35
Haigh JD (2001) Climate variability and the influence of the sun. Science 294:2109–2111
Halfman JD, Johnson TC, Finney BP (1994) New AMS dates, stratigraphic correlations and decadal climatic cycles for the past 4 ka at Lake Turkana, Kenya. Palaeogeogr Palaeoclim Palaeoecol 111:83–98
Hall RI, Smol JP (2010) Diatoms as indicators of lake eutrophication. In: Smol JP, Stoermer EF (eds) The diatoms: applications for the environmental and earth sciences, 2nd edn. Cambridge University Press, Cambridge, pp 122–151
Hodell DA, Brenner M, Curtis JH, Guilderson T (2001) Solar forcing of drought in the Maya lowlands. Science 292:1367–1370
Juggins S, Birks HJB (2012) Chapter 14: Quantitative environmental reconstructions from biostratigraphical data. In: Birks HJB, Lotter AF, Juggins S, Smol JP (eds) Tracking environmental change using lake sediments. Volume 5: Data handling and numerical techniques. Springer, Dordrecht
Laird KR, Fritz SC, Maasch KA, Cumming BF (1996) Greater drought intensity and frequency before AD 1200 in the northern Great Plains, USA. Nature 384:552–554
Laird KR, Fritz SC, Cumming BF (1998) A diatom-based reconstruction of drought intensity, duration, and frequency from Moon Lake, North Dakota: a sub-decadal record of the last 2300 years. J Paleolimnol 19:161–179
Laird KR, Cumming BF, Wunsam S, Rusak JA, Oglesby RJ, Fritz SC, Leavitt PR (2003) Lake sediments record large-scale shifts in moisture regimes across the northern prairies of North America during the past two millennia. Proc Natl Acad Sci USA 100:2483–2488
Legendre P, Birks HJB (2012) Chapter 8: From classical to canonical ordination. In: Birks HJB, Lotter AF, Juggins S, Smol JP (eds) Tracking environmental change using lake sediments. Volume 5: Data handling and numerical techniques. Springer, Dordrecht
Legendre P, Dutilleul P (1992) Introduction to the analysis of periodic phenomena. In: Ali MA (ed) Rhythms in fishes. Volume A-236: NATO ASI series. Plenum Press, New York, pp 11–25
Legendre P, Fortin M-J (1989) Spatial pattern and ecological analysis. Vegetatio 80:107–138
Legendre P, Vaudor A (1991) The R Package: multidimensional analysis. Spatial analysis. Département de sciences biologiques, Université de Montréal, Montréal
Lotter AF, Pienitz R, Schmidt R (2010) Diatoms as indicators of environmental change in subarctic and alpine regions. In: Smol JP, Stoermer EF (eds) The diatoms: applications for the environmental and earth sciences, 2nd edn. Cambridge University Press, Cambridge, pp 231–248
Moran PAP (1950) Notes on continuous stochastic phenomena. Biometrika 37:17–23

Moy CM, Seltzer GO, Rodbell DT, Anderson DM (2002) Variability of El Niño/Southern Oscillation activity at millennial timescales during the Holocene epoch. Nature 162:162–165
Ord JK (1988) Time-series. In: Kotz S, Johnson NL (eds) Encyclopedia of statistical sciences. Volume 9. Wiley, New York, pp 245–255
Pienitz R, Lotter AF, Newman L, Kiefer T (eds) (2009). Advances in paleolimnology. PAGES News 17:89–136
Priestley MB (1981) Spectral analysis and time-series. Academic Press, London
Rind D (2002) The sun's role in climate variations. Science 296:673–677
Rittenour TM, Brigham-Grette J, Mann ME (2000) El Niño-like climate teleconnections in New England during the late Pleistocene. Science 288:1039–1042
SAS Institute Inc. (1999) The SAS system for Windows, v. 8. SAS Institute, Cary
Schulz M, Stattegger K (1997) SPECTRUM: spectral analysis of unevenly spaced palaeoclimatic time-series. Comput Geosci 23:929–945
Schuster A (1898) On the investigation of hidden periodicities with application to a supposed 26-day period in meteorological phenomena. Terr Mag Atmos Elect 3:13–41
Smol JP, Cumming BF (2000) Tracking long-term changes in climate using algal indicators in lake sediments. J Phycol 36:986–1011
Stager JC, Cumming BF, Meeker L (1997) A high-resolution 11,400-yr diatom record from Lake Victoria, East Africa. Quaternary Res 47:81–89
Stager JC, Ryves DS, Chase BM, Pausata FSR (2011) Catastrophic drought in the Afro-Asian monsoon region during Heinrich Event 1. Science 331:1299–1302
Tardif J, Dutilleul P, Bergeron Y (1998) Variations in periodicities of the ring width of black ash (*Fraxinus nigra* Marsh.) in relation to flooding and ecological site factors at Lake Duparquet in northwestern Québec. Biol Rhythm Res 29:1–29
Upton GJG, Fingleton B (1985) Spatial data analysis by example: point pattern and quantitative data. Wiley, Chichester
Verschuren D, Laird KR, Cumming BF (2000) Rainfall and drought in equatorial East Africa during the past 1,100 years. Nature 403:410–414
Wilson SE, Cumming BF, Smol JP (1996) Assessing the reliability of salinity inference models from diatom assemblages: an examination of a 219-lake dataset from western North America. Can J Fish Aquat Sci 53:1580–1594
Yu Z, Ito E (1999) Possible solar forcing of century-scale drought frequency in the northern Great Plains. Geology 27:263–266

Part IV
Case Studies and Future Developments in Quantitative Palaeolimnology

Chapter 17
Introduction and Overview of Part IV

H. John B. Birks

Abstract This introduction to Part IV emphasises how palaeolimnology and its research focus has changed from a qualitative and rather academic subject to a quantitative and strongly applied subject. Three of the chapters in Part IV present contrasting palaeolimnological case-studies, all of which have used numerical techniques as important research tools. The case studies consider responses of lakes to external forcing factors at inter-annual to decadal time-scales; the reconstruction and evaluation of surface-water acidification and the use of palaeolimnology in providing information on reference conditions as a guide for lake restoration; and the use of palaeolimnology to reconstruct climate change over the time-scale of the Holocene (last 11,700 years). These chapters highlight that numerical techniques are tools to help answer research questions and that they are not ends in themselves but are means to an end. The last chapter discusses areas of research that represent future challenges in quantitative palaeolimnology. These areas will, as in the initial development of the subject, require active two-way interaction and collaboration with applied statisticians.

Keywords Acidification • Applied palaeolimnology • Climate change • Eutrophication • Future challenges • Human impact • Hypothesis testing • Lake restoration • Models • Reference conditions • Research collaboration

H.J.B. Birks (✉)
Department of Biology and Bjerknes Centre for Climate Research, University of Bergen, PO Box 7803, N-5020 Bergen, Norway

Environmental Change Research Centre, University College London, Pearson Building, Gower Street, London, WC1E 6BT, UK

School of Geography and the Environment, University of Oxford, Oxford, OX1 3QY, UK
e-mail: john.birks@bio.uib.no

H.J.B. Birks et al. (eds.), *Tracking Environmental Change Using Lake Sediments*, Developments in Paleoenvironmental Research 5, DOI 10.1007/978-94-007-2745-8_17,

Overview

The pioneering studies in palaeolimnology were primarily qualitative and considered academic questions in limnology such as lake ontogeny, changes in lake productivity, processes and rates of lake infilling, the balance between within-lake (autochthonous) and outside-lake (allochthonous) organic sources, and the time and impact of catchment disturbances on lake development (e.g., Deevey 1955; Livingstone 1957; Frey 1964, 1969; Mackereth 1966; Wright 1966; Hutchinson 1970; Pennington et al. 1972; Likens and Davis 1975; Oldfield 1977; Likens 1985). The now very out-dated chapter in Birks and Birks (1980) on palaeolimnology reflects the nature of palaeolimnology as it was in the late 1970s. With the increasing realisation of major environmental problems facing society in the late 1970s and early 1980s, such as surface-water acidification, eutrophication, atmospheric contamination by heavy metals and persistent organic pollutants, and in the early 1990s global warming, palaeolimnologists quickly responded to the challenges faced by these problems and transformed their subject from a primarily descriptive and academic discipline into a quantitative and strongly applied subject. In a period of 20 years, palaeolimnology has become a highly societal relevant, quantitative, and rigorous subject, as reviewed in detail by Smol (2008) (see also Smol and Douglas 2007; Pienitz et al. 2009; Battarbee and Bennion 2011; Bennion et al. 2011a; Dearing et al. 2011; Smol et al. 2012: Chap. 1). Palaeolimnologists now employ a huge range of analytical techniques, including an arsenal of numerical and statistical techniques, many of which are described and discussed in Parts II and III of this book. However, numerical methods are not ends in themselves but are means to ends, and their role in palaeolimnology is to help in the elucidation of lake history, the understanding and quantification of the role of external impacts on limnic systems, the quantification of lake-water changes in response to external stressors, and the separation of internal dynamics and external factors on lake change and development. These, and several other problems in palaeolimnology, require quantitative approaches for reconstructing past environments and for testing multiple working hypotheses (Birks and Birks 2006).

The chapters in this Part IV provide three contrasting palaeolimnological case-studies, all of which have used numerical methods as important research tools. Lotter and Anderson (2012: Chap. 18) consider the responses of lakes and their biota to external forcing factors at inter-annual to decadal time-scales. Their case-studies illustrate the power of constrained ordination techniques and their partial relatives such as partial redundancy analysis and variation partitioning analysis (see Legendre and Birks 2012: Chap. 8) to test specific hypotheses about external factors on lake dynamics. Simpson and Hall (2012: Chap. 19) review the use of quantitative palaeolimnological techniques in reconstructing and evaluating surface-water acidification and eutrophication. The numerical techniques illustrated include constrained ordination and variation partitioning analysis (see Legendre and Birks

2012: Chap. 8), calibration functions and quantitative environmental reconstructions (see Juggins and Birks 2012: Chap. 15), and modern analogue matching (see Simpson 2012: Chap. 15). Chapter 19 also shows that palaeolimnology can provide a means not only of reconstructing past lake-water status but also of providing guidelines in lake restoration and in assessing causes of change (see also Battarbee and Bennion 2011; Battarbee et al. 2011a,b; Bennion and Simpson 2011; Bennion et al. 2011a,b). Cumming and colleagues (2012: Chap. 20) discuss the use of palaeolimnology in reconstructing Holocene (last 11,700 years) climate change from assemblages of aquatic biota preserved in lake sediments. They show the value of ordinations and variation partitioning (Legendre and Birks 2012: Chap. 8) and calibration functions (Juggins and Birks 2012: Chap. 14) in reconstructing past climatic changes and in the validation of such reconstructions and in assessing the influence of climate on lake history.

These three case-studies illustrate the central role that numerical techniques play in palaeolimnological research and also highlight the amazing advances that have occurred in palaeolimnology in the last 20 years. Smol (2008) discusses several additional environmental problems that palaeolimnologists are addressing – persistent organic pollutants, mercury, erosion, species invasions, biomanipulations, extirpations, ozone depletion, UV-B effects, and fish-stock declines. Many of these problems, when studied in a palaeolimnological context, are amenable to numerical analysis including new modelling approaches (e.g., Anderson et al. 2006; Boyle 2007, 2008; Dearing 2008). The important thing about numerical techniques is that they are designed to be used and that they are usable. The challenge for palaeolimnologists is to use techniques appropriate to the research questions at hand. There is also a challenge to applied statisticians and numerical palaeolimnologists to ensure that the methods being used are appropriate and optimal for particular research questions and to develop new methods to contribute solutions to the future challenges faced by palaeolimnological research (Smol 2008).

Chapter 21 discusses eight areas of research that represent future challenges in the numerical and statistical analysis of palaeolimnological data such as model selection, modelling of biological dynamics, and Bayesian inference. This chapter outlines some of the challenges that palaeolimnologists will face as they explore their data in new and potentially exciting ways and as they attempt to answer previously unexplored research questions (Birks 2012: Chap. 21). These challenges highlight the need for renewed collaboration between palaeolimnologists and applied statisticians, following the valuable collaborations that resulted in the publications by ter Braak and van Dam (1989), Birks et al. (1990), and ter Braak and Juggins (1993), all of which involved palaeolimnologists working closely with applied statisticians and contributed to the rapid development of palaeolimnology as a quantitative and applied subject in the early 1990s.

Acknowledgements I am indebted to John Smol for helpful comments and to Cathy Jenks for invaluable help in the preparation of this chapter. This is publication A349 from the Bjerknes Centre for Climate Research.

References

Anderson NJ, Bugmann H, Dearing JA, Gaillard M-J (2006) Linking palaeoenvironmental data and models to understand the past and to predict the future. Trends Ecol Evol 21:696–704

Battarbee RW, Bennion H (2011) Palaeolimnology and its developing role in assessing the history and extent of human impact on lake ecosystems. J Paleolimnol 45:399–404

Battarbee RW, Morley D, Bennion H, Simpson GL, Hughes N, Bauere V (2011a) A palaeolimnological meta-database for assessing the ecological status of lakes. J Paleolimnol 45:405–414

Battarbee RW, Simpson GL, Bennin H, Curtis C (2011b) A reference typology of low alkalinity lakes in the UK based on pre-acidification diatom assemblages from lake sediment cores. J Paleolimnol 45:489–505

Bennion H, Simpson GL (2011) The use of diatom records to establish reference conditions for UK lakes subject to eutrophication. J Paleolimnol 45:469–488

Bennion H, Battarbee RW, Sayer CD, Simpson GL, Davidson TA (2011a) Defining reference conditions and restoration targets for lake ecosystems using palaeolimnology: a synthesis. J Paleolimnol 45:533–544

Bennion H, Simpson GL, Anderson NJ, Clarke G, Dong X, Hobæk A, Guilizzoni P, Marchetto A, Sayer CD, Thies H, Tolotti M (2011b) Defining ecological and chemical reference conditions and restoration targets for nine European lakes. J Paleolimnol 45:415–431

Birks HH, Birks HJB (2006) Multi-proxy studies in palaeolimnology. Veg Hist Archaeobot 15:235–251

Birks HJB (2012) Chapter 21: Conclusions and future challenges. In: Birks HJB, Lotter AF, Juggins S, Smol JP (eds) Tracking environmental change using lake sediments. Volume 5: Data handling and numerical techniques. Springer, Dordrecht

Birks HJB, Birks HH (1980) Quaternary palaeoecology. Arnold, London

Birks HJB, Line JM, Juggins S, Stevenson AC, ter Braak CJF (1990) Diatoms and pH reconstruction. Philos Trans R Soc Lond B 327:263–278

Boyle JF (2007) Simulating long-term weathering loss of primary silicate minerals from soil using ALLOGEN: comparison with soil chronosequence, lake sediment, and river solution flux data. Geomorphology 83:121–135

Boyle JF (2008) Climate and surface water acidity: development and application of a generalized predictive model. Holocene 18:69–81

Cumming BF, Laird KR, Fritz SC, Verschuren D (2012) Chapter 20: Tracking Holocene climatic change with aquatic biota from lake sediments: case studies of commonly used numerical techniques. In: Birks HJB, Lotter AF, Juggins S, Smol JP (eds) Tracking environmental change using lake sediments. Volume 5: Data handling and numerical techniques. Springer, Dordrecht

Dearing JA (2008) Landscape change and resilience theory: a palaeoenvironmental assessment from Yunnan, SW China. Holocene 18:117–127

Dearing JA, Dotterweich M, Foster T, Newman L, von Gunten L (eds) (2011) Integrative paleoscience for sustainable management. PAGES News 19:42–92

Deevey ES (1955) The obliteration of the hypolimnion. Mem Ist Ital Idrobiol 8(suppl):9–38

Frey DG (1964) Remains of animals in Quaternary lake and bog sediments and their interpretation. Arch Hydrobiol Beih 2:1–114

Frey DG (1969) The rationale of paleolimnology. Mitt Int Verein Limnol 17:7–18

Hutchinson GE (1970) Ianula: an account of the history and development of the Lago di Monterosi, Latium, Italy. Trans Am Philos Soc 60:1–178

Juggins S, Birks HJB (2012) Chapter 14: Quantitative environmental reconstructions from biological data. In: Birks HJB, Lotter AF, Juggins S, Smol JP (eds) Tracking environmental change using lake sediments. Volume 5: Data handling and numerical techniques. Springer, Dordrecht

Legendre P, Birks HJB (2012) Chapter 8: From classical to canonical ordination. In: Birks HJB, Lotter AF, Juggins S, Smol JP (eds) Tracking environmental change using lake sediments. Volume 5: Data handling and numerical techniques. Springer, Dordrecht

Likens GE (ed) (1985) An ecosystem approach to aquatic ecology. Mirror lake and its environment. Springer, New York

Likens GE, Davis MB (1975) Post-glacial history of Mirror Lake and its watershed in New Hampshire, USA: an initial report. Verh Int Verein Limnol 19:982–993

Livingstone DA (1957) On the sigmoid growth phase in the history of Linsley Pond. Am J Sci 255:364–373

Lotter AF, Anderson NJ (2012) Chapter 18: Limnological responses to environmental changes at inter-annual to decadal time-scales. In: Birks HJB, Lotter AF, Juggins S, Smol JP (eds) Tracking environmental change using lake sediments. Volume 5: Data handling and numerical methods. Springer, Dordrecht

Mackereth FJH (1966) Some chemical observations on post-glacial sediments. Philos Trans R Soc Lond B 250:165–213

Oldfield F (1977) Lakes and their drainage basins as units of sediment-based ecological study. Prog Phys Geog 1:460–504

Pennington W, Haworth EY, Bonny AP, Lishman JP (1972) Lake sediments in northern Scotland. Philos Trans R Soc Lond B 264:191–294

Pienitz R, Lotter AF, Newman L, Kiefer T (eds) (2009) Advances in paleolimnology. PAGES News 17:89–136

Simpson GL (2012) Chapter 15: Modern analogue techniques. In: Birks HJB, Lotter AF, Juggins S, Smol JP (eds) Tracking environmental change using lake sediments. Volume 5: Data handling and numerical techniques. Springer, Dordrecht

Simpson GL, Hall RI (2012) Chapter 19: Human impacts – applications of numerical methods to evaluate surface-water acidification and eutrophication. In: Birks HJB, Lotter AF, Juggins S, Smol JP (eds) Tracking environmental change using lake sediments. Volume 5: Data handling and numerical methods. Springer, Dordrecht

Smol JP (2008) Pollution of lakes and rivers: a paleoenvironmental perspective, 2nd edn. Wiley-Blackwell, Oxford

Smol JP, Douglas MSV (2007) From controversy to consensus: making the case for recent climate change in the Arctic using lake sediments. Front Ecol Environ 5:466–474

Smol JP, Birks HJB, Lotter AF, Juggins S (2012) Chapter 1: The march towards the quantitative analysis of palaeolimnological data. In: Birks HJB, Lotter AF, Juggins S, Smol JP (eds) Tracking environmental change using lake sediments. Volume 5: Data handling and numerical techniques. Springer, Dordrecht

ter Braak CJF, Juggins S (1993) Weighted averaging partial least squares regression (WA-PLS): an improved method for reconstructing environmental variables from species assemblages. Hydrobiologia 269:485–502

ter Braak CJF, van Dam H (1989) Inferring pH from diatoms: a comparison of old and new calibration methods. Hydrobiologia 178:209–223

Wright HE (1966) Stratigraphy of lake sediments and the precision of the paleoclimatic record. In: Sawyer JS (ed) World climate from 8000 to 0 BC. Royal Meteorological Society, London, pp 157–173

Chapter 18
Limnological Responses to Environmental Changes at Inter-annual to Decadal Time-Scales

André F. Lotter and N. John Anderson

Abstract Lake responses to environmental change are complex and occur at a variety of time-scales. Three case studies are presented and discussed to illustrate how numerical and statistical methods can be used to answer critical palaeolimnological questions about lake responses to environmental change at the inter-annual or sub-decadal time-scales. These all involve lakes with annually-laminated sediments. They concern the comparison of annual sediment records with instrumental data to study the effects of nutrients and climate on the Swiss lake of Baldeggersee; the comparison of diatom stratigraphical data and tree-ring based climate at Kassjön, a lake in central-northern Sweden; and the assessment of the effects of volcanic tephra deposition and climatic change on pollen and diatom stratigraphical assemblages at Holzmaar in west-central Germany. These case studies highlight the need for careful definition of the research question or hypothesis to be tested, the selection of ecologically relevant response proxies, the critical choice of predictor variables as proxies for stress and environmental change, and the selection of appropriate numerical and statistical techniques.

Keywords Climate change • Constrained ordination • Eutrophication • Hypothesis testing • Laacher See tephra • Land use • Limnological responses • Partitioning • Tree rings • Variation • Varves • Volcanic tephra

---

A.F. Lotter (✉)
Laboratory of Palaeobotany and Palynology, Department of Biology, University of Utrecht, Budapestlaan 4, NL-3584 CD Utrecht, The Netherlands
e-mail: a.f.lotter@uu.nl

N.J. Anderson
Department of Geography, Loughborough University, Loughborough LE11 3TU, UK
e-mail: n.j.anderson@lboro.ac.uk

H.J.B. Birks et al. (eds.), *Tracking Environmental Change Using Lake Sediments*, Developments in Paleoenvironmental Research 5, DOI 10.1007/978-94-007-2745-8_18,

Introduction

Lake response to environmental change is complex and occurs at a variety of time-scales. Both contemporary monitoring and whole-lake manipulations have furthered our understanding of community and productivity effects of varying environmental stress at intra-annual to decadal time-scales. Some disturbance processes are subtle and insidious (e.g., acidification, eutrophication) and, because of the strong inter-annual variability, some effects can be difficult to identify from short-term monitoring (Anderson and Battarbee 1994; Lotter and Psenner 2004). Similarly, for ecologically-rapid changes, cause-and-effect can be difficult to identify at remote sites because of the lack of suitable physical or chemical background data against which ecological change can be measured.

Environmental stress can be categorised as either internal (autogenic) or external (allogenic) and covers a range of time-scales and intensities. These range from abrupt (i.e., a pulse disturbance or pulse experiment in manipulative ecological studies) to chronic, long-term stress (10^1–10^2 years; analogous to a so-called press ecological experiment with a 'sustained alteration of species densities' *sensu* Bender et al. 1984). Pulse experiments measure the resilience of the system to the experimental treatment, whereas press experiments measure the resistance of the system to the experimental treatment (Gotelli and Ellison 2004). Furthermore, both allogenic and autogenic forcing of lake ecosystems can be divided again into natural (e.g., pedogenesis, fire, volcanic eruptions) and anthropogenic factors (e.g., cultural eutrophication, acidification, introduction of alien species).

Long-term autogenic processes are, for example, the development of internal nutrient loading, food-web ontogeny associated with species immigration, and changing trophic interactions (e.g., Smol 2008). Long-term allogenic forcing is more commonly associated with landscape development (soil, climate, and vegetation development), (e.g., Korsman et al. 1994; Engstrom et al. 2000) and land-management changes (i.e., cultural eutrophication associated with increased nutrient loading). Allogenic short-term forcing tends to be catastrophic (e.g., volcanic eruptions, hydrological flash flooding). Short-term autogenic effects are, for example, winter anoxia that may lead to fish kills. An example of intermediate-term autogenic change is fish introductions where the response is more gradual and moderated by the development and ageing of the fish population (Carpenter et al. 1987). Lake response to climate forcing covers a range of time-scales (see e.g., Fritz 2003, 2008; Cumming et al. 2012: Chap. 20).

Lake response to disturbance is complex partly because multiple stressors can influence lake communities simultaneously (Carpenter 1988; see also Fig. 18.1) but also because the response of one group of organisms to a given stress can be confounded by another stress (Hall et al. 1999). Moreover, the trajectory of aquatic systems may be influenced by hysteresis, multiple stable states (Scheffer 1998), and in-lake processes with species interactions masking or enhancing the effect of an external disturbance. Given the complexity of ecological response of lake ecosystems to internal and external forcing, it is important that the interpretation of sediment records is approached in a critical manner. There is a range of numerical

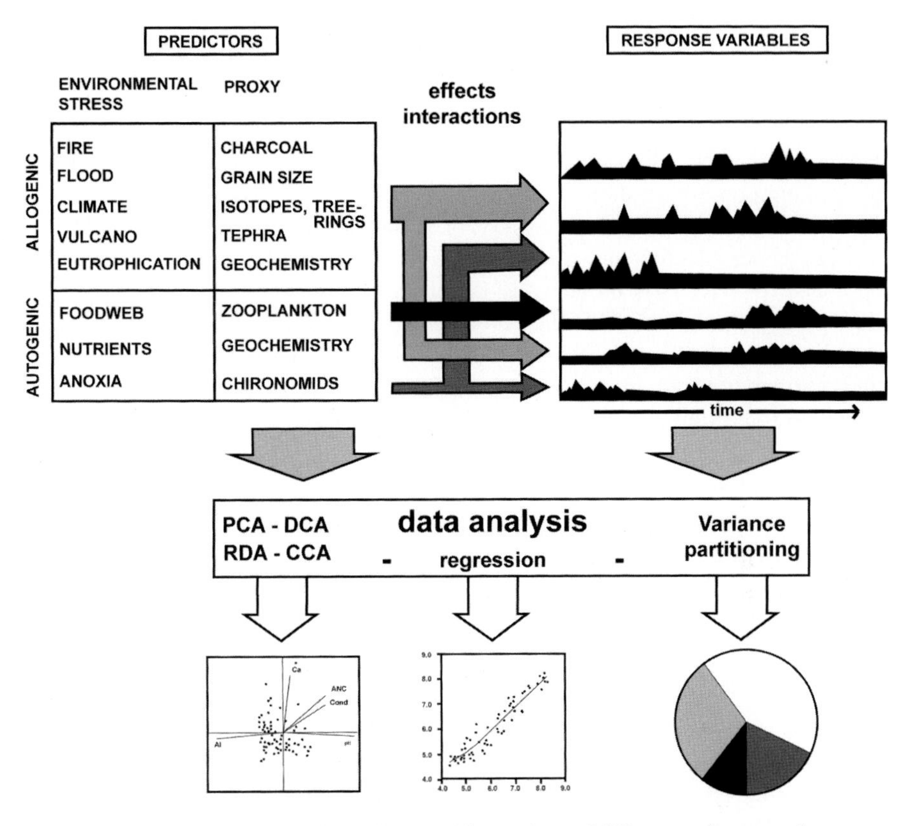

Fig. 18.1 Schematic overview of the effects and interactions of different environmental stressors on palaeolimnological response variables with different possibilities of numerical and statistical data analysis

and statistical tools that can be used to assist in the interpretation of sediment records, to separate between the signals originating from internal and external forcing, and to test palaeolimnological hypotheses (see e.g., Birks 1998).

Palaeolimnology has provided unambiguous evidence for long-term trends in lake development, e.g., catchment processes altering surface water and hence lake acidity that lead to changes in lake biota (e.g., Renberg 1990). The combination of independent predictors and response variables at sub-decadal time-scales in sediment studies are, however, rare and represent an important challenge for the subject. High-resolution sediment records are at the interface of ecology and palaeoecology (Kitchell et al. 1988; Smol 1991), in that there is a direct link to and sometimes an overlap with contemporary monitoring data. Furthermore, sediment records can be critical in providing background data for ecosystem changes where its onset predates lake-monitoring programmes. Sediment-trap data (Flower 1990; Lotter and Bigler 2000) and high-resolution sediment records (e.g., Simola et al. 1990; Anderson 1995a; Lotter 1998) can resolve seasonal assemblage changes, thereby providing information on the life-cycle of aquatic organisms (e.g., the pioneering work of Nipkow 1927). The selection of appropriate sampling strategies and the selection of proxies suitable for recording particular environmental stresses are, however, critical (Anderson 1995b).

In this chapter we use three case studies to illustrate how bi- and multivariate numerical methods outlined earlier in this book can be used to tackle critical palaeolimnological questions, especially when using independent proxies derived from sediment records, documentary or instrumental data, and other sources. Moreover, we identify and discuss some of the problems associated with this approach. Our focus is on annually-laminated sediments because they represent the optimal means to address the problems we are concerned with at inter-annual to sub-decadal time-scales (Brauer et al. 2009). However, with good time control and contiguous sampling, non-varved sediments can also be used to address similar questions (e.g., Hall et al. 1999).

Disturbance and Response Dynamics in Lakes: Three Case Studies

In this section we present, as three case studies, examples of how numerical methods such as constrained canonical ordination and variation partitioning (Legendre and Birks 2012: Chap. 8) can be used to test specific hypotheses about limnological responses to environmental changes. We discuss environmental disturbance and the response dynamics of lakes on time-scales that link ecology and palaeoecology. We explicitly neglect centennial to millennial scale ontogeny of aquatic systems commonly addressed in palaeolimnology (e.g., Deevey 1984; Anderson 1995a) and concentrate on time-scales of annual to decadal resolution that allow a direct comparison with real-time observations and the results of contemporary limnological experiments. However, the approach outlined here is also well suited for the study of external forcing of lakes and their biota at Holocene time-scales (e.g., vegetation and soil processes and the subsequent aquatic response; Engstrom et al. 2000; Lotter and Birks 2003; Bradshaw et al. 2005).

Predictor and Response Variables

The quantitative analysis of the relationship between environmental stress and biotic response in sediment sequences requires the identification of independent proxies for the predictors (explanatory variables) that are being used in the analyses. Proxies are required because it is very rare that an explanatory process will have a direct analogue or variable that can be derived from the sediment record. The environmental predictors can be extracted from documentary or instrumental data sources (e.g., meteorological, land-use, limnological monitoring data), independent natural archives (e.g., tree-rings, glaciers) or derived from sediment sources (Fig. 18.1 and Table 18.1). It is important that the variables chosen have a temporal variability

Table 18.1 Lake internal and external stressors, and examples of palaeolimnological proxies that may be used as response variables and independent predictors or explanatory variables to help test the effects of these stressors

Type of forcing	Stressor, perturbation	Dependent response variables	Independent predictor or explanatory variable
Autogenic (internal)	Acidification	Cladocerans	Measured overturn pH, DI pH
	Eutrophication	Varve thickness	Measured phosphorus, DI TP
	Lake ontogeny (infilling)	Diatoms	Sediment depth, cladoceran-inferred water depth
	Anoxia, meromixis	Biomarkers	Fossil pigments, geochemistry, benthic organisms
Allogenic (external)	Climate change	Chironomids	Instrumental and historical record, tree rings, $\delta^{18}O$
	Land-use change	Diatoms	Pollen, fungal and bacterial spores
	Vegetation change	Geochemistry	Pollen, plant macrofossils, biomarkers
	Forest fire	Pollen	Charcoal particles, DI pH
	Volcanic eruption	Pollen, diatoms	Tephra layer

DI diatom-inferred, *TP* total phosphorus

similar to the temporal resolution of the sediment sequence under investigation, but also that they are conceptually, ecologically, and mechanistically relevant to the questions being asked. While it is important that predictor selection is done critically, imagination and lateral thinking are also vital.

Generally, increased variability in the composition of biotic assemblages may be regarded as an indicator of environmental stress and expressed as changes in the diversity of living and hence also subfossil assemblages. To assess limnological response to past environmental perturbation, albeit natural or anthropogenic, different numerical techniques may be used (Fig. 18.1). However, the most crucial point in any such palaeolimnological study is to define carefully the hypothesis to be tested, as this will eventually determine the numerical methods to be used. Also of importance is the careful selection of predictor variables (proxies for perturbation or stress, explanatory variables) and response variables (e.g., biotic or abiotic proxies, see Fig. 18.1 and Table 18.1).

Comparison of Annual Sediment Records with Instrumental Data: The Effects of Nutrients and Climate on Baldeggersee

Anthropogenic nutrient enrichment with its global ecological and socio-economic impact is one of the major current issues affecting aquatic ecosystems (see Smol 2008; Simpson and Hall 2012: Chap. 19). Eutrophication of freshwater systems is commonly a gradual process that is often only apparent in biotic assemblages when certain nutrient thresholds are passed. Palaeolimnological studies show that in most regions extreme eutrophication started during the late nineteenth century and peaked during the twentieth century (see Anderson 1997 and references therein).

The palaeolimnological study of the Swiss lake Baldeggersee (47°10′N; 8°17′E, 464 m asl), a medium-sized (5.2 km^2, 66 m maximum depth), nutrient-rich hard-water lake in central Switzerland, is a good example to illustrate the dynamics of eutrophication and the interaction of nutrient enrichment and climate change on lake development. Baldeggersee is located in a region of intensive agriculture and farming. Since 1885 the 66-m deep lake has had an anoxic hypolimnion, thus preserving annually laminated sediments. Since 1982 a lake restoration programme with artificial oxygenation and circulation has been running (Wehrli et al. 1997). However, as oxygen availability in the deepest part of the basin did not improve, bioturbation is absent and annual layers are still preserved. In 1993 the annually-laminated sediments were sampled by *in situ* freezing. The freeze-cores were sampled for individual varves (Lotter et al. 1997b) and subsequently varve measurements on petrographic thin-sections, and sedimentological and diatom analyses were carried out (Lotter et al. 1997c; Lotter 1998).

Concurrent to the significant anthropogenic increase in nutrient loads over the past 100 years, air temperatures in Switzerland rose by about 1°C (Lister et al. 1998). Therefore, an important issue for lake development is whether the influence of climate and trophic state on both the varve formation and the composition of diatom assemblages can be separated, thereby permitting an estimation of the effect of these two factors on the biotic and abiotic aquatic system independently. This question was tackled by carrying out a series of partial constrained ordinations, where predictors (explanatory variables) are used as covariables (see Legendre and Birks 2012: Chap. 8). In this way the effect of each predictor on the response variables can be allowed for in a statistical sense and thus eliminated in the analysis. The aim is to determine the variation in the response variables that is uniquely attributable to a particular set of predictors by partialling out the effects of other predictors in the ordination (Borcard et al. 1992).

An initial detrended canonical correspondence analysis (DCCA) (detrending by segments, non-linear rescaling: ter Braak 1986) with sample age as the only external constraining variable was used to estimate the gradient length of the stratigraphical data-sets to decide whether linear-based redundancy analysis (RDA) or unimodal-based canonical correspondence analysis (CCA) partial ordination methods need to be used (Birks 1995). RDA was used for the partitioning of the variation in the varve measurements because of the linear or at least monotonic relationship between the response and the predictor variables, whereas for the diatoms the floristic compositional gradient was >2 standard deviations, suggesting that the use of unimodal methods such as CCA (ter Braak and Prentice 1988) would be appropriate.

The diatom sediment data-set consists of 91 samples and percentage values for 75 taxa (Lotter 1998), whereas the varve data-set consists of 91 matching samples analysed for light (calcite), dark (organic), and total layer thickness (Fig. 18.2;

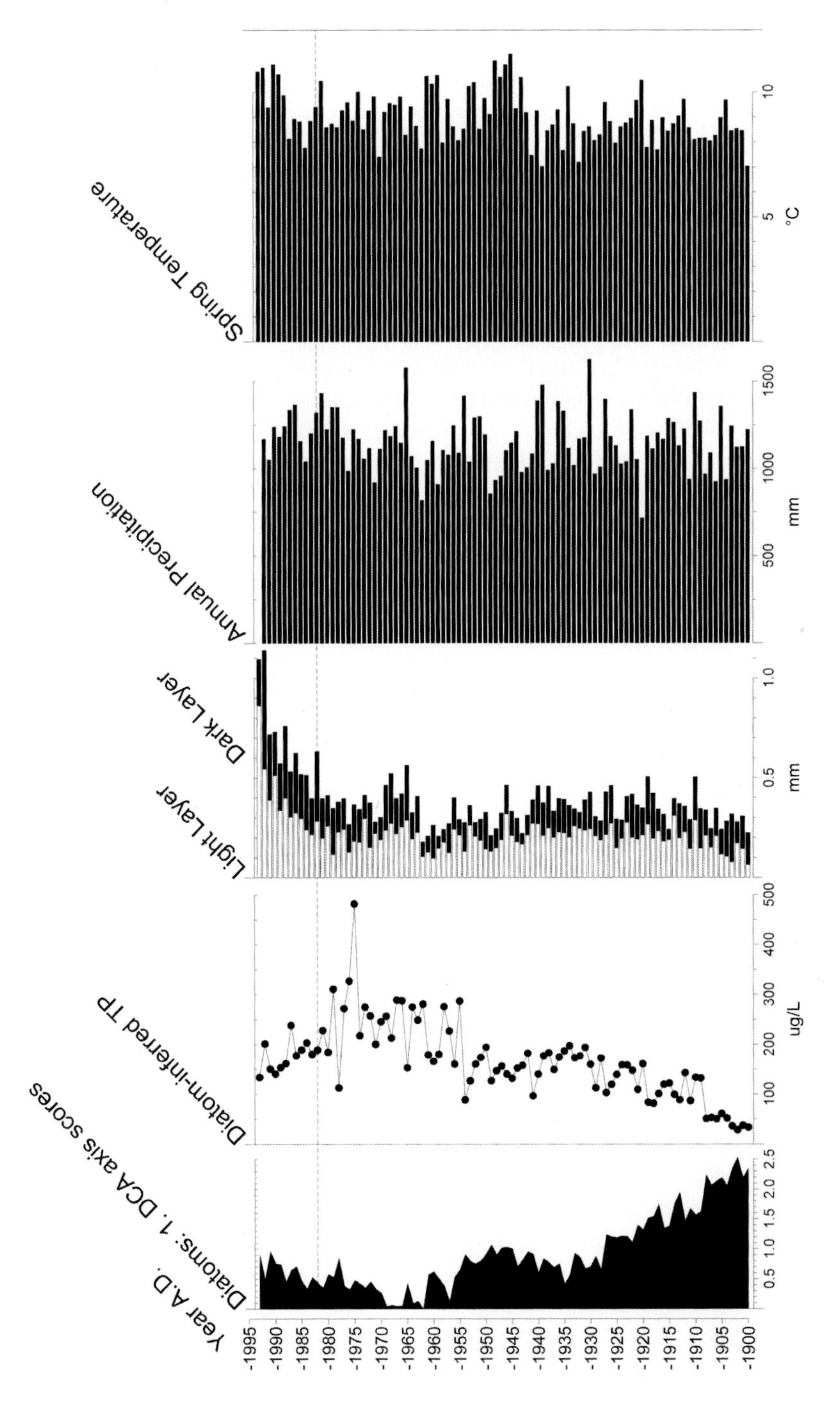

Fig. 18.2 Comparison of diatom sample scores on the first detrended correspondence analysis (DCA) axis as a measure of compositional turnover, epilimnetic diatom-inferred total phosphorus concentrations, and varve measurements between AD 1993 and 1900 from Baldeggersee, with regional meteorological measurements for the same period that were used to partition the variation in the varve data

Lotter et al. 1997c). The total variation in these two data-sets was partitioned into the four following components:

1. *Variation uniquely explained by lake trophic state, independent of climate.* This was modelled by total phosphorus concentration (TP) and by the presence or absence of the lake restoration programme since 1982. The latter was modelled as a dummy (0/1) variable, whereas for TP, spring-circulation water column measurements back to 1958 were used. For the older samples diatom-inferred TP concentrations (Lotter 1998) were used (see Fig. 18.2). To prevent circularity, TP concentrations have been assumed to decrease linearly by 2 $\mu g\ L^{-1}\ year^{-1}$ before 1958 (i.e., the onset of water column TP measurements) for the diatom data-set. Furthermore, an interaction term between TP and lake restoration was introduced for the diatom data-set because the changed mixing regime is likely to have had an influence on water circulation and nutrient cycling.
2. *Variation uniquely explained by climate, independent of trophic state.* For the diatom data-set mean spring air temperature was used to model climate as only spring air temperature accounted for a statistically significant amount of the total variation in CCAs with seasonal air temperatures as the sole constraining variables. For the varve data-set the variables summer air temperature and annual precipitation (see Fig. 18.2) explained the highest amount of variation among the climate data and were therefore used as predictors.
3. *Covariation between trophic state and climate.*
4. *Variation unexplained by these models.*

Because of the strong temporal autocorrelation in the data the two time-series were detrended by partialling out sample age as a covariable in all analyses. In this way the major effects of time-dependent ecological and environmental processes were removed. Statistical significance of the different RDAs and CCAs was assessed using restricted Monte Carlo permutation tests for temporally structured samples (ter Braak and Šmilauer 2002).

Disentangling the effects of climate and trophic state on the composition of annual diatom assemblages and the formation of annual layers in Baldeggersee for a period when both predictors changed is important for the assessment of the effects of both. Here we are in the fortunate situation in having a well-dated sediment record of seasonal or annual resolution that may be compared directly to instrumental meteorological and water chemistry measurements of the same temporal resolution.

The fact that the models used in the above partial ordinations are rather simple becomes evident by the amount of unexplained variation (85.5% for the diatoms and 63.4–66.8% for the varves, see Table 18.2). This is especially the case for the noisy diatom data that consist of many taxa and contain many zero values. Nevertheless, these models give some indications of the importance of the forcing mechanisms that may have influenced the biotic and abiotic systems in Baldeggersee.

Of the 14.5% of the variation in the diatom data explained by trophic state and climate (Table 18.2), a major but marginally non-significant part (12.4%, $p = 0.06$) is explained by TP and lake restoration, whereas only a small, statistically insignificant part of the diatom variation (1.4%, $p = 0.08$) can be explained by

Table 18.2 Results of variation partitioning using climate (mean spring air temperature for diatoms; mean summer air temperature and annual precipitation for varves, see Fig. 18.2) and trophic state (total phosphorus lake-water concentration, restoration, and their interaction) as predictors after sample age was fitted as a covariable

	Diatoms 1992–1902	Varves 1992–1902	Varves 1980–1920
Unexplained variation	85.5%	63.4%	66.8%
Climate effects, independent of trophic state	1.4%	17.6%	28.1%
Effects of trophic state, independent of climate	12.4%	17.6%	5.8%
Covariation between the effect of climate and trophic state, after the effects of age have been allowed for	0.7%	1.3%	−0.7%

climate. Despite indications of an empirical relationship between diatoms and temperature (e.g., Pienitz et al. 1995; Lotter et al. 1997a) the composition of diatom assemblages along such a short temperature gradient (~1°C in 100 years) seems to be more dependent on limiting nutrients such as Si and P.

The analysis of the varve data was done for two different time-slices (Lotter and Birks 1997): the period between 1992 and 1902 where regional instrumental meteorological records are available and for the period between 1980 and 1920. The difference in results between the whole time-series (1992–1902) and the one without the restoration period (1980–1920), as well as the significance of the restoration variable (Table 18.2), shows that lake restoration has had a substantial effect on sediment formation. In the long time-series (1992–1902), climate and trophic state explain the same amount of variation, whereas in the pre-restoration period climate was more important. This implies that the formation of calcite varves depends strongly on climate and that they may therefore potentially be used as climate proxies (e.g., Livingstone and Hajdas 2001).

Comparison of Two Independent Proxy-Archives with Comparable Temporal Resolution: Diatoms and Tree-Ring Based Climate at Kassjön

The possibility of comparing the individual laminae from a varved record with both contemporary limnological data and meteorological data represents an ideal opportunity for linking palaeolimnology and limnology. However, for many lakes and for time periods that pre-date routine monitoring and instrumental measurements, such an approach is not possible. An alternative approach and one that maintains the high temporal resolution is the use of an independent natural archive with annual resolution, e.g., tree-rings.

Varved lake sediments are common over much of central-northern Sweden (Petterson 1996). A considerable amount of work has focused on Kassjön, (63° 55′N; 20° 01′E), a small (23 ha, 12 m maximum depth) dimictic lake situated in boreal forest, some 30 km from Umeå (Segerström 1990b; Anderson et al. 1995). An isolation lake, formed by isostatic rebound about 6300 years BP, it has a continuous record of annual laminations characterised by a minerogenic layer deposited by the spring melt-water (Renberg 1981). At Kassjön agriculture started c. AD 1350 and peaked in the late nineteenth century, but today only about 20% of the catchment is used for grazing and hay production, while the abandoned fields are being re-colonised by birch scrub (Segerström 1990b; Anderson et al. 1995).

The varved sediments at Kassjön have been used to reconstruct vegetation history via pollen analysis (Segerström 1990a) and to estimate sediment yield at inter-annual resolution through image analysis (Petterson 1999; Petterson et al. 1999). Anderson et al. (1995) used RDA to assess whether catchment vegetation changes associated with the start of agriculture resulted in significant responses in the diatom flora. Three pollen types (*Salix, Juniperus,* and *Rumex* as identified by forward selection in RDA) explained 23% of the variation in the diatom data over a 500-year period. Linear regression revealed a significant relationship between quantitative estimates of diatom production (cells cm^{-2} $year^{-1}$) and increasing field area (derived from old maps and parish data) during the seventeenth to nineteenth centuries (Anderson et al. 1995). Spores of the bacteria *Thermoactinomyces vulgaris* were also shown to be highly and significantly correlated to anthropogenic pollen indicators over a 1200-year period: pollen explained 83% of the variation in the mean annual accumulation rates of these endospores (Nilsson and Renberg 1990).

Although a clear example of quantitative palaeolimnology, these studies from Kassjön were at a rather coarse temporal resolution (2–41 years per sample, mean 18 years). The sediment sequence was sampled contiguously but had been cut to distinctive marker varves, hence the irregular sampling interval. Such an approach makes conventional time-series analysis problematical (see Dutilleul et al. 2012: Chap. 16). Furthermore, no attempt was made to separate the possible confounding effects of long-term climatic variability from catchment land-use change.

As a result of these limitations, in a subsequent study, Anderson et al. (1996) used tree-ring based summer temperatures from northern Sweden (Briffa et al. 1990) as an independent measure of climatic variability to assess the role of climate on both lake (diatoms) and catchment vegetation (pollen). The same diatom data as in Anderson et al. (1995) were used, but sediment samples were re-analysed for pollen since the original pollen work had even coarser sampling intervals than those used by Anderson et al. (1995). Tree-ring based temperature anomalies (expressed as a deviation from the mean of 1950–1980; see Briffa et al. 1990) were averaged to match the same time periods covered by the diatom and pollen samples (see below). As detrended correspondence analysis indicated short compositional gradients in the pollen and diatom data they were analysed using redundancy analysis. Variable time-lags (i.e., adding a different number of years to the age of each diatom and pollen sample) were applied to the diatom data and tree-ring based temperatures to determine the maximum fit (as the maximum correlation coefficient obtained).

The pollen assemblages showed no significant response to climate forcing as indicated by tree-ring inferred summer temperatures, whereas 5% of variation in the diatom assemblages (partial RDA; $p = 0.014$) were explained with a 20-year lag (Anderson et al. 1996). A major decrease in diatom richness, as estimated by rarefaction analysis (with a base count of 180 diatoms, see Birks and Line 1992), was attributed to effects of the Little Ice Age cooling (Anderson et al. 1996). A combination of summer temperature and a pollen taxon (*Juniperus,* as shown by forward selection) explained over 34% of the variation in diatom richness. The presence of juniper is indicative of cultural activity in Sweden and these results possibly indicate a role of climate (see below) and land-use changes in structuring diatom communities in boreal forest systems.

Anderson et al. (1996) concluded that the relationship between temperature and diatom richness was confounded by nineteenth to twentieth century land-use changes. In boreal-forest catchments such as Kassjön, catchment land-use is in part a reflection of climate forcing as well as cultural activity. Separating the indirect effects of climate on a lake, i.e., those that are mediated through forest dynamics, from direct climate effects (lake-stratification, lake ice-cover) and anthropogenic drivers is very difficult using proxy data. It was decided, therefore, in a more recent study of Kassjön (I Renberg, G Petterson, and NJ Anderson, unpublished data) to remove the possibly complicating effects of land-use change by examining a period (AD 540–1015) that pre-dates the start of agriculture in the area. Furthermore, given the largely synchronous links observed between lakes and external meteorological forcing observed in contemporary systems over recent decades (e.g., North Atlantic Oscillation (NAO); Livingstone and Dokulil 2001), it was argued that a time-lag was not ecologically realistic. As a result, the relationship between the two time-series (sediment-core data and tree rings) in the latest study was analysed after adding 20 years to the varve chronology. Previously, a 20-year adjustment was found to maximise the fit between tree-ring temperatures and diatoms and it was interpreted as a lag-effect (Anderson et al. 1996). Compared to varves, tree-ring chronologies are virtually error-free (Briffa 2000). Given the errors associated with varve counting, it was decided in the most recent analysis to treat the 20-year offset as an error and not as a lag.

The varve sequence from AD 476–1015 analysed by Petterson et al. (1999) was sub-sampled at contiguous 5-year intervals (540 varves in 108 samples) and analysed for dry mass accumulation rate (DMAR), carbon content, biogenic silica, and grey scale (Petterson et al. 1999). The sub-sampling technique was replicated and shown to be highly repeatable: the DMAR for each 5-year section had a standard error of 13%. Through the analysed period there is a substantial change in DMAR, which averaged 30 mg cm^{-2} $year^{-1}$ prior to AD 730 and reached a minimum of ~15 mg cm^{-2} $year^{-1}$ between AD 700 and 850 (Fig. 18.3). Because of the strong dilution effects of the minerogenic inputs to the lake on concentrations of sediment variables, the carbon data were expressed as accumulation rates, which vary between 2 and 4 mg C cm^{-2} $year^{-1}$. As the central aim of the study was to identify the response of the lake to climate forcing, tree-ring based summer temperatures were again used as the independent, predictor variable. Lake response was indicated by organic carbon, biogenic silica, and dry mass accumulation rates. Prior to data analysis, the sediment sequences were detrended by fitting a running

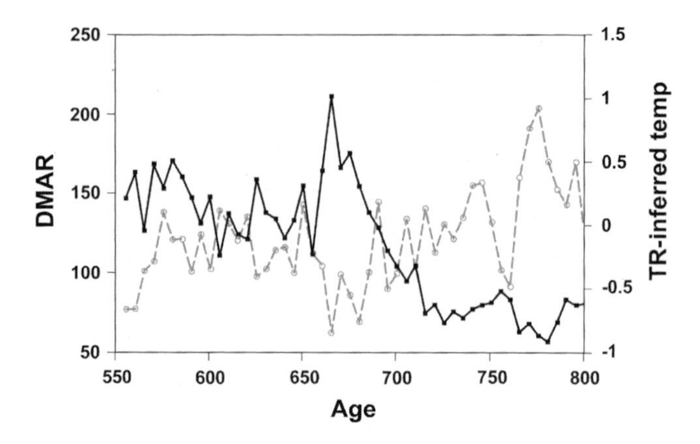

Fig. 18.3 Dry mass accumulation rates (DMAR; *black squares*) from Kassjön for the period AD 550–800 compared with tree-ring (TR) based temperature (as anomalies from present-day; *grey circles*)

mean over 21 samples (105 years), which effectively removed variation with a frequency >100 years. Observed values were subtracted from the running mean and the final analyses were undertaken on the residuals.

There is a strong inverse relationship between the lake-sediment proxies and tree-ring based temperatures in the Kassjön study (Fig. 18.3). The strength of the relationship is variable over the analysed time-periods, partly due to the inherent natural variability in lake-climate linkages (see Livingstone and Dokulil 2001; Straile et al. 2003) but also due to the varying chronological control. The inverse relationship between temperature and lake productivity (as indicated by organic carbon, $r = -0.56$; $p \leq 0.05$) can be attributed to two possible mechanisms. First, increased catchment-derived nutrient input to the lake associated with increased runoff, particularly during spring meltwater input. Presumably, cooler years have greater snow cover and hence enhanced spring meltwater flux. It has been shown that nutrient input to aquatic systems in northern Sweden today is highest during snowmelt, while Petterson and others (unpublished data) have argued that varve structure (measured as grey scale) primarily reflects minerogenic inputs during the spring meltwater flux. The strong inverse relationship between DMAR and tree-ring based summer temperatures (Fig. 18.3) lends some credence to this hypothesis. However, as neither pigment nor diatom data are available for this period at the same temporal resolution as the other proxies, it is not possible to partition the variation in the biological responses in relation to the independent predictors (as indicated in Fig. 18.1) as was undertaken at Baldeggersee.

An equally plausible mechanism may be the hydrological control of allochthonous carbon inputs to Kassjön from the catchment. Many lakes in northern Sweden have been shown to be net heterotrophic in terms of their carbon budgets (system respiration > primary production) with the imbalance being attributed to carbon inputs from the catchment (Algesten et al. 2003). Kassjön has relatively high dissolved organic carbon (DOC) concentrations today (~10 mg L^{-1}) and a large catchment-to-lake ratio. Increased discharge has been shown to increase DOC export from boreal-forest catchments. It is possible that the climate-lake linkages proposed for this site do not involve productivity *per se* but an indirect link where climate controls the hydrological supply of allochthonous carbon from the catchment to the lake. Given the chronological uncertainties and the lack of suitable predictors (e.g., phosphorus flux data), it is difficult to test these competing hypotheses critically.

The studies at Kassjön illustrate the development of a research project over a number of years in response to a changing research environment. The original aims (reconstructing vegetation history and diatom productivity) were quite different to those that are being pursued now, such as assessing the variability of aquatic systems under natural climate forcing. However, the work at Kassjön illustrates some of the major problems that have to be tackled when working with sediment records and independent proxies at high resolution: uncertainties in the chronologies and in the correlation and variable chronological control among different environmental archives. Tree-ring chronologies are considered to be highly reliable, whereas varve sequences may have cumulative errors approaching a decade or more for periods covering thousands of years (see Maher et al. 2012: Chap. 6). But more importantly, this error is not constant and it is difficult to determine the agreement between two chronologies at any one point. It is unlikely that the uncertainty in the varve chronology at Kassjön (corrected by adding 20 years) is actually constant over time, even within the short period covered by the analysis shown in Fig. 18.3. Given the problems with matching two independent proxies at high resolution, a better way to proceed is to use palaeoecological proxies as predictors, which are derived from the same sample level as the response variable. This approach is outlined in the next case study.

Sediment-Derived Proxies and Hypothesis Testing: The Effects of Tephra Deposition and Climatic Change at Holzmaar

The further back in time we go, there are fewer instrumental and documentary records available. Furthermore, as is often the case in multi-proxy studies, the correlation between independent proxy records (e.g., tree-rings, ice-core data, marine data) and palaeolimnological time-series becomes a major problem, especially at annual to decadal time-scales (Lotter 2003). As shown in the example of Kassjön, a precise correlation of records is of utmost importance if the emphasis of a study lies not only in the detection of correlations but also in explaining causalities and mechanisms. The use of independent proxies derived from the same sediment samples helps solve the dilemma of different chronologies and matching of time-series. In palaeolimnological studies, independent proxies (see Table 18.1) such as stable isotopes (e.g., ^{18}O as a proxy for climate change: Lotter et al. 1992), charcoal (as a proxy for forest fire: Korsman and Segerström 1998), pollen and plant macrofossils (as a proxy for catchment vegetation: Korsman et al. 1994; Hausmann et al. 2002; Lotter and Birks 2003) or sedimentology (e.g., grain-size distribution as a proxy for flooding events: Nesje et al. 2001; Parris et al. 2010) may be used as predictors to examine the effects of environmental change on aquatic biota such as diatoms, chironomids, cladocerans, or chrysophyte cysts.

Here, we present an example of an environmental perturbation caused by an abrupt event, namely the eruption of the Laacher See volcano in the Eifel Mountains of western-central Germany and the subsequent deposition of the Laacher See Tephra (LST) 12,880 cal. year BP (Bittmann et al. 2002). Within several weeks this highly explosive plinian eruption produced more than 6 km^3 of maphic phonolite magma, corresponding to more than 20 km^3 of ejected tephra, and released at least 2 megatons of sulphur into the atmosphere that remained as sulphuric acid aerosols in the stratosphere for years (Schmincke et al. 1999). Different studies have examined the effects of past volcanic eruptions on terrestrial and aquatic systems (e.g., Lotter and Birks 1993; Birks and Lotter 1994; Barker et al. 2000). However, as the effects of aerosols on climate as well as the effects of the tephra deposition on ecosystems are assumed to be of short duration (i.e., several years to one or two decades) high-resolution studies are needed to test the hypothesis that this eruption had no significant effect on terrestrial pollen or aquatic diatom assemblages. Lotter et al. (1995) analysed 40 contiguous sediment samples of known volume and number of varves from Holzmaar, a well-studied crater lake 60 km to the south-west of the origin of the LST (Zolitschka et al. 2000). The sediment sequence includes 475 varves with a 78-mm thick LST layer as well as the transition from the Allerød (AL) to the Younger Dryas (YD) biozone (Ammann and Lotter 1989). The sediments above the LST show an immediate increase in grass pollen that lasted for 17 years (two samples), whereas the diatom assemblages are characterised by 17 years of increased *Asterionella formosa* percentages. To test whether the observed changes in the pollen and diatom assemblages are statistically significantly different from random variation a set of partial-RDAs was carried out. The following predictors were used: biozones (AL, YD) as dummy variables (1/0) to represent broad-scale late-glacial climatic change; sample age to represent long-term unidirectional change (e.g., succession, soil development, lake infilling); LST impacts modelled as an exponential decay function (see Fig. 18.4). When all these explanatory variables are used the model is statistically significant for pollen and diatoms both for percentages and accumulation rates (Table 18.3). When the effects of long-term temporal and climatic change are partialled out as covariables, both pollen (% and accumulation rates) and diatoms (only accumulation rates, not %) show a statistically significant response in relation to the LST (Table 18.3). These results suggest that the eruption of the Laacher See volcano and/or the deposition of its tephra had substantial effects on pollen assemblages in terms of relative composition and absolute rates, and hence also on the composition and density of the vegetation in the catchment of Holzmaar. The predominant birch woodland in the catchment was affected for a period of about two decades. As the relative diatom composition showed no significant effect, the eruption and tephra deposition presumably did not alter the water chemistry of the lake permanently. However, the

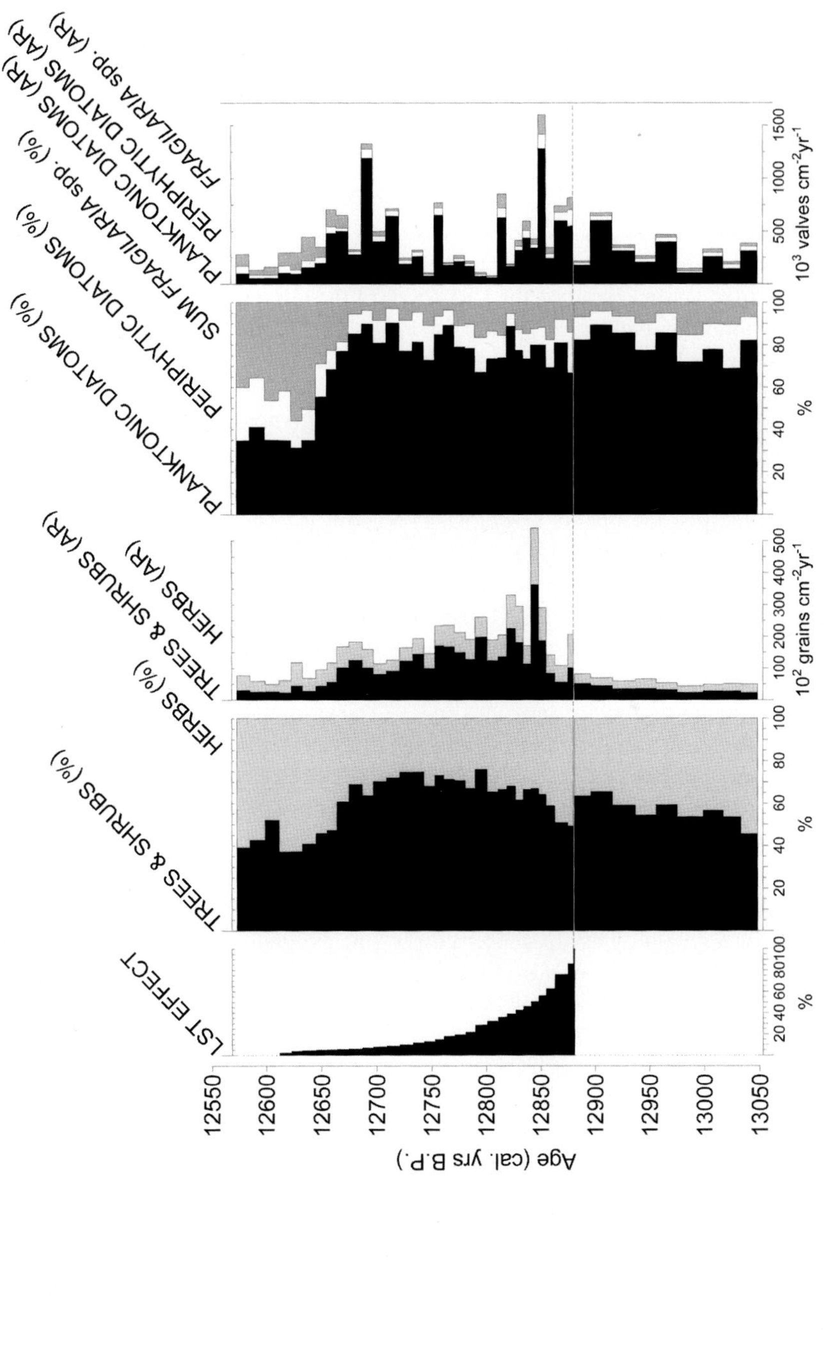

Fig. 18.4 Pollen and diatom relative abundances and accumulation rates from Holzmaar in relation to the deposition and effect of the Laacher See Tephra (LST)

Table 18.3 The impact of the Laacher See Tephra (LST) (see Fig. 18.4) on pollen and diatom assemblages at Holzmaar

Model	Covariable	Pollen %	Pollen AR	Diatoms %	Diatoms AR
Biozone + Age + LST	–	0.01*	0.01*	0.01*	0.01
LST	Biozone+Age	0.04*	0.01*	0.14	0.01*

The *p*-values of (partial) redundancy analysis (RDA) of percentage (%) and accumulation rates (*AR*) were assessed by 99 restricted Monte Carlo permutations. A variance-covariance matrix was used, all data were log-transformed prior to RDA and double centring was applied
*significant at $p < 0.05$

diatom productivity increased substantially after the LST deposition, most likely as an effect of higher silica input by the LST (60% of the ash consists of SiO_2).

Altered pollen assemblages, mainly an increase in non-arboreal pollen types (NAP) that occurred about 216 varves after the deposition of the LST, indicate the transition from the AL to the YD biozone, which according to Greenland ice-core records took place within several decades. Already 205 varves after the LST deposition, diatom assemblages show a transition from plankton- (*Stephanodiscus*) to periphyton-dominated (*Fragilaria*) assemblages that took place within 30 varves (Fig. 18.4). Despite the diatom assemblages reacting 11 varves (i.e., one sample) earlier, cross-correlation analysis shows the highest correlation between the sum of *Fragilaria* spp. and the sum of NAP ($r = 0.74$), as well as between the first principal component analysis (PCA) axes of pollen and diatom percentages at a lag of 0 ($r = 0.79$; see Lotter et al. 1995). Given this synchronous reaction between the two biota, we might want to know how well the diatom data can be explained by external predictors such as biozones (AL, YD), temporal effects (sample age), LST, or regional vegetation (summarised by the first four pollen PCA axes). A RDA of the diatom percentage data with forward selection of variables was used to find the minimal set of statistically significant predictors ($p \leq 0.05$; 999 restricted Monte Carlo permutations with Bonferroni correction for multiple simultaneous tests). The YD biozone (41%, $p = 0.001$) and pollen PCA axis 1 (5%, $p = 0.001$) explained the highest statistically significant part of the variation, suggesting a strong response of the diatom composition to the Younger Dryas environmental change and to the associated vegetation change (as represented by PCA axis 1). The strong relationship between the aquatic system and the terrestrial vegetation suggests climatic change to be the common underlying forcing factor.

Discussion

Sediment samples used in palaeolimnological and palaeoecological studies encompass, for the most part, time-spans of several years to decades. Such studies generally provide data-series covering time periods of a few centuries to millennia.

Over such time-scales palaeoecological investigations may help to assess the natural variability of biotic and abiotic systems and provide information on the long-term reaction and recovery to different perturbations. Given good time-control, dynamics, cyclicities, and extent of disturbances can be estimated and palaeolimnological studies can thus improve the understanding of long-term ecological processes (Schoonmaker and Foster 1991; Anderson et al. 2006; Willis et al. 2010).

Contemporary limnologists and ecologists are working on time-scales of weeks to decades. Due to the higher sample density and the better temporal precision in contemporary ecological studies, there has traditionally been little overlap with palaeolimnological and palaeoecological data (Kitchell et al. 1988). Although palaeolimnological investigations can provide a continuous and long-term record of environmental change where no long-term monitoring data or experimental results are available, only high-resolution studies can provide data of a temporal resolution and precision comparable to contemporary limnological studies (i.e., one or few years per sample: Anderson 1995a). To trace environmental change and limnological response at ecologically relevant time-scales, annual to decadal time resolution is, therefore, absolutely necessary.

In this chapter we have concentrated on biogenic varves. However, varves from glacio-lacustrine systems have also been used to relate varve thickness and sediment yield to air temperature (e.g., Leemann and Niessen 1994; Zolitschka 1996). Well-dated non-laminated sediments with high accumulation rates may also be suitable for high-resolution studies. However, if the influence of external stressors is to be compared with sediment-derived data on annual to decadal time-scales, the need for reliable chronological control cannot be over-emphasised. Low sediment accumulation rates in oligotrophic lakes combined with a low temporal resolution, even at contiguous 2-mm sediment samples, make it difficult to compare non-varved sediment records with, for example, meteorological data (see e.g., Lotter et al. 2002; Sorvari et al. 2002).

The study at Kassjön highlights the problem of floating chronologies without a fixed reference point when sediment sequences are to be compared to independent time-series. At Baldeggersee, the core surface as well as other independent dates (^{137}Cs, historical data: see Lotter et al. 1997c) were used as reference points. At Holzmaar, both response variables (diatoms, pollen) and predictors (LST, biozones, age) are derived from the same sediment samples. Therefore, they have a matching time-resolution and thus circumvent problems of matching independent time-series.

Once the problem of matching the time-resolution of response and predictor variables is solved, several approaches to analyse the data numerically can be taken (Fig. 18.1). Yet, there are often several different stressors affecting the biotic or abiotic response variable. As in the example of Baldeggersee, it might be important to assess the importance of the different stressors through decomposition of the signal using variation partitioning approaches. However, the choice of predictors is crucial. The analysis can often be supported by using constrained ordination with forward selection of predictor variables that explain a significant amount of the variation in the response variables (ter Braak and Šmilauer 2002). Eventually, this should enable us to choose the minimum adequate model (Birks 2012: Chap. 2) that will allow an assessment of the influence of stressors on biotic assemblages and, hence, to gain further insights into those mechanisms controlling change in aquatic systems.

An alternative approach is to formulate hypotheses that, as in the case of Holzmaar, may be tested using sediment data. Given the problems of covariation in inference models and the difficulties in making reliable inferences of a single variable from a matrix of changing environmental stressors and responses (Anderson 2000), the use of constrained ordination and carefully selected response and predictor variables to address species-environment interactions over time represents a promising, yet under-rated and only occasionally used approach in palaeolimnology.

The need for independence in data analysis is an increasingly recognised component of quantitative palaeolimnology. This need will press researchers to use independent proxies, either 'external' or those derived from the sediment itself. With the latter approach, one of the most difficult aspects is to decide which variable to use as the predictor and which to use as the response, because there is a high level of interdependence within the data.

Conclusions

There is a wealth of powerful numerical tools that may help us to analyse complex palaeolimnological data-sets and detect patterns and relationships in the data. Correlations between predictors and response variables do not necessarily imply a cause-and-effect relationship and common sense as well as understanding of limnological and ecological processes are thus indispensable. Of great importance is the definition of the research question or the hypothesis to be tested, the selection of ecologically relevant response proxies, the critical choice of predictors as proxies for stress and change, and the selection of numerical analysis tools (e.g., partial ordination, cross-correlation, forward selection of potentially important explanatory variables). The results and conclusions are, as in all analyses, heavily dependent on the quality of the data. Palaeolimnological analyses of disturbances require, therefore, critical site selection, sub-sampling at relevant temporal scales (Anderson 1995a), the application of rigorous data analysis and statistical testing (Birks 1998), and finally the application of ecological and limnological sense.

References

Algesten G, Sobek S, Bergström A-K, Agren A, Tranvik LJ, Jansson M (2003) Role of lakes for organic carbon cycling in the boreal zone. Global Change Biol 10:141–147

Ammann B, Lotter AF (1989) Late-Glacial radiocarbon- and palynostratigraphy on the Swiss Plateau. Boreas 18:109–126

Anderson NJ (1995a) Temporal scale, phytoplankton ecology and palaeolimnology. Freshw Biol 34:367–378

Anderson NJ (1995b) Using the past to predict the future: lake sediments and the modelling of limnological disturbance. Ecol Model 78:149–172

Anderson NJ (1997) Reconstructing historical phosphorus concentrations in rural lakes using diatom models. In: Tunney H, Carton OT, Brookes PC, Johnston AE (eds) Phosphorus loss from soil to water. CAB International, Oxford, pp 95–118

Anderson NJ (2000) Diatoms, temperature and climatic change. Eur J Phycol 35:307–314

Anderson NJ, Battarbee RW (1994) Aquatic community persistence and variability: a palaeolimnological perspective. In: Giller PS, Hildrew AG, Raffaelli D (eds) Aquatic ecology: scale patterns and processes. Blackwell Scientific Publications, Oxford, pp 233–259

Anderson NJ, Renberg I, Segerström U (1995) Diatom production response to the development of early agriculture in a boreal forest lake-catchment (Kassjön, northern Sweden). J Ecol 83: 809–822

Anderson NJ, Odgaard BV, Segerström U, Renberg I (1996) Climate-lake interactions recorded in varved sediments from a Swedish boreal forest lake. Global Change Biol 2:399–405

Anderson NJ, Bugmann H, Dearing JA, Gaillard M-J (2006) Linking palaeoenvironmental data and models to understand the past and to predict the future. Trends Ecol Evol 21:696–704

Barker P, Telford RJ, Merdaci O, Williamson D, Taieb M, Vincens A, Gibert E (2000) The sensitivity of a Tanzanian crater lake to catastrophic tephra input and four millennia of climate change. Holocene 10:303–310

Bender EA, Case TJ, Gilpin ME (1984) Perturbation experiments in community ecology: theory and practice. Ecology 65:1–13

Birks HJB (1995) Quantitative palaeoenvironmental reconstructions. In: Maddy D, Brew JS (eds) Statistical modelling of Quaternary science data. Quaternary Research Association, Cambridge, pp 161–254

Birks HJB (1998) Numerical tools in palaeolimnology – progress, potentialities, and problems. J Paleolimnol 20:307–332

Birks HJB (2012) Chapter 2: Overview of numerical methods in palaeolimnology. In: Birks HJB, Lotter AF, Juggins S, Smol JP (eds) Tracking environmental change using lake sediments. Volume 5: Data handling and numerical techniques. Springer, Dordrecht

Birks HJB, Line JM (1992) The use of rarefaction analysis for estimating palynological richness from Quaternary pollen-analytical data. Holocene 2:1–10

Birks HJB, Lotter AF (1994) The impact of the Laacher See volcano (11000 yr B.P.) on terrestrial vegetation and diatoms. J Paleolimnol 11:313–322

Bittmann F, Weninger B, Wiethold J (2002) Impact of the Late Glacial eruption of the Laacher See volcano, central Rhineland, Germany. Quaternary Res 58:273–288

Borcard D, Legendre P, Drapeau P (1992) Partialling out the spatial component of ecological variation. Ecology 73:1045–1055

Bradshaw EG, Rasmussen P, Odgaard BV (2005) Mid- to late-Holocene land-use change and lake development at Dallund Sø, Denmark: synthesis of multiproxy data, linking land and lake. Holocene 15:1152–1162

Brauer A, Dulski P, Mangili C, Mingram J, Liu J (2009) The potential of varves in high-resolution paleolinological studies. PAGES News 17:96–98

Briffa KR (2000) Annual climate variability in the Holocene: interpreting the message of ancient trees. Quaternary Sci Rev 19:87–105

Briffa KR, Bartholin TS, Eckstein D, Jones PD, Karlen W, Schweingruber FH, Zetterberg P (1990) A 1,400-year tree-ring record of summer temperature in Fennoscandia. Nature 346:434–439

Carpenter SR (ed) (1988) Complex interactions in lake communities. Springer, New York

Carpenter SR, Kitchell JF, Hodgson JR, Cochran PA, Elser JJ, Elser M, Lodge DM, Kretchmer D, He X, von Ende CN (1987) Regulation of lake primary productivity by food web structure. Ecology 68:1863–1876

Cumming BF, Laird KR, Fritz SC, Verschuren D (2012) Chapter 20: Tracking Holocene climatic change with aquatic biota from lake sediments: case studies of commonly used numerical

techniques. In: Birks HJB, Lotter AF, Juggins S, Smol JP (eds) Tracking environmental change using lake sediments. Volume 5: Data handling and numerical techniques. Springer, Dordrecht
Deevey ES (1984) Stress, strain, and stability of lacustrine ecosystems. In: Haworth EY, Lund JWG (eds) Lake sediments and environmental history. University of Leicester Press, Leicester, pp 203–229
Dutilleul P, Cumming BF, Lontoc-Roy M (2012) Chapter 16: Autocorrelogram and periodogram analysis of palaeolimnological temporal series from lakes in central and western North America to assess shifts in drought conditions. In: Birks HJB, Lotter AF, Juggins S, Smol JP (eds) Tracking environmental change using lake sediments. Volume 5: Data handling and numerical techniques. Springer, Dordrecht
Engstrom DR, Fritz SC, Almendinger JE, Juggins S (2000) Chemical and biological trends during lake evolution in recently deglaciated terrain. Nature 408:161–166
Flower RJ (1990) Seasonal changes in sedimenting material collected by high aspect ratio sediment traps operated in a holomictic eutrophic lake. Hydrobiologia 214:311–316
Fritz SC (2003) Lacustrine perspectives on Holocene climate. In: Mackay AW, Battarbee RW, Birks HJB, Oldfield F (eds) Global change in the Holocene. Edward Arnold, London, pp 227–241
Fritz SC (2008) Deciphering climatic history from lake sediments. J Paleolimnol 39:5–16
Gotelli NJ, Ellison AM (2004) A primer of ecological statistics. Sinauer Associates, Sunderland
Hall RI, Leavitt PR, Quinlan R, Dixit AS, Smol JP (1999) Effects of agriculture, urbanization, and climate on water quality in the northern Great Plains. Limnol Oceanogr 44:739–756
Hausmann S, Lotter AF, van Leeuwen JFN, Ohlendorf C, Lemcke G, Grönlund E, Sturm M (2002) Interactions of climate and land use documented in the varved sediments of Seebergsee in the Swiss Alps. Holocene 12:279–289
Kitchell JF, Bartell SM, Carpenter SR, Hall DJ, McQueen DJ, Neill WE, Scavia D, Werner EE (1988) Epistemology, experiments, and pragmatism. In: Carpenter SR (ed) Complex interactions in lake communities. Springer, New York, pp 263–280
Korsman T, Segerström U (1998) Forest fire and lake-water acidity in a northern Swedish boreal area: Holocene changes in lake-water quality at Makkassjön. J Ecol 86:113–124
Korsman T, Renberg I, Anderson NJ (1994) A palaeolimnological test of the influence of Norway spruce (*Picea abies*) immigration on lake-water acidity. Holocene 4:132–140
Leemann A, Niessen F (1994) Varve formation and the climatic record in an Alpine proglacial lake: calibrating annually-laminated sediments against hydrological and meteorological data. Holocene 4:1–8
Legendre P, Birks HJB (2012) Chapter 8: From classical to canonical ordination. In: Birks HJB, Lotter AF, Juggins S, Smol JP (eds) Tracking environmental change using lake sediments. Volume 5: Data handling and numerical techniques. Springer, Dordrecht
Lister GS, Livingstone DM, Ammann B, Ariztegui D, Haeberli W, Lotter AF, Ohlendorf C, Pfister C, Schwander J, Schweingruber F, Stauffer B, Sturm M (1998) Alpine paleoclimatology. In: Cebon P, Dahinden U, Davies HC, Imboden D, Jaeger CC (eds) Views from the Alps. Regional perspectives on climate change. Massachusetts Institute of Technology Press, Cambridge, pp 73–170
Livingstone DM, Dokulil MT (2001) Eighty years of spatially coherent Austrian lake surface temperatures and their relationship to regional air temperature and the North Atlantic Oscillation. Limnol Oceanogr 46:1220–1227
Livingstone DM, Hajdas I (2001) Climatically relevant periodicities in the thicknesses of biogenic carbonate varves in Soppensee, Switzerland (9740–6870 calendar yr BP). J Paleolimnol 25: 17–24
Lotter AF (1998) The recent eutrophication of Baldeggersee (Switzerland) as assessed by fossil diatom assemblages. Holocene 8:395–405
Lotter AF (2003) Multi-proxy climatic reconstructions. In: Mackay AW, Battarbee RW, Birks HJB, Oldfield F (eds) Global change in the Holocene. Edward Arnold, London, pp 373–383
Lotter AF, Bigler C (2000) Do diatoms in the Swiss Alps reflect the length of ice-cover? Aquat Sci 62:125–141

Lotter AF, Birks HJB (1993) The impact of the Laacher See Tephra on terrestrial and aquatic ecosystems in the Black Forest, southern Germany. J Quaternary Sci 8:263–276
Lotter AF, Birks HJB (1997) The separation of the influence of nutrients and climate on the varve time-series of Baldeggersee, Switzerland. Aquat Sci 59:362–375
Lotter AF, Birks HJB (2003) The Holocene palaeolimnology of Sägistalsee (1935 m asl) and its environmental history - a synthesis. J Paleolimnol 30:333–342
Lotter AF, Psenner R (2004) Global change impacts on mountain waters: lessons from the past to help define monitoring targets for the future. In: Lee C, Schaaf M (eds) Global environmental and social monitoring. UNESCO, Paris, pp 102–114
Lotter AF, Eicher U, Birks HJB, Siegenthaler U (1992) Late-glacial climatic oscillations as recorded in Swiss lake sediments. J Quaternary Sci 7:187–204
Lotter AF, Birks HJB, Zolitschka B (1995) Late-glacial pollen and diatom changes in response to two different environmental perturbations: volcanic eruption and Younger Dryas cooling. J Paleolimnol 14:23–47
Lotter AF, Birks HJB, Hofmann W, Marchetto A (1997a) Modern diatom, Cladocera, chironomid, and chrysophyte cyst assemblages as quantitative indicators for the reconstruction of past environmental conditions in the Alps. I. Climate. J Paleolimnol 18:395–420
Lotter AF, Renberg I, Hansen H, Stöckli R, Sturm M (1997b) A remote controlled freeze corer for sampling unconsolidated surface sediments. Aquat Sci 59:295–303
Lotter AF, Sturm M, Teranes JL, Wehrli B (1997c) Varve formation since 1885 and high-resolution varve analysis in hypertrophic Baldeggersee (Switzerland). Aquat Sci 59:304–325
Lotter AF, Appleby PG, Bindler R, Dearing JA, Grytnes J-A, Hofmann W, Kamenik C, Lami A, Livingstone DM, Ohlendorf C, Rose N, Sturm M (2002) The sediment record of the past 200 years in a Swiss high-alpine lake: Hagelseewili (2339 m a.s.l.). J Paleolimnol 28:111–127
Maher LJ, Heiri O, Lotter AF (2012) Chapter 6: Assessment of uncertainties associated with palaeolimnological laboratory methods and microfossil analysis. In: Birks HJB, Lotter AF, Juggins S, Smol JP (eds) Tracking environmental change using lake sediments. Volume 5: Data handling and numerical techniques. Springer, Dordrecht
Nesje A, Dahl SO, Matthews JA, Berrisford MS (2001) A 4500 yr record of river floods obtained from a sediment core in Lake Atnsjøen, eastern Norway. J Paleolimnol 25:329–342
Nilsson M, Renberg I (1990) Viable endospores of *Thermoactinomyces vulgaris* in lake-sediments as indicators of agricultural history. Appl Environ Microbiol 56:2025–2028
Nipkow F (1927) Über das Verhalten der Skelette planktischer Kieselalgen im geschichteten Tiefenschlamm des Zürich- und Baldeggersees. Neue Beiträge zur Biologie der Planktondiatomeen und zur Biomorphose der subalpinen Seen. ETH Zürich Nr. 455
Parris AS, Bierman PR, Noren AJ, Prins MA, Lini A (2010) Holocene paleostorms identified by particle size signatures in lake sediments from the northeastern United States. J Paleolimnol 43:29–49
Petterson G (1996) Varved sediments in Sweden: a brief review. In: Kemp AE (ed) Palaeoclimatology and palaeoceanography from laminated sediments. Geological Society of London Special Publication, pp 73–77
Petterson G (1999) Image analysis, varved lake sediments and climate reconstruction. PhD Thesis, University of Umeå, Umeå
Petterson G, Odgaard BV, Renberg I (1999) Image analysis as a method to quantify sediment components. J Paleolimnol 22:443–455
Pienitz R, Smol JP, Birks HJB (1995) Assessment of freshwater diatoms as quantitative indicators of past climatic change in the Yukon and Northwest Territories, Canada. J Paleolimnol 13: 21–49
Renberg I (1981) Formation, structure and visual appearance of iron-rich, varved lake sediments. Verhandlungen Internationale Vereinigung für Limnologie 21:94–101
Renberg I (1990) A 12600 year perspective of the acidification of Lilla Öresjön, southwest Sweden. Palaeolimnology and lake acidification. Philos Trans R Soc Lond B 327:357–361
Scheffer M (1998) Ecology of shallow lakes. Chapman & Hall, London

Schmincke H-U, Park C, Harms E (1999) Evolution and environmental impacts of the eruption of Laacher See volcano (Germany) 12,900 a BP. Quaternary Int 61:61–72

Schoonmaker PK, Foster DR (1991) Some implications of paleoecology for contemporary ecology. Bot Rev 57:204–245

Segerström U (1990a) The natural Holocene vegetation development and the introduction of agriculture in northern Norrland, Sweden. Studies of soil, peat and especially varved lake sediments. PhD thesis, University of Umeå, Umeå

Segerström U (1990b) The post-glacial history of the vegetation and agriculture in the Luleälv river valley. Archaeol Environ 7:9–80

Simola H, Hanski I, Liukkonen M (1990) Stratigraphy, species richness and seasonal dynamics of plankton diatoms during 418 years in Lake Lovojärvi, South Finland. Ann Bot Fenn 27: 241–259

Simpson GL, Hall IR (2012) Chapter 19: Human impacts – applications of numerical methods to evaluate surface-water acidification and eutrophication. In: Birks HJB, Lotter AF, Juggins S, Smol JP (eds) Tracking environmental change using lake sediments. Volume 5: Data handling and numerical techniques. Springer, Dordrecht

Smol JP (1991) Are we building enough bridges between paleolimnology and aquatic ecology? Hydrobiologia 214:201–206

Smol JP (2008) Pollution of lakes and rivers: a paleoenvironmental perspective, 2nd edn. Wiley-Blackwell, Oxford

Sorvari S, Korhola A, Thompson R (2002) Lake diatom response to recent Arctic warming in Finnish Lapland. Global Change Biol 8:153–163

Straile D, Livingstone DM, Weyhenmeyer GA, Gerorge DG (2003) The response of freshwater ecosystems to climate variability associated with the North Atlantic Oscillation. In: Hurrell JW, Kushnir Y, Ottersen G, Visbeck M (eds) The North Atlantic Oscillation. Climatic significance and environmental impact, Geophysical Monograph. American Geophysical Union, Washington, DC, pp 263–279

ter Braak CJF (1986) Canonical correspondence analysis: a new eigenvector technique for multivariate direct gradient analysis. Ecology 67:1167–1179

ter Braak CJF, Prentice IC (1988) A theory of gradient analysis. Adv Ecol Res 18:271–317

ter Braak CJF, Šmilauer P (2002) CANOCO reference manual and CanoDraw for Windows user's guide: software for Canonical Community Ordination (version 4.5). Microcomputer Power, Ithaca

Wehrli B, Lotter AF, Schaller T, Sturm M (1997) High-resolution varve studies in Baldeggersee (Switzerland): project overview and limnological background data. Aquat Sci 59:285–294

Willis KJ, Bailey RM, Bhagwat SA, Birks HJB (2010) Biodiversity baselines, thresholds and resilience: testing predictions and assumptions using palaeoecological data. Trends Ecol Evol 25:583–591

Zolitschka B (1996) Recent sedimentation in a high arctic lake, northern Ellesmere Island, Canada. J Paleolimnol 16:169–186

Zolitschka B, Brauer A, Negendank JFW, Stockhausen H, Lang A (2000) Annually dated late Weichselian continental paleoclimate record from the Eifel, Germany. Geology 28:783–786

Chapter 19
Human Impacts: Applications of Numerical Methods to Evaluate Surface-Water Acidification and Eutrophication

Gavin L. Simpson and Roland I. Hall

Abstract In this chapter we review the contributions that numerical techniques have made in answering key questions in applied palaeolimnology relating to studies of lake acidification and eutrophication. Palaeoecological data and calibration functions in particular provided some of the key observations implicating acid emissions from industrial and power-generation sources as the major cause of the recent acidification of lakes in northern Europe and North America. Sedimentary records and subsequent quantitative analyses play a similar role in understanding the eutrophication of lakes, and today are being used widely to inform management of enriched lakes and to set restoration targets for recovery.

Keywords Acidification • Analogues • Calibration functions • Constrained ordination • Diatoms • Ecosystem restoration • Environmental management • Eutrophication • Gaussian logistic (= logit) regression • Maximum likelihood calibration • Ordination • pH • Phosphorus loadings • Reference conditions • Total phosphorus • Variation partitioning • Weighted-averaging partial least squares regression and calibration

---

G.L. Simpson (✉)
Environmental Change Research Centre, University College London, Pearson Building, Gower Street, London WC1E 6BT, UK
e-mail: gavin.simpson@ucl.ac.uk

R.I. Hall
Department of Biology, University of Waterloo, 200 University Avenue West, Waterloo, ON N2L 3G1, Canada
e-mail: rihall@uwaterloo.ca

H.J.B. Birks et al. (eds.), *Tracking Environmental Change Using Lake Sediments*, Developments in Paleoenvironmental Research 5, DOI 10.1007/978-94-007-2745-8_19,

Introduction

The science of palaeolimnology has advanced considerably during the last 30 years as new questions are asked of lake-sediment records and as techniques have been developed to help answer them. The application of palaeolimnology to questions regarding recent environmental change and lake management are two areas of the science that have witnessed considerable advances (e.g., Battarbee 1994, 1999; Smol 2008; Pienitz et al. 2009). Palaeolimnology is particularly suited to tackling questions of recent environmental change where contemporary monitoring data are generally only available for the period following the initial identification of the disturbance. In many cases little information concerning background, or pre-disturbance, reference conditions, including the range of natural variation, exists in any form.

Acid deposition and the discharge of nutrients to surface waters have had considerable impact upon many freshwater ecosystems globally. The widespread losses of fish from acidified lakes and streams (e.g., Beamish and Harvey 1972; Harriman et al. 1987), as well as the loss of aquatic macrophytes, increase in phytoplankton, and loss of biodiversity in systems receiving increased nutrient loads are the most visible changes associated with anthropogenic acidification and eutrophication. In many cases whole ecosystem shifts have occurred as the direct result of acid deposition or eutrophication due to the alteration of biogeochemical cycling and the loss of sensitive organisms (Smol 2008).

In this chapter we review the use of numerical techniques in palaeolimnological studies and examine some methods that can be applied to palaeolimnological data in order to answer questions central to the study of recent human-induced environmental change. Specifically, we present case studies to illustrate how numerical approaches are used to track the degradation and recovery of lake ecosystems, to identify the causes of change, and to guide restoration of impacted surface waters.

Reconstructing and Evaluating Past Trends in Surface-Water Acidity and Nutrient Status

Quantitative Reconstruction of Lake-Water Chemistry

Biologists have long recognised that species respond to changes along environmental gradients (e.g., Whittaker 1956, 1967), and that groups of species can be identified which exhibit similar patterns of occurrence along these gradients. It is this motivation that has led to the transfer-function or calibration-function paradigm in modern palaeolimnology (ter Braak and Prentice 1988; ter Braak and van Dam 1989; Birks et al. 1990a, 2010). Calibration functions have been referred to as the “paleolimnologist’s Rosetta Stone” (Smol 2008) and have been widely used to reconstruct past environmental changes from lake-sediment archives, not least because of the development of user-friendly computer software which implements these methods (ter Braak and Šmilauer 2002; Juggins 2003; Lepš and Šmilauer 2003). Like all numerical techniques, calibration-function methodologies should be applied with care, interrogated by critical analysis of the model output, and, most importantly, interpreted in ecologically meaningful ways (see also Juggins and Birks 2012: Chap. 14; Lotter and Anderson 2012: Chap. 18; Cumming et al. 2012: Chap. 20).

Acidification

As an example, we apply several pH calibration-function models to a ca. 10,000 year diatom sequence (RLGH3) from The Round Loch of Glenhead (RLGH, Jones et al. 1989; Birks and Jones 2012: Chap. 3), an acidified loch in the Galloway region of south-west Scotland, using 138 modern samples from the Surface Waters Acidification Programme (SWAP) diatom-pH training-set (Stevenson et al. 1991) and provide some evaluation of one of these models for use in reconstructing past changes in pH.

Detrended correspondence analysis (DCA) of the SWAP data suggests that moderately long compositional gradient lengths (axis 1 = 3.4 Hill's standard deviation (SD) units of compositional turnover, axis 2 = 2.66 SD units – see Legendre and Birks 2012b: Chap. 8) are present in the diatom abundances, and a detrended canonical correspondence analysis (DCCA) (ter Braak 1986; Birks 1995) with pH as the sole constraining external variable indicates that the diatoms respond in a non-linear way to pH (DCCA axis 1 = 2.56 SD units). A useful rule-of-thumb (ter Braak and Prentice 1988) is that where DCCA gradient lengths are greater than 2 SD units, several taxa in the training-set have their optima located within the gradient and unimodal-based methods are appropriate. Where gradient lengths are less than 2 SD units, species responses are generally monotonic along the gradient and linear-based methods would potentially be more appropriate (ter Braak and Juggins 1993; Birks 1995, 1998). Therefore, constructing a calibration-function model for pH using non-linear or unimodal methods of regression and calibration would be appropriate given the properties of the training-set (Juggins and Birks 2012: Chap. 14).

Weighted-averaging partial least squares (WAPLS) and maximum likelihood (Gaussian logistic or logit) regression and calibration (ML) are two methods for constructing a calibration function, where parameters for m species are estimated from the modern training-set for the environmental variable of interest and are subsequently used to reconstruct the past environment from these same m species in fossil samples. An alternative to these species-based techniques is to use the modern analogue technique (MAT) (see Juggins and Birks 2012: Chap. 14; Simpson 2012: Chap. 15), an assemblage approach where the k most similar assemblages from the modern training-set are selected for each fossil sample and the environmental reconstruction for each fossil sample is then simply a (weighted) average of the environmental variable(s) of interest over the k closest modern analogues. The main difference between the two approaches is that in MAT it is the assemblage of species that determines the reconstructed values, whereas in WAPLS and ML the reconstructed values are based on individual species' responses to the environmental variable in question, not the assemblages in the training-set from which the species data were derived (Birks et al. 2010). We fit all three methods to the SWAP training-set and the RLGH3 core to permit a comparison and assessment of the reconstructions. If all the reconstructions are similar then we can be more confident about the inferred values for the environmental variable than if the various models produced different reconstructions. A further advantage of fitting more than a single model is the ability to use the results from all models to derive a consensus reconstruction (Birks 1998).

Table 19.1 shows the results of the WAPLS model when applied to the SWAP training-set to produce calibration-function models for estimating past lake-water pH. Listed are the root mean squared error (RMSE), r^2, and selected bias statistics for each of the first five WAPLS components computed using the C2 software of Juggins (2003). These provide overly optimistic assessments of the error, r^2, and bias in the calibration function because they have been derived by using the same data to construct and then test the model (Birks 1995). A better indication of these model-performance measures is given by cross-validation (Birks 1995). In this example we have used 999 bootstrap re-sampling cycles to derive pseudo-independent indicators of model performance. Table 19.2 shows the bootstrap-derived performance statistics for the three calibration-function methods fitted to the SWAP diatom-pH training-set. The RMSEP values indicate that the error in predicting pH from new data is of the order ±0.3 pH units for all three methods.

A key concept in statistics is that of the parsimonious or minimal adequate model (Birks 2012a: Chap. 2), which states that simple models are to be preferred over more complex models and that to accept a more complex model over a simple model, the complex model must offer significant improvements in explanatory power. A useful rule-of-thumb, when selecting a parsimonious WAPLS model is to choose the model that gives the lowest RMSEP and results in a 5% or greater reduction in the RMSEP over the previous model (ter Braak and Juggins 1993; Birks 1998; Juggins and Birks 2012: Chap. 14). In the SWAP example, the increase in complexity from a one- to a two-component WAPLS model reduces the RMSEP by 0.00423 ($0.31741 - 0.31318 = 0.00423$) which is less than 5% of the RMSEP of the one-component model ($0.31741 \times 0.05 = 0.01587$) and as such does not represent a sufficient improvement over the one-component model to accept the additional complexity of a two-component model. An alternative, statistically-based, approach to model selection is the randomisation t-test (van der Voet 1994; see Juggins and Birks 2012: Chap. 14). It is worth noting that a one-component WAPLS model is the equivalent of a simple weighted averaging (WA) model using an inverse deshrinking regression (ter Braak and Juggins 1993), if untransformed percentage biological data are used (see Juggins and Birks 2012: Chap. 14).

Table 19.2 shows the bootstrap-derived estimates of model performance for the one-component WAPLS (WAPLS(1)), ML, and MAT models. The RMSEP for all

Table 19.1 Weighted-averaging partial least squares (WAPLS) apparent model performance statistics

	1 Comp.	2 Comp.	3 Comp.	4 Comp.	5 Comp.
RMSE	0.27565	0.23212	0.19359	0.17248	0.15314
r^2	0.87168	0.90901	0.93670	0.94976	0.96040
Average Bias	0.00158	0.00055	−0.00015	0.00058	0.00049
Maximum Bias	0.19303	0.15359	0.14203	0.11548	0.09889

Comp component, *RMSE* root mean squared error for the 138-lake Surface Waters Acidification Programme (SWAP) diatom-pH training-set

Table 19.2 Bootstrap-derived performance statistics for the various calibration-function methods fitted to the Surface Waters Acidification Programme (SWAP) diatom-pH training-set

	RMSEP	S_1	S_2	Average bias	Maximum bias
WAPLS(1)	0.317	0.064	0.311	0.016	0.430
ML	0.344	0.133	0.318	0.004	0.348
MAT	0.328	0.123	0.304	0.047	0.444

RMSEP root mean squared error of prediction, S_1 and S_2 are components of RMSEP (see Birks et al. 1990a; Juggins and Birks 2012: Chap. 14; Simpson 2012: Chap. 15 for details)
WAPLS(1) one-component weighted-averaging partial least squares model, *ML* maximum likelihood regression and calibration, *MAT* modern analogue technique

three models is similar, with WAPLS(1) having the lowest RMSEP and ML the highest with MAT roughly in between these two. The ML model has the lowest average and maximum bias (0.004 and 0.348, respectively) of the three models, with MAT being the worst performer of the three in terms of bias.

As with many statistical methods, it is important to examine the fitted values and the residuals from the calibration-function model to look for systematic differences between the fitted and the observed values. Figure 19.1 shows two such views of the model from the WAPLS(1) reconstruction for the RLGH3 core described above. In Fig. 19.1 we follow the recommendation of Racca and Prairie (2004) and plot the fitted or predicted pH values on the x-axis and the observed values on the y-axis. The dashed line is a 1:1 line and if the model perfectly predicted the observed values all the points would lie on this line. The vertical distance between the 1:1 line and each point is the residual variance in pH not explained by the model. The solid line is a simple linear regression of the form:

$$\mathrm{pH}_{\mathrm{observed}} \sim \mathrm{b}_0 + \mathrm{b}_1 \mathrm{pH}_{\mathrm{fitted}}$$

The difference between this fitted regression line and the 1:1 line is a measure of the tendency for the model to over- or under-predict the response (the observed pH), which is very low in the RLGH3 example, as also indicated by the average-bias statistic shown in Table 19.2. Figure 19.1b shows a plot of the residuals against the fitted values (Racca and Prairie 2004). Overlain on this plot are two lines: (i) the solid, horizontal line shows the average bias (i.e., the mean of the

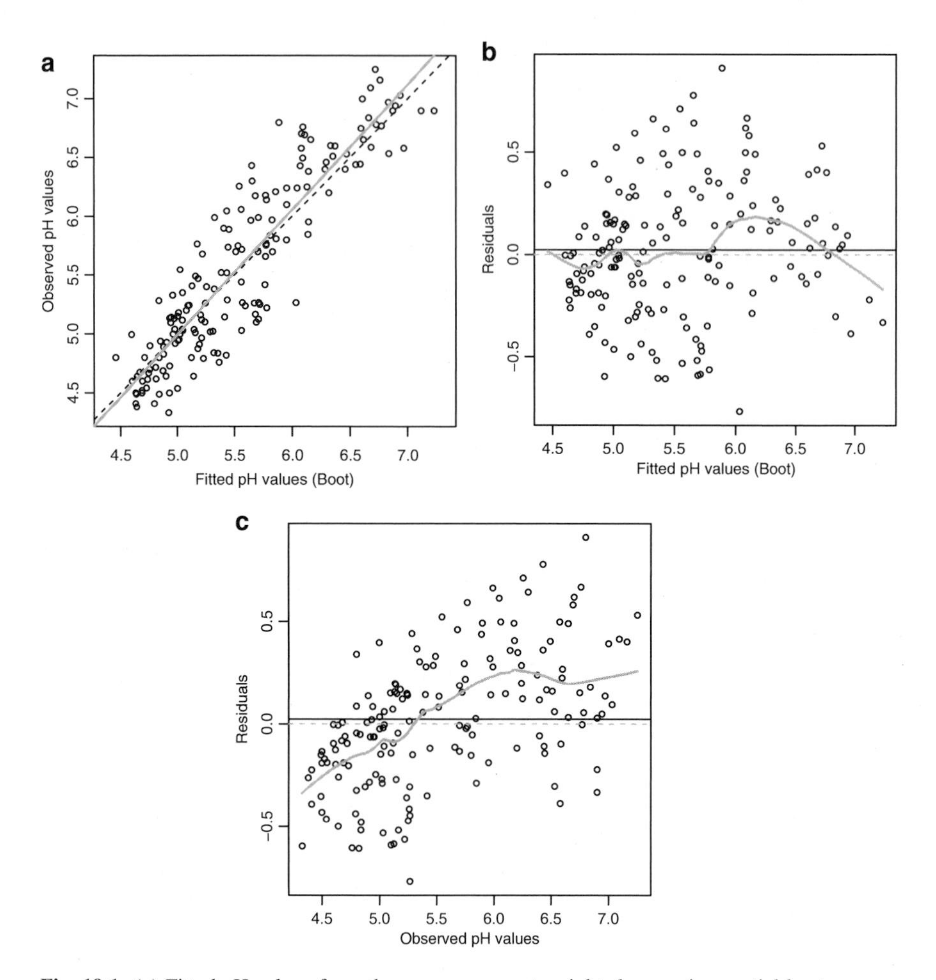

Fig. 19.1 (**a**) Fitted pH values from the one-component weighted-averaging partial least squares model (WAPLS(1)) plotted against the observed pH values for the Surface Waters Acidification Programme (SWAP) diatom-pH training-set. The difference between the 1:1 line (*solid*) and the fitted line of a regression of the fitted values on the observed values (*dashed*) is the tendency for the model to over- or under-predict. (**b**) Fitted values from the WAPLS(1) model plotted against the residuals. The solid horizontal line illustrates the average bias in the fitted values (i.e., the mean of the residuals) and a LOESS smoother is fitted to the points using a span of 0.3. The LOESS smoother highlights the presence of systematic patterns in the fitted values from the model. (**c**) Observed values of lake-water pH plotted against residuals from the WAPLS(1) model, as before showing average bias and a LOESS smoother. This presentation of model performance highlights the presence of strong under-prediction at low pH values, a pattern that is not observed in (**b**) (see text for explanation)

residuals) in the WAPLS(1) model and (ii) the solid, irregular line shows a LOESS smoother (Cleveland 1979; see Birks 2012a: Chap. 2; Juggins and Telford 2012: Chap. 5) that is used to highlight any pattern in the residuals which might indicate problems in model formulation, or locations along the environmental gradient being reconstructed where the model is performing poorly. For the RLGH3 example, the residuals from the WAPLS(1) model suggest that overall the model is well fitted with no major pattern in the residuals. The LOESS smoother suggests that in the interval pH 6–6.5 the model tends to under-estimate lake-water pH by about 0.2 pH units and that above pH 7.0 the model over-predicts by 0.2–0.3 pH units. We should take this into account when evaluating reconstructions from this model, but for the RLGH3 core the model is likely to perform well as the loch is naturally acidic and the reconstructed values are lower than pH 6.0, which is where there is little evidence of bias in the predicted values in the residual plot. Figure 19.1c shows the residuals plotted against the observed pH values. The LOESS smoother now shows a clear tendency to over-predict pH values at the low pH end of the gradient. This apparent contradiction between the two forms of display is discussed by Juggins and Birks (2012: Chap. 14). It is clear that Fig. 19.1c provides the more informative display in this case, alerting us to the tendency to over-predict at low pH.

The WAPLS(1) reconstruction for the full Holocene history of the RLGH is illustrated in Fig. 19.2a along with bootstrap-derived sample-specific errors. The reconstruction highlights three main periods of change: (i) the slow acidification of the loch between ~10,000 and 4000 radiocarbon year before present (BP), as organic soils developed in the catchment; (ii) the rapid fluctuations in pH between 4000 and 2000 radiocarbon year BP, thought to be due to the spread of blanket mire and a decline in tree cover in the catchment (~4000 radiocarbon year BP), and the subsequent erosion of peat (~3000 radiocarbon year BP; Jones et al. 1989); and (iii) the rapid acidification of the RLGH that begins in the early 1800s when pH declines to pH 4.75. When placed in the longer-term context it is clear that the recent history of the RLGH has seen an unparalleled rate and magnitude of change, which when linked with other proxy data (e.g., pollution indicators such as spheroidal carbonaceous particles and heavy metals) clearly demonstrates the deleterious effects of industrial air pollution on the water quality of the RLGH (Battarbee et al. 1989; Renberg and Battarbee 1990).

The reconstruction described above (Fig. 19.2a) suggests a change in pH of 0.75 units from pH 5.5 to 4.75 in the RLGH since the mid nineteenth century. The average annual pH of the loch measured when the core was taken in 1984 was 4.7 with an annual range (1981–82) of 4.6–5.0 (Jones et al. 1989; Birks and Jones 2012: Chap. 3), which shows good agreement with the reconstructed values of 4.8–5.0 for samples in the upper 1 cm of the core. The WAPLS(1) pH reconstruction also demonstrates that the pH of the RLGH, far from being stable since the onset of the Holocene, has fluctuated between 5.85 and 5.0, but we can see that at no time in the history of the loch did the pH fall to the extremely low levels seen in the twentieth Century.

The uppermost sample in the RLGH3 core (Fig. 19.2) suggests the beginnings of a recovery from acidification at the site with the reconstruction indicating that pH rose from pH 4.8 to 5.0 for this sample. Great care must be taken when interpreting small changes in reconstructed values like these where the values fall well within the prediction error of the model and where there is no pattern in diatom composition

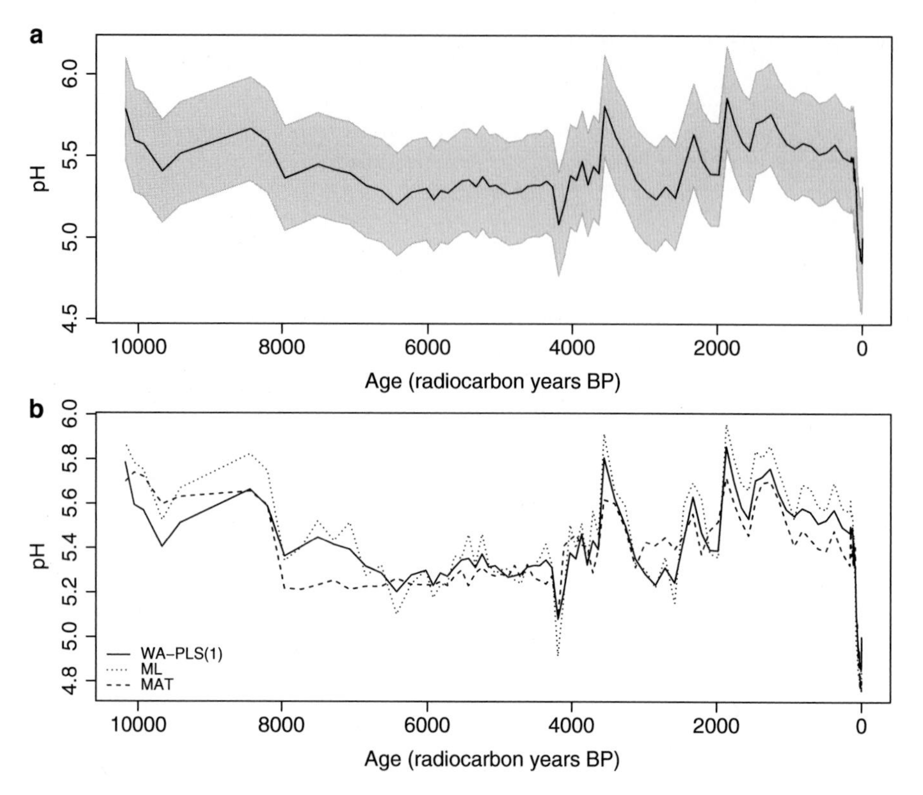

Fig. 19.2 (**a**) The reconstructed pH values for the samples of The Round Loch of Glenhead (RLGH3) core (*solid line*) plotted with sample-specific bootstrap-derived errors (*shaded area*) using the one-component weighted-averaging partial least squares model (WAPLS(1)) and the Surface Waters Acidification Programme (SWAP) training-set. (**b**) Comparison of the reconstructed pH for samples of the RLGH3 core from the WAPLS(1) (*solid line*), maximum likelihood (ML) (*dotted line*), and modern analogue technique (MAT) (*dashed line*) models

that could be used to suggest a trend towards higher pH values even if we are not confident about the precision of the fitted values (see Juggins and Birks 2012: Chap. 14). The same can be said of the entire reconstruction and it is useful to know if both the pattern in the reconstruction and the actual predicted values show some consistency across reconstruction methods. Figure 19.2b shows the inferred values from the three reconstructions. The reconstructions show good agreement in the patterns of pH change throughout the post-glacial history of the RLGH, which imparts a degree of confidence in the reconstructed values for a single method. There are some departures in the reconstructed values across the three methods, particularly in the period of the core representing >10,000–6000 radiocarbon year BP, which is perhaps related to the ways the three methods deal with samples close to or outside the extremes of the modern training-set. Generating a consensus reconstruction using non-parametric smoothers, such as LOESS (Cleveland 1979; see also Birks 2012a, b: Chaps. 2 and 11; Juggins and Telford 2012: Chap. 5), from the recon-

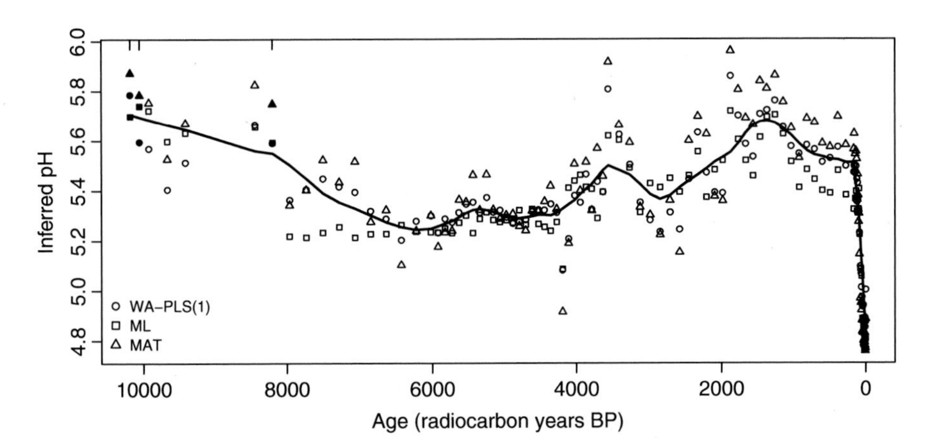

Fig. 19.3 Consensus reconstruction based upon the reconstructed pH values for the fossil samples in The Round Loch of Glenhead (RLGH3) core of all three reconstruction methods (∘ = one-component weighted averaging partial least squares model (WAPLS(1)), □ = maximum likelihood (ML) and Δ = modern analogue technique (MAT)). The consensus reconstruction has been generated using a LOESS smoother fitted to the inferred pH values as a function of sample age with a span of 0.1

structed values of several reconstructions is a useful way of highlighting the signal or pattern from noisy reconstructions (Birks 1998, 2003). However, where individual reconstructions are strongly divergent, the consensus reconstruction may take values that are not supported by any of the individual reconstructions and must be regarded with scepticism. Figure 19.3 shows a consensus reconstruction for the RLGH3 core, using a LOESS smoother to highlight the main patterns in the reconstruction.

Birks (1998) suggests four criteria for evaluating reconstructions: (i) sample-specific RMSEPs; (ii) 'goodness-of-fit' statistics from a constrained or canonical ordination; (iii) analogue measures between fossil and training-set samples; and (iv) the percentages of taxa in fossil samples (a) not represented or (b) poorly represented in the training-set (see also Juggins and Birks 2012: Chap. 14). All are equally important in assessing the potential reliability of the individual reconstructed values, but of the four criteria, the 'goodness-of-fit' statistics are rarely presented in research studies. We discuss this technique here in some detail. Goodness-of-fit is assessed by passively fitting the fossil samples into a constrained ordination of the training-set with the sole constraint being the environmental variable being reconstructed. The fossil samples are positioned as supplementary samples within the ordination space by means of transition formulae (equations 6.20 and 6.21 in ter Braak and Šmilauer 2002: p. 161) which determine a score for each fossil sample by taking an abundance weighted sum (redundancy analysis, RDA) or a weighted average (canonical correspondence analysis, CCA) of the species scores extracted from the ordination of the training-set samples (ter Braak and Šmilauer 2002). In this way, fossil samples are positioned within the ordination without influencing the underlying ordination based on modern data only. The squared residual length (SqRL) in the CANOCO solution file contains the squared residual distance between each sample point and its fitted position on the first constrained axis (entries in the AX1 column).

The distribution of the squared residual distances for the training-set samples is then calculated and the squared distances that equate to the 95th and 90th percentiles of that distribution are calculated. Any fossil sample that has a squared residual fit greater than the 95th percentile distance for the training-set samples is very poorly fitted and any that lie between the 90th and 95th percentiles are poorly fitted within the calibration-function model (Birks et al. 1990a). This is exemplified in Fig. 19.4 using the SWAP training-set and the RLGH3 core samples. The upper panel shows the distribution of the SqRL for samples in the SWAP 138-lake training-set. The vertical dotted lines correspond to the values of the 90th, 95th, and 99th percentiles of this distribution. Two samples (S21 and S261) in the SWAP training-set are extremely poorly fitted in the CCA constrained by pH, lying beyond the 99th percentile. A further seven samples are very poorly fitted and another eight samples are poorly fitted. The lower panel of Fig. 19.4 shows the distribution of the squared distances for the fossil samples in the RLGH3 core, showing that only three samples are not well fitted to pH in the modern CCA model; one sample is very poorly fitted (224.5 cm) with a further two samples poorly fitted (254.5, 256.5 cm). The analysis of the squared residual distance for the fossil samples to pH suggests that the majority of samples are well fitted to pH within the unimodal response-model framework and that we can therefore be confident of the pH reconstructed values (see also Telford and Birks 2011).

Eutrophication

Eutrophication of surface waters has a myriad of adverse ecological effects, including stimulation of excessive plant growth, depletion of oxygen in deep-water habitats, alteration of biogeochemical cycles and biological communities, and loss of biodiversity (Hutchinson 1973; Harper 1992; Scheffer et al. 1993; Smol 2008). For most aquatic systems, however, monitoring records are too short or too sparse to define pre-eutrophication conditions, detect trends, or provide sufficient information about changes in specific habitats (e.g., littoral or profundal habitats), as required to establish restoration targets, quantify the magnitude of the problem, or identify the causes of eutrophication and factors promoting (or delaying) ecosystem recovery. Remediation programmes have often employed reductions in nutrient loads as a method of controlling the effects of eutrophication, but they rarely assess whether such actions restore biotic communities or conditions in deep-water and littoral habitats (Dillon et al. 1978; Carvalho et al. 1995). For example, oxygen availability in deep-water habitats is rarely considered, although it plays a critical role in many invertebrate and commercially valuable fish communities and in regulating internal phosphorus loads (Nürnberg 1995).

Fortunately, quantitative palaeolimnological methods can provide a useful and rigorous scientific approach to address some of the above shortcomings. For example, a number of calibration functions are available that can quantify limnological

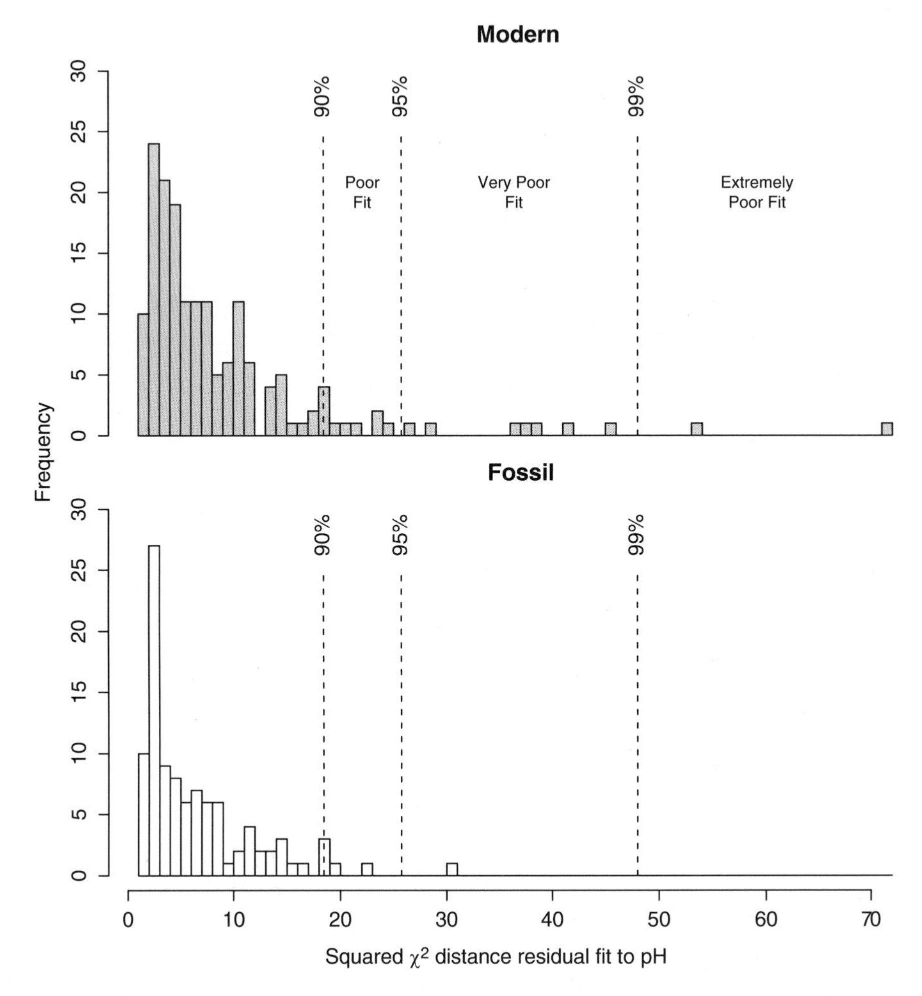

Fig. 19.4 Frequency of the squared residual fit to pH for the Surface Waters Acidification Programme (SWAP) diatom-pH training-set (*upper panel*) and the fossil samples from The Round Loch of Glenhead (RLGH3) core (*lower panel*) derived from passively overlaying the fossil samples on to a canonical correspondence analysis ordination of the SWAP training-set samples. The labelled dashed lines are for the 90th, 95th, and 99th percentiles of the distributions of the two sets of squared residual lengths. Samples lying beyond the 99th percentile are extremely poorly fitted to pH, those between the 95th and 99th percentiles are very poorly fitted, and those samples between the 90th and 95th percentiles are poorly fitted to pH

changes related to lake eutrophication (or trophic status). Most commonly, calibration functions have been developed to estimate epilimnetic total phosphorus concentration (TP) of surface waters (see Table 19.3) because phosphorus is the nutrient that often limits aquatic productivity (Sas 1989) and strongly regulates communities (Schindler 1971), although total nitrogen calibration functions also exist (Christie and Smol 1993; Siver 1999).

Table 19.3 Summary of many of the currently available calibration functions for reconstructing surface-water total phosphorus concentration based on diatom, chydorid, or chironomid remains in aquatic sediments

Range (μg L^{-1}) (mean)	Season	Model	r^2	$r_{jack/boot}^2$	RMSEP	Variable transformation	No. of sites	Location	Special	References
Diatoms										
30-550(71)	Annual	WA	n.a.	0.82	0.12	log(x)	43	S.E. China	Yangtze River floodplain	Yang et al. (2008)
0-675 (33)		*WAPLS(2)*	0.67	n.a.	n.a.		73	Ireland		Taylor et al. (2006)
6-49	Spring	WA	n.a.	0.57	0.20	log(x)	30	S.E. Ontario, Canada	Polymictic, alkaline	Werner & Smol (2005)
2-171 (14.4)	Annual	WA_{tol}	n.a.	0.50	0.20	log(x)	53	New Zealand		Reid (2005)
7-451 (64)	Annual	*WAPLS(2)*	0.94	0.74	0.23	log(x)+1	31	S.E. Australia	Reservoirs	Tibby (2004)
3–89 (38)	Autumn	WA	n.a.	0.76	0.16	log(x)	61	S. Finland		Kaupilla et al. (2002)
3–52 (10.3)	Summer	*WAPLS(3)*	0.92	0.51	3.2	none	75	Quebec, Canada	Oligotrophic	Philibert and Prairie (2002)
9–1,687 (96)	Annual	WA_{tol}	0.86	n.a.	0.44	ln(x)	69	N.E. Germany	Only littoral diatoms	Schönfelder et al. (2002)
24–1,145 (164)	Annual	*WAPLS(2)*	0.86	0.37	0.28	log(x)	29	Denmark	Shallow, eutrophic	Bradshaw et al. (2002)
24–1,145 (164)	Annual	WA_{tol}	0.62	0.23	0.32	log(x)	29	Denmark	Only planktonic diatoms; shallow, eutrophic	Bradshaw et al. (2002)

7–370	Annual	WA	0.75	0.47	0.24	log(x)	43	Sweden		Bradshaw and Anderson (2001)
4–54 (14.1)	Summer	WA	0.64	0.47	10	none	64	S.E. Ontario, Canada	Alkaline	Reavie and Smol (2001)
3–83	Summer	WA	0.77	0.52	0.23	log(x)	51	Alaska, USA	Single water samples	Gregory-Eaves et al. (1999)
0–8,740	Summer	WA	n.a.	0.55	0.79	ln(x)	238	N.E. USA		Dixit et al. (1999)
6–520 (73)	Spring	*WAPLS(2)*	0.93	0.79	0.19	log(x)	72	Switzerland		Lotter et al. (1998)
3–24 (7.6)	Spring over-turn	WA	0.62	0.41	4.2	none	54	Ontario, Canada	Oligotrophic	Hall and Smol (1996)
5–1,190 (104)	Annual	*WAPLS(2)*	0.91	0.82	0.21	log(x)	152	N.W. Europe	Small, shallow, eutrophic	Bennion et al. (1996)
6–42	Spring-autumn	WA	0.73	0.46	0.48	none	59	W. Canada	Single water samples	Reavie et al. (1995a)
2–266 (115)	Summer	WA_{tol}	0.57	0.31	0.35	log(x)	86	Alpine (Austria, Bavaria, Italy)	Oligo-mesotrophic	Wunsam and Schmidt (1995)
25–646	Annual	WA	0.79	n.a.	0.28	log(x)	30	S.E. England	Shallow, eutrophic	Bennion (1994)

(continued)

Table 19.3 (continued)

Range (μg L^{-1}) (mean)	Season	Model	r^2	$r_{jack/boot}^2$	RMSEP	Variable transformation	No. of sites	Location	Special	References
11–800		WA	0.75	n.a.	0.23	log(x)	49	Northern Ireland		Anderson et al. (1993)
1–51	July	WA	0.73	n.a.	0.41*	ln(x + 1)	41	Michigan, USA	Single water samples	Fritz et al. (1993)
5–28	Spring-autumn	WA	0.86	n.a.	0.25*	ln(x + 1)	37	W. Canada	Single water samples	Hall and Smol (1992)
Chydorids										
16–765	Summer mean	WA	n.a.	0.79	0.24	log(x)1	32	Denmark		Brodersen et al. (1998)
6–930 (11.5)	Spring	WA	0.62	0.49	0.28	log(x)	69	Switzerland		Lotter et al. (1998)
Chironomids										
5.3–1,162 (>200)	Annual mean	WA	n.a.	0.60	0.34	log(x)	44	English Midlands & Wales	Eutrophic	Brooks et al. (2001)
10–102 (28)	Spring	*WAPLS(2)*	0.84	0.68	0.14	log(x)	48	Switzerland		Lotter et al. (1998)

Asterisk indicates apparent RMSE (root mean squared error) values, estimated without bootstrapping or jackknifing (leave-one-out cross-validation) *RMSEP* root mean squared error of prediction, *WA* weighted averaging, *WAPLS(2)/(3)* = two/(three)-component weighted-averaging partial least squares model, *n.a.* not available

Calibration functions for reconstructing TP are developed using the same regional training-set approach as outlined above for pH (see also Juggins and Birks 2012: Chap. 14). Most commonly, TP calibration functions are based on diatoms, but TP calibration-functions have also been developed using chydorid and chironomid assemblages (Table 19.3). The majority of diatom-TP calibration functions are based on all taxa in sedimentary assemblages, including planktonic, epiphytic, and benthic taxa. However, a few studies (e.g., Siver 1999) have attempted to improve calibration-function performance statistics (r^2, RMSEP) by focussing on planktonic taxa only, based on the concept that errors increase because epiphytic and benthic taxa experience nutrient concentrations that differ from those of the epilimnion. But, in practice they tend to provide little or no improvement (Siver 1999; Bradshaw et al. 2002; Philibert and Prairie 2002), perhaps because of errors due to incorrect assignment of taxa to a habitat type and because many taxa may grow in more than one habitat.

Calibration functions also exist that can quantify changes in other important limnological conditions affected by eutrophication; including deep-water oxygen availability from chironomid remains (e.g., hypolimnetic anoxia) as the Anoxic Factor (Quinlan et al. 1998), end-of-summer volume-weighted hypolimnetic oxygen concentration (Quinlan and Smol 2001), and changes in the density of planktivorous fish from zooplankton microfossils (Jeppesen et al. 1996). As an example, Little et al. (2000) used quantitative multi-proxy palaeolimnological methods (diatom-TP, chironomid-anoxia factor calibration functions) to demonstrate that reduced nutrient loads and restoration of epilimnetic TP concentration to pre-disturbance values cannot be assumed to translate into recovery of benthic communities or greater deep-water oxygen availability over the multi-decadal time-scales of remediation programmes. Additionally, analysis of sedimentary pigments can quantify changes in the standing crops of major algal groups and photosynthetic bacteria (e.g., Hall et al. 1999).

Using a TP calibration-function developed for the Swiss Alps (Lotter et al. 1998), Lotter (1998) tracked the eutrophication of Baldeggersee during 109 years (1885–1993) using diatoms preserved in annually laminated sediments. This study provides an elegant example of applying a TP calibration function to quantify lake eutrophication and recovery because Baldergersee's varved sediments provide an annual record of the eutrophication trends and allowed Lotter to assess the reliability of the inferred diatom-TP by comparison with epilimnetic TP measurements recorded between 1957 and 1993 (see also Lotter and Anderson 2012: Chap. 18). The diatom-TP calibration function is based on a training-set of 78 lakes spanning a surface-water spring TP gradient of 5–520 μg L^{-1} (median = 31 μg TP L^{-1}) and a two-component WAPLS model. Performance statistics of this calibration function are stronger than other diatom-TP calibration functions currently available (jack-knifed $r^2 = 0.79$, RMSEP = 0.19 $\log_{10}$ μg TP L^{-1}; Table 19.3), and so this example most likely represents a 'best-case scenario' in TP reconstructions.

The diatom TP calibration-function estimated relatively consistent epilimnetic spring TP during the period AD 1885–1909, with a range of 25–50 μg L^{-1} (Fig. 19.5). Due to the long history of human activity at Balderggersee, epilimnetic

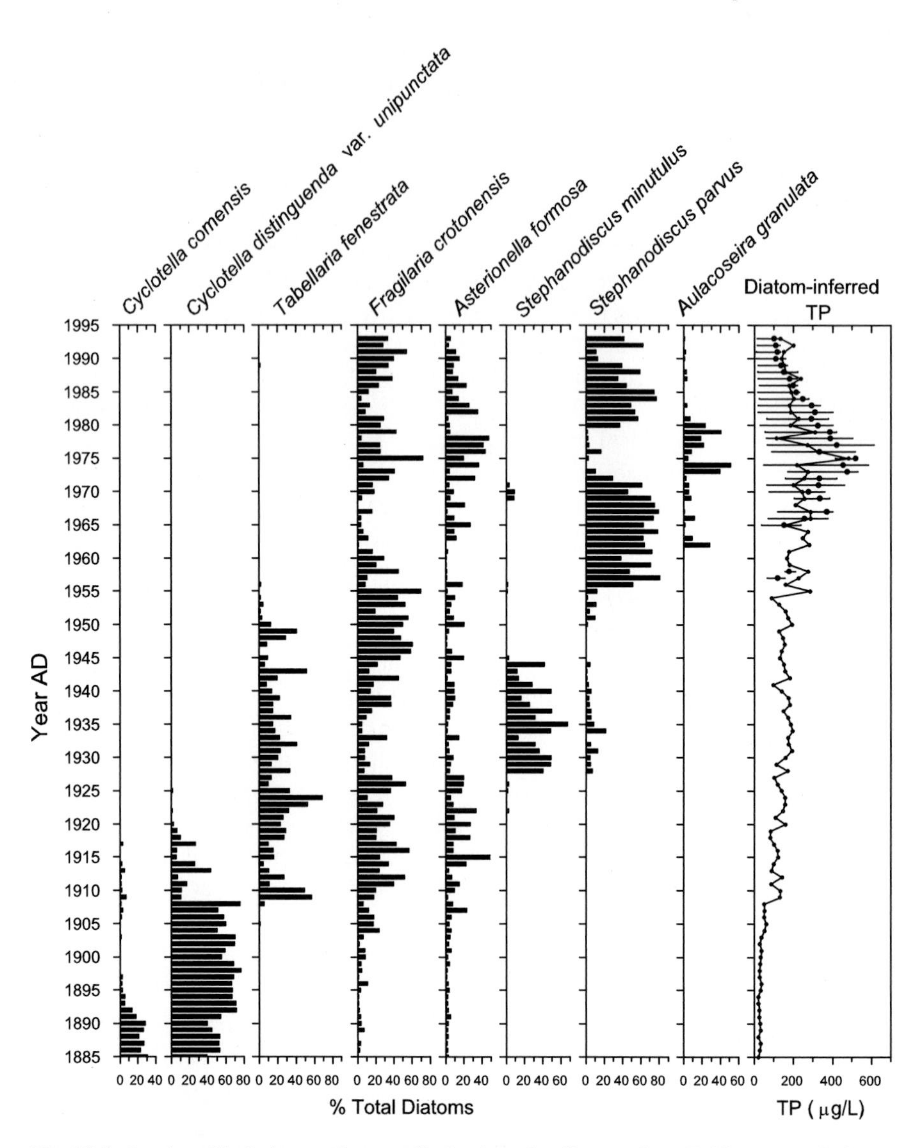

Fig. 19.5 Stratigraphical changes in annually-laminated sediments from Baldeggersee, Switzerland, between AD 1885 and 1993, obtained with freeze core BA93-C. Percent abundances of selected diatom taxa are shown. The right-hand panel shows diatom-inferred total phosphorus (*TP*) concentrations (*small solid dots* joined by a *solid line*), and compares this with measured spring circulation TP concentrations for the uppermost 15 m of the water column (*large solid dots*) along with the annual TP range measured in the uppermost 15 m (*horizontal lines*) (Modified from Lotter 1998)

TP of 25–50 μg L^{-1} is unlikely to represent the pre-impact, or pristine, state of the lake. But, it does provide a useful benchmark to guide lake-restoration programmes by indicating that nutrient-load reductions could achieve a marked improvement from the current hyper-eutrophic conditions to a meso-eutrophic state. During 1909–1910, diatom assemblages switched suddenly from those dominated by *Cyclotella comensis* Grun. and *C. distinguenda* var. *unipunctata* (Hust.) Håkansson & Carter, to domination by *Tabellaria fenestrata* (Lyngb.) Kütz. and *Fragilaria crotonensis* Kitton, and diatom-inferred TP (DI-TP) jumped to >100 $\mu g\ L^{-1}$, indicating a marked shift to hyper-eutrophy. Subsequent increases in *Asterionella formosa* Hassall and *Stephanodiscus minutulus* (Kütz.) Cleve & Euler during 1910–1956 resulted in a trend of increasing DI-TP from 100 to 200 $\mu g\ L^{-1}$. After 1956, DI-TP increased further to between 200 and 300 $\mu g\ L^{-1}$, and reached a maximum of nearly 500 $\mu g\ L^{-1}$ in 1977. The 1977 peak in DI-TP coincided with the highest epilimnetic TP ever measured in the lake, and subsequent declines in DI-TP closely track measured TP (~100 $\mu g\ L^{-1}$ in 1993).

Overall, the diatom-TP calibration function appears to have performed well in Balderggersee in that it demonstrated a transition of the lake from a meso-eutrophic to hyper-eutrophic state at 1909 due to increased nutrient loading from expanding agricultural activity and human sewage, as well as identifying the peak TP concentration in 1977 and subsequent water-quality improvements following the onset of a lake-restoration programme (Lotter 1998). Moreover, DI-TP corresponded reasonably closely with spring TP measurements since 1956, including rising TP until 1977, and subsequent declines (Fig. 19.5). However, the calibration function estimated greater inter-annual variability of TP than the measured spring values, probably due to the greater inherent variability of diatom assemblages compared to chemical conditions (Lotter 1998), and it consistently under-estimated measured spring TP during the period of highest values (Fig. 19.5). Under-estimation at high TP probably occurs, at least in part, because the modern calibration-set (with a median TP of 31 $\mu g\ L^{-1}$) is biased towards lakes with low and medium TP.

Several other studies have shown that diatom calibration functions tend to under-estimate measured TP at high values and over-estimate it under less productive conditions (e.g., Anderson et al. 1993; Anderson and Rippey 1994; Bennion 1994; Bennion et al. 1995). Several factors can impair the ability of organism-based calibration-functions to quantify surface-water nutrient concentrations under both highly productive and oligotrophic conditions. For example, calibration functions based on WA (and the related WAPLS) suffer from so-called 'edge-effects' where the optima of eutrophic or oligotrophic taxa are poorly estimated due to truncation of their response curves at the extreme ends of the gradient (see Oksanen et al. 1988; Birks et al. 1990a, 2010; Birks 1995, 1998; Juggins and Birks 2012: Chap. 14). Also, constant species composition among highly productive lakes (Anderson et al. 1993; Bennion 1994; Reavie et al. 1995a) or across broad TP gradients (Bennion and Appleby 1999; Bennion et al. 2001; Sayer 2001), and the absence of analogues in modern training-sets for fossil assemblages (Bennion et al. 1995, 1996; Reavie et al. 1995a, b), can weaken inferences based on relative percentage diatom compositional data. Fortunately, fossil assemblages which lack modern analogues or which consist of high proportions of taxa showing poor relationship to TP (i.e., poor fit to TP) can be identified using numerical methods (see the above section on Acidification; Birks 1995; Juggins and Birks 2012: Chap. 14). Furthermore, high intra-annual variability in surface-water TP may introduce errors into the estimates of taxon optima because the optima must be calculated from a single value that summarises the TP concentration of a lake (e.g., annual or seasonal averages of empirical TP measurements), rather than from the precise concentrations that existed during the period of growth of each taxon (Anderson and Odgaard 1994; Anderson and Rippey 1994; Bennion et al. 1995; Bennion and Smith 2000; Bradshaw et al. 2002). This problem is particularly acute in eutrophic lakes, because they exhibit the greatest annual variability of TP (Gibson et al. 1996).

Perhaps the greatest obstacle to accurate inferences of surface-water TP is the observation that biological communities respond to many limnological factors and that TP is only one of many factors (e.g., light, habitat availability, dissolved organic carbon, other nutrients (N, Si, etc.), food-web structure) which change during eutrophication. For example, comparison of published CCA biplots suggests that TP reconstructions based on diatoms can be confounded by changes in total nitrogen (Christie and Smol 1993; Fritz et al. 1993; Jones and Juggins 1995), lake depth (Dixit and Smol 1994; Hall and Smol 1992, 1996), pH (Fritz et al. 1993), and transparency (Fritz et al. 1993; Dixit and Smol 1994; Reavie et al. 1995a). Consequently, diatom-based reconstructions (and those based on other biota) are potentially affected by past changes in covariates that are weakly related or even unrelated to alterations in TP. However, despite the numerous sources of error and noise, several validation studies show there is often good agreement between DI-TP and measured records (Anderson and Rippey 1994; Anderson 1995; Bennion et al. 1995; Hall et al. 1997; Lotter 1998; Bradshaw and Anderson 2001) in most lakes. Calibration functions for reconstructing TP certainly are not as precise as for pH, but they are sufficient to distinguish between oligotrophic, mesotrophic, eutrophic, and hyper-eutrophic states and to identify restoration targets to a higher precision than remediation efforts can predict they will achieve. Thus, they provide a highly useful tool for aquatic resource managers and provide quality data that cannot be readily achieved by other means.

Detection of Long-Term Trends

Training-sets used in the generation of calibration functions can also be used in ordination-based approaches to detect trends and patterns of change. It is possible to add fossil samples passively to canonical correspondence analysis (CCA) or redundancy analysis (RDA) biplots so that changes in the assemblage composition in a sediment core can be related to the species-environment relationships described by CCA or RDA models created using only the modern training-set data (ter Braak and Šmilauer 2002; see also Legendre and Birks 2012b: Chap. 8).

Figure 19.6 shows a so-called time-track (trajectory) plot of the RLGH3 core samples projected passively into the ordination space of a CCA of the SWAP 138 data-set (Birks et al. 1990b; Stevenson et al. 1991). We have used the SWAP 138 data-set as this reduced training-set has a more complete set of hydrochemical

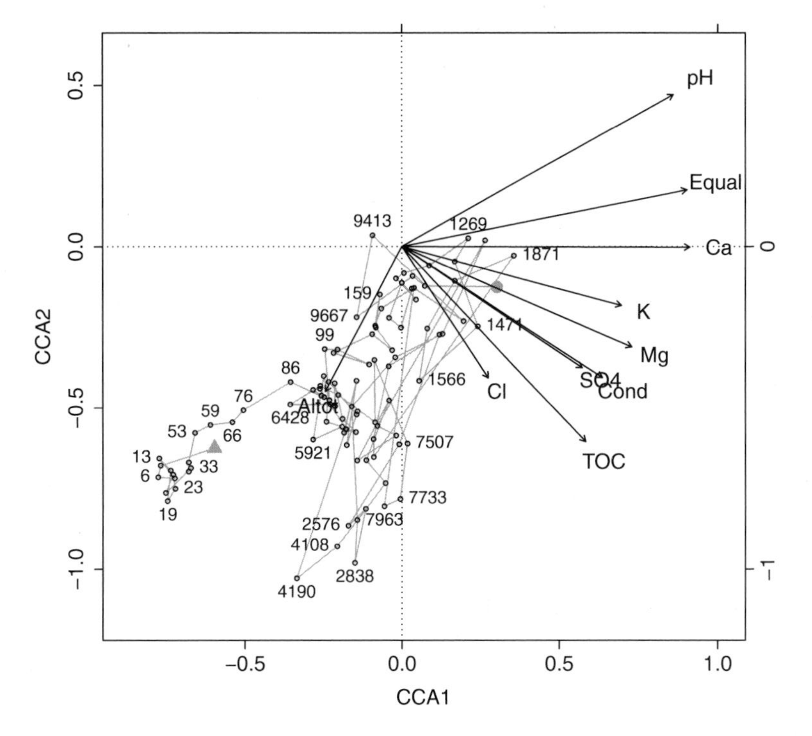

Fig. 19.6 Time-track plot of The Round Loch of Glenhead (RLGH3) core samples plotted passively into the ordination space of a canonical correspondence analysis of the Surface Waters Acidification Programme (SWAP) 138 diatom and water chemistry training-set (Stevenson et al. 1991). The start of the RLGH3 record (i.e., the oldest sediments) is indicated by the filled circle and the surface-sample of the core is indicated by the filled triangle. The labelled points are the interpolated ^{210}Pb or ^{14}C ages in years BP. *Equal* equivalent alkalinity, *Ca* calcium, *K* potassium, *Mg* magnesium, *SO_4* sulphate, *Cond* conductivity, *TOC* total organic carbon, *Cl* chloride, *Altot* total aluminium

parameters than the full SWAP training-set (Birks and Jones 2012: Chap. 3). The time-track plot illustrates the possible changes in hydrochemistry through time based on the assemblage composition of the fossil samples and the modern species scores obtained from the SWAP training-set samples. There is a lot of noise in the diagram, just as there is in the calibration-function reconstructions (Fig. 19.4), but the general patterns can be discerned. The time-track plot indicates that the RLGH acidified and total organic carbon (TOC) concentrations increased up to approximately 8000–7500 year BP, when TOC concentrations appear to fluctuate over the next 1500–2000 years. The RLGH has experienced a number of acidification events, illustrated in the time-track plot by the samples located at the bottom of the diagram at ~4000 and ~2800–2500 radiocarbon year BP. During these acidification periods, the diatom assemblage is somewhat different to that found previously in the loch, denoted by the sample points being located away from the cluster of samples on the plot. From ~150 year BP the strong anthropogenic acidification indicated by the calibration-function models can be seen in the time-track plot, with later samples following parallel, but in an opposite direction, to the projected vector for pH. It is also interesting to note that this anthropogenic acidification is associated with diatom assemblages that are very dissimilar to those seen previously in the history of the RLGH, including the previous acidification phases. The plot also suggests that the RLGH had higher TOC concentrations than those indicated for the recent acidification period.

Allott et al. (1992) made extensive use of time-track plots studying recovery from acidification at the RLGH. High-resolution diatom analyses on a number of cores from different parts of the RLGH basin were undertaken and the results indicated that the diatom assemblages had responded to an increase in pH of 0.2 units between AD 1978 and 1989, with time-track plots for some of the cores analysed showing clear signs of a reversal in the diatom assemblages in the recent sediment samples. However, the results also indicated that this floristic recovery was only apparent in those cores with accumulation rates greater than 0.7 mm $year^{-1}$ (Allott et al. 1992), a pattern that was effectively illustrated using CCA time-track plots.

Detecting trends and identifying change in sediment records can also be undertaken using rate-of-change analysis, in which the sediment record is amalgamated into groups of samples, each group encompassing the same time unit; 25 or 50 years for example (see also Birks 2012: Chap. 11). The amount of assemblage change between consecutive time units is quantified using a suitable dissimilarity coefficient, such as chi-square distance or squared chord (=Hellinger) distance (Overpeck et al. 1985; see Legendre and Birks 2012b: Chap. 8) between the fossil assemblages of each slice. Birks (1997) used this approach at the RLGH, and showed significant rates of change (as assessed by restricted permutation tests) at the onset of the Holocene (~9750 years ago) and unprecedented rates of assemblage change in the last 150 years, brought about as a result of anthropogenic acidification. The use of rate-of-change analysis convincingly demonstrated the difference in the rate of change associated with natural long-term acidification processes at the RLGH and that linked to anthropogenic acidification.

Identifying Patterns of Change

An important step in evaluating the effect of human disturbance on aquatic ecosystems is the process of scaling up from site-specific palaeolimnological evidence to regional assessments of change. An example of such an approach is that of Cumming et al. (1994), who used cluster analysis and canonical variates analysis (CVA) (=multiple discriminant analysis; ter Braak and Šmilauer 2002; Birks 2012a: Chap. 2) to determine when acid-sensitive lakes in the Adirondack Park (New York, USA) began to acidify.

Diatom-based pH reconstructions were performed on chysophyte profiles of ^{210}Pb-dated sediment cores from 20 low-alkalinity lakes in the Adirondack Park

(Cumming et al. 1994). Following this, 17 pH values per core were generated by interpolation and extrapolation of the inferred-pH profiles to a common set of dates (at 10-yearly intervals between 1850 and 1970, and at roughly 5-yearly intervals between 1970 and 1988). A cluster analysis was then performed on the 20 sites using the pH values in each of the 17 time-slices at each lake as variables in the analysis, which identified four types of pH profile: (i) lakes with no or little evidence of acidification; (ii) lakes with pre-industrial pH between 5 and 6 that began acidifying ca. 1900; (iii) 'naturally' acidic lakes that had acidified further since 1900; and (iv) lakes with pre-industrial pH around 6 that acidified ca. 1930–1950 (Cumming et al. 1994). CVA of the groupings of pH profiles with measured physical, chemical, and 1850- and present-day inferred-pH as explanatory variables showed that 1850- and present-day inferred-pH and lake elevation were the three variables that could be best discriminated between the four pH-profile groups. By overlaying the remaining measured variables passively onto the CVA biplot, Cumming et al. (1994) were able to show that lakes which had acidified by ~1900 were associated with higher precipitation and hence atmospheric deposition than lakes with little or no acidification.

Assessing Causes of Change

In many cases, palaeolimnological data provide the only long-term evidence available to determine the extent of human disturbance on aquatic ecosystems, and the timing and causes of these changes. Good experimental design is required for hypothesis-testing studies and needs to be coupled with appropriate data analysis to answer the questions at hand (Birks 1998, 2010).

The case of recent lake acidification stimulated a highly charged scientific and political debate both in northern Europe and North America over the cause of increased lake acidity and the loss or decline of fish stocks from sensitive lakes on a regional scale. A number of causes were proposed, including natural long-term acidification processes (Pennington 1984), land-use changes (Rosenqvist 1977, 1978), afforestation, and increased acid deposition (Odén 1968) resulting from the burning of fossil fuels. In Europe, the Surface Waters Acidification Programme (SWAP) palaeolimnology project (Renberg and Battarbee 1990) set about assessing these various hypotheses by carefully selecting sites to study and by designing suitably tailored experimental designs, whilst in North America, the Paleoecological Investigation of Recent Lake Acidification (PIRLA) I (Charles and Whitehead 1986) and PIRLA II (Charles and Smol 1990) projects attempted to tackle similar questions regarding the cause of the observed changes.

Quantitative reconstructions of pH using calibration functions can play a role in determining the causes of change as well as being used to demonstrate that change has taken place. At a number of sites where full Holocene sediment-records had been analysed for diatom remains, the results of quantitative pH reconstructions showed that lakes did acidify over long time scales as a result of the leaching of base cations from developing soils in the early Holocene, but never to the extremely low pH seen in many acidic lakes today (e.g., Renberg 1990: Renberg et al. 1993). Indeed, these longer term perspectives on acidification convincingly illustrated that the severe acidification affecting many acid-sensitive lakes in the Northern Hemisphere is a relatively recent phenomenon (see the earlier example for the RLGH; Fig. 19.4).

Kreiser et al. (1990) compared two moorland lakes with two lakes in recently (since the mid-1900s) afforested catchments. One pair of moorland/afforested sites was located in an area of low acid deposition and the other pair in an area receiving high levels of deposition. By carefully selecting their study sites and applying the SWAP diatom-pH calibration function to well-dated sediment cores from each of the lakes, Kreiser et al. (1990) showed that acidification began prior to afforestation, that non-afforested sites in areas of high acid deposition had acidified, and finally that, following afforestation at the high deposition site, acidification was intensified. The lack of historical data on emissions of acid-forming compounds from industrial sources meant that SWAP and PIRLA had to use robust experimental designs, evidence from multiple proxy records, and quantitative pH-reconstructions to answer questions concerning the causes of change.

Where historical data are available a wider suite of quantitative techniques can be employed than indicated in the examples above to attempt to address directly the causes of limnological changes. For example, with the availability of historical records for the period 1920–1994, Hall et al. (1999) were able to use variation partitioning analysis (VPA) on palaeolimnological indicators preserved in sediments of eutrophic lakes of the Qu'Appelle Valley (Saskatchewan, Canada) to determine the relative importance of climatic variability, resource-use, and urbanisation as controls of aquatic communities. VPA is a numerical technique based on constrained and partial canonical ordinations that was first developed by Borcard et al. (1992), and provides an effective method to estimate the fraction of variation in assemblage composition that can be explained by categories of measured variables (see Legendre and Birks 2012b: Chap. 8).

Prior to using VPA, Hall et al. (1999) identified three distinct biological assemblages since ca. 1775 from sedimentary analyses of diatoms, pigments, and chironomids at Pasqua Lake (Saskatchewan); the first lake in a series of six lakes situated along the Qu'Appelle River and the first to receive nutrients in sewage from the main urban centres of Regina and Moose Jaw. Before the onset of agriculture ~1890, the lake was naturally eutrophic with abundant cyanobacterial carotenoids, diatoms indicative of productive waters, and anoxia-tolerant chironomids. Distinct assemblages formed ~1930–1960 that were characterised by elevated algal biomass (inferred as β-carotene), nuisance cyanobacteria, eutrophic *Stephanodiscus hantzschii* Grun., and a low abundance of deep-water zoobenthos. Sedimentary assemblages deposited after ~1977 were variable and indicated that water quality had not improved despite a three-fold reduction in P loading due to tertiary sewage treatment.

Hall et al. (1999) used constrained and partial canonical ordinations (ter Braak 1986) for a three-category VPA to determine the relative importance of climate

Table 19.4 Results of variation partitioning analysis of diatom, pigment, and chironomid assemblages from sediments deposited 1920–1993 in Pasqua Lake, Saskatchewan, Canada

	Diatoms (%)	Pigments (%)	Chironomids (%)
Climate effects, independent of resource-use and urban factors (C)	10.4	4.2	19.2
Resource-use effects, independent of urban and climatic factors (R)	24.7	13.9	18.7
Urban effects, independent of resource-use and climatic factors (U)	9.4	10.8	18
Covariation between effects of climatic and resource-use factors (independent of urban; CR)	22.1	3.5	5.2
Covariation between effects of climatic and urban factors (independent of resource-use; CU)	6.8	8.3	0
Covariation between effects of resource-use and urban effects (independent of climate; RU)	4.1	13.6	0
Covariation among effects of climatic, resource-use and urban factors (CRU)	13.2	33.1	16.6
Unexplained variation	9.3	12.6	22.3

Modified from Hall et al. (1999). With permission from the American Society of Limnology and Oceanography

(C), resource-use (R), and urban activity (U) on fossil assemblages in Pasqua Lake deposited between AD 1920 and 1997, the period for which reliable continuous historical records of 83 potential explanatory variables were available (Table 19.4). Redundancy analysis was used to partition the variation in the fossil assemblages at Pasqua Lake because exploratory DCA suggested that fossil assemblages varied along environmental gradients in a linear rather than unimodal fashion (ter Braak 1986; ter Braak and Prentice 1988). RDA was performed with percent abundance of diatom and chironomid taxa or pigment concentrations after smoothing with an unweighted three-point running mean. Because the research interests were to investigate controls of long-term biological changes rather than inter-annual variability, historical data were also smoothed using a 3-year running mean.

Variation partitioning requires similar numbers of variables within each explanatory category to avoid bias in the analyses (Borcard et al. 1992). Consequently, Hall et al. (1999) developed and used objective, *a priori* criteria to ensure similar numbers of statistically significant variables per category. Only explanatory variables accounting for significant amounts of variation in the fossil data ($\alpha = 0.05$) were selected for inclusion in the VPA computations, based on a series of RDAs constrained to a single explanatory variable at a time (ter Braak and Šmilauer 2002). Significant variables were then assigned to one of the explanatory categories (C, R, or U). Finally, a series of RDAs were performed on each category, sequentially eliminating the explanatory variable with the highest variance inflation factor (VIF) until all VIFs were less than 20. This step reduced multicollinearity among variables within each category (ter Braak and Šmilauer 2002) and resulted in similar numbers of variables per category.

Five steps were required to partition variation in the fossil data among the C, R, and U categories. First, canonical ordination with no covariables was used to measure the total amount of variation (as the sum of canonical eigenvalues) in the fossil assemblages attributable to all explanatory variables (C + R + U + CR + CU + RU + CRU) and the total unexplained variation (Total variation – [C + R + U + CR + CU + RU + CRU]). Second, a series of partial canonical ordinations was used to calculate the variation explained by the unique effects of each category (C, R, or U). In this step, ordinations of individual explanatory categories were run with the remaining two categories as covariables. Third, a series of partial canonical ordinations were used to calculate the unique effects plus covariation between pairs of explanatory categories (C + CR, C + CU, R + CR, R + RU, U + CU, U + RU). In each analysis, one category was used as explanatory variables with one of the remaining categories acting as covariables. The third category was not included in the analysis. Fourth, 'first-order' covariation terms (CR, RU, CU) were calculated by subtracting appropriate terms generated during steps 2 and 3 (e.g., CU = [C + CU]–C). Finally, the variation explained by covariation among all three categories (CRU) was calculated as the difference between 100% and the sum of variation captured in the first, second, and fourth steps (CRU = 100–C–R–U–CR–CU–RU–unexplained).

At Pasqua Lake, VPA captured 78–91% of the variation in fossil assemblage composition using only 11–13 significant environmental variables. Resource-use (crop-land area, livestock biomass) and urbanisation (nitrogen in sewage) were stronger determinants of algal and chironomid assemblage change than were climatic factors (temperature, evaporation, river discharge) (Table 19.4). Covariation among resource-use and urban activities (R, U, RU), independent of climate, accounted for 27–38% of the total variation in the fossil assemblages since 1920. In particular, analysis of algal assemblage change since 1920 demonstrated that the long-term influence of resource-use on algae was mediated mainly through changes in terrestrial practices involving livestock or crops (Hall et al. 1999). While the effects of climatic variables on biotic assemblages were also important, climate impacts were mediated by human activity, as demonstrated by the high proportion of variation attributable to covariation with resource-use and urban activities (i.e., CU, CR, CRU), rather than to unique effects (C):

The use of VPA and information contained in century-long sediment cores allowed Hall et al. (1999) to formulate specific recommendations for prairie-lake managers.

1. Fossil analyses demonstrated that Pasqua Lake was naturally eutrophic, and so should not be managed for low productivity.
2. Despite high baseline production, recent water quality in Pasqua Lake was considerably worse than before European settlement, indicating that water-quality improvements are possible.

3. Management strategies should further investigate the role of sewage inputs, agriculture, and reservoir hydrology on water quality, because variables reflecting nutrient export from Regina's sewage (Regina population, TP and TN fluxes, TN:TP), crop-land area, livestock biomass, and discharge volume from the Lake Diefenbaker reservoir consistently accounted for significant variation in the fossil pigments, diatoms, and chironomids.
4. Nutrient abatement programmes should reduce N inputs to Qu'Appelle lakes because palaeolimnological data showed that inferred algal abundance and water quality had not substantially improved in Pasqua Lake since 1977 despite tertiary sewage treatment which reduced P loading to levels of the 1930s.

VPA provides a single measure for the amount of variation in species composition explained by groups of explanatory variables. It cannot indicate where in the sediment sequence (time) this effect is observed. Passively placing core samples within an ordination of a training-set can address this issue to some extent, although the inferred effect of one or more variables is global and generally linear. Simpson and Anderson (2009) address these issues by modelling the relationship between species composition and one or more sedimentary proxies of disturbance using additive regression techniques enhanced to account for potential autocorrelation in the stratigraphical observations. Using data from Loch Coire Fionnaraich, north-west Scotland (Pla et al. 2009), additive models incorporating smooth functions of temperature and atmospheric deposition were fitted to the first principal component of the Hellinger transformed diatom data to investigate the relative roles of climate (temperature) and atmospheric deposition as drivers of change in diatom composition over the past ~200 years (Simpson and Anderson 2009). Using this approach, Simpson and Anderson (2009) were able to demonstrate (i) that atmospheric deposition, not climatic variability, was the major driver of diatom compositional change in Loch Coire Fionnaraich; and (ii) that temperature was able to explain some of the inter-sample variability about the trend due to atmospheric deposition. Their approach is computationally and data demanding but it provides a sophisticated means of modelling the effects of explanatory variables through time on species composition, and can address issues of non-linear and relative effects and the timing of effects.

Quantitative Palaeolimnology and Lake Restoration

The restoration of acidified and eutrophic lakes is another area where palaeolimnology is having an important influence on the management of freshwater systems such as in determining targets for recovery and setting critical loads for pollutants. For example, Battarbee et al. (1996) used palaeolimnological records from 41 UK lakes to determine whether they had acidified or not. They then used the ratio of calcium to sulphur (S) deposition to discriminate between the acidified and non-acidified sites using logistic regression (Birks 2012a: Chap. 2). A ratio of 94:1 was identified as the optimal ratio that discriminated between the acidified and non-acidified sites. They were then able to use this critical ratio to set critical loads for S, which have been used to assess the potential benefits of emission reduction policies such as the Oslo and Gothenburg protocols.

Flower et al. (1997) and Simpson et al. (2005) used an approach similar to the MAT (see Simpson 2012: Chap. 15) to determine recovery targets for acidified lakes. Using diatom (Flower et al. 1997) and diatom and cladoceran (Simpson et al. 2005) remains, respectively, they identified sites from within modern training-sets that were biologically similar to pre-acidification sediment samples from a number of acidified lakes in the UK using dissimilarity coefficients such as the squared chi-square or chord (=Hellinger) distances (Overpeck et al. 1985; see Legendre and Birks 2012b: Chap. 8). The identified modern analogues can be used as reference conditions or recovery targets for the acidified lakes because the modern analogues can be surveyed to gather information on species groups that do not leave reliable assemblages in lake sediments. Simpson et al. (2005) applied the technique to identify modern analogues for ten lakes that form part of the United Kingdom Acid Waters Monitoring Network (UKAWMN; Moneteith and Evans 2000) using a training-set of diatom and cladoceran remains enumerated from the surface-sediment samples of 83 upland, acid-sensitive lakes in the United Kingdom. Close modern analogues were identified for eight of the ten UKAWMN lakes (Simpson et al. 2005). These analogue sites were then assessed in terms of their hydrochemistry, aquatic macrophyte flora, and macro-invertebrate fauna as to their suitability for use in defining wider hydrochemical and biological reference conditions for the studied UKAWMN lakes. The modern analogues identified for individual UKAWMN lakes showed a close degree of similarity in terms of their hydrochemical characteristics and aquatic macrophyte flora, and, to a lesser extent, in the macro-invertebrate fauna. The results of the surveys of the modern-analogue lakes indicated that the reference conditions of the acidified UKAWMN lakes are inferred to be less acidic than today, and to support a wider range of acid-sensitive aquatic macrophyte and macro-invertebrate taxa than were recorded in the UKWAMN lakes during monitoring since 1988. At two UKAWMN sites where biological recovery from acidification is in progress, acid-sensitive species, which were predicted from the modern analogues to have been present in the lakes before acidification, have been recorded in the most recent monitoring surveys (Battarbee 2010; Kernan et al. 2010), further validating the technique (Simpson et al. 2005).

An alternative palaeolimnological approach for defining and assessing reference conditions was presented by Bennion et al. (2004) where DCA and chord distances were applied to dated sediment cores from 26 Scottish freshwater lochs to establish the amount of diatom assemblage change at each site, demonstrating significant nutrient enrichment in 18 of the lochs. TWINSPAN (see Legendre and Birks 2012a: Chap. 7) was then used to classify the pre-disturbance diatom assemblages of the 26 lochs to characterise the reference-condition diatom assemblages of the different lake types. Diatom-TP calibration functions were applied to each loch to determine the reference-condition TP concentrations. In this way ecological and chemical reference conditions, as required by recent legislation, such as the US Clean Water Act (Barbour et al. 2000) and the European Union Water Framework Directive (European Union 2000), can be defined using biological communities for a range of lake types. The use of quantitative palaeolimnological approaches in defining reference conditions and assessing the ecological status of lakes is presented in detail by Battarbee and Bennion (2011), Battarbee et al. (2011a, b), Bennion and Simpson (2011), and Bennion et al. (2011a, b).

A different approach was developed by Rippey and Anderson (1996) and Jordan et al. (2001) that combines information from diatom-TP calibration functions, P sedimentation rates, and lake-flushing rates to quantify past changes in TP loading. This approach can be used to quantify changes over time in TP loads from diffuse sources such as agriculture, which are notoriously difficult to measure using alternate methods (Jordan et al. 2001). Also, it can be used to quantify how changes in external TP loads are variably directed as fluxes to the sediments, the outflow, or lake storage (Rippey and Anderson 1996). Moreover, the approach can identify when fertiliser applications began to exceed threshold soil P concentrations at which point soluble P in runoff increases, and thus can substantially improve our understanding of mechanisms promoting lake eutrophication in rural catchments (Jordan et al. 2001). The approach is powerful because few lakes in the world have P loading data for more than a decade or two (Rippey 1995) and it can provide a rigorous method to establish critical P loading limits above which eutrophication exceeded acceptable standards in individual lakes.

To reconstruct P loads, the above approach involves the application of lake-chemistry mass-balance assumptions. TP inputs are assumed to be the sum of TP losses via the outflow and TP losses to the sediments, and are quantified by rearranging the Vollenweider (1975) steady-state lake P model with an extra term to account for periods of non-steady state changes in lake storage (Jordan et al. 2001):

$$\mathrm{L}_i = (\mathrm{TP}_i \mathrm{z}\rho_i + \mathrm{TP}_i \mathrm{z}\sigma_i) + [(\mathrm{TP}_i \mathrm{z} - \mathrm{TP}_{i-1}\mathrm{z})\,/(i - (i-1))] \tag{19.1}$$

where, L_i is the external TP loading on the lake (g m^{-2} year^{-1}) in a time interval i, $\mathrm{TP}_i\mathrm{z}\rho_i$ is the TP loss through the outflow (g m^{-2} year^{-1}), $\mathrm{TP}_i\mathrm{z}\sigma_i$ is the accumulation of TP in the sediments (g m^{-2} year^{-1}), and $[(\mathrm{TP}_i\mathrm{z}{-}\mathrm{TP}_{i-1}\mathrm{z})/(i{-}(i{-}1)]$ is a term to account for non-steady state periods of changing lake TP storage between adjacent time intervals, expressed relative to the lake surface. Past TP outflow losses and periods of changing TP storage require an estimate of lake-water TP concentration, which can be provided by application of diatom-TP calibration functions to fossil diatom assemblages (or TP calibration functions based on microfossils of other biota). Mean depth (z, in metres) is estimated from basin morphometry. Changes in flushing rates (ρ, year^{-1}) can be estimated from precipitation records, and the whole-basin mean sedimentary TP accumulation rate ($\mathrm{TP}_i\mathrm{z}\sigma_i$) can be determined from dated multiple sediment cores analysed for TP concentration.

The above approach was used effectively by Rippey and Anderson (1996) to reconstruct phosphorus loading to a small lake in Northern Ireland (Augher Lough)

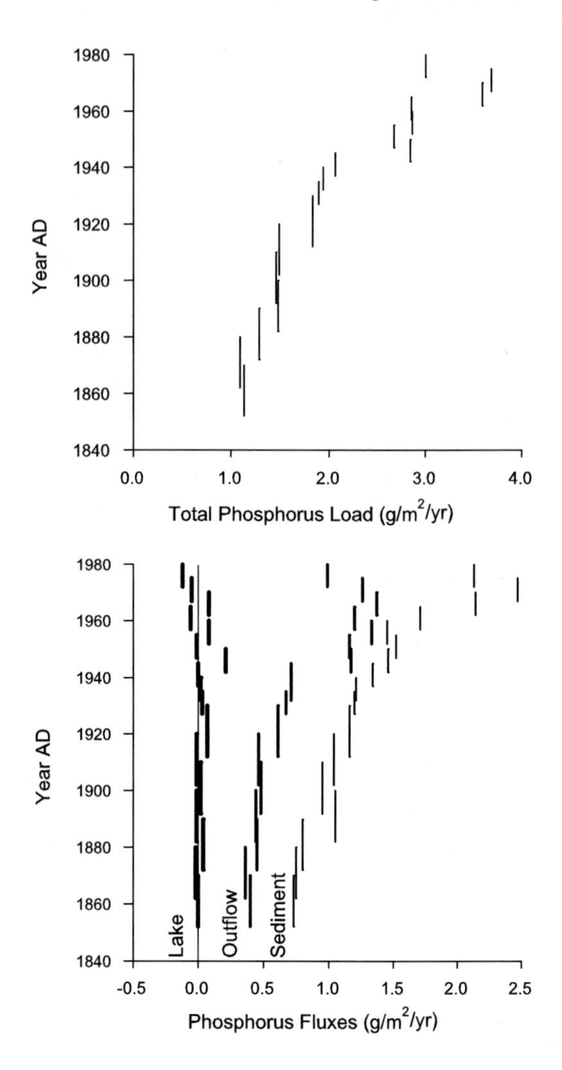

Fig. 19.7 Total phosphorus loading to Augher Lough from 1850 to 1980 reconstructed from the sedimentary record (*upper panel*). The phosphorus fluxes via the outflow, into the sediments and changes due to lake storage are shown in the *lower panel* (Modified from Rippey and Anderson 1996. With permission from the Chemical Society of America)

between ~1850 and 1980 (Fig. 19.7, upper panel). External TP loading rose steadily ~1850–1945, with more rapid increases for most of the period after ~1945. Interestingly, the approach was able to identify an important change in lake P dynamics during the period ~1945–1950, characterised by elevated internal P loading to the water column, due to anoxia at the sediment-water interface, which caused non-steady-state water column P storage (Fig. 19.7, lower panel). The approach was extended by Rippey et al. (1997) to elucidate complex interactions that can occur among climatic variability (drought) and human stressors in eutrophic lakes.

Conclusions

The development of palaeoecological quantitative techniques has seeded a revolution in the field, one that has allowed increasingly complex ecological problems and a wider range of human impacts to be addressed in a rigorous manner. The calibration-function approach in particular has had a profound impact on our ability to identify the nature of change and to reconstruct reference conditions suitable for restoration targets. The importance of calibration functions for palaeoecology is hard to overestimate.

In the 25 or so years since the development of modern calibration-function techniques the science has seen a maturation of these methods. As this maturation has occurred the standard techniques have become commonplace in research publications, which is, at least in part, due to the ready availability of powerful yet user-friendly computer software implementations. Much applied palaeolimnological work related to lake acidification and eutrophication is focussed on the development or application of calibration functions to reconstruct changes in pH or TP from a wide array of proxies. Smol's "Rosetta Stone" (Smol 2008), which promised much and most certainly delivered, could be in danger of becoming a victim of its own success. Those of us in this field must guard against complacency and not allow the standard techniques to become a hindrance to the pursuit of novel research questions and innovation (Birks 2012c: Chap. 21).

Palaeoecologists of late have striven to reconstruct the environment from the biology, yet all too often the counterpoise question, of how has the biology responded to the environment, remains unasked. Above, we have discussed the use of variation partitioning and careful study design as two approaches towards addressing the causes of change, and we briefly mentioned how modern regression methods could also be usefully employed in such situations. We hope that wider usage of these types of approaches will become part of the standard toolbox of techniques in the palaeoecologist's arsenal (see Birks 2012c: Chap. 21). Placing a greater emphasis on the biology will allow us to ask questions of palaeoecological data that tackle key questions in the fields of ecological restoration, aquatic biodiversity, and ecological theory, amongst others. Just as the environmental problems posed over 25 years ago required the development and application of new numerical techniques to answer them, addressing these new questions will require us to adopt novel data analysis techniques and improved study designs.

Acknowledgements We thank André Lotter for providing data from Baldeggersee, Viv Jones for her RLGH diatom data and for assisting in assembling the chronology for the RLGH3 core, Brian Rippey and John Anderson for their data from Lough Augher, and John Smol for useful comments. Tammy Karst-Riddoch prepared Figs. 19.5 and 19.7.

References

Allott TEH, Harriman R, Battarbee RW (1992) Reversibility of lake acidification at The Round Loch of Glenhead, Galloway, Scotland. Environ Pollut 77:219–225

Anderson NJ (1995) Using the past to predict the future: lake sediments and the modelling of limnological disturbance. Ecol Model 78:149–172

Anderson NJ, Odgaard BV (1994) Recent palaeoecology of three shallow Danish lakes. Hydrobiologia 275(276):411–422

Anderson NJ, Rippey B (1994) Monitoring lake recovery from point-source eutrophication: the use of diatom-inferred epilimnetic total phosphorus and sediment chemistry. Freshw Biol 32: 625–639

Anderson NJ, Rippey B, Gibson CE (1993) A comparison of sedimentary and diatom-inferred phosphorus profiles: implications for defining pre-disturbance nutrient conditions. Hydrobiologia 253:357–366

Barbour MT, Swietlik WF, Jackson SK, Courtemanch DL, Davies SP, Yoder CO (2000) Measuring the attainment of biological integrity in the USA: a critical element of ecosystem integrity. Hydrobiologia 422(423):453–464

Battarbee RW (1994) Palaeolimnology. In: Maitland PS, Boon PJ, McLusky DS (eds) The fresh waters of Scotland: a national resource of international significance. Wiley, Chichester, pp 113–130

Battarbee RW (1999) The importance of palaeolimnology to lake restoration. Hydrobiologia 395/396:149–159

Battarbee RW (2010) Are our acidified upland waters recovering? Freshw Biol Assoc News 52:5–6

Battarbee RW, Bennion H (2011) Palaeolimnology and its developing role in assessing the history and extent of human impact on lake ecosystems. J Paleolimnol 45:399–404

Battarbee RW, Stevenson AC, Rippey B, Fletcher C, Natkanski J, Wik M, Flower RJ (1989) Causes of lake acidification in Galloway, south-west Scotland: a palaeoecological evaluation of the relative roles of atmospheric contamination and catchment change for two acidified sites with non-afforested catchments. J Ecol 77:651–672

Battarbee RW, Allott TEH, Juggins S, Kreiser AM, Curtis C, Harriman R (1996) Critical loads of acidity to surface waters: an empirical diatom-based palaeolimnological model. Ambio 25: 366–369

Battarbee RW, Morley D, Bennion H, Simpson GL, Hughes M, Bauere V (2011a) A palaeolimnological meta-database for assessing the ecological status of lakes. J Paleolimnol 45:405–414

Battarbee RW, Simpson GL, Bennin H, Curtis C (2011b) A reference typology of low alkalinity lakes in the UK based on pre-acidification diatom assemblages from lake sediment cores. J Paleolimnol 45:489–505

Beamish J, Harvey HH (1972) Acidification of LaCloche Mountain lakes, Ontario, and resulting fish mortalities. J Fish Res Board Can 29:1331–1143

Bennion H (1994) A diatom-phosphorus transfer-function for shallow, eutrophic ponds in southeast England. Hydrobiologia 275(276):391–410

Bennion H, Appleby P (1999) An assessment of recent environmental change in Llangorse Lake using palaeolimnology. Aquat Conserv Mar Freshw Ecosyst 9:361–375

Bennion H, Simpson GL (2011) The use of diatom records to establish reference conditions for UK lakes subject to eutrophication. J Paleolimnol 45:469–488

Bennion H, Smith MA (2000) Variability in the water chemistry of shallow ponds in southeast England, with special reference to the seasonality of nutrients and implications for modelling trophic status. Hydrobiologia 436:145–158

Bennion H, Wunsam S, Schmidt R (1995) The validation of diatom-phosphorus transfer functions: an example from Mondsee, Austria. Freshw Biol 34:271–83

Bennion H, Juggins S, Anderson NJ (1996) Predicting epilimnetic phosphorus concentrations using an improved diatom-based transfer function and its application to lake management. Environ Sci Technol 30:2004–2007

Bennion H, Appleby P, Phillips GL (2001) Reconstructing nutrient histories in the Norfolk Broads, UK: implications for the role of diatom-total phosphorus transfer functions in shallow lake management. J Paleolimnol 26:181–204

Bennion H, Fluin J, Simpson GL (2004) Assessing eutrophication and reference conditions for Scottish freshwater lochs using subfossil diatoms. J Appl Ecol 41:124–138

Bennion H, Battarbee RW, Sayer CD, Simpson GL, Davidson TA (2011a) Defining reference conditions and restoration targets for lake ecosystems using palaeolimnology: a synthesis. J Paleolimnol 45:533–544

Bennion H, Simpson GL, Anderson NJ, Clarke G, Dong X, Hobæk A, Guilizzoni P, Marchetto A, Sayer CD, Thies H, Tolotti M (2011b) Defining ecological and chemical reference conditions and restoration targets for nine European lakes. J Paleolimnol 45:415–431

Birks HJB (1995) Quantitative palaeoenvironmental reconstructions. In: Maddy D, Brew JS (eds) Statistical modelling of Quaternary science data, Technical guide 5. Quaternary Research Association, Cambridge, pp 161–254

Birks HJB (1997) Environmental change in Britain – a long-term palaeoecological perspective. In: Mackay AW, Murlis J (eds) Britain's natural environment: a state of the nation review. ENSIS Ltd, London, pp 23–28

Birks HJB (1998) Numerical tools in palaeolimnology – progress, potentialities, and problems. J Paleolimnol 20:307–332

Birks HJB (2003) Quantitative palaeoevironmental reconstructions from Holocene biological data. In: Mackay AW, Battarbee RW, Birks HJB, Oldfield F (eds) Global change in the Holocene. Arnold, London, pp 107–123

Birks HJB (2010) Numerical methods for the analysis of diatom assemblage data. In: Smol JP, Stoermer EF (eds) The diatoms – applications for the environmental and earth sciences. Cambridge University Press, Cambridge, pp 23–54

Birks HJB (2012a) Chapter 2: Overview of numerical methods in palaeolimnology. In: Birks HJB, Lotter AF, Juggins S, Smol JP (eds) Tracking environmental change using lake sediments. Volume 5: Data handling and numerical techniques. Springer, Dordrecht

Birks HJB (2012b) Chapter 11: Analysis of stratigraphical data. In: Birks HJB, Lotter AF, Juggins S, Smol JP (eds) Tracking environmental change using lake sediments. Volume 5: Data handling and numerical techniques. Springer, Dordrecht

Birks HJB (2012c) Chapter 21: Conclusions and future challenges. In: Birks HJB, Lotter AF, Juggins S, Smol JP (eds) Tracking environmental change using lake sediments. Volume 5: Data handling and numerical techniques. Springer, Dordrecht

Birks HJB, Jones VJ (2012) Chapter 3: Data-sets. In: Birks HJB, Lotter AF, Juggins S, Smol JP (eds) Tracking environmental change using lake sediments. Volume 5: Data handling and numerical techniques. Springer, Dordrecht

Birks HJB, Line JM, Juggins S, Stevenson AC, ter Braak CJF (1990a) Diatoms and pH reconstructions. Philos Trans R Soc Lond B 327:263–278

Birks HJB, Juggins S, Line JM (1990b) Lake surface-water chemistry reconstructions from palaeolimnological data. In: Mason BJ (ed) The Surface Waters Acidification Programme. Cambridge University Press, Cambridge, pp 301–313

Birks HJB, Heiri O, Seppä H, Bjune AE (2010) Strengths and weaknesses of quantitative climate reconstructions based on late-Quaternary biological proxies. Open Ecol J 3:68–110

Borcard D, Legendre P, Drapeau P (1992) Partialling out the spatial component of ecological variation. Ecology 73:1045–1055

Bradshaw EG, Anderson NJ (2001) Validation of a diatom-phosphorus calibration-set for Sweden. Freshw Biol 46:1035–1048

Bradshaw EG, Anderson NJ, Jensen JP, Jeppesen E (2002) Phosphorus dynamics in Danish lakes and the implications for diatom ecology and palaeoecology. Freshw Biol 47:1963–1975

Brodersen KP, Whiteside MC, Lindegaard C (1998) Reconstruction of trophic state in Danish lakes using subfossil chydorid (Cladocera) assemblages. Can J Fish Aquat Sci 55:1093–1103

Brooks SJ, Bennion H, Birks HJB (2001) Tracing lake trophic history with a chironomid-total phosphorus inference model. Freshw Biol 46:513–533

Carvalho L, Bekelioglu M, Moss B (1995) Changes in a deep lake following sewage diversion - a challenge to the orthodoxy of external phosphorus control as a restoration strategy? Freshw Biol 34:399–410
Charles DF, Smol JP (1990) The PRILA II project: regional assessment of lake acidification trends. Verh Internat Verein Limnol 24:474–480
Charles DF, Whitehead DR (1986) The PIRLA project: paleoecological investigations of recent lake acidification. Hydrobiologia 143:13–20
Christie CE, Smol JP (1993) Diatom assemblages as indicators of lake trophic status in southeastern Ontario lakes. J Phycol 29:575–586
Cleveland WS (1979) Robust locally weighted regression and smoothing scatterplots. J Am Stat Assoc 74:829–836
Cumming BF, Davey KA, Smol JP, Birks HJB (1994) When did acid-sensitive Adirondack lakes (New York, USA) begin to acidify and are they still acidifying? Can J Fish Aquat Sci 51: 1550–1568
Cumming BF, Laird KR, Fritz SC, Verschuren D (2012) Chapter 20: Tracking Holocene climatic change with aquatic biota from lake sediments: case studies of commonly used numerical techniques. In: Birks HJB, Lotter AF, Juggins S, Smol JP (eds) Tracking environmental change using lake sediments. Volume 5: Data handling and numerical techniques. Springer, Dordrecht
Dillon PJ, Nicholls KH, Robinson GW (1978) Phosphorus removal at Gravenhurst Bay, Ontario: an 8 year study on water quality changes. Verh Internat Verein Limnol 20:263–271
Dixit SS, Smol JP (1994) Diatoms as indicators in the Environmental Monitoring and Assessment Program – Surface Waters (EMAP-SW). Environ Monit Assess 31:275–306
Dixit SS, Smol JP, Charles DF, Hughes RM, Paulsen SG, Collins GB (1999) Assessing water quality changes in the lakes of the northeastern United States using sediment diatoms. Can J Fish Aquat Sci 56:131–152
European Union (2000) Directive 2000/60/ec of the European Parliament and the Council of 23 October 2000 establishing a framework for community action in the field of water policy. Off J Eur Commun L327:1–72
Flower RJ, Juggins S, Battarbee RW (1997) Matching diatom assemblages in lake sediment cores and modern surface sediment samples: the implications for lake conservation and restoration with special reference to acidified systems. Hydrobiologia 344:27–40
Fritz SC, Kingston JC, Engstrom DR (1993) Quantitative trophic reconstructions from sedimentary diatom assemblages: a cautionary tale. Freshw Biol 30:1–23
Gibson CE, Foy RH, Bailey-Watts AE (1996) An analysis of the total phosphorus cycle in some temperate lakes: the response to enrichment. Freshw Biol 35:525–532
Gregory-Eaves R, Smol JP, Finney BP, Edwards ME (1999) Diatom-based transfer functions for inferring past climatic and environmental changes in Alaska, USA. Arct Antarct Alp Res 31:353–365
Hall RI, Smol JP (1992) A weighted-averaging regression and calibration model for inferring total phosphorus concentration from diatoms in British Columbia (Canada) lakes. Freshw Biol 27:417–434
Hall RI, Smol JP (1996) Paleolimnological assessment of long-term water-quality changes in south-central Ontario lakes affected by cottage development and acidification. Can J Fish Aquat Sci 53:1–17
Hall RI, Leavitt PR, Smol JP, Zirnhelt N (1997) Comparison of diatoms, fossil pigments and historical records as measures of lake eutrophication. Freshw Biol 38:401–417
Hall RI, Leavitt PR, Quinlan R, Dixit AS, Smol JP (1999) Effects of agriculture, urbanization, and climate on water quality in the northern Great Plains. Limnol Oceanogr 44:739–756
Harper D (1992) Eutrophication of freshwaters: principles, problems and restoration. Chapman and Hall, London, 327 pp
Harriman R, Morrison BRS, Caines LA, Collen P, Watt AW (1987) Long-term changes in fish populations of acid streams and lochs in Galloway south west Scotland. Water Air Soil Pollut 32:89–112
Hutchinson GE (1973) Eutrophication. Am Sci 61:269–279

Jeppesen E, Madsen EA, Jensen JP (1996) Reconstructing the past density of planktivorous fish and trophic structure from sedimentary zooplankton fossils: a surface sediment calibration data-set from shallow lakes. Freshw Biol 36:115–127

Jones VJ, Juggins S (1995) The construction of a diatom-based chlorophyll *a* transfer function and its application at three lakes on Signy Island (maritime Antarctic) subject to differing degrees of nutrient enrichment. Freshw Biol 34:433–445

Jones VJ, Stevenson AC, Battarbee RW (1989) Acidification of lakes in Galloway, south west Scotland: a diatom and pollen study of the post-glacial history of The Round Loch of Glenhead. J Ecol 77:1–23

Jordan P, Rippey B, Anderson NJ (2001) Modelling diffuse phosphorus loads from land to freshwater using the sedimentary record. Environ Sci Technol 35:815–819

Juggins S (2003) C2 user guide. Software for ecological and palaeoecological data analysis and visualisation. University of Newcastle, Newcastle–upon–Tyne, UK. http://www.campus.ncl.ac.uk/staff/Stephen.Juggins/software/C2Home.htm

Juggins S, Birks HJB (2012) Chapter 14: Quantitative environmental reconstructions from biological data. In: Birks HJB, Lotter AF, Juggins S, Smol JP (eds) Tracking environmental change using lake sediments. Volume 5: Data handling and numerical techniques. Springer, Dordrecht

Juggins S, Telford RJ (2012) Chapter 5: Exploratory data analysis and data display. In: Birks HJB, Lotter AF, Juggins S, Smol JP (eds) Tracking environmental change using lake sediments. Volume 5: Data handling and numerical techniques. Springer, Dordrecht

Kaupilla T, Moisio T, Salonen V-P (2002) A diatom-based inference model for autumn epilimnetic total phosphorus concentration and its application to a presently eutrophic Boreal lake. J Paleolimnol 27:261–273

Kernan M, Battarbee RW, Curtis CJ, Monteith DT, Shilland EM (eds) (2010) Recovery of lakes and streams in the UK from acid rain. The United Kingdom acid waters monitoring network 20 year interpretative report. ECRC Research Report 141. Report to the Department for Environment, Food and Rural Affairs (Contract EPG 1/3/160)

Kreiser AM, Appleby PG, Natkanski J, Rippey B, Battarbee RW (1990) Afforestation and lake acidification: a comparison of four sites in Scotland. Philos Trans R Soc Lond B 327:377–383

Legendre P, Birks HJB (2012a) Chapter 7: Clustering and partitioning. In: Birks HJB, Lotter AF, Juggins S, Smol JP (eds) Tracking environmental change using lake sediments. Volume 5: Data handling and numerical techniques. Springer, Dordrecht

Legendre P, Birks HJB (2012b) Chapter 8: From classical to canonical ordination. In: Birks HJB, Lotter AF, Juggins S, Smol JP (eds) Tracking environmental change using lake sediments. Volume 5: Data handling and numerical techniques. Springer, Dordrecht

Lepš J, Šmilauer P (2003) Multivariate data analysis using CANOCO. Cambridge University Press, Cambridge

Little JL, Hall RI, Quinlan R, Smol JP (2000) Past trophic status and hypolimnetic anoxia during eutrophication and remediation of Gravenhurst Bay, Ontario: comparison of diatoms, chironomids, and historical records. Can J Fish Aquat Sci 57:333–341

Lotter AF (1998) The recent eutrophication of Baldeggersee (Switzerland) as assessed by fossil diatom assemblages. Holocene 8:395–405

Lotter AF, Anderson NJ (2012) Chapter 18: Limnological responses to environmental changes at inter-annual to decadal time-scales. In: Birks HJB, Lotter AF, Juggins S, Smol JP (eds) Tracking environmental change using lake sediments. Volume 5: Data handling and numerical techniques. Springer, Dordrecht

Lotter AF, Birks HJB, Hofmann W, Marchetto A (1998) Modern diatom, Cladocera, Chironomid, and Chrysophyte cyst assemblages as quantitative indicators for the reconstruction of past environmental conditions in the Alps. II. Nutrients. J Paleolimnol 19:443–463

Moneteith DT, Evans CD (2000) UK acid waters monitoring network: 10 year report. ENSIS Publishing, London

Nürnberg GK (1995) The anoxic factor, a quantitative measure of anoxia and fish species richness in central Ontario lakes. Trans Am Fish Soc 124:677–686

Odén S (1968) The acidification of air precipitation and its consequences in the natural environment. Energy Committee Bulletin, 1. Swedish Natural Sciences Research Council, Stockholm
Oksanen J, Läärä E, Huttunen P, Meriläinen J (1988) Estimation of pH optima and tolerances of diatoms in lake sediments by the methods of weighted averaging, least squares and maximum likelihood, and their use for the prediction of lake acidity. J Paleolimol 1:39–49
Overpeck JT, Webb T III, Prentice IC (1985) Quantitative interpretation of fossil pollen spectra – dissimilarity coefficients and the method of modern analogs. Quaternay Res 23:87–108
Pennington W (1984) Long-term natural acidification of upland sites in Cumbria: evidence from post-glacial lake sediments. Freshw Biol Assoc Ann Rep 52:28–46
Philibert A, Prairie Y (2002) Is the introduction of benthic species necessary for open-water chemical reconstruction in diatom-based transfer functions? Can J Fish Aquat Sci 59:938–951
Pienitz R, Lotter AF, Newman L, Kiefer T, (eds) (2009) Advances in paleolimnology. PAGES News 17:89–136
Pla S, Monteith DT, Flower RJ (2009) Assessing the relative influence of climate warming and atmospheric contamination on the diatom and chysophyte communities of a remote Scottish loch. Freshw Biol 54:505–523
Quinlan R, Smol JP (2001) Chironomid-based inference models for estimating end-of-summer hypolimnetic oxygen from south-central Ontario shield lakes. Freshw Biol 46:1529–1551
Quinlan R, Hall RI, Smol JP (1998) Quantitative inferences of past hypolimnetic anoxia in south-central Ontario lakes using fossil midges (Diptera: Chironomidae). Can J Fish Aquat Sci 55:587–596
Racca JMJ, Prairie YT (2004) Apparent and real bias in numerical transfer functions in Palaeolimnology. J Paleolimnol 31:117–124
Reavie ED, Smol JP (2001) Diatom-environmental relationships in 64 alkaline southeastern Ontario (Canada) lakes: a diatom-based model for water quality reconstructions. J Paleolimnol 25:25–42
Reavie ED, Hall RI, Smol JP (1995a) An expanded weighted-averaging model for inferring past total phosphorus concentrations from diatom assemblages in eutrophic British Columbia (Canada) lakes. J Paleolimnol 14:49–67
Reavie ED, Smol JP, Carmichael B (1995b) Post-settlement eutrophication histories of six British Columbia (Canada) lakes. Can J Fish Aquat Sci 52:2388–2401
Reid M (2005) Diatom-based models for reconstructing past water quality and productivity in New Zealand lakes. J Paleolimnol 33:13–38
Renberg I (1990) A 12 600 year perspective of the acidification of Lilla Öresjön, southwest Sweden. Philos Trans R Soc Lond B 327:357–361
Renberg I, Battarbee RW (1990) The SWAP palaeolimnology programme: a synthesis. In: Mason BJ (ed) The Surface Waters Acidification Programme. Cambridge University Press, Cambridge, pp 281–300
Renberg I, Korsman T, Anderson NJ (1993) A temporal perspective of lake acidification in Sweden. Ambio 22:264–271
Rippey B (1995) Lake phosphorus models. In: Patrick ST, Anderson NJ (eds) Ecology and palaeoecology of Lake Eutrophication. Geological Survey of Denmark DGU Service Report no. 7, Copenhagen, Denmark, pp 58–60
Rippey B, Anderson NJ (1996) Reconstruction of lake phosphorus loading and dynamics using the sedimentary record. Environ Sci Technol 30:1786–1788
Rippey B, Anderson NJ, Foy RH (1997) Accuracy of diatom-inferred total phosphorus concentrations, and the accelerated eutrophication of a lake due to reduced flushing and increased internal loading. Can J Fish Aquat Sci 54:2637–2646
Rosenqvist IT (1977) Acid soil – acid water. Ingeniørforlaget, Oslo
Rosenqvist IT (1978) Alternative sources for acidification of river water in Norway. Sci Total Environ 10:39–49
Sas H (1989) Lake restoration by reduction of nutrient loading: expectations, experiences, extrapolations. Academia Verlag, St. Augustin

Sayer C (2001) Problems with the application of diatom-total phosphorus transfer functions: examples from a shallow English lake. Freshw Biol 46:743–757

Scheffer M, Hosper SM, Meijer ML, Moss B, Jeppesen E (1993) Alternative equilibria in shallow lakes. Trends Ecol Evol 8:275–279

Schindler DW (1971) Carbon, nitrogen and phosphorus and the eutrophication freshwater lakes. J Phycol 7:321–329

Schönfelder I, Gelbrecht J, Schönfelder J, Steinberg CEW (2002) Relationships between littoral diatoms and their chemical environment in northeastern German lakes and rivers. J Phycol 38:666–82

Simpson GL (2012) Chapter 15: Modern analogue techniques. In: Birks HJB, Lotter AF, Juggins S, Smol JP (eds) Tracking environmental change using lake sediments. Volume 5: Data handling and numerical techniques. Springer, Dordrecht

Simpson GL, Anderson NJ (2009) Deciphering the effect of climate change and separating the influence of confounding factors in sediment records using additive models. Limnol Oceanogr 54:2529–2541

Simpson GL, Shilland EM, Winterbottom J, Keay J (2005) Defining reference conditions for acidified waters using a modern analogue approach. Environ Pollut 137:119–133

Siver PA (1999) Development of paleolimnological inference models for pH, total nitrogen and specific conductivity based on planktonic diatoms. J Paleolimnol 21:45–59

Smol JP (2008) Pollution of lakes and rivers: a paleoenvironmental perspective, 2nd edn. Blackwell, Oxford

Stevenson AC, Juggins S, Birks HJB, Anderson DS, Anderson NJ, Battarbee RW, Berge F, Davis RB, Flower RJ, Haworth EY, Jones VJ, Kingston JC, Kreiser AM, Line JM, Munro MAR, Renberg I (1991) The Surface Waters Acidification Project Palaeolimnology Programme: modern diatom/lake-water chemistry data-set. ENSIS Publishing, London

Taylor D, Dalton C, Leira M, Jordan P, Chen G, Leon-Vintro L, Irvine K, Bennion H, Nolan T (2006) Recent histories of six productive lakes in the Irish Ecoregion based on multiproxy palaeolimnological evidence. Hydrobiologia 571:237–259

Telford RJ, Birks HJB (2011) A novel method for assessing the statistical significance of quantitative reconstructions inferred from biotic assemblages. Quaternary Sci Rev 30: 1272–1278

ter Braak CJF (1986) Canonical correspondence analysis: a new eigenvector technique for multivariate direct gradient analysis. Ecology 67:1167–1179

ter Braak CJF, Juggins S (1993) Weighted averaging partial least squares regression (WAPLS): an improved method for reconstructing environmental variables from species assemblages. Hydrobiologia 269(270):485–502

ter Braak CJF, Prentice IC (1988) A theory of gradient analysis. Adv Ecol Res 18:271–317

ter Braak CJF, Šmilauer P (2002) CANOCO reference manual and CanoDraw for Windows user's guide: software for canonical community ordination (version 4.5). Microcomputer Power, Ithaca

ter Braak CJF, van Dam H (1989) Inferring pH from diatoms: a comparison of old and new calibration methods. Hydrobiologia 178:209–223

Tibby J (2004) Development of a diatom-based model for inferring total phosphorus concentration in southeastern Australian water storages. J Paleolimnol 31:23–36

van der Voet H (1994) Comparing the predictive accuracy of models using a simple randomization test. Chemometr Intell Lab Syst 25:313–323

Vollenweider RA (1975) Input–output models with special reference to the phosphorus loading concept. Schweiz Z Hydrol 37:58–83

Werner P, Smol JP (2005) Diatom–environmental relationships and nutrient transfer functions from contrasting shallow and deep limestone lakes in Ontario, Canada. Hydrobiologia 533:145–173

Whittaker RH (1956) Vegetation of the Great Smoky Mountains. Ecol Monogr 26:1–80
Whittaker RH (1967) Gradient analysis of vegetation. Biol Rev Camb Philos Soc 49:207–264
Yang X, Anderson NJ, Dong X, Shen JI (2008) Surface sediment diatom assemblages and epilimnetic total phosphorus in large, shallow lakes of the Yangtze floodplain: their relationships and implications for assessing long-term eutrophication. Freshw Biol 53:1273–1290
Wunsam S, Schmidt R (1995) A diatom-phosphorus transfer function for alpine and pre-alpine lakes. Mem Ist Ital Idrobiol 53:85–99

Chapter 20
Tracking Holocene Climatic Change with Aquatic Biota from Lake Sediments: Case Studies of Commonly used Numerical Techniques

Brian F. Cumming, Kathleen R. Laird, Sherylyn C. Fritz, and Dirk Verschuren

Abstract It is now widely recognised that reliable long-term climatic data are required to evaluate the impact of human activities on climate. Lake-sediment records are an important source of such palaeoclimatic information, on time-scales from years to millennia. However, unequivocal interpretation of biological climate-proxy data preserved in lake sediments can be very challenging. Here we review the different numerical approaches that are used to evaluate the sensitivity and reliability of species assemblages of aquatic biota (algae and invertebrates) extracted from lake-sediment records as proxies of past climatic conditions. The most common techniques used to assess this relationship between these proxies and climate include calibration functions that model the relationship across modern lake environments between species composition in the indicator group and particular climate-influenced aspects of their aquatic habitat, and assessments of the main directions of variation in species composition in relation to independent climatic data. Other statistical techniques, such as variation partitioning analysis, are used to assess the relative importance of climate versus other factors in influencing limnological changes seen in the sedimentary record. These techniques show that in climate-sensitive lake systems, the sedimentary remains of aquatic biota can be

B.F. Cumming (✉) • K.R. Laird
Paleoecological Environmental Assessment and Research Laboratory (PEARL), Department of Biology, Queen's University, Kingston, ON K7L 3N6, Canada
e-mail: brian.cumming@queensu.ca; lairdk@queensu.ca

S.C. Fritz
Department of Earth and Atmospheric Sciences, University of Nebraska, Lincoln, NE 68588-0340, USA
e-mail: sfritz2@unl.edu

D. Verschuren
Limnology Unit, Department of Biology, Ghent University, KL Ledeganckstraat 35, B-9000, Ghent, Belgium
e-mail: dirk.verschuren@ugent.be

H.J.B. Birks et al. (eds.), *Tracking Environmental Change Using Lake Sediments*, Developments in Paleoenvironmental Research 5, DOI 10.1007/978-94-007-2745-8_20,

sensitive and trustworthy proxies, permitting quantitative reconstructions of past climatic conditions with high temporal resolution.

Keywords Calibration functions • Chironomids • Climatic change • Diatoms • Invertebrates • Modern analogues • Ordination • Palaeoclimate • Rate-of-change analysis • Reconstruction validation • Variation partitioning

Introduction

With recent concerns about global change, the need for reliable long-term palaeoclimatic data is now universally recognised by governments and international organisations. Such records are necessary to provide information on the magnitude and patterns of past climatic change against which recent changes can be assessed. Furthermore, prognoses of future climatic conditions will be enhanced when we better understand the temporal and spatial patterns of natural climatic variability.

Historically, a great amount of information on long-term climatic variability has been provided by reconstructions of past vegetation changes from pollen and plant macrofossils preserved in lake sediments (e.g., Webb 1986; Prentice et al. 1991) and from studies applying geomorphological, sedimentological, and biostratigraphical methods to transects of sediment cores to reconstruct changes in water level (e.g., Harrison and Digerfeldt 1993). However, as noted by many authors (e.g., Ritchie 1987), pollen-based studies have certain limitations, which include the inability to identify many pollen taxa below the family level; the broad dispersal of many wind-blown pollen grains; the differential response of various tree species to climatic change associated with variation in topography, soils, life-cycle characteristics, and anthropogenic factors; and the general scarcity of pollen and macrofossils in some environments (e.g., regions at high altitudes and/or latitudes). None-the-less, contributions from palynological and lake-level studies have elucidated the broad-scale temporal and spatial template of climatic change in many regions, and in some regions have the potential to provide records of climatically-driven vegetation change with high temporal resolution (e.g., Clark et al. 2002; Lamb et al. 2003; Gajewski 2008; Kröpelin et al. 2008; St. Jacques et al. 2008a, b).

Many lakes exhibit physical, chemical, and/or biological responses to changes in climatic conditions, and evidence of these changes are often preserved in their sediment record (Battarbee 2000; Smol and Cumming 2000; Fritz 2008). Analyses of the remains of aquatic algae (e.g., diatoms, chrysophytes, pigments) and invertebrates (e.g., chironomids, cladocerans, ostracods) in sediment cores from such climatically-sensitive lakes can potentially provide records of climatic and environmental change at a higher resolution than most palynological and many lake-level studies. Enhanced sensitivity of aquatic biological indicators (proxies) to environmental change is due to our ability to identify many of these organisms to species level, their short life-span, and their fast dispersal, which guarantees rapid colonisation of newly available habitat and thus avoids significant lags in their response to climatic change. Thus, examination of changes in the species composition of aquatic biota in sediment cores can provide valuable insights into how climatic conditions have varied in the past that may not be readily apparent from terrestrial indicators. However, the relative importance of climatic change can be complex, even in apparently simple aquatic systems (e.g., Anderson et al. 2008; Fritz 2008).

In this chapter, we provide an overview of the different numerical techniques that have been used to study the linkages between changes in biotic assemblages preserved in lake sediments and climatic change during the Holocene. It is important to note that these approaches are simply tools that are useful to assess, simplify, and visualise past environmental changes in a repeatable and somewhat objective manner. Many studies make limited use of numerical approaches but still yield valuable insights into past climatic conditions (e.g., Baker et al. 2001; Bennett et al. 2001; Dean 2002; Dean et al. 2002; Spooner et al. 2002). Finally, although the focus of this chapter is on approaches that have been used to interpret climatic signals in fossil assemblages of lake biota, equally important palaeoclimatic information is derived from physical (e.g., Lamoureux 2000; Noon et al. 2001; Dean et al. 2002; Pienitz et al. 2009) and geochemical proxies (e.g., Dean 2002; Tierney et al. 2008; Pienitz et al. 2009; Toney et al. 2010). The latter increasingly include stable-isotope signatures extracted from lake biota, such as ostracods (e.g., von Grafenstein et al. 1999), chironomids (e.g., Verbruggen et al. 2010), and diatoms (e.g., Hernandez et al. 2010).

Biological Proxies of Past Climatic Conditions

Proxy data based on biotic assemblages preserved in lake sediments have long been known to be sensitive and reliable indicators of past climatic change, because of the strong influence of temperature and rainfall variations on the structure and function of their aquatic habitats. Lake characteristics that are sensitive to climatic changes can be both physical (e.g., temperature, ice cover, lake depth, river discharge, mixing regime, light transparency) and chemical (e.g., changes in alkalinity, pH, nutrients, salinity) (e.g., Walker et al. 1991; Walker 1995; Vinebrooke et al. 1998; Korhola 1999; Battarbee 2000; Smol and Cumming 2000; Korhola and Rautio 2001; Brodersen and Anderson 2002; Fritz 2008; Lotter et al. 2010; Wolin and Stone 2010). However, establishing an unequivocal connection of these limnological changes to climate is often difficult. The challenge is to demonstrate that the biotic proxy indicators are connected to climate in a systematic, predictable fashion. In some systems, this is exceedingly difficult because of the non-linear responses of the limnological system to climatic change and subsequent sedimentological complications, such as variable sedimentation rates and/or mixing (Verschuren 1999a, b). Such complexities are related to basin hydrology, as well as to the physical, chemical, and biological characteristics of individual lake basins (Smol and Cumming 2000; Schwalb and Dean 2002; Fritz 2008). Additionally, because different groups of biological proxies have different life-history strategies, changes in climate that influence aquatic systems can impact various proxies in different ways (Battarbee et al. 2002a; Heegaard et al. 2006). Therefore, it is not surprising that the response of different lakes, and of different proxies within a single lake, to climatic change can vary tremendously (Smol and Cumming 2000; Fritz 2008).

Successful palaeoclimatic interpretation of in-lake biological proxies depends on a good understanding of the climatic response of a lake to changes in climate. Not surprisingly, selecting and then demonstrating the climatic sensitivity of a site is not a simple procedure. Smol and Cumming (2000) identified lakes that are thought to be especially sensitive. These include lakes from high latitudes, lakes that are close to ecotonal boundaries (e.g., near tree-line, near the forest-prairie boundary), and lakes in arid to semi-arid regions with limited groundwater inputs and, ideally, large catchments (i.e., amplifier lakes: Olaka et al. 2010). In addition to geographical position, sites sensitive to climatic changes can be identified from measurements of limnological change (e.g., changes in lake-water chemistry, transparency, declines in water levels, and/or changes in salinity) and from instrumental and/or historical records (Fritz 1990; Schindler et al. 1996; Verschuren 1996, 2003; Webster et al. 2000; Doran et al. 2002). However, data of sufficient duration and frequency are relatively rare, even in populated regions. Additional information on the hydrological responsiveness to changes in climatic conditions can be obtained by careful examination of temporal sequences of aerial photographs and satellite images (Donovan et al. 2002; Smith et al. 2005).

Once climate-sensitive lakes have been identified, it is becoming increasingly common to assess the strength of the climatic impact on biological proxies or inferred limnological variables in the sediment record by comparison of the proxy records with instrumental meteorological data (temperature, precipitation, drought indices), historical data of lake response to climate (e.g., lake-level changes, chemical constituents), or with independent evidence of climatic change (e.g., documentary, other palaeoclimatic proxies). These comparisons may involve simple visual matching of pattern similarities in the palaeolimnological and meteorological records, calculation of correlation coefficients between lake and climatic variables, or using multivariate numerical methods (e.g., variation partitioning: Borcard et al. 1992; Legendre and Birks 2012: Chap. 8).

Comparison of Inferred Limnological Variables and Instrumental Climate Data

In arid to sub-humid regions, long-term (decade-scale and longer) variations in the balance of rainfall and evaporation can drive significant changes in lake level, which in turn may influence the physical, chemical, and biological environment of lakes to the extent that their signatures are recorded in the sedimentary record (Fritz 1996). In other regions, the influence of temperature is evident in the sedimentary record,

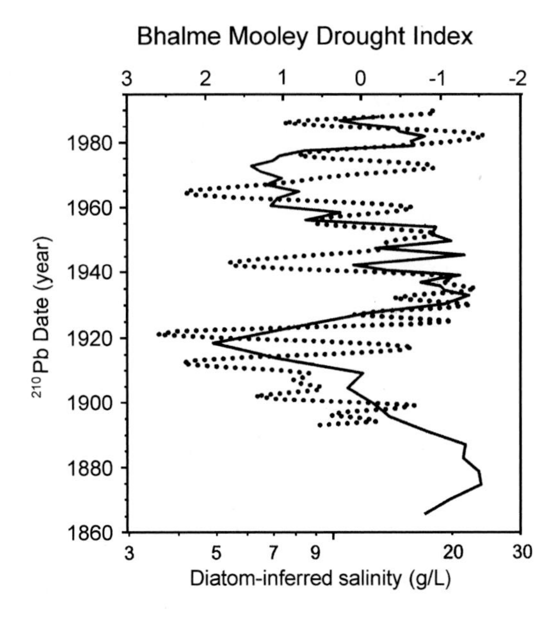

Fig. 20.1 Diatom-inferred salinity estimates from Moon Lake (North Dakota, USA) (*solid line*) compared to a smoothed Bhalme-Mooley Drought Index (*dotted line*) based on nearby instrumental precipitation records over the last 100 years; $r = 0.49$, $p < 0.01$ (Modified from Laird et al. 1998a)

either directly or indirectly, via changes in catchment hydrology, vegetation, ice cover, or mixing regimes. One way to validate proxy-based climatic inferences is to compare quantitatively or qualitatively the proxy-based reconstructions of environmental change (e.g., lake-water salinity, lake level, temperature) in the recent past with historical time-series of instrumental meteorological data. To date, the majority of such direct, quantitative comparisons have used approaches based on calibration functions (see Juggins and Birks 2012: Chap. 14).

To evaluate the climatic sensitivity of Moon Lake (Northern Great Plains, USA), Laird et al. (1996a) assessed the correlation between diatom-inferred (DI) salinity through time in a ^{210}Pb-dated sediment core and instrumental time-series of effective moisture (precipitation minus evapotranspiration or ET). Diatom-inferred salinity showed major increases coincident with the major droughts of the 1930s and the 1890s, with a statistically significant relationship between DI salinity and effective moisture (Laird et al. 1996a). Because ET can be difficult to estimate, Laird et al. (1998a) replaced estimates of effective moisture with the Bhalme and Mooley drought index (BMDI), derived from nearby precipitation records (Fig. 20.1). The BMDI was chosen as a measure of drought, because it is based on precipitation records from local stations, as opposed to regional precipitation averages used in some other indices, such as the Palmer Drought Severity Index. In this analysis, the BMDI was smoothed using a four-point Fourier transform filter to achieve approximately equal time resolution between the sediment record and the instrumental data. A modest correlation between diatom-inferred salinity and BMDI (Fig. 20.1) can be attributed to several factors, including the low time resolution in core data from before 1940, the further distance of the meteorological station from the lake for the early instrumental period, and limnological changes

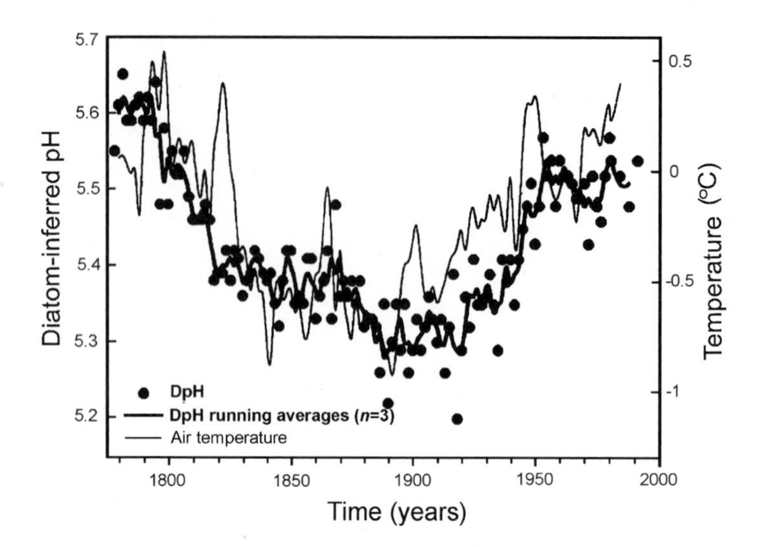

Fig. 20.2 Diatom-inferred lake-water pH from Schwarzsee ob Sölden in the Austrian Alps since ~1780 compared with smoothed annual mean air temperature from 20 regional weather stations; $r = 0.68$, $p < 0.001$ (From Sommaruga-Wögrath et al. 1997. With permission)

that are potentially attributable to settlement and agricultural practices that may have impacted lake chemistry independent of climate since the early 1900s. Comparison of the Moon Lake record with regionally recognised climatic anomalies prior to the instrumental record (e.g., a tree-ring inferred wet period from 1825 to 1838 and the extreme 1890s drought) suggest, however, that the diatom-inferred salinity record does capture the general drought trends characteristic for the Northern Great Plains (Laird et al. 1996a, 1998a, b). Consequently, this site was chosen to reconstruct drought intensity, duration, and frequency throughout the Holocene at both centennial and sub-decadal scales (Laird et al. 1996a, b, 1998a, b).

In a study from a mountain lake in the Austrian Alps, comparison between diatom-inferred pH and instrumental air temperature between 1778 and 1991 revealed a significant positive correlation ($r = 0.68$, $p < 0.001$) (Fig. 20.2) (Sommaruga-Wögrath et al. 1997), suggesting that diatoms may be an important proxy of past temperatures. In two other alpine lakes, diatom-inferred pH and temperature were correlated during the entire 19th century, but the relationship weakened during the 20th century with the onset of acidic deposition (Psenner and Schmidt 1992). The mechanism that produced the pH increases during warming events and declines during cooler periods is likely to be related to changes in the duration and extent of ice-cover because these would alter light, temperature, and nutrient regimes, all of which affect diatom production and could influence in-lake production of alkalinity, as well as lake-water pH (Sommaruga-Wögrath et al. 1997). Periods of prolonged ice-cover would also reduce CO_2 exchange with the atmosphere and temporarily eliminate atmospheric inputs of base cations into the lake (Psenner and Schmidt 1992), both of which would affect the acid–base status in poorly-buffered lakes.

Bigler and Hall (2003) and Larocque and Hall (2003) compared chironomid and diatom-inferred mean-July air temperatures from northern Swedish sediment cores over the last century to measured values from nearby meteorological stations. In all four lakes studied, chironomid-inferred July temperatures were significantly but weakly correlated to the instrumental record ($r = 0.35–0.39$). The relationship between diatom-inferred and instrumental July temperatures was weaker, and no correlation coefficients or levels of significance were reported.

Many studies have validated climate proxies through a qualitative comparison of reconstructed environmental change to historically documented changes in climatically-sensitive limnological variables, an approach first used in comparisons of diatom-inferred salinity and ostracod trace-metal content with documented changes in the depth and salinity of Devils Lake (Northern Great Plains, USA) (Fritz 1990; Engstrom and Nelson 1991). One particular series of studies (Verschuren 1999a, b, 2001; Verschuren et al. 1999a, b, 2000a) investigated the sedimentological, geochemical, and biological signatures of climatically driven lake-level changes in a system of four contrasting lakes in the Eastern Rift Valley of Kenya, which because of their hydrological inter-connectedness have a common recorded lake-level history spanning the last 120 years (Verschuren 1996). Together these studies illustrate the complexity of chemical and biological responses to climatically-driven changes in water balance at sub-decadal time scales (Verschuren et al. 1999a, b, 2000a), and how the signatures of this response in the sediment record are affected by sedimentation dynamics and taphonomy (Verschuren 1999a; Verschuren et al. 1999a). For example, although reconstructed salinity and lake-level changes in Lake Oloidien displayed the expected inverse relationship, this was modulated by delayed dilution of residual salts following a modest lake-level rise in the late 1950s and early 1960s (Fig. 20.3) (Verschuren et al. 2000a). In nearby Lake Sonachi, diatoms quickly responded to a major 1890s lake-level rise which created stable density stratification of the water column. In the sediment record, however, their response is delayed by almost a decade because previously buried shallow-water diatoms are re-deposited in deep-water sediments during the transgression (Verschuren et al. 1999b). In the three main groups of aquatic invertebrates present (ostracods, cladocerans, chironomids), only some species responded directly to salinity change (via the physiological impact of osmotic stress). Most species responded more strongly to substrate changes associated with the lake-level change itself or to changes in the distribution of aquatic vegetation (Verschuren et al. 2000a). While lake depth, salinity, and substrate quality are all important factors in structuring aquatic invertebrate communities, it is sobering to realise that at the time-resolution of many modern palaeoclimatic studies, the intuitive co-variation between these factors can be strongly modulated by transient system dynamics. None-the-less, invertebrate communities do remain useful proxies of past hydrological change, particularly in combination with diatoms and non-biological climatic proxies. Other qualitative studies (e.g., Legesse et al. 2002) have used both biological and sedimentological proxies to reveal fluctuations in water depth and salinity that are broadly consistent with available instrumental and historical records.

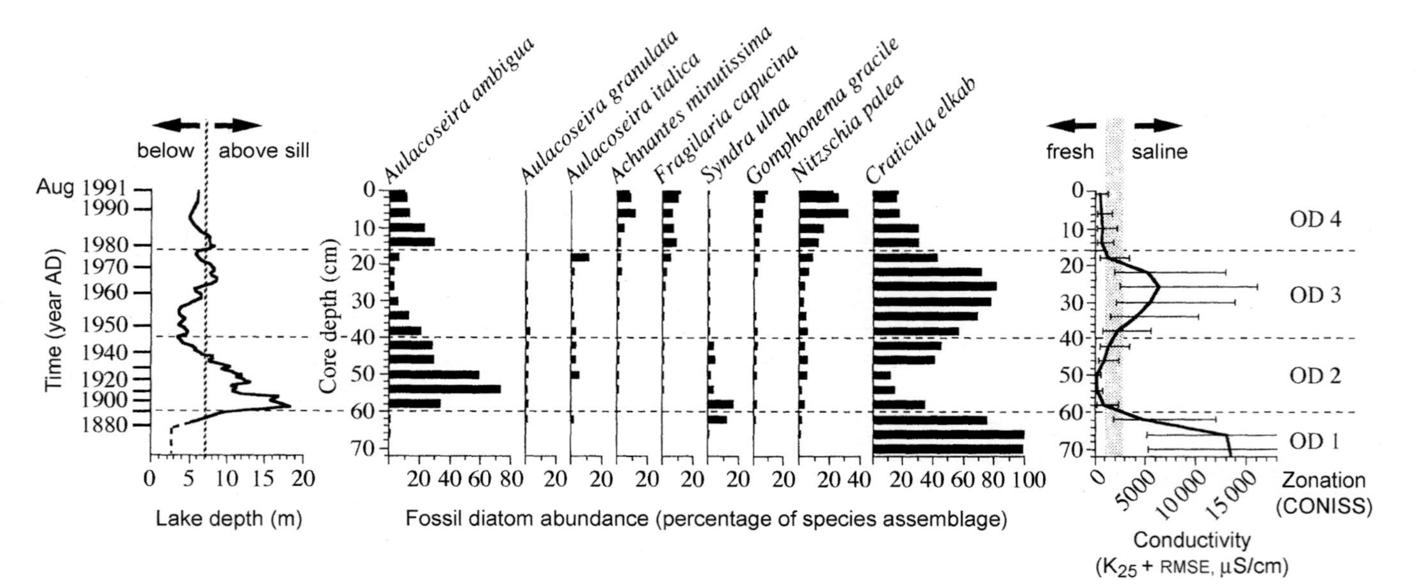

Fig. 20.3 Changes in dominant diatom taxa and diatom-inferred conductivity in Lake Oloidien (Kenya) since ~AD 1860 compared with the instrumental lake-level record. 'Below/above sill' refers to the elevation of the sill between the hydrologically-closed Lake Oloidien and the nearby hydrologically-open Lake Naivasha. When the two lakes are confluent at high lake-level (~AD 1885–1938, and much of the period 1963–1984), exchange of water between them removes dissolved salts from Lake Oloidien (From Verschuren et al. 2000a. With permission)

Although direct comparison of palaeoclimatic reconstructions with meteorological data has yielded many valuable insights into the response of specific lakes to climatic variation, other approaches may be used. These include comparisons between the main directions of variation in fossil assemblages and instrumental climatic records (Sorvari et al. 2002), the use of variation partitioning to assess the role of climate in proxy records over the instrumental period (Lotter and Birks 1997; Hall et al. 1999; Quinlan et al. 2002), and comparisons of within-lake proxy records with independent estimates of climatic change (Anderson et al. 1996, 2008; Stager et al. 1997; Cumming et al. 2002; Wolfe 2003; Enache and Cumming 2009; Stone et al. 2011).

Direct Comparison of Biological Data to Instrumental Climatic Data

Ordination techniques (see Legendre and Birks 2012: Chap. 8) are now widely used to quantify and summarise the main directions of variation in biological assemblages (Cumming and Moser 2007), which can then be correlated directly to instrumental climatic records. For example, Sorvari et al. (2002) examined the relationship between species-composition changes in the diatom communities of five lakes with varying characteristics in Finnish Lapland over the last 200 years using principal component analysis (PCA), a common ordination technique (Fig. 20.4).

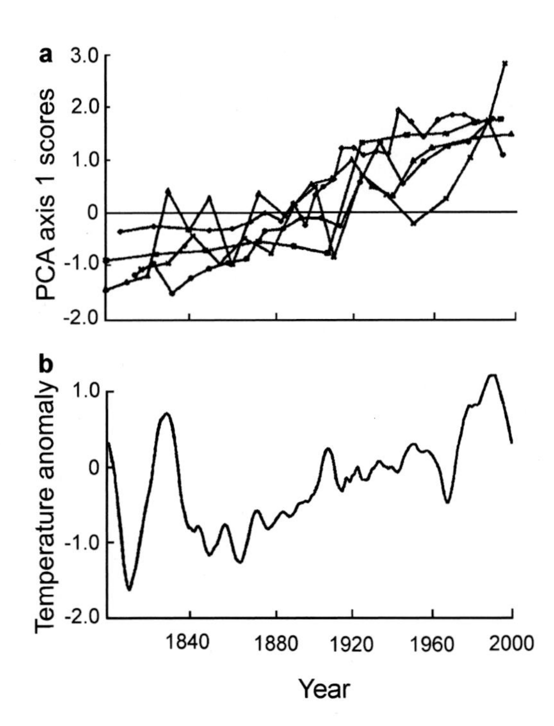

Fig. 20.4 (**a**) Trends in principal component analysis (PCA) axis-1 sample scores of fossil diatom floras in five lakes from Finnish Lapland over the last 200 years in relation to (**b**) smoothed spring temperature anomalies (°C) (Modified from Sorvari et al. 2002)

The PCA sample scores from each lake record were highly correlated to each other ($r = 0.74$–0.95), as well as being significantly correlated to instrumental records of yearly temperature ($r = 0.39$–0.62 in three of the five lakes investigated). Interestingly, the highest correlations ($r = 0.52$–0.70) were between the main direction of variation in the diatom assemblages and spring temperatures. Sorvari et al. (2002) suggested that the strong coherence of sedimentary signatures in these five lakes, and their significant correlation with temperature change over the last ~200 years, indicate that diatom floras did indeed respond to regional temperature rise, possibly through decreased ice-cover, a longer growing-season, or enhanced thermal stratification. This example illustrates that arctic as well as alpine lakes are sensitive locations to search for ecosystem changes in response to recent climatic change (Battarbee et al. 2002a; Heiri and Lotter 2005; Smol et al. 2005: Lotter et al. 2010). The response of these systems to climate is not limited to direct temperature influences, but may also include responses to changes in physical and chemical aspects of the aquatic environment, which themselves are affected by changes in temperature or precipitation. For example, such changes may include the duration of ice-cover, microhabitat availability, and changes in dissolved organic carbon (DOC) or acid–base status (Smol and Cumming 2000; Smol and Douglas 2007; Lotter et al. 2010).

In a large interdisciplinary study involving seven European high-elevation lakes, the Mountain Lake Research (MOLAR) project, researchers sought to establish links between various biotic and abiotic proxies in lake sediments and instrumental temperature records (summarised by Battarbee et al. 2002a). The biological proxy indicators included assemblages of diatoms, chrysophytes, chironomids, and cladocerans. Potential explanatory variables included: mean summer temperature (June to August and July to September), mean winter temperature (December to February), an index of continentality (difference between June to August and December to February), and mean annual temperature (average of all 12 months). Because instrumental data series directly linked to these remote high-elevation sites were not long enough for comparison with core data, homogenised lowland air-temperature records were transformed to values appropriate for each high-elevation site using the contemporary relationship between lowland temperatures and recently established on-site automatic weather stations (Agusti-Panareda and Thompson 2002). Prior to regression analysis, the variation in each of the biotic proxy data-sets was numerically reduced to its main direction of variation by calculating sample scores along PCA axes, following a square-root transformation of the species percent-abundance data. Correlations between the main direction of variation in the biological data and the instrumental records were very low, with few clear or consistent patterns. The highest correlations were found using the greatest smoothing of the instrumental data, suggesting that the biological proxy data may capture only the broadest, long-term changes in these records. Planktonic diatom assemblages (at most sites) and chrysophytes and chironomids (at a few sites) showed the highest correlation with the instrumental data (Battarbee et al. 2002b; Koinig et al. 2002). The generally weak correlations were mainly attributed to inaccurate dating and/or the limited sensitivity of some proxies to modest recent climatic forcing, as well as confounding forcing factors, such as air pollution and earthquakes, at some sites. These results are perhaps not surprising, because different species thrive at different locations within the lake as well as at different times in the season (e.g., Interlandi et al. 1999; Bradbury et al. 2002; Catalan et al. 2002). Consequently, the use of yearly averages or average temperature in any given season will likely fail to represent accurately the ecological characteristics of some of the species to temperature. The low correlations found in this study in comparison to the five sites in Finnish Lapland may also suggest that the lakes in the MOLAR project were not as sensitive to recent climatic changes or that the degree of climatic change has been less in the MOLAR lakes in comparison to those in Lapland. Continued investigation of the response of varied proxies and European lakes to climate is ongoing in the large interdisciplinary European Millennium Project (e.g., von Gunten et al. 2008), which is focused on reconstructing regional climates of the last 1000 years.

Assessing the Influence of Climate on Lake History Using Variation Partitioning and Linear Mixed-Effects Models

Variation partitioning analysis (VPA) has been used increasingly to assess the relative influence of climate versus other factors in driving limnological changes reconstructed from the sedimentary record. Variation partitioning (Borcard et al. 1992; Legendre and Birks 2012: Chap. 8; Lotter and Anderson 2012: Chap. 18; Simpson and Hall 2012: Chap. 19) uses direct gradient analysis or canonical ordination to estimate which fraction of the total variation in assemblage composition can be explained by specific categories of measured environmental variables, and which fraction is shared between different categories of variables. For example, Lotter (1998) used VPA to estimate the amount of variance in diatom data from a 100+-year varved sediment record from Baldeggersee (Switzerland) that could be explained independently by trophic state and climate (also see Lotter and Anderson 2012: Chap. 18). The variation in the diatom data was partitioned into: (i) variation due to changes in lake trophic state only (modelled by measured and extrapolated total phosphorus, lake restoration, and their interaction); (ii) variation due to climatic change only (modelled by mean spring air temperatures); (iii) covariation between (i) and (ii); and (iv) variation unexplained by the model. Because of strong temporal autocorrelation in the data, the effect of time was removed prior to the statistical analyses. Trophic state and climate together accounted for only 14.2% of the total variation in the diatom data ($p = 0.01$), with 12.6% being explained by trophic state ($p = 0.06$), and only 1.1% by climate ($p = 0.08$). Hence, Lotter (1998) concluded that changes in climatic variables were relatively unimportant in comparison with trophic-state variables in explaining changes in the diatom assemblages. Lotter and Birks (1997) used a similar approach to disentangle the relative importance of trophic state and climate on varve thickness in the same lake. Heiri and Lotter

(2005) used variance partitioning to examine the influence of co-variation between summer temperature and trophic state on calibration-function development in the Swiss Alps. Their analyses suggest that independent calibration functions can be developed for diatoms and benthic cladocerans but caution should be used in interpreting total phosphorus (TP) reconstructions from chironomids.

Hall et al. (1999) used VPA to determine how much of the down-core variation in diatom, pigment, and chironomid assemblages from Pasqua Lake (Saskatchewan, Canada) could be accounted for solely by climate (C), resource-use (R), urbanisation (U), and by covariations between them (i.e., CR, CU, RU, and CRU). Hall et al. (1999) performed separate VPAs on the percent abundances of diatoms and chironomids and a 3-year running mean of pigment concentrations for periods beginning in 1890, 1920, 1950, and 1970, so that the relative importance of climate, resource-use, and urbanisation on the biological records could be evaluated through time. All categories, both independently and in combination, explained significant amounts of variation in the biological data, although their relative importance depended on the fossil group and the period considered. Changes in urban population, sewage characteristics, livestock biomass, crop area, and temperature and precipitation proved to be consistently important explanatory variables in the history of this lake (Hall et al. 1999). However, the most consistently important explanatory variable was resource-use, accounting for between 14% and 25% of the variation in the fossil assemblages according to the VPAs. Variation attributable solely to climate was consistently low (between 4% and 10%), but the covariation between climate and resource-use and urbanisation (CR, CU, or CRU) were typically quite large, at times accounting for more than 50% of the total variation. In a similar study from the Canadian prairies involving eight lakes, Quinlan et al. (2002) found that VPA identified climate (specifically winter temperature) as the category that accounted for the largest amount of variation in past chironomid assemblages (on average ~25%), whereas changes in resource-use and urbanisation only accounted for ~7% and ~4% of variation, respectively. In summary, one of the benefits of VPA is that it potentially allows the relative importance of competing influences to be evaluated and quantified (Birks 1998).

Eggermont et al. (2010a) assessed whether the Rwenzori mountain lakes (Uganda/DR Congo) are sensitive to climate-driven environmental change of the same order of magnitude as that expected to result from current and future anthropogenic global warming. This was done by comparing the species assemblages of larval chironomid remains deposited recently in lake sediments with those deposited at the base of the short cores (dated to within or shortly after the Little Ice Age) in 16 lakes. Chironomid-based temperature reconstructions were made using temperature-inference models (Eggermont et al. 2010b) with calibration functions based on weighted averaging, weighted-averaging partial least squares, or a weighted modern analogue technique (see Juggins and Birks 2012: Chap. 14; Simpson 2012: Chap. 15). Excluding one atypical mid-elevation lake, Eggermont et al. (2010a) found a three-to-one ratio of sites with inferred warming against inferred cooling. The chironomid-inferred temperature changes mostly fell within the error range of the inference models, but a generalised linear mixed-effects analysis (SAS Inc. 2004; Birks 2012: Chap. 11) of the combined result nevertheless indicated significantly warmer mean annual air temperature (on average $+0.38 \pm 0.11$°C) at present compared to between ~85 and ~645 years ago. This result was independent of whether lakes were located in glaciated or non-glaciated Rwenzori catchments, and of basal core age, suggesting that at least part of the signal is due to relatively recent, anthropogenic warming.

Qualitative Validation of Biological Proxies Using Independent Evidence of Climatic Change

Support for the climatic sensitivity of lake systems is not limited to comparisons with instrumental and other historical data. Lake response to climate also can be assessed by comparisons (qualitative or quantitative) of in-lake climate-proxy indicators with other known climatic proxies. For example, Pienitz et al. (1999) assessed changes in diatom species composition in a sediment core from Queen's Lake (Northwest Territories, Canada), located north of the tree-line. Sharp increases in the abundances of black spruce (*Picea mariana*) pollen and sediment organic matter between ~5000 and 3000 years ago (Fig. 20.5), in conjunction with isotopic changes, strongly suggested regional climatic warming during this interval (MacDonald et al. 1993; Wolfe et al. 1996). Coincident with these changes were abrupt changes in both the concentration and species composition of diatoms in the sediment profiles (Fig. 20.5). By using a diatom-based calibration function for dissolved organic carbon (DOC), developed from a regional surface-sediment calibration-set (Pienitz et al. 1995), Pienitz et al. (1999) showed that the diatom changes were consistent with changes in lake-water transparency caused by changes in soils as trees moved into the watershed during climatic warming.

Another example of palynological data being used to support inferences of past changes in diatom species composition comes from Lake Victoria, East Africa (Stager et al. 1997). In this study, changes in diatom species composition over the last ~14,000 years were summarised into two main directions of variation by correspondence analysis (CA) (see Legendre and Birks 2012) (Fig. 20.6). The two main directions of variation were interpreted to represent relative water column stability (or conversely mixing, an index of wind strength) and the ratio of precipitation to evaporation (or effective moisture, assumed to be reflected in lake depth) based on the known habitat requirements of the dominant diatom species found in the fossil record. In conjunction with pollen records of past changes in terrestrial vegetation within the Lake Victoria catchment from an earlier study from a nearby location (Kendall 1969), the diatom data permitted the delimitation of four major periods of limnological/climatic change over the past 14,000 years: (1) a period of aridity, from ~13,400 to 11,400 cal year. BP, represented by extremely high abundances of *Fragilaria* and other benthic diatoms coincident with the abundant occurrence of grass pollen and phytoliths; (2) an early-Holocene humid phase characterised

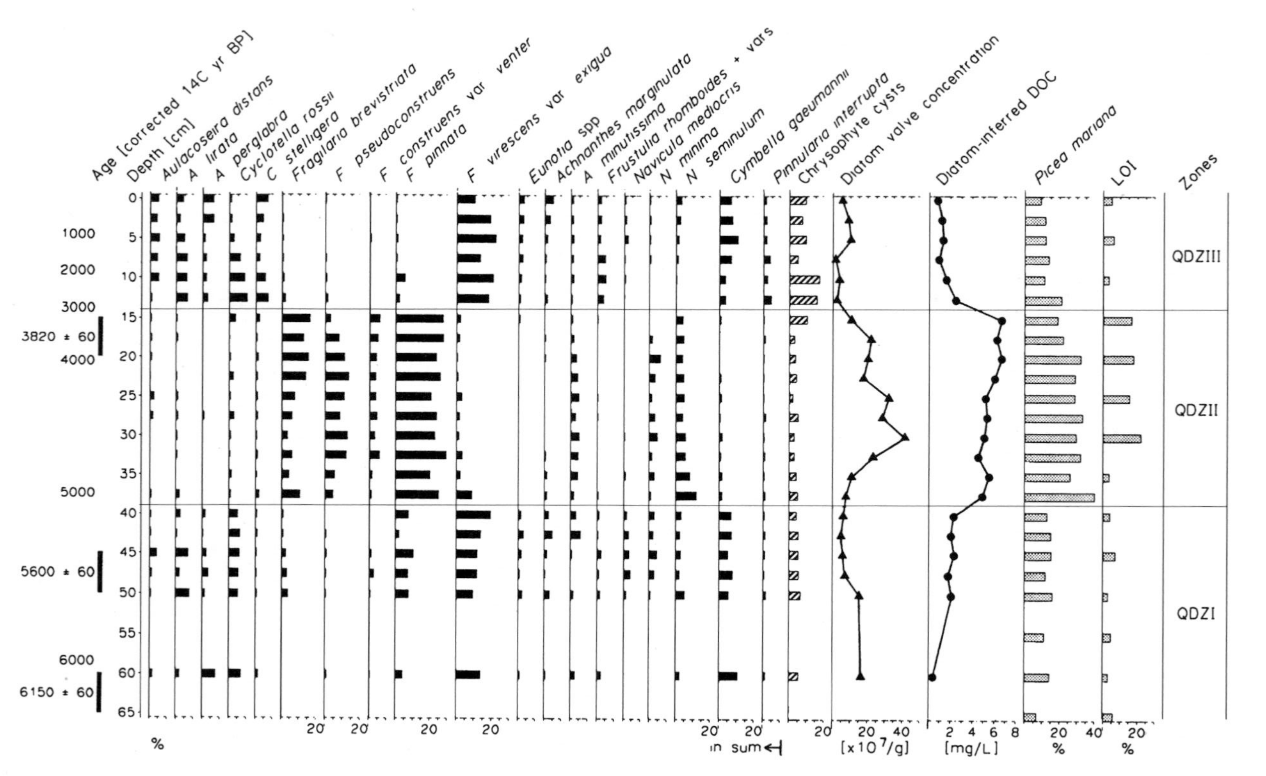

Fig. 20.5 Changes in the percent abundance of diatom taxa and diatom-inferred dissolved organic carbon (DOC) over the past 6000 years from a tundra lake near the arctic tree-line (Northwest Territories, Canada). The abrupt changes in diatom composition and productivity between ~5000 and 3000 years ago is attributed to changes in DOC resulting from the advance and retreat of black spruce (*Picea mariana*) in this region during mid-Holocene climatic warming (From Pienitz et al. 1999. With permission)

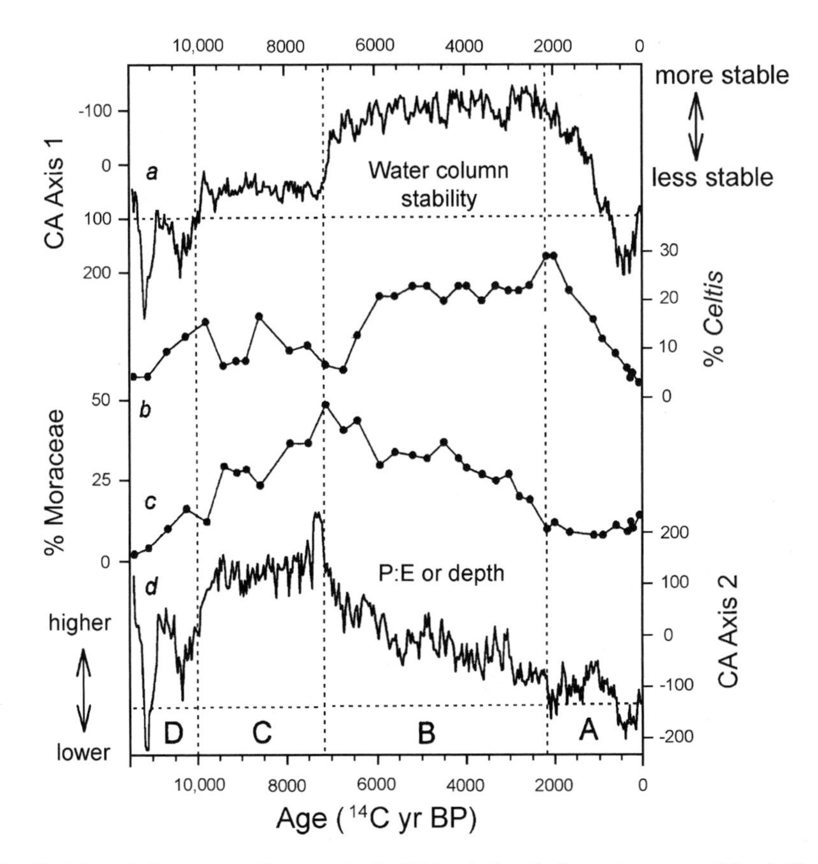

Fig. 20.6 Trends in correspondence analysis (CA) axis-1 and -2 sample scores of fossil diatom assemblages from Lake Victoria's Damba Channel (East Africa) over the past 11,000 years, in relation to changes in the regional abundance of *Celtis* (reflecting seasonally dry forest) and Moraceae (reflecting moist forest) as reconstructed at a nearby location in Lake Victoria from fossil pollen (From Stager et al. 1997. With permission)

by high abundances of *Aulacoseira* species, that prefer well-mixed deep water, coincident with high abundances of pollen from humid forest trees (e.g., Moraceae); (3) a period of increased seasonality beginning ~7900 cal year BP characterised by high abundances of taxa that thrive under enhanced thermal stratification, including *A. nyassensis* and *Nitzschia* taxa, coincident with evidence of seasonally dry forests (e.g., *Celtis*); and (4) a late-Holocene period (from ~2300 cal year BP) of increasing aridity characterised by a return to higher abundances of *Fragilaria* taxa, benthic diatoms, and grasses and phytoliths (Stager et al. 1997). Virtually identical interpretations have been presented for two other cores from Lake Victoria, one from the centre of the lake (Stager and Johnson 2000) and one from a shallow peripheral basin (Stager et al. 2003).

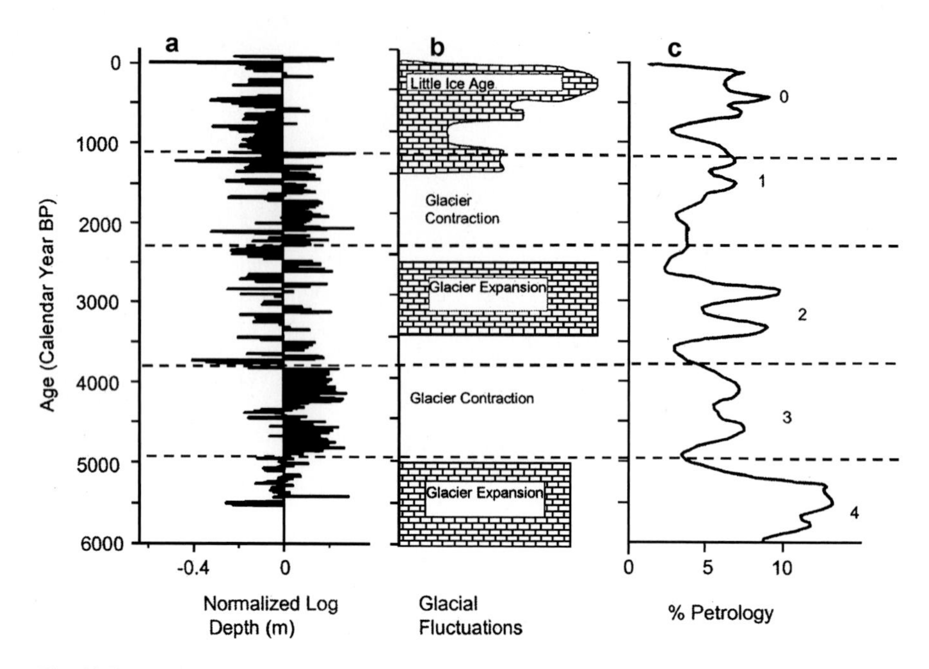

Fig. 20.7 Relationship between (**a**) diatom-inferred changes in lake depth at Big Lake (British Columbia, Canada) over the past 5500 years and (**b**) a summary of world-wide glacier fluctuations (Denton and Karlen 1973), and (**c**) a composite record of ice-rafted-debris (IRD) in the North Atlantic Ocean (Bond et al. 2001) (Modified from Cumming et al. 2002)

In a study of multi-decadal to millennium-scale climate dynamics in British Columbia (Canada) since the renewal of glacial activity ~5500 years ago, Cumming et al. (2002) recorded major shifts in diatom assemblages and diatom-inferred depth and salinity, corresponding well with millennial-scale variations in the expansion/recession of continental glaciers in the region, as well as with ice-rafting events in the North Atlantic Ocean (Fig. 20.7). This result provides supporting evidence for the sensitivity of such lake systems to broad-scale climatic forcing. However, it has been suggested that reconstructions of lake depth from biological assemblages based on regional calibration-sets, as was done in this study (Fig. 20.7), may need to be interpreted with caution because the specific morphometric features of the study lake may not be adequately represented by a regional reference data-set. Birks (1998) suggested that calibration models for lake depth may be best developed on the basis of surface-sediment samples from different depth transects within the lake under investigation. Laird et al. (2011) provide a summary of how quantitative techniques and selection of sensitive near-shore coring locations can be exploited to produce high-resolution palaeoclimatic reconstructions. This approach can yield valuable insights, but has an increased risk of non-analogue situations, in which the dominant species in certain sections of the core profile are not represented in the local surface-sediments today. Consequently, in many instances the reconstruction of a limnologically important variable, such as lake depth, can benefit from modern calibration-sets that include both surface-sediment samples from within the lake under study (i.e., a transect across depth) as well as in many lakes across the modern landscape. In some cases, quantitative reconstructions of lake-level change using biotic indicators can be constrained by geophysical or geomorphic data that provide clear physical evidence of lake stage at given points in time. This approach has been used in studies of Quaternary lake-level variation in Lake Malawi in Africa (Stone et al. 2011), where a multi-proxy reconstruction of lake level was compared with seismic data; and in a comparison of diatom-inferred conductivity as a proxy for precipitation minus evapotranspiration in West Greenland with dated Holocene palaeoshorelines (Aebly and Fritz 2009).

Finally, several studies from Europe have compared quantitative reconstructions of temperature with other regional reconstructions of temperature change. Korhola et al. (2002) used both a Bayesian multinomial model and a weighted-averaging partial least squares model (see Juggins and Birks 2012: Chap. 14) to reconstruct temperature based on chironomid assemblages over the Holocene. The results generally agreed with inferences from the Greenland ice-cores and marine sediments, as well as with previous reconstructions from diatoms and pollen (Korhola et al. 2002). Ampel et al. (2010) used a diatom calibration-function at a site in eastern France to quantify the magnitude of summer temperature change associated with millennial events during the late Pleistocene and compared it with other proxy data from the same site, other regional records, and the patterns of variation manifested in the Greenland ice core.

Quantitative Validation of Biological Proxies Using Independent Evidence of Climatic Change

A number of studies have attempted a quantitative assessment of how much of the variation in biological climate-proxy assemblages can be attributed to climatic forcing, using ordination and partial ordination techniques including VPA. For example, in an 1100-year lake record from northern Sweden, Anderson et al. (1996) (see also Lotter and Anderson 2012: Chap. 18) found a weak relationship between changes in diatom species assemblages and tree-ring inferred estimates of summer temperature. However, this relationship could only explain 5.2% of the variation in the diatom assemblages, and only after the temperature data were lagged by 20 years. Interestingly, a stronger relationship ($\sim$10% of the observed variation) was found between species richness and temperature. Other investigators have used VPA to compare the relative importance of climate and other factors contributing to changes in biological assemblages. For example, Birks and Lotter (1994) and Lotter et al. (1995) attempted to disentangle the impacts of volcanic tephra and climatic change on terrestrial and aquatic ecosystems (see Lotter and Anderson 2012: Chap. 18). Similarly, Barker et al. (2000) used VPA to investigate the relative importance of climate (represented by the ratio between tree and grass pollen), catchment disturbance (represented by magnetic properties of the sediments), and tephra deposition on a 4100-year diatom record from Lake Massoko (southern Tanzania). The effects of time-dependent ecological and environmental processes were removed by using sample age as a covariable in partial constrained ordination (see Legendre and Birks 2012: Chap. 8). Anderson et al. (2008) used independent climatic data (e.g., Greenland ice-core data) to assess the degree to which climate versus in-lake processes explained the community structure of chironomids and other proxies. In this study, catchment changes or biotic relationships explained more variation than Holocene climate, with the exception of the early lake development. VPA also has been used to demonstrate that changes in land-use have been more important than climate in driving changes in the diatom flora of Seebergsee (Switzerland) over the last 2600 years (Hausmann et al. 2002). A recurring theme in many of these studies is that, although climatic forcing has accounted for a significant fraction of the reconstructed biological variation, the amount of variation explained by climate alone, over the time scales examined, has been relatively small.

Other Numerical Techniques Used in Lake-Based Studies of Climatic Change

As is clear from the material discussed above, calibration functions, ordination, and regression techniques are the methods most commonly used to track palaeoclimatic signals in biological and other proxies extracted from lake sediments. Calibration functions are continually being tested and refined (e.g., Köster et al. 2004; Eggermont et al. 2006; Battarbee et al. 2008; Birks et al. 2010), and new numerical approaches continue to be developed (e.g., Racca et al. 2003; Hübener et al. 2008). Many examples of these techniques are presented above, and/or are covered in detail in Parts II–IV of this book. Methods based on aquatic algae have also been summarised in Smol and Cumming (2000), as well as in more detailed accounts of palaeoclimatic techniques for semi-arid regions (Bradbury 1999; Fritz 2008; Fritz et al. 2010) and arctic and alpine regions (Smol and Douglas 2007; Douglas and Smol 2010; Lotter et al. 2010). Additionally, there have been substantial developments in the application of chironomids and cladocerans as quantitative indicators of past climatic conditions (e.g., Lotter et al. 1997; Larocque et al. 2001; Korhola et al. 2002; Palmer et al. 2002; Heiri et al. 2004; von Gunten et al. 2008; Eggermont et al. 2008; Kröpelin et al. 2008). For example, Palmer et al. (2002) provided a midge-based consensus reconstruction of Holocene mean July air temperatures for southern British Columbia (Canada) based on sediment cores from four lake sites.

Although the number of lake-based quantitative palaeoenvironmental reconstructions has increased rapidly in the past decade, for some time relatively little attention was given to a thorough evaluation and validation of these reconstructions (Birks 1998). This has now improved with the availability of a number of simple approaches to assess the basic reliability of environmental reconstructions (Telford and

Birks 2011). For example, Laird et al. (1998a) assessed the reliability of the Moon Lake salinity inferences over the past 2300 years (Laird et al. 1996a) (Fig. 20.8) by (1) calculating how well a 'fossil' sample is represented by modern assemblages (Fig. 20.8b), and (2) assessing the overall 'goodness-of fit' of the environmental reconstruction (Fig. 20.8c). The lower the dissimilarity coefficient or the less often a sample is in the extreme of the distribution of the 'goodness-of-fit', the more confidence one can place in a reconstructed value, because the 'fossil' assemblages are well represented in the modern data-set (Birks et al. 1990). For their palaeosalinity reconstructions based on African chironomid assemblages, Verschuren et al. (2004) and Eggermont et al. (2006) additionally assessed to what extent numerical differences between calibration functions affected reconstructed salinity trends through time. Initially this was deemed necessary because, lacking a traditional calibration data-set of faunal composition based on surface-sediment assemblages, Verschuren et al. (2000b) had performed a hybrid procedure in which a weighted-averaging (WA) inference model calibrated with presence/absence distributional records from the literature was applied to abundance-weighted fossil data. This helped ensure that changes in the relative abundance between taxa, not just their new appearance or complete disappearance, would generate a proxy climatic signal. Verschuren et al. (2004) justified this hybrid procedure by proposing that calibration based on presence/absence data is not dissimilar to an abundance-weighted calibration with down-weighting of rare taxa. Using (semi-) independent palaeohydrological reconstructions based on sediment stratigraphy and fossil diatom assemblages (Verschuren et al. 2000b) as a reference framework, coupled with a consideration of modern-day benthic community dynamics in shallow closed-basin African lakes, Eggermont et al. (2006) documented significant variation among calibration functions in reconstructed salinity trends, mainly reflecting their different sensitivity to the presence or relative abundance of certain key taxa. Specifically, WAPLS and maximum likelihood (ML) techniques statistically 'camouflage' the step change in chironomid faunal composition near the freshwater-saline boundary, resulting in less robust reconstructions. These authors concluded that selection of the most appropriate calibration function for environmental reconstruction should not solely optimise the statistical performance of the resulting inference model, but carefully consider whether it produces ecologically meaningful reconstruction results with high signal content (see Juggins and Birks 2012: Chap. 14).

Confidence in palaeoclimatic inferences will also further increase as broad-scale patterns are replicated in space and time. Examples of coherence are starting to emerge at a number of sites, over different temporal and spatial scales, for example, the studies by Sorvari et al. (2002) and Palmer et al. (2002) discussed above. Similarly, several studies (Fritz et al. 2000; Yu et al. 2002; Laird et al. 2003; Schmieder et al. 2011) have presented strong evidence of synchronous changes at various sites in the Great Plains of North America over the last 1500–2000 years.

A number of studies of climatic change have begun to address ecologically interesting questions that attempt to assess changes in stability and diversity (e.g., Anderson et al. 1996; Laird et al. 1998a, b; Wolfe 2003; Stone and Fritz 2006; Hobbs et al. 2010). For example, Laird et al. (1996a, 1998a) found a large switch in inferred

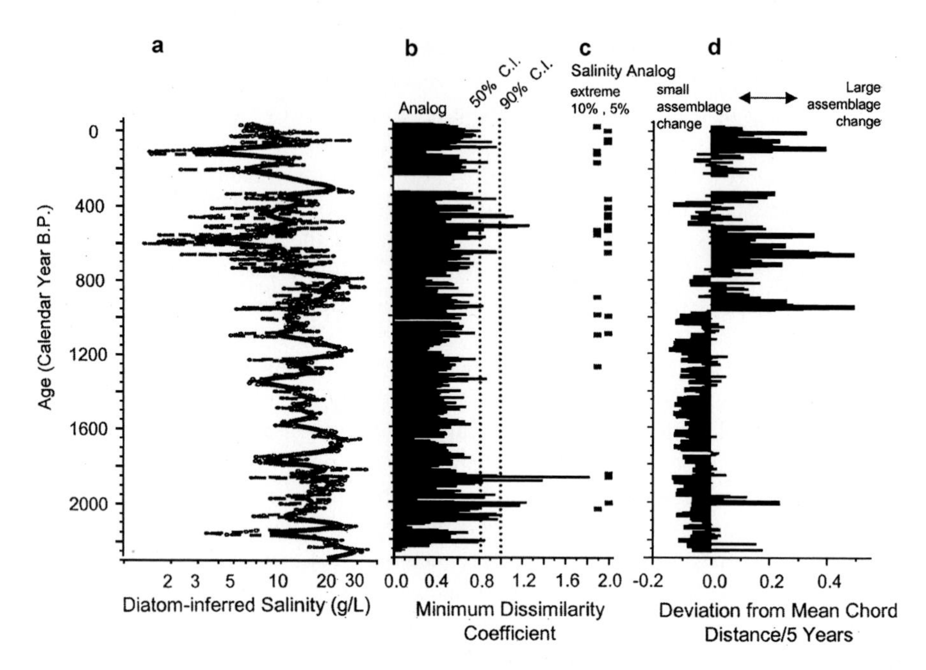

Fig. 20.8 (**a**) Diatom-inferred salinity changes at Moon Lake (North Dakota, USA) over the last 2200 years. (**b**) Analogue analysis and (**c**) 'goodness-of-fit' analyses based on comparisons between dissimilarities between core and modern diatom assemblages, and diagnostics from a constrained ordination, respectively (see text for details). (**d**) The results of a rate-of-change analysis used to assess changes in community stability (Modified from Laird et al. 1998a)

climatic conditions in the record from Moon Lake (North Dakota, USA) indicated by a distinct change in inferred salinity before and after AD 1200 (Fig. 20.8a). The differences in the environmental stability of Moon Lake before and after ~AD 1200 were assessed by estimating the rate-of-change in the diatom assemblages, based on a dissimilarity coefficient between adjacent samples over a fixed time interval (see Birks 2012: Chap. 11). The results of this analysis suggested that the diatom assemblages in Moon Lake were significantly more stable prior to AD 1200 (during a period of inferred prolonged aridity) than for the last ~800 years, which were relatively wet (Fig. 20.8d). This approach has also been employed to assess changes in limnological stability between the early- and late-Holocene (Laird et al. 1998b; Clark et al. 2002), and through periods of Holocene cooling and warming (Anderson et al. 1996; Palmer et al. 2002; Wolfe 2003). Heegaard et al. (2006) used cumulative rate-of-change and cumulative relative rate-of-change to assess the detection of aquatic ecotones along an altitudinal gradient in the Swiss Alps for cladocera, chironomids, and diatoms. Assessing changes in biological richness and diversity has become a topic of renewed interest in palaeolimnological studies. For example, Sorvari et al. (2002) noticed a decline in richness (estimated as Hill's (1973) N2, a measure of richness of common species) coincident with the recent inferred warming of lakes in Finnish Lapland. A significant negative correlation between diatom richness and temperature was also found in northern Sweden (Anderson et al. 1996). Conversely, Wolfe (2003) showed that richness increased following climatic deterioration from the climatic optimum, which started ~2000 years BP.

In the complex system of the Mackenzie River and Slave River delta systems, species richness and 'indicator' taxa were used to distinguish between three hydrologically distinct lake types (Hay et al. 2000; Sokal et al. 2008). Ordination methods were first used to assess the distribution of diatom assemblages in the three lake categories. Sokal et al. (2008) then used analysis of similarities (ANOSIM: Clarke and Warwick 2001) to test whether the diatom assemblages differed significantly among the three lake categories in the Slave River Delta. Finally, similarity percentage tests (SIMPER: Clarke and Gorley 2006) were performed to identify specific diatom taxa that accounted for these differences and canonical variates analysis was used to assess whether 'indicator' taxa significantly discriminate the hydrological lake categories (Sokal et al. 2008). These studies indicate that numerical methods can be used to assess the hydrological and climatic variability within complex delta systems.

Finally, it is becoming increasingly evident that the temporal structure of many palaeolimnological records is composed of periodic components. Correct identification of these periodicities and how they vary over time is becoming increasingly important in palaeolimnological studies. Aspects of this critical topic are covered in Dutilleul et al. (2012: Chap. 16).

Acknowledgement We would like to thank HJB Birks for suggesting that we write this review. His interest and involvement in a wide array of ecological, palaeoecological, and statistical issues continue to serve as a source of inspiration.

References

Aebly F, Fritz SC (2009) Palaeohydrology of Kangerlussuaq (Søndre Strømfjord), West Greenland during the last 8000 years. Holocene 19:91–104

Agusti-Panareda A, Thompson R (2002) Reconstructing air temperatures at eleven remote alpine and arctic lakes in Europe from 1781 to 1997 AD. J Paleolimnol 28:7–23

Ampel L, Bigler C, Wolfarth B, Risberg J, Lotter AF, Veres D (2010) Modest summer temperature variability during DO cycles in western Europe. Quaternary Sci Rev 29:1322–1327

Anderson NJ, Odgaard BV, Segerström U, Renberg I (1996) Climate-lake interactions recorded in varved sediments from a Swedish boreal forest lake. Glob Change Biol 2:399–405

Anderson NJ, Brodersen KP, Ryves DB, McGowan S, Johansson LS, Jeppesen E, Leng MJ (2008) Climate versus in-lake processes as controls on the development of community structure in a low-arctic lake (south-west Greenland). Ecosystems 11:307–324

Baker PA, Seltzer GO, Fritz SC, Dunbar RB, Grove MJ, Tapia PM, Cross SL, Rowe HD, Broda JP (2001) The history of South American tropical precipitation for the past 25,000 years. Science 291:640–643

Barker P, Telford RJ, Merdaci O, Williamson D, Taieb M, Vincens A, Gilbert E (2000) The sensitivity of a Tanzanian crater lake to catastrophic tephra input and four millennia of climate change. Holocene 10:303–310

Battarbee RW (2000) Palaeolimnological approaches to climate change, with special regard to the biological record. Quaternary Sci Rev 19:107–124

Battarbee RW, Grytnes J-A, Thompson R, Appleby PG, Catalan J, Korhola A, Birks HJB, Heegaard E, Lami A (2002a) Comparing palaeolimnological and instrumental evidence of climate change for remote mountain lakes over the last 200 years. J Paleolimnol 28:161–179

Battarbee RW, Thompson R, Catalan J, Grytnes J-A, Birks HJB (2002b) Climate variability and ecosystem dynamics of remote alpine and arctic lakes: the MOLAR project. J Paleolimnol 28:1–6

Battarbee RW, Monteith DT, Juggins S, Simpson GL, Shilland EW, Flower RJ, Kreiser AM (2008) Assessing the accuracy of diatom-based transfer functions in defining reference conditions for acidified lakes in the United Kingdom. Holocene 18:57–67

Bennett JR, Cumming BF, Leavitt PR, Chiu M, Smol JP, Szeicz J (2001) Diatom, pollen, and chemical evidence of postglacial climatic change at Big Lake, south-central British Columbia, Canada. Quaternary Res 55:332–343

Bigler C, Hall RI (2003) Diatoms as quantitative indicators of July temperature: a validation attempt at century-scale with meteorological data from northern Sweden. Palaeogeogr Palaeoclim Palaeoecol 189:147–160

Birks HJB (1998) Numerical tools in palaeolimnology – progress, potentialities, and problems. J Paleolimnol 20:307–332

Birks HJB (2012) Chapter 11: Analysis of stratigraphical data. In: Birks HJB, Lotter AF, Juggins S, Smol JP (eds) Tracking environmental change using lake sediments. Volume 5: Data handling and numerical techniques. Springer, Dordrecht

Birks HJB, Lotter AF (1994) The impact of the Laacher See Volcano (11000 years BP) on terrestrial vegetation and diatoms. J Paleolimnol 11:313–322

Birks HJB, Line JM, Juggins S, Stevenson AC, ter Braak CJF (1990) Diatoms and pH reconstructions. Philos Trans R Soc Lond B 327:263–278

Birks HJB, Heiri O, Seppä H, Bjune AE (2010) Strengths and weaknesses of quantitative climate reconstructions based on late-Quaternary biological proxies. Open Ecol J 3:68–110

Bond G, Kromer B, Beer J, Muscheler R, Evans MN, Showers W, Hoffmann S, Lotti-Bond R, Hajdas I, Bonani G (2001) Persistent solar influence on north Atlantic climate during the Holocene. Science 294:2130–2136

Borcard D, Legendre P, Drapeau P (1992) Partialling out the spatial component of ecological variation. Ecology 73:1045–1055

Bradbury JP (1999) Continental diatoms as indicators of long-term environmental change. In: Smol JP, Stoermer EF (eds) The diatoms: applications for the environmental and earth sciences. Cambridge University Press, Cambridge, pp 169–182

Bradbury P, Cumming B, Laird K (2002) A 1500-year record of climatic and environmental change in Elk Lake, Minnesota III: measures of past primary productivity. J Paleolimnol 27:321–340

Brodersen KP, Anderson NJ (2002) Distribution of chironomids (Diptera) in low Arctic west Greenland lakes: trophic conditions, temperature and environmental reconstruction. Freshw Biol 47:1137–1157

Catalan J, Plas S, Rieradevall M, Felip M, Ventura M, Buchaca T, Camarero L, Brancelj A, Appleby PG, Lami A, Grytnes J-A, Agusti-Panareda A, Thompson R (2002) Lake Redó ecosystem response to an increasing warming in the Pyrenees during the twentieth century. J Paleolimnol 28:129–145

Clark JS, Grimm EC, Donovan JJ, Fritz SC, Engstrom DR, Almendinger JE (2002) Drought cycles and landscape responses to past aridity on prairies of the northern Great Plains, USA. Ecology 83:595–601

Clarke KR, Gorley RN (2006) PRIMER v6: User manual/tutorial. PRIMER-E, Plymouth

Clarke KR, Warwick RM (2001) Change in marine communities: an approach to statistical analysis and interpretation, 2nd edn. PRIMER-E, Plymouth

Cumming BF, Moser KA (2007) Applications of commonly used numerical techniques in diatom-based paleoecology. Paleontol Soc Pap 13:37–56

Cumming BF, Laird KR, Bennett JR, Smol JP, Salomon AK (2002) Persistent millennial-scale shifts in moisture regimes in western Canada during the past six millennia. Proc Natl Acad Sci USA 99:16117–16121

Dean WE (2002) A 1500-year record of climatic and environmental change in Elk Lake, Clearwater County, Minnesota II: geochemistry, mineralogy, and stable isotopes. J Paleolimnol 27:301–319

Dean WE, Anderson RY, Bradbury JP (2002) A 1500-year record of climatic and environmental change in Elk Lake, Minnesota - I: Varve thickness and gray-scale density. J Paleolimnol 27:287–299

Denton GH, Karlen W (1973) Holocene climatic variations – their pattern and possible cause. Quaternary Res 3:155–205

Donovan JJ, Smith AJ, Panek VA, Engstrom DR, Ito E (2002) Climate-driven hydrologic transients in lake sediment records: calibration of groundwater conditions using 20th century drought. Quaternary Sci Rev 21:605–624

Doran PT, Priscu JC, Lyons WB, Walsh JE, Fountain AG, McKnight DM, Moorhead DL, Virginia RA, Wall DH, Clow GD, Fritsen CH, McKay CP, Parsons AN (2002) Antarctic climate cooling and terrestrial ecosystem response. Nature 415:517–520

Douglas MSV, Smol JP (2010) Freshwater diatoms as indicators of environmental change in the high arctic. In: Smol JP, Stoermer EF (eds) The diatoms: applications for the environmental and earth sciences, 2nd edn. Cambridge University Press, Cambridge, pp 249–266

Dutilleul P, Cumming BF, Lontoc-Roy M (2012) Chapter 16: Autocorrelogram and periodogram analysis of palaeolimnological temporal series from lakes in central and western North America to assess shifts in drought conditions. In: Birks HJB, Lotter AF, Juggins S, Smol JP (eds) Tracking environmental change using lake sediments. Volume 5: Data handling and numerical techniques. Springer, Dordrecht

Eggermont H, Heiri O, Verschuren D (2006) Fossil Chironomidae (Insecta: Diptera) as quantitative indicators of past salinity in African lakes. Quaternary Sci Rev 25:1966–1994

Eggermont H, Verschuren D, Fagot M, Rumes B, Van Bocxlaer B, Kröpelin S (2008) Aquatic community response in a groundwater-fed desert lake to Holocene desiccation of the Sahara. Quaternary Sci Rev 27:2411–2425

Eggermont H, Verschuren D, Audenaert L, Lens L, Russell J, Klaassen G, Heiri O (2010a) Limnological and ecological sensitivity of Rwenzori mountain lakes to climate warming. Hydrobiologia 648:123–142

Eggermont H, Heiri O, Russell J, Vuille M, Audenaert L, Verschuren D (2010b) Paleotemperature reconstruction in tropical Africa using fossil Chironomidae (Insecta: Diptera). J Paleolimnol 43:413–435

Enache MD, Cumming BF (2009) Extreme fires under warmer and drier conditions inferred from sedimentary charcoal morphotypes from Opatcho Lake, central British Columbia, Canada. Holocene 19:835–846

Engstrom DR, Nelson S (1991) Palaeosalinity from trace metals in fossil ostracods compared with observational records at Devils Lake, North Dakota, USA. Palaeogeogr Palaeoclim Palaeoecol 83:295–312

Fritz SC (1990) Twentieth-century salinity and water-level fluctuations in Devils Lake, North Dakota: Test of a diatom-based transfer function. Limnol Oceanogr 35:1171–1781

Fritz SC (1996) Palaeolimnological records of climate change in North America. Limnol Oceanogr 41:882–889

Fritz SC (2008) Deciphering climatic history from lake sediments. J Paleolimnol 39:5–16

Fritz SC, Ito E, Yu Z, Laird KR, Engstrom DR (2000) Hydrologic variation in the northern Great Plains during the last two millennia. Quaternary Res 53:175–184

Fritz SC, Cumming BF, Gasse F, Laird KR (2010) Diatoms as indicators of hydrologic and climatic change in saline lakes. In: Smol JP, Stoermer EF (eds) The diatoms: applications for the environmental and earth sciences, 2nd edn. Cambridge University Press, Cambridge, pp 186–208

Gajewski K (2008) The global pollen database in biogeographical and palaeoclimatic studies. Progr Phys Geog 32:379–402

Hall RI, Leavitt PR, Quinlan R, Dixit AS, Smol JP (1999) Effects of agriculture, and climate on water quality in the northern Great Plains. Limnol Oceanogr 44:739–756

Harrison SP, Digerfeldt G (1993) European lakes as palaeohydrological and palaeoclimatic indicators. Quaternary Sci Rev 12:233–248

Hausmann S, Lotter AF, van Leeuwen JFN, Ohlendorf C, Lemcke G, Gronlund E, Strum M (2002) Interactions of climate and land use documented in the varved sediments of Seebergsee in the Swiss Alps. Holocene 12:279–289

Hay MB, Michelutti N, Smol JP (2000) Ecological patterns of diatom assemblages from Mackenzie Delta lakes, Northwest Territories, Canada. Can J Bot 78:19–33

Heegaard E, Lotter AF, Birks HJB (2006) Aquatic biota and the detection of climate change: are there consistent aquatic ecotones? J Paleolimnol 35:507–518

Heiri O, Lotter AF (2005) Holocene and lateglacial summer temperature reconstruction in the Swiss Alps based on fossil assemblages of aquatic organisms: a review. Boreas 34:506–516

Heiri O, Tinner W, Lotter AF (2004) Evidence for cooler European summers during periods of changing meltwater flux to the North Atlantic. Proc Natl Acad Sci USA 101:15285–15288

Hernandez A, Giralt S, Bao R, Saez A, Leng MJ, Barker PA (2010) ENSO and solar activity signals from oxygen isotopes in diatom silica during late glacial-Holocene transition in Central Andes (18° S). J Paleolimnol 44:413–429

Hill MO (1973) Diversity and evenness: a unifying notation and its consequences. Ecology 54: 427–431

Hobbs WO, Telford R, Birks HJB, Saros J, Hazewinkel R, Perren B, Saulnier E, Wolfe A (2010) Quantifying recent ecological changes in remote lakes of North America and Greenland using sediment diatom assemblages. PLoS One 5:e10026

Hübener T, Dreßler M, Schwarz A, Langner K, Adler S (2008) Dynamic adjustment of training sets (‘moving-window’ reconstruction) by using transfer functions in paleolimnology – a new approach. J Paleolimnol 40:79–95

Interlandi SJ, Kilham SS, Theriot EC (1999) Responses of phytoplankton to varied resource availability in large lakes of the Greater Yellowstone Ecosystem. Limnol Oceanogr 44: 668–682

St. Jacques J-M, Cumming BF, Smol JP (2008a) A pre-European settlement pollen-climate calibration set for Minnesota, USA: developing tools for palaeoclimatic reconstructions. J Biogeogr 35:306–324

St. Jacques J-M, Cumming BF, Smol JP (2008b) A 900-year pollen-inferred temperature and effective moisture record from varved Lake Mina, west-central Minnesota, USA. Quaternary Sci Rev 27:781–796

Juggins S, Birks HJB (2012) Chapter 14: Quantitative environmental reconstructions from biological data. In: Birks HJB, Lotter AF, Juggins S, Smol JP (eds) Tracking environmental change using lake sediments. Volume 5: Data handling and numerical techniques. Springer, Dordrecht

Kendall RL (1969) An ecological history of the Lake Victoria basin. Ecol Monogr 39:121–176

Koinig KA, Kamenik C, Schmidt R, Agusti-Panareda A, Appleby PG, Lami A, Prazakova M, Rose N, Schnell OA, Tessadri R, Thompson R, Psenner R (2002) Environmental changes in an alpine lake (Gossenkollesee, Austria) over the last two centuries the influence of air temperature on biological parameters. J Paleolimnol 28:147–160

Korhola A (1999) Distribution patterns of Cladocera in subarctic Fennoscandian lakes and their potential in environmental reconstruction. Ecography 22:357–373

Korhola A, Rautio M (2001) Cladocera and other small brachiopods. In: Smol JP, Birks HJB, Last WM (eds) Tracking environmental change using lake sediments. Volume 4: Zoological indicators. Kluwer Academic Publishers, Dordrecht, pp 5–41

Korhola A, Vasko K, Toivonen HTT, Olander H (2002) Holocene temperature change in northern Fennoscandia reconstructed from chironomids using Bayesian modelling. Quaternary Sci Rev 21:1841–1860

Köster D, Racca JMJ, Pienitz R (2004) Diatom-based inference models and reconstructions revisited: methods and transformations. J Paleolimnol 32:233–246

Kröpelin S, Verschuren D, Lézine AM, Eggermont H, Cocquyt C, Francus P, Cazet J-P, Fagot M, Rumes B, Russell JM, Conley S, Schuster M, von Suchodoletz H, Engstrom DR (2008) Climate-driven ecosystem succession in the central Sahara: the last 6000 years. Science 320:765–768

Laird KR, Fritz SC, Grimm EC, Mueller PG (1996a) Century-scale palaeoclimatic reconstruction from Moon Lake, a closed-basin lake in the northern Great Plains. Limnol Oceanogr 41: 890–902

Laird KR, Fritz SC, Maasch KA, Cumming BF (1996b) Greater drought intensity and frequency before AD 1200 in the northern Great Plains, USA. Nature 384:552–554

Laird KR, Fritz SC, Cumming BF (1998a) A diatom-based reconstruction of drought intensity, duration, and frequency from Moon Lake, North Dakota: a sub-decadal record of the last 2300 years. J Paleolimnol 19:161–179

Laird KR, Fritz SC, Cumming BF (1998b) Early-Holocene limnological and climatic variability in the northern Great Plains. Holocene 8:275–286

Laird KR, Cumming BF, Wunsam S, Rusak J, Oglesby RJ, Fritz SC, Leavitt PR (2003) Large-scale shifts in moisture regimes from lake records across the northern prairies of North America during the past two millennia. Proc Natl Acad Sci USA 100:2483–2488

Laird KR, Kingsbury MV, Lewis CFM, Cumming BF (2011) Diatom-inferred depth models in 8 Canadian boreal lakes: inferred changes in the benthic:planktonic depth boundary in northwestern Ontario over the Holocene. Quaternary Sci Rev 30:1201–1217

Lamb HF, Darbyshire I, Verschuren D (2003) Vegetation response to rainfall variation and human impact in central Kenya during the past 1100 years. Holocene 13:285–292

Lamoureux S (2000) Five centuries of interannual sediment yield and rainfall-induced erosion in the Canadian high arctic recorded in lacustrine varves. Water Res 36:309–318

Larocque I, Hall RI (2003) Chironomids as quantitative indicators of mean July air temperature: validation by comparison with century-long meteorological records from northern Sweden. J Paleolimnol 29:475–493

Larocque I, Hall RI, Grahn E (2001) Chironomids as indicators of climate change: a 100 lake training set from a subarctic region of northern Sweden (Lapland). J Paleolimnol 172:133–142

Legendre P, Birks HJB (2012) Chapter 8: From classical to canonical ordination. In: Birks HJB, Lotter AF, Juggins S, Smol JP (eds) Tracking environmental change using lake sediments. Volume 5: Data handling and numerical techniques. Springer, Dordrecht

Legesse D, Gasse F, Radakovitch O, Vallet-Coulomb C, Bonnefille R, Verschuren D, Gibert E, Barker P (2002) Environmental changes in a tropical lake (Lake Abiyata, Ethiopia) during recent centuries. Palaeogeogr Palaeoclim Palaeoecol 187:233–258

Lotter AF (1998) The recent eutrophication of Baldeggersee (Switzerland) as assessed by fossil diatom assemblages. Holocene 8:95–405

Lotter AF, Anderson NJ (2012) Chapter 18: Limnological responses to environmental changes at inter-annual to decadal time-scales. In: Birks HJB, Lotter AF, Juggins S, Smol JP (eds) Tracking environmental change using lake sediments. Volume 5: Data handling and numerical techniques. Springer, Dordrecht

Lotter AF, Birks HJB (1997) The separation of the influence of nutrients and climate on the varve time-series of Baldeggersee, Switzerland. Aquat Sci 59:362–375

Lotter AF, Birks HJB, Zolitschka B (1995) Late-glacial pollen and diatom changes in response to two different environmental perturbations: volcanic eruption and Younger Dryas cooling. J Paleolimnol 14:23–47

Lotter AF, Birks HJB, Hofmann W, Marchetto A (1997) Modern diatom, cladocera, chironomid and chrysophyte cyst assemblages as quantitative indicators for the reconstruction of past environmental conditions in the Alps. I. Climate. J. Paleolimnol 18:395–420

Lotter AF, Pienitz P, Schmidt R (2010) Diatoms as indicators of environmental change in subarctic and alpine regions. In: Smol JP, Stoermer EF (eds) The diatoms: applications for

the environmental and earth sciences, 2nd edn. Cambridge University Press, Cambridge, pp 231–248
MacDonald GM, Edwards TWD, Moser KA, Pienitz R, Smol JP (1993) Rapid response of treeline vegetation and lakes to past climate warming. Nature 361:243–246
Noon PE, Birks HJB, Jones VJ, Ellis-Evans JC (2001) Quantitative models for reconstructing catchment ice-extent using physical-chemical characteristics of lake sediments. J Paleolimnol 25:375–392
Olaka LA, Odada EO, Trauth MH, Olago DO (2010) The sensitivity of East African rift lakes to climate fluctuations. J Paleolimnol 44:629–644
Palmer S, Walker I, Heinrichs M, Hebda R, Scudder G (2002) Postglacial midge community change and Holocene palaeotemperatures reconstructions near treeline, southern British Columbia (Canada). J Paleolimnol 28:469–490
Pienitz R, Smol JP, Birks HJB (1995) Assessment of freshwater diatoms as quantitative indicators of past climatic-change in the Yukon and Northwest Territories, Canada. J Paleolimnol 13: 21–49
Pienitz R, Smol JP, MacDonald GM (1999) Paleolimnological reconstruction of Holocene climatic trends from two boreal treeline lakes, Northwest Territories, Canada. Arct Antarct Alp Res 31:82–93
Pienitz R, Lotter AF, Newman L, Kiefer T (eds) (2009) Advances in paleolimnology. PAGES News 17:89–136
Prentice IC, Bartlein PJ, Webb T (1991) Vegetation and climate change in eastern North America since the last glacial maximum. Ecology 72:2038–2056
Psenner R, Schmidt R (1992) Climate-driven pH control of remote alpine lakes and effects of acid deposition. Nature 356:781–783
Quinlan R, Leavitt PR, Dixit AS, Hall RI, Smol JP (2002) Landscape effects of climate, agriculture, and urbanization on benthic invertebrate communities of Canadian prairie lakes. Limnol Oceanogr 47:378–391
Racca JMJ, Wild M, Birks HJB, Prairie YT (2003) Separating wheat from chaff: diatom taxon selection using an artificial neural network pruning algorithm. J Paleolimnol 29:123–133
Ritchie JC (1987) Postglacial vegetation of Canada. Cambridge University Press, Cambridge
SAS Inc (2004) Online Doc 9.1.3. SAS Institute Inc., Cary
Schindler DW, Bayley SE, Parker BR, Beaty KG, Cruikshank DR, Fee EJ, Stainton MP (1996) The effect of climatic warming on the properties of boreal lakes and streams at the Experimental Lakes Area, northwestern Ontario. Limnol Oceanogr 41:1004–1017
Schmieder J, Fritz SC, Swinehart J, Shinneman AC, Wolfe AP, Miller G, Daniels N, Jacobs K, Grimm EC (2011) A regional-scale climate reconstruction of the last 4000 years from lakes in the Nebraska Sand Hills, USA. Quaternary Sci Rev 30:1797–1812
Schwalb A, Dean WE (2002) Reconstruction of hydrological changes and response to effective moisture variations from North-Central USA lake sediments. Quaternary Sci Rev 21: 1541–1554
Simpson GL (2012) Chapter 15: Analogue methods in palaeolimnology. In: Birks HJB, Lotter AF, Juggins S, Smol JP (eds) Tracking environmental change using lake sediments. Volume 5: Data handling and numerical techniques. Springer, Dordrecht
Simpson GL, Hall IR (2012) Chapter 19: Human impacts – applications of numerical methods to evaluate surface-water acidification and eutrophication. In: Birks HJB, Lotter AF, Juggins S, Smol JP (eds) Tracking environmental change using lake sediments. Volume 5: Data handling and numerical techniques. Springer, Dordrecht
Smith LC, Sheng Y, MacDonald GM, Hinzman LD (2005) Disappearing arctic lakes. Science 308:1429
Smol JP, Cumming BF (2000) Tracking long-term changes in climate using algal indicators in lake sediments. J Phycol 36:986–1011
Smol JP, Douglas MSV (2007) From controversy to consensus: making the case for recent climatic change in the arctic using lake sediments. Front Ecol Environ 5:466–474

Smol JP, Wolfe AP, Birks HJB, Douglas MSV, Jones VJ, Korhola A, Pienitz R, Ruhland K, Sorvari S, Antoniades D, Brooks SJ, Fallu MA, Hughes M, Keatley BE, Laing TE, Michelutti N, Nazarova L, Nyman M, Paterson AM, Perren B, Quinlan R, Rautio M, Saulnier-Talbot E, Siitoneni S, Solovieva N, Weckstrom J (2005) Climate-driven regime shifts in the biological communities of arctic lakes. Proc Natl Acad Sci USA 102:4397–4402

Sokal MA, Hall RI, Wolfe BB (2008) Relationships between hydrological and limnological conditions in lakes of the Slave River Delta (NWT, Canada) and quantification of their roles on sedimentary diatom assemblages. J Paleolimnol 39:533–550

Sommaruga-Wögrath S, Koinig KA, Schmidt R, Sommaruga R, Tessadri R, Psenner R (1997) Temperature effects on the acidity of remote alpine lakes. Nature 387:64–67

Sorvari S, Korhola A, Thompson R (2002) Lake diatom response to recent Arctic warming in Finnish Lapland. Glob Change Biol 8:171–181

Spooner IS, Mazzucchi D, Osborn G, Gilbert R, Larocque I (2002) A multi-proxy Holocene record of environmental change from the sediments of Skinny Lake, Iskut region, northern British Columbia, Canada. J Paleolimnol 28:419–431

Stager JC, Johnson TC (2000) A 12,400 C-14 yr offshore diatom record form east central Lake Victoria, East Africa. J Paleolimnol 23:373–383

Stager JC, Cumming BF, Meeker L (1997) A high-resolution 11,400-yr diatom record from Lake Victoria, East Africa. Quaternary Res 47:81–89

Stager JC, Cumming BF, Meeker L (2003) A 10,000 year high-resolution diatom record from Pilkington Bay, Lake Victoria, East Africa. Quaternary Res 59:172–181

Stone JR, Fritz SC (2006) Multidecadal drought and Holocene climate instability in the Rocky Mountains. Geology 34:409–412

Stone JR, Westover KS, Cohen A (2011) Late Pleistocene paleohydrography and diatom paleoecology of the central basin of Lake Malawi, Africa. Palaeogeog Palaeoclim Palaeoecol 303:51–70

Telford RJ, Birks HJB (2011) A novel method for assessing the statistical significance of quantitative reconstructions inferred from biotic assemblages. Quaternary Sci Rev 30: 1272–1278

Tierney JE, Russell JM, Huang Y, Sinninghe Damsté JS, Hopmans EC, Cohen AS (2008) Northern hemisphere controls on tropical southeast African climate during the past 60,000 years. Science 322:252–255

Toney J, Huang Y, Fritz SC, Baker PA, Nyren P, Grimm E (2010) Climatic and environmental controls on the occurrence and distribution of long-chain alkenones in lakes of the interior United States. Geochim Cosmchim Acta 74:1563–1578

Verbruggen F, Heiri O, Reichart GJ, Lotter AF (2010) Chironomid δ^{18}O as a proxy for past lake water δ^{18}O: a Late-glacial record from Rotsee (Switzerland). Quaternary Sci Rev 29: 2271–2279

Verschuren D (1996) Comparative palaeolimnology in a system of four shallow, climate-sensitive tropical lake basins. In: Johnson TC, Odada E (eds) The limnology, climatology and palaeoclimatology of the east African lakes. Gordon and Breach, Newark, pp 559–572

Verschuren D (1999a) Sedimentation controls on the preservation and time resolution of climate-proxy records from shallow fluctuating lakes. Quaternary Sci Rev 18:821–837

Verschuren D (1999b) Influence of depth and mixing regime on sedimentation in a small fluctuating tropical soda lake. Limnol Oceanogr 44:1103–1113

Verschuren D (2001) Reconstructing fluctuations of a shallow East African lake during the past 1800 yrs from sediment stratigraphy in a submerged crater basin. J Paleolimnol 25:297–311

Verschuren D (2003) Lake-based climate reconstruction in Africa: progress and challenges. Hydrobiologia 500:315–330

Verschuren D, Tibby J, Leavitt PR, Roberts CN (1999a) The environmental history of a climate-sensitive lake in the former 'White Highlands' of central Kenya. Ambio 28:494–501

Verschuren D, Cocquyt C, Tibby J, Roberts CN, Leavitt PR (1999b) Long-term dynamics of algal and invertebrate communities in a small, fluctuating tropical soda lake. Limnol Oceanogr 44:1216–1231

Verschuren D, Tibby J, Sabbe K, Roberts N (2000a) Effects of depth, salinity, and substrate on the invertebrate community of a fluctuating tropical lake. Ecology 81:164–182

Verschuren D, Laird K, Cumming BF (2000b) Rainfall and drought in equatorial East Africa during the past 1,100 years. Nature 403:410–414

Verschuren D, Cumming BF, Laird KR (2004) Quantitative reconstruction of past salinity variations in African lakes using fossil midges (Diptera: Chironomidae): assessment of inference models in space and time. Can J Fish Aquat Sci 61:986–998

Vinebrooke RD, Hall RI, Leavitt PR, Cumming BF (1998) Fossil pigments as indicators of phototrophic response to salinity and climatic changes in lakes of western Canada. Can J Fish Aquat Sci 55:668–681

von Grafenstein U, Erlenkeuser H, Brauer A, Jouzel J, Johnsen SJ (1999) A mid-European decadal isotope-climate record from 15,500 to 5000 years BP. Science 284:1654–1657

von Gunten L, Heiri O, Bigler C, Casty C, Lotter AF, Sturm M (2008) Seasonal temperatures for the past 400 years reconstructed from diatom and chironomid assemblages in a high-altitude lake (Lej da la Tscheppa, Switzerland). J Paleolimnol 39:283–299

Walker IR (1995) Chironomids as indicators of past environmental change. In: Armitage PD, Cranston PS, Pinder LCV (eds) The chironomidae. The biology and ecology of nonbiting midges. Chapman & Hall, London, pp 405–422

Walker IR, Smol JP, Engstrom DR, Birks HJB (1991) An assessment of Chironomidae as quantitative indicators of past climatic change. Can J Fish Aquat Sci 28:975–987

Webb T (1986) Is vegetation in equilibrium with climate? How to interpret late-Quaternary pollen data. Vegetatio 67:75–91

Webster KE, Soranno PA, Baines SB, Kratz TK, Bowser CJ, Dillon PJ, Campbell P, Fee EJ, Hecky RE (2000) Structuring features of lake districts: landscape controls on lake chemical responses to drought. Freshw Biol 43:499–515

Wolfe AP (2003) Diatom community responses to late-Holocene climatic variability, Baffin Island, Canada: a comparison of numerical approaches. Holocene 13:29–37

Wolfe BB, Edwards TWD, Aravena R, MacDonald GM (1996) Rapid Holocene hydrologic changes along boreal treeline revealed by $\delta^{13}C$ and $\delta^{18}O$ in organic lake sediments, Northwest Territories, Canada. J Paleolimnol 15:171–181

Wolin JA, Stone JR (2010) Diatoms as indicators of water-level change in freshwater lakes. In: Smol JP, Stoermer EF (eds) The diatoms: applications for the environmental and earth sciences, 2nd edn. Cambridge University Press, Cambridge, pp 174–185

Yu Z, Ito E, Engstrom DR, Fritz SC (2002) A 2100-year trace-element and stable-isotope record at decadal resolution from Rice Lake in the northern Great Plains, USA. Holocene 12: 605–617

Chapter 21
Conclusions and Future Challenges

H. John B. Birks

Abstract Quantitative palaeolimnology has made great advances in the last 20 years. The subject is not static, however, and as more and more demanding questions are asked of palaeolimnology in the future, there will be more and more future numerical challenges to be addressed and subjects to be explored. These include the problems of model selection, trait analysis, data-mining, time-warp analysis, quantile regression, additive modelling, new techniques for temporal-series analysis, and increasing use of Bayesian inference. The practical problems of computing and of available software are also discussed and it is clear that future developments in quantitative palaeolimnology will depend on researchers becoming proficient in the use of R and its innumerable packages relevant to palaeolimnological data-analysis.

Keywords Additive modelling • Additive monotone regression splines • Bayesian hierarchical models • Bayesian inference • Biological dynamics • Computing • Constrained Gaussian ordination • Data-mining • Ecological modelling • Mechanistic modelling • Model selection • Multi-proxy studies • Palaeolimnology • Principal response curves • Quadratic reduced-rank vector-based generalised linear and additive models • Quantile regression • R • Software • Statistical modelling • Temporal-series analysis • Time-warp analysis • Trait analysis • Uncertainty

---

H.J.B. Birks (✉)
Department of Biology and Bjerknes Centre for Climate Research, University of Bergen, PO Box 7803, N-5020 Bergen, Norway

Environmental Change Research Centre, University College London, Pearson Building, Gower Street, London WC1E 6BT, UK

School of Geography and the Environment, University of Oxford, Oxford OX1 3QY, UK
e-mail: john.birks@bio.uib.no

H.J.B. Birks et al. (eds.), *Tracking Environmental Change Using Lake Sediments*, Developments in Paleoenvironmental Research 5, DOI 10.1007/978-94-007-2745-8_21,

Introduction

Palaeolimnology has witnessed many important advances in the past 25–30 years with the study of an increasing number of biological, chemical, and physical proxies preserved in lake sediments (Last and Smol 2001a, b; Smol et al. 2001a, b; Cohen 2003; Smol 2008; Pienitz et al. 2009) and with ever-increasing attention to project design, site selection, sampling protocols, analytical procedures, and data interpretation. Numerical and statistical analyses of palaeolimnological data have contributed to the rapid development of palaeolimnology as a rigorous, quantitative branch of environmental science. Considerable advances have been made in the quantitative analysis of palaeolimnological data since the pioneering attempts in the early 1970s and the seminal publications of ter Braak (1986), ter Braak and Prentice (1988), and ter Braak and van Dam (1989). These advances have contributed primarily to the descriptive and narrative phases (*sensu* Ball 1975) of palaeolimnological studies. In the descriptive phase, basic patterns are detected, assessed, described, and grouped and the relevant numerical techniques for data collection, assessment, and summarisation such as error estimation (Maher et al. 2012: Chap. 6), exploratory data analysis (Juggins and Telford 2012: Chap. 5), clustering and partitioning (Legendre and Birks 2012a: Chap. 7), classical and constrained ordination (Legendre and Birks 2012b: Chap. 8), classification and regression trees and other tree-based and statistical-learning techniques (Simpson and Birks 2012: Chap. 9), self-organising maps (Simpson and Birks 2012: Chap. 9), principal curves (Simpson and Birks 2012: Chap. 9), constrained clustering and partitioning (Birks 2012b: Chap. 11), age-depth modelling (Blaauw and Heegaard 2012: Chap. 12), and core correlation by sequence-slotting (Thompson et al. 2012: Chap. 13). In the narrative phase, inductively based explanations, reconstructions, and generalisations are derived from the observed patterns as, for example, environmental reconstructions based on regression and calibration (Juggins and Birks 2012: Chap. 14), modern analogues using the modern analogue technique (Simpson 2012: Chap. 15), and temporal patterns such as periodicities (Dutilleul et al. 2012: Chap. 16). These numerical approaches all fall under the general category of data analysis (Birks 2012a: Chap. 2) where particular numerical characteristics are estimated from palaeolimnological stratigraphical data.

Statistical methods, such as regression analysis and statistical modelling (Birks 2012a: Chap. 2), and constrained ordination techniques, such as canonical correspondence analysis, redundancy analysis, and their partial relatives, and associated Monte Carlo permutation tests and variation partitioning (Birks 2012a: Chap. 2; Legendre and Birks 2012b: Chap. 8) provide one approach to data interpretation and to the analytical phase (*sensu* Ball 1975) in palaeolimnology (Birks 1998) where testable and falsifiable hypotheses are proposed, evaluated, tested, and rejected (Dutilleul et al. 2012: Chap. 16; Lotter and Anderson 2012: Chap. 18; Simpson and Hall 2012: Chap. 19; Cumming et al. 2012: Chap. 20).

Quantitative palaeolimnologists have every reason to be proud of their achievements in the last ~20 years but no vibrant science like palaeolimnology is ever static and more detailed and more demanding questions are being asked of palaeolimnology and palaeolimnological data (e.g., Dearing 2008; Leavitt et al. 2009; Saros 2009; Dearing et al. 2010, 2011; Downes 2010; Sayer et al. 2010; Smol 2010). These and related questions present many new challenges to the numerical and statistical analysis of palaeolimnological data. What are these future challenges? What will be the new directions of study in quantitative palaeolimnology?

Model Selection

A recurring theme that runs through many of the chapters in this book is the question of model selection in data analysis and statistical modelling. Model selection arises in, for example, age-depth modelling (Blaauw and Heegaard 2012: Chap. 12), environmental reconstructions and calibration functions (Juggins and Birks 2012: Chap. 14), partitioning of data (Birks 2012a, b: Chaps. 2 and 11; Legendre and Birks 2012a: Chap. 7), sequence-slotting of stratigraphical sequences (Thompson et al. 2012: Chap. 13), classification and regression trees and artificial neural networks (Birks 2012b: Chap. 11; Simpson and Birks 2012: Chap. 9), the modern analogue technique (Simpson 2012: Chap. 15), ordinations both classical and constrained (Legendre and Birks 2012b: Chap. 8), and temporal-series analysis (Dutilleul et al. 2012: Chap. 16). Despite its importance in so many aspects of quantitative palaeolimnology, model selection is a topic that has received surprisingly little attention in palaeolimnology (Birks 1998). Exceptions include Birks et al. (1990), ter Braak and Juggins (1993), Cottingham et al. (2000), Racca et al. (2001, 2003), Korhola et al. (2002), Paterson et al. (2002), Köster et al. (2004), Telford et al. (2004), Telford and Birks (2005, 2009), Ferguson et al. (2008), and Simpson and Anderson (2009). In contrast, model selection has received comparably much more attention in ecology and evolutionary biology (Hilborn and Mangel 1997; Johnson and Omland 2004; Mangel 2006; Whittingham et al. 2006; DR Anderson 2008). See Wolf and Mangel (2008) for an example of careful model selection within the framework of multiple hypothesis testing.

The key aspects of model selection in the context of modern statistical modelling (Burnham and Anderson 2002; DR Anderson 2008) are that the approach of testing a null hypothesis is replaced by model selection as a means of making inferences (Stephens et al. 2007). In the model-selection approach, several equally plausible minimally adequate models, each representing one hypothesis, are simultaneously evaluated in terms of support from the observed data. Models can be ranked and assigned weights, providing a quantitative measure of relative support for each hypothesis. When different models have similar levels of support, model averaging can be used to make robust parameter estimates and predictions (Kass and Raftery 1995; Volinsky et al. 1997; Hoeting et al. 1999; Johnson and Omland 2004; Wang et al. 2004; Callaghan and Ashton 2008).

There are well developed numerical approaches to model selection (Burnham and Anderson 2002). These include the adjusted coefficient of determination (r_{adj}^2 or R_{adj}^2) (Sokal and Rohlf 1995) as a measure of fit where the number of parameters included in the model is included in estimating R^2; likelihood-ratio tests (Sokal and Rohlf 1995); Akaike Information Criterion (AIC) (see Birks 2012: Chap. 2), and the small sample unbiased AIC (AIC_c) (Burnham and Anderson 2002) that involve measures of fit and model complexity and with a bias-correction term for small sample-size in AIC_c; and Schwarz's criterion (Schwarz 1978) (also known as the Bayesian Information Criterion (BIC) – see Birks 2012a: Chap. 2) where model fit, complexity, and sample size are all considered.

Shrinkage techniques, such as principal components regression, partial least squares regression (van der Voet 1999), ridge regression, the lasso, and the elastic net (Dahlgren 2010; Hastie et al. 2011; Simpson and Birks 2012: Chap. 9), can help with model selection and the very real problem with biological and environmental data of collinearity problems. In such problems, the explanatory or predictor variables are related by a linear function, thereby making the estimation of regression coefficients impossible or, at least, unstable. These and related techniques (Hastie et al. 2011) are an important addition to the palaeolimnologist's numerical tool-kit, because as research questions in palaeolimnology become more refined and demanding, model-selection and variable procedures will play an increasing part in quantitative palaeolimnology involving statistical modelling, model development, environmental inferences, and evaluating competing hypotheses.

There are currently three philosophically different paradigms used to make statistical inferences and to select between statistical models (Alexander et al. 2011). These are the null-hypothesis approach of classical inferential statistics, the information-theory based approach involving AIC and BIC (Ramsey and Schafer 1997), and Bayesian inference. At present there is no consensus view and palaeolimnologists should keep an open and critical mind in model selection (see Murtaugh 1998; Anderson and Burnham 2002; Guthery et al. 2005; Stephens et al. 2005; Hobbs and Hilborn 2006; Hoeting et al. 2006; Lukacs et al. 2007; Sauerbrei et al. 2007; Raffalovich et al. 2008; Ward 2008 for contrasting views on model selection).

In model selection, as in *all* numerical data-analyses, it is important to favour simplicity (Occam's (or Ockham's) razor or the principle of parsimony) and avoid the danger that Murtaugh (2007) quotes as "I can easily test the hypothesis by simple *t*-tests, but want something more 'elegant' that will fit well with a 'better' journal". There are several examples in the recent palaeolimnological literature where basic biology, limnology, ecology, and statistics appear to have been sacrificed for 'fancy' statistical techniques in so-called 'better' journals (cf. Chamberlain 2008).

Modelling of Biological Dynamics

Palaeolimnology is the study of lakes in the past using the physical, chemical, and biological information archived in lake sediments (Smol 2008; Smol et al. 2012: Chap. 1). In practice, it has mainly been concerned with the reconstruction of past biota, populations, communities, environments, and ecosystems. In such reconstructions, often all the available palaeolimnological data—biotic and abiotic––are used. With the recent upsurge of interest and activity in multi-proxy studies (Lotter 2003; Birks and Birks 2006) where a range of biotic and abiotic proxies is studied on the same core or set of correlated cores, palaeolimnology can better understand a lake's ecology and its dynamics in the past.

There are at least two major approaches to studying the past ecology of lakes. First, we study the responses of limnic organisms in the past preserved in sediments to environmental change but the palaeoenvironmental record is **not** based on the fossil group of interest. Instead it is based on independent palaeoenvironmental records such as stable isotopes (e.g., Wolfe et al. 2001, 2007; Leng and Marshall 2004; Wooller et al. 2004, 2008; Leng 2006; Heiri et al. 2009; Verbruggen et al. 2010), lake-level changes (Jackson and Booth 2002; Shuman et al. 2004), sediment geochemistry (Boyle 2001; Grosjean et al. 2009; Virah-Sawmy et al. 2009a, b; Weijers et al. 2009; Trachsel et al. 2010; Jeffers et al. 2011a, b), or sedimentary parameters (e.g. Francus 2004; Blass et al. 2007). This approach is very much in the scientific philosophy and methodology of using "the geological record of ecological dynamics" and "the geological record as an ecological laboratory" presented by Flessa and Jackson (2005a, b).

Second, in a multi-proxy study (e.g., Lotter 2003; Birks and Birks 2006), one or more biological proxies can be used to provide an independent palaeoenvironmental reconstruction using calibration functions (e.g., chironomids as a temperature proxy (Brooks and Birks 2000) or plant macrofossils as a catchment-vegetation proxy (Wick et al. 2003)). The responses in other proxies to the inferred environmental changes can be studied and questions of response times, lags, rates of turnover, and ecosystem drivers can be studied and quantified (Lotter and Birks 2003; Birks and Birks 2008; Lotter and Anderson 2012: Chap. 18).

In both approaches the end result is a temporal series of biological assemblages (e.g., diatoms) and an independent temporal series of past environmental change (e.g., lake-water temperature, regional climate, catchment vegetation, available nitrogen in the catchment). Millennial-scale ecological dynamics, as reconstructed from palaeolimnological records, often show non-linear shifts in the trajectory of drivers of biotic change such that discrete disturbance, stress, and climate regimes can be observed on either side of a breakpoint or threshold in the temporal-series data (Dearing 2008; Willis et al. 2010; Jeffers et al. 2011a; Seddon et al. 2011). Such thresholds can set off cascading changes through ecosystems including shifts between alternative stable states in population and community dynamics (Scheffer and Carpenter 2003) but they are often difficult to predict (Scheffer et al. 2009). Can such palaeolimnological temporal-series be used to assess how interactions between biota respond to abrupt environmental change: do they show gradual responses or sharp shifts between alternative stable states (Dearing 2008; Jeffers et al. 2011a)?

Jeffers et al. (2011a, b) used model-fitting and model-selection analyses of a series of dynamic models to generate predicted changes in biotic interactions and abundances over time and across abrupt change in climate, fire, ungulate herbivore density, and nitrogen availability. Relatively simple mechanistic models of ecological dynamics were used and AIC weights were used to assess the relative amount of support for each mechanistic model, as discussed in the section on Model Selection above. This approach has considerable potential in palaeolimnology as it may allow the distinction between chaos or non-linear deterministic dynamics from noise or random effects in stratigraphical data and to test the ideas of Deevey (1984) about stress, strain, and stability in lacustrine systems. There is increasing evidence for a wide range of complex non-linear dynamic behaviour in ecology that can arise from simple deterministic systems (Stone and Ezrati 1996). In some fine-resolution palaeolimnological time-series from annually laminated sediments (Simola et al. 1990; McQuoid and Hobson 1997), there are hints of erratic and explosive population changes and other features possibly associated with deterministic chaos.

Mechanistic modelling using ecological dynamic models (e.g., Cottingham et al. 2000; Clark 2007; Otto and Day 2007; Bolker 2008; Soetaert and Herman 2009; Stevens 2009) with fine-resolution palaeolimnological data (biotic responses and environmental predictors) is an important future challenge for quantitative palaeolimnology (Birks 1998). It has the potential to assess the relative importance of multiple 'stressors' (Ormerod et al. 2010) on lake systems by careful model building, model selection, and model assessment. For example, Clark and McLachlan (2003) use pollen-stratigraphical data to test competing ideas about stabilising mechanisms and neutral dynamics in forest systems. The same approach could usefully be applied and extended to palaeolimnological data-sets.

Related to possible approaches to modelling ecological dynamics in limnological systems, there is considerable interest in detecting and interpreting ecological community thresholds and assessing abrupt changes and regime shifts in lake ecosystems (e.g., Smol et al. 2005; Manly and Chotkowski 2006; Rodionov 2006; Andersen et al. 2008; Carpenter and Lathrop 2008; Dakos et al. 2008; Ficetola and Denoël 2009; Scheffer et al. 2009; Baker and King 2010; Gal and Anderson 2010; Hastings and Wysham 2010; Seddon et al. 2011). These numerical approaches can usefully be applied to fine-resolution palaeolimnological temporal-series to test ideas about regime shifts, potential drivers, resilience, and thresholds in a long-term perspective (Dearing 2008; Gil-Romera et al. 2010: Willis et al. 2010; Allen et al. 2011; Seddon et al. 2011). There is a need to integrate such models with data in limnology, ecology, palaeolimnology, and palaeoecology, as discussed by Dearing et al. (2010, 2011) and Peng et al. (2011).

As computing power increases and algorithms become better, ecological modelling techniques can now fit models of bewildering complexity (Lavine 2010). There is an increasing tension between accurate characterisation of ecological patterns and processes and the need for accessible models that provide novel inferences and robust and valid predictions (LaDeau 2010; Lele 2010) and thus for bridging the gap between mathematical modelling and statistical modelling in ecology (Waller 2010). Palaeolimnologists should be aware of the limitations and assumptions of both statistical models and mathematical models in ecology and limnology and thus in palaeoecology and palaeolimnology, so as to avoid "living dangerously with big fancy models" (Lavine 2010).

Trait Analysis

A potentially useful but, as far as I know, largely untried approach to the interpretation of biological palaeolimnological data is to use life-history and other ecological traits and to link these traits to past environmental conditions through the stratigraphical fossil record. Trait analysis is a very active research area in modern ecology (e.g., Weiher et al. 1999; Westoby et al. 2002; Reich et al. 2003; Choler 2005; Violle et al. 2007; Shurin et al. 2009; Shipley 2010; Carpenter et al. 2011; Diamond et al. 2011; Keller et al. 2011; Pakeman 2011). Special numerical techniques that can simultaneously link the three matrices of palaeoenvironmental data (**R**), species traits (**Q**), and fossil data (**L**), so-called **RLQ** analysis (Dolédec et al. 1996) are required for this type of trait analysis. Fourth-corner analysis (Legendre et al. 1997; Dray and Legendre 2008) is also useful for quantifying and testing the relationships between **R** and **L** (see Brind'Amour et al. 2011 for a limnological example of RLQ and fourth-corner analyses). Lacourse (2009) illustrates the use of these three-matrix numerical procedures in a palaeoecological study in British Columbia where she demonstrates that climate is the ultimate control on Holocene forest composition and species abundance but that long-term community assembly is also constrained through inter-specific differences in plant traits. Bhagwat and Willis (2008) use traits to explore biological attributes of plants and animals that survived the last glacial maximum in refugia in southern or northern Europe.

Trait analysis has considerable potential in palaeolimnology (e.g., Hering et al. 2010; Kernan et al. 2010; Allen et al. 2011). The relevant numerical tools exist but appropriate trait data are lacking for very many aquatic organisms.

Data-Mining

There is increasing interest and activity in the compilation, merging, and analysis of large paleolimnological data-sets (typically over 1000 samples) (e.g., Smol et al. 2005; Rühland et al. 2008; Vanormelingen et al. 2008; Verleyen et al. 2009; Battarbee et al. 2010; Bennett et al. 2010; Smol and Stoermer 2010; Stomp et al. 2011). Given such data-sets, it is important to detect the major patterns of composition and variation within such data-sets, to study aspects of biodiversity, especially beta diversity or compositional heterogeneity (Legendre et al. 2005; Legendre 2008; Jurasinski et al. 2009; Tuomisto 2010a, b; Anderson et al. 2011) and, at the same time, to summarise the data as simple groups in terms of overall composition and the relationships of such groups to the environment and to other biota. There are now several specialised techniques for analysing large and heterogeneous data-sets, so-called data-mining techniques (Raymond et al. 2005; Witten and Frank 2005; Hastie et al. 2011; Torgo 2011). These include generalised additive models (see Birks 2012a: Chap. 2), classification, regression, and related decision trees (see Simpson and Birks 2012: Chap. 9), artificial neural networks and self-organising maps (see Birks 2012a: Chap. 2; Simpson and Birks 2012: Chap. 9), clustering and partitioning (see Legendre and Birks 2012a: Chap. 7), and a range of kernel methods, additive trees, and support vector machines (Hastie et al. 2011). Data-mining is defined by Everitt (2002) as "The nontrivial extraction of implicit, previously unknown, and potentially useful information from data. It uses expert systems, statistical and graphical techniques to discover and present knowledge in a form which is easily comprehensible to humans". However, this definition is almost identical to one's general idea of data exploration, exploratory data analysis, and data summarisation. Data-mining refers more specifically to the analysis of large data-sets (Hand 1998; Raymond et al. 2005; Witten and Frank 2005; Torgo 2011). Hand (1998) defines it as "the process of secondary analysis of large databases aimed at finding unsuspected relationships which are of interest or value to the database owners". One of the most important issues in data-mining is thus size. With the widespread use of computer technology and information systems, the amount of data available for exploration (e.g., biodiversity data – see GBIF Global Biodiversity Information Facility 2008) has increased exponentially in the last decade. This poses major challenges to the standard numerical and statistical techniques for data analysis. Data-mining considers issues like computing efficiency, limited memory resources, and interfaces to data-bases. Data-mining is thus a highly interdisciplinary subject involving not only typical data analyses such as clusterings, partitionings, and ordinations, but also interfacing with data-bases, providing rapid data visualisation in many dimensions using interactive graphics, devising rigorous and rapid outlier detection, etc. (see Torgo 2011).

Within the data-mining research community, enthusiasts of data-mining propose that with enough data, traditional data analysis and hypothesis testing will no longer be necessary (CR Anderson 2008). It is suggested that correlations will reveal mechanisms in comprehensive statistical models that encompass all possible data (Gotelli 2008). Exciting developments in computer science are leading to 'reverse-engineering' algorithms that can help to uncover the functional form of relationships among correlated variables (Gotelli 2008). These new iterative methods use data partitioning, automated probing, and snipping to modify sequentially and test underlying dynamic non-linear functions with data-rich time-series (Gotelli 2008; Petris et al. 2009).

Within the standard current array of data-mining techniques (Witten and Frank 2005; Hastie et al. 2011; Torgo 2011), classification and regression trees (CART) and associated decision-tree procedures (Fielding 2007; Simpson and Birks 2012: Chap. 9) have considerable potential in data-mining large palaeolimnological data-sets. Use of these methods is rapidly increasing in the related fields of ecology, biogeography, conservation biology, and applied ecology, and their wider use in palaeolimnology in the near future can be confidently expected (see Davidson et al. 2010a, b for examples of their use in palaeolimnology). The necessary software is available in various R packages (see e.g., Borcard et al. 2011; Wehrens 2011; Simpson and Birks 2012: Chap. 9 for details).

The analysis of large palaeolimnological data-sets through data-mining techniques will provide important links with macroecology where dynamics over ecological time-scales are being addressed (e.g., Gotelli 2008; Fisher et al. 2010; Stomp et al. 2011), and with meta-community ecology (e.g., Telford et al. 2006; Allen et al. 2011). There is great scope for linking palaeolimnology with macro-ecology, biogeography, meta-community ecology, and population dynamics in the future (see below).

Potentially Useful Numerical Tools in Palaeolimnology

There are several well-developed numerical tools discussed by, for example, Venables and Ripley (2002), Fielding (2007), Zuur et al. (2007), Manly (2009), Hastie et al. (2011), and Wehrens (2011) that are potentially useful tools in the numerical analysis of palaeolimnological data. These include time-warp analysis, quantile regression, additive models, and hierarchical partitioning.

Time-warp analysis (Giorgino 2009; Wehrens 2011) is a technique for comparing ordered series of values with each other. Dynamic time-warping (Giorgino 2009) is most used in natural sciences. The rationale is to stretch or compress two temporal or stratigraphical series locally in order to make one resemble the other as much as possible. Time warping optimally deforms one of the two series onto the other. The distance between the two series is computed, after stretching, by summing the distances of individual aligned elements. This distance is insensitive to local compression (Giorgino 2009). Time-warp analysis has some links with sequence-slotting (Thompson et al. 2012: Chap. 13) but it is a more flexible way of aligning two temporal series.

Quantile regression (Scharf et al. 1998; Koenker and Hallock 2001; Cade and Noon 2003; Hao and Naiman 2007) is a method for estimating functional relationships between variables for all portions of a probability distribution. Quantiles are divisions of a probability distribution or frequency distribution into equal, ordered subgroups, e.g., quartiles or percentiles. Statistical distributions of ecological or palaeolimnological data often have unequal variation due to the complex interactions between the factors that influence organisms that cannot be measured or accounted for in a statistical model. Unequal variation implies that there is more than a single slope (rate of change) that describes the relationship between a response variable and predictor variables measured on a subset of these factors. Quantile regression estimates multiple slopes from the minimum to the maximum response, thereby providing a more complex and fuller picture of the relationships between variables than can be obtained by other regression models (Cade and Noon 2003; Yee 2004a). The ecological concept of limiting factors as constraints on organisms often focuses on rates of change in quantiles near the maximum response (optimum) when only a subset of limiting factors are measured (Cade et al. 1999). Ecological applications of quantile regression include Cade and Guo (2000), Knight and Ackerly (2002), Brown and Peet (2003), Cade et al. (2005), Schröder et al. (2005), MJ Anderson (2008), Vaz et al. (2008), Ricotta et al. (2010), and Cade (2011), whereas limnological examples include Lancaster and Belyea (2006) and

Kelly et al. (2008). Quantile regression and its relative quantile splines (Koenker and Schorfheide 1994; MJ Anderson 2008) are increasingly used in climatology for the analysis of trends in climate data (e.g., Elsner et al. 2008). These techniques may be important tools in quantitative palaeolimnology in the near future, in particular at looking at complex biological responses along environmental gradients (cf. Knight and Ackerly 2002; Schröder et al. 2005; MJ Anderson 2008).

Additive models (Fox 2000; Wood 2006) are a non-parametric form of regression in which the predictor or explanatory variables have an additive effect on the response variable. If, for example, predictor variable A has an effect of size a on some response variable and variable B has an effect of size b on the same response variable, then in an assumed additive model for A and B their combined effect is $a + b$. In additive models (Fox 2000; Ferguson et al. 2008; Simpson and Anderson 2009) the sum of regression coefficients $\times$ predictor variables of a linear regression (see Birks 2012a: Chap. 2), is replaced by a sum of arbitrary smooth functions of the predictor variable. This allows the shape of the relationship between the response and the predictor variables to be determined from the data themselves (Birks 2012a: Chap. 2), rather than being assigned a prescribed functional form such as a linear or quadratic function (Simpson and Anderson 2009). Additive models are thus able to model local features of the relationship between the response and predictor variables. Simpson and Anderson (2009) applied additive models (with a serial correlation structure to model residual temporal autocorrelation) as a tool to model statistically palaeolimnological core records in relation to possible forcing factors. They reduced the biological assemblage data to principal component analysis (PCA) axes that capture the major patterns in the data through time (Birks 2012a: Chap. 2; Legendre and Birks 2012b: Chap. 8). Each set of PCA sample scores were then modelled using an additive model where the predictor variables represented potential forcing factors such as tree-ring inferred temperatures or proxies for atmospheric deposition and, where necessary, periodic components. The effect of the predictor variables on assemblage composition through time was determined from the contribution that each predictor variable made to the fitted additive model. These contributions were then used to separate the effects due to competing forcing variables such as climate and atmospheric nitrogen deposition. Additive modelling when applied to palaeolimnological temporal series is a powerful approach to tease apart the potential importance of two or more forcing variables (Ives et al. 2003; Hampton et al. 2006; Hobbs et al. 2010). It warrants further use in palaeolimnology.

Hierarchical partitioning (Chevan and Sutherland 1991; Olea et al. 2010) allows the contribution of each predictor variable to the total explained variance of a multiple regression model, both independently and in conjunction with the other predictor variables, to be estimated for all possible candidate regression models. The independent contribution of predictor variable x_i is calculated by comparing the fit of all models that include x_i with their reduced model, namely its exact same model but with x_i omitted within each hierarchical level. The improvement in model fit for each hierarchical level that considers x_i is then averaged across all hierarchies, giving the independent contribution of x_i (Quinn and Keogh 2002). Hierarchical partitioning does not produce a predictive model. Instead it allows identification of the predictor variables that explain most variance independently of the other predictors, thereby helping to minimise the problems of multicollinearity between predictors in regression models (MacNally 2002). It has been used by Steve Juggins (unpublished) to assess the amount of variance explained by different environmental variables as a guide to discovering which environmental variables can potentially be reconstructed from fossil assemblages using modern species-environment calibration functions (Juggins and Birks 2012: Chap. 14).

Many current techniques for data analysis and data interpretation in palaeolimnology assume a symmetric Gaussian unimodal species response to environmental variables (ter Braak and Prentice 1988) that can be approximated by the simple weighted averaging algorithm (WA) (see Birks 2012a: Chap. 2; Juggins and Birks 2012: Chap. 14). With the great increase in computing power and in sophisticated algorithms, Yee (2004b) has returned to ter Braak's (1986) idea of constrained Gaussian ordination (CGO) with estimation by maximum likelihood rather than by the simple WA algorithm used in canonical correspondence analysis (CCA) and its close partial relative (partial CCA). Quadratic reduced-rank vector-based generalised linear models (GLMs) implement CGO but assume symmetric unimodal species responses (cf. ter Braak and Verdonschot 1995). Yee (2006) has extended this approach to use generalised additive models (GAMs – see Birks 2012a: Chap. 2) where no species response model is assumed. Instead the data determine the response model as in simple GAMs. These techniques are still under development as they are "fragile with dirty data" (Yee 2006). Further developments will hopefully give palaeolimnologists powerful ordination and constrained ordination techniques within the GLM/GAM theoretical framework incorporating a mixture of linear, quadratic, and flexible smooth responses (e.g., Zhu et al. 2005). This general approach can, in theory, be extended to regression or prediction and to calibration and environmental reconstructions (Yee 2006), leading to a significant extension to ter Braak and Prentice's (1988) theory of gradient analysis.

There are several other statistical techniques of potential value in quantitative palaeolimnology and aquatic ecology. Additive monotone regression splines (de Boer et al. 2002) provide a means of implementing non-linear regression to estimate thresholds in dose-effect relations. Principal response curves (PRC) (van den Brink and ter Braak 1998, 1999; van den Brink et al. 2003, 2009; Timmerman and ter Braak 2008) are a very effective tool in detecting trends over time in relation to an internal reference (e.g., overall mean, reference sample, or experimental control) or external references (e.g., preferred water quality or reference site). Principal response curves are a form of redundancy analysis (see ter Braak and Šmilauer 2002; Lepš and Šmilauer 2003; Legendre and Birks 2012b: Chap. 8) and are a potentially useful tool in assessing recent changes in limnological systems, as well as in terrestrial systems (e.g., Heegaard and Vandvik 2004; Vandvik 2004; Vandvik et al. 2005).

It is clearly a challenge for palaeolimnologists to keep up with the many important advances being made in applied statistics that are of potential relevance to the analysis of palaeolimnological data. As these methods and approaches are increasingly complex and computationally demanding, it is important that palaeolimnologists extend their 'statistical fluency' (*sensu* Ellison and Dennis 2010) and develop close collaborations with applied statisticians and quantitative palaeolimnologists who understand not only the numerical methods but also the nature of palaeolimnology *and* the numerical problems posed by palaeolimnological data.

Numerical Methods to Be Developed

It would be very misleading to convey to the palaeolimnological research community that no new numerical methods need to be developed. Despite the wide range of techniques presented in this book, there are many questions in data analysis to which we do not have the answers. Some of these have been mentioned in particular chapters, such as how to interpret confidence intervals on fossil taxon counts in stratigraphical data (Maher et al. 2012: Chap. 6) or to interpret sample-specific errors of prediction in environmental reconstructions in the presence of the strong temporal autocorrelation that is a basic property of all palaeolimnological stratigraphical data and resulting environmental reconstructions (Juggins and Birks 2012: Chap. 14).

Despite considerable advances in, for example, calibration-function methodology and cross-validation procedures, our abilities to interpret and compare temporal-series of palaeoenvironmental reconstructions and of palaeolimnological data in general have hardly developed beyond visual comparison of temporal-series (Bennett 2002). The work presented by Dutilleul et al. (2012: Chap. 16) is particularly important in that it provides new robust procedures for analysing and comparing palaeolimnological data-sets and environmental reconstructions. There is a great need for applying and extending robust procedures for comparing different types of temporal-series (e.g., Burnaby 1953; Malmgren 1978; Schuenemeyer 1978; Malmgren et al. 1998; Hammer and Harper 2006; Manly 2007; Tian et al. 2011), including comparisons with random-walk simulations of palaeolimnological data (Blaauw et al. 2010).

Another area of palaeolimnological research where new statistical techniques are needed is in the analysis of the large amounts of data from multi-proxy palaeolimnological studies (e.g., Birks et al. 2000; Birks and Birks 2006). The rigorous analysis of such data is still in its infancy.

Bayesian Inference

There is an increasing awareness and interest in applying the approach of Bayesian inference in various topics within quantitative palaeolimnology, for example age-depth modelling (Blaauw and Heegaard 2012: Chap. 12) and environmental reconstructions (Juggins and Birks 2012: Chap. 14). It has considerable potential in topics such as model selection (see above) (Kass and Raftery 1995; Hoeting et al. 1999; Johnson and Omland 2004; Wang et al. 2004), modelling of biological dynamics (see above) (Jeffers et al. 2011a, b), trait analysis (see above), some areas of regression analysis and statistical modelling (Birks 2012a: Chap. 2; Simpson and Birks 2012: Chap. 9), and the assessment of uncertainties in palaeolimnological data (Maher et al. 2012: Chap. 6).

Bayesian inference consists of four major stages (Everitt 2002; Christensen et al. 2011; Kruschke 2011; Simpson and Birks 2012: Chap. 9):

1. Obtain the likelihood or conditional probability $f\ (\mathbf{X}|\boldsymbol{\theta})$ describing the process giving rise to the data $\mathbf{X}$ in terms of the unknown parameters $\boldsymbol{\theta}$
2. Obtain the prior distribution $f\ (\boldsymbol{\theta})$ expressing what is known about $\boldsymbol{\theta}$ prior to observing the data
3. Apply Bayes' theorem to derive the posterior distribution $f\ (\boldsymbol{\theta}|\mathbf{X})$ expressing what is known about $\boldsymbol{\theta}$ after observing the data
4. Derive appropriate inference statements from the posterior distribution, such as point estimates, interval estimates, or probabilities of hypotheses.

Bayesian inference differs from the classical form of so-called frequentist inference in several ways, particularly in the use of a prior distribution which is absent in classical inference. It represents the researcher's knowledge about the parameters before seeing the data. Classical statistics only uses the likelihood, whereas in Bayesian inference every problem is potentially unique as it is characterised by the researcher's ideas about the parameters expressed in the prior distribution for the specific analysis (Everitt 2002). Bayesian statistical analysis thus combines the data likelihood with a prior distribution using Bayes' theorem. A key task is then to summarise the posterior distribution, for example by the mean, the covariance, or percentiles of individual parameters. When this summarisation cannot be implemented by analytical means or analytical approximation, simulation methods such as Markov chain Monte Carlo techniques (Clark and Gelland 2006a; Christensen et al. 2011; Kruschke 2011) have to be used to generate a sample from the posterior distribution. The desired summary of the posterior distribution is then obtained from the sample. The posterior distribution is typically multi-dimensional (ter Braak and Vrugt 2008).

One of the attractions of the Bayesian approach to statistical modelling is that it allows the evaluation of how well multiple working hypotheses fit data. Instead of rejecting a null hypothesis, the result of a Bayesian analysis is an index of confidence in each of several hypotheses. The Bayesian approach lends itself elegantly to evaluating alternatives and has been shown to be an ideal tool for assessing alternative ecological models and hypotheses (e.g., Hilborn and Mangel 1997; Fabricius and De'ath 2004; Gotelli and Ellison 2004; Johnson and Omland 2004; Wolf and Mangel 2008). Bayesian approaches are widely used in a range of archaeological, environmental, and ecological problems (e.g., Buck et al. 1996; Dennis 1996; Ellison 1996, 2004; van Hoef 1996; Fabricius and De'ath 2004; Clark 2005; McCarthy and Masters 2005; Clark and Gelland 2006b; Golicher et al. 2006). Tjelmeland and Lund (2003) develop Bayesian modelling of spatial compositional (proportional) data to analyse sedimentary data (sand, silt, clay) (Coakley and Rust 1968) from 39 locations within an Arctic lake. Useful introductions to Bayesian statistical inference include Iversen (1984), Gotelli and Ellison (2004), Clark and Gelland (2006a), Clark (2007), McCarthy (2007), Webb and King (2009), Kéry (2010), Christensen et al. (2011), and Kruschke (2011), whereas Albert (2007) discusses the very real problems in Bayesian computation and Beaumont (2010) reviews approximate Bayesian computation in evolution and ecology.

Within palaeolimnology and related disciplines, Bayesian approaches have primarily been used in age-depth modelling and radiocarbon calibration (see Buck and Millard 2004; Blaauw and Heegaard 2012: Chap. 12) and in environmental reconstructions (see Juggins and Birks 2012: Chap. 14). Several age-depth studies show the advantage of adopting a Bayesian inference in developing robust age-depth models (Blockley et al. 2004, 2007, 2008; Wohlfarth et al. 2006; Yeloff et al. 2006; Bronk Ramsey 2008).

In the field of environmental reconstructions, Oehlert (1988) pioneered the use of Bayesian inference to derive uncertainty estimates for lake-water pH reconstructions based on diatoms. Robertson et al. (1999) presented a simple kernel-based Bayesian approach to climate reconstruction from tree-rings. Toivonen et al. (2001) presented a Bayesian model with a conditional probability based on the Gaussian unimodal response model. This was extended by Vasko et al. (2000) and Erästö and Holmström (2006) to include the more realistic but more complex multinomial Gaussian response model (see ter Braak and van Dam 1989; Birks 1995; Juggins and Birks 2012: Chap. 14). This so-called Bummer model has been applied to derive a chironomid-based temperature reconstruction by Korhola et al. (2002). Haslett et al. (2006) have developed further these ideas to more generalised modelling of modern pollen-climate response surfaces (see Birks et al. 2010) and Holden et al. (2008) provide a computationally efficient approach based on probability weighting of diatom-pH response curves. Haslett and Challenor (2010) present a very readable account of the current 'state-of-the-art' of palaeoenvironmental reconstructions within the Bayesian inference framework.

As an approach to environmental reconstruction, Bayesian inference has several potential advantages. It does not rely on an explicit model of the relationship between two sets of variables (e.g., diatoms and pH) (Haslett et al. 2006) but on the modification of some prior belief about the specific value of a variable on the basis of some additional information (Robertson et al. 1999). This prior information or probability can be refined with additional measured information by a modern calibration data-set to give the likelihood of conditional probability. Once this has been estimated, it can be combined with the prior probability function to provide the posterior probability density function using Bayes' theorem (Robertson et al. 1999).

Salonen et al. (2012) compare the results of using the Bayesian Bummer model of Vasko et al. (2000) with results from weighted averaging (WA) and weighted-averaging partial least squares (WAPLS) regression and calibration (see Juggins and Birks 2012: Chap. 14) in developing a pollen-based temperature calibration function in north-eastern Europe. The Bayesian model shows a significantly improved performance in leave-one-out cross-validation compared to WA or WAPLS and it is little affected by spatial autocorrelation. However, when the down-core reconstructions are compared with independent palaeoclimate records, there are some clear biases in the Bayesian reconstruction. They show that the prior parameters significantly influence not only the Bayesian model's performance statistics in cross-validation but also the resulting reconstruction. They conclude that a major future challenge in Bayesian inferences is the identification of ecologically and environmentally realistic prior parameters in relation to the time-scale and geographical setting of each reconstruction. This is a major challenge not only ecologically but also computationally because even with 2011 computer power, predicting the temperature of just one lake took about 3.6 h CPU time on an Intel Core DUOE 6750 computer and a leave-one-out cross-validation of the modern calibration set of 113 samples required 17 days computing.

A largely unexplored area of Bayesian inference in palaeoecology is its ability to provide an appropriate framework to integrate various types of data (Saros 2009), for example combining cosmogenic, stratigraphical, and palaeomagnetic data in a chronological tool (Muzikar and Granger 2006). In addition, palaeolimnologists have always been aware that there are many sources of uncertainty associated with their data (see Maher et al. 2012: Chap. 6; Juggins and Birks 2012: Chap. 14). Modern statistical frameworks involving Bayesian inference allow these uncertainties to be incorporated explicitly into numerical analyses (Clark and Bjørnstad 2004). Bayesian hierarchical models (Clark and Gelland 2006a) have considerable potential in the integration of different types of palaeolimnological data and estimating the uncertainties in such an integration. Li et al. (2010a) have used Bayesian hierarchical modelling to integrate tree-ring, borehole temperature, and pollen-stratigraphical data to provide a reconstruction of past climate along with realistic uncertainty estimates. This pioneering approach has stimulated comment and discussion (e.g., Cressie and Tingley 2010; Li et al. 2010b; Smith 2010; Wahl et al. 2010).

Although the uncertainties for Bayesian environmental reconstruction models are generally of a similar magnitude to those for non-Bayesian approaches, they have the major advantage that they provide a coherent and explicit handling of uncertainty (Juggins and Birks 2012: Chap. 14). Bayesian models offer an elegant solution for modelling multiple sources of evidence and their associated uncertainties within a single integrated framework and are thus an important priority for future work (Haslett and Challenor 2010).

Despite its great potential, Bayesian inference and modelling have not been widely used in palaeolimnology. One reason is that the computing demands are huge when moderate-sized or larger data-sets are analysed. Hopefully, this situation will change in the future. Salonen et al. (2012) argue that given the total effort in data collection, if a Bayesian model works well and has superior model performance to simple WA or WAPLS models, it is worth the additional heavy computational burden. Better simulation techniques, suitable model approximations, and improved software availability (Haslett et al. 2006; Beaumont 2010) may contribute to making Bayesian inference and modelling more accessible to the palaeolimnological research community. Its full exploitation is certainly a major challenge for the future.

Computing

A future challenge facing palaeolimnologists wanting to apply appropriate state-of-the-art numerical and statistical techniques to their data is implementing these techniques. Numerical analysis of palaeolimnological data requires several essential components – research questions to be answered or hypotheses to be tested; careful site selection; coring; laboratory analyses; high-quality data; appropriate numerical techniques; and computer software to implement the numerical procedures. Since the beginning of the application of quantitative techniques in palaeolimnology in the late 1980s (see Smol et al. 2012: Chap. 1), there has been dedicated, relatively easy-to-use software for the major numerical methods used by palaeolimnologists at that time. Programs like WACALIB for environmental reconstructions written by John Line, Cajo ter Braak, and John Birks (Line and Birks 1990; Line et al. 1994), CALIB, WAPLS, and GLR for environmental reconstructions and Gaussian logit regression written by Steve Juggins and Cajo ter Braak (www.staff.ncl.ac.uk/staff/stephen.juggins/software.htm), CANOCO for classical and constrained ordinations and variation partitioning with associated Monte Carlo permutation tests written by Cajo ter Braak (ter Braak 1987), TILIA and TILIA-GRAPH written by Eric Grimm for data-handling and graphical display (http://www.ncdc.noaa.gov/paleo/tiliafaq.html), ZONE for clustering or partitioning stratigraphical data written by Steve Juggins, Allan Gordon, and John Birks (www.staff.ncl.ac.uk/staff/stephen.juggins/software.htm), and PSIMPOLL written by Keith Bennett for handling, analysing, and displaying palaeoecological data (http://chrono.qub.ac.uk/psimpoll/psimpoll.html) were widely used in the 1980s–1990s, almost exclusively running under DOS.

With the rapid widespread of the Microsoft Windows® operating system and the increasing obstacles that Microsoft make in limiting or preventing the use of DOS software as each new version of Windows is released, the choice of software for palaeolimnologists is becoming more restricted with C2 written by Steve Juggins for environmental reconstructions and graphical display (Juggins 2007), CANOCO 4.5 and CanoDraw for Windows written by Cajo ter Braak and Petr Šmilauer for ordination and graphical display (ter Braak and Šmilauer 2002), TILIA and TG-View for Windows written by Eric Grimm for data-handling and graphical display (http://www.ncdc.noaa.gov/paleo/tiliafaq.html), and PSIMPOLL written by Keith Bennett (http://chrono.qub.ac.uk/psimpoll/psimpoll.html) that is platform-independent. These programs do not implement the most recent numerical developments in quantitative palaeolimnology and applied statistics such as additive modelling, multivariate regression trees, random forests, tests for spatial autocorrelation, statistical testing of environmental reconstructions, etc.

Today almost all new techniques are being developed as freely available, open-source packages for R distributed under the GNU General Public Licence version 2 (R Development Core Team 2010). The R language provides a single, extremely powerful environment for statistical computing, including data-set compilation and screening, exploratory data analysis, regression analysis and statistical modelling, environmental reconstructions and evaluation, ordination and many other techniques in multivariate data analysis, time-series analysis, graphical display, interactive graphics, and much more.

The following R packages are particularly relevant in quantitative palaeolimnology:

- vegan (ordination and cluster analysis, dissimilarity measures, permutation tests, etc.) (Oksanen et al. 2011)
- analogue (weighted averaging, modern analogue technique) (Simpson and Oksanen 2009)
- rioja (environmental reconstructions, zonation, plotting of stratigraphical data, etc.) (Juggins 2009)
- paltran (environmental reconstructions) (Adler 2010)
- fossil (palaeoecological data analysis) (Vavrek 2010)
- palaeoSig (testing the statistical significance of environmental reconstructions) (Telford 2011)
- paleoMAS (modern analogue technique, palaeoecological data analysis) (Correa-Metrio et al. 2010)
- simba (dissimilarity coefficients, diversity indices, permutation tests) (Jurasinski 2009)
- BiodiversityR (analysis of ecological and palaeoecological data) (Kindt 2009)

In addition there are R packages for generalised linear modelling, generalised additive modelling, mixed-effects modelling, age-depth modelling, classification and regression trees, random forests, boosted trees, artificial neural networks, time-series analysis, interactive graphical display, and very much more. Fox (2002), Venables and Ripley (2002), Faraway (2005, 2006), Murrell (2006), Wood (2006), Crawley (2007), Zuur et al. (2007, 2009), Rizzo (2008), Sawitzki (2009), Everitt and Hothorn (2010), Maindonald and Braun (2010), and Qian (2010) outline part of the range of statistical, numerical, and graphical techniques freely available in R.

In addition to the books mentioned in Chap. 1 of this volume that provide excellent introductions to the use of R in statistical data handling and computing, there are many books in the Use R! series that cover topics of direct relevance to quantitative palaeolimnologists now and in the future. These include numerical ecology (Borcard et al. 2011), spatial data analysis (Bivand et al. 2008), data manipulation (Spector 2008), time-series analysis (Cowpertwait and Metcalf 2009), population ecology (Stevens 2009), chemometrics and multivariate data analysis (Wehrens 2011), multivariate data analysis (Everitt and Hothorn 2011), non-linear regression (Ritz and Streibig 2008), wavelets (Nason 2008), Monte Carlo methods (Robert and Casella 2010), interactive graphics (Cook and Swayne 2007), Bayesian statistics (Albert 2007), and graphics for multivariate data (Sarkar 2008), and for univariate and bivariate data (Whickham 2009). There is inevitably a steep learning-curve in the initial stages of using R, just as there is in any other programming environment, but the final rewards are well worth the initial effort. Almost all applied statisticians use R today because it is good, it is international, it is versatile, it has a vast library of packages and functions, and it is free. Almost all recent advances in numerical and statistical techniques in palaeolimnology are only available as R packages. Using R and these techniques is thus a challenge for the palaeolimnological community now and in the immediate future.

Conclusions

The various chapters in this book show that palaeolimnologists now have an impressive array of powerful numerical and statistical techniques to help in the analysis of palaeolimnological data. The use of these methods has contributed to some of the exciting developments in palaeolimnology that have occurred in the last two decades (see Smol 2008, 2010; Pienitz et al. 2009 for details of these developments). As the recent interest in the development of modern calibration data-sets and the use of these data-sets to make down-core environmental reconstructions swings to more fine-resolution stratigraphical studies and more multi-disciplinary stratigraphical studies, there will be an increasing need for further numerical techniques and developments to help summarise, analyse, and decipher the extraordinary diversity of environmental and biological proxies preserved in lake sediments. It is, however, necessary to emphasise the importance of 'expert knowledge' that comes from a sound training and experience in the many disciplines that contribute to palaeolimnology. Without such 'expert knowledge' it is easy for a researcher to be misled by results from a particular numerical analysis or statistical model. It is essential to remember that numerical methods are only tools to help palaeolimnologists address and hopefully answer interesting and challenging questions about lake history, biotic response, and environmental change.

Acknowledgements In concluding this chapter and this book as a whole, I wish to acknowledge my great debt to the applied statisticians and quantitative ecologists and palaeoecologists with whom I have collaborated or who have greatly influenced my ideas about quantitative data-analysis, in particular, Cajo ter Braak, Allan Gordon, Steve Juggins, Gavin Simpson, Richard Telford, Petr Šmilauer, Einar Heegaard, Pierre Legendre, John Line, Mark Hill, John Imbrie, Daniel Borcard, Lou Maher, Svante Wold, Keith Bennett, Thompson Webb, Maarten Blaauw, Eric Grimm, and Ed Cushing. I am also indebted to the many palaeolimnologists with whom I have worked, for sharing ideas, data, and insights, in particular, Brigitta Ammann, John Anderson, Rick Battarbee, Helen Bennion, the late Frode Berge, Christian Bigler, Hilary Birks, John Boyle, Emily Bradshaw, Steve Brooks, Nigel Cameron, Don Charles, Brian Cumming, Tom Davidson, Roger Flower, Roland Hall, Oliver Heiri, Ulrike Herzschuh, Will Hobbs, Viv Jones, the late John Kingston, Atte Korhola, Tom Korsman, Jorunn Larsen, Andy Lotter, Anson Mackay, Reinhard Pienitz, Ingemar Renberg, Neil Rose, Peter Rosén, Carl Sayer, Heikki Seppä, John Smol, Tony Stevenson, Gaute Velle, Kathy Willis, Alex Wolfe, and Barbara Zeeb.
I am very grateful to Gavin Simpson, Richard Telford, Steve Juggins, Rick Battarbee, John Smol, Alistair Seddon, and Cajo ter Braak for helpful discussions and to Cathy Jenks for help in preparing this chapter. This is publication number A350 from the Bjerknes Centre for Climate Change.

References

Adler S (2010) paltran: WA, WA-PLS, MW for paleolimnology. http://cran.r-project.org/web/packages/paltran/index.html

Albert J (2007) Bayesian computation with R. Springer, New York

Alexander KA, Blackburn JK, Frimpong EA (2011) Buffalo and Maslow's hammer. Front Ecol Environ 9:302–303

Allen MR, van Dyke JN, Cáceres CE (2011) Metacommunity assembly and sorting in newly formed lake communities. Ecology 92:269–275

Andersen T, Carstensen J, Hernández-García E, Duarte CM (2008) Ecological thresholds and regime shifts: approaches to identification. Trends Ecol Evol 24:49–57

Anderson C (2008) The end of theory: the data deluge makes the scientific method obsolete. Wired Magazine. www.wired.com/science/discoveries/magazine/16-07/pb_theory. Accessed 8 August 2011

Anderson DR (2008) Model based inference in the life sciences. A primer on evidence. Springer, New York

Anderson DR, Burnham KP (2002) Avoiding pitfalls when using information-theoretic models. J Wildl Manage 66:912–918

Anderson MJ (2008) Animal-sediment relationships re-visited: characterising species' distributions along an environmental gradient using canonical analysis and quantile regression splines. J Exp Mar Biol Ecol 366:16–27

Anderson MJ, Crist TO, Chase JM, Velland M, Inouye BD, Freestone AL, Sanders NJ, Cornell HV, Comita LS, Davies KF, Harrison SP, Kraft NJB, Stegen JC, Swenson NG (2011) Navigating the multiple meanings of β diversity: a roadmap for the practicing ecologist. Ecol Lett 14:19–28

Baker ME, King RS (2010) A new method for detecting and interpreting biodiversity and ecological community thresholds. Method Ecol Evol 1:25–37

Ball IR (1975) Nature and formulation of biogeographic hypotheses. Syst Zool 24:407–430

Battarbee RW, Charles DF, Bigler C, Cumming BF, Renberg I (2010) Diatoms as indicators of surface-water acidity. In: Smol JP, Stoermer EF (eds) The diatoms: applications for the environmental and earth science, 2nd edn. Cambridge University Press, Cambridge, pp 98–121

Beaumont MA (2010) Approximate Bayesian computation in evolution and ecology. Ann Rev Ecol Evol Syst 41:379–406

Bennett KD (2002) Comment: the Greenland 8200 cal. yr BP event detected in loss-on-ignition profiles in Norwegian lacustrine sediment sequences. J Quaternary Sci 17:97–99

Bennett JR, Cumming BF, Ginn BK, Smol JP (2010) Broad-scale environmental response and niche conservatism in lacustrine diatom communities. Global Ecol Biogeogr 19:724–732

Bhagwat S, Willis KJ (2008) Species persistence in northerly glacial refugia of Europe: a matter of chance or biogeographical traits? J Biogeogr 35:464–482

Birks HH, Birks HJB (2006) Multi-proxy studies in palaeolimnology. Veg Hist Archaeobot 15:235–251

Birks HH, Battarbee RW, Birks HJB (2000) The development of the aquatic ecosystem at Kråkenes Lake, western Norway, during the late glacial and early Holocene – a synthesis. J Paleolimnol 23:91–114

Birks HJB (1995) Quantitative palaeoenvironmental reconstructions. In: Maddy D, Brew JS (eds) Statistical modelling of Quaternary science data, Technical guide 5. Quaternary Research Association, Cambridge, pp 161–254

Birks HJB (1998) Numerical tools in palaeolimnology – progress, potentialities, and problems. J Paleolimnol 20:307–332

Birks HJB (2012a) Chapter 2: Overview of numerical methods in palaeolimnology. In: Birks HJB, Lotter AF, Juggins S, Smol JP (eds) Tracking environmental change using lake sediments. Volume 5: Data handling and numerical techniques. Springer, Dordrecht

Birks HJB (2012b) Chapter 11: Analysis of stratigraphical data. In: Birks HJB, Lotter AF, Juggins S, Smol JP (eds) Tracking environmental change using lake sediments. Volume 5: Data handling and numerical techniques. Springer, Dordrecht

Birks HJB, Birks HH (2008) Biological responses to rapid climate changes at the Younger Dryas-Holocene transition at Kråkenes, western Norway. Holocene 18:19–30

Birks HJB, Line JM, Juggins S, Stevenson AC, ter Braak CJF (1990) Diatoms and pH reconstruction. Philos Trans R Soc Lond B 327:263–278

Birks HJB, Heiri O, Seppä H, Bjune AE (2010) Strengths and weaknesses of quantitative climate reconstructions based on late-Quaternary biological proxies. Open Ecol J 3:68–110

Bivand RS, Pebesma EJ, Gómez-Rubio V (2008) Applied spatial data analysis with R. Springer, New York

Blaauw M, Heegaard E (2012) Chapter 12: Estimation of age-depth relationships. In: Birks HJB, Lotter AF, Juggins S, Smol JP (eds) Tracking environmental change using lake sediments. Volume 5: Data handling and numerical techniques. Springer, Dordrecht

Blaauw M, Bennett KD, Christen JA (2010) Random walk simulations of fossil proxy data. Holocene 20:645–649

Blass A, Bigler C, Grosjean M, Sturm B (2007) Decadal-scale autumn temperature reconstruction back to AD 1580 inferred from the varved sediments of Lake Silvaplana (southeastern Swiss Alps). Quaternary Res 68:184–195

Blockley SPE, Lowe JJ, Walker MJC, Asioli A, Trincardi F, Coope GR, Donahue RE, Pollard AM (2004) Bayesian analysis of radiocarbon chronologies: examples from the European late-glacial. J Quaternary Sci 19:159–175

Blockley SPE, Blaauw M, Bronk Ramsey C, van der Plicht J (2007) Building and testing age models for radiocarbon dates in lateglacial and early Holocene sediments. Quaternary Sci Rev 26:1915–1926

Blockley SPE, Bronk Ramsey C, Lane CS, Lotter AF (2008) Improved age modelling approaches as exemplified by the revised chronology for the Central European varved lake Soppensee. Quaternary Sci Rev 27:61–71

Bolker BM (2008) Ecological models and data in R. Princeton University Press, Princeton

Borcard D, Gillet F, Legendre P (2011) Numerical ecology with R. Springer, New York

Boyle JF (2001) Inorganic geochemical methods in paleolimnology. In: Last WM, Smol JP (eds) Tracking environmental change using lake sediments. Volume 2: Physical and geochemical methods. Kluwer, Dordrecht, pp 83–141

Brind'Amour A, Boisclair D, Dray S, Legendre P (2011) Relationships between species feeding traits and environmental conditions in fish communities: a three-matrix approach. Ecol Appl 21:363–377

Bronk Ransey C (2008) Deposition models for chronological records. Quaternary Sci Rev 27:42–60

Brooks SJ, Birks HJB (2000) Chironomid-inferred late-glacial and early-Holocene mean July air temperatures for Kråkenes Lake, western Norway. J Paleolimnol 23:77–89

Brown RL, Peet RK (2003) Diversity and invasibility of southern Appalachian plant communities. Ecology 84:32–39

Buck CE, Millard AR (2004) Tools for constructing chronologies. Springer, New York

Buck CE, Cavanagh WG, Litton CD (1996) Bayesian approach to interpreting archaeological data. Wiley, Chichester

Burnaby TP (1953) A suggested alternative to the correlation coefficient for testing the significance of agreement between pairs of time series and its application to geological data. Nature 172:210–211

Burnham KP, Anderson DR (2002) Model selection and multimodel inference: a practical information-theoretic approach, 2nd edn. Springer, New York

Cade BS (2011) Estimating equivalence with quantile regression. Ecol Appl 21:281–289

Cade BS, Guo Q (2000) Estimating effects of constraints on plant performance with regression quantiles. Oikos 91:245–254

Cade BS, Noon B (2003) A gentle introduction to quantile regression for ecologists. Front Ecol Environ 1:412–420
Cade BS, Terrell JW, Schroeder RL (1999) Estimating effects of limiting factors with regression quantiles. Ecology 80:311–323
Cade BS, Noon BR, Flather CH (2005) Quantile regression reveals hidden bias and uncertainty in habitat models. Ecology 86:786–800
Callaghan DA, Ashton PA (2008) Knowledge gaps in bryophyte distribution and prediction of species-richness. J Bryol 30:147–158
Carpenter SR, Lathrop RC (2008) Probabilistic estimate of a threshold for eutrophication. Ecosystems 11:601–613
Carpenter SR, Cole JJ, Pace ML, Batt R, Brock WA, Cline T, Coloso J, Hodgson JR, Kitchell JF, Seekell DA, Smith L, Weidel B (2011) Early warnings of regime shifts: a whole-ecosystem experiment. Science 332:1079–1082
Chamberlain MJ (2008) Are we sacrificing biology for statistics? J Wildl Manage 72:1057–1058
Chevan A, Sutherland M (1991) Hierarchical partitioning. Am Stat 45:90–96
Choler P (2005) Consistent shifts in alpine plant traits along a mesotopographical gradient. Arct Antarct Alp Res 37:444–453
Christensen R, Johnson WO, Branscum AJ, Hanson TE (2011) Bayesian ideas and data analysis. Chapman & Hall/CRC, Boca Raton
Clark JS (2005) Why environmental scientists are becoming Bayesians. Ecol Lett 8:2–14
Clark JS (2007) Models for ecological data – an introduction. Princeton University Press, Princeton
Clark JS, Bjørnstad ON (2004) Population time series: process variability, observation errors, missing values, lags, and hidden states. Ecology 85:3140–3150
Clark JS, Gelland AE (eds) (2006a) Hierarchical modelling for the environmental sciences. Oxford University Press, Oxford
Clark JS, Gelland AE (2006b) A future for models and data in environmental science. Trends Ecol Evol 21:375–380
Clark JS, McLachlan JS (2003) Stability of forest biodiversity. Nature 423:635–638
Coakley JP, Rust BR (1968) Sedimentation in an Arctic lake. J Sediment Pet 38:1290–1300
Cohen AS (2003) Palaeolimnology: the history and evolution of lake systems. Oxford University Press, Oxford
Cook D, Swayne DF (2007) Interactive and dynamic graphics for data analysis. Springer, New York
Correa-Metrio A, Urrego DH, Cabrera KR, Bush MB (2010) paleoMAS: paleoecological analysis. http://cran.r-project.org/web/packages/paleoMAS/index.html
Cottingham KL, Rusak JA, Leavitt PR (2000) Increased ecosystem variability and reduced predictability following fertilization: evidence from palaeolimnology. Ecol Lett 3:340–348
Cowpertwait PSP, Metcalf AV (2009) Introductory time series with R. Springer, New York
Crawley MJ (2007) The R book. Wiley, Chichester
Cressie N, Tingley MP (2010) Comment: hierarchical statistical modelling for paleoclimate reconstruction. J Am Stat Assoc 105:895–900
Cumming BF, Laird KR, Fritz SC, Verschuren D (2012) Chapter 20: Tracking Holocene climatic change with aquatic biota from lake sediments: case studies of commonly used numerical techniques. In: Birks HJB, Lotter AF, Juggins S, Smol JP (eds) Tracking environmental change using lake sediments. Volume 5: Data handling and numerical techniques. Springer, Dordrecht
Dahlgren JP (2010) Alternative regression methods are not considered in Murtaugh (2009) or by ecologists in general. Ecol Lett 13:E7–E9
Dakos V, Scheffer M, van Nes EH, Brovkin V, Petoukhov V, Held H (2008) Slowing down as an early warning signal for abrupt climate change. Proc Natl Acad Sci USA 105:14308–14312
Davidson TA, Sayer CD, Langdon PG, Burgess A, Jackson M (2010a) Inferring past zooplanktivorous fish and macrophyte density in a shallow lake: application of a new regression tree model. Freshw Biol 55:584–599

Davidson TA, Sayer CD, Perrow M, Bramm M, Jeppesen E (2010b) The simultaneous inference of zooplanktivorous fish and macrophyte density from sub-fossil cladoceran assemblages: a multivariate regression tree approach. Freshw Biol 55:546–564

de Boer WJ, den Besten PJ, ter Braak CJF (2002) Statistical analysis of sediment toxicity by additive monotone regression splines. Ecotoxicology 11:435–450

Dearing JA (2008) Landscape change and resilience theory: a palaeoenvironmental assessment from Yunnan, SW China. Holocene 18:117–127

Dearing JA, Braimoh AK, Reenberg A, Turner BL, van der Leeuw S (2010) Complex land systems: the need for long time perspectives to asses their future. Ecol Soc 15:21

Dearing JA, Dotterweich M, Foster T, Newman L, Gunter T (eds) (2011) Integrative paleoscience for sustainable management. PAGES News 19:42–92

Deevey ES (1984) Stress, strain and stability of lacustrine ecosystems. In: Haworth EY, Lund JWG (eds) Lake sediments and environmental history. Leicester University Press, Leicester, pp 203–229

Dennis B (1996) Discussion: should ecologists become Bayesians? Ecol Appl 6:1095–1103

Diamond SE, Frame AM, Martin RA, Buckley LB (2011) Species' traits predict phenological response to climate change in butterflies. Ecology 92:1005–1012

Dolédec S, Chessel D, ter Braak CJF, Champely S (1996) Matching species traits to environmental variables: a new three-table ordination method. Environ Ecol Stat 3:143–166

Downes BJ (2010) Back to the future: little-used tools and principles of scientific inference can help disentangle effects of multiple stressors on freshwater ecosystems. Freshw Biol 55:60–79

Dray S, Legendre P (2008) Testing the species traits-environment relationships: the fourth-corner problem revisited. Ecology 89:3400–3412

Dutilleul P, Cumming BF, Lontoc-Roy M (2012) Chapter 16: Autocorrelogram and periodogram analyses of palaeolimnological temporal series from lakes in central and western North America to assess shifts in drought conditions. In: Birks HJB, Lotter AF, Juggins S, Smol JP (eds) Tracking environmental change using lake sediments. Volume 5: Data handling and numerical techniques. Springer, Dordrecht

Ellison AM (1996) An introduction to Bayesian inference for ecological research and environmental decision-making. Ecol Appl 6:1036–1046

Ellison AM (2004) Bayesian inference in ecology. Ecol Lett 7:509–520

Ellison AM, Dennis B (2010) Paths to statistical fluency for ecologists. Front Ecol Environ 8: 362–370

Elsner JB, Kossin JP, Jagger TH (2008) The increasing intensity of the strongest tropical cyclones. Nature 455:92–95

Erästö P, Holmström L (2006) Selection of prior distributions and multiscale analysis in Bayesian temperature reconstructions based on fossil assemblages. J Paleolimnol 36:69–80

Everitt BS (2002) The Cambridge dictionary of statistics, 2nd edn. Cambridge University Press, Cambridge

Everitt BS, Hothorn T (2010) A handbook of statistical analyses using R, 2nd edn. CRC Press, Boca Raton

Everitt BS, Hothorn T (2011) An introduction to applied multivariate analysis with R. Springer, New York

Fabricius KE, De'ath G (2004) Identifying ecological change and its causes: a case study on coral reefs. Ecol Appl 14:1448–1465

Faraway JJ (2005) Linear models with R. CRC Press, Boca Raton

Faraway JJ (2006) Extending the linear model with R. Generalized linear, mixed effects and nonparametric regression. Chapman & Hall, Boca Raton

Ferguson CA, Carvalho L, Scott EM, Bowman AW, Kirika A (2008) Assessing ecological responses to environmental change using statistical models. J Appl Ecol 45:193–203

Ficetola GF, Denoël M (2009) Ecological thresholds: an assessment of methods to identify abrupt changes in species-habitat relationships. Ecography 32:1075–1084

Fielding AH (2007) Cluster and classification techniques for the biosciences. Cambridge University Press, Cambridge

Fisher JAD, Frank KT, Leggett WC (2010) Dynamic macroecology on ecological time-scales. Glob Ecol Biogeogr 19:1–15
Flessa KW, Jackson ST (2005a) Forging a common agenda for ecology and paleoecology. Bioscience 55:1030–1031
Flessa KW, Jackson ST (eds) (2005b) The geological record of ecological dynamics. Understanding the biotic effects of future environmental change. National Research Council of the National Academies, Washington, DC
Fox J (2000) Multiple and generalized nonparametric regression. Sage, Thousand Oaks
Fox J (2002) An R and S-PLUS® companion to applied regression. Sage, Thousand Oaks
Francus P (ed) (2004) Image analysis, sediments and paleoenvironments. Springer, Dordrecht
Gal G, Anderson W (2010) A novel approach to detecting a regime shift in a lake ecosystem. Method Ecol Evol 1:45–52
Gil-Romera G, López-Merino L, Carríon JS, González-Sampériz P, Martín-Peurtas C, López Sáez JA, Fernández S, Carcía-Antón M, Stefanova V (2010) Interpreting resilience through long-term ecology: potential insights in western Mediterranean landscapes. Open Ecol J 3:43–53
Giorgino T (2009) Computing and visualizing dynamic time warping alignments in R: the dtw package. J Stat Softw 31:1–23
GBIF Global Biodiversity Information Facility (2008) http://www.gbif.org/press/factsheet
Golicher DJ, O'Hara RB, Ruiz-Montoya L, Cayuela L (2006) Lifting a veil on diversity: a Bayesian approach to fitting relative-abundance models. Ecol Appl 16:202–212
Gotelli NJ (2008) Hypothesis testing, curve fitting, and data mining in macroecology. Int Biogeogr Soc Newsl 6:3–7
Gotelli NJ, Ellison AM (2004) A primer of ecological statistics. Sinauer Associates, Sunderland
Grosjean M, von Gunten L, Trachsel M, Kamenik C (2009) Calibration-in-time: transforming biogeochemical lake sediment proxies into quantitative climate variable. PAGES News 17: 108–110
Guthery FS, Brennan LA, Peterson MJ, Lusk JJ (2005) Information theory in wildlife science: critique and viewpoint. J Wildl Manage 69:457–465
Hammer Ø, Harper DAT (2006) Paleontological data analysis. Blackwell, Oxford
Hampton SE, Scheuerell MD, Schindler DE (2006) Coalescence in the Lake Washington story: interaction strength in a planktonic food web. Limnol Oceanogr 51:2042–2051
Hand DJ (1998) Data mining: statistics and more? Am Stat 52:112–118
Hao L, Naiman DQ (2007) Quantile regression. Sage, Thousand Oaks
Haslett J, Challenor P (2010) Palaeoclimate histories. Insights from the Institute of Advanced Study, Durham University, Durham
Haslett J, Whiley M, Bhattacharya S, Salter-Townsend M, Wilson SP, Allen JRM, Huntley B, Mitchell FJG (2006) Bayesian palaeoclimate reconstruction. J R Stat Soc A 169:395–438
Hastie TJ, Tibshirani RJ, Friedman J (2011) The elements of statistical learning. Data mining, inference, and prediction, 2nd edn. Springer, New York
Hastings A, Wysham DB (2010) Regime shifts in ecological systems can occur with no warning. Ecol Lett 13:464–472
Heegaard E, Vandvik V (2004) Climate change affects the outcome of competitive interactions—an application of principal response curves. Oecologia 139:459–466
Heiri O, Wooller MJ, van Hardenbroek M, Wang YV (2009) Stable isotopes in chitinous fossils of aquatic vertebrates. PAGES News 17:100–102
Hering D, Haidekker A, Schmidt-Kloiber A, Barker T, Buisson L, Graf W, Grenouillet G, Lorenz A, Sandin L, Stendera S (2010) Monitoring the responses of freshwater ecosystems to climate change. In: Kernan M, Battarbee RW, Moss B (eds) Climate change impacts on freshwater ecosystems. Wiley-Blackwell, Chichester, pp 84–118
Hilborn R, Mangel M (1997) The ecological detective – confronting models with data. Princeton University Press, Princeton
Hobbs NY, Hilborn R (2006) Alternatives to statistical hypothesis testing in ecology: a guide to self teaching. Ecol Appl 16:5–19

Hobbs WO, Telford RJ, Birks HJB, Saros JE, Hazewinkel RRO, Perren BB, Saulnier-Talbot É, Wolfe AP (2010) Quantifying recent ecological changes in remote lakes of North America and Greenland using sediment diatom assemblages. PLoS One 5:e10026
Hoeting JA, Madigan D, Raftery AE, Volinsky CT (1999) Bayesian model averaging: a tutorial. Stat Sci 14:382–417
Hoeting JA, Davis RA, Merton AA, Thompson SE (2006) Model selection for geostatistical models. Ecol Appl 16:87–98
Holden PB, Mackay AM, Simpson GL (2008) A Bayesian palaeoenvironmental transfer function model for acidified lakes. J Paleolimnol 39:551–566
Iversen GR (1984) Bayesian statistical inference. Sage, Newbury Park
Ives AR, Dennis B, Cottingham KL, Carpenter SR (2003) Estimating community stability and ecological interactions from time-series data. Ecol Monogr 73:301–330
Jackson ST, Booth RK (2002) The role of late Holocene climate variability in the expansion of yellow birch in the western Great Lakes region. Divers Dist 8:275–284
Jeffers ES, Bonsall MB, Brooks SJ, Willis KJ (2011a) Abrupt environmental changes drive shifts in tree–grass interaction outcomes. J Ecol 99:1063–1070
Jeffers ES, Bonsall MB, Willis KJ (2011b) Stability in ecosystem functioning across a climatic threshold and contrasting forest regimes. PLoS One 6:e16134
Johnson JB, Omland KS (2004) Model selection in ecology and evolution. Trends Ecol Evol 19:101–108
Juggins S (2007) C2 version 1.5 user guide. Software for ecological and palaeoecological data analysis and visualisation. University of Newcastle, Newcastle-upon-Tyne
Juggins S (2009) rioja: Analysis of Quaternary science data. http://cran.r-project.org/web/packages/rioja/index.html
Juggins S, Birks HJB (2012) Chapter 14: Quantitative environmental reconstructions from biological data. In: Birks HJB, Lotter AF, Juggins S, Smol JP (eds) Tracking environmental change using lake sediments. Volume 5: Data handling and numerical techniques. Springer, Dordrecht
Juggins S, Telford RJ (2012) Chapter 5: Exploratory data analysis and data display. In: Birks HJB, Lotter AF, Juggins S, Smol JP (eds) Tracking environmental change using lake sediments. Volume 5: Data handling and numerical techniques. Springer, Dordrecht
Jurasinski G (2009) simba: a collection of functions for similarity analysis of vegetation data. http://cran.r-project.org/web/packages/simba/index.html
Jurasinski G, Retzer V, Beierkuhnlein C (2009) Inventory, differentiation, and proportional diversity: a consistent terminology for quantifying species diversity. Oecologia 159:15–26
Kass RE, Raftery AE (1995) Bayes factors. J Am Stat Assoc 90:773–795
Keller RP, Kocev D, Dzeroski S (2011) Trait-based risk assessment for invasive species: high performance across diverse taxonomic groups, geographic ranges and machine learning/statistical tools. Divers Dist 17:451–461
Kelly M, Juggins S, Guthrie R, Pritchard S, Jamieson J, Rippey B, Hirst H, Yallop M (2008) Assessment of ecological status in UK rivers using diatoms. Freshw Biol 53:403–422
Kernan M, Battarbee RW, Moss B (eds) (2010) Climate change impacts on freshwater ecosystems. Wiley-Blackwell, Chichester
Kéry M (2010) Introduction to WinBUGS for ecologists. Academic Press, Burlington
Kindt R (2009) BiodiversityR: GUI for biodiversity and community ecology analysis. http://cran.r-project.org/web/packages/BiodiversityR/index.html
Knight CA, Ackerly DD (2002) Variation in nuclear DNA content across environmental gradients: a quantile regression analysis. Ecol Lett 5:66–76
Koenker R, Hallock KF (2001) Quantile regression. J Econ Perspect 15:143–156
Koenker R, Schorfheide F (1994) Quantile spline models for global temperature change. Clim Chang 28:395–404
Korhola A, Vasko K, Toivonen HTT, Olander H (2002) Holocene temperature changes in northern Fennoscandia reconstructed from chironomids using Bayesian modelling. Quaternary Sci Rev 21:1841–1860

Köster D, Racca JMJ, Pienitz R (2004) Diatom-based inference models and reconstructions revisited: methods and transformations. J Paleolimnol 32:233–245
Kruschke J (2011) Doing Bayesian analysis. A tutorial with R and BUGS. Academic Press, Burlington
Lacourse T (2009) Environmental change controls postglacial forest dynamics through inter-specific differences in life-history traits. Ecology 90:2149–2160
LaDeau S (2010) Advances in modeling highlight a tension between analytical accuracy and accessibility. Ecology 91:3488–3492
Lancaster J, Belyea LR (2006) Defining the limits to local density: alternative views of abundance–environment relationships. Freshw Biol 51:783–796
Last WM, Smol JP (eds) (2001a) Tracking environmental change using lake sediments. Volume 1: Basin analysis, coring, and chronological techniques. Kluwer, Dordrecht
Last WM, Smol JP (eds) (2001b) Tracking environmental change using lake sediments. Volume 2: Physical and geochemical methods. Kluwer, Dordrecht
Lavine M (2010) Living dangerously with big fancy models. Ecology 91:3487
Leavitt PR, Fritz SC, Anderson NJ, Baker PA, Blenckner T, Bunting L, Catalan J, Conley DJ, Hobbs WO, Jeppesen E, Korhola A, McGowan S, Rühland K, Rusak JA, Simpson GL, Solovieva N, Werne J (2009) Paleolimnological evidence of the effects on lakes of energy and mass transfer from climate and humans. Limnol Oceanogr 54:2330–2348
Legendre P (2008) Studying beta diversity: ecological variation partitioning by multiple regression and canonical analysis. J Plant Ecol 1:3–8
Legendre P, Birks HJB (2012a) Chapter 7: Clustering and partitioning. In: Birks HJB, Lotter AF, Juggins S, Smol JP (eds) Tracking environmental change using lake sediments. Volume 5: Data handling and numerical techniques. Springer, Dordrecht
Legendre P, Birks HJB (2012b) Chapter 8: From classical to canonical ordination. In: Birks HJB, Lotter AF, Juggins S, Smol JP (eds) Tracking environmental change using lake sediments. Volume 5: Data handling and numerical techniques. Springer, Dordrecht
Legendre P, Galzin R, Harmelin-Vivien M-L (1997) Relating behaviour to habitat: solutions to the fourth-corner problem. Ecology 78:547–562
Legendre P, Borcard D, Peres-Neto PR (2005) Analyzing beta diversity: partitioning the spatial variation of community composition data. Ecol Monogr 75:435–450
Lele SR (2010) Model complexity and information in the data: could it be a house built on sand? Ecology 91:3493–3496
Leng MJ (ed) (2006) Isotopes in palaeoenvironmental research. Springer, Dordrecht
Leng MJ, Marshall JD (2004) Paleoclimate information of stable isotope data from lake sediment archives. Quaternary Sci Rev 23:811–831
Lepš J, Šmilauer P (2003) Multivariate analysis of ecological data using CANOCO. Cambridge University Press, Cambridge
Li B, Nychka DW, Amman CM (2010a) The values of multiproxy reconstruction of past climate. J Am Stat Assoc 105:883–895
Li B, Nychka DW, Amman CM (2010b) Rejoinder. J Am Stat Assoc 105:910–911
Line JM, Birks HJB (1990) WACALIB 2.1 a computer program to reconstruct environmental variables from fossil assemblages by weighted averaging. J Paleolimnol 3:170–173
Line JM, ter Braak CJF, Birks HJB (1994) WACALIB version 3.3 - a computer program to reconstruct environmental variables from fossil assemblages by weighted averaging and to derive sample- specific errors of prediction. J Paleolimnol 10:147–152
Lotter AF (2003) Multi-proxy climatic reconstructions. In: Mackay AW, Battarbee RW, Birks HJB, Oldfield F (eds) Global change in the Holocene. Arnold, London, pp 373–383
Lotter AF, Anderson NJ (2012) Chapter 18: Limnological responses to environmental changes at inter-annual to decadal time-scales. In: Birks HJB, Lotter AF, Juggins S, Smol JP (eds) Tracking environmental change using lake sediments. Volume 5: Data handling and numerical techniques. Springer, Dordrecht
Lotter AF, Birks HJB (2003) The Holocene palaeolimnology of Sägistalsee and its environmental history – a synthesis. J Paleolimnol 30:333–342

Lukacs PM, Thomson WL, Kendall WL, Gould WR, Doherty PF, Burnham KP, Anderson DR (2007) Concerns regarding a call for pluralism of information theory and hypothesis testing. J Appl Ecol 44:456–460
MacNally R (2002) Multiple regression and inference in ecology and conservation biology: further comments on identifying important predictor variables. Biodiv Cons 11:1397–1401
Maher LJ, Heiri O, Lotter AF (2012) Chapter 6: Assessment of uncertainties associated with palaeolimnological laboratory methods and microfossil analysis. In: Birks HJB, Lotter AF, Juggins S, Smol JP (eds) Tracking environmental change using lake sediments. Volume 5: Data handling and numerical techniques. Springer, Dordrecht
Maindonald J, Braun WJ (2010) Data analysis and graphics using R, 3rd edn. Cambridge University Press, Cambridge
Malmgren BA (1978) Comparison of visual and statistical correlation in time series curves. Math Geol 10:103–106
Malmgren BA, Winter A, Chen D (1998) El-Niño-Southern Oscillation and North Atlantic Oscillation control of climate in Puerto Rico. J Climate 11:2713–2717
Mangel M (2006) The theoretical biologist's toolbox. Cambridge University Press, Cambridge
Manly BFJ (2007) Randomization, bootstrap, and Monte Carlo methods in biology, 3rd edn. Chapman & Hall/CRC, London
Manly BFJ (2009) Statistics for environmental science and management, 2nd edn. CRC, Boca Raton
Manly BFJ, Chotkowski M (2006) Two new methods for regime change analysis. Archiv für Hydrobiologie 167:593–607
McCarthy MA (2007) Bayesian methods in ecology. Cambridge University Press, Cambridge
McCarthy MA, Masters P (2005) Profiting from prior information in Bayesian analyses of ecological data. J Appl Ecol 42:1012–1019
McQuoid MR, Hobson AL (1997) A 91-year record of seasonal and interannual variability of diatoms from laminated sediments in Sanich Inlet, British Columbia. J Plankton Res 19: 173–194
Murrell P (2006) R graphics. Chapman & Hall/CRC, Boca Raton
Murtaugh PA (1998) Methods of variable selection in regression modeling. Commun Stat Simul 27:711–734
Murtaugh PA (2007) Simplicity and complexity in ecological data analysis. Ecology 88:56–62
Muzikar P, Granger D (2006) Combining cosmogenic, stratigraphic, and paleomagnetic information using a Bayesian approach: general results and an application to Sterkfontein. Earth Planet Sci Lett 243:400–408
Nason GP (2008) Wavelet methods in statistics with R. Springer, New York
Oehlert GW (1988) Interval estimates for diatom-inferred pH histories. Can J Stats 16:51–60
Oksanen J, Blanchet FG, Kindt R, Legendre P, O'Hara RB, Simpson GL, Solymos P, Stevens MHM, Wagner H (2011) vegan: Community Ecology Package. R package version 1.17-8 http://cran.r-project.org/package=vegan
Olea PP, Mateo-Tomás P, de Frutos Á (2010) Estimating and modelling bias of the hierarchical partitioning public-domain software: implications in environmental management and conservation. PLoS One 5:e11698
Ormerod SJ, Dobson M, Hildrew AG, Townsend CR (2010) Multiple stressors in freshwater ecosystems. Freshw Biol 55:1–4
Otto SP, Day T (2007) A biologists guide to mathematical modelling in ecology and evolution. Princeton University Press, Princeton
Pakeman RJ (2011) Multivariate identification of plant functional response and effect traits in an agricultural landscape. Ecology 92:1353–1365
Paterson AM, Cumming BF, Dixit SS, Smol JP (2002) The importance of model choice on pH inferences from scaled chrysophyte assemblages in North America. J Paleolimnol 27:379–391
Peng C, Guiot J, Wu H, Jiang H, Luo Y (2011) Integrating models with data in ecology and palaeoecology: advances towards a model-data fusion approach. Ecol Lett 14:522–536
Petris G, Petrone S, Campagnoli P (2009) Dynamic linear models with R. Springer, New York

Pienitz R, Lotter AF, Newman L, Kiefer T (eds) (2009) Advances in paleolimnology. PAGES News 17:89–136
Qian SS (2010) Environmental and ecological statistics with R. CRC Press, Boca Raton
Quinn GP, Keogh MJ (2002) Experimental design and data analysis for biologists. Cambridge University Press, Cambridge
R Development Core Team (2010) R: a language and environment for statistical computing. R Foundation for Statistical Computing, Vienna. www.r-project.org/
Racca JMJ, Philibert A, Racca R, Prairie YT (2001) A comparison between diatom-based pH inference models using artificial neural networks (ANN), weighted averaging (WA) and weighted averaging partial least squares (WA-PLS) regressions. J Paleolimnol 26:411–422
Racca JMJ, Wild M, Birks HJB, Prairie YT (2003) Separating wheat from chaff: diatom taxon selection using an artificial neural network pruning algorithm. J Paleolimnol 29:123–133
Raffalovich L, Deane GD, Armstrong D, Tsao H-S (2008) Model selection procedures in social research: Monte Carlo simulation results. J Appl Stat 35:1093–1114
Ramsey FL, Schafer DW (1997) The statistical sleuth – a course in methods of data analysis. Duxbury Press, Belmont
Raymond B, Watts DJ, Burton H, Bonnice J (2005) Data mining and scientific data. Arct Antarct Alp Res 37:348–357
Reich PB, Wright IJ, Cavender-Bares J, Craine JM, Oleksyn J, Westoboy M, Walters MB (2003) The evolution of plant functional variation: trait, spectra, and strategies. Int J Plant Sci 164(suppl3):S143–S164
Ricotta C, Godefroid S, Rocchini D (2010) Invasiveness of alien plants in Brussels is related to their phylogenetic similarity to native species. Divers Dist 16:655–662
Ritz C, Streibig JC (2008) Nonlinear regression with R. Springer, New York
Rizzo ML (2008) Statistical computing with R. Chapman & Hall/CRC, Boca Raton
Robert CP, Casella G (2010) Introductory Monte Carlo methods with R. Springer, New York
Robertson I, Lucy D, Baxter L, Pollard AM, Aykroyd RG, Barker AC, Carter AHC, Sirotsur VR, Waterhouse JS (1999) A kernel-based Bayesian approach to climatic reconstruction. Holocene 9:495–500
Rodionov S (2006) A brief overview of the regime shift detection methods. In: Velikova V, Chipev N (eds) Large-scale disturbances (regime shifts) and recovery in aquatic ecosystems. Challenges for management toward sustainability. UNESCO-ROSTE/BAS Workshop on Regime Shifts, Varna, Bulgaria, pp 68–72
Rühland K, Paterson AM, Smol JP (2008) Hemispheric-scale patterns of climate-related shifts in planktonic diatoms from North American and European lakes. Glob Change Biol 14:2470–2475
Salonen JS, Ilvonen L, Seppä H, Holmström L, Telford RJ, Gaidamavicius A, Stancikaite M, Subetto D (2012) Inverse multivariate regression (WA and WA-PLS) and Bayesian modelling: comparing two transfer function techniques for quantitative palaeoclimate reconstruction. Holocene. doi: 10.1177/095968361142554
Sarkar D (2008) Lattice – multivariate data visualization with R. Springer, New York
Saros JE (2009) Integrating neo- and paleolimnological approaches to refine interpretations of environmental change. J Paleolimnol 41:243–252
Sauerbrei W, Royston P, Binder H (2007) Selection of important variables and determination of functional form for continuous predictors in multivariate model building. Stat Med 26: 5512–5528
Sawitzki G (2009) Computational statistics: an introduction to R. CRC Press, Boca Raton
Sayer CD, Davidson TA, Jones JI, Langdon PG (2010) Combining contemporary ecology and palaeolimnology to understand shallow lake ecosystem change. Freshw Biol 55:487–499
Scharf FS, Juanes F, Sutherland M (1998) Inferring ecological relationships from the edges of scatter diagrams: comparison of regression techniques. Ecology 79:448–460
Scheffer M, Carpenter SR (2003) Catastrophic regime shifts in ecosystems: linking theory to observation. Trends Ecol Evol 18:648–650

Scheffer M, Bascompte J, Brock WA, Brovkin V, Carpenter SR, Dakos V, Held H, van Nes EH, Rietkerk M, Sugihara G (2009) Early-warning signals for critical transitions. Nature 461:53–59
Schröder HK, Andersen HE, Kiehl K (2005) Rejecting the mean: estimating the response of fen plant species to environmental factors by non-linear quantile regression. J Veg Sci 16:373–382
Schuenemeyer JH (1978) Reply to comparison of visual and statistical correlation in time series curves. Math Geol 10:106–108
Schwarz G (1978) Estimating the dimensions of a model. Ann Stat 6:461–464
Seddon AWR, Froyd CA, Leng MJ, Milne GA, Willis KA (2011) Ecosystem resilience and threshold response in the Galápagos coastal zone. PLoS One 6:e22376
Shipley B (2010) From plant traits to vegetation structure. Cambridge University Press, Cambridge
Shuman B, Newby P, Huang YS, Webb T (2004) Evidence for the close climatic control of New England vegetation history. Ecology 85:1297–1310
Shurin JB, Cottenie K, Hillebrand H (2009) Spatial autocorrelation and dispersal limitation in freshwater organisms. Oecologia 159:151–159
Simola H, Hanski I, Liukkonen M (1990) Stratigraphy, species richness and seasonal dynamics of planktonic diatoms during 418 years in Lake Lovajärvi, south Finland. Ann Bot Fenn 27:241–259
Simpson GL (2012) Chapter 15: Analogue methods in palaeolimnology. In: Birks HJB, Lotter AF, Juggins S, Smol JP (eds) Tracking environmental change using lake sediments. Volume 5: Data handling and numerical techniques. Springer, Dordrecht
Simpson GL, Anderson NJ (2009) Deciphering the effect of climate change and separating the influence of confounding factors in sediment core records using additive models. Limnol Oceanogr 54:2529–2541
Simpson GL, Birks HJB (2012) Chapter 9: Statistical learning in palaeolimnology. In: Birks HJB, Lotter AF, Juggins S, Smol JP (eds) Tracking environmental change using lake sediments. Volume 5: Data handling and numerical techniques. Springer, Dordrecht
Simpson GL, Hall IR (2012) Chapter 19: Human impacts – applications of numerical methods to evaluate surface-water acidification and eutrophication. In: Birks HJB, Lotter AF, Juggins S, Smol JP (eds) Tracking environmental change using lake sediments. Volume 5: Data handling and numerical techniques. Springer, Dordrecht
Simpson GL, Oksanen J (2009) analogue: a palaeoecological data analysis package for R. http://analogue.r-forge.r-project.org/
Smith RL (2010) Comment. J Am Stat Assoc 105:905–910
Smol JP (2008) Pollution of lakes and rivers: a paleoenvironmental perspective, 2nd edn. Blackwell, Oxford
Smol JP (2010) The power of the past: using sediments to track the effects of multiple stressors on lake ecosystems. Freshw Biol 55:43–59
Smol JP, Stoermer EF (eds) (2010) The diatoms: applications for the environmental and earth sciences, 2nd edn. Cambridge University Press, Cambridge
Smol JP, Birks HJB, Last WM (eds) (2001a) Tracking environmental change using lake sediments. Volume 3: Terrestrial, algal, and siliceous indicators. Kluwer, Dordrecht
Smol JP, Birks HJB, Last WM (eds) (2001b) Tracking environmental change using lake sediments. Volume 4: Zoological indicators. Kluwer, Dordrecht
Smol JP, Wolfe AP, Birks HJB et al. (2005) Climate-driven regime shifts in the biological communities of Arctic lakes. Proc Natl Acad Sci USA 102:4397–4402
Smol JP, Birks HJB, Lotter AF, Juggins S (2012) Chapter 1: The march towards the quantitative analysis of palaeolimnological data. In: Birks HJB, Lotter AF, Juggins S, Smol JP (eds) Tracking environmental change using lake sediments. Volume 5: Data handling and numerical techniques. Springer, Dordrecht
Soetaert K, Herman PMJ (2009) A practical guide to ecological modelling. Springer, New York
Sokal RR, Rohlf FJ (1995) Biometry – the principles and practice of statistics in biological research. WH Freeman, New York
Spector P (2008) Data manipulation with R. Springer, New York

Stephens PA, Buskik SW, Hayward GD, del Rio CM (2005) Information theory and hypothesis testing: a call for pluralism. J Appl Ecol 42:4–12
Stephens PA, Buskirk SW, Martínez del Rio C (2007) Inference in ecology and evolution. Trends Ecol Evol 22:192–197
Stevens MHH (2009) A primer of ecology with R. Springer, New York
Stomp M, Huisman J, Mittelbach GG, Litchman E, Klausmeir CA (2011) Large-scale biodiversity patterns in freshwater phytoplankton. Ecology 92:2096–2107
Stone L, Ezrati S (1996) Chaos, cycles and spatiotemporal dynamics in plant ecology. J Ecol 84:279–291
Telford RJ (2011) palaeoSig: significance tests for palaeoenvironmental reconstructions. http://cran.r-project.org/web/packages/palaeoSig/index.html
Telford RJ, Birks HJB (2005) The secret assumption of transfer functions: problems with spatial autocorrelation in evaluating model performance. Quaternary Sci Rev 24:2173–2179
Telford RJ, Birks HJB (2009) Design and evaluation of transfer functions in spatially structured environments. Quaternary Sci Rev 28:1309–1316
Telford RJ, Heegaard E, Birks HJB (2004) All age-depth models are wrong: but how badly? Quaternary Sci Rev 23:1–5
Telford RJ, Vandvik V, Birks HJB (2006) Dispersal limitations matter for microbial morphospecies. Science 312:015–1015
ter Braak CJF (1986) Canonical correspondence analysis: a new eigenvector technique for multivariate direct gradient analysis. Ecology 67:1167–1179
ter Braak CJF (1987) CANOCO – a FORTRAN program for CANOnical Community Ordination by [partial] [detrended] [canonical] correspondence analysis, principal components analysis and redundancy analysis (version 2.1). TNO Institute of Applied Computer Science, Wageningen
ter Braak CJF, Juggins S (1993) Weighted Averaging Partial Least-Squares Regression (WA-PLS) - an improved method for reconstructing environmental variables from species assemblages. Hydrobiologia 269:485–502
ter Braak CJF, Prentice IC (1988) A theory of gradient analysis. Adv Ecol Res 18:271–317
ter Braak CJF, Šmilauer P (2002) CANOCO reference manual and CanoDraw for Windows user's guide: Software for canonical community ordination (version 4.5). Microcomputer Power, Ithaca
ter Braak CJF, van Dam H (1989) Inferring pH from diatoms - a comparison of old and new calibration methods. Hydrobiologia 178:209–223
ter Braak CJF, Verdonschot PFM (1995) Canonical correspondence analysis and related multivariate methods in aquatic ecology. Aq Sci 57:255–289
ter Braak CJF, Vrugt JA (2008) Differential evolution Markov chain with snooker updater and fewer chains. Stat Comp 18:435–446
Thompson R, Clark RM, Boulton GS (2012) Chapter 13: Core correlation. In: Birks HJB, Lotter AF, Juggins S, Smol JP (eds) Tracking environmental change using lake sediments. Volume 5: Data handling and numerical techniques. Springer, Dordrecht
Tian J, Nelson DM, Hu FS (2011) How well do sediment indicators record past climate? An evaluation using annually laminated sediments. J Paleolimnol 45:73–84
Timmerman ME, ter Braak CJF (2008) Bootstrap confidence intervals for principal response curves. Comp Stat Data Anal 52:1837–1849
Tjelmeland H, Lund KV (2003) Bayesian modelling of spatial compositional data. J Appl Stat 30:87–100
Toivonen HTT, Mannila H, Korhola A, Olander H (2001) Applying Bayesian statistics to organism-based environmental reconstructions. Ecol Appl 11:618–630
Torgo L (2011) Data mining with R. Learning with case studies. CRC, Boca Raton
Trachsel M, Grosjean M, Laroque-Tobler I, Schwikowsko M, Blass A, Sturm B (2010) Quantitative summer temperature reconstruction derived from a combined biogenic Si and chironomid record from varved sediments of Lake Silvaplana (south-eastern Swiss Alps) back to AD 1177. Quaternary Sci Rev 29:2719–2730

Tuomisto H (2010a) A diversity of beta diversities: straightening up a concept gone awry. Part 1: Defining beta diversity as a function of alpha and gamma diversity. Ecography 33:2–22
Tuomisto H (2010b) A diversity of beta diversities: straightening up a concept gone awry. Part 2: Quantifying beta diversity and related phenomena. Ecography 33:23–45
van den Brink PJ, ter Braak CJF (1998) Multivariate analysis of stress in experimental ecosystems by principal response curves and similarity analysis. Aquat Ecol 32:161–178
van den Brink PJ, ter Braak CJF (1999) Principal response curves: analysis of time dependent multivariate responses of a biological community to stress. Environ Toxic Chem 18:138–148
van den Brink PJ, van den Brink NW, ter Braak CJF (2003) Multivariate analysis of ecotoxicological data using ordinations: demonstrations of utility on the basis of various examples. Aust J Ecotoxic 9:141–156
van den Brink PJ, den Besten PJ, bij de Vaate A, ter Braak CJF (2009) Principal response curves technique for the analysis of multivariate biomonitoring time series. Environ Monit Assess 152:271–281
van der Voet H (1999) Pseduo-degrees of freedom for complex predictive models: the example of partial least squares. J Chemometr 13:195–208
van Hoef JM (1996) Parametric empirical Bayes methods for ecological applications. Ecol Appl 6:1047–1055
Vandvik V (2004) Gap dynamics in perennial subalpine grasslands: trends and processes change during secondary succession. J Ecol 92:86–96
Vandvik V, Heegaard E, Måren IE, Aarrestad PA (2005) Managing heterogeneity: the importance of grazing and environmental variation on post-fire succession in heathlands. J Appl Ecol 42:139–149
Vanormelingen P, Verleyen E, Vyverman W (2008) The diversity and distribution of diatoms: from cosmopolitanism to narrow endemism. Biodiv Cons 17:393–405
Vasko K, Toivonen HTT, Korhola A (2000) A Bayesian multinomial Gaussian response model for organism-based environmental reconstructions. J Paleolimnol 24:243–250
Vavrek M (2010) fossil: palaeoecological and palaeogeographical analysis tools. http://cran.r-project.org/web/packages/fossil/index.html
Vaz S, Martin CS, Eastwood PD, Ernande B, Carpentier A, Meaden GJ, Coppin F (2008) Modelling species distributions using regression quantiles. J Appl Ecol 45:204–217
Venables WN, Ripley BD (2002) Modern applied statistics with S, 4th edn. Springer, New York
Verbruggen F, Heiri O, Reichart G-J, Lotter AF (2010) Chironomid $\delta^{18}O$ as a proxy for past lake water $\delta^{18}O$: a Lateglacial record from Rotsee (Switzerland). Quaternary Sci Rev 29:2271–2279
Verleyen E, Vyverman W, Sterken M, Hodgson DA, De Wever A, Juggins S, van de Vijver B, Jones VJ, Vanormelingen P, Roberts D, Flower R, Kilroy C, Souffreau C, Sabbe K (2009) The importance of dispersal-related and local factors in shaping the taxonomic structure of diatom metacommunities. Oikos 118:1239–1249
Violle C, Navas M-L, Vile D, Kazakou E, Fortunel C, Hummel I, Garnier E (2007) Let the concept of trait be functional! Oikos 116:882–892
Virah-Sawmy M, Gillson L, Willis KJ (2009a) How does spatial heterogeneity influence resilience to climatic change? Ecological dynamics in southeast Madagascar. Ecol Monogr 79:557–574
Virah-Sawmy M, Willis KJ, Gillson L (2009b) Threshold response of Madagascar's littoral forest to sea-level rise. Global Ecol Biogeogr 18:98–110
Volinsky CT, Madigan D, Raftery AE, Kronmal RA (1997) Bayesian model averaging in proportional hazard models: assessing the risk of a stroke. J R Stat Soc Ser C 46:433–448
Wahl E, Schoelzel C, Williams J, Tigrek S (2010) Comment. J Am Stat Assoc 105:900–905
Waller LA (2010) Bridging gaps between statistical and mathematical modeling in ecology. Ecology 91:3500–3502
Wang D, Zhang W, Bakhai A (2004) Comparison of Bayesian model averaging and stepwise methods for model selection in logistic regression. Stat Med 23:3451–3467
Ward EJ (2008) A review and comparison of four commonly used Bayesian and maximum likelihood model selection tools. Ecol Model 211:1–10

Webb JA, King EL (2009) A Bayesian hierarchical trend analysis finds strong evidence for large-scale temporal declines in stream ecological condition around Melbourne, Australia. Ecography 32:215–225
Wehrens R (2011) Chemometrics with R. Springer, New York
Weiher E, van der Werf A, Thompson K, Roderick M, Garnier E, Eriksson O (1999) Challenging Theophrastus: a common core list of plant traits for functional ecology. J Veg Sci 10:609–620
Weijers JWH, Blaga CI, Werne JP, Sinninghe Damsté JS (2009) Microbial membrane lipids in lake sediments as a paleothermometer. PAGES News 17:102–104
Westoby M, Falster DS, Moles AT, Vesk PA, Wright IJ (2002) Plant ecological strategies: some leading dimensions of variation between species. Ann Rev Ecol Evol Syst 33:125–159
Whickham H (2009) ggplot2. Springer, New York
Whittingham MJ, Stephens PA, Bradbury RB, Freckleton RP (2006) Why do we still use stepwise modelling in ecology and behaviour? J Anim Ecol 75:1182–1189
Wick L, van Leeuwen JFN, van der Knaap WO, Lotter AF (2003) Holocene vegetation development in the catchment of Sagistalsee (1935 m asl), a small lake in the Swiss Alps. J Paleolimnol 30:261–272
Willis KJ, Bailey RM, Bhagwat SA, Birks HJB (2010) Biodiversity baselines, thresholds, and resilience: testing predictions and assumptions using palaeoecological data. Trends Ecol Evol 25:583–581
Witten IH, Frank E (2005) Data mining: practical machine learning tools and techniques. Morgan Kaufmann/Elsevier, Amsterdam
Wohlfarth B, Blaauw M, Davies SM, Andersson M, Wastegård S, Hornes A, Possnert G (2006) Constraining the age of lateglacial and early Holocene pollen zones and tephra horizons in southern Sweden using Bayesian probability methods. J Quaternary Sci 21:321–334
Wolf N, Mangel M (2008) Multiple hypothesis testing and the declining-population paradigm in Steller Sea Lions. Ecol Appl 18:1932–1955
Wolfe BB, Edwards TWD, Elgood RJ, Benning KRM (2001) Carbon and oxygen isotope analysis of lake sediment cellulose: methods and applications. In: Last WM, Smol JP (eds) Tracking environmental change using lake sediments. Volume 2: Physical and geochemical methods. Kluwer, Dordrecht, pp 373–400
Wolfe BB, Falcone MD, Clogg-Wright KP, Mongeon CL, Yi Y, Brock BE, St Amour NA, Mark WA, Edwards TWD (2007) Progress in isotope paleohydrology using lake sediment cellulose. J Paleolimnol 37:221–231
Wood SN (2006) Generalized additive models. An introduction with R. Chapman & Hall, Boca Raton
Wooller MJ, Francis D, Fogel ML, Miller GH, Walker IR, Wolfe AP (2004) Quantitative paleotemperature estimates from $\delta^{18}O$ of chironomid head capsules preserved in Arctic lake sediments. J Paleolimnol 31:267–274
Wooller MJ, Wang Y, Axford Y (2008) A multiple stable isotope record of late Quaternary limnological changes and chironomid paleoecology from north-eastern Iceland. J Paleolimnol 40:63–77
Yee TW (2004a) Quantile regression via vector generalized additive models. Stat Med 23:2295–2315
Yee TW (2004b) A new technique for maximum-likelihood canonical Gaussian ordination. Ecol Monogr 74:685–701
Yee TW (2006) Constrained additive ordination. Ecology 97:203–213
Yeloff D, Bennett KD, Blaauw M, Mauquoy D, Sillasoo Ü, van der Plicht J, van Geel B (2006) High precision ^{14}C dating of Holocene peat deposits: a comparison of Bayesian calibration and wiggle-matching approaches. Quaternary Geochron 1:222–235
Zhu M, Hastie TJ, Walther G (2005) Constrained ordination analysis with flexible response functions. Ecol Model 187:524–536
Zuur AF, Ieno EN, Smith GM (2007) Analyzing ecological data. Springer, New York
Zuur AF, Ieno EN, Walker NJ, Savelier AA, Smith GM (2009) Mixed effect models and extensions in ecology with R. Springer, New York

Glossary

χ^2 See chi-square test, chi-square criterion, and chi-square distance coefficient.

Abscissa The horizontal x axis on a graph.

Accelerator mass spectrometry (AMS) A method for radiocarbon dating that involves directly counting the number of ^{12}C, ^{13}C, and ^{14}C atoms in the material being dated using a large-scale, high-voltage mass spectrometer.

Accumulation rate The rate at which sediment accumulates, estimated from an age-depth model or a varved sediment sequence.

Accuracy The degree to which the measured value approaches the true value of what is being measured.

Additive model A model where the combined effect of the explanatory variables and their interactions equals the sum of their separate effects.

Additive monotone regression splines Splines that guarantee strictly monotonically increasing functions by the use of regression splines and as a result thresholds can be modelled in, for example, ecotoxicology.

ade4 Software in R for a range of ordination and graphical techniques.

Adjusted coefficient of determination ($r_{adj}{}^2$ or $R_{adj}{}^2$ or $R_a{}^2$) A recommended modification of R^2 to adjust for the number of parameters fitted by a regression model. For large sample sizes, $R_{adj}{}^2$ is approximately equal to R^2, the coefficient of determination.

AEM See asymmetric eigenvector maps.

Age-depth model A numerical model of the relationship between the age of sediment (typically determined using radiometric dating such as ^{210}Pb or ^{14}C) and its stratigraphical depth. The model is usually defined by a small number of dated points and then used to provide estimates of the age of intermediate levels by interpolation.

Agglomerative clustering Methods of cluster analysis that begin with each individual object in a separate group and then, in a series of steps, combine objects and, later, clusters into new, larger clusters until all objects are members of a single group.

AIC See Akiake information criterion.

H.J.B. Birks et al. (eds.), *Tracking Environmental Change Using Lake Sediments*, Developments in Paleoenvironmental Research 5, DOI 10.1007/978-94-007-2745-8,

Akiake information criterion (AIC) An index that aids choosing between competing statistical models by opting for the model that minimises a likelihood-ratio goodness-of-fit statistic.

Aliasing effect In signal processing, the effect that causes different signals to be indistinguishable when sampled.

Algorithm A stated procedure consisting of a series of steps, often repetitive, for solving a problem.

Allochthonous Derived from outside the system under study, e.g., terrestrial plant material that is preserved in lake sediments. Often used synonymously with allogenic.

Allogenic Sediment which originated away from the area of sedimentation and has been transported to the site. Often used synonymously with allochthonous.

AMS See accelerator mass spectrometry.

ANALOG Software for analogue matching and modern analogue technique (MAT) for palaeoenvironmental reconstructions.

analogue Package for R statistical software for analogue matching and palaeoenvironmental reconstruction.

Analogue matching A methodology and group of numerical procedures for identifying modern analogues of fossil biological assemblages.

Analysis of covariance (ANCOVA) Analysis of variance that uses a mixture of continuous random variables and qualitative variables. ANCOVA models can be thought of as multiple regression with some dummy variables. It is an extension of analysis of variance that allows for the possible effects of continuous explanatory variables on the response variable, in addition to the effects of the factor or treatment variables.

Analysis of similarities (ANOSIM) Numerical procedure for testing whether there is a statistically significant difference between two or more groups of sampling units.

Analysis of variance (ANOVA) The attribution of variation in a variable to variations in one or more explanatory variables, where each explanatory variable can take one of a small number of values.

Analyte A substance that is the subject of chemical analysis.

ANCOVA See analysis of covariance.

ANN See artificial neural networks.

ANOSIM See analysis of similarities.

ANOVA See analysis of variance.

ape Software in R for principal coordinate analysis.

Area under the curve Means of summarising information from a series of measurements. In palaeolimnology, used as a measure of the degree to which differences in species composition discriminate between analogue and non-analogue samples. Equivalent to the Mann-Whitney statistic U.

Artificial neural networks (ANNs) A family of numerical models that learn and predict from a set of data by mimicking the way the human neural network processes information.

Asymmetric eigenvector maps A technique used in the analysis of spatial data to model spatial distributions of biological data generated by hypothesised directional physical processes such as migrations in river networks and currents in water bodies.

Asymptotic Approaching a value or curve arbitrarily closely.

Authigenic A diagenetic mineral or sedimentary deposit formed *in situ*. Also referred to as autochthonous or autogenic.

Autochthonous Material (e.g., plants, sediment) which originated within the system under study. For example, algal microfossils are autochthonous, whilst charcoal particles are not. Also referred to as autogenic or authigenic.

Autocorrelation Internal correlation of the observations in a time series, usually expressed as a function of the time lag between observations. Also the internal correlation of observations in space (spatial autocorrelation).

Autocorrelogram See correlogram.

Autogenic See authigenic.

Automated probing Iterative tool in data-mining to modify sequentially and test underlying dynamic non-linear functions in complex time-series.

Autoregressive process A model used in time-series analysis in which an observation at time t is postulated to be a linear function of previous values of the time-series. A first-order process only considers the preceding sample in the series. The process can extend to order p where p is less than t.

Back-propagation A learning algorithm for training artificial neural networks.

Backward elimination A method for selecting a subset of explanatory variables based on sequential removal of variables.

Bacon R software package for Bayesian age-depth modelling using the model of piece-wise linear accumulation.

Bagging See bootstrap aggregating.

Bar-chart A graphical representation of data grouped into a series of (usually unordered) categories. Equal-width rectangular bars are drawn over each category with height equal to the observed frequencies of the categories.

Basis functions Used in multivariate adaptive regression splines (MARS), they are defined by a single knot location and take the value of 0 on one side of the knot and a linear function on the opposite side. Each such basis function has a reflected partner where the 0-value region and the linear-function region are reversed.

Bayes classifier A simple probabilistic classifier based on applying Bayes theorem.

Bayes theorem A procedure for determining inverse probabilities, that is finding the conditional probability of A given B from the conditional probability of B given A.

Bayesian approach An approach concerned with the modifying of previous beliefs as a result of receiving new data.

Bayesian belief networks See Bayesian networks.

Bayesian hierarchical model A means of combining data of different types into a single coherent model, by having different basic components at different levels in a hierarchy. The data model occupies one level, the process model resides below it, and a third hierarchical level contains the statistical models or priors for unknown parameters. The levels are formally generated by a series of conditional steps where one level is conditioned on knowledge of the levels below it.

Bayesian inference An approach concerned with the consequences of modifying previous beliefs as a result of receiving new data.

Bayesian information criterion (BIC) A way of choosing between competing statistical models that usually results in a simpler model being selected. Also known as Schwartz's criterion.

Bayesian multinomial model An environmental inference model based on the multinomial logistic response model of species and their environment and involving Bayesian inference. It is presently very demanding computationally. The relevant software is bummer. The multinomial logistic model for proportional species data estimates species parameters simultaneously with the constraint that the abundance of all species sums to 1. It is difficult to fit and its parameters are often difficult to interpret.

Bayesian networks An expert system in which uncertainty is handled using conditional probabilities and Bayes theorem.

BCal Online software for radiocarbon calibration using Bayesian methods.

Bchron Package for R statistical software for age-depth modelling using Bayesian methods.

BDK See bi-directional Kohonen network.

Belief networks See Bayesian networks.

Bernoulli trial An experiment or trial that has exactly two possible results, usually classified as 'success' or 'failure'.

Bernstein polynomial A polynomial that is a linear combination of Bernstein basis polynomials.

Bézier curve A parametric curve constructed as a sequence of cubic segments used to model smooth curves that appear reasonably smooth at all scales.

Bhalme and Mooley drought index (BMDI) An index of drought intensity calculated from monthly precipitation measurements from individual meteorological stations.

Bias See systematic error.

BIC See Bayesian information criterion.

Bi-directional Kohonen network (BDK) A type of supervised Kohonen network in which the input and output maps are updated in an alternating way.

Binary divisive analysis A partitioning procedure that divides a set of observations into 2 groups, 4 groups, 8 groups, etc., until the groups are too small for further division or their numerical properties suggest further division is not warranted.

Binary splitting zonation method A method of sequentially dividing a stratigraphical sequence into a desired number of zones, with each partition found to minimise an overall sum-of-squares or information criterion.

Binomial distribution The discrete probability distribution of the number of successes in a sequence of n independent yes/no experiments, each of which yields success with a constant probability.

biodiversityR R package for analysing ecological and palaeoecological data.

Bioturbation The mixing of sediment by organisms.

Biplot An ordination diagram of two kinds of entities (e.g., samples and environmental variables) which can be interpreted by the biplot rule. Interpretation proceeds by projecting points onto directions defined by arrows in the biplot.

Bivariate kernel-density estimate A non-parametric estimate of the joint probability density function of two variables.

Bivariate statistics Statistical procedures used to describe the relationship between two variables.

BMDI See Bhalme and Mooley drought index.

Bonferroni comparison A simultaneous test of whether the means of three or more populations are equal.

Bonferroni correction A multiple-comparison correction used when several statistical tests are being performed simultaneously to guard against inflated Type 1 errors.

Boosting A machine-learning algorithm for iteratively improving and combining a number of weak classifiers into a single strong classifier.

Bootstrap aggregating (bagging) A machine-learning algorithm for improving classification and regression models by combining the output from multiple randomly generated training-sets created by bootstrap re-sampling.

Bootstrap cross-validation A method of cross-validation in which multiple random training-sets are generated by bootstrap re-sampling.

Bootstrap re-sampling A computer intensive statistical resampling procedure that randomly generates 'new' data-sets (e.g., 1000), with replacement, that are the same size as the original data-set. The predictive ability of a model is based on estimates derived on samples when they do not form part of the randomly generated data-set.

Bootstrapping A simulation method for statistical inference based on bootstrap re-sampling.

Box-Cox transformations A way of converting a general set of n observations into a set of n independent observations from a normal distribution with constant variance.

Box-plot See box-whisker plot.

Box-whisker plot A graphical tool for displaying the important features of a set of observations in terms of the median, inter-quartile range (box part) and the 'whiskers' extending to include the minimum and maximum but not the outside observations which are displayed separately.

Bpeat Software for Bayesian age-depth modelling of peat cores.

Bray-Curtis (Odum or Steinhaus) distance A dissimilarity coefficient used to quantify the biological compositional dissimilarity between two different sites. It is equivalent to the ratio between the turnover of species between the two sites and the total species richness over the two sites.

Broken-stick model A model based on the broken-stick distribution where a set of objects is taken as equivalent to a stick of unit length that is broken randomly into a number of pieces.

BSiZer Method and software for constructing SiZer maps using Bayesian inference.

C2 Software for palaeoecological data analysis.

CA See correspondence analysis.

CALIB Software for calibrating radiocarbon dates available as an on-line procedure or as a computer program.

Calibration In statistics, a procedure that enables a series of easily obtainable measurements to be used to provide an estimate of a quantity of interest. In palaeolimnology, used to refer to the range of procedures used to express values of an environmental variable (e.g., pH) as a function of species data (e.g., diatom assemblages). Calibration differs from regression because the causal relationships between species and environment are reversed and asymmetric.

Calinski-Harabasz criterion A variance ratio criterion used to estimate the number of clusters in a set of sampling units.

CANOCO Software for canonical community ordination by [partial] [detrended] [canonical] correspondence analysis, principal components analysis, and redundancy analysis, written by CJF ter Braak.

CanoDraw Graphical software for drawing ordination plots. It directly interfaces with CANOCO. It also implements a range of other graphical and statistical tools (e.g., GAM, GLM, LOESS) Written by P Šmilauer.

Canonical analysis of principal coordinates (CAP) A constrained ordination technique involving many predictor and many explanatory variables. It differs from RDA and CCA in that any dissimilarity measure between response variables can be used as in db-RDA.

Canonical correlation analysis (CCorA) Method of assessing the linear relationship between two groups of variables.

Canonical correspondence analysis (CCA) A constrained ordination technique that uses a weighted-average algorithm, in which ordination axes are constrained to be linear combinations of the supplied environmental variables.

Canonical gradient analysis See direct gradient analysis.

Canonical ordination In ecology, a set of ordination methods for relating species assemblages to their environment.

Canonical variates (analysis) (CVA) Canonical variates are linear combinations of variables that maximise the ratio of the between-group variance to the within-group variance. Also known as multiple discriminant analysis.

CAP See canonical analysis of principal coordinates.

CART See classification and regression trees.

Cartesian space Euclidean space defined by Cartesian coordinates in two or three dimensions.

CCA See canonical correspondence analysis.

cclust Software package in R to implement K-means clustering.

CCorA See canonical correlation analysis.

CEM See certified reference material.

Centroid The centre of gravity of a cluster of sampling units.

Centroid rule In ordination plots for CA, CCA, and DCA, if the scaling is chosen that species scores are weighted mean sample scores, each species point will be at the centre of its niche in the plot. Their interpretation is thus based on the centroid principle.

Certified reference material (CRM) Reference standards used to check the accuracy of analytical instruments.

CGO See constrained Gaussian ordination.

Chaos The behaviour of complex deterministic systems in which small changes in initial conditions lead to divergent outcomes, making long-term prediction impossible.

Chemometrics The science of extracting information from chemical systems by data-driven means.

Chi-square (χ^2) criterion A statistic having, at least approximately, a chi-square distribution. A simple example is the test statistic used to assess the independence of two variables forming a contingency table.

Chi-square (χ^2) distance coefficient A coefficient that estimates the dissimilarity between assemblages (presence/absence, ranks, or quantitative) in two samples that is central to correspondence analysis. It is similar to Euclidean distance but it compensates for different relative frequencies or probabilities of occurrence. There is also a chi-square metric.

Chi-square (χ^2) test A statistical test in which the sampling distribution of the test statistic is a chi-square distribution.

Chord distance (dissimilarity) A coefficient that estimates the dissimilarity between quantitative assemblages in two samples. It is similar to Euclidean distance but it uses square-root transformed percentage or proportional data.

Chronological clustering A clustering method that takes account of the temporal sequence of sampling (see also zonation).

Cladistic Pertaining to a clade, where members of the group share closer ancestry with one another than with taxa of other clades.

clam R software for classical age-depth modelling including calibration of radiocarbon dates. Error estimation for the various models are based on Monte Carlo sampling from the calibrated-age distributions.

Classical gradient analysis See indirect gradient analysis.

Classification and regression trees (CART) An alternative to regression techniques for determining subsets of explanatory variables most important in the prediction of the response variable. Rather than fitting a model to the data, a tree structure is generated by dividing the data recursively into a number of groups, each division being chosen so as to maximise some measure of the difference in the response variable in the resulting two groups. If the response variables are presence/absence, a classification tree is generated. If the response variables are quantitative, a regression tree is generated.

cluster Software package in R for Ward's agglomerative clustering, other clustering procedures, and calculation of dissimilarity measures.

Cluster analysis A set of methods for deriving a classification of a set of data using variables measured on each individual.

Co-correspondence analysis An ordination-based procedure to relate two sets of biological data (e.g., beetles and plants) for the same samples.

Co-plot A powerful graphical tool for studying how a response depends on an explanatory variable given the values of other explanatory variables. The plot consists of a number of panels, one of which (the 'given' panel) shows the values of a particular explanatory variable divided into a number of intervals, while the others (the 'dependence' panels) show the scatter-plots of the response variable and another explanatory variable corresponding to each interval in the given panel.

cocorresp R software to implement co-correspondence analysis.

Codebook vector An important component in self-organising maps that map all data to a set of discrete locations organised in a regular grid. Associated with every location is a prototypical object, called a codebook vector, representing part of the space of the data. The complete set of codebook vectors is a concise summary of the original multivariate data.

Coefficient of determination (R^2, r^2) In regression, the square of the correlation coefficient between two variables. It gives a measure of the proportion of variation in one variable accounted for by the other.

Coefficient of variation A measure of spread for a data-set defined as 100 x (standard deviation / mean).

Collinearity See multicollinearity.

COMBINE Software for the statistical analysis of concentration data.

Combined path length (CPL) In core correlation using sequence slotting, CPL is the sum of the distance between consecutive samples in the pooled sequence and is a measure of the total discordance between the two sequences.

Complete linkage agglomerative clustering A method for grouping multivariate data into clusters where the distance between two clusters is defined as the greatest distance between an item in one cluster and an item in another cluster.

Compositional data Data consisting of a set of variables, each represented by proportions, and which sum to one.

Compositional turnover Amount of difference in assemblage composition and/or abundance along a known environmental or temporal gradient or along the major direction of variation in a data-set.

Comprehensive R archive network (CRAN) A network of ftp and web servers for distributing the R language and environment for statistical computing and graphics.

CONCENTR Software for the statistical analysis of concentration data.

Confidence intervals See confidence limits.

Confidence limits A range of values, calculated from the samples, which are believed, with a particular probability, to contain the true but unknown parameter value. A 95% confidence interval implies that when the estimation is repeated many times, then 95% of the calculated intervals would be expected to contain the true parameter value. Also referred to as confidence intervals.

CONIIC Constrained information cluster analysis – see constrained incremental sum-of-squares cluster analysis.

CONISS See constrained incremental sum-of-squares cluster analysis.

Conjoint coding A simple means of substituting a quantitative continuous variable with several qualitative variables (e.g., pseudospecies). Used in two-way indicator species analysis. An advantage of conjoint coding is that if a species' abundance shows a unimodal response along a gradient, each qualitative variable (pseudospecies) also shows a unimodal response curve. If the response curve for abundance is skewed, then the pseudospecies response curves will differ in their optimum.

CONSLINK See constrained single-link cluster analysis.

const.clust Software package in R for constrained clustering.

Constrained cluster analysis A type of cluster analysis in which an external constraint such as temporal ordering or geographic location is used to constrain group formation.

Constrained Gaussian ordination (CGO) A Gaussian ordination in which the ordination axes of biological data are constrained to be linear combinations of external environmental variables.

Constrained gradient analysis See direct gradient analysis.

Constrained incremental sum-of-squares cluster analysis (CONISS) A stratigraphically constrained cluster analysis where groups of adjacent samples or groups of samples are grouped together so as to minimise the within-group sum-of-squares and thus maximise the between-group sum-of-squares. Used for zonation of stratigraphical data. Constrained information cluster analysis (CONIIC) is similar but uses information content rather than sum-of-squares criterion.

Constrained ordination An ordination of assemblage data but where the ordination axes are constrained to be linear combinations of the external environmental variables. Also known as canonical ordination.

Constrained single-link cluster analysis (CONSLINK) A variant of single-link agglomerative clustering in which clusters are constrained to consist of adjacent samples or group.

Contingency table A table to display the frequency of each combination of two or more variables.

Convex hull The edges of a convex polygon that bounds a set of observations in variable space.

Cophenetic correlation A measure of how well the results of a cluster analysis match the original data, calculated as the correlation between observed values in the dissimilarity (or similarity) matrix and the corresponding fusion levels in the dendrogram. Also known as matrix correlation.

Core correlation Correlation of two or more sedimentary sequences. In palaeolimnology, core correlation usually involves matching individual cores from different parts of a basin (e.g., littoral vs profundal) on the basis of palaeomagnetic, geochemical, and/or biological data.

Correlation biplot A type of ordination diagram that is optimal for displaying correlations between variables.

Correlation coefficient A measure of the linear dependence of one random variable on another. It has a value between −1 and +1 with 0 indicating no correlation between the variables.

Correlation matrix A square symmetric matrix with rows and columns corresponding to variables, in which the off-diagonal elements are correlations between pairs of variables, and elements on the diagonal are unity. Often the lower half below the diagonal is presented.

Correlogram A plot of the values of the autocorrelation against the time lag. Also known as an autocorrelogram.

Correspondence analysis (CA) An ordination technique that may use a weighted average algorithm to maximise the dispersion of species or sites in low dimensional space.

Covariable Often used as an alternative name for explanatory or predictor variable, but in the context of ordination the term refers more specifically to a variable that is not of primary interest but whose effects need to be included and allowed for in the analysis. Also known as a concomitant or background variable. It often corresponds to an incidental or nuisance parameter.

Covariance The expected value of the product of the deviations of two variables, x and y, from their respective means, μ_x and μ_y, namely cov $(x, y) = \mathrm{E}\,(x - \mu_x)(y - \mu_y)$. The corresponding sample statistic is

$$\mathrm{cov}(x, y) = \frac{1}{n}\sum_{i=1}^{n}(x_i - \bar{x})(y_i - \bar{y})$$

where n is the number of objects and $\bar{x}$ and $\bar{y}$ are the respective means of x and y.

Covariate Often used as an alternative name for explanatory or predictor variables but perhaps most specifically to refer to variables that are not of primary interest but are measured because it is thought that they may affect the response variable and consequently should be included in regression analysis and model building, for example in analysis of covariance (ANCOVA), partial ordination, and partial constrained (or canonical) ordination.

Covariation The difference between the expected value of the product of two random variables and the product of their separate expected values.

CPL See combined path length.

CPLSlot Windows®-based program for sequence slotting.

CRAN See comprehensive R archive network.

CRM See certified reference material.

Cross-classified data Data in which the observations have been independently grouped into two or more categorical variables.

Cross-correlogram A plot of the lagged correlations between two time series.

Cross-spectral analysis A technique to estimate the frequencies and periodicities shared between two variables in time-series data. Each time series is decomposed into an infinite number of periodic components and the contributions of these components in the two variables in certain ranges of frequency (spectrum) are estimated to derive a cross-spectral function.

Cross-validation (CV) Division of data into a training-set that is used to estimate the parameters of the model of interest and a test-set that is used to assess whether the model with these parameters fits or predicts well. There are several ways of implementing cross-validation including leave-one-out cross-validation.

Cubic spline A continuous smooth curve consisting of piecewise third-order polynomials passing through a set of control points.

Curse of dimensionality The tendency for some numerical problems to become intractable as the number of variables increases.

Curvilinear trend A relationship between variables that is not linear but appears as a curve when the relationship is graphed.

CV See cross-validation.

CVA See canonical variates analysis.

DAGs See directed acyclic graphs.

Data-mining The processing of large amounts of data to extract new useful information from them based on patterns and relationships.

db-RDA See distance-based redundancy analysis.

DCA See detrended correspondence analysis.

DCCA See detrended canonical correspondence analysis.

Decision tree A graphical tool to display a tree-like graph or model of decisions and their consequences or outcomes.

Degrees of freedom A parameter in some probability distributions giving the number of independent pieces of information concerning the variance.

Degree of smoothing See span.

Delauney triangulation A way of dividing a plane into tessellating triangles.

Dendrogram A diagram used in cluster analysis to show the steps of aggregation that form the clusters.

Density estimation procedure A way of estimating the population probability density function from a sample of observations.

Density function A curve described by a mathematical formula that specifies, by way of areas under the curve, the probability that the variable of interest falls within a particular interval, e.g., normal distribution. Also called probability function or probability density.

Determinand That which is to be determined.

Detrended canonical correspondence analysis (DCCA) Canonical correspondence analysis but where the first and later axes are detrended to remove any curvature or 'horseshoe' structure that may be an artifact for a particular data-set. If a particular scaling is used, the analysis provides a convenient estimate of compositional change or turnover along particular environmental gradients.

Detrended correspondence analysis (DCA) The detrended form of correspondence analysis. Detrending is a mathematical technique used to remove the 'arch' or 'horseshoe effect' on the second axis, which is a mathematical artifact.

Detrending by segments An algorithm used in detrended correspondence analysis for removing the arch effect.

Deviance A measure of the extent to which a particular model differs from the saturated or full model for a data-set.

Deviant index A quantitative measure of how different an individual object (e.g., fossil) is to the mean, median, or mode of reference material. Used in identifying fossil objects numerically.

Digital moments Morphometric measure of the shape of an individual.

Direct gradient analysis The analysis of assemblage data in which samples are positioned not only on the basis of their assemblages but also in relation to their position along one or more known environmental gradients. If one species is considered, regression analysis is used; if two or more species are considered, constrained ordination is used.

Directed acyclic graphs (DAGs) The formal name for Bayesian networks or Bayesian belief networks where the nodes represent random variables and the linkages between nodes represent the conditional dependencies between the joined nodes. The graph is acyclic, meaning that there are no loops or feedbacks in the network structure and is directed because the relationships between nodes have a stated direction – A causes B.

Dirichlet probability distribution A family of continuous multivariate probability distributions often used as prior distributions in Bayesian statistics.

Discordance Where members of a group do not share a particular trait.

Discrete-time stochastic process A sequence of random variables observed at $T = [0, 1, 2, \ldots]$.

DISCRIM Software developed by CJF ter Braak to derive simple discriminant functions based on external predictor variables for a pre-existing classification (e.g., from TWINSPAN) based on multivariate assemblage data of response variables. This DOS program and its data-preparation program are available from HJB Birks.

Discriminant analysis A form of supervised pattern recognition used to derive rules (discriminant functions) for allocating individuals to *a priori* defined groups on the basis of a set of measured attributes.

Discriminant function Linear regression model that attempts to predict group membership based on a linear combination of predictor variables.

Disjoint coding Coding of nominal variables as a series of 1 / 0 dummy variables, one for each category of each nominal variable. After being formed into categories, quantitative variables can be coded in the same way. Used in two-way species analysis and simple discriminant functions.

Dispersion The distribution or scatter of observations or values about the mean or central value.

Dispersion matrix See variance-covariance matrix.

Dissimilarity coefficient or function or index An index for quantifying the difference between two observations in a set of multivariate data.

Dissimilarity matrix A matrix of dissimilarity coefficients expressing pair-wise dissimilarities between all observations in a set of data.

Distance-based redundancy analysis A form of redundancy analysis where any dissimilarity coefficient can be used in place of the Euclidean distance coefficient that is implicit in redundancy analysis.

Distance biplot A type of ordination plot that is optimal for displaying ecological distances between sites.

Distance coefficient A dissimilarity coefficient that satisfies the inequality that the dissimilarity between two points i and j is less than or equal to the sum of their dissimilarities from a third point.

Distance matrix A matrix of distance coefficients expressing pair-wise distances between all observations in a set of data.

Diversity A characterisation of species composition of a habitat or sample that may include richness (number of species) and/or evenness (their relative abundance) of species.

Dot-plot A more effective display than, for example, pie-charts or bar-charts, for displaying quantitative data that can be clearly labelled.

Double square-root transformation A transformation involving square roots twice, sometimes applied to biological count data to stabilise variance.

Down-weighting In correspondence analysis, detrended correspondence analysis, and detrended canonical correspondence analysis, an algorithm to down-weight the influence of rare species on the analysis.

Dummy variable A variable that can take the value of either 0 or 1 to indicate the absence or presence of some categorical effect that may be expected to alter the outcome.

earth R software package to implement multivariate adaptive regression splines (MARS).

Edge effect Refers to the bias introduced into estimates of species optima when surveys do not sample the entire range of environmental conditions that a species inhabits. Edge-effects refer to truncated species responses at the ends of environmental gradients (e.g., at very low or high values).

Effective number of taxa A measure of the degree to which proportional abundances are distributed among the taxa.

Effective precipitation Net precipitation after losses by evaporation and transpiration.

Eigenanalysis The search for a coordinate system that provides a simplification of the problem at hand.

Eigenvalue decomposition The redescription or decomposition of a square symmetric matrix of dissimilarities or distances into eigenvalues and their associated eigenvectors. The eigenvalues are selected to satisfy particular mathematical criteria.

Eigenvalues and eigenvectors If $\mathbf{A}$ is a square matrix, $\mathbf{x}$ is a column vector not equal to 0, and λ is a scalar so that $\mathbf{Ax} = \lambda\mathbf{x}$, then $\mathbf{x}$ is an eigenvector of $\mathbf{A}$ and λ is the corresponding eigenvalue. In CA or PCA, the eigenvalue of each axis reflects the proportion of the total variance accounted for by that axis and is a measure of the importance of an ordination axis. The eigenvectors ($\mathbf{x}$) define the linear function of the variables in the above relation.

Elastic net A compromise in regression shrinkage techniques between the rather weak lasso penalty that selects predictors via shrinkage and the ridge penalty that tends to shrink coefficients of correlated variables towards each other. The elastic net combines the two via a weighted combination to form the elastic-net penalty.

Entropy A measure of the amount of information received or outputted by a system, usually measured in bits.

ESS Error sum-of-squares – see residual sum-of-squares.

Euclidean distance A measure of dissimilarity between pairs of samples calculated by extending Pythagoras' theorem from two dimensions to the full dimensionality of multivariate data.

Euclidean space The Euclidean plane and three-dimensional space of Euclidean geometry, and their generalisations to higher dimensions.

Evapotranspiration Water lost as vapour from both soil or open water (evaporation) and from the surface of plants (transpiration).

Evenness A measure of the similarity in numbers of organisms of each species in a habitat or sample.

Explanatory variable A variable on which the response variable is assumed to depend. Also called the predictor variable.

Exploratory data analysis An approach to data analysis that emphasises the use of informal data summarisation and graphical procedures not based on prior assumptions about the data structure.

***F*-ratio test** See *F*-test.

***F*-statistic** The result of a test of equality of two or more variances.

***F*-test** A test of equality of the variances from samples of two populations that have normal distributions.

Factorial design A way of investigating the effects of several explanatory variables (factors) on a single response variable.

FD R package for measuring functional diversity from multiple traits, and other tools for functional ecology.

Feed-forward network A type of artificial neural network in which information moves in one direction only, from the input, through one or more hidden layers to the output.

Fidelity The degree of restriction of a taxon to a particular situation, community, or assemblage.

Floating chronology A chronology without a fixed reference point.

Forward selection A method for selecting a 'good' (but not necessarily the 'best') subset of explanatory or predictor variables in regression analysis, including constrained ordinations that are, in reality, multivariate regressions. The criterion used for assessing if a variable should be added to an existing model is the change in the residual sum-of-squares produced by the addition of the variable.

fossil R package for analysing palaeoecological and palaeontological data.

Fourier frequencies Used in harmonic analysis that determines the period of the cyclical term in a time-series. Fourier frequencies result from the decomposition of the periodic function into its constituent sine and cosine terms.

Fourier spectral analysis The analysis of data subject to a fast Fourier transformation based on Fourier's theorem that proposes that any periodic function can be reduced to a series of sine and cosine terms, each represented by an amplitude and a phase.

Fourier transform filter The simplest way to estimate the power spectrum of a time-series is to find the inner products (proportional to correlation coefficients) between the discrete time-series and a harmonic series of sines and cosines. This is the discrete Fourier transform. Its filter is used to achieve equal time resolution with the series.

Fourier transformed infra-red spectroscopy (FTIRS) Similar to near infra-red spectroscopy (NIRS) but uses longer wavelengths (4000–400 cm^{-1}) than NIRS (12,500–4000 cm^{-1}) and directly monitors molecular vibrations. It thus allows more detailed structural and compositional analysis of both organic and inorganic compounds than is possible with NIRS.

Fourth-corner analysis Synonym for RLQ analysis

Freedman and Diaconis's rule A rule for calculating the appropriate number of classes or bins to use in the construction of a histogram.

Freeze-corer A device that freezes lake sediment *in situ* to enable retrieval of undisturbed material.

Frequentist inference Where probability is viewed as being equal to the limiting relative frequency as the sample size increases.

FTIRS See Fourier transformed infra-red spectroscopy.

Gabor transformation In signal processing, a special case of the short-time Fourier transform used to determine the sinusoidal frequency and phase content of local sections of a signal as it changes over time.

Gabriel graph A subgraph of Delaunay triangulation showing the proximity of a set of points.

GAM See generalised additive model.

Gamma distribution A two-parameter family of continuous probability distributions.

GARP See genetic algorithm for rule-set prediction.

Gaussian distribution A normal distribution.

Gaussian filter A filter whose impulse function is a Gaussian function designed to give no overshoot to a step function input whilst minimising the rise and fall time. It modifies the input signal by convolution with a Gaussian function. Also used in smoothing time-series in which the smoothed value is the average of k points around the central value but with each point weighted according to the value of the appropriate Gaussian or normal density function.

Gaussian logistic regression Mathematical technique that attempts to fit either a Gaussian (bell shaped) regression curve or an increasing or decreasing monotonic curve to species abundance data generated from a training-set.

Gaussian logit regression See Gaussian logistic regression

Gaussian ordination An ordination procedure based on Gaussian unimodal species-response curves that aims to construct one or more latent variables so that these curves optimally fit the species data.

gbm Package using R software for gradient boosting in boosted regression trees.

GCV See generalised cross-validation.

Generalised additive model (GAM) An extension of the generalised linear model in which the link function of the expected value of the response is modelled as the sum of smooth functions of the explanatory variables, rather than the variables themselves.

Generalised cross-validation (GCV) A computationally efficient alternative to leave-one-out cross-validation.

Generalised linear mixed-effects analysis An analysis involved in generalised linear modelling but extended to include random effects in the linear predictor.

Generalised linear model (GLM) A generalisation of the ordinary or general linear model in which the response is related to the explanatory variables via a link function and by specifying the form of the variance of the response variable. Includes linear, logistic, and Poisson regression.

Genetic algorithm In machine learning, an optimisation procedure motivated by a biological analogy.

Genetic algorithm for rule-set prediction (GARP) A computer program based on a genetic algorithm for deriving rules that describe a set of ecological niches for species.

Gini index or Gini coefficient A coefficient for measuring the inequality of a distribution. Used as a measure or node impurity in regression trees.

glmnet R package for fitting a range of shrinkage methods (ridge regression, lasso, elastic net, etc.).

GLR See Gaussian logistic (logit) regression. Also software for Guassian logistic regression and for estimating optima and tolerances of taxa using maximum-likelihood estimation.

Goodness of fit Measure of the agreement between a set of observations and the corresponding values predicted from some model of interest.

Gower's coefficient of similarity A similarity coefficient suitable when the variables are mixed, consisting of continuous quantitative variables and categorical variables (including presence/absence variables).

Gower's general dissimilarity coefficient See Gower's coefficient of similarity.

Gradient analysis A method used in community ecology to relate the abundances of various species in a biological community to one or more environmental gradients, usually by ordination or weighted averaging.

Group-average sorting A method for collecting multivariate data into clusters.

***h*-block cross-validation** A method of cross-validation that preserves the autocorrelation structure in spatially- or temporally-ordered data.

Hellinger distance A measure of the distance between populations with multivariate distributions having two probability density functions. In ordination, a Euclidean distance between two samples where the abundance values are first divided by the sample total abundance and then square-root transformed (Hellinger transformation). See chord distance (dissimilarity).

Helmert's contrasts (= orthogonal dummy variables) A coding system for categorical variables in analysis of variance, in which each level of a factor is tested against the average of the remaining levels.

Heteroscedasticity The property of a set of random variables that have different variances.

Heuristic A general recommendation based on statistical evidence.

Hexagonal binning A form of bivariate histogram for displaying large data-sets, in which the original data are grouped or binned, and the binned data plotted as a bivariate scatterplot using differently sized or shaded hexagons.

Hierarchical classification or clustering Clustering techniques that combine objects into groups which are arranged in a hierarchy, with similarities between different groups displayed at different levels in the hierarchy.

Hierarchical partitioning A numerical technique that allows the contribution of each predictor or explanatory variable to the total explained variance of a regression model, both independently and in conjunction with the other predictors, to be estimated for all possible candidate regression models.

Highest posterior density Largest value of the posterior density, used in problems of Bayesian inference.

Hill's diversity measures N0 is the number of all taxa in the sample regardless of their relative abundance; N1 is the number of abundant taxa in the sample; and N2 is the number of very abundant taxa in the sample. The effective number of taxa is a measure of the number of taxa in the sample where each taxon is weighted by its abundance.

Hill's index of similarity A measure based on information theory for assessing the similarity between two clusterings or partitionings of the same sets of objects but based on different types of variables.

Hill's scaling Method of scaling CA, DCA, and CCA ordination axes in Hill's standard deviation units.

Hill's standard deviation units of compositional turnover The length of a CA, DCA, CCA, or DCCA ordination axis (range of sample scores) expressed in standard deviation units of compositional turnover. The tolerance of the species' curves along the axis is close to 1 after rescaling, and each curve therefore rises and falls over about 4 standard deviations. A gradient of more than four standard deviations can thus be expected to have no species in common.

Histogram A graph of a frequency distribution in the form of rectangles whose base coincides with the class interval and whose area is proportional to the class frequency.

HOF Software for the HOF modelling of species responses to an environmental gradient, written by J Oksanen and PR Minchin.

HOF modelling Huismann, Olff, and Fresco modelling of species responses to an environmental gradient using a hierarchical set of generalised linear models (skewed, symmetric, plateau, monotonic, null).

Holocene The name of a geological epoch of the Quaternary period, covering the last ~11,700 years. It means 'entirely recent' (*holos* = whole and *kainos* = new in Greek), because it represents modern times.

Homogeneity in the mean A term used in statistics to indicate the equality of some quantity of interest (e.g., mean) in a number of different groups, etc. Very relevant in time-series analysis.

Homogeneity test A test for the homogeneity or equality of some quantities of interest (most often variance) in different groups.

Hutchinsonian niche A view of the niche as *n*-dimensional hyper-volume which describes the environmental conditions in which a species can survive.

Hypergeometric distribution A discrete probability distribution that describes sampling with replacement from a finite population.

Hysteresis Delay in the adjustment of a process as a result of a change in an associated process.

IKFA Imbrie and Kipp factor analysis – see Q-mode factor analysis.

indicspecies Software for R that extends the IndVal approach.

Indirect gradient analysis A generic term for ordination methods that only analyse assemblage data and represent the data in a low-dimensional graphical form where the axes are selected to capture the variation in the data as effectively as possible according to an assumed underlying species response model. Interpretation is often aided by a *post-hoc* regression on external variables.

IndVal Software for undertaking indicator species analysis.

Interactive graphic display Modern graphical analysis that is interactive and allows rotations, colour coding, selective labelling, etc.

Internal block length In sequence slotting when the sediment properties in part of either sequence being correlated do not vary much, the optimum slotting often contains blocks of consecutive sediment samples. In most cases it can be useful to impose the constraint of a maximum internal block length (e.g., 2 or 3 samples).

Internal cross-validation Cross-validation using some form of data splitting, as opposed to external cross-validation using an independent test set.

Interpolation (interpolate) The process of determining the value of a function between two known values.

Interquartile range A measure of spread given by the differences between the first and third quartiles of a set of values for a variable.

Inverse regression A means of calibration for inferring environmental variables. The environmental variable is used as the response variable and the responses of the species are the predictor variables. The resulting regression equation is then the calibration or transfer function used in reconstruction.

Isolation lake or isolation basin A freshwater lake that was once connected to the sea but has become isolated as a result of relative sea-level change.

Isostatic rebound Used to refer to the state of gravitational equilibrium where land masses that were previously depressed by an enormous weight (typically ice sheets or rock) subsequently undergo uplifting after the weight is removed.

Iterative methods Methods that repeatedly use a series of operations until a good fit is obtained.

Jaccard coefficient A measure of the similarity in species composition between two communities using only presence/absence data. Can also be expressed as a distance.

Joint plot An ordination diagram of two kinds of entities which can be interpreted by the centroid rule.

'Jump' approach A heuristic method for selecting *k*, the number of close analogues in the modern analogue technique.

***K*-means partitioning** Method of cluster analysis in which a group of observations is partitioned into K groups.

***k*-nearest neighbours** Method of discriminant analysis based on studying the training-set subjects most similar to the subject to be classified. Classification is based on the k nearest neighbours where k is >1.

Kendall's non-parametric tau statistic (= Kendall's rank correlation coefficient) A method for determining the significance of association between two variables whose values have been replaced by ranks within their respective samples.

Kernel density A method for the estimation of probability density functions involving a window width or bandwidth and a kernel function. A powerful tool in exploratory data analysis.

Kernel function See kernel density.

Kernel matrix A similarity matrix between objects and codebook vectors in a self-organising map (SOM) weighted by kernel functions.

Kernel methods See kernel density.

Kernel regression A distribution-free method for smoothing data.

Kernel smoother A smoother produced by kernel regression smoothing, a distribution-free method for smoothing data. In one dimension, the method consists of estimating $f(x)$ in the relation $y_i = f(x) + \varepsilon_i$ where ε_i is assumed to be symmetric errors with zero means. There are several ways of estimating f, for example by averaging the y_i values that have x_i close to x.

Knot The tie points in a spline function.

kohonen Package of R software to create self-organising maps.

Kronecker's delta A measure to describe the presence (=1) or absence (=0) of information relating to a variable. Used in Gower's general similarity coefficient.

Kurtosis The departure of a frequency distribution from a normal distribution.

labdsv R software for analysis of ecological data including indicator species analysis using the function `indval()`.

lars R software for least angle regression used to compute the entire lasso path from no predictors to the full least-squares solution.

Lasso See least absolute shrinkage and selection operator.

Latent variable A variable that cannot be measured directly but is assumed to be related to a number of observed samples. It is selected to 'best' explain the data according to an assumed response model and in ordinations it is the first major ordination axis.

Least absolute shrinkage and selection operator (lasso) A technique for shrinkage and variable selection in regression modelling. The lasso is a general technique that has been applied to generalised linear models and is used as a shrinkage technique in boosted trees.

Least-squares criterion The criterion of minimising the sum-of-squared residuals in fitting a model to a set of data.

Leave-one-out cross-validation A method of cross-validation for determining the error rate in which each observation in turn is omitted from the data, the prediction model recalculated, and the error rate calculated from the left-out observations.

Likelihood-ratio test A statistical test used to compare the fit of two models.

Linear discriminant analysis See discriminant function analysis.

Linear discriminant function See discriminant function.

Linear regression A simple linear regression model involving a response variable that is a continuous variable and a single predictor variable related by an intercept and slope term.

Linear transformation A transformation of q variables $x_1, x_2, \ldots, x_q$ given by the equations $y_1 = a_{1\,1}\, x_1 + a_{1\,2}\, x_2 + \cdots + a_{1\,q}\, x_q$ to $y_{p\,1} = a_{p\,1}\, x_1 + a_{p\,2}\, x_2 + \cdots + a_{p\,q}\, x_q$. Such a transformation is the basis of principal component analysis (PCA).

Linkage clustering A family of clustering methods in which objects are assigned to clusters when a user-determined proportion of the similarity links has been realised.

Local discordance Used in core correlation by sequence slotting. It is the dissimilarity or distance between any two samples in either sequence being correlated.

Locally weighted scatterplot smoothing (LO(W)ESS) A method of regression analysis in which polynomials of degree one (linear) or two (quadratic) are used to approximate the regression function in particular 'neighbourhoods' of the space of the predictor variables. It uses weighted least squares with local subsets of the data so as to pay less attention to distant points. It assumes no predetermined model for the entire data-set and therefore provides no explicit formula for the fitted curve.

LOESS See locally weighted scatterplot smoothing (also known as LOWESS).

Log-normal distribution The probability distribution of a random variable whose logarithm is normally distributed.

Log transformation A transformation that replaces original values by their logarithms. Often useful for transforming chemical concentration data which often exhibit a log-normal distribution.

Logistic regression A model that can be used to predict the probability of an event occurring by fitting data to a logit function.

Logit Logistic transformation or logit of a proportion; logit $(p) = \log_e p \,/\, (1 - p)$. As p tends to 0, logit (p) tends to $-\infty$ and as p tends to 1, logit (p) tends to ∞. The function logit (p) is a sigmoid curve that is symmetric about $p = 0.5$.

Logit or logistic model A linear model in which the dependent variable is a logit and the explanatory variables are categorical.

Logit regression See logistic regression.

LOI See loss-on-ignition.

Loss-on-ignition (LOI) Organic matter content of the sediment estimated by measuring weight loss in sub-samples after burning at selected temperatures (typically 550°C for 5 h). Often used as a first-order estimate of organic carbon content of sediments and widely used in lake sediment studies.

Low-pass filter In time-series analysis, a filter that passes low-frequency signals but reduces the amplitude of high-frequency variation.

LOWESS See locally weighted scatterplot smoothing (also known as LOESS).

Macroecology Biogeographical studies of populations and species interactions on a broad rather than local scale, and which considers both geography and history in understanding the abundance, distribution, and diversity of species.

Magma A molten fluid formed within or below the Earth's crust which may consolidate to form igneous rock.

Manhattan distance Also known as the city block distance. A distance or dissimilarity measure that is the sum of the differences, irrespective of sign, between values for object i and j for all k variables. It gets its name from the distance travelled by a taxi around blocks in a city with an orthogonal plan like Manhattan.

Mann-Whitney U statistic The statistic used in a Mann-Whitney test, a distribution free, non-parametric alternative to Student's t-test.

MANOVA See multivariate analysis of variance.

Maphic Ferromagnesian minerals (silicates containing iron and/or magnesium)

Markov chain Monte Carlo (MCMC) methods Powerful but computer-intensive methods for indirectly simulating random observations from complex, and often high dimensional probability functions. Used widely in application of Bayesian inference.

Markov property A Markov chain and its equation imply that to make predictions about the future behaviour of the system of interest, it suffices to consider only its present state and not its history.

MARS See multivariate adaptive regression spline.

MART See multiple additive regression trees.

MASS R software for a huge range of modern applied statistical procedures.

MAT See modern analogue technique.

MATLAB Software package for statistical, graphical, and numerical analysis with many add-on functions and libraries and programming facilities.

Matrix correlation See cophenetic correlation.

MATTOOLS R Software for the modern analogue technique (MAT).

Maximum likelihood (regression and calibration) An estimation procedure involving maximisation of the likelihood or the log-likelihood with respect to the parameters. It is commonly used in generalised linear models.

MCMC See Markov chain Monte Carlo methods.

Mean A measure of location or central value for a continuous quantitative variable estimated as the sum of the variable of interest in all samples divided by the number of samples.

Mean squared error The square of the differences between an estimated and true value of a parameter.

Mean squares The ratio of the sum-of-squares to the corresponding number of degrees of freedom in ANOVA.

Measurand A physical quantity, property, or condition which is measured.

Median The value of a variable in an ordered array that has an equal number of observations above and below it (= second quartile).

MEMs See Moran's eigenvector maps.

Meta-data Data concerning the collection of data and data-bases, e.g., site location and description, date of sampling, investigator.

Metric ordination or metric scaling Any ordination method (e.g., PCA, CA, PCoA) that approximates a linear relationship between the configuration of points in ordination space and the original input distances.

MFPA See multi-frequential periodogram analysis.

Minimum-variance clustering (= Ward's method) An agglomerative clustering algorithm that constructs clusters which have minimum within-cluster variance and maximum between-cluster variance.

Minitab Software package for general statistical analysis.

Mixed-effect model A class of models that contain both fixed and random effects. They are often used where repeated measures are made on the same sampling units.

Modal Of a value that occurs most frequently in a distribution or set of data.

Model I regression Ordinary least-squares regression in which the sum of squared errors in y are minimised, and assuming no errors in x.

Model II regression Regression when there is error in both the dependent and independent variables. Includes techniques such as major axis (MA), standard major axis (SMA), and ranged major axis (RMA) regression.

Model-fitting analysis A general approach in statistics that attempts to find the simplest model for a set of observations that provides an adequate fit to the data under the principle of parsimony.

Modern analogue matching See analogue matching.

Modern analogue technique (MAT) A technique developed in palaeoecology in which fossil assemblages are compared, in turn, using an appropriate dissimilarity measure, with each sample in a modern assemblage data-set to identify modern analogues as an aid in interpretation of the fossil assemblage in terms of past communities or past environments.

MODPOL Software for modern analogue matching.

Monothetic divisive classification method Methods of hierarchical cluster analysis that recursively partition the original data-set into groups using a single variable to determine the split at each level of the hierarchy.

Monte Carlo permutation test A procedure for determining statistical significance directly from data without reference to a particular sampling distribution. It is similar to a randomisation test except that the permutations are restricted in some way to take account of, or maintain, the sampling design of the observed data (e.g., time series, line transects). For example, in a study involving the comparison of two groups, the data are divided (permuted) repeatedly between groups and for each division (permutation) the relevant test statistic is calculated to estimate the proportion of the data permutations that provide as large a test statistic as that calculated form the observed data.

Monte Carlo sampling A method of generating random samples from a probability density function.

Moran's eigenvector maps (MEMs) Formerly called principal coordinates of neighbour matrices, Moran's eigenvector maps are a more general procedure for spatial partitioning and modelling.

Moran's *I* statistic A measure of spatial autocorrelation using information from specified pairs of spatial observations.

MOSITEST Software that tests whether microfossil counts of taxa from two or more samples are likely to have come from the same population.

MOSLIMIT Software for calculating confidence limits for proportional data.

MRT See multivariate regression tree.

MS See mean squares.

Multicollinearity In regression analysis, the situation in which two or more explanatory variables are highly correlated, leading to regression coefficients with high variance.

Multi-frequential periodogram A statistical method for detecting periodic components in temporal series with unequally spaced samples. The periodic components are detected through the estimation of the corresponding frequencies in a step-wise procedure. All frequencies are re-estimated at each step and a significance test is performed at each step. It can determine if a periodicity is present or absent in portions of the temporal series.

Multifurcations In cluster analysis or phylogeny, a node in a tree that connects more than three branches.

Multi-layer perceptron A feed-forward artificial neural network model that maps a set of input data to an output.

Multi-modal A frequency distribution containing more than one mode.

Multinomial distribution A generalisation of the binomial distribution to situations where *r* outcomes can occur in each trial, e.g., when there are more than two classes.

Multinomial logistic regression A generalisation of logistic regression that allows more than two discrete outcomes.

Multinomial logit regression See multinomial logistic regression.

Multinormal distribution (= multivariate normal distribution) A generalisation of the univariate normal distribution to higher dimensions.

Multiple additive regression trees (MART) A form of boosted regression tree involving gradient boosting.

Multiple correlation coefficient The correlation between the observed values of the response variable in a multiple regression and the values predicted by the estimated regression coefficient. It is often used as an indicator of how useful the predictor variables are in predicting the response. Its square gives the proportion of variance of the response variable that is accounted for by the predictor variables.

Multiple discriminant analysis An alternative term for canonical variates analysis.

Multiple regression A method to relate one dependent or response variable to several independent explanatory or predictor variables.

Multiple simultaneous tests In regression modelling, it is possible to perform multiple significance tests simultaneously, for example in forward selection procedures. In such instances, it is important to guard against an increase in the probability of a Type I error. To maintain the probability of a Type I error at some selected value of α, each of the m tests to be performed is judged against a significance level of α/m.

Multivariate adaptive regression spline (MARS) A flexible, non-parametric regression method.

Multivariate analysis of covariance An extension of analysis of covariance to include more than one dependent variable.

Multivariate analysis of variance (MANOVA) A procedure for testing the equality of the mean vectors of more than two groups for a multivariate response variable. It is directly analogous to analysis of variance of univariate data except that the groups are compared on all response variables simultaneously.

Multivariate data Data in which an observation is characterised by more than one independent variable.

Multivariate regression tree A regression tree in which the response is multivariate (e.g., a matrix of biological abundance data).

mvpart R software for multivariate regression trees.

***n*-fold leave-out cross-validation** Similar to leave-one-out cross-validation, except the data-set is divided into n groups and each group is left-out in turn.

NCAP See non-linear canonical analysis of principal coordinates.

Naïve Bayes classifier A probabilistic classifier that uses Bayes theorem and is the optimal supervised statistical-learning method if the predictors are independent (uncorrelated) given the classes. If its assumptions are met, it is guaranteed to produce the most accurate predictions.

Near infra-red spectroscopy (NIRS) A rapid non-destructive technique for estimating the concentration of chemical constituents of organic materials (e.g., carbon, nitrogen, phosphorus, lignin) and for obtaining proxy environmental data (e.g., lake-water pH, nitrogen, phosphorus, total organic carbon) from lake sediments or peat.

Negative binomial distribution A mathematical distribution used as a model of an aggregated or contagiously distributed population in which the presence of an individual at any given point increases the probability of another individual occurring nearby and in which the variance is greater than the mean.

Nelder-Mead simplex algorithm A method to find the maximum or minimum of a function.

Newman-Keuls test A multiple comparison test where the equality of three or more population means can be simultaneously tested.

NIRS See near infra-red spectroscopy.

NMDS See non-metric multidimensional scaling.

Node impurity An important concept in classification and regression trees as splits are chosen on the basis of how much they reduce the node impurity of the resulting tree. For regression trees, residual sum-of-squares about the child groups or nodes or residual sums of absolute deviations from the child-node medians are commonly used. There are other measures of node impurity. The overall node impurity evaluated for all possible splits is the sum of the impurities of the two groups formed by the split.

Non-linear canonical analysis of principal coordinates (NCAP) An extension of CAP that considers intrinsically non-linear relationships between biotic assemblages (responses) and non-linear environmental (predictor) variables. It combines canonical ordination of RDA and CAP to a GLM through a link function and non-linear optimisation procedures as in GLM.

Non-linear deterministic dynamics A mathematical non-linear model in which all the relationships are fixed and the concept of probability is not involved, so that a given input produces one exact prediction as an output, in contrast to a stochastic model.

Non-linear rescaling Used in CA, DCA, and DCCA. Non-linear rescaling of an ordination axis attempts to equalise the breadth of species response curves along the axis by means of equalising the within-sample variances of the species scores.

Non-metric multidimensional scaling (NMDS) An ordination or indirect gradient analysis method in which only the ranks of the dissimilarity or similarity coefficients are used to produce a low-dimensional representation of the data.

Non-parametric regression A regression in which the predictor effect is assumed only to be 'smooth' rather than of some specific linear or non-linear form as in GLM. Examples are LOESS (LOWESS) and regression splines.

Normal distribution A bell-shaped curve that is symmetrical about the mean (= Gaussian curve) representing the distribution of results from a normal sample population.

Normalised data Data transformed so it approximates a normal distribution.

Normalising transformations Data transformed to zero mean and unit variance (see standardisation).

NULCONC Software for the statistical analysis of concentration data.

Null hypothesis The hypothesis that no real difference or association exists between two populations, i.e., that an observed difference is due to chance alone. The null hypothesis is usually expressed as the converse of the expected results.

Ochiai coefficient A similarity coefficient to quantify the similarity between pairs of samples using presence/absence data.

Odum distance See Bray-Curtis distance.

OOB See out-of-bag samples.

Optimal splitting zonation method A method of dividing a stratigraphical sequence into a desired number of zones (n), without regard to the pattern of splits into $n-1$ zones, based on minimising the sum-of-squares for each value of n.

Ordinate The vertical y axis on a graph.

Ordination A collective term for numerical techniques that attempts to arrange sites in low-dimensional space based on their species composition (e.g., a data-set consisting of many sites and species can effectively be summarised by one or more ordination axes).

Out-of-bag samples (OOB) Samples not selected for a bootstrap sample and used as a test-set in deriving bagged regression trees and sample-specific errors of prediction in environmental reconstructions.

Outlier An observation that appears to deviate markedly from the other samples in which it occurs.

Over-fitting The fitting of statistical models that contain more unknown parameters than can be justified by the data.

OxCal Software for calibrating radiocarbon dates and analysing archaeological and palaeoenvironmental chronological information.

packfor R software for forward selection.

PAIRS Software for calculating the confidence intervals for two samples and for the combined data.

palaeoSig R package to test the significance of environmental reconstructions.

Paleoecological Investigation of Recent Lake Acidification (PIRLA) A co-ordinated study in North America to assess the extent, timing, and causes of recent lake acidification. It consisted of two projects PIRLA I and PIRLA II and ran from the mid 1980s to the mid 1990s.

paleoMAS R package for palaeoecological data analysis.

paleoNet R package for artificial neural networks in palaeolimnology.

PaleoToolBox R package for palaeoenvironmental reconstructions.

Palmer drought severity index A standardised drought index based on time-series data including precipitation, air temperature, and soil moisture.

paltran R package for palaeoenvironmental reconstructions.

PALYHELP A software package including the programs COMBINE, CONCENTR, MOSITEST, MOSLIMIT, and PAIRS.

Parsimony A general principle in statistics that among competing models, all of which provide an adequate fit for a set of data, the model with the fewest parameters is to be preferred.

Partial constrained (or canonical) ordination Constrained ordination (e.g., CCA, RDA) in which the statistical effects of some environmental variables, referred to as covariables, are removed or 'partialled out'.

Partial least squares (PLS) A statistical technique related to principal components regression (PCR), but whereas PCR finds latent variables that maximise variance in a set of explanatory variables **X**, and uses these as predictors of **Y** in a separate multiple regression, PLS directly maximises the covariance between **X** and **Y**, and finds latent variables that are often more efficient in predicting **Y**. It is often used in multivariate calibration where there is multicollinearity in the predictor variables.

Partial ordination Unconstrained ordination (e.g., PCA, CA, DCA) in which the statistical effects of some environmental variables, referred to as covariables, are removed or 'partialled out'.

Partialling out Performing partial ordination or partial constrained or canonical ordination.

Partitioning Dividing a set of samples into a series of non-overlapping groups. Starting with all samples and progressively dividing them according to some specified mathematical criterion.

PC-ORD Software for the analysis of ecological data including ordination and clustering techniques.

PCA See principal component analysis.

PCNM See principal coordinates of neighbour matrices.

PCNM R software for computing principal coordinates of neighbour matrices or Moran's eigenvectore maps.

PCoA See principal coordinate analysis.

PCR See principal components regression.

Pearson chi-square statistic A simple goodness-of-fit test which is the sum of a set of terms, each term being the quotient of the squared difference between an observed frequency and the corresponding expected frequency divided by the expected frequency.

Pearson's product-moment coefficient or Pearson's *r* linear correlation coefficient A measure of the strength of the linear relationship between two variables which can take on the values from −1.0 to +1.0, where −1.0 is a perfect negative (inverse) correlation, 0.0 is no correlation, and +1.0 is a perfect positive correlation.

Peeled hull Also known as convex-hull trimming. A procedure that can be applied to a set of bivariate data to allow robust estimation of Pearson's product-moment correlation coefficient. The points defining the convex hull of the observations are deleted before calculating the correlation coefficients. It eliminates isolated outliers without disturbing the general bivariate relationship.

Percentile A division in a cumulative frequency graph. The 50th percentile is equivalent to the median.

PeriodMod Software available from Pierre Dutilleul for multi-frequential periodogram analysis (MFPA).

Periodogram A graphical representation of the results of a harmonic or spectral analysis that determines the period of the cyclical terms in a time series.

Permutation test A simple type of hypothesis test where the observed data values are randomly redistributed amongst the experimental units. A test statistic is calculated for each redistribution and its significance is determined by the proportion of permutations that lead to values greater than or equal to it. All possible permutations can be made or a random selection used.

Phonolite A fine-grained igneous rock containing little silica.

Phylogenetic Pertaining to evolutionary relationships within and between groups.

Pie-chart Graphical method for displaying the relative frequencies of a categorical variable, consisting of a circle sub-divided into sectors, with each sector proportional to the percentage they represent.

Piece-wise regression A regression procedure where 'broken stick' models with two or more lines are fitted at unknown point(s) (so-called break-points) representing abrupt change in the data-set of interest.

PIRLA See Paleoecological Investigation of Recent Lake Acidification.

Plinian eruption A violent volcanic eruption where gas and ash extend into the stratosphere and the ejection of large amounts of pumice.

Point estimate The value of an estimated parameter for a particular sample.

Poisson distribution A probability distribution used to model events which have a discrete number of outcomes (i.e., counted as integers) and in which the probability of occurrence reduces as the integer count of the event increases (e.g., biological count data). A model of randomly distributed populations in which the presence of an individual at any given point does not alter the probability of another individual occurring nearby and in which the variance is approximately equal to the mean.

Poisson process Description of a situation where events occur randomly in time or space in such a way that for each small interval of time or region of space, the probability that it contains exactly one event is proportional to the size of the interval or region.

PolPal Software for constrained single-link cluster analysis.

Polynomial function A function of x in which x is raised to the power of one or more non-negative integers.

Polynomial regression A form of linear regression in which the relationship between the independent variable x and the dependent variable y is modelled as an nth order polynomial.

Polynomial_RDACCA Software package for polynomial redundancy analysis and canonical correspondence analysis. Available from Pierre Legendre.

Polythetic divisive procedure A method of cluster analysis that begins with all objects in a single group and sequentially divides the groups until all groups contain a single object or a stopping criterion is reached. Polythetic methods use a combination of attributes that are used to define the divisions at each stage, whereas monothetic methods use a single attribute.

Power function A function of the form $f(x) = cx^a$, where c and a are constant real numbers and x is a variable.

Power spectral analysis Analysis of the frequencies and periodicities in time-series data.

Power spectrum In time-series analysis, a function that defines the amount of 'power' or contribution to the total variance of the series made by particular frequencies.

PRC See principal response curves.

PrCoord Software for principal coordinates analysis distributed with the CANOCO and CanoDraw package.

Precision The closeness of repeated measurements of the same quantity.

Predictor variable A variable on which the response variable is assumed to be dependent. Also called an explanatory or independent variable.

Press experiment An experiment that measures the resistance of a system to an experimental treatment.

primer Software for similarity percentage tests (SIMPER) and other multivariate techniques. Widely used in marine ecology.

Principal component analysis (PCA) An ordination method for indirect gradient analysis that finds the best-fitting linear combination of variables (latent variables) to minimise the total residual sum-of-squares and finds subsequent linear combinations that are uncorrelated to previous combinations. It assumes a linear species response model to these latent variables. In the process it transforms original variables in the multivariate data into new composite variables that are uncorrelated and account for decreasing proportions of the variance of the data. The main aim of the method is to reduce the dimensionality of the data and to find the latent variables.

Principal components regression (PCR) A statistical technique for multivariate calibration that addresses the problem of multicollinearity among the predictor variables by summarising these in a small number of principal components and using these components as predictors in a multiple regression.

Principal coordinate analysis (PCoA) An ordination method for indirect gradient analysis in which the required coordinate values for the axes are found from the eigenvectors of a matrix of inner products of any dissimilarity or similarity coefficient. The object points are mapped onto the resulting low-dimensional ordination space so that the distances between the objects are as close as possible to the original dissimilarity (= distance) or transformed similarity. Also known as classical or metric scaling.

Principal coordinates of neighbour matrices (PCNM) An ordination method for exploring patterns in ecological or other data at multiple spatial scales.

Principal curves A generalisation of principal component analysis that fits smooth, one-dimension curves that pass through the middle of the data-set, providing a non-linear summary of the data.

Principal response curves (PRC) A multivariate method for the analysis of repeated measurement designs that tests and displays treatment effects across time in experimental or monitoring studies. It is based on redundancy analysis (= reduced rank regression) that is adjusted for changes across time in the control treatment or control site. In some instances, space can be substituted for time in ecological and limnological studies. It focuses on the time-dependent (or space-dependent) effects.

Probability density See density function.

Probability function See density function.

Procrustes analysis A method of comparing alternative geometrical ordination representations or solutions of a set of multivariate data. The two solutions are compared using a residual sum-of-squares criterion which is minimised by allowing the co-ordinates corresponding to one solution to be rotated, reflected, and translated relative to the other.

Proportional-link agglomerative clustering A clustering algorithm, also known as intermediate linkage clustering, where the fusion criterion of an object or a cluster with another cluster is considered satisfied when a given proportion of the total possible number of similarity links is reached.

PROTEST Software for Procrustes analysis with associated permutation tests to assess the statistical significance of the Procrustes goodness-of-fit statistic.

PSIMPOLL Software to plot and analyse pollen stratigraphical data.

Pseudospecies See conjoint coding.

Pulse experiment An experiment that measures the resilience of a system to an experimental treatment.

Q-mode factor analysis A type of principal components analysis based on distances between observations rather than relationships between variables. Often includes a varimax rotation of the axes to so-called 'simple structure'.

Quadratic A function of predictor variables in a regression model where the variable x_i is combined with x_i^2.

Quadratic reduced-rank vector-based generalised linear and additive models Vector-based generalised linear and additive models include all known generalised linear and additive models and thus are very general and includes as many distributions and models as possible. The introduction of reduced-rank regression allows for dimension reduction and the use of quadratic terms extends the approach to model Gaussian unimodal species responses.

Qualitative variable A nominal or categorical variable.

Quantile-quantile plot A plot for comparing two probability distributions, where the coordinates are the quantiles for different values of the cumulative probabilities. Quantiles are values that divide a frequency distribution or probability distribution into equal ordered subgroups such as quartiles or percentiles.

Quantile regression A type of regression analysis in which the result estimates the median or other quantile of the response, rather than the mean as in least-squares regression.

Quartiles The values that divide a frequency distribution or a probability distribution into four equal parts.

R Package A software package for multivariate and spatial analysis created by Casgrain and Legendre in 1984 that is distinct from the free software environment R and its associated packages and libraries (http://www.bio.umontreal.ca/casgrain/en/labo/R/index.html)

R **software** R is a free software environment for statistical computing and graphics. It is a collection of user-generated code that enhances R with specialised statistical techniques and graphics, import/export capabilities, etc. (hhtp://www.r-project.org)

R^2 or r^2 See coefficient of determination.

${R_{adj}}^2$ or ${r_{adj}}^2$ or ${R_a}^2$ See adjusted coefficient of determination.

RAM See revised analogue method.

RAM98 Software for the revised analogue method (RAM).

Rand index An index that measures the similarity between two data clusterings.

Random error A deviation for which the magnitude and direction cannot be predicted.

Random forest In statistical learning, a classifier whose prediction is based on the modal value of an ensemble of decision trees. The method combines the idea of bagging and random feature selection.

Randomisation *t*-test A statistical test for assessing the significance of components in PLS and related models.

Range The numerical difference between the largest and the smallest values for a variable in a data-set.

Ranging A transformation of quantitative data where the data values for a variable first have the minimum value of the variable subtracted from the observed values and are then divided by the range of the values for the variable of interest (for interval-scale variables such as temperature in °C). For relative-scale variables such as calcium concentrations, the transformation is simply division of the observed values by their maximum value.

RARECEP Software for reformatting data for use in the program RAREPOLL.

Rarefaction A means of standardising assemblage samples for estimating taxonomic richness to a common sample size and for estimating the number of taxa that would be expected in samples if all the samples had the same count size.

RAREFORM Software for reformatting data for use in the program RAREPOLL.

RAREPOLL Software for performing rarefaction analysis of ecological and palaeoecological data.

Rate-of-change analysis A method for estimating the rate of compositional change in stratigraphical assemblage data. Compositional dissimilarity between adjacent pairs of samples is calculated and this is standardised to units of dissimilarity or difference per unit time.

RDA See redundancy analysis.

Realised niche That part of the fundamental niche actually occupied by a species in the presence of other species.

Receiver operating characteristic (ROC) curve A graphical plot of the sensitivity (true positive rate) of a test versus one minus the specificity (false positive rate). It is often used for choosing between competing tests.

Recursive partitioning regression Synonym for classification and regression trees (CART).

REDFIT Software for estimating red-noise spectra directly from unevenly spaced time series.

Reduced rank multivariate regression A way of constraining the multivariate linear regression model so that the rank of the matrix of regression coefficients is less than full. This constraint allows such models to be applied to data-sets with more variables than observations.

Redundancy analysis (RDA) A constrained ordination technique based on principal component analysis in which the ordination axes are constrained to be linear combinations of the environmental variables. Being based on principal component analysis, it assumes a linear species-environment response model. Also known as reduced-rank regression.

Regression (i) A statistical technique that describes the dependence of one variable on another (cf. correlation, which assesses the relationship between two variables); (ii) When used in the context of training-sets, the 'regression' step refers to the estimation of species parameters (e.g., optima, tolerance) from the species abundances in the training-set; (iii) Retreat of the sea from a land area.

Regression tree See classification and regression trees.

Relative neighbourhood graph In computational geometry, a triangulation of a set of points in two dimensions.

Repeatability An estimate of the precision of replicate measurements performed on independent samples of the substance of interest by the same analyst, the same equipment, using the same conditions of use, at the same location, and over a short period of time.

Reproducibility The precision of measurements performed on subsamples of the substance of interest by the same method but under changing conditions such as with different analysts and equipment, at a different location, or at a different time.

Resemblance coefficient In ecology, another name for similarity coefficient.

Residual The difference between the observed value and the value fitted or predicted by some model of interest.

Residual sum-of-squares (RSS) The sum of the squared residuals, a measure of the difference between the data and a fitted model.

Residual sums of absolute deviations (RSAD) The sum of the absolute values of residuals in a fitted model. This value is minimised in some forms of robust regression.

Residual variation Variation in a response not accounted for by a fitted model.

Resilience The ability of a system to recover to its original equilibrium or stable state following a perturbation or disturbance.

Resistance The ability of a system to remain unaffected by a perturbation or disturbance.

Response variable A variable that is presumed to be dependent on one or more other variables (called predictor or explanatory variables). Also known as dependent variable.

Restricted permutation tests A type of permutation test in which the range of possible permutations is restricted to take account of any spatial, temporal, or other structure in the data that results when observations are not sampled randomly and independently from one another.

Reversal Reversals in the order of fusion in a dendrogram can occur in certain clustering methods involving an unweighted centroid or weighted centroid algorithm. A reversal may be interpreted as nearly equivalent to a trichotomy in the hierarchical dendrogram.

Reverse-engineering algorithms Techniques in data-mining that help to uncover the functional form of relationships among correlated variables.

Revised analogue method A variant of the modern analogue technique that uses gridded interpolated assemblages in addition to the original training-set samples.

Ridge regression A regression method designed to overcome the problem of multicollinearity among the explanatory variables.

rioja An R package for the analysis of Quaternary science data.

RLQ analysis A special numerical technique that can simultaneously link the three matrices of palaeoenvironmental data (R), species traits (Q), and fossil data (L).

RMSE See root mean squared error.

RMSEP See root mean squared error of prediction.

ROC See receiver operating characteristic curve.

ROCKIT Software for receiver operating characteristic curve estimation.

Root mean squared error (RMSE) The square root of the sum of the differences between the observed and estimated values for a variable squared and divided by the number of objects. The square root of the mean squared error.

Root mean squared error of prediction (RMSEP) The root mean squared error derived under cross-validation or using an independent test data-set.

rpart Package using R software for fitting classification trees.

RSAD See residual sums of absolute deviations.

RSS See residual sum-of-squares.

Rug-plot A tool for displaying graphically a sample of values on a continuous variable by indicating their position on a horizontal line.

Running mean A method of smoothing a time series to reduce the effects of random variation and to reveal any underlying trend or seasonality. Weights can be used in the averaging.

S-PLUS Software package for statistical data analysis and graphical display with several add-on functions and libraries and programming facilities.

SAS Software package for statistical data analysis and graphical display (Statistical Analysis System).

Scaling (i) A general term for a low-dimensional geometric representation of multivariate data (e.g., by PCA, CA, NMDS, PCoA), (ii) Scatter or ordination plots can be drawn using different combinations of the matrix scalings used in the computations of, for example, CA, PCA, RDA, and CCA.

Scatter-plot A graph plotting pairs of values as points with the ability to incorporate other categorical variables by changing the shape or colour of the points.

Schuster's periodogram Uni-frequential periodogram used in time-series analysis. Schuster's test is available for assessing non-zero periodic ordinates in a periodogram.

Schwartz's criterion Equivalent to the Bayesian information criterion (BIC).

Self-organising map (SOM) A type of artificial neural network for unsupervised machine learning that is trained to produce a low-dimensional representation of the input variables.

Sensitivity A statistical measure of the performance of a binary classifier that measures the proportion of true positives which are correctly identified.

Separability A problem that can arise in logistic regression and modern analogues when there is no overlap in the dissimilarity for analogues (1) and non-analogues (0). This problem arises when the regression model does too good a job and perfect predictions can be made. In such circumstances the maximum-likelihood estimates of the regression coefficients may not exist or be subject to large uncertainties.

Sequence slotting A numerical procedure for correlating two or more stratigraphical sequences by identifying common patterns in a set of variables that allows sequences to be combined, or slotted together, into a single sequence.

Sequence splitting A numerical procedure for making independent splits in the records of individual variables in a stratigraphical sequence. The approach involves first distinguishing effectively non-zero parts of the sequences for zero portions, and then further splitting the non-zero sequences quantitatively using a maximum likelihood method.

Serial correlation See autocorrelation.

SHEFFPOLL Software for agglomerative zonation of stratigraphical data using total within group sum-of-squares or information content.

Shrinkage The phenomenon that can occur when a regression from, for example, a multiple regression, is applied to a new data-set in which the model predicts much less well than in the original sample. In particular, the value of the multiple correlation coefficient becomes less, i.e., it 'shrinks'.

Significance of zero crossing of the derivative (SiZer) A graphical tool for use in association with smoothing methods in the analysis of temporal series that helps to answer which observed features are 'real' as opposed to being spurious sampling artefacts.

Significant non-stationarities Stationarity is a term applied to time-series or spatial data to describe their equilibrium behaviour. The key aspect of stationarity is the invariance of their joint to a common translation in time or space. Non-stationarity is the absence of stationarity and significant non-stationarity is of sufficient extent that it can be the result of stochastic processes.

simba R package for dissimilarity coefficients, diversity indices, and permutation tests.

SIMCA See soft modelling of class analogy.

Similarity coefficient or function or index An index used to quantify the similarity of two observations in terms of a number of attributes or character states.

Similarity matrix A square matrix containing similarity measures, taken pairwise, among a set of observations.

Similarity percentage test (SIMPER) A method for assessing which taxa are responsible for observed differences between groups of samples.

SIMMAX A variation of the modern analogue technique that inversely weights modern analogues according to their geographical distance from the fossil samples.

SIMMAX Software to implement the SIMMAX variation of the modern analogue technique.

SIMPER See similarity percentage test.

Single-factor analysis of variance This is the simplest form of analysis of variance (ANOVA). The groups of samples are classified by only a single criterion. Also known as single classification analysis of variance.

Single-link agglomerative clustering A method to group multivariate data into clusters where the distance between two clusters is defined as the least distance between an item in one cluster and an item in the other cluster.

Singular value decomposition A method of matrix decomposition into eigenvalues and eigenvectors that underlies a number of multivariate methods (e.g., PCA, CA).

SiNos See significant non-stationarities.

SiZer See significance of zero crossings of the derivative.

Skeletonisation procedure A pruning algorithm for artificial neural networks where variables are progressively removed in an attempt to simplify the model and to minimise over-fitting without drastically affecting model performance. It is analogous to backward-elimination in regression modelling.

Skewness The lack of symmetry in a probability or frequency distribution. A distribution is said to have positive skewness when it has a long tail to the right and to have negative skewness when it has a long tail to the left.

Smoothing The averaging of data in time or space to compensate for random errors or variations on a scale smaller than that presumed significant to a given problem.

Smoothing spline A curve consisting of a sequence of cubic polynomials with no discontinuities drawn through a set of data.

Snipping An iterative procedure in data-mining to modify and test underlying dynamic non-linear functions in data-rich time-series.

Soft independent modelling of class analogy (SIMCA) A battery of techniques developed in chemometrics consisting of four levels. Level 1 is devoted to developing mathematical rules for each of a number of pre-set groups (called classes in SIMCA) in a training-set by fitting separate PCA models to each of them. Level 2, the prediction phase, uses these rules to assign new observations to any of the classes. Levels 3 and 4 implement quantitative predictions of one or several variables through PLS. In level 3 the predictions are for one variable, in level 4 the predictions are for two or more variables.

Softmax function In artificial neural networks, a function that calculates a layer's output from the inputs.

SOM See self-organising map.

Sørensen coefficient of similarity (Sørensen's index) In ecology, an index used to quantify the similarity of two samples.

spacemakeR Software in R for computing PCNM and MEM spatial eigenfunctions.

SpaceMaker2 Stand-alone package to compute PCNM eigenfunctions. Available from Pierre Legendre.

Span The proportion of a set of observations used in LOESS (= LOWESS) in which a locally weighted regression is fitted as part of deriving LOESS curves.

Spatial autocorrelation The correlation of a variable with itself through space. Positive spatial autocorrelation violates the assumption that the values of observations in each sample are independent and may invalidate some statistical tests.

Spatial contiguity A constraint imposed on spatially distributed data derived from the spatial connections between neighbouring points on a regular grid or irregularly-spaced points.

Spatial contiguity matrix A matrix depicting the spatial contiguities of objects, with 1 representing the spatial contiguity between two neighbouring objects and 0 representing the lack of spatial contiguity between two objects.

Spatial eigenfunctions Functions of the geographical co-ordinates of a set of objects derived from a principal coordinates analysis of neighbour matrices. These spatial eigenfunctions represent a spectral decomposition of the spatial relationships among the objects.

Spearman's (rank coefficient) rho A non-parametric correlation coefficient based on ranks that is used to measure the association between two variables. It is equivalent to Pearson's product-moment correlation coefficient between the rankings of the two variables.

Species loading Value of a taxon on a PCA or RDA ordination axis. It is an eigenvector coefficient. The comparable entity in CA, DCA, CCA, or DCCA is the centre of the taxon curve.

Species optimum The value of an environmental variable where a species reaches its maximum abundance. Often calculated as the mean of environmental values across sites where a species occurs, weighted by the species abundance at each site.

Species tolerance The range of a species around its optimum. Often calculated as the standard deviation of environmental values across sites where the species occurs, weighted by the abundance of the species at each site.

Specificitiy Used in indicator species analysis to estimate the extent that a species is only found in a particular group or cluster.

Spectral density The population counterpart of a periodogram which is a graphical tool in time-series analysis where the time series is decomposed into an infinite number of periodic components. Estimates of the contributions of these components in certain ranges of frequency are termed the spectrum of the series.

SPECTRUM Software for spectral analysis of time-series data with unequally spaced observations in time that uses the Welch-Overlapping-Segment-Averaging method.

Spiked samples Samples in which a known concentration of an analyte has been added for calibration purposes.

Spline A smoothly joined piecewise polynomial.

SPLITLSQ Method for constrained binary divisive partitioning using the sum-of-squares criterion. Also name of earlier software to implement the method.

SPSS Statistical Package for the Social Sciences.

SqRL See squared residual length.

Squared chord distance A simple measure of dissimilarity used, for example, to assess floristic dissimilarity between diatom samples.

Squared residual length (SqRL) The distance squared between an observed value and the predicted value of a variable or sample in a statistical model. Least-squares estimation attempts to minimise these distances.

Stable state Remaining unaltered for an extended period of time.

Standard deviation The most commonly used measure of the spread of a set of observations. It is equal to the square root of the variance.

Standard error The standard deviation of the sampling distribution of a statistic. The standard error of the mean of n observations is $\sigma/\sqrt{n}$ where σ^2 is the variance of the original observation.

Standard scores See z-scores.

Standardisation Converting a random variable or data-set into another variable with mean of 0 and variance of 1.

Standardised residual A residual divided by an estimate of its standard deviation.

Stationarity When the expected value at all time points in a time-series is the same and the correlation between two values depends only on the lag.

Statistical learning A general term applied to supervised and unsupervised machine learning for problems of data mining, inference, and prediction.

stats R package which has a function to implement Ward's agglomerative clustering and to calculate a range of dissimilarity coefficients.

Steinhaus distance See Bray-Curtis distance.

Stochastic process In probability theory, a series of random variables, indexed, for example, against time. More generally, a process involving chance.

Stratigraphically constrained cluster analysis A type of cluster analysis in which groups are constrained to consist of stratigraphically adjacent observations.

Student's *t* distribution An estimation of the deviation from the mean for values in a small data-set.

Sturges' formula A rule for calculating the appropriate number of classes or bins to use in the construction of a histogram.

Supervised classification The process of deriving a classification function or rule from a training data-set in which group membership is known *a priori*. The function can then be used to classify unknown objects.

Supervised learning In machine learning, the task of inferring a function or set of rules from a training data-set.

Support vector machine In a set of supervised learning methods, related to artificial neural networks, for solving regression and classification problems.

Surface Waters Acidification Programme (SWAP) An international research programme in the late 1980s and early 1990s designed to establish the cause and timing of recent lake acidification in Norway, Sweden, and Britain.

SWAP See Surface Waters Acidification Programme.

SYN-TAX A general purpose program for the analysis of ecological and taxonomic data.

Synthetic variables See latent variables.

Systematic error A bias in the measurement of a variable such that the mean of many separate measurements differs significantly from the actual values.

***t*-test** A method to determine the significance of the difference between two means when the samples are small and drawn from a normally distributed population with an unknown standard deviation. Common versions test whether the mean of a single population takes a specific value, or whether the means of two populations have the same value.

Tanimoto distance See Jaccard distance.

Taphonomic process A process that affects an organism after death.

tb-PCA See transformation-based principal component analysis.

tb-RDA See transformation-based redundancy analysis.

Tephra Volcanic ash.

TESS See total error sum-of-squares.

TGView Software for plotting pollen and other biostratigraphical diagrams.

TILIA Software for manipulating and plotting pollen and other biostratigraphical data.

TILIA-GRAPH Software for the plotting of stratigraphical diagrams. Now superceded by TILIA for Windows.

Time-series A series of observations over a long period of time, usually at regular intervals, of a random variable. The observed movement and fluctuations of many such series consist of four components – secular trend, seasonal variation, cyclical variation, and irregular variation.

Time-warp analysis A method for correcting for small shifts in two data-series, for example mass chromatograms. Small shifts in peak positions may occur because of external factors such as pH. Correction of such shifts is known as time-warping that originated in speech processing. It is widely used in chemometrics and biological dynamics.

Tolerance In community ecology, the range of an environmental variable which an organism or population can survive or one standard deviation of a Gaussian unimodal response model estimated by Gaussian logistic regression or as a weighted standard deviation estimated by weighted averaging regression.

Tolerance down-weighting In weighted averaging, the process of down-weighting taxa with large tolerances, based on the assumption that those with narrow tolerances are better environmental indicators.

Total discordance Used in core correlation by sequence slotting. It is the sum of the distances between consecutive samples in the pooled sequence created by sequence slotting. Synonymous with combined path length.

Total error sum-of-squares The sum of squared distances of each object to its cluster centre. Used in cluster analysis as a measure of the tightness of the clusters.

Training-set A data-set used to develop a predictive regression or classification model.

Trait analysis Analysis of characters or properties of organisms (e.g., body size, reproductive mode, fecundity).

Transformation A numerical change applied to a variable to simplify the analysis or to meet assumptions of a statistical test.

Transformation-based principal component analysis (tb-PCA) A principal component analysis (PCA) with prior transformation of the input-data to yield an ordination based on e.g., chord or chi-square distance rather than Euclidean distance.

Transformation-based redundancy analysis (tb-RDA) A redundancy analysis (RDA) with prior transformation of the input data to yield an ordination based on a distance e.g., chord or chi-square distance other than Euclidean distance.

Transformed data Data that have been transformed to aid analysis.

Tree complexity A tuning parameter in boosting that affects the learning rate required to yield a large ensemble of trees and determines the types of interactions that can be fitted by the final model.

Triangle inequality A property of dissimilarity or distance measures (D) between objects a and b, namely $D\ (a, b) + (b, c) \geq D\ (a + c)$. The sum of two sides of a triangle drawn in Euclidean space is necessarily equal to or larger than the third side.

Triplot An ordination diagram with three kinds of entities of which all pairs can form biplots. Examples are RDA and CCA triplots that consist of sample, species, and environmental variables. Sometimes called erroneously biplots.

True negative fraction A measure of performance of a receiver operating characteristic (ROC) curve used in modern analogue matching. Known as specificity, it is the proportion of true non-analogues out of all non-analogues for the same critical value of the chosen dissimilarity measure's critical threshold.

True positive fraction Also known as sensitivity in receiver operating characteristic (ROC) curves in modern analogue matching. It is the proportion of true analogues out of all analogues for a given critical value of the dissimilarity coefficient used.

Trueness A measure of whether the average of replicate measurements accurately reflects the true value of a measurand.

Tukey test Usually refers to Tukey's HSD (honest significant difference) test, a multiple comparison test usually used in conjunction with ANOVA to determine which means are significantly different from one another.

Turbidites Sediments deposited by currents containing a slurry of sediment and water.

Turnover The fraction of an assemblage that is exchanged or lost per unit time or per unit of an environmental variable or gradient.

TWINSPAN See two-way indicator species analysis.

TWINSPAN MS-DOS software to implement TWINSPAN (two-way indicator species analysis).

Two-way indicator species analysis A method for partitioning large data-sets of assemblages into groups using the first correspondence analysis axis as a basis for division. The algorithm continues to produce 2, 4, 8, 16, etc. groups unless a resulting group is too small to justify further division. Species are then grouped on the basis of their indicator value in relation to the groups of samples.

Two-way weighted averaging (regression and classification) Given a data-set of species abundances and environmental values, the process of applying weighted averaging twice, first in a regression step to estimate species scores, or optima, and second, in a calibration step, to estimate sample scores or environmental estimates from the species scores.

Type I error The rejection of a null hypothesis when it is true and should have been accepted.

Type II error The acceptance of a null hypothesis when it is false and should have been rejected.

UKAWMN See United Kingdom Acid Waters Monitoring Network.

Uni-frequential peridogram See Schuster's periodogram.

Uniformitarianism principle The concept that processes that operate in the present also operated in the past and produced the same results.

Unimodal response The expected non-linear response of a biological species to an environmental variable along an environmental gradient. The abundance of a species is expected to be at its maximum at the centre of its range.

United Kingdom Acid Waters Monitoring Network (UKAWMN) A monitoring scheme of 23 sites in the UK involving epilithic diatoms, aquatic macrophytes, aquatic invertebrates, and water chemistry. It has run for 20 years and is co-ordinated by University College London.

Univariate data Data involving a single quantity measured on a set of observations.

Unsupervised classification A clustering or partitioning technique that involves some form of cluster analysis. It is used in situations when there is little or no *a priori* information about group structure within the data.

Unsupervised learning In machine or statistical learning, the problem of finding structure or pattern in a data-set without *a priori* information. Approaches to unsupervised learning include clustering and indirect ordination.

Unweighted arithmetic average clustering See unweighted pair-group method using arithmetic averages.

Unweighted centroid clustering (UPGMC) A clustering algorithm that fuses groups on the basis of the distance between the centroids of each group.

Unweighted pair-group method using arithmetic averages (UPGMA) A clustering algorithm that fuses groups on the basis of the mean distance between objects in each group.

UPGMA See unweighted pair-group method using arithmetic averages.

UPGMC See unweighted centroid clustering.

Variable-barriers approach A method for zonation of stratigraphical data that allows for the delimitation of clearly defined zones and variable transitional periods.

Variance A measure of the variability in the values of a random variable. It is estimated by the squared difference between the variable and its mean.

Variance inflation factor (VIF) An indicator of the effect other predictor values have on the variance of a regression coefficient of a particular predictor. It is the reciprocal of the square of the multiple correlation coefficient of the variable with the other variables.

Variance-covariance matrix A square symmetrical matrix in which the elements on the main diagonal are variances and the remaining elements are covariances.

Variation partitioning analysis The partitioning of the variation in a data-set into the variation uniquely explained, in a statistical sense, by two or more sets of predictor variables, into the covariance between variable sets, and into the unexplained component. It most commonly uses (partial) CCA or (partial) RDA.

Varve A pair of contrasting sediment laminae representing accumulation during two seasons of a single year (e.g., typically a light summer layer and a dark winter one).

vegan An R package for the analysis of community ecology data.

VIF See variance inflation factor.

VPA See variation partitioning analysis.

WA See weighted averaging.

WACALIB Software for weighted-averaging regression and calibration, with bootstrap sample-specific error estimation.

WAPLS See weighted-averaging partial least squares. Also DOS software for weighted-averaging partial least squares regression and calibration and estimation of sample-specific errors.

Ward's agglomerative clustering See minimum-variance clustering.

Weighted arithmetic average clustering See weighted pair-group method using arithmetic averages.

Weighted average An average that attaches greater importance (adds weight) to some observations than others.

Weighted averaging (WA) A technique used to estimate either (i) the optimum of a taxon (weighted-averaging regression) based on measured values of environmental variables from the lakes in a training-set, where the weight is proportional to the species abundance; or (ii) an environmental variable from the species composition of a sample, based on estimates of species parameters (optima) from a training-set, where species are weighted relative to their abundance (weighted-averaging calibration).

Weighted-averaging partial least squares (WAPLS) An extension of weighted-averaging regression where partial least squares regression is used to find components within the modern assemblage data that will maximise the covariance between the species' weighted averages and the environmental variable of interest. It uses residual structure in the species data to improve the estimates of the species' parameters in the final prediction model. A WAPLS first component is the same as a two-way weighted averaging model that uses an inverse deshrinking regression. Prediction or reconstruction (calibration) of an environmental variable from a fossil assemblage is done by weighted averaging of the species' parameters, as in weighted averaging calibration.

Weighted centroid clustering (WPGMC) A clustering algorithm that fuses groups on the basis of the distance between the centroids of each group, weighted by group size.

Weighted mean See weighted average.

Weighted pair-group method using arithmetic averages (WPGMA) A clustering algorithm that fuses groups on the basis of the mean distance between objects in each group, weighted by group size.

Welch overlapping segment averaging (WOSA) (Welch's method) A method for estimating the power spectrum of a time-series based on averaging the periodograms derived from overlapping segments of the original series.

Wiggle-matching A way of aligning two curves by comparing the overall shape of each curve.

WinTWINS TWINSPAN for Windows® computer program.

WOSA See Welch overlapping segment averaging.

WPGMA See weighted pair-group method using arithmetic averages.

WPGMC See weighted centroid clustering.

WynKyst Software for non-metric multidimensional scaling distributed witht eh CANOCO and CanoDraw package.

X-Y fused Kohonen network A type of supervised Kohonen network in which the input and output maps are linked by a fused similarity map.

z-scores Variable values transformed to zero mean and unit variance. Also known as standard scores.

Zonation The distribution of organisms in distinctive areas or the process of partitioning a temporally or spatially ordered sequence of observations into groups or zones of similar character.

Zone A geographical or stratigraphical area or subdivision that has a cohesive character.

Zone MS-DOS zonation software developed by Steve Juggins that also runs under Windows® that combines CONSLINK, CONISS, binary splitting by least-squares and information content, optimal splitting by least-squares, and the variable barriers approach. Available from Stephen Juggins.

Acknowledgement HJBB wishes to acknowledge his great debt to Cathy Jenks, Steve Juggins, and John Smol for their invaluable help in compiling this glossary. It has been a challenging task with entries not only from palaeolimnology, ecology, limnology, and environmental science but also from applied statistics and computing.

Sources

Allaby M (2008) Oxford dictionary of earth sciences, 3rd edn. Oxford University Press, Oxford
Allaby M (2010) Oxford dictionary of ecology, 4th edn. Oxford University Press, Oxford
Everitt BS (2002) The Cambridge dictionary of statistics, 2nd edn. Cambridge University Press, Cambridge
Jongman RHG, ter Braak CJF, van Tongeren OFR (1987) Data analysis in community and landscape ecology. Pudoc, Wageningen
Legendre P, Legendre L (1998) Numerical ecology, 2nd English edn. Elsevier, Amsterdam
Lincoln RJ, Boxshall GA (1987) The Cambridge illustrated dictionary of natural history. Cambridge University Press, Cambridge
Lincoln RJ, Boxshall GA, Clark PF (1982) A dictionary of ecology, evolution and systematics. Cambridge University Press, Cambridge
Lincoln RJ, Boxshall GA, Clark PF (1998) A dictionary of ecology, evolution and systematics, 2nd edn. Cambridge University Press, Cambridge
Matthews JA (ed) (2001) The encyclopaedic dictionary of environmental change. Arnold, London
Park C (2008) Oxford dictionary of environment and conservation. Oxford University Press, Oxford
Smol JP (2008) Pollution of lakes and rivers – a paleoenvironmental perspective, 2nd edn. Blackwell, Oxford
ter Braak CJF, Šmilauer P (2002) CANOCO reference manual and CanoDraw user's guide – software for Canonical Community Ordination (version 4.5). Microcomputer Power, Ithaca
Upton G, Cook I (2006) Oxford dictionary of statistics, 2nd edn. Oxford University Press, Oxford
Whitten DGA, Brooks JRV (1972) The Penguin dictionary of geology. Penguin Books, London

Index

H.J.B. Birks et al. (eds.), *Tracking Environmental Change Using Lake Sediments*, Developments in Paleoenvironmental Research 5, DOI 10.1007/978-94-007-2745-8,

B

D

N

P

Q

R

S

T